I0811096

PALEOAMERICAN ORIGINS

Peopling of the Americas Publications

Sponsored by the Center for the Study of the First Americans
General editors: Michael R. Waters and Ted Goebel

PALEOAMERICAN ORIGINS: BEYOND CLOVIS

Edited by Robson Bonnichsen, Bradley T. Lepper,
Dennis Stanford, and Michael R. Waters

Texas A&M University Press • College Station

PALEOAMERICAN ORIGINS: BEYOND CLOVIS

Library of Congress Cataloging-in-Publication Data
Paleoamerican origins: beyond Clovis / edited by Robson Bonnichsen . . .[et al.].
p. cm.—(Peopling of the Americas publication)
Includes bibliographical references and index.
ISBN 13: 978-1-58544-540-0 (cloth)
ISBN 13: 978-1-60344-812-3 (paper)
1. Paleo-Indians—North America. 2. Clovis culture. 3. Paleoecology—North American—Pleistocene. 4. North America—Antiquities.
I. Bonnichsen, Robson. II. Title. III. Series.
E77.9.P35 2005
970.0'1—dc22 2005054293

©2005 Center for the Study of the First Americans. All rights reserved. No part of this book may be reproduced, projected, stored in a retrieval system, or transmitted, for whatever purpose, in any form or by any means, whether electronic, mechanical, magnetic, photographic, laser, or otherwise, without the prior written permission of the Center for the Study of the First Americans, Department of Anthropology, Texas A&M University, College Station, TX 77843-4352

Second printing, 2013

ROBSON BONNICHSEN
1940–2004

Founder and First Director, Center for the Study of the First Americans

Rob Bonnichsen was a unique individual who played many roles in the field of archaeology. Rob was first and foremost a scholar. Rob had a distinguished career in archaeology and was a guiding light in the study of the first Americans. Rob was a visionary, who established the **Center for the Study of the First Americans** in 1981 and served as its Director for 24 years. Rob was a dedicated teacher who encouraged and helped students become scholars. Rob was a story teller who was committed to sharing the excitement of the past with the public. Rob was a fighter who advocated for the rights of all archaeologists to pursue their research into the past. And finally, Rob was a friend to all—always helping and encouraging colleagues and students. Rob was truly a one-of-a-kind. His passing is a great loss and we will miss our friend and colleague.

Contents

Foreword

Michael R. Waters
Director, Center for the Study of the First Americans

Paleoamerican Origins: Beyond Clovis is an outgrowth of the Clovis and Beyond Conference held in Santa Fe, New Mexico, in 1999. Included in this volume are a number of articles written by the foremost authorities in first Americans research. These papers deal with the archaeological evidence for Paleoamerican sites in North and South America, and the genetic and skeletal record of the first Americans.

Rob Bonnichsen had been compiling this volume for some time and was working on the final chapter for this book at the time of his death. It had taken Rob a long time to get to this point because he had been distracted by other research, teaching, **Center** business, and the challenges of the Kennewick Man case. During the year prior to his death, he dedicated as much time as he could to finishing the last chapter of this volume. Rob wanted to wrap up the conference and all the papers in a comprehensive concluding chapter. While much of the manuscript was completed at the time of his death, much remained to be done—editing remained to be completed, sections needed to be added, illustrations needed to be finalized, and references had to be inserted. Dennis Stanford took up this challenge and brought the final chapter to completion in the summer of 2005. I was tasked with the final organization of the chapters and other book tasks. I have done my best to bring this book to publication as Rob would have wished.

This book is a tribute to Rob. He was the visionary and the spirit behind Clovis and Beyond, and he moved forward with his usual zeal to assemble the authoritative and provocative chapters in this volume. Rob felt that the Clovis and Beyond Conference was a turning point in our understanding of the first Americans and signaled a shift from the dogma of "Clovis first." This is the last time Rob will speak to us directly, but his spirit and enthusiasm for new knowledge will live on forever and his contributions will continue to ignite the imaginations of future scholars and fuel the great debate over the first Americans.

I. Paleoamerican Origins

An Introduction to Paleoamerican Origins

Robson Bonnichsen

The future of First Americans Studies is at a two-way crossroad between science and public policy. The traditional scientific explanation used to explain the peopling of the Americas, known as the Clovis-first model, does not adequately account for the nature of the Paleoamerican archaeological, skeletal, and genetic records. A challenge that lies before members of the scientific community is how to fill this void left by the Clovis-first model. New, more appropriate models must be constructed using appropriate methods for gathering new data to create new more adequate explanations of how the Americas were peopled. Additionally, First Americans Studies is also at a public policy crossroad. In the recent past, government decision makers frequently used the Public Trust Doctrine to guide decisions regarding the use and study of cultural resources. Some agency decision makers are now attempting to replace this doctrine with a new Exclusive Rights Doctrine. Whichever public policy prevails will dramatically affect the development of new knowledge about America's earliest cultural and biological heritage.

Traditionally, archaeologists have benefited from government decision makers' use of the Public Trust Doctrine. This doctrine, backed by laws such as the American Antiquity Act and the Archaeological Resources Protection Act, supported and encouraged the scientific community to use new scientific methods to explore public lands with the goal of expanding public understanding of the past. Under this doctrine, empirical scientific research on the initial peopling of the Americas was done in the name of all people. As shall be seen in this volume, scientific research in the field of First Americans studies benefited from this receptive political climate.

In contrast, the new Exclusive Rights Doctrine contends that archaeological resources belong not to society as a whole, but to various ill-defined segments of society. Nowhere is this new perspective more clear than in the Native Americans Grave Protection and Repatriation Act of 1990 (NAGPRA) and recent state laws in California, Montana, and Maine. This new public policy perspective raises fundamental questions, including who owns the past and who has the right to interpret ancient cultural and biological remains. Rather than aiding scientific research by qualified specialists, new and restrictive laws limit or deny access to ancient collections and to private, state, and federal lands. Furthermore, federal decision makers are now making decisions about very ancient remains and sites based on non-empirical criteria such as oral history and geography of modern groups where ancient remains are found. The emergence of a politically motivated doctrine to interpret ancient archaeological and skeletal remains is a very real threat to the future of scientific investigation of the past.

Scientific Research Directions

Until the late 1990s, most specialists thought that the Americas were peopled only once during late Ice Age times. New discoveries and better scientific methods are changing long-held perceptions about how the initial peopling of the Americas occurred.

During the 1960s specialists began to experiment with scientific model building as a means for explaining specific historical events and processes responsible for the peopling of the Americas. One of the first and most important has been called the Clovis-first model. The fundamental premises of this model, to be described in detail shortly, led many to the conclusion that the Americas were only peopled once by a single founding population from Northeast Asia approximately 11,500 RCYBP. The implication of this model is that the colonists and their immediate descendants gave rise to all subsequent New World cultures, languages, and biological populations. The evidentiary record provided by new early archaeological sites, early cranial remains that differ from modern Native Americas, and genetic evidence

Director, Center for the Study of the First Americans, deceased December 2004. Address all inquiries to Michael R. Waters, Director, Center for the Study of the First Americans, Department of Anthropology, Texas A&M University, College Station, TX 77843-4352; e-mail: mwaters@tamu.edu suggests key propositions of the Clovis-first model can no longer be regarded as correct. Specialists are searching for a new paradigm to replace the Clovis-first model. Gradually a new multi-origin theory that envisions the Americas were colonized multiple times during the late Ice Age and more recent times is beginning to develop. It should be emphasized, however, that detailed specific models to be embraced by this general theory are still in an embryonic stage of development.

Developing an understanding of the initial peopling of the Americas is of fundamental importance for understanding the contributions that the Americas have made to world prehistory. First Americans Studies is a relatively small subfield of archaeology united by a common interest in Paleoamerican origins. The goal of First Americans Studies is to develop an adequate theory for explaining the peopling of the Americas. Such an explanation has both significant scientific and humanistic values. From a humanistic perspective, most people want to know where they and other people came from. Such knowledge is important for how people feel about themselves and others. From a scientific perspective, First Americans Studies provides scientific information about our past and enhances understanding of the dispersion, cultural, and biological processes and patterns that led to the peopling and settlement of the New World.

Solving the history of the peopling of the Americas is an enormous problem involving events on three continents: Eurasia, North America and South America. Events that occurred in one area can have implications for interpreting events on other continents. Because of the enormous scale of the overall history of the peopling of the Americans, the relative paucity of data relating to this early period, and numerous lines of evidence that must be considered, constant review and synthesis of new discoveries is required. Each of the four subfields of anthropology (i.e., archaeology, physical anthropology, linguistics, and cultural anthropology) seeks to explain similarities and differences found in its subject material, e.g., American cultures, languages, and biological populations. Ultimately, specialists in each subfield draw on Paleoamerican origin theories, since they provide the beginning foundation on which all other developmental theories are based. When there is a paradigm shift in First Americans Studies, subfields that rely on outdated information must be adjusted accordingly

Models

Models are, in a sense, theories that provide a scenario to explain a specific event such as the peopling of the Americas. What distinguishes scientific model building from mere speculation is that models are based in part on the surviving evidentiary record. Models are constructed to explain and integrate biological, cultural, and environmental information that documents the presence and activities of past peoples and their environments. In constructing such models, investigators attempt to find a fit between processes supplied by the model and available archaeological, biological, linguistic, and environmental data.

The Clovis-first model, soon after its introduction during the early 1960s, became a controlling theory in the field of First America Studies. Proponents of the model rejected other possible alternatives. Not surprisingly, this posture became a focal point of a long debate in the professional community regarding the timing of the initial peopling of the Americas.

Although the Clovis-first model (Haynes 1964, 1966; Martin 1973, 1984; Mosimann and Martin 1975) originally was created to explain the extinction of Pleistocene megamammals, weighing more than 100 kg, and the apparently simultaneous appearance of Clovis fluted projectile points, it was widely adopted by prehistorians to explain the initial peopling of the Americas.

The Clovis-first model envisions that a band of Siberian big-game hunters numbering no more than 150 people came from northeastern Siberia to interior North America approximately 11,500 RCYBP. Modelers envisioned that these people walked across the Bering Land Bridge near the end of the last glacial period when lower sea levels exposed the Bering Land Platform connecting Asia and North America. The band of hunters quickly found their way across eastern Beringia, the unglaciated portion of Alaskan and Yukon Territory, and then down the Ice-free Corridor between the Cordilleran and Laurentide Ice sheets onto what is now the Great Plains. These hunters allegedly found a virgin continent teeming with large and unsuspecting game animals. Using a sophisticated tool kit and a new weapon system, they were able to slaughter mammoths and other unsuspecting large mammals. Life was good, and these groups prospered, multiplied, and rapidly expanded throughout North and South America. In the wake of their expansion, the colonists and their immediate descendants are said to have exterminated 33 genera of large mammals in North America and over 55 genera in South America in about 1,000 years (Martin 1984).

The archaeological signature for this Clovis big-game hunter expansion is said to include fluted projectile points, bifaces, scrapers, gravers, blade cores, blade tools, use of large bone and ivory tools, and red ocher (Haynes 1980). Clovis sites with stratigraphic context have consistently yielded radiocarbon ages between 11,700 and 10,800 RCYBP. Where preservation occurs, Clovis artifacts are often associated with extinct fauna and especially the mammoth. Armed with an efficient tool kit, these hunters and their descendants proliferated in a continent devoid of human populations. Clovis descendants are said to have settled all of the Americas by 10,600 RCYBP. With the demise of the mammoth, the Clovis cultural pattern ended as abruptly and mysteriously as it started. The widespread distribution of Clovis points in every state in the continental United States and the apparent lack of evidence of earlier occupations led many regional American experts to conclude that Clovis was the basal culture in North America and South America.

The Clovis-first colonization scenario is based on several testable propositions. It predicts: (1) where the Clovis colonization event originated; (2) when colonization occurred; (3) the route taken to the Americas; (4) the nature of adaptations used by the early hunters; and (5) the direction of movement from Alaska to South America. It also implies that the initial founding group shared a common culture, language, and gene pool that gave rise to all the diverse languages, cultures, and biological populations found in the Americas at the time of European contact.

There has long been a small but vocal minority who have argued that the peopling of the Americas occurred well before 11,500 RCYBP (e.g., Bryan 1986; Gruhn, this volume, Guidon 1994). Only recently has attention turned from a focus on chronology to systematically testing propositions on which colonization models are based (Bonnichsen and Schneider 1999; Dillehay 1997, 2000; Gruhn, this volume) as means for evaluating model validity.

Various alternatives to the Clovis-first model were advanced to explain the initial peopling of the Americas. For example, Fladmark (1978, 1979, 1983) has proposed the Americans were settled by populations moving along the Pacific Northwest coast. Greenberg et al. (1966) proposed the three-wave model that draws on linguistic, archaeological, tooth, and genetic data in postulating three separate colonization events. Others, Torroni (1992), Torroni et al. (1993), and Wallace and Torroni (1992), draw on genetic evidence to formulate multiple colonization events. Steele and Powell (1999) consider skeletal evidence, and Nichols (1991) uses linguistic evidence in support of her view that the Americans were colonized several times. Despite these modeling efforts, the Clovis-first model dominated archaeological inquiry over the last four decades, until quite recently (Dillehay 2000; Gruhn this volume).

Many investigators are now seeking to raise the bar on the standards of evidence used in model building. Most agree models that seek to explain the peopling of the Americans need to be grounded in evidence. Interdisciplinary teams of specialists who study a variety of different kinds of evidence such as landforms, sediments, soils, plant and animal remains, artifacts, features, sites, and human remains collect and analyze evidence required to explain what happened in the past. Modelers use data to test model premises of how the Americas were colonized. Model construction should involve as many independent lines of evidence as possible. Models that link many lines of independent data are stronger than those based on only one or a few lines of evidence.

The shift in perception about the initial peopling of the Americas is a consequence of new discoveries and advances in scientific methodology. These discoveries and advances demonstrate that the Western Hemisphere was peopled earlier than anticipated by the predictions of the Clovis-first model. In addition, there are increasing data that indicate that multiple groups from different areas of Eurasia were involved in colonizing the Americas.

A new but still embryonic theory is beginning to fill the void created by the collapse of the Clovis-first model. Basic propositions of the new multi-origin theory are as follows. First, the Americas were peopled before 11,500 RCYBP. Second, groups with different cultures and languages and from different genetic stocks peopled the Americas. Third, once these groups were in the Americas, a complex interaction occurred among peoples with different origins to create new peoples, cultures, and adaptive systems that were uniquely American. The new perception does not exclude the possibility that the appearance of Clovis could have represented a distinct colonization event. It does, however, hold that Clovis colonists were not the first and only colonizers.

The multi-origin theory challenges the research community to create new and more accurate models to trace the origin, spread, rise, and demise of past cultural systems and biological populations through time and across space from Eurasia to the Americas. It sets the formidable task of conducting comparative studies of New and Old World collections to better understand the processes responsible for dispersion of humans to the New World, how they adapted to American conditions, and to understand how early colonizing populations interacted with one another. It also challenges biological anthropologists to develop phylogenetic models of biological population descent throughout time. To answer these other questions will require new research agendas focused not only on American data but also comparative research between American and Eurasia data sets.

Public Policy

A forward-looking approach to First Americans Studies in North America would be incomplete without considering changes in the legal framework that surrounds archaeological research in the United States. Scientific research occurs within a societal context. Changes in public policy can affect research goals and directions. Of particular importance, not only for First Americans studies but American archaeology in general, was the passage of the Natives American Graves Protection and Repatriation Act of 1990 (NAGPRA) and similarly restrictive state laws. These laws place new constraints on research and limit access to sites on private, state, and Federal lands. It appears as if neither the scientific community nor the public fully appreciates the implications of these laws.

The congressional intent of NAGPRA was to correct perceived errors of the past. It did so by giving Native Americans the legal right to obtain from museums and government agencies human remains, funerary objects, and items of cultural patrimony to which they are culturally affiliated or have certain types of geographic connections. Most archaeologists and First American specialists are sympathetic to the intent of this law. Many, however, are troubled by how it is being applied in specific situations.

Differences over interpreting the law came to a head with the discovery of Kennewick Man. In late July of 1996, human skeletal remains were found next to a bank of the Columbia River in Kennewick, Washington. Preliminary studies by James Chatters, commissioned by the Benton County coroner, suggested that Kennewick Man is one of the most complete and significant skeletons ever recovered in the Americas.

Upon learning of the antiquity of the Kennewick skeleton, the Umatilla Tribe of Oregon contacted the Army Corps of Engineers and claimed Kennewick Man was one of their lineal descendants. Soon, the Yakima, Colville, and Nez Perce tribes and the Wanapum Band joined the Umatilla in requesting that these ancient skeletal remains be turned over for repatriation. The Army Corps of Engineers accepted the tribes' claims and published a notice of intent to transfer the skeleton to their possession. Eight scientists asked for the right to study these remains before transfer. The Army Corps refused to respond to written inquiries from the scientists and proceeded with its plan to give the remains to the tribal claimants.

The eight scientists (R. Bonnichsen, C. L. Brace, G. Gill, C. V. Haynes, R. Jantz, D. Owsley, D. Stanford, and D. G. Steele) understood that without study, it would never be possible to determine from a scientific perspective whether Kennewick Man was or was not related to modern tribal peoples. The scientists were faced with a choice. They could either accept the Army Corps decision or file a lawsuit to study. Acting on their behalf, Alan L. Schneider and Paula Barran, attorneys, filed a lawsuit in federal District Court in Portland, Oregon. In June 1997, Federal Magistrate John Jelderks set aside the Army Corps initial decision to give the skeleton to the tribal claimants. The Army Corps was ordered to conduct new administrative proceedings on the fate of the skeleton and the scientists' study requests. The Department of the Interior subsequently joined the lawsuit, and in September 2000 ruled that geography and tribal origin accounts were sufficient evidence to demonstrate that the skeleton is culturally affiliated to the claiming tribes. Concurrent with this ruling, the Department of the Army denied the scientists' study requests. These decisions have been challenged in court.

One can only wonder how the views of government decision makers and those of scientists who study Paleoamerican prehistory became so polarized? Prior to the passage of NAGPRA in 1990, agency decision makers followed what is called the Public Trust Doctrine (Knudson and Keel 1995). Under this doctrine, the government was supportive of scientific research by trained personnel. Cultural resources were to be researched, conserved, and preserved for the benefit of all people. The government's role was as a steward of the past. Following the passage of NAGPRA, some agency decision makers have begun to follow what can best be described as the "Exclusive Benefit Doctrine." This doctrine holds that Native American tribes are the exclusive owners of all ancient human remains, grave goods, and items of cultural patrimony that date earlier than 1492, regardless of their age and regardless of whether they can or cannot be culturally affiliated to a modern tribe. This exclusive benefit policy has been extended to decisions concerning use of archaeological sites. The scientific and humanistic values of these remains, objects, and sites to non-tribal groups are in most cases excluded from consideration.

Science offers the only objective approach available for learning about very ancient human remains, objects, and sites, and their value for understanding Paleoamerican origins. Yet those who should be speaking out for the importance of scientific perspectives for making rational NAGPRA decisions generally are not doing so (but see Funk's recent letter [to the SAA newsletter] for important exceptions). The changes occurring in our understanding of Paleoamerican origins have important implications for NAGPRA decision makers. In respect to the repatriation of very ancient human remains, some decision makers have apparently assumed that there is an inevitable relationship between ancient human remains thousands of years old and contemporary Native American tribes. This need not, however, be the case. Some early groups may have gone extinct and lack modern tribal descendants. It would seem that study is the only fair and objective way to resolve repatriation issues involving Paleoamerican skeletal remains, associated funerary items, and items of cultural patrimony.

Public policy regarding First Americans cultural and biological resources are out of sync with where science is at today. There are fewer than a dozen known human skeletons from the United States older than 8000 RCYBP. These remains are 400 to 600 or more generations removed from the present, and most do not look like modern Native Americans (Steele and Powell 1999). When very ancient human remains are buried, we all lose information regarding our common human heritage.

Available scientific evidence suggests that late Ice Age and early-Holocene human remains and sites from the first Americans may not be directly related to modern-day tribal peoples. Scientists are being asked to suspend scientific rules of empirical evidence in favor of non-empirical political agendas. These are important issues that should come to the forefront of discussion among specialists who seek to understand the past and should as well be open to public discussion. As citizens and scientists, it behooves us to have clarity in our public policy choices and to make our preferences known to government decision makers.

It was in the context of these two disparate themes in First Americans Studies that the decision was made to convene the Clovis and Beyond Conference 29–31 October 1999 in Santa Fe, New Mexico. On the one hand, important scientific advances were being made that were changing our understanding of Paleoamerican origins and are important to public policy governing America's past. On the other hand, a new restrictive U.S. political environment was directly impacting the ability of scientists to learn about the American past.

Clovis and Beyond

The conference organizers included the Center for the Study of the First Americans, the Smithsonian Institution Paleoindian Program, *Discovering Archaeology* magazine, Forrest Fenn, Mark Mullins, and Kenneth Tankersley. The Clovis and Beyond Conference was organized in the spirit of furthering communication among all parties with a stake in Paleoamerican prehistory. Development of the conference was guided by four educational goals. First, we wanted to create an exhibit of late Ice Age artifacts so that the conference participants and public could develop an appreciation of the nature of the pre-Clovis and Clovis archaeological record from the United States. Artifacts from Clovis sites are widely dispersed and held in both private collections and professional institutions. In fact, many of the most important collections had never been assembled in one place before. Conference organizers Mullins, Fenn, and Tankersley made arrangements with institutions and collectors for the conference exhibit. A second conference goal was to have leading specialists present the latest archaeological, skeletal, and genetic evidence that is contributing to our changing perceptions of the peopling of the Americas. Dennis Stanford, Bradley Lepper, and Kenneth Tankersley assisted in choosing and inviting colleagues to make 40-minute presentations. D. Gentry Steele took a leadership role in selecting specialists with an emphasis in skeletal biology and genetics to cover new developments in paleobiology. As our third goal, we wanted to provide participants with an overview of current public policy issues in First Americans studies. Jo Ann Harris organized a panel discussion with representatives from the Society for American Archaeology and the scientific, legal, government, and Native American communities. As our fourth goal, we decided to have a poster session to accommodate anyone who wanted to share new information on the peopling of the Americas. All the hard work and organization that took place over a two-year period paid off. The conference was a resounding success. The speakers, poster presenters, and collection exhibitors outdid themselves. Over 1,400 people came to hear speakers, see the collections, witness the public policy panel discussion, and view the poster sessions during the three-day conference. Abstracts for the conference are posted on the web at www.clovisandbeyond.org and published in the conference abstract volume *Clovis and Beyond, Abstracts, Collections, and Exhibits* (Bonnichsen et al. 1999). Additionally, *Discovering Archaeology* in February 2000 published brief essays by many of the participants of Clovis and Beyond.

Publication

Collectively, the conference presenters developed a significant amount of new scientific information for the conference. In all, featured speakers, panel discussants, artifact exhibitors, and poster presenters made 97 presentations. In consulting with colleagues who became the publication committee—Ruth Gruhn, Dennis Stanford, D. Gentry Steele, Bradley Lepper, and Claude N. Warren—during the last day of the conference, we concluded that significant advances had been made in the field and that this new evidence merited publication.

The conference contributors were offered two venues for publication: they could submit short 2,000-word concept papers, journal-length articles, or both short concept papers and full-length articles. Ruth Gruhn agreed to serve as series editor, and other members of the editorial team refereed papers in their areas of expertise.

Bradley Lepper agreed to serve as the editor for *New Perspectives on the First Americans*, a volume of short concept papers. I took on the responsibility of organizing *Paleoamerican Origins: Beyond Clovis* and was assisted by a very able editorial team. Collectively, contributions in these two volumes provide an overview of past, present, and future directions of First Americans studies.

This volume is organized into three major sections: Paleoamerican Origins, Paleoamerican Prehistory, and Conclusions. Many of the manuscripts included present a foundation for a new research paradigm. Bonnichsen and Lepper (this volume) present overviews of the archaeology and paleobiology sections to give the reader a quick grasp of issues covered by contributors' papers. In the concluding section, A. L. Schneider and R. Bonnichsen provide a wake-up call regarding science politics and public policy issues that are affecting First Americans research. The final manuscript discusses future Directions in First American Studies. It focuses on the development of the multi-origin theory and requirement for developing Bering Land Bridge/Ice-free Corridor, Pacific Rim, the North Atlantic Rim and South Pacific entry models for explaining the peopling of the Americas.

References Cited

Bonnichsen, R., M. Punke, C. Allred, K. B. Tankersley, F. Fenn, and M. Mullins
1999 Clovis and Beyond: A Peopling of the Americas Conference: Abstracts, Collections and Exhibits. Published by Clovis and Beyond, Santa Fe, NM.

Bonnichsen, R., and A. L. Schneider
1999 Breaking the Impasse on the Peopling of the Americas. In *Ice Age Peoples of North America*, edited by R. Bonnichsen and K. L. Turnmire, pp. 497–519. Oregon State University Press, Corvallis, OR.

Bryan, A. L.
1986 Paleoamerican Prehistory as Seen from South America. In *New Evidence for the Pleistocene Peopling of the Americas*, edited by A. L. Bryan, pp. 1–14. Center for the Study of Early Man, University of Maine, Orono, ME.

Dillehay, T. D., Editor
1997 *Monte Verde: A Late Pleistocene Settlement in Chile I: The Archaeological Context*. Smithsonian Institution Press, Washington, D.C.
2000 *The Settlement of the Americas: A New Prehistory*. Basic Books, New York, NY.

Fladmark, K. R.
1978 The Feasibility of the Northwest Coast as a Migration Route for Early Man. In *Early Man in America from a Circum-Pacific Perspective*, edited by A. L. Bryan, pp. 119–128. Department of Anthropology, Occasional Paper 1, Edmonton, Alberta.
1979 Routes: Alternative Corridors for Early Man in North America. *American Antiquity* 44:55–69.
1983 Times and Places: Environmental Correlates of Mid-to-late Wisconsinan Human Populations Expansion in North America. In *Early Man in the New World*, edited by R. Shutler, pp. 13–41. Sage Publications, Beverly Hills, CA.

Guidon, N., A. M. Pessis, F. Parenti, M. Fontugue, and C. Guerin
1994 Nature and Age of the Deposits in Pedra Furada: Reply to Meltzer, Adovasio & Dillehay. *Antiquity* 70:408–21.

Greenberg, J. H., C. G. Turner II, and S. L. Zegura.
1966 The Settlement of the Americas: A Comparison of the Linguistic, Dental, and Genetic Evidence. *Current Anthropology* 27(5):477–97.

Haynes, C. V., Jr.
1964 Fluted Projectile Points: Their Age and Dispersion. *Science* 145:1408–13.
1966 Elephant-hunting in North America. *Scientific American* 214:104–12.
1980 The Clovis Culture. *Canadian Journal of Anthropology* 1:115–21.

Knudson, R., and B. C. Keel (editors)
1995 *The Public Trust and the First Americans*. Oregon State University Press, Corvallis, OR.

Martin, P. S.
1967 Prehistoric Overkill. In *Pleistocene Extinctions: The Search for a Cause*, edited by P. S. Martin and H. E. Wright, Jr., pp. 75–120. Yale University Press, New Haven, CT.
1973 The Discovery of America. *Science* 117:969–73.
1984 Prehistoric Overkill: The Global Model. In *Quaternary Extinctions: A Prehistoric Revolution*, editors P. S. Martin and R. G. Klein, pp. 354–403. University of Arizona Press, Tucson, AZ.

1987 Clovisia the Beautiful! *Natural History* pp. 10–13.

Mosimann, J. F., and P. S. Martin
1975 Simulating Overkill by Paleoindains. *American Scientist* 63(3):304–13.

Nichols, J.
1991 Linguistic Diversity and the First Settlement of the New World. *Language* pp. 261–73.

Steele, D. G., and J. F. Powell
1999 Peopling of the Americas: A Historical and Comparative Perspective. In *Who Were the First Americans? Proceedings of the 58th Annual Biology Colloquium, Oregon State University*, edited by R. Bonnichsen, pp. 97–126. Center for the Study of the First Americans, Oregon State University, Corvallis, OR.

Torroni, A.
1992 Native American Mitochondrial DNA Analysis Indicates that the Amerind and the Nadene Populations Were Founded by Two Independent Migrations. *Genetics* 130:153–62.

Torroni, R., I. Sukernik, T. G. Schurr, Y. B. Starkikovskaya, M. F. Cabell, M. H. Crawford, A. G. Comuzzie, and D. C. Wallace
1993 mtDNA Variation of Aboriginal Siberians Reveals Distinct Genetic Affinities with Native Americans. *American Journal of Human Genetics* 130:53:591–608.

Wallace, D. C., and A. Torroni
1992 American Indian Prehistory as Written in the Mitochondrial DNA: A View. *Human Biology* 64(3):403–16.

Changing Perceptions of Paleoamerican Prehistory

Robson Bonnichsen[1] and Bradley T. Lepper[2]

The goal of this essay is to provide an overview of new information presented by the contributors to this volume relating to Paleoamerican origins and the initial peopling of the Americas. No attempt is made here to provide comprehensive coverage of every important newly investigated site in each geographical area. Rather, our focus is placed on sites and research that are important for testing fundamental propositions of the Clovis-first model originally formulated by Haynes (1964, 1966) and Martin (1967, 1973, 1984, 1987); and Mosimann and Martin (Martin 1975) and which Haynes (this volume) and others (Fiedel this volume) still champion. Until recently, the Clovis-first model was the quintessential theory that dominated scientific thinking regarding the initial colonization of the Americas (see the Introduction or Gruhn in this volume for brief overviews of the Clovis-first model).

From a historical perspective, the Clovis-first model represented an important departure from the cultural historical approach that preceded it. Rather than organizing data into static stages, the Clovis-first model focused on scientific model building. In seeking to develop an answer to one of the truly great watersheds of world prehistory, specialists embarked on a research agenda of setting forth a model composed of propositions that could be tested through the study of empirical phenomena. The Clovis-first model was proposed as an explanation for the apparently rapid demise of Ice Age mammals at the end of the Pleistocene that seemed to have coincided with the equally sudden and simultaneous appearance of a new element in the ecosystem—human big-game hunters from Northeast Asia. According to the tenets of the Clovis-first model, these hunters, armed with a new and supposedly more efficient killing technology—most particularly, the Clovis projectile point—brought about the rapid demise of over 33 genera of large game animals in North America and more than 55 genera in South America in under 1,000 years. As the initial colonists and their descendants dispersed rapidly from Alaska throughout the North and South American continents, they met no resistance from prior immigrants for, in this model, there were none. In their wake, they left behind the elements of their tool kits, including fluted points, blade cores, tools made on blades (such as scrapers and spokeshaves), spurred gravers, large bifaces, mammoth bone and ivory foreshafts, shaft straighteners, and points, as well as red ocher found in archaeological sites (Haynes 1980).

The Clovis-first model is an elegant articulation of a series of propositions that have both explanatory and predictive power. The model specifies where earliest colonists came from, the route taken as well as the rate of colonization, what archaeologists should expect to find in the most basal level of the American archaeological record, and the earliest possible age for human occupation. Although many authors have questioned the validity of the Clovis-first model (Bonnichsen and Schneider 1999; Dillehay 2000; Gruhn this volume), others have argued in support of it (Dincauze 1984; Fiedel this volume; Lynch 1990, 2001).

In our view, the long-standing impasse between the Clovis and pre-Clovis camps has been broken. With revelations brought about through the discovery of new archaeological sites older than 11,500 RCYBP (Adovasio et al. this volume; Anderson this volume; Bonnichsen and Turnmire 1999; Dillehay 1997, 2000; Goodyear this volume; Gruhn this volume; Johnson this volume; and Overstreet this volume), and integration of new skeletal and genetic evidence (Jantz and Owsley; Owsley and Jantz, Schurr, and Smith et al. this volume), many specialists have rejected the Clovis-first model. Some have gone back to the field to gather more data;

[1]Director, Center for the Study of the First Americans, deceased December 2004. Address all inquiries to Michael R. Waters, Director, Center for the Study of the First Americans, Department of Anthropology, Texas A&M University, College Station, TX 77843-4352; e-mail: mwaters@tamu.edu

[2] Curator of Archaeology, Ohio Historical Society, 1982 Velma Avenue, Columbus, OH 43211-2497. e-mail: blepper@ohiohistory.org The opinions expressed in this paper are those of the author and do not necessarily reflect those of the Ohio Historical Society.

others have gone back to the drawing board to create a new generation of colonization models to account for the peopling of the Americas.

As indicated in the Introduction, what is emerging is a new theory that envisions the Americas were not peopled once, but were colonized in several waves beginning before the end of the last Ice Age by groups from different areas of Eurasia (Bonnichsen 1999). Using new and existing data, we review the empirical archaeological evidence to examine the validity of testable propositions from which the Clovis-first model is constructed.

Our discussion, which parallels the organization of this volume, focuses on the following topical areas: regional archaeological records, Clovis, non-Clovis traditions, pre-Clovis, and the South American Paleoamerican record. We consider how the widespread acceptance of the Clovis-first model has influenced the development of Paleoamerican prehistory and reasons why research on both the northern and southern continents suggests this once popular and useful model is no longer useful or relevant. It is time to search for a replacement.

Regional Archaeological Records

The importance of interdisciplinary research in First American studies cannot be overemphasized. Scientific data developed through the study of paleoenvironmental, paleobiological, and archaeological records provide the evidentiary foundation for constructing models that seek to explain the peopling of the Americas. Dated stratigraphic records of sediments and soils, plant and animal remains, and artifacts and features allow specialists to create models at the local, regional, and global scales to explain the processes and events responsible for climate change, landscape evolution, changes in plant and animal communities, and corresponding changes in human lifeways. This focus, which is known as environmental archaeology, is now widely accepted by First American specialists and by contributors to this volume in particular. This holistic approach fosters a contextual understanding of the past and a foundation of information from which prehistorians can construct an array of settlement, subsistence, lithic procurement, adaptation, colonization, and descent models.

Recent archaeological research in the southeastern United States dispels the Clovis-first proposition that no human occupation occurred in the Americas prior to Clovis times. A series of important new sites of pre-Clovis age have yielded cultural material associated with pre-12,000 RCYBP dates (Bonnichsen and Schneider 2001). The most important include: Coats-Hines, Tennessee (Anderson this volume); Cactus Hill, Virginia (McAvoy and McAvoy 1997); Saltville, Virginia (MacDonald 2000); Page-Ladson, Florida (Dunbar 2001); Sloth Hole, Florida (Dunbar 2001; Hemmings 1997); and Topper, South Carolina (Anderson this volume and Goodyear this volume). These sites along with the rich record of sites of Clovis and post-Clovis age have led Anderson to suggest new timelines, based on calendar years before present, are warranted for the Southeast. Anderson observes that radiocarbon ages compress the time scale, making it difficult to understand or appreciate the rapid changes that occurred during the Paleoamerican period. He proposes that three sub-periods are in order: Early (prior to 11,700 RCYBP or 13,450 CALYBP), Middle (11,700–10,800 RCYBP or 13,450–12,900 CALYBP, and Late (10,800–10,000 RCYBP or 12,900–11,450 CALYBP). These periods correlate with pre-Clovis, Clovis, and post-Clovis occupations in the Southeast. This timeline may be useful for other areas, as they seem to correlate well with global climatic events and periods of rapid culture change.

New discoveries at single sites can change understanding of regional cultural sequences. Dunbar et al. (this volume) present such a case in their discussion of new discoveries from the Ryan/Harley site 8JE1004, Florida. The site is 1 m below low-water stage of the Wacissa River. It represents the first occurrence of an in situ Suwannee complex site with stone tools and associated faunal remains. A remarkable aspect of this site is the rich faunal record of all five classes of vertebrates including mammalian species now extinct or that no longer reside in Florida. The remains of horse and tapir were found in place. However, the remains of the giant tortoise (*Hesperotestudo crassicutata*), giant armadillo (*Holmesina septentrionalis*), sloth (*Paramylodon harlani*), and American mastodon (*Mammut americanum*) were not in primary depositional context in the riverbed. Mammal bones found in context exhibit green-bone breaks that suggest processing when fresh. Of particular interest is the beveled end of a mastodon ivory foreshaft found out of context. A total of 193 stone tools found in place include a Suwannee preform with flute-like basal thinning, bifaces, oval and conical scrapers, tools made on blades, end scrapers, and debitage that is consistent with late-stage tool production and maintenance. Dunbar et al. (this volume) conclude that the tool kit is similar to Clovis tool kits. Efforts to date the site more precisely so far have failed; organic silts replaced bone collagen that was dated to about 4500 RCYBP. Dunbar et al. suggest Pleistocene megafauna survived beyond the Allerød boundary into the Younger Dryas (after 10,800 RCYBP or 11,450 CALYBP). They speculate that the Suwannee people are descendants of Clovis, and unlike the Clovis upland big-game hunters, the Suwannee folk hunted wetland species including tapir, turtles, alligators, fish, birds, and amphibians to fill their nutritional requirements. Another possibility, in contrast to the proposition that Suwaneee derives from Clovis, is that Suwannee and Clovis may have shared a common ancestor and the Suwannee complex represents a co-tradition contemporaneous with Clovis or may be earlier than Clovis. Secure dates will be required to resolve this issue.

Dansie and Jerrems's innovative contribution (this volume) to regional studies is their documentation of region-wide lake level fluctuations triggered by climate change that would have dramatically affected human adaptive strategies. Pleistocene Lake Lahontan was a vast mountain-rimmed freshwater Pleistocene lake in the northwestern Great Basin (Chandler 2001). At high stand, the lake was about 900 ft deep. It filled several linked basins and covered about 8,665 square miles at 13,070 RCYBP. The rapidly changing nature of this complex hydrological system is vital to interpreting late-Pleistocene and early-Holocene archaeological and paleontological records of the Great Basin.

Using approximately 200 radiocarbon dates, Dansie and Jerrems constructed a graph of lake elevations through time. It establishes a temporal framework for integrating geomorphic, paleontological, and cultural remains. Between 12,600 RCYBP and 12,080 RCYBP Lake Lahontan crashed. It receded 536 ft in just 170 years. By dating extinct faunal remains held in museum collections from localities contained within the Lahontan system, it was possible to determine that Pleistocene fauna survived the crash. The two earliest known cultural traditions of the Lake Lahontan system, Clovis and Western Stemmed Point traditions, appear to correlate with the "low stand" of Lake Lahontan. The importance of museum collections for future study is highlighted by surprisingly recent megafauna dates, 10,350 RCYBP for mammoth and 9960 RCYBP for short-faced bear. If additional dates support these late ages on megafauna, further doubt would be cast on the abrupt demise of megafauna called for by the Clovis-first over-kill model.

Clovis

Paleoamerican-origin models commonly draw on data compiled in several regions. Such is the case with the study of Clovis origins. The sudden and widespread appearance of Clovis and its equally sudden disappearance are as much a mystery today as when first recognized during the early part of the last century. Four authors consider ways of explaining Clovis. Haynes and Fiedel each present a "Revised View" of the Clovis-first model. Haynes concedes that there may have been other cultures in the Americas as early as Clovis, but he otherwise holds to positions adopted earlier in his career. He maintains that the Nenana complex of Alaska is a direct ancestor to Clovis, and, by implication, that big-game hunters from Northeast Asia walked across the Bering Land Bridge and became the people in Alaska who created the Nenana complex. From this, it could be argued that they traveled southeast down through the Ice-free Corridor, dramatically changing tool types and hunting patterns and becoming Clovis by the time they arrived in the continental United States. While this scenario has coherence and a certain elegance, it does not solve the persistent problem of a lack of a clear Clovis antecedent in Northeast Asia, the presumed homeland of Clovis. Moreover, proposed ancestral relationship between Nenana and Clovis is speculative at best, given the absence of detailed comparative technological analyses of Nenana and Clovis assemblages. Nonetheless, Haynes deserves credit for his attempt to integrate the latest and most refined paleoecological/paleoclimatological data and use it as a causal force responsible for the rise and demise of the Clovis complex.

In seeking to explain the spread of Clovis, Fiedel uses the dispersion of Thule culture as an analog. The details of the analogy can be debated, but the demonstration that people can sometimes move rapidly is well taken. Fiedel uses this analogy in support of the Clovis-first model and does not take seriously the evidence of pre-Clovis cultures or contemporaneous co-traditions that would complicate the analogy.

Collins (this volume) suggests that debate over the origins of "Clovis" is needed, but he calls attention to the importance of carefully considering the "rules of evidence." How do we judge varying degrees of historical relatedness (homology) versus convergent cultural evolution (analogy)? Collins makes the excellent point that "the peopling of the Americas is a complex problem for which a solution is dependent upon evidence from much of the world over the last 20,000 years or so."

So where does this leave us in our efforts to explain the *origins* of Clovis? The search for such an explanation currently revolves around several competing models that attempt to weave together the various strands of data into a coherent framework. The Revised Clovis-first model (Haynes and Fiedel this volume) seeks to keep the door open to the possibilities that: (1) Clovis had its roots in the far north; (2) the Ice-free Corridor was a viable route for Clovis colonists; (3) Clovis was the first major archaeological culture in North America. West (1997) supports a variant of this model and proposes that Clovis originated on the Bering Land Bridge now submerged by the Bering Sea.

In contrast, Bryan (1986) and others propose that human populations were in the New World before Clovis times. In their view, the emergence of Clovis was precipitated by rapid environmental change at the end of the last Ice Age (Bonnichsen et al. 1987; Collins 1999; Frison and Bonnichsen 1996; and Lepper 2002). Global climate change disrupted environments and existing adaptive systems. Human groups responded by devising new and more efficient technological organizations to respond more effectively to rapidly changing environmental circumstances. During a period of environmental change important technological concepts caught on and rapidly spread from group to group and from region to region. This model envisions Clovis as a diffusion pattern and not a monolithic culture.

Several researchers (Anderson and Gillam 2000; Chandler 2001; Stanford and Bradley 2000) observe that more fluted-point sites occur in the southeastern United States than in the Southwest, which may imply Clovis originated in the Southeast and spread west and north. Stanford and Bradley propose that the ancestors of Clovis were descendants of the Solutrean populations from the Iberian Peninsula who used small watercraft to cross the North Atlantic to the East coast of North America during the Last Glacial Maximum (Chandler 2001, Stanford and Bradley 2000; cf. Straus 2000). By contrast, Dixon (1999, 2000) envisions that early human populations from Asia moved along the North Pacific Rim using small watercraft. Once they were south of the Cordilleran Ice Sheet along the West coast of North America, these foragers adapted to local conditions and moved inland up rivers into interior North America, subsequently giving rise to the Clovis complex.

Haynes's call (this volume) for using multiple working hypotheses is being answered. We now have a number of alternative hypotheses to explain Clovis origins. To move our thinking forward regarding the Clovis origins enigma, we need more than hypotheses; we need a definitive test that can be addressed through comparative research to sort out which, if any, of the current generation of Clovis origins models is correct. We need standards or agreed upon "rules of evidence" that will allow us to translate variation in stone tool types into an understanding of human adaptive patterns as a means to track Paleoeurasians and

Paleoamericans across time and through space and to distinguish among migration, diffusion, and in situ development. Although these are very old anthropological issues, they have yet to be adequately resolved.

Four of the most fundamental premises of the Clovis-first model are: (1) Clovis represents a single culture; (2) Clovis hunters first entered the New World about 11,500 RCYBP; (3) the Clovis cultural complex disappeared with (and probably caused) the extinction of large megafauna about 10,800 RCYBP; and (4) the Clovis culture is basal to and gave rise to all other archaeological traditions (see also Willig 1994) in North and South America.

Bonnichsen et al. (this volume) propose, to the contrary, that: (1) Clovis does not represent the first human group to enter the New World; (2) the Americas were peopled several different times by different human groups; (3) several North American archaeological co-traditions are as old as, if not older than Clovis; (4) Paleoamerican groups responded to global environmental change using strategies that made sense at the local level, giving rise to regional diversification by Clovis times; and (5) Clovis is not ancestral to all North and South American populations (Gruhn this volume; Dillehay 2000). In support of such a multilinear model of Paleoamerican prehistory that calls for early regional diversification, Bonnichsen and his coauthors summarize radiocarbon dates from stratified archaeological sites, document tool assemblages, and outline possible alternative adaptive strategies. They identify Nenana, Sluice Way, Western Stemmed Lanceolate, and Goshen/Plainview fishtail traditions as some of the most prominent co-traditions that are coeval with Clovis.

Pre-Clovis

Evidence of pre-Clovis cultures has increased substantially during recent years. Gradually, the quantity and quality of interdisciplinary site investigations have improved. Although stratigraphic, dating, and artifact problems persist at some locations, there are now many sites with pre-11,600 RCYBP radiocarbon ages, stratigraphy, and associated archaeological remains. The most important of these include: Blue Fish Caves, Yukon Territory (Cinq-Mars and Morlan 1999); Old Crow Flats (Bonnichsen 1979); Morlan (1980); Manis, Washington (Gustafson et al. 1979); Wilson Butte Cave, Idaho (Gruhn 1995); Pendejo Cave, New Mexico (Chrisman and MacNeish 1996; MacNeish 1991; MacNeish 1993; MacNeish and Libby 2003); Lamb Spring, Colorado (Rancier et al. 1981); Dutton, Colorado (1979, 1981); Selby, Colorado (1979); Schaefer, Wisconsin (Overstreet this volume); Mud Lake, Wisconsin (Overstreet this volume); Fenske, Wisconsin (Overstreet this volume); Hebior, Wisconsin (Overstreet this volume), Schaefer, Wisconsin (Bonnichsen and Turnmire 1999; Holen 1994, 1995); La Sena, Shaffert and Jensen, Nebraska (Bonnichsen and Turnmire 1999; Holen 1994, 1995; May and Holen 1993); Lovewell, Kansas (Holen 1996); Big Eddy, Missouri (Lopinot et al. 1998, 2000); Burnham, Oklahoma (Wyckoff et al. 2003); Sand Creek, Texas (Johnson et al. 1994); Meadowcroft Rockshelter, Pennsylvania (Adovasio and Pedler this volume); Coats-Hines, Tennessee (Anderson this volume); Cactus Hill, Virginia (McAvoy and McAvoy 1997); Saltville, Virginia (MacDonald 2000); Topper, South Carolina (Goodyear this volume); Page-Ladson, Florida (Dunbar 2001); Sloth Hole, Florida (Dunbar 2001; Hemmings 1998); Little Salt Spring, Florida (Clausen et al. 1979). There are at least an additional 20 pre-Clovis sites reported from Mexico and South America (See Dillehay 2000; Gruhn this volume).

Contributors to this volume collectively make the point that our perception of human occupation in the mid-continent and eastern North America has significantly shifted over the last decade. From the 1970s through most of the 1990s, the archaeological record from Meadowcroft Rockshelter, Pennsylvania, was characterized by advocates of the Clovis-first model as an anomaly (Adovasio et al. this volume). Critics wondered how pre-Clovis ages reported from Meadowcroft could possibly be correct, since no other early sites had been found with artifacts in stratigraphic context that were of pre-Clovis age. Consequently, Meadowcroft received more than its "fair share" of criticism by those unwilling to accept evidence of a pre-Clovis colonization of the Americas.

Adovasio and Pedler (this volume) present an essentially sociological review of the profession's reaction to a site investigation that found itself in the middle of (and, indeed, helped to initiate) a paradigm shift over the last 25 years. The work by Adovasio and his team at Meadowcroft Rockshelter has the distinction of being the first to present stratigraphic evidence, radiocarbon chronology, and associated clear and unambiguous artifacts in context, all of which were said to be necessary proof for sites of pre-Clovis age. In their efforts to meet these standards, Adovasio's team challenged the premises of the Clovis-first model; see Goodyear (this volume) for a very brief summary. In their unique paper, Adovasio and Pedler review reactions to evidence as well as lack of evidence that has been part of the paradigm conflict and controversy surrounding work at Meadowcroft. The authors conclude that Meadowcroft provides one of the best examples of a pre-Clovis site in the eastern United States.

One of the "standards" long demanded by advocates of the Clovis-first model is that more than one site of pre-Clovis age needs to be found in an area to prove the existence of pre-Clovis populations. Their reasoning is that if humans reside in a region, there should be multiple sites with similar records. Thus, multiple sites with replicate ages and artifacts would prove the presence of a pre-Clovis occupation in the Americas. This standard has now been met at several locations.

Overstreet's discussion (this volume) focuses on human-mammoth interaction at four southeastern Wisconsin sites dated from 13,500 to 12,500 RCYBP. His research provides insights into human adaptations along the southern margin of the Laurentide Ice sheet. Investigations of the Schaefer, Hebior, Mud Lake, and Fenske sites provide insights into how local populations adapted to ice-marginal conditions. Mammoth bones from all four sites show modifications, which Overstreet (this volume) and Johnson (this volume) interpret as butchering marks, but chipped-stone artifacts were recovered only from Schaefer and Hebior. The lithic assemblage appears to be part of an unfluted biface tradition.

Overstreet proposes that these ice-marginal peoples used points like Hi-Lo (previously known, but presumed to be late Paleoamerican). These points are similar to the lanceolate points from Meadowcroft and Cactus Hill. Overstreet envisions that the early Paleoamerican occupation occurred in two "waves," one at 13,500 RCYBP followed by a climate change (Port Huron advance), then another at 12,400 RCYBP.

Johnson's study, "Late-Wisconsinan Mammoth Procurement patterns in the North American Grasslands" (this volume), provides additional evidence of the widespread occurrence of mammoth procurement in the North American grassland ecosystem well before the appearance of Clovis. Her work provides important insights into human subsistence practices before the appearance of the Clovis pattern as well as during Clovis times. The Clovis/pre-Clovis dichotomy traditionally has emphasized the abrupt change and sudden appearance of Clovis in conformity with the Clovis-first model. Yet, with respect to mammoth procurement, Johnson's data support a continuity model envisioning late-Pleistocene peoples who relied upon mammoth meat and materials that were found throughout the American grasslands beginning about 18,000 RCYBP. These practices are said to have persisted until the extinction of the mammoth about 10,800 RCYBP.

Johnson's review of mammoth bone modification relies on 13 study sites that included the presence of culturally modified mammoth bone from dated stratified deposits in the Mexican, Southern, and Northern grasslands. Sites studied included: Mud Lake (ca. 13,530 RCYBP), Wisconsin; Fenske (ca. 13,530 RCYBP), Wisconsin; Schaefer (ca. 12,300 RCYBP), Wisconsin; Lubbock Lake (ca. 11,000 RCYBP), Texas; Sand Creek (13,450 RCYBP), Texas; Blackwater Draw locality 1, gray sand, (ca. 11,630–11,040), New Mexico; Blackwater Draw locality 1, sand wedge (ca. 10,800 RCYBP), New Mexico; Miami (ca. 11,400–10,350 RCYBP), Texas; Domebo (ca. 11,200 and 11,045 RCYBP), Oklahoma; Duewall-Newberry (terminal Pleistocene), Texas; Bonfire Shelter, stratum E (12,400–10,200 RCYBP), Texas; and Tocuila (11,188 RCYBP), Mexico.

Using a taphonomic approach for analyzing cutmarks, push marks, and fractures inferred to have been produced by cutting, levering, and high-velocity impacts produced by hammerstones, Johnson distinguishes quarrying, scavenging, and butchering activities. She concludes that late-Wisconsin grassland peoples had a broad-spectrum mammoth meat and materials subsistence economy based on encounter hunting of mammoths. Prior to ca. 13,000 RCYBP mammoth bones were quarried without direct evidence of butchering marks (cutmarks). The lack of cutmarks suggests to Johnson that scavenging may have occurred prior to 13,000 RCYBP and that the earliest mammoth hunting did not begin until about 12,000 RCYBP. Alternative interpretations are possible given our incomplete understanding of mammoth procurement and utilization practices during the last glacial cycle.

Other taphonomic studies of late-Pleistocene fauna in the Yukon Territory (Bonnichsen 1979; Cinq-Mars and Morlan 1999; Morlan 1980) and the Great Plains (Stanford 1978, 1981) underscore the importance of a taphonomic approach for recognizing the presence of humans at pre-11,600 RCYBP sites. Collectively, these studies suggest early populations with repertoires for modifying large mammal bones extended from the northern Yukon Territory to the Valley of Mexico prior to the appearance of Clovis.

Many early sites yielding bones altered by humans often have few or no flaked-stone tools and are regarded by some as suspect. "What if" arguments proposing nonhuman agencies are commonly used to counter human alteration arguments (Bonnichsen and Schneider 1999). Although the human versus natural modification issues will continue to be prominent in Paleoamerican studies, new evidence is emerging that supports the case for an early bone technology. There are now several examples of radiocarbon-dated pre-Clovis-age sites with good stratigraphy and with which modified bones and flaked-stone tools are associated (e.g., Blue Fish Caves, and the Hebior, Schaefer, and Saltville sites). Examples of Clovis-age sites with flaked mammoth bone include Owl Cave, Idaho (Miller 1989), Lange/Ferguson, South Dakota (Hannus 1983, 1989), and Hiscock, New York (Tomenchuck and Laub 1995). There is another group of sites of pre-Clovis age with flaked mammoth bones modified by humans that lack stone tools. These include La Sena, Jensen, and Shaffert sites of Nebraska (Holen 1994 1995, May and Holen 1993); the Lovewell site of Kansas (Holen 1996); and the Duewall-Newberry site of Texas (Steele and Carlson 1989).

Some specialists cite a protocol of standards that they insist a site must meet to qualify as a legitimate human pre-Clovis-age site. The list varies slightly from author to author (Sellars 1999; Toth 1990; Waters 2000). Such lists almost always include clear stratigraphy, radiocarbon age determinations, diagnostic artifacts, and an unambiguous association between specimens and radiocarbon age. Goodyear (this volume) and Anderson (this volume) synthesize research at four pre-Clovis-age sites (Meadowcroft Rockshelter, Pennsylvania; Saltville, Virginia; Cactus Hill, Virginia; and Topper, South Carolina) that appear to meet these traditional standards. Of these, Cactus Hill, located along the Nottoway River near Richmond, meets an additional standard. It provides archaeological evidence of a pre-Clovis occupation in deposits that occur stratigraphically below a Clovis level.

Cactus Hill (Goodyear, this volume) is a stratified multicomponent site in a sand dune overlooking the Nottoway River on the interior Coastal Plain of Virginia excavated by teams of specialists led by Joseph McAvoy (McAvoy and McAvoy 1997) and Michael Johnson. In an excavation of several hundred square meters, a well-documented and -dated archaeological sequence was established in the upper 1.0 m of dune deposit. The site contains a conventional sequence beginning with Clovis (10,920 ± 250 RCYBP) and ending with historic material. An archaeological component that occurs below the Clovis horizon contains an assemblage of quartzite flakes, prismatic blades, and polyhedral cores. McAvoy suggests that two separate sub-Clovis lithic horizons are present: the lower contains prismatic blades followed by a zone of smaller blades. Two basally thinned trianguloid-to-lanceolate bifacially thinned points were found associated with the upper horizon. Pine charcoal from this level yielded ^{14}C ages of 15,070 ± 70 RCYBP and 16,670 ± 730 RCYBP. An additional sample collected in 1997–1998 yielded an age of 16,940 ± 50 RCYBP. Radiocarbon dating of charcoal, sedimentology, and optically stimulated thermoluminescence (OSL) dating of quartz sands all support an in situ stratigraphic sequence.

A second locality that contains evidence of a non-Clovis occupation below a Clovis level is the Topper site (Goodyear, this volume). It is a stratified multicomponent site situated on a Pleistocene terrace of the Savannah River and located adjacent to a chert outcrop and quarry area. The artifact-bearing sediments are represented in two stratigraphic units. The upper 1.0 to 1.4 m of sands are colluvial slopewash and contain a nearly complete archaeological sequence beginning with Clovis and ending with material from the 18th century. From a depth of 1.4 m and continuing to 2.2 m below surface are sands that are alluvial in origin. Unusual fractured chert and quartz specimens that may be artifacts occur primarily in the lower half of the unit. Goodyear characterizes these materials as a microlithic industry. Elements of this assemblage are small chert flakes, burin spalls, microblades, and microblade cores. The upper portion of the alluvial sand has been dated to between 15,000 and 16,000 CALYBP. Analysis of Topper is in progress.

Excavation by Jerry MacDonald (MacDonald 2000) at the Saltville site, located in the Saltville Valley of Virginia, exposed a detailed biostratigraphic record in a colluvial and lacustrine section approximately 1.0 m thick, which is supported by multiple ^{14}C dates. In the lower part of the deposits occur at least three discrete dated horizons ranging in age from 14,510 to 13,000 RCYBP, and each contains archaeological evidence of human occupation. From the oldest horizon came a musk ox tibia tool (*Bootherium bombifrons*) shaped by percussion flaking and altered by use-wear polish. The tool yielded a ^{14}C AMS age of 14,510 RCYBP. Additional evidence from this part of the deposit includes proboscidean bone, possibly from mastodon, that shows evidence of butchering and burning. Additionally, a partially worn chert flake and chunk of chert with possible use-wear patterns are associated. The middle horizon yielded 12 pieces of microdebitage that could be from bifacial retouch. The most recent level contained midden-like deposits with over 200 shells of clam, 500 pieces of small vertebrate bones and teeth, charcoal, and 125 pieces of microdebitage, including chert flakes. MacDonald infers from the faunal remains that people were harvesting clams and collecting fish and amphibians in ponds during periods of low water levels.

From the research summarized above, it is becoming difficult to argue that humans were not present in the eastern half of North America well before the appearance of Clovis. During an August 1999 workshop at the Smithsonian Institution organized by David Matson and Dennis Stanford, artifacts from Topper, Meadowcroft, and Cactus Hill, as well as Siberian and several North American Paleoamerican sites were exhibited by site principal investigators. This occasion allowed conference participants to make visual comparative inspections of important pre-Clovis specimens for the first time. Although there are differences between Meadowcroft and Cactus Hill artifacts, most participants were impressed with the similarities between the projectile points, core forms, and blade-like flakes, which led some to suggest a cultural connection between the peoples who produced the Meadowcroft and Cactus Hill specimens.

South America

Gruhn (this volume) provides an overview of the Paleoamerican archaeological record from South America and its implication for the Clovis-first model. Also, see Dillehay (2000) for a complementary perspective. Gruhn indicates that several corollary restrictions followed from the imposition of the North American model of early prehistory upon South American early archaeological remains. The Clovis-first model proposed that specialized big-game hunters from the United States spread all the way to the Straits of Magellan and would have arrived there approximately 10,600 RCYBP. The South America record provides two important tests of the Clovis-first model: (1) sites older than 10,700 RCYBP challenge the assumption that Clovis people were the initial colonists of South America; (2) the North American evolutionary model of a basal Paleoindian stage followed by an Archaic stage is not applicable to the South American archaeological record.

Early South American sites suggest people had arrived in South America before 35,000 RCYBP. One is Monte Verde I, which is located on an older and higher terrace that occurs stratigraphically above the 12,500 RCYBP Monte Verde II occupation (Dillehay 1997). In a 1985 test pit, charcoal (possibly from a hearth) found associated with 26 lithic artifacts yielded radiocarbon ages of 33,000 RCYBP. Another important early South American site is Pedra Furada (Guidon 1989, 1990; Guidon and Delibras 1990; Guidon et al. 1996). It is a large stratified rockshelter near the town of São Raimundo Nonato in the state of Piaui in northeast Brazil. Throughout the lower 4 m of deposit, quartz flake and pebble tools were recovered in association with large burned areas (possibly hearths) with charcoal concentrations. A series of finite radiocarbon dates from these deposits yielded dates ranging from 32,000 to 17,000 RCYBP. The lowest deposits yielded infinite dates greater than 40,000 RCYBP. Meltzer et al. (1994) have challenged the work of Guidon and her team by constructing "what if" scenarios, including rock fall, to account for the flaked pebble and cobble tools. Additionally, they propose forest fires to account for the charcoal. These arguments have led many to regard Pedra Furada as equivocal, and it has been largely relegated to a footnote in the ongoing discussion about the peopling of the Americas. Gruhn concludes that the antiquity of Monte Verde I and the early components at Pedra Furada suggest humans were in South America prior to 35,000 RCYBP. This antiquity provides sufficient time depth to allow for gradual population growth giving rise to the diversity of cultural traditions found in each major environmental area by the end of the Pleistocene.

Gruhn (this volume) observes that by the time Clovis appeared in North America, populations in South America already had highly diversified economies and technologies. For example,

Pedra Pintada, located in the Amazon rain forest of Brazil, was occupied by the last millennium of the Pleistocene (Roosevelt et al. 1996). At Quebrada Jaguay (Sandweiss et al. 1998), located on the southern coast of Peru, site occupants had developed a specialized maritime economy by 11,000 RCYBP. Less well known is research in southern Argentina by Laura Miotti and her colleagues (Miotti and Salemme this volume) at the Piedra Museo locality. The occurrence of a small paleolake system, extensive grasslands, and good toolstone drew people to a well-stratified rockshelter (Alero El Puesto). Unit 6, the lowest stratigraphic unit, produced a radiocarbon age of 12,890 ± RCYBP. Other dates from the lower part of Unit 6 were on charcoal (11,000 ± 65 RCYBP) and bone (10,925 ± 65 RCYBP). The artifact assemblage is composed of flaking debitage, including bifacial thinning flakes and retouched flakes, and a polished bone point. Flaked-stone projectile points were absent. Pleistocene species include a small camelid (*Lama gracilis* sp.), horse (*Hippidion saldiasi*), and sloth (mylodontids). These remains suggest to Miotti and colleagues that these archaeological remains represent the initial occupation of the Patagonia grasslands. Moving from Argentina to the upland interior of eastern Brazil, a region of broad valleys covered by savanna-parkland rich in edible plants and game, researchers have reported occupations at half a dozen early rockshelters. Lapa do Boquete is the best described of the lot and has yielded five radiocarbon dates ranging in age from 11,440 to 12,070 RCYBP. Remains from the basal level include small-size game, fish, freshwater mussels, land snails, and palm nuts associated with an assemblage of unifacial knives, scrapers, and limaces. These remains from basal level VIII indicate a generalized foraging economy. In central Peru, at an elevation of 4,300 m on the Puna Grassland, the basal occupation of Pachamachay Cave produced a radiocarbon age of 11,800 ± 930 RCYBP. Apparently humans were hunting camelids at high elevations soon after deglaciation.

Several other important South American pre-Clovis sites provide insight into how sites dated earlier than Clovis have been dismissed out of hand. Following up on the work of Cruxent, Bryan and Gruhn initiated excavations at Taima-taima, Venezuela, during the 1970s (Ochsenius and Gruhn 1979). In their excavations, they found the remains of a juvenile mastodon in a sealed stratigraphic context with two flaked artifacts, a median section fragment of a quartzite El Jobo point, and a flake in direct association with the skeleton. Rodofo Casamiquella, project paleontologist, observed modifications he interpreted as cutmarks on several ribs and on a humerus as well as evidence of purposeful disarticulation. Additionally, masses of masticated fragments of twigs, thought to be digesta or stomach contents of the mastodon, yielded four radiocarbon dates of ca. 13,000 RCYBP. Although Taima-taima fits the North American big-game hunting model, the projectile points are not Clovis and the radiocarbon dates are earlier than predicted by the Clovis-first model. Although the evidence from Taima-taima appeared to refute the Clovis-first model, critics developed a series of "what if" scenarios, including the suggestions that the spring had caused stratigraphic disturbance and that the leaf-shaped Lerma-like points came from a higher stratigraphic level. Gruhn discounts such speculations as unfounded and believes that Taima-taima has been unfairly relegated to the "equivocal and controversial" category.

At the Tibitó site, located in an ancient lake basin in the highlands of Colombia, Gonzalo Correal excavated fragmented mastodon, horse, and white-tailed deer remains. The associated lithic assemblage consisting of 120 lithic artifacts included small amorphous chunks of chert with minimal edge retouch; this assemblage is known as the Abriense industry in Colombia. The Tibitó site produced a radiocarbon age of 11,740 ± 110 RCYBP. Although the radiocarbon age is earlier than predicted by the Clovis-first model, it has not been challenged. At least one critic (Lynch 1990:17) indicated that the lack of projectile points was due to sampling error and that the Abriense industry should be Archaic in age. The Tibitó site did not fulfill the expectations of North American archaeologists, even though it produced a simple uniface industry associated with extinct species. It failed to match the predictions of the Clovis-first model and therefore was dismissed as uncertain.

Monte Verde is one of the most important and controversial sites in the Clovis-first debate (Dillehay 1989, 1997). It is this southern Chile site that was the final "straw that broke the back" of the Clovis-first model. Here, organic materials preserved under a later peat deposit provided the first glimpse into many aspects of past lifeways not previously seen at other Paleoamerican sites. Perishables include wooden frameworks for dwellings secured by stakes and cordage, fragments of hide cover, an array of wooden artifacts such as digging sticks, mortars with remains of edible plants, a fire drill, and hafts for stone tools, and a variety of plant remains from forest, wetland, Andean, and coastal environments. Moreover, modified tusk and bone remains from seven individual mastodons, several bifacial implements, and two long lanceolate projectile points indicate the presence of a non-fluted-point technology. Flaked-stone tools are not common, and many stone tools were naturally faceted or sharp-edged stones that show evidence of use. Organic materials from the site provided a suite of radiocarbon dates that average 12,500 RCYBP.

The exhaustive state-of-the-art interdisciplinary research conducted at Monte Verde involved 40 specialists who presented more lines of evidence than from any previous Paleoamerican site (Dillehay 1997). In January 1997, just before the second Monte Verde volume was scheduled to be released, a delegation of experts, which included staunch supporters of the Clovis-first model, examined artifacts in Lexington, Kentucky, and then traveled to Monte Verde for a site visit. By the end of the trip, there was a consensus that Monte Verde was valid and provided conclusive evidence of the failure of the Clovis-first model (Adovasio and Pedler 1997; Meltzer 1997; Meltzer et al. 1997).

Shortly before the 1997 Monte Verde announcement, geological work in the Richardson and MacKenzie Mountains (Catto 1996; Duk-Rodkin and Hughes 1991; Szeicz et al 1995; Jackson and Duk-Rodkin (1996); and Lemenon et al (1994) was reported that indicates the northern portion of the Ice-free Corridor route was blocked by glacial ice between 30,000 and 11,500 RCYBP. This work challenges the primacy of the Ice-free Corridor as a viable route for sites earlier than 11,500 RCYBP. Mandryk et al. (2001) note that biological productivity was low in the corridor during full glacial times and would not have sustained human or large-mammal populations. Burns's (2001) study of radiocarbon-dated extinct fauna from the Alberta portion of the Corridor supports Mandryk's contention and shows a clear gap in the faunal record from 11,600 to 21,000 RCYBP. The Ice-free Corridor could have been a colonization route that funneled vegetation, animals, and people between the far north and southern North America before its closure at 30,000 RCYBP. An Arctic Steppe fauna, including mammoth, horse, and bison, both preceded and followed the Late Glacial Maximum in the Alberta portion of the Corridor (Burns 1996).

Human presence in the Arctic Steppe biome is suggested by modification of bones from fauna reported from the Old Crow Basin (Bonnichsen 1979; Morlan 1980) and Blue Fish drainage (Cinq-Mars and Morlan 1999) and from ice marginal contexts in Wisconsin (Overstreet this volume) and possibly Nebraska and Kansas (Holen 1994, 1995; May and Holen 1993). Until recently, the earliest dated evidence from the Ice-free Corridor region came from the Vermillion Lake site in Banff, Alberta, where two dates on the earliest component of the site average 10,770 ± 175 RCYBP (Fedje 1985, 1986; Fedje et al. 1995:90); and Charlie Lake, British Columbia (Fladmark et al. 1988), with ages of 10,450 ± 160 (SFU 300), 10,380 ± 160 (SFU 378) 10,770 ± 120 (SFU 454) RCYBP. Kooyman's et al. (2001) work at the Corridor at the Wally's Beach site (DhPg-8) on the St. May's Reservoir, which occurs immediately south of the southern end of the Ice-free Corridor, provides the closest linkage between Arctic Steppe fauna and early human populations. This linkage was established through the discovery of horse blood residues on three Clovis fluted points. Although none of the points were found in situ, they were collected from an area producing footprints from Arctic Steppe fauna including mammoth, horse, bison, musk ox, and caribou. Collagen dates on Arctic Steppe fauna from this locality yielded ages from 11,000 to 11,350 RCYBP. At this juncture, it is not clear whether humans were moving south as postulated by both the Clovis-first model and the "revised Clovis-first model" (Haynes, this volume) or whether they were moving north from what is now the continental United States (Clark 1991).

Realizing that the Ice-free Corridor entry route will not adequately accommodate the earliest archaeological record from the Americas, many specialists are now searching for alternative routes for bringing the peoples to the Americas. Ideas originally formulated during the 1970s by Fladmark (1978, 1979, 1983) and rejected as implausible for many years are undergoing new consideration. Fladmark speculated that Northeast Asian colonists with a maritime adaptation using small boats coast-hopped along the Pacific Rim from Asia to the Americas during the terminal phase of the last Ice Age. Unlike interior routes that were covered by glacial ice, coastal routes were open between 14,000 and 11,500 RCYBP. In fact, Dixon (1999) and Mandryk et al. (2001) suggest that the Bølling warming interval may have facilitated such a move through increased productivity in the littoral zone and may have provided an "optimal interval" for humans and animals to migrate to the New World during the critical late-Pleistocene period. This does not, however, exclude the possibility that colonization by Eurasian populations could have occurred at earlier times. For example, in their North Atlantic model, Stanford and Bradley propose that populations moved along the southern edge of the North Atlantic ice shelf during full glacial times. Early Eurasian populations using small watercraft also could have moved along the edge of the Beringian Ice Shelf (Grosswald 1999), finding dry land areas between piedmont glaciers along the coast of the North Pacific. Coastal and ice marginal fringes are localities of high biological productivity. It is possible that humans could have moved along high-latitude coastal areas during glacial and interstadial times perhaps as early as 40,000 years ago, as has been suggested by Gruhn (1994).

Critics of the Pacific Rim model have long argued there is no direct evidence that people came to the New World via watercraft. The most difficult problem is that most coastal areas where humans might have resided have since been drowned by rising sea levels. Also, no early boats have been found and are not likely to be found owing to the perishable nature of organic materials used to construct boats.

Despite these obstacles, Margaret Jodry (this volume) makes a compelling case for the use of Pleistocene watercraft using indirect evidence. Ethnographic studies indicate that boats and water transport technology have a near worldwide distribution. Archaeological remains from Australia, an "island continent," date at least to 40,000 RCYBP, and obsidian from an island off the coast of Honsu Island, Japan, comes from an occupation site dated to 30,000 RCYBP. These data imply the presence of boats in the Southern and Northern Hemispheres by mid-Wisconsinan times. The occurrence of obsidian on islands from mainland sources along the North Pacific coast; the presence of chert from the lower levels of Daisy Cave, San Miguel Island, California; and an occupation at Monte Verde, Chile, dating to 12,500 RCYBP (a time when the Ice-free Corridor was closed) are indicators boats were used to colonize the Americas. Not only were boats important for group mobility, they would have enabled early populations to exploit marine, estuarine coastal environments, riverine, and lake environments in the interior. Collectively, the data and interpretations demonstrate that boats are likely to be far more significant to understanding the expansion of our species through maritime colonization than previously recognized. In the past, proponents of the Clovis-first model assumed that the Americas were peopled by pedestrian big-game hunters. This assumption was integrated into local and regional models of how humans related to the land. Jodry suggests that too much emphasis has been placed on pedestrian models and that a new emphasis should explore the role of watercraft in transport, communication, and paleoeconomic patterns in our search for understanding past human adaptive systems.

The articles presented in this volume make it abundantly clear that there is now a substantial evidentiary record from First American sites in North and South America supporting the view that the Americas were peopled before the appearance of Clovis. The Clovis-first model productively guided research for more than three decades, but it now has been falsified. The demise of the Clovis-first model leaves a void that challenges model builders to construct new Paleoamerican origin models that will encompass the wealth of archaeological and paleobiological data presented by the contributors to this volume. This is an exciting time to be studying the peopling of the Americas.

References Cited

Adovasio, J. M., and D. R. Pedler
1997 Monte Verde and the Antiquity of Humankind in the Americas. *Antiquity* 71:573–80.

Anderson, D. G., and J. C. Gillam
1999 Paleoindian Colonization of the Americas: Implications from an Examination of Physiograpy, Demography, and Artifact Distribution. *American Antiquity* 65(1):43–66.

Bonnichsen, R.
1979 Pleistocene Bone Technology in the Beringian Refugium. *National Museum of Man Mercury Series, Archaeological Survey of Canada* No. 89, 1–297. Ottawa.

Bonnichsen, R. (Editor)
1999 *Who Were the First Americans?* Center for the Study of the First Americans, Oregon State University, Corvallis, OR.

Bonnichsen, R., and K. L. Turnmire
1999 An Introduction to the Peopling of the Americas. In *Ice Age Peoples of North America*, edited by R. Bonnichsen and K. L. Turnmire, pp. 1–26. Oregon State University Press, Corvallis, OR.

Bonnichsen, R., and A. L. Schneider
1999 Breaking the Impasse on the Peopling of the Americas. In *Ice Age Peoples of North America*, edited by R. Bonnichsen and K. L. Turnmire, pp. 497-519. Oregon State University Press, Corvallis, OR.

2001 The Case for a Pre-Clovis People. *American Archaeology* 5(4):35–39.

Bonnichsen, R., D. Stanford, and J. L. Fastook
1987 Environmental Change and Developmental History of Human Adaptive Patterns: The Paleoindian Case. In *North American and Adjacent Oceans During the Last Deglaciation*, edited by W. F. Ruddiman and H. W. Wright, Jr., pp. 403–24. *The Geology of North America, DNAG, V. K-3*. Geological Society of American, Boulder, CO.

Bryan, A. L
1986 Paleoamerican Prehistory as Seen from South America. In *New Evidence for the Pleistocene Peopling of the Americas*, edited by A. L. Bryan, pp. 1–14. Center for the Study of Early Man, University of Maine, Orono, ME.

Burns, J. A.
1996 Vertebrate Paleontology and the Alleged Ice-Free Corridor: The Meat of the Matter. *Quaternary International* 32:107–112.

Catto, N. R.
1996 Richardson Mountains. Yukon-Northwest Territories: The Northern Portal of the Postulated "Ice-free Corridor." *Quaternary International* 32:3–19.

Chandler, J.
2001 Pleistocene Lake Lahontan: Filling in More Blanks. *Mammoth Trumpet* 17(1):1–3, 17–20.

2001 Immigrants from the Other Side? *Mammoth Trumpet* 17(1):11-16.

Cinq-Mars, J., and R. E. Morlan
1999 Bluefish Caves and Old Crow Basin: A New Rapport. In *Ice Age Peoples of North America*, edited by R. Bonnichsen and K. L. Turnmire, pp. 200–12. Oregon State University Press, Corvallis, OR.

Chrisman, D., and R. S. MacNeish
1996 Late Pleistocene Human Friction Skin Prints from Pendejo Cave, New Mexico: *American Antiquity* 61:357–76.

Clark, D. W.
1991 The Northern (Alaska-Yukon) Fluted Points. In *Clovis: Origins and Adaptations* edited by R. Bonnichsen and K. L. Turnmire, pp. 15–34. Center for the Study of the First Americans, Oregon State University, Corvallis, OR.

Clausen, O. J. C., A. D. Cohen, C. Emmilani, J. A. Holman, and J. J. Stipp
1979 Little Salt Spring, Florida: A Unique Underwater Site. *Science* 203:609–13.

Collins, M. B.
1999 *Clovis Blade Technology*. University of Texas Press, Austin, TX.

Dillehay, T. D., Editor
1989 *Monte Verde: A Late Pleistocene Settlement in Chile I: Paleoenvironment and Site Context*. Smithsonian Institution Press, Washington, D.C.

1997 *Monte Verde: A Late Pleistocene Settlement in Chile I: The Archaeological Context*. Smithsonian Institution Press, Washington, D.C.

2000 *The Settlement of the Americas: A New Prehistory*. Basic Books, New York.

Dincauze, D. F.
1984 An Archeo-Logical Evaluation of the Case for Pre-Clovis Occupations. *Advances in World Archaeology* 3:275–323.

Dixon, J. A.
1998 *Bones, Boats, & Bison: Archaeology of Western North America*. The University of New Mexico Press, Albuquerque, NM.

1999 Coastal Navigators. *Discovering Archaeology* January/February 34–35.

Driver, J. C., M. Handly, K. R. Fladmark, D. E. Nelson, K. G. M. Sullivan, and R. Preston
1996 Stratigraphy, Radiocarbon Dating, and Culture History of Charlies Lake Cave, British Columbia. *Arctic* (3):265–77.

Duk-Rodkin, A., and O. L. Hughes
1990 Age Relationships of Laurentide and Montane Glaciations. Mackenzie Mountains, Northwest Territories. *Géographie Physique et Quaternarire* 45:78–90.

Dunbar, J.
2000 Aucilla River Prehistory Project. http://www.flmnh.ufl.edu/natsci/vertpaleo/arpp.htm.

Fedje, D.
1985 *Archaeological Investigations in Banff National Park*. Parks Canada Research Bulletin No. 236.

1986 Banff Archaeology 1983–1985. In *Eastern slopes Prehistory: Selected Papers*, edited by B. M. Ronaghan, pp. 25–62. Archaeological Survey of Alberta Occasional Paper No. 30.

Fedje, D. W., J. M. White, M. C. Wilson, D. E. Nelson, J. S. Vogel, and J. E. Southon
1995 Vermilion Lakes Site: Adaptations and Environments in the Canadian Rockies During the Latest Pleistocene and Early Holocene. *American Antiquity* 60(1):81–108.

Fladmark, K. R.
1978 The Feasibility of the Northwest Coast as a Migration Route for Early Man. In *Early Man in America from a Circum-Pacific*

Perspective, edited by A. L. Bryan, pp. 119–128. Department of Anthropology, Occasional Paper 1, Edmonton, Alberta.

1979 Routes: Alternative Corridors for Early Man in North America. *American Antiquity* 44:55–69.

1983 Times and Places: Environmental Correlates of Mid-to-late Wisconsinan Human Populations Expansion in North America. In *Early Man in the New World*, edited by R. Shutler, pp. 13–41. Sage Publications, Beverly Hills, CA.

Fladmark, K. R., J. C. Driver, and D. Alexander
1987 The Paleoindian Component at Charlie Lake Cave (HbRf 39), British Columbia. *American Antiquity* Vol. 53 (2):375.

Frison, G. C., and R. Bonnichsen
1996 The Pleistocene-Holocene Transition of the Plains and Rocky Mountains of North America. In *Humans at the End of the Ice-Age: The Archaeology of Pleistocene-Holocene Transition*, edited by L. G. Straus, B. V. Eriksen, J. M. Erlandson, and D. R. Yesner, pp. 303–18. Plenum Press, New York, NY.

Gruhn, R.
1995 Results of New Excavations at Wilson Butte Cave, Idaho. *Current Research in the Pleistocene* 12:16–17.

1994 The Pacific Coast Route of Initial Entry: An Overview. In *Method and Theory for Investigating the Peopling of the Americas*, edited by R. Bonnichsen and D. G. Steele, pp. 249–56. Center for the Study of the First Americans, Oregon State University, Corvallis, OR.

Guidon, N.
1989 On Stratigraphy and Chronology at Pedra Furada. *Current Anthropology* 30 (5):641–42.

1990 The Chrononogy of the New World: Two Faces of One Reality. *World Archaeology* 23(2):167–78.

Guidon, N., and G. Delibrias
1991 Carbon-14 Dates Point to Man in the Americas 32,000 Years Ago. *Nature* 321:769–71.

Guidon, N., A. M. Pessis, F. Parenti, M. Fontugue, and C. Guerin
1995 Nature and Age of the Deposits in Pedra Furada: Reply to Meltzer, Adovasio and Dillehay. *Antiquity* 70:408–21.

Gustafson, C. E, D. Gilbow, and R. D. Daugherty
1979 The Manis Mastodon Site: Early Man on the Olympic Peninsula. *Canadian Journal of Archaeology* 3:157–64.

Hall, D. G.
1995 Bones of Nebraska Mammoths Imply Early Human Presence. *Mammoth Trumpet* 10(1):1, 4–7.

Hannus, L. A.
1983 *Lange/Ferguson (39SH33): A Clovis Mammoth Kill/butchery Site.* Ph.D. Dissertation, Department of Anthropology, University of Utah. Salt Lake City.

1989 Flaked Mammoth Bone from the Lange/Ferguson Site, South Dakota. In *Bone Modification*, edited by R. Bonnichsen and M. H. Sorg, pp. 381–94. Center for the Study of the First Americans, University of Maine, Orono.

Haynes, C. V., Jr.
1964 Fluted Projectile Points: Their Age and Dispersion. *Science* 145:1408–13.

1966 Elephant-hunting in North America. *Scientific American* 214:104–12.

1980 The Clovis Culture. *Canadian Journal of Anthropology* 1:115–21.

Hemmings, C. A.
1997 Probable Association of Paleoindian Artifacts and Mastodon Remains from Sloth Hole, Aucilla River, North Florida. *Current Research in the Pleistocene* 15:16–18.

Holen, S.
1994 Did Someone Eat the La Sena Mammoth? In Cellars of Time. Paleontology and Archaeology in Nebraska. *Nebraskaland Magazine* 72(1):88.

1995 Evidence of the First Humans in Nebraska, pp. 1–6. *Museum Notes* 90, University of Nebraska State Museum, Lincoln.

1996 The Lovewell Mammoth: A Late Wisconsinan Site in North-Central Kansas. *Current Research in the Pleistocene* 13:69–70.

Jackson, L. E., Jr., A. Duk-Rodkin
1997 Quaternary Geology of the Ice-Free Corridor: Glacial Controls on the Peopling of the New World. In *Prehistoric Mongoloid Dispersersals*, edited by T. Akazawa and E. J. E. Szathmary, pp. 214–27. Oxford University Press, Oxford.

Johnson, E., L. Litwinionek, and V. T. Holliday
1994 The Sand Creek Mammoth Site, Llano Estacado of Texas. *Current Research in the Pleistocene* 11:60–72.

Kooyman, B., M. E. Newman, C. Cluney, M. Lobb, S. Toman, P. McNeil, and L. V. Hill
2001 Identification of Horse Exploitation by Clovis Hunters Based on Protein Analysis. *American Antiquity* 66(4):686–91.

Lemmon, D. S., A. Duk-Rodkin, and J. N. Bednarski
1995 Late Glacial Drainage Systems along the Northwestern Margin of the Laurentide Ice Sheet. *Quaternary Science Review* 13:805–28.

Lepper, B. T.
2002 Reading between the Lines: Drawing the boundaries of Paleoindian Culture Areas in Midcontinental North America. In *The Paleoindian Period in Pennsylvania,* edited by K. Carr, Pennsylvania Archaeological Council.

Lopinot, N. H., J. H. Ray, and M. D. Conner
1998 *The 1997 Excavations at the Big Eddy Site (23CE426) in Southwest Missouri.* Center for Archaeological Research, Southwest Missouri State University, Special Publication No. 2, Springfield, MO.

2000 *The 1999 Excavations at the Big Eddy Site (23CE426)*. Center for Archaeological Research, Southwest Missouri State University Special Publication, No. 3, Springfield, MO.

Lynch, T. F
1990 Glacial-Age Man in South America: A Critical Review. *American Antiquity* 55(1):12–36.
2001 Commentary: On the Road Again . . . Reflections on Monte Verde. *The Review of Archaeology* 22(1):39–43.

Mandryk, C. A. S, H. Josenhans, D. W. Fedje, and R. W. Mathewes
2002 Late Quaternary Paleoenvironments of Northwestern North America: Implications for Inland versus Coastal Migration Routes. *Quaternary Sciences Reviews* 20:301–14.

May, D. W., and S. R. Holen
1992 Radiocarbon Ages of Soils and Charcoal in Late Wisconsinan Loess, South-Central Nebraska. *Quaternary Research* 39:55–58.

McDonald, J. N.
2003 An Outline of the Pre-Clovis Archeology of SV-2, Saltville, Virginia, with Special Attention to a Bone Tool Dated 14,510 yr B.P. *Jeffersonia: Contributions from the Virginia Museum of Natural History* 9:1–59.

MacNeish, R. S.
1991 The Fort Bliss Archaeological Project by Afar Excavation of Pintada and Pendejo Caves Near Orogrande, New Mexico. February 3–May 11, 1–34. 1991b. Andover MA, Andover Foundation for Archaeological Research.

1996 Pendejo Pre-Clovis Proofs and Their Implications. Fumdhamentos, Proceedings of the International Meeting on the Peopling of the Americas São Raimundo Nonato, Piaui, Brasil (1993). Revista da Fundacao Museu do Homenm Americano, V.I. No. 1, São Raimundo Nonato Brasil.

MacNeish, R. S. and J. G. Lilly, editors
2003 *Pendejo Cave.* University of New Mexico Press, Albuquerque.

Martin, P. S.
1967 Prehistoric Overkill. In *Pleistocene Extinctions: The Search for a Cause*, edited by P. S. Martin and H. E. Wright, Jr., pp. 75–120. Yale University Press, New Haven, CT.
1973 The Discovery of America. *Science* 117:969–73.
1984 Prehistoric Overkill: The Global Model. In *Quaternary Extinctions: A Prehistoric Revolution*, editors P. S. Martin and R. G. Klein, pp. 354–403.University of Arizona Press, Tucson.
1988 Clovisia the Beautiful! *Natural History* pp. 10–13.

McAvoy, J. M., and L. D. McAvoy
1996 *Archaeological Investigations of Site 44SX202, Cactus Hill, Sussex County Virginia.* Research Report No. 8, Commonwealth of Virginia, Department of Historic Resources, Richmond.

Meltzer, D. J.
1997 Monte Verde and the Pleistocene Peopling of the Americas. *Science* 276:276–755.

Meltzer, D. J., J. M. Adovasio, and T. D. Dillehay
1993 On a Pleistocene Human Occupation at Pedra Furada, Brazil. *Antiquity* 68:695–714.

Meltzer, D. J., D. K. Grayson, G. Ardila, A. W. Barker, D. F. Dincause, C. V. Haynes, F. Mena, L. Nunez, and D. J. Stanford
1997 On Pleistocene Antiquity of Monte Verde, Southern Chile. *American Antiquity* 62(4):659–63.

Miller, S. J.
1989 Characteristics of Mammoth Bone Reduction at Owl Cave, the Wasden Site, Idaho. In *Bone Modification*, edited by R. Bonnichsen and M. H. Sorg, pp. 381–94. Center for the Study of the First Americans, University of Maine, Orono.

Morlan, R. E.
1980 *Taphonomy and Archaeology in the Upper Pleistocene of the Northern Yukon Territory: A Glimpse of the Peopling of the New World.* National Museum of Man Mercury Series, Archaeological Survey of Canada Paper No. 94. Ottawa.

Mosimann, J. F., and P. S. Martin
1975 Simulating Overkill by Paleoindains. *American Scientist* 63(3):304–313.

Ochsenius, C., and R. Gruhn (editors)
1979 *Taima-taima: A Late Pleistocene Paleo-Indian Kill Site in Northernmost South America—Final Report of 1976 Excavation.* South American Quaternary Documentation Program, Universidad Francisco de Miranda, Coro, Venezuela. Printed in 1992 by the Center for the Study of the First Americans, Oregon State University, Corvallis.

Rancier, J, G. Haynes, and D. Stanford
1981 Investigations of Lamb Springs. *Southwestern Lore* 48(2):1–17.

Roosevelt, A. C., M. Lima da Costa, C. Lopes Machado, M. Michab, N. Mercier, H. Valladas, J. Feathers, W. Barnett, M. Imazio da Silveria, A. Henderson, J. Sliva, B. Chernoff, D. S. Reese, J. A. Holan, N. Toth, and K. Schick
1994 Paleoindian Cave Dwellers in the Amazon: The Peopling of the Americas. *Science* 272:373–84.

Sandweiss, D. H., H. McInnis, R. L. Burger, A. Cano, B. Ojeda, R. Parades, M. C. Sandweiss, and M. D. Glasscock
1998 Quebrada Jaguay: Early South American Maritime Adaptations. *Science* 281:1830–1832.

Sellars, V. S.
1999 The Accuracy of AMS 14C Dates on Human Bone and Associated Vertebrate Remains—Determining Chronologies for Late Pleistocene Human Occupation of the New World. In *Clovis and Beyond: A Peopling of the Americas Conference, Abstracts, Collections, and Exhibits*, edited by R. Bonnichsen, M. Punke, and C. Allred, Abstract 65. Center for the Study of the First Americans, Oregon State University, Corvallis.

Stanford, D.
1978 Archeological Investigations of the Selby and Dutton Mammoth Kill Sites, Yuma, Colorado. *National Geographic Society Research Reports* pp. 519–541.
1982 Archaeological Investigation of the Lamb Spring Site. *Southwestern Lore* 47:14–27.

Stanford, D., and B. Bradley
2000 The Solutrean Solution. *Discovering Archaeology* January/February: 54–55.

Steele, D. G, and D. L. Carlson
1989 Excavation and Taphonomy of Mammoth Remains from the Duewall-Newberry Site, Brazos County Texas. In *Bone Modification*, edited by R. Bonnichsen and M. H. Sorg, pp. 413–430. Center for the Study of the First Americans, University of Maine, Orono.

Straus, L. G.
2000 Soultrean Settlement of North America? *American Antiquity* (65(2):29–226.

Szeicz, J. M.
1995 Late Quaternary Vegetation History of the Central Mackenzie Mountains, Northwest Territories, Canada. *Palaeography, Palaeoclimatology, Palaeocology* 113:351–71.

Tomenchuck, J., and R. S. Laub
1996 New Insights into the Late Pleistocene Bone Technology at the Hiscock Site, Western New York State. *Current Research in the Pleistocene* 12:71–74.

Toth, N.
1991 The Material Record. In *The First Americans: Search and Research*, edited by T. Dillehay and D. J. Meltzer, pp. 53–76. CRC Press, Boca Raton, LA.

Waters, M. R.
2000 Proving Pre-Clovis. *Scientific American Discovering Archaeology* January/February, pp. 47–49.

West, F. H.
1997 The Archaeological Evidence. In *American Beginnings*, edited by F. H. West, pp.537–560. University of Chicago Press, Chicago, IL.

Willig, J. A
1989 Clovis Technology and Adaptation in Far Western North America: Regional Pattern and Environmental Context. In *Clovis: Origins and Adaptations*, edited by R. Bonnichsen and K. L. Turnmire, pp. 91–118. Center for the Study of the First Americans, Oregon State University, Corvallis.

Wycoff, D. G., J. L. Theler, and B. J. Carter
2004 The Burnam site in Northwestern Oklahoma: Glimpses beyond Clovis? Sam Noble Oklahoma Museum of Natural History, Oklahoma Anthropological Society, Memoir 9.

II. Paleoamerican Prehistory

Archaeological Evidence, North America

A Long View of Deep Time at Meadowcroft Rockshelter

J. M. Adovasio[1] and D. R. Pedler[2]

Abstract

Since the initiation of excavations in 1973, Meadowcroft Rockshelter has borne levels of visibility and notoriety far out of proportion with its relatively modest physical dimensions. By turns praised, vilified, and ignored in scholarly debate, Meadowcroft Rockshelter nonetheless remains one of the best candidates for the oldest locus of human occupation in North America and the longest record of human occupation in the Western Hemisphere. Whatever its antiquity, the site is widely viewed as the most carefully excavated and extensively documented site of its kind in North America. This contribution summarizes the history of work at Meadowcroft and assesses the implications of that work for the archaeology of Pennsylvania and North America.

Scholarly Perspectives on Meadowcroft Rockshelter

"It is unlikely that any other site in North America has been more carefully excavated or more extensively documented." (Custer 1996:92)

"Impeccable excavation procedures uncovered a long sequence of human occupation." (Lepper 1999:366)

"Whatever the precise date of first occupation, Meadowcroft is one of the earliest well documented archaeological sites in North America." (Fagan 1995:80)

"Although many candidates for pre-Clovis have not weathered the intense scrutiny to validate their claims, others have survived the test. Questions still arise about . . . Meadowcroft in Pennsylvania but [its] validity cannot be challenged on grounds of faulty data collection or analysis." (Frison 2000:42)

"The careful work at Meadowcroft has set a standard." (Fagan 1990:20)

[1,2]Mercyhurst Archaeological Institute, Mercyhurst College, Erie, PA 16546.

"The painstaking and complete research you are engaged in is without equal in the world." (Gary Haynes, pers. comm. 1976)

"I can find no holes in your data." (C. Vance Haynes, pers. comm. 1974)

"The Meadowcroft conclusions are not fully supported by the published data." (Dincauze 1989:141)

"The faunal remains in [Stratum] IIa are very limited and . . . uninformative about environments" (Dincauze 1989:139)

"There are few identifiable bones from the early levels and thus it is not surprising that no boreal or extinct animal species were recognized." (Lynch 1983:88)

"The strata from II to XI contain . . . fauna and flora of completely Holocene character." (Haynes 1980:582)

"The recovered fauna and flora are discordant with other local Wisconsinan-age fossils." (Mead 1980:579)

"The grand total (i.e., of wood, charcoal, seeds, and fruits) from the glacial age levels was a pitiful 11.9 grams. That's a hopelessly inadequate basis on which to judge whether the anomaly is real." (Meltzer 1993:73)

"A thin band of totally vitrinized wood of Pennsylvanian age is present at the rockshelter. . . . The 'thin band' of vitrinite is synonymous with a coal seam." (Tankersley et al. 1987:319)

"Unfortunately, the question of contamination in the lowest levels has not been resolved." (Tankersley et al. 1987:324)

"Arguments that the radiocarbon dates were contaminated by 'old' carbon from coal are not supported by the evidence." (Gowlett 1992:141)

"The fact that coal . . . decomposes and can form humates that are subsequently transported by groundwater remains a noteworthy point." (Tankersley and Munson 1992:323)

"Groundwater can transport humates for several kilometers." (Tankersley and Munson 1992:322)

"More significantly in terms of the Meadowcroft dating controversy, we see no evidence of the groundwater saturation of these sediments, nor were particulate coal materials visible in any sample." (Goldberg and Arpin 2000)

"Even the lowest [IIa] lithic artifact assemblage contains typical Early Holocene microblades, bifacial thinning flakes, biface fragments, flake knives and a typical Plano-type projectile point." (Haynes 1980:582)

"The cultural assemblage from lower Stratum IIa has been plausibly argued to represent a pre-Clovis Upper Paleolithic technology. There are no 'anomalous' later Paleoindian or Archaic artifacts associated with these dated early levels that might suggest substantial mixing or contamination." (Lepper 1999:366)

"I think that most archaeologists would now argue that critics have run out of objections and we must regard the Meadowcroft Rockshelter as a bona fide archaeological site predating 12,000 years ago." (Custer 1996:93)

"The age of Stratum IIa at Meadowcroft still remains unresolved." (Haynes 2001 [this volume])

"After 25 years since its first publication, Meadowcroft Rockshelter appears to be what the investigators originally claimed." (Goodyear 2001 [this volume])

Introduction

As these often diametrically opposed comments reveal, since the initiation of excavations in 1973, Meadowcroft Rockshelter has borne levels of visibility, notoriety, and rancorous controversy far out of proportion with its relatively modest physical dimensions. For those few happy souls still unaware of Meadowcroft or the debate surrounding it, herewith are a few details about the site and the research conducted there.

Meadowcroft Rockshelter is a deeply stratified multicomponent site located 48.3 km (30 mi) southwest of Pittsburgh, Pennsylvania, and 4 km (2.5 mi) northwest of Avella, Washington County, Pennsylvania. It is situated on the north bank of Cross Creek, a small tributary of the Ohio River (Figure 1), some 12.2 km (7.6 mi) east of the creek's confluence with that river. The site is a typical immature sandstone reentrant oriented roughly east-west with a southern exposure (Donahue and Adovasio 1990). It is located 15.1 m (49.4 ft) above Cross Creek and ca. 259.9 m (852.5 ft) above mean sea level (msl). The rockshelter's extant overhang covers an area of ca. 65 m^2 (699.4 ft^2) and stands ca. 13 m (42.6 ft) above the modern surface of the site.

Meadowcroft Rockshelter was originally discovered (and subsequently protected) by the late Albert Miller, whose family has owned the property continuously since 1795. The site was brought to the attention of J. M. Adovasio through the efforts of Miller and the late P. Jack, California State College, California, Pennsylvania. It became the focal point of a long-term multidisciplinary project, the most intensive field phase of which began in the summer of 1973 and concluded in the fall of 1978. Additional field work was conducted in 1982, 1983, and 1987, with major re-excavation of the site's Holocene deposits occurring in 1994–1995. The analysis and publication phase is ongoing and to date has generated some 85 articles, book chapters, monographs, and papers. The final report is in preparation. Owing to the unique multidisciplinary expertise brought to bear on every aspect of the research, facilitated in very large part by abundant financial support, the Meadowcroft excavations are widely considered to represent state-of-the-art closed-site excavations (cf. Custer 1996; Fagan 1990, 1995; Feder 1996).

The 11 colluvially and attritionally emplaced strata identified at Meadowcroft, which include Stratum I (the culturally sterile shale "basement") and 10 overlying lithostratigraphic units (i.e., Strata II–XI) of widely varying thickness and composition, extend to a maximum depth of 4.6 m (15.1 ft) and have produced what is presently the longest intermittent occupational sequence in the New World. These sediments yielded some 20,000 artifacts (principally flaked stone), over 150 fire pits (final tallies from the 1994–1995 excavation season are not yet available), 33 fire floors, 52 ash and charcoal lenses, 16 specialized activity areas, and 21 refuse/storage pits. The site's massive ecofactual yield includes nearly 1 million faunal remains and ca. 1.4 million plant remains.

The rockshelter excavations were complemented by archaeological reconnaissance of the Cross Creek drainage, which covered ca. 14,164.5 ha (35,000 acres) and identified 236 additional prehistoric sites. All these sites were surface collected, 22 were tested, and 2 were extensively excavated. The resultant corpus of data is as detailed a prehistoric record as currently exists for any comparable-size study area anywhere in the Americas, if not the world.

The Great Debate

Since the publication of the first radiocarbon dates from Meadowcroft Rockshelter (Adovasio et al. 1975), the site has been close to the epicenter of the debate about the timing of the arrival of the first migrants to the New World. It is one of two sites (the other being Monte Verde in Chile) most frequently cited as evidence that the 57-year-old "Clovis-first" colonization scenario is simply wrong. As with its South American contemporary, Meadowcroft has been the focus of concerted criticism, which, until recently, had addressed virtually every aspect of the site *except* the context and association of the dated material. For reasons detailed in many other publications (Adovasio 1993, Adovasio et al., 1979–1980a, 1979–1980b, 1980, 1984, 1985, 1988, 1989, 1990, 1992; Carlisle and Adovasio 1982), most of the criticism has been directed at the 13 radiocarbon dates older than ca. 12,800 ± 870 RCYBP, with the repeated assertion that all may be contaminated with particulate or non-particulate "dead" carbon (Dincauze 1981, 1984, 1989; Haynes 1980, 1991; Mead 1980; Tankersley and Munson 1992; Tankersley et al. 1987.

Forests of paper have been devoted to refuting these criticisms, the most recent examples of which (Adovasio et al. 1999; Goldberg and Arpin 2000) essentially close the book on the contamination issue. But characteristically, perhaps, the debate has not ended. It has now been suggested that the dates are, indeed, correct after all. The most recently derived problem with Meadowcroft instead lies with the Pleistocene artifacts and ecofacts from the site's deeper levels, which are out of context because the stratigraphy is hopelessly mixed—a fact somehow overlooked by the excavators and missed by the hundreds of archaeologists who have visited the site over the past 25 years.

Rather than respond to this latest assertion, we wish to offer a suggestion for why there is a continuing debate over Meadowcroft,

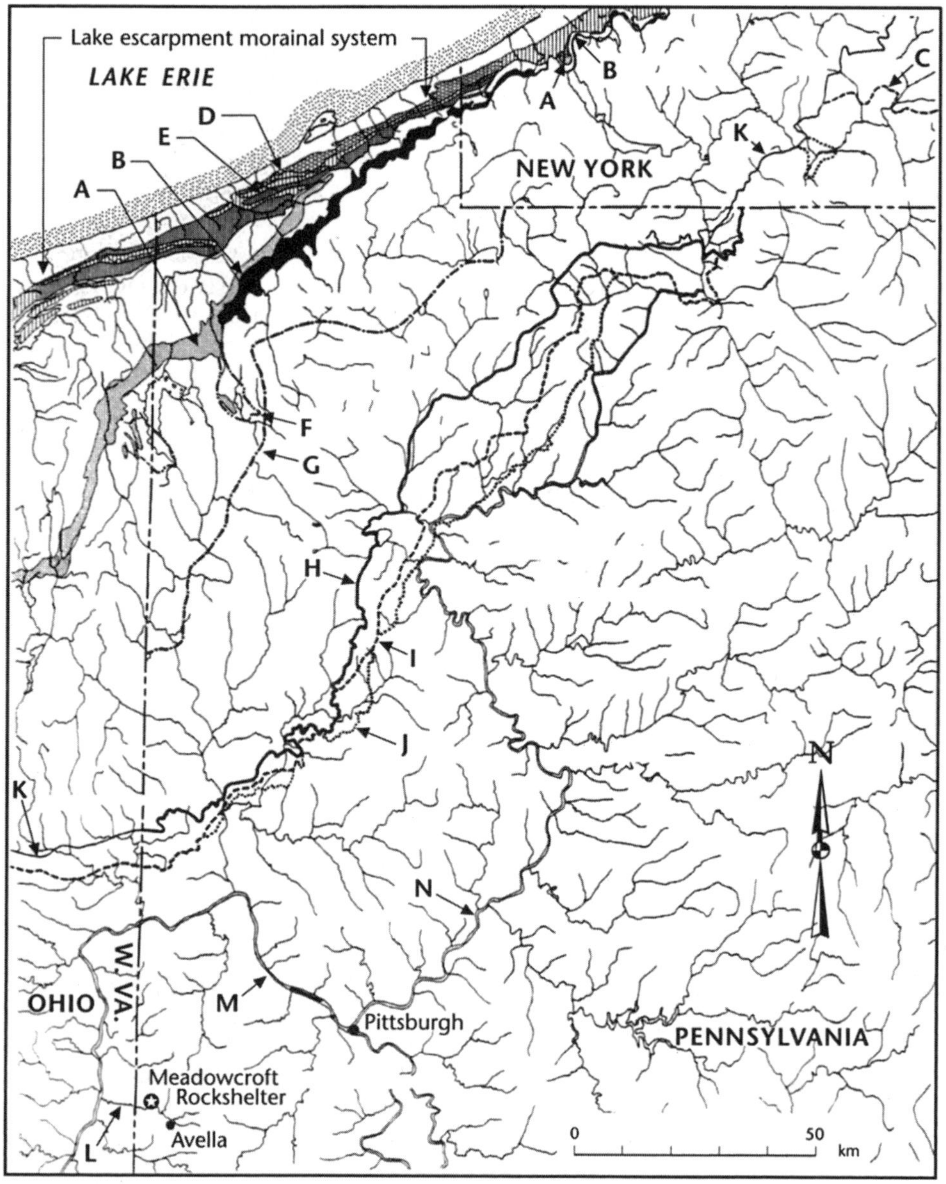

Figure 1. The upper Ohio River valley and contiguous regions, showing the maximum southward extension of the Wisconsinan glacial advance (after Johnson 1981). **A,** Defiance moraine; **B,** Lavery moraine; **C,** Olean till; **D,** Girard moraine; **E,** Ashtabula moraine; **F,** Hiram till maximum; **G,** Lavery till (outer discontinuous border); **H,** Kent till maximum; **I,** Titusville till maximum; **J,** Mapledale till maximum; **K,** "mapped" Kent moraine; **L,** Cross Creek; **M,** Ohio River; **N,** Allegheny River.

Monte Verde, and the broader concept of pre-Clovis—a debate, incidentally, that is viewed as ludicrous in Europe and laughable to our colleagues in South America. Perhaps the fundamental reason for the continued playing out of this archaeological farce is not the script (i.e., the data) but, rather, the actors—many of whom have invested so heavily in their own canonical views that the acknowledgment of a "new" story somehow threatens the integrity of their life's work. This situation is either comic or tragic, depending on your perspective, but it has *never* been science.

Given the manner in which this debate has unfolded, it is doubtful that the remaining spear-carriers will improvise on their appointed lines. But before the curtain falls, we would like to provide several "bottom line" observations pertaining to the Meadowcroft/Cross Creek project.

The Bottom Lines

1. There is not now, nor has there ever been a coal lens, coal seam, or coal outcrop within 800 m (2,625 ft) of Meadowcroft Rockshelter, let alone in the rockshelter.

2. Despite extensive investigations by four different radiocarbon laboratories, including the Oxford AMS Facility and detailed research by micromorphologists, no evidence of particulate contamination has ever been detected in any of the Meadowcroft radiocarbon samples.

3. Despite 20 years of fanciful scenarios and wishful thinking, no evidence of any groundwater movement—and, hence, no vehicle for non-particulate dissolved contaminants—exists at Meadowcroft Rockshelter.

4. Applying the most conservative interpretation of the 52-date chronology, even if only the youngest assay from upper middle Stratum IIa is valid, the minimum age for the presence of human population in this portion of Pennsylvania is on the order of 10,600–12,000 yr B.P.
5. If the six deepest dates unequivocally associated with cultural material are averaged—a procedure with which these writers are uncomfortable—then humans were definitely present at this site (and by implication, throughout much and perhaps all of the Americas) sometime between 13,955 and 14,555 radiocarbon years ago.
6. Contrary to the received wisdom and consistent with the topographic setting of the rockshelter and the distance (ca. 150 km [93 mi] south) of Meadowcroft from the waning glacial front, the earliest inhabitants of the site were operating in a mosaic environment fully consistent with data now emerging from elsewhere in eastern North America.
7. The oldest archaeological manifestation identified at Meadowcroft and several other loci in the Cross Creek drainage is the Miller complex. This complex appears to represent the pioneer population in the upper Ohio Valley and, possibly, the Northeast.
8. The flaked stone artifact inventory from lower and middle Stratum IIa at Meadowcroft Rockshelter contains small, prismatic blades that were detached from small, prepared cores. Although complete cores were not recovered at Meadowcroft, the artifact assemblage from the nearby and apparently contemporaneous Krajacic site contains a great variety of the distinctive Meadowcroft-style blade implements and several small, cylindrical polyhedral cores. Recovered after the initial study of the Meadowcroft lithic assemblage had been undertaken in 1975, the Krajacic cores precisely parallel the core-reduction strategy previously posited for the Meadowcroft blades.
9. Collectively, these data suggest that the first inhabitants of eastern North America employed a technologically standardized and sophisticated small polyhedral core- and blade-based industry of decidedly Eurasiatic, upper-Paleolithic aspect. Not surprisingly, although this assemblage is presently unique in northeastern North America, it reflects precisely the sort of lithic reduction strategy that one would think *should* be evidenced at this time. Moreover, this assemblage is not at variance technologically or chronologically with its possible northeast Asian prototypes. Significantly, a very similar technology is also represented at Cactus Hill, Virginia (McAvoy and McAvoy 1997), with dates comparable to those from lower and middle Stratum IIa at Meadowcroft.
10. Despite unfounded observations to the contrary, nothing in the Miller complex lithic suite occurs in or is apparently related to any later cultural manifestations. Its unique blademaking technology, in particular, is unknown in later contexts. Additionally, though partially coeval at least in its later stages with Clovis, few definitive connections between these two early cultures are presently apparent.
11. In the long view, and based on the meager but informative ecofactual remains from lower and middle Stratum IIa, it appears that the Miller complex populations can be tentatively characterized as generalized hunter-foragers rather than specialized hunters and, further, that despite their currently geographically circumscribed distribution, they represent the baseline—as of 2001—for all subsequent cultural developments in the upper Ohio Valley and, perhaps, eastern North America.

Discussion

The statements outlined above are not offered as collective proof of the existence of pre-Clovis, nor are they tendered as a definitive refutation of the long-standing "Clovis-first" model. Rather, they simply reiterate certain seminal facts concerning more than 25 years of research at Meadowcroft Rockshelter and provide observations about the site's apparent relationships with broader themes in the North American late Pleistocene. Perhaps these statements will reinforce the site's supporters, sway agnostics, and even achieve a majority opinion among North American archaeologists, but it is virtually certain that they will not change the minds of Clovis-first's most adamant proponents.

Despite the continued refutations, protestations, and retrenchments of this last group of archaeologists and geologists, however, it nonetheless appears that the majority opinion concerning the antiquity of humans in the New World has changed—or at least has become open to new, emerging possibilities. Instead of singular, fleeting "flavor of the week" pre-Clovis localities, there are now at least five New World sites widely recognized as bona fide entities. Monte Verde's antiquity is now widely accepted, there is similar majority opinion on the ultimate disposition of the Meadowcroft radiocarbon sequence, and three more recently excavated sites (i.e., Saltville and Cactus Hill in Virginia and Topper in South Carolina) are changing the way scholars think about Clovis antecedents and the late-Pleistocene landscape. It appears, then, that pre-Clovis sites are on the way to satisfying the criteria of replicability and high visibility (here defined as "standardized" artifacts and toolkits occurring in contexts with extensive regional or continent-wide analogues). These criteria have long impeded—either rightly or wrongly—the wide acceptance of pre-Clovis sites by the majority of North American scholars.

Whether a result of the growing collaboration between New World and European scholars, the Monte Verde site visit, the Beyond Clovis Conference, or a vague discomfort with an exceptionally tidy Clovis-first model that abides no exceptional cases, it is now apparent that the acceptance of pre-Clovis has reached what social psychologists would call its "tipping point" (Gladwell 2000). That is, it is now widely considered possible and even probable that humans occupied the New World prior to 11,500 RCYBP, despite fanciful, celeritous colonization scenarios to the contrary. The theory of the tipping point requires that at some juncture we—as consumers of information—are compelled to "reframe the way we think about the world" (Gladwell 2000:257). It is our contention that sufficient data exist for archaeologists to continue with the project of reframing the way they think about the world and how humans moved across its surface over 11,500 years ago.

Acknowledgments

The excavations and attendant analyses at Meadowcroft Rockshelter and the Cross Creek drainage were conducted under the auspices of the former Archaeological Research Program (subsequently the Cultural Resource Management Program) of the Department of Anthropology, University of Pittsburgh. Recent re-excavations in 1994 and 1995 were conducted under the aegis of the Mercyhurst Archaeological Institute, Mercyhurst College, Erie, Pennsylvania. The initial 1973 field project and the 1977–1978 field seasons were directed by J. M. Adovasio. The 1974–1976 field seasons were co-directed by J. M. Adovasio and J. D. Gunn. The 1994 and 1995 field projects were directed by J. M. Adovasio. Analysis of all materials recovered during the 1973–1995 field seasons is under the ultimate direction of J. M. Adovasio. The late Albert Miller, owner of the site, is acknowledged in particular for his long-standing support of the project and his warm humanitarianism. Generous financial and logistic support for the 1973–1978 excavations and analyses was provided by the University of Pittsburgh, the Meadowcroft Foundation, the National Geographic Society, the National Science Foundation, the Alcoa Foundation, the Buhl Foundation, the Leon Falk Family Trust, and Messrs. John and Edward Boyle of Oil City, Pennsylvania. Support for the 1994 excavations was provided by the Western Pennsylvania Historical Society and the Pennsylvania Historic and Museum Commission. The 1995 excavations were partially underwritten by the National Geographic Society.

References Cited

Adovasio, J. M.
1993 The Ones That Will Not Go Away: A Biased View of Pre-Clovis Populations in the New World. In *From Kostenki to Clovis: Upper Paleolithic–Paleo-Indian Adaptations*, edited by O. Soffer and N. D. Praslov, pp. 199–218. Plenum Press, New York.

Adovasio, J. M., A. T. Boldurian, and R. C. Carlisle
1988 Who Are Those Guys? Some Biased Thoughts on the Initial Peopling of the New World. In *Americans Before Columbus: Ice Age Origins*, edited by R. C. Carlisle, pp. 45–61. Ethnology Monographs No. 12. Department of Anthropology, University of Pittsburgh, Pittsburgh.

Adovasio, J. M., R. C. Carlisle, K. A. Cushman, J. Donahue, J. E. Guilday, W. C. Johnson, K. Lord, P. W. Parmalee, R. Stuckenrath, and P. W. Wiegman
1985 Paleoenvironmental Reconstruction at Meadowcroft Rockshelter, Washington County, Pennsylvania. In *Environments and Extinctions: Man in Late Glacial North America*, edited by J. I. Mead and D. J. Meltzer, pp. 73–110. Peopling of the Americas Edited Volume Series, Center for the Study of Early Man, University of Maine, Orono.

Adovasio, J. M., J. Donahue, R. C. Carlisle, K. Cushman, R. Stuckenrath, and P. Wiegman
1984 Meadowcroft Rockshelter and the Pleistocene/Holocene Transition in Southwestern Pennsylvania. In *Contributions in Quaternary Vertebrate Paleontology: A Volume in Memorial to John E. Guilday,* edited by H. H. Genoways and M. R. Dawson, pp. 347–69. Special Publication No. 8. Carnegie Museum of Natural History, Pittsburgh.

Adovasio, J. M., J. Donahue, and R. Stuckenrath
1990 The Meadowcroft Rockshelter Radiocarbon Chronology 1975–1990. *American Antiquity* 55:348–54.

1992 Never Say Never Again: Some Thoughts on Could Haves and Might Have Beens. *American Antiquity* 57:327–31.

Adovasio, J. M., J. Donahue, R. Stuckenrath, and R. C. Carlisle
1989 The Meadowcroft Rockshelter Radiocarbon Chronology 1975–1989: Some Ruminations. Paper presented at the First World Summit Conference in the Peopling of the Americas, University of Maine, Orono, May 24–28.

Adovasio, J. M., J. D. Gunn, J. Donahue, and R. Stuckenrath
1975 Excavations at Meadowcroft Rockshelter, 1973–1974: A Progress Report. *Pennsylvania Archaeologist* 45(3):1–30.

Adovasio, J. M., J. D. Gunn, J. Donahue, R. Stuckenrath, J. Guilday, and K. Lord
1979–80a Meadowcroft Rockshelter—Retrospect 1977: Part 1. *North American Archaeologist* 1(1):3–44.

1979–80b Meadowcroft Rockshelter—Retrospect 1977: Part 2. *North American Archaeologist* 1(2):99–137.

Adovasio, J. M., J. D. Gunn, J. Donahue, R. Stuckenrath, J. Guilday, and K. Volman
1980 Yes Virginia, It Really is That Old: A Reply to Haynes and Mead. *American Antiquity* 45:588–95.

Adovasio, J. M., D. Pedler, J. Donahue, and R. Stuckenrath
1999 No Vestige of a Beginning nor Prospect for an End: Two Decades of Debate on Meadowcroft Rockshelter. In *Ice Age Peoples of North America: Environments, Origins, and Adaptations of the First Americans,* edited by R. Bonnichsen and K. L. Turnmire, pp. 416–31. Oregon State University Press, Corvallis.

Carlisle, R. C., and J. M. Adovasio (editors)
1982 *Meadowcroft: Collected Papers on the Archaeology of Meadowcroft Rockshelter and the Cross Creek Drainage.* Department of Anthropology, University of Pittsburgh, Pittsburgh.

Custer, J. F.
1996 *Prehistoric Cultures of Eastern Pennsylvania.* Anthropological Series Number 7. Commonwealth of Pennsylvania, Pennsylvania Historical and Museum Commission, Harrisburg.

Dincauze, D. F.
1981 The Meadowcroft Papers. *Quarterly Review of Archaeology* 2:3–4.

1984 An Archaeo-Logical Evaluation of the Case for Pre-Clovis Occupations. *Advances in World Archaeology* 3:275–324.

1989 The Meadowcroft Papers. *The Review of Archaeology* 10(1):137–45.

Donahue, J., and J. M. Adovasio
1990 The Evolution of Sandstone Rockshelters in Eastern North America; A Geoarchaeological Approach. In *Archaeological Geology of North America,* edited by N. P. Lasca and J. Donahue, pp. 231–251. Geological Society of America Centennial Special Volume 4. Geological Society of America, Boulder.

Fagan, B. M.
1990 Tracking the First Americans. *Archaeology* November/December 1990:14–20.

1995 *Ancient North America: The Archeology of a Continent.* Thames and Hudson, second edition, London.

Feder, K. L.
1996 *The Past in Perspective.* Mayfield Publishing Company, Mountain View, California.

Frison, G. C.
2000 Progress & Challenges: Intriguing Questions Linger after 75 Years of Answers. *Discovering Archaeology* February 2000:40–42.

Gladwell, M.
2000 *The Tipping Point: How Little Things Can Make a Big Differ2ence.* Little, Brown and Company, New York.

Goldberg, P., and T. L. Arpin
2000 Micromorphological Analysis of Sediments from Meadowcroft Rockshelter: Implications to the Radiocarbon Dates. *Journal of Field Archaeology*, in press.

Goodyear, A. C.
2001 Evidence of Pre-Clovis Sites in the Eastern United States. [*contribution to this volume*]

Gowlett, J. A.
1992 *Ascent to Civilization: The Archaeology of Early Humans.* 2nd edition. McGraw-Hill, Inc., New York.

Haynes, C. V.
1980 Paleoindian Charcoal from Meadowcroft Rockshelter: Is Contamination a Problem? *American Antiquity* 45(3):582–87.
1991 More on Meadowcroft Rockshelter Radiocarbon Chronology. *The Review of Archaeology* 12(1):8–14.
2001 Clovis, Pre-Clovis, Climate Change, and Extinction. [*contribution to this volume*]

Johnson, W. C.
1981 *Archaeological Review Activities in Survey Region IV, Northwestern Pennsylvania: Year End Report of the Regional Archaeologist for the Period September 1979 through August 1980.* Cultural Resource Management Program, University of Pittsburgh, Pittsburgh. Submitted to the Pennsylvania Historical and Museum Commission, Harrisburg, Pennsylvania. Service Purchase Contract 645987.

Lepper, B. T.
1999 Pleistocene Peoples of Midcontinental North America. In *Ice Age Peoples of North America: Environments, Origins, and Adaptations of the First Americans,* edited by R. Bonnichsen and K. L. Turnmire, pp. 362–394. Oregon State University Press, Corvallis.

Lynch, T. F.
1983 The Paleo-Indians. In *Ancient South Americans,* edited by J. D. Jennings, pp. 87–138. W. H. Freeman and Company, New York.

McAvoy, J. M., and L. D. McAvoy
1997 *Archaeological Investigations of Site 44SX202, Cactus Hill, Sussex County Virginia.* Virginia Department of Historic Resources No. 8 and Nottoway River Survey Archaeological Research Report No. 2. Sandston, Virginia.

Mead, J. I.
1980 Is It Really that Old? A Comment about the Meadowcroft Rockshelter "Overview." *American Antiquity* 45(3):579–81.

Meltzer, D. J.
1993 *Search for the First Americans.* St. Remy Press, Montreal, and Smithsonian Books, Washington, D.C.

Tankersley, K. B., and C. A. Munson
1992 Comments on the Meadowcroft Rockshelter Radiocarbon Chronology and the Recognition of Coal Contaminants. *American Antiquity* 57(2):321–326.

Tankersley, K. B., C. A. Munson, and D. Smith
1987 Recognition of Bituminous Coal Contaminants in Radiocarbon Samples. *American Antiquity* 52(2):318–29.

Pleistocene Human Occupation of the Southeastern United States: Research Directions for the Early 21st Century

David G. Anderson

Abstract

Paleoindian research effort in the Southeast and beyond in the years to come should be directed, as much as possible, to primary data collection, both through fieldwork and laboratory analyses; absolute and relative dating of artifacts and assemblages; the use of calibrated or calendar dates; the increasing adoption of multidisciplinary research approaches; the development of well-grounded models; and the full publication of the results of current and past fieldwork. The reward structure of the archaeological profession, in academia and beyond, should be refocused to emphasize the production of comprehensive site reports and interpretive analyses employing large data sets, not the fragmentation of results into a myriad of preliminary papers or journal articles. New research and researchers should be welcomed and encouraged by members of the existing regional professional community working on the Paleoindian era. The Southeast has long prided itself on maintaining a tradition of careful fieldwork, sound linkages between primary data and interpretations and models based on that data, and an openness and hospitality among the members of its research community. These characteristics are ideally suited to furthering Paleoindian research in the 21st century, both in the Southeast and throughout the Americas.

Introduction

The southeastern United States has a rich and varied archaeological record encompassing the period prior to 11,425 CALYBP/10,000 RCYBP, during what has traditionally been called the Paleoindian or late-Pleistocene era. This record has been summarized at length in a number of recent publications (Anderson 1990, 2001, 2003; Anderson and Sassaman, eds., 1996; Dincauze 1993a; Ellis et al. 1998; Goodyear 1999a; Lepper and Meltzer 1991; Mason 1962; Morse et al. 1996; Williams and Stoltman 1965). In this paper, rather than repeat this existing information, I look at the state of current research and offer suggestions about what we should be doing to better understand the initial human occupation of the region. These suggestions are intended to guide and not straitjacket or constrain research in the Southeast in years to come. I have no doubt that exciting discoveries will occur and that new models and approaches will be developed that are completely unanticipated here. That is as it should be, and is, after all, what makes archaeology such an exciting field. Just as we often don't know what we will find when we dig, so too are we frequently surprised at where our analyses and modeling take us. Those measures proposed here, for the most part, are things that regional Paleoindian specialists are well aware need doing. My purpose, however, is to emphasize how wide open the opportunities are and to foster greater interest and participation in local research. Many of the observations advanced here for the Southeast, furthermore, are likely to prove equally valid in other areas.

First and foremost, increased effort needs to be devoted to primary data collection, both through fieldwork directed to the discovery of early sites as well as analyses directed to examining and recording existing collections. Second, the absolute and relative dating of early artifacts and assemblages from this period should be emphasized. Third, when referring to time, calibrated or calendar dates should be universally adopted and incorporated into models of culture change. Fourth, when evidence for early occupations is detected, multidisciplinary research teams need to be brought to bear in their interpreting and dating. The research programs undertaken at pre-Clovis–era sites like Cactus Hill and Topper, or at later Paleoindian sites like Dust Cave and Big Eddy, serve as models of this kind of multidisciplinary approach. Fifth, developing models of what the Paleoindian archaeological record means in terms of human behavior and adaptation should be encouraged; but above all, these models should

Department of Anthropology, 250 South Stadium Hall, Knoxville, TN 37996-0720; e-mail: dander19@utk.edu or jmuse@sc.tds.net be well grounded, that is, tied to existing data. Finally, full publication of the results of both field and laboratory studies should be encouraged; and where earlier work has not been adequately reported, every effort should be made to see that this is done.

Enlarging the Primary Database

More effort needs to be directed to locating and collecting primary data across the region through surveying, excavating, and analyzing existing collections, both from earlier investigations and in private hands. We need to dig deeper and in places informed by geoarchaeological research if we are to find well-preserved assemblages, particularly those dating to the early Paleoindian or pre-Clovis era (Goodyear 1999a). Submerged and wet-site archaeology should receive increasing attention, both as a means of determining the archaeological record on the continental shelf (Faught 1996, 2004; Faught and Donoghue 1997; Faught et al. 1992) as well as in the hope of recovering perishable artifacts and human remains, as has occurred in Florida at the Little Salt Spring and Page-Ladson sites (e.g., Clausen et al. 1979; Dunbar et al. 1988).

Fluted-point surveys across the region need to be established or reinvigorated. While compilations of primary data exist from every Southern state, only in Virginia at present is there an active Paleoindian artifact recording project in which primary data are regularly and systematically published. This is, of course, the fluted-point survey project initiated by McCary (1984, 1991) in the late 1940s and carried on in recent years by Johnson and Pearsall (e.g., 1993, 1995, 1996) and Hranicky (2004). To date, attribute data (e.g., metric measurements, raw material type, color, presence or absence of grinding, etc.) on over 1,000 fluted points from Virginia have been published, a figure that encompasses just under 10 percent of all the fluted points known from the entire country at present, at least as presently compiled (Anderson and Faught 1998, 2000; see Shott 2002 for an excellent critique of the strengths and weaknesses of these kinds of compilations). If the professional or avocational communities in every state were as conscientious as they are in Virginia, we would have a much better understanding of the range of variation within these early artifacts and their location on the landscape.

More work on compiling and recording data on primary artifacts is needed in the Southeast; ideally, work should go beyond fluted points to encompass first all Paleoindian point forms, and then other tool forms and assemblage characteristics. Fortunately, data on Paleoindian artifacts, typically fluted points, are presently being conscientiously recorded by individuals in many Southeastern states. In a few states, notably Georgia, Mississippi, and Tennessee, these surveys encompass all Paleoindian point types. Only rarely, however, has any of this primary data been fully published, although all current researchers are very generous in sharing their primary data records. Publishing primary data should be encouraged, however, to avoid the loss of this information in the future. Interestingly, when such reports or papers appear they tend to become instant classics, since they typically provide some of the only readily accessible primary data on Paleoindian materials from the state in question (e.g., Anderson et al. 1990; Dunbar 1991; Dunbar and Waller 1983; Gagliano and Gregory 1965; Goodyear et al. 1990; McGahey 1996; Rolingson 1964; Rolingson and Swartz 1966; Tankersley 1990; Wittkofski and Reinhart, eds., 1989).

The reanalysis, or in some cases initial analysis, of older excavation assemblages is another area where important research should be accomplished. Some classic Paleoindian sites in the region have either been minimally reported or were reported so long ago that reanalysis can prove rewarding. This has been demonstrated with materials from the late-Paleoindian Hardaway site in North Carolina, for example, which was excavated from the 1940s through the 1980s. Materials from this site helped define the basic late-Paleoindian through Archaic regional cultural sequence (Coe 1964) and, more recently, the associated toolkits and possible settlement strategies of the site's early occupants (Daniel 1998). Reanalyzing materials from sites such as Stanfield-Worley Bluff Shelter in Alabama (DeJarnette et al. 1962), Silver Springs in Florida (Neill 1958), or Parrish Village in Kentucky (Webb 1951), to cite other possible examples, would unquestionably yield new insights.

The early assemblages found at Macon Plateau in the 1930s by New Deal–era archaeologists, including the first Clovis fluted point found in stratified context in the Southeast and recognized for what it was, have never been systematically described; only a few pages in a preliminary report on the investigations discuss these early materials (Kelly 1938:2–5). The collections from Macon Plateau are maintained at the Southeast Archeological Center, however, and have been undergoing re-cataloguing for several years now. From casual inspection of the materials in that laboratory, I know that dozens of Dalton and early side-notched points are present in the collections, as well as hundreds of unifacial and bifacial tools, including at least one Dalton adze. The provenience data for these materials are not the best in some cases—for many artifacts, intrasite provenience data are unknown or ambiguous, and some early artifacts occurred in later Mississippian-era mound fill and hence are clearly redeposited. Nonetheless, the collection is a superb example of the kind of early assemblages that can occur in a Fall Line setting in this part of the region.

Increasing the Use of Absolute Dating Procedures

Compared with the cost of excavating, absolute dating is relatively inexpensive. A variety of procedures can be used, including conventional radiocarbon dating, accelerator mass spectrometry (AMS) radiocarbon dating, thermoluminescent dating (TL), optically stimulated luminescence dating (OSL), and a comparatively new procedure, oxidized carbon ratio dating (OCR; see Frink 1992, 1994, 2004; but see also Killick et al. 1999). As recent research at Topper, Coats-Hines, Cactus Hill, and Saltville has shown, use of multiple dating techniques can be an invaluable means of assessing the integrity and age of site deposits when collected as part of a multidisciplinary research effort. Three of these sites contain early-Paleoindian (i.e., pre-Clovis) remains, and the use of multiple dating techniques increases confidence that the deposits and associated artifacts date to this period (Goodyear 2004).

At the Topper site in South Carolina, for example, Goodyear has used AMS and OSL dating to bracket the age of a possible pre-Clovis lithic assemblage, which occurred in alluvial sands situated above a gray clay and below a thick layer of colluvium. Two AMS dates obtained on humic acids from beneath the gray clay were 19,280 ± 140 RCYBP/22,470–23,261 CALYBP (CAMS–59593) and 20,860 ± 90 RCYBP/beyond current calibration curve CALYBP (CAMS–58432) (Goodyear 1999b:10), while OSL dates ranged from ca. 13,000 to 14,000 CALYBP at the base of the colluvium, and from 15,000 to 16,000 CALYBP in the top of the alluvial sand layer (Goodyear 2000, 2004). These dates suggest the archaeological remains encountered in the alluvial sands date to at least 15,000–16,000 CALYBP, appreciably pre-Clovis, and possibly earlier than ca. 23,000 CALYBP, the age of the gray clay. The samples were collected as part of an extensive program of geoarchaeological and paleoenvironmental research that complements the archaeological investigations.

At Coats-Hines (40Wm31), an apparent kill site in Tennessee, 10 chert tools and 24 flakes were found with the remains of a disarticulated mastodon. Butchering marks and other evidence of human modification are apparent on a number of bones, and the tip of a bone projectile point was found between the ribs of the mastodon (Breitburg et al. 1996; J. Broster pers. comm. 2000). Four OCR dates have yielded an age around 13,000 CALYBP, comparable to calibrated radiocarbon dates for Clovis. A radiocarbon date of 27,050 ± 200 RCYBP/beyond current calibration curve CALYBP (Beta–80169) was obtained from the base of the deposits at the site, below the cultural level. A second date, on material from within the dental cusps of the mastodon, was 6530 ± 70 RCYBP/7334–7554 CALYBP (Beta–75403) (Breitburg et al. 1996:7). An AMS date on materials from the bone bed yielded a date of 12,030 ± 40 RCYBP/13,823–14,290 CALYBP (Beta–125350) (J. Broster pers. comm. 2000).

At Cactus Hill in Virginia, a number of conventional and AMS radiocarbon determinations have been run on material from pre-Clovis deposits located in a sand dune setting (McAvoy and McAvoy 1997:124, 167, 169). Small blades, polyhedral blade cores, retouched flakes, and abrading stones have been found at the site stratigraphically below a well-defined Clovis occupation. The assemblage was documented in two separate areas, in excavations by two different teams of researchers, led by Joseph and Lynn McAvoy and Michael F. Johnson (Johnson 1997; McAvoy and McAvoy 1997). Two unfluted lanceolate/triangular bifaces, which McAvoy and McAvoy (1997:136) have called early Triangular, occurred in the pre-Clovis deposits. Seven quartzite flakes and three quartzite blade cores were found in and near an amorphous hearth-like scatter of white pine charcoal that yielded an AMS radiocarbon determination of 15,070 ± 70 RCYBP/17,743–18,298 CALYBP (Beta–81590) (McAvoy and McAvoy 1997:167). Three additional early dates (16,670 ± 730 RCYBP/18,968–20,756 CALYBP [Beta–97708], 16,940 ± 50 RCYBP/19,854–20,492 CALYBP [Beta–128330], and 19,700 ± 130 RCYBP/22,976–23,717 CALYBP [Beta–128331]) and two anomalously recent dates, 9250 ± 60 RCYBP/10,277–10,501 CALYBP (Beta–93899) and 10,160 ± 60 RCYBP/11,571–12,060 CALYBP (Beta–92923), have also been obtained on charcoal from the pre-Clovis levels (McAvoy et al. 2000). The 16,670 ± 730 and 16,940 ± 50 RCYBP dates are from hearth areas, while the 19,700 ± 130 RCYBP date is near the base of the dune, below the cultural levels. The overlying Clovis assemblage is well defined, with numerous points and tools, and a hearth-like scatter of Southern pine charcoal from the same level has been radiocarbon dated to 10,920 ± 250 RCYBP/12,436–13,183 CALYBP (Beta–81589) (McAvoy and McAvoy 1997:124, 167, 169).

AMS and conventional radiocarbon dating have also been used with great success in the pre-Clovis levels at the Saltville site in Virginia (McDonald 2000:8, 37–46). Three artifact-bearing horizons were recognized, dating from roughly 14,500, 13,900, and 13,000 RCYBP/17,365, 16,675, and 15,636 CALYBP (McDonald 2000; see also Goodyear n.d.). Collagen from a fractured and apparently use-worn tibia of a probable musk ox (*Bootherium bombifrons*) was AMS dated to 14,510 ± 80 RCYBP/17,112–17,640 CALYBP (Beta–117541) from the lowest level; a second date on wood from the same level is nearly identical, at 14,480 ± 300/16,916–17,767 CALYBP (Beta–5701) (McDonald 2000:8, 37–46). The middle horizon at the Saltville site included a cluster of pebbles and cobbles from a small depression, the uppermost stratum of which yielded 12 pieces of microdebitage and some fish bones. Twigs collected from a sand lens from within the block yielded a radiocarbon date of 13,950 ± 70/16,486–16,978 CALYBP RCYBP (Beta–65209); two other dates from the same stratum were similar in age, 13,460 ± 420/15,563–16,771 CALYBP (SI–641) on tusk and 13,130 ± 330/14,533–16,371 CALYBP (A–2985) on wood (McDonald 2000:8, 33). The upper horizon, currently undated, was a feature in an eroded rill into the middle horizon that contained a midden-like concentration containing over 200 clam shells, over 500 pieces of small vertebrate teeth and bones, and 125 pieces of chert microdebitage, some of which appear to be intentionally produced flakes. McDonald (2000:34–36) suggests the debris was formed by people harvesting shellfish and small animals from the lake during periods of low water. If created by human action, the feature would be the oldest shell midden in the New World.

Other absolute dating procedures also warrant consideration. Dendrochronology may ultimately prove a useful procedure for precisely dating early sites, provided wood with enough growth rings can be found either to tie in to existing reconstructions or, as is more likely for the foreseeable future, to develop fairly precise dates through wiggle-matching radiocarbon-dating procedures (e.g., Stuiver et al. 1998; see also Christen and Litton 1995 and Goslar and Madry 1998). Developing dendrochronological reconstructions for the Pleistocene may prove extremely difficult, but having annual scale proxy climate data for this era would give us great insight into the world that early Southeastern peoples experienced. In the deep South, subfossil cypress logs dating back into the Pleistocene may be preserved in stream channels, sinkholes, or marsh deposits; these may prove a source of annual scale climate data, as well as provide a means of dating (Stahle et al. 1985). Archaeomagnetic dating may also eventually prove useful, if sufficient samples can be found to develop a calibration curve (Eighmy and Sternberg, eds., 1990). Even if absolute dates cannot be immediately obtained using this procedure, it may provide information on the relative dates of closely spaced features.

Finally, increased effort should be paid to existing artifact collections from across the region to see if new diagnostic point and tool forms are present that can help identify components of the various Paleoindian subperiods. The early Triangulars found at Cactus Hill may be present in any number of surface collections from the region, for example, as may points comparable to the Miller Lanceolate type found at Meadowcroft (Adovasio et al. 1999:28). These points would likely be called middle- or late-Paleoindian lanceolates if found in surface context. The early Triangulars from Cactus Hill might even be considered preforms for early-Archaic notched points; now we believe them to be much older. Literally millions of projectile points are maintained in collections around the region, both in museums and other formal curatorial repositories and in the hands of amateur collectors (e.g., Charles 1981). These data, when properly examined, can yield important knowledge about the past and just may contain many more early-Paleoindian diagnostics. Over the past several decades, furthermore, measurement data from thousands of Paleoindian points have been collected from across the region and beyond. It is time to start seriously examining these data to understand the meaning of the variability present within them.

Using Calendar Dates When Reporting Time

If we are to understand what happened in the Southeast prior to 11,425 CALYBP/10,000 RCYBP, we must routinely employ calendar ages when discussing sites and assemblages and when modeling culture change. The development of an extended radiocarbon calibration encompassing upwards of 20,000 years in recent years (Stuiver et al. 1998) and the increased use of dating procedures like OSL or TL that yield calendar ages on Southeastern Paleoindian sites have enabled researchers to use calendar dates with confidence when discussing early occupations. Given the great difference between uncalibrated radiocarbon and calendar ages at this time level, at a minimum some 1,400 years, and the fact that plateaus, jumps, and even reversals are evident in the radiocarbon record, using calendar time is absolutely essential to understanding the regional archaeological record (Anderson 2001:143–44; Fiedel 1999, 2000). We now know, for example, that the incredible diversification of sites and point forms that occurred between ca. 10,500 and 10,000 RCYBP (12,568–11,425 CALYBP) did not occur within a mere 500 years, as we once thought, but over ca. 1,100 years, given the discovery of a major plateau in the radiocarbon calibration during this interval.

Accordingly, a new timeline encompassing the Paleoindian era in the region is advanced that incorporates calendar ages (Table 1). Three subperiods are proposed, the early, middle, and late Paleoindian encompassing pre-Clovis, Clovis, and post-Clovis Paleoindian occupations (see also Anderson 2001:153–156, 2004). When the first people actually arrived in the Southeast is unknown, but it is assumed to have been prior to 13,500 CALYBP (i.e., >11,500 RCYBP), the earliest date currently accepted for the inception of the Clovis tradition. There appears to be little argument that pre-Clovis remains exist in the Americas at present, although appreciable discussion may focus on the dating of specific sites and assemblages. A number of early-Paleoindian sites are currently reported in the Southeast, including Cactus Hill, Topper, Little Salt Springs, and Saltville. While the cultures represented by these assemblages remain somewhat enigmatic and are known by little more than their presence and the dating of some of these sites remains somewhat controversial, some or all appear to be genuinely pre-Clovis in age (Goodyear 2004). One of three explanations for these sites appears likely, that they (1) represent the remains of small groups that arrived in the region but died without issue, so-called "failed migrations"; (2) represent the remains of early continuous settlement by such few numbers of people that the archaeological record they produced is nearly invisible; or (3) indicate that a combination of the first two options may have occurred in different parts of the region. That is, people may have successfully settled in some areas and died out in others during this period. While not all these sites are universally accepted as early Paleoindian in age, most researchers accept that pre-Clovis occupations are increasingly probable.

The middle-Paleoindian subperiod encompasses the interval when Clovis and related fluted-point assemblages are assumed to have been common over the region, from roughly 13,500 to 12,825 CALYBP (i.e., ca. 11,500–10,800 RCYBP). While few sites have actually been dated to this interval in the Southeast, the large numbers of Clovis points that have been reported and the existence of a number of sites with dense assemblages, such as Carson-Conn-Short (Broster and Norton 1996:290–293; Broster et al. 1994, 1996; Nami et al. 1996) and Wells Creek Crater (Dragoo 1973) in Tennessee, Adams (Sanders 1988, 1990) in Kentucky or Williamson (McCary 1951, 1975) in Virginia, suggest a fairly substantial human presence, at least when compared with what came before. While Clovis assemblages in North America are for the most part dated somewhat later in time, to between ca. 13,172 and 12,982 CALYBP/ca. 11,200–10,900 RCYBP (Fiedel 1999, 2000; Taylor et al. 1996), a starting date of 13,500 CALYBP/11,500 RCYBP is advanced here, since two dates from the Aubrey site in Texas are in the ca. 11,500–11,600 RCYBP range/ca. 13,500–13,625 CALYBP (Ferring 1995), and since 11,500 RCYBP is a number that has been used for the onset of Clovis in the region for about a decade now in Southeastern timelines (e.g., Anderson 1990).

When Clovis is more precisely dated, the starting point for this subperiod may have to be adjusted either forward or backward in time. There are hints that the starting date may move backward, perhaps to around 14,000 CALYBP or ca. 12,000 RCYBP. A date of 11,900 ± 80 RCYBP/13,636–14,094 CALYBP (AA–27486) was obtained from a level containing a Clovis assemblage at the Big Eddy site in Missouri, although the other

Table 1. A Southeastern Paleoindian Chronology (calibrated dates obtained using the Calib 4.4 program, and averaging the one-sigma range; adapted from Stuiver et al. 1998).

	Approximate Calendar age (CALYBP)	Radiocarbon age (RCYBP)	Culture complex	Climatic event
Early Archaic				
	8892	8000	Bifurcate	
	10,197	9000		Boreal
	10,865	9500	Corner-notched widespread	
	11,243	9900	Side-notched widespread	
Late Paleoindian				
	11,425	10,000		
	11,586	10,100		Younger Dryas ends/Preboreal
	11,886	10,200	Early Side-notched	
	12,568	10,500	Dalton/Suwannee, Quad/Beaver Lake	
	12,826	10,800	Cumberland/Folsom	
Middle Paleoindian				
	12,982	10,900		Younger Dryas begins
	13,039	11,000		
	13,077	11,100		Inter-Allerød Cold Period ends
			Clovis widespread	
	13,172	11,200		
	13,316	11,400		Inter-Allerød Cold Period begins
	13,497	11,500		
Clovis beginnings??				
	13,675	11,750		Allerød
	13,456	11,950		Older Dryas ends
	14,044	12,000	Little Salt Springs/Page-Ladson	
Early Paleoindian				
	14,342	12,100		Older Dryas begins
	14,808	12,500	Monte Verde	
	14,897	12,600		Bolling begins
	19,091	16,000	Cactus Hill (?)	
	21,392	18,000	Initial Colonization (?)	Glacial maximum

dates from the same level fall into the more expected range of from ca. 11,400 to 10,700 RCYBP/13,316 to 12,770 CALYBP (Lopinot et al., eds., 1998:92–93, 218). Three dates from the Johnson site in central Tennessee—11,700 ± 980 RCYBP/14,039 CALYBP (TX–7000); 11,980 ± 110 RCYBP/13,996 CALYBP (TX–7454); and 12,660 ± 970 RCYBP/15,048 CALYBP (TX–6999) on hearths with associated fluted preforms (Broster and Barker 1992; Broster and Norton 1996:292–294; Broster et al. 1991; Goodyear 1999a:448–449)—when coupled with the date from Big Eddy suggest Clovis may originate earlier than traditionally thought in the Southeast.

The late-Paleoindian subperiod is dated from roughly 12,826 to 11,425 CALYBP or 10,800–10,000 RCYBP, a range that closely corresponds to the Younger Dryas climate interval, a pronounced return to cold and highly variable climatic conditions worldwide that lasted from roughly 12,900 until about 11,650 CALYBP/10,850–10,100 RCYBP. During the late-Paleoindian era, Clovis technology was replaced by local assemblages whose distribution was restricted to portions of the larger region. The emergence of these subregional cultural traditions is thought to have been brought about by rising population levels and the extinction of megafauna, which would have reduced the need for long-distance movement and interaction to obtain subsistence and maintain a mating network. The onset of the Younger Dryas, which brought a dramatic change in global conditions, and the ensuing period of increased climatic variability are thought to have played a role in the changes observed over this interval.

A range of fluted and unfluted lanceolate forms occur early in the late-Paleoindian subperiod, followed by notched points toward the end of the era. Specific types that are thought to date early in the subperiod, from ca. 12,826 to 12,568 CALYBP/10,800–10,500 RCYBP, include Beaver Lake, Clovis Variant, Cumberland, Quad, Suwannee, and Simpson, as well as Plains forms in the western part of the region such as Folsom, Plainview, Midland, and later in time, Angostura. Around 12,500 CALYBP Dalton points become common over much of the region (Goodyear 1982), with distinct variants occurring in different areas, such as Colbert, Greenbrier, Hardaway, Nucholls, and San Patrice *vars.* Hope and St. Johns. In Florida, Dalton points are comparatively rare; Suwannee points are thought to have continued in use, possibly as a local substitute, until they were replaced by side-notched points. In the western part of the region, Plains Paleoindian forms continue to occur. By ca. 11,886 CALYBP/10,200 RCYBP, side-notched point forms appear, as suggested by dates at both the Dust Cave and Page-Ladson sites; and by ca. 11,425 CALYBP/10,000 RCYBP, side-notched variants occur widely over the region. The onset of the Holocene climate interval, and the Archaic cultural period, is traditionally dated to 10,000 RCYBP/11,425 CALYBP, shortly after the end of the Younger Dryas.

To reiterate, in the future, calendar dates should always be used in discussing Paleoindian occupations in the Southeast. When radiocarbon dates are cited, calibrated calendar dates should typically be given as well. The ca. 1,425-year offset between calendar and radiocarbon time at the end of the Paleoindian period (i.e., 10,000 RCYBP/11,425 CALYBP) increases to over 2,000 years at 12,000 RCYBP/14,044 CALYBP and to almost 3,500 years at 18,000 RCYBP/21,392 CALYBP (Table 1). When coupled with the existence of extended plateaus, such as that from 10,500 to 10,100 RCYBP/ca. 12,568–11,586 CALYBP, using radiocarbon dates alone to discuss events during the Paleoindian era gives a highly erroneous picture.

Emphasizing Multidisciplinary Research Programs

Some of the finest Paleoindian research projects currently underway in the Southeast are taking place at sites like Cactus Hill, Coats-Hines, Dust Cave, Page-Ladson, and Topper, where many different specialists are working together to understand and interpret the assemblages. Geology, geoarchaeology, botany, ethnobotany, palynology, paleontology, and zooarchaeology are all disciplines routinely brought to bear on Paleoindian sites in the region. Not only are collaborative research and the mixing of different research perspectives exciting, they invariably lead to new insights.

Lithic raw-material sourcing studies involving archaeologists and geologists are particularly important, since the information can be used to evaluate directly where raw materials originated and to learn indirectly how early peoples themselves moved over the landscape. Appreciable research has been directed to delimiting lithic raw-material sources in the Southeast using data on trace elements, petrography, or fossil microfauna (e.g., Anderson et al. 1982; Banks 1990; Daniel and Butler 1991; Goodyear and Charles 1984; Upchurch 1984). Since many lithic raw materials in the Southeast are difficult or impossible to distinguish macroscopically, such studies are extremely important. In Georgia, for example, there are cherts in the Piedmont that are identical in appearance to cherts from the Coastal Plain over 100 km away (Ledbetter et al. 1981). The Piedmont cherts lack the microfossil inclusions ubiquitous in cherts from the Coastal Plain, however, and a microscope is needed to differentiate the materials.

Even information on variations in material quality within an outcrop can be important. A multiyear research program at the Allendale chert quarries in South Carolina, for example, has shown that assemblage composition can vary markedly, detectable by fairly subtle variations in topography and local raw-material quality (Goodyear 1992, 1999a, 1999b, 2000, 2004; Goodyear and Charles 1984; Goodyear et al. 1985). Paleoindian populations apparently focused on chert boulders freshly exposed in stream beds rather than on more weathered materials on nearby hillsides. Differences of no more than a few meters' vertical elevation and a few tens of meters' horizontal distances separate Paleoindian period quarrying, workshop, and possible habitation areas within the locality.

Collaborative research also means open research. Visits by other professional archaeologists are routinely encouraged in the Southeast, a by-product of the "tradition of hospitality" and courtesy researchers in the region have routinely extended to one another as far back as the New Deal era (Brown 1994). Over 100 different professional archaeologists, for example, have visited Al Goodyear's investigations at the Topper site in recent years, including many of the top Paleoindian researchers in the country. (I know this for a fact, since I throw a barbecue for the project visitors and crew each year at my home—which is located just 30 miles from Al's site—at the end of the field season, and have kept a guest book.) Specialists with widely varying positions on the existence of pre-Clovis deposits in the Americas have been present. Besides assisting with the digging and examining the artifacts and profiles, they commonly engage in freewheeling discussions round the clock. This is exactly the way science should proceed, through the collegial exchange of ideas. Similar activities characterize the ongoing Dust Cave and Cactus Hill projects, which have become something of in-the-field meeting places.

Encouraging Model Building

While primary data collecting is critical, interpreting the information that comes in is just as important. Indeed, our theoretical perspective often shapes the kind of data we collect. As Albert Goodyear has often noted, until the late 1990s and his work at Topper, he typically never dug below the Clovis levels at the sites he was examining because he "knew" there were no earlier peoples in the Americas; as he put it, "You don't look for what you don't believe in" (Marshall 2001:1730). He was not alone in this perspective. Twenty years ago I opened three 1-by-1-m units 80 cm below the Paleoindian and early-Archaic levels at the Rucker's Bottom site in Georgia, but I did so "to document the absence of artifacts below the level of the remainder of the block" (Anderson and Schuldenrein, eds., 1985:289). A few small flakes were found that were attributed to intrusion from above; but in light of what we know now, I can only wonder if they really were intrusive, or perhaps were evidence of an earlier occupation. Although we mentioned their presence, since we were unable to persuade the funding agency to provide funds to explore the site's well-documented Paleoindian and early-Archaic remains further, much less the many later-period components (we dug almost twice the area contracted for anyway, using volunteers), I know what the reaction would have been to a request to look for possible pre-Clovis remains on the basis of a few flakes. I would like to think that such a request today, in light of recent findings, would receive more serious consideration.

The point, of course, is that we are always trying to explain the archaeological record; and we do so in terms of what makes sense or seems reasonable, given the data at hand and expectations from our knowledge of anthropological and archaeological theory. Accordingly, greater attention should be directed to well-grounded, theoretically informed analyses and model-building activity. For example, while a number of probable pre-Clovis sites have been found in the Southeast and across the Americas, archaeologists are struggling to interpret what these remains mean. If people really were present 15,000 or 20,000 years ago, why is the archaeological record prior to the middle-Paleoindian Clovis subperiod beginning some 13,450 years ago so spotty and varied? Why didn't earlier peoples in the Americas quickly reproduce themselves into greater visibility, as hunter-gatherer demographic theory suggests they should (e.g., Hassan 1981:137–42, 201–203)? Several answers come to mind, including that these earlier remains represent "failed migrations" or groups that died out; that the pre-Clovis archaeological record is actually right in front of us, but we haven't recognized it as such (i.e., points resembling early Triangulars and Miller Lanceolates are not all that atypical to anyone who has handled tens of thousands of Southeastern projectile points); or that early populations were present in some areas and settings but not in others, such as in (now submerged) coastal areas, or on riverine terrace margins long since scoured away (e.g., Butzer 1991; Faught 1996, 2004; Goodyear 1999a).

Likewise, since we know that hunter-gatherer technological organization and mobility strategies are shaped by resource structure, physiography, and effective temperature (e.g., Binford 1980; Kelly 1983, 1995), it is reasonable to expect that Paleoindian adaptations over a region as large and diverse as the Southeast varied appreciably over both time and space. Meltzer (1984, 1988) has explored this idea on a larger scale in examining Paleoindian assemblages in the northern and southern parts of eastern North America. Northern populations were inferred to be economic specialists, exploiting primarily caribou, and leaving behind dense kill and occupation sites. A more generalized foraging adaptation was proposed for the South, with occupations in most areas of fairly short duration. In a similar modeling analysis, Cable and Claggett (Cable 1982, 1996; Claggett and Cable, eds., 1982) argued that the dramatic changes in global temperature at the end of the Pleistocene would have produced differing adaptations over time as temperatures warmed, and over space from lower to higher latitudes. They evaluated the model with data from the deeply stratified Haw River sites in North Carolina and demonstrated that a fairly dramatic shift from curated to expedient tools apparently occurred by the initial part of the early Archaic, around 11,425 CALYBP/10,000 RCYBP (Cable 1982:686–687; see Shott 1996 for a discussion of curated versus expedient technology and how these concepts can be used in interpreting Paleoindian assemblages). This technological shift was thought to reflect the replacement of logistically organized "collector" adaptations (after Binford 1980) by residentially mobile foraging adaptations.

Thus, given differences in climate and biota, contemporaneous Paleoindian occupations in Florida likely differed from those in Kentucky or Virginia, or in the Great Lakes and the Northeast. Likewise, Paleoindian occupations and settlement systems in the same locality likely differed over time, as climate and biota changed. To date there has been little exploration of these possibilities, aside from Meltzer's (1984, 1988) broad regional comparative analyses and other more focused intersite comparisons (e.g., Sanders 1990:65–69; Tankersley 1998). Instead, Paleoindian settlement and mobility strategies tend to be subsumed under somewhat monolithic models that emphasize generalized foraging (Meltzer and Smith 1986) or more focused large-animal procurement (e.g., Kelly and Todd 1988). Indeed, the high-technology forager (HTF) model proposed by Kelly and Todd (1988) explains how highly mobile populations could exploit a wide range of environments with a standardized technology and organizational strategy. The HTF model, in fact, helps to explain why Clovis assemblages appear so similar over large areas. But are they really? Can we really say that the sites and assemblages found in the Tennessee River valley are identical in composition to those in Florida or South Carolina? More comparative analyses of site and assemblage data are clearly needed, and more subregional or locality specific settlement models need to be proposed.

Likewise, Paleoindian ceremonialism needs greater consideration. The existence of a formal Dalton cemetery has been pretty conclusively documented at the Sloan site in northeast Arkansas, where some 200 tiny, weathered human bone fragments were found amid a remarkable assemblage of Dalton points and other tools that were apparently interred with the dead (Morse 1975; Morse, ed., 1997). Sloan is located well away from contemporaneous Dalton sites, on a sand dune that saw only minor use in later prehistory. The reason we know so little about Southeastern Paleoindian ceremonialism, Sloan demonstrates, is because we may be looking in the wrong places. Given this, claims that Paleoindian caching behavior—like that which produced sites like Richey-Roberts/East Wenatchee (Gramly 1993) and Anzick (Lahren and Bonnichsen 1974)—is absent in the East would appear premature. These types of assemblages may be located in relatively isolated parts of the landscape; or, as the human remains found at Little Salt Springs and the later-Archaic Windover site suggest, submerged in bogs or marshes (Clausen et al. 1979; Doran et al. 1988).

Ceremonialism and interaction were clearly linked in the later-Paleoindian Southeast. Over 1,000 late-Paleoindian Dalton sites have been reported in the central Mississippi River valley, from northeast Arkansas to south-central Illinois, in an area that was almost certainly fabulously rich in natural resources and was likely settled quite early. The scattered occurrence of large, exquisitely chipped Sloan points over this area, in caches and as isolated finds, many of them made on Burlington chert from the Crescent Quarries near the Missouri-Mississippi confluence, has been used to infer the existence and extent of a possible late-Paleoindian prehistoric ceremonial and alliance network aptly called the "Cult of the Long Blade" (Walthall and Koldehoff 1998). The existence of adzes in Dalton toolkits has long been thought to reflect appreciable woodworking skills, including the manufacture of dugout canoes (Morse and Goodyear 1973); watercraft are inferred to have been a primary way these groups were linked together (Walthall and Koldehoff 1998:261; see also Engelbrecht and Seyfort 1994 and Jodry 1999 for discussions of evidence of the probable use of watercraft by Paleoindian populations). Whether and how ceremonialism linked Paleoindian groups across the region is a subject that has seen little examination. Many Paleoindian sites occur in dramatic settings on the landscape, suggesting locations had a sacred aspect while also serving as meeting places between different groups.

Exemplary craftsmanship was a major and widely shared aspect of middle- and late-Paleoindian culture, even in producing stone tools used for routine tasks. Was this Paleoindian fascination with high-quality lithic raw materials solely due to the need to maintain a highly curated toolkit by mobile populations, as Goodyear (1979, 1989) has argued? Or was it also shaped by the ceremonial role of artifacts made from these materials, as exemplified by the presence of elaborate specimens in caches and burials (e.g., Walthall and Koldehoff 1998)? In my opinion the effort and care that went into the production of the Paleoindian toolkit, which includes some of the finest chipped-stone artifacts ever produced in the New World, had to go beyond functional considerations. Visiting quarry areas accordingly may have been as much about promoting ceremony and interaction as about replenishing high-quality stone if groups knew they could find others at these locations at certain times of the year (Daniel 1998:194–95; 2001). By procuring high-quality stone and fashioning it into exquisite tools, these early peoples were likely reinforcing basic cultural values and, not unexpectedly, increasing their chances of surviving. The fact that aspects of the Paleoindian toolkit were retained for thousands of years after the demise of the conditions that supposedly caused them to come about in the first place further suggests this was a highly important, and conservative, aspect of these people's culture. Paleoindian use of stone, in my opinion, at least at some times and in some areas had as much to do with ideology as it did with technological and functional considerations.

Existing Paleoindian settlement models advanced for the region should be evaluated and refined wherever possible. These include, to list a few, Morse (1977) and Gillam's (1996, 1999) thoughts on Dalton settlement patterns in northeast Arkansas, and how they are influenced by drainage features and lithic outcrops as well as by potential for interaction; Walthall's (1998) ideas on why rockshelters were not occupied until the late-Paleoindian era; Morrow's (1996) ideas on Paleoindian technological organization and mobility; Daniel (1998) and Gardner's (1983, 1989) views on the importance of quarry/raw material source areas in shaping settlement systems; Dunbar's (1991) ideas on how physiography, drainage conditions, and the occurrence of knappable stone shaped Paleoindian occupations in Florida; Hubbert's (1989) views on Paleoindian site variability in the central Tennessee River Valley; various perspectives on the subsistence orientation of these early populations (e.g., Meltzer 1988; Meltzer and Smith 1986; Walker 2000); ideas on possible movement directions or corridors by colonizing populations, based on geographic and demographic modeling or raw material distributions (Anderson and Gillam 2000, 2001; Moore and Moseley 2001; Steele et al. 1988; Tankersley 1991, 1994); ideas Paleoindian mortuary behavior, ceremonialism, and interaction (Anderson 1995; Morse, ed., 1997; Walthall and Koldehoff 1998); or my own ideas on Paleoindian staging areas or possible loci of original settlement from which expansion over larger areas could be launched (Anderson 1990, 1996; see Dincauze 1993b and Cable 1996:144–145 for examples of extensions and challenges to this model). Fortunately, in the Southeast we do not appear to be suffering from any dearth of ability to think about and interpret Paleoindian data, although much more could obviously be done.

Emphasizing the Importance of Thorough Publishing of Research

The publication of research conducted at Paleoindian sites and with Paleoindian assemblages in the Southeast needs to be encouraged. Full publication includes the thorough description of what was done, what was found (by provenience), and what measurements were taken or results were obtained during associated analyses. It also includes the reporting of all samples collected and processed in specialized analyses, including absolute dating, as well as the inclusion of sufficient photographs, floor plans, and profiles so that researchers can understand the assemblage context. Finally, when interpretive claims are made, these should be backed up with analyses that can be replicated using the collected data.

Comparatively few Southeastern Paleoindian sites can be considered adequately reported by these standards, but those that have been serve as excellent role models. The only work at an early-Paleoindian site that has been thoroughly reported to date is at Cactus Hill through the 1996 field season in an outstanding monograph summarizing the work of two separate teams of researchers (Johnson 1997; McAvoy and McAvoy 1997). Excellent descriptive summaries are also available for the work at Saltville (McDonald 2000) and Little Salt Spring (Clausen et al. 1979). At Topper only limited reporting has occurred, which is fully justifiable since work in the pre-Clovis levels was initiated in 1998 and is ongoing (Goodyear 1999b, 2000, 2004).

Examples of thoroughly published middle-Paleoindian Clovis sites are rare, at least among those yielding extensive assemblages; Adams in Kentucky (Sanders 1988, 1990) and Williamson in Virginia (McCary 1975) are perhaps the best known. Examples of late-Paleoindian sites meeting most or all the reporting standards suggested above (and with fairly extensive assemblages) include Stanfield-Worley Bluff Shelter in Alabama (DeJarnette et al. 1962), Brand and Sloan in northeast Arkansas (Goodyear 1974; Morse 1975; Morse, ed., 1997), Harney Flats in Florida (Daniel and Wisenbaker 1987); early work at Hester in Mississippi (Brookes 1979), Big Eddy in Missouri (Lopinot et al., eds., 1998, 2001), and the Haw River and Hardaway sites in North Carolina (Claggett and Cable, eds., 1982; Daniel 1998).

Many more Paleoindian sites are, of course, known from the Southeast. Some have been the subject of exceptional summary papers or preliminary reports, such as Dust Cave in Alabama (Driskell 1994, 1996; Goldman-Finn and Driskell 1994); Page-Ladson (Dunbar et al. 1988), and Wakulla Springs Lodge in Florida (Jones and Tesar 2000); John Pearce in Louisiana (Webb et al. 1971); Taylor in South Carolina (Michie 1996); the Pierce (Broster 1982), Twelkemeier (Broster and Norton 1990), and Wells Creek Crater (Dragoo 1973) sites in Tennessee; or the Flint Run site complex in Virginia (Gardner, ed., 1974), to cite a few of the many possible examples. In the case of smaller site assemblages encompassing limited surface or excavation data, such papers accomplish the goal of full publication. Such reporting efforts deserve our commendation, since they ensure that site data get out well ahead of final reports.

Survey projects, and not just excavation activity, also warrant thorough publication. An exceptional report on a site survey focusing on Paleoindian assemblages is that by McAvoy (1992), encompassing portions of southern Virginia in the vicinity of the Nottoway River and its tributaries. Using extensive and well-controlled surface and excavation data from over 100 sites, including the well-known Williamson site, the author advanced a series of detailed observations about culture change over time, differing site types, the activities that likely occurred on these sites, patterns of settlement movement within the study locality, and possible group territories/ranges in the larger region of southern Virginia and northern North Carolina. The importance of this study lies in its extensive presentation of primary data, its use of well-documented avocational collections for serious research, and in highlighting the variability in local Paleoindian site assemblages. A second well-documented survey effort is that by Goodyear and Charles (1984) of sites of all periods in the Allendale chert quarry area in South Carolina. Finally, the primary data Michael K. Faught and I have used to generate fluted-point distribution maps has been available on request for over a decade; these data have been posted on the Web for the past several years (Anderson 1990; Anderson and Faught 1998, 2000; Faught et al. 1994).

For larger sites or survey projects, especially those in which extensive data collection over large areas or many seasons occurred, however, short papers or preliminary reports are no substitute for monograph-length treatments or extended papers presenting the full range of data described above. Unfortunately, final reports on large-scale Paleoindian archaeological research projects remain comparatively rare in the Southeast. No final reports exist for many classic sites, and in some cases the original researchers have passed on. Fortunately, if the collections and original records are well maintained, upcoming generations of researchers will be able to salvage some information from this work; graduate students in particular can write theses and dissertations with these materials. We all must take great care when collecting primary data to ensure that as much information as possible is recorded about what we did and found so that someone down the line will be able to write up what was done, even if we are unable to do so ourselves.

Archaeologists should be encouraged to publish detailed site reports and be rewarded when they do so (conversely, when they fail to publish perhaps they should be discouraged from conducting new fieldwork or should be encouraged to team up with people who can write). Within academia, in my opinion, those capable of producing sound descriptive and theoretically informed interpretive site reports or research syntheses should receive the highest priority for promotion and tenure, while those producing trendy theoretical papers with brief intellectual half lives should receive far less consideration. Unfortunately, today's academic world encourages the latter kind of publication and promotes fragmenting research results into a myriad of publication outlets, since what appears to be typically rewarded is a high number of publications and only rarely a solid book or monograph of long-term intellectual quality. This must change if we are to advance as a profession.

Authors of good site reports should thus be cherished as exemplars of responsible archaeological behavior, for they are producing records that will last far beyond their lifetime and will be referred to for generations to come. When we dig, write, and curate our collections and records properly, and do not merely dig and through inaction destroy, we are upholding the highest ethics of our profession, meeting our responsibility to the archaeological record and hence to the people of the past whose lives our work has touched and brought back to life.

Conclusions

The Southeastern Paleoindian archaeological record is among the richest in the New World in terms of numbers of sites and artifacts. The community of scholars working with these assemblages, dedicated professional and avocational archaeologists alike, are generating a wealth of important information. Most of us feel somewhat overwhelmed with what needs to be done, however; and if there is a logical conclusion to a paper pointing out new research directions, it is that more people, avocational and professional archaeologists alike, need to get involved in Paleoindian/late-Pleistocene archaeology in the region.

At present, only rarely do more than one or two people in any given Southeastern state focus on Pleistocene archaeological remains, and even they are typically involved in a host of other duties or research areas. Much of the Paleoindian-period archaeological reporting in Mississippi in recent years, for example, has been accomplished by state archaeologist Sam McGahey (1981, 1987, 1993, 1996) and U.S. Forest Service archaeologist Sam Brookes (1979). To the best of my knowledge, not a single person in the region is able to work full time on the subject. In fact, the vast majority of the primary data and literature produced in the region over the past quarter century reflects the work of about 40 or 50 people. The miracle is that we know as much as we do. The field is thus wide open, and I think I speak for all those working in the Southeast in saying we would welcome with open arms anyone interested in working on Paleoindian archaeology in the region.

Acknowledgments

I wish to thank David S. Brose, Dena Dincauze, J. Christopher Gillam, Albert C. Goodyear, Robert Grumet, Michael K. Faught, Michael Shott, and two anonymous reviewers for useful comments and suggestions that helped shape the contents of this ms. Particular thanks go to Robson Bonnichsen for asking me to participate and for valuable comments throughout the process of writing and revision. All radiocarbon dates were calibrated using the Calib 4.4 program, using a one-sigma date range. Unless given as a range, the date reported reflects an average of the endpoints of the ranges. As new information is obtained, calibrated dates may likewise change, explaining differences from earlier reporting (cf. Anderson 2001, 2004; see also Fiedel 2004:78–80 for a discussion of this effect).

References Cited

Adovasio, J. M., D. Pedler, J. Donahue, and R. Stuckenrath
1999 No Vestiges of a Beginning nor Prospect for an End: Two Decades of Debate on Meadowcroft Rockshelter. In *Ice Age Peoples of North America,* edited by R. Bonnichsen and K. L. Turnmire, pp. 416–31. Center for the Study of the First Americans, Oregon State University, Corvallis.

Anderson, D. G.
1990 The Paleoindian Colonization of Eastern North America: A View from the Southeastern United States. In *Early Paleoindian Economies of Eastern North America,* edited by K. B. Tankersley and B. L. Isaac, pp. 163–216. Research in Economic Anthropology, Supplement 5. JAI Press, Greenwich, Connecticut.

1995 Paleoindian Interaction Networks in the Eastern Woodlands. In *Native American Interaction: Multiscalar Analyses and Interpretations in the Eastern Woodlands,* edited by M. S. Nassaney and K. E. Sassaman, pp. 1–26. University of Tennessee Press, Knoxville.

1996 Models of Paleoindian and Early Archaic Settlement in the Southeastern United States. In *The Paleoindian and Early Archaic Southeast,* edited by D. G. Anderson and K. E. Sassaman, pp. 29–57. University of Alabama Press, Tuscaloosa.

2001 Climate and Culture Change in Prehistoric and Early Historic Eastern North America. *Archaeology of Eastern North America* pp. 143–186.

2003 Southeast Context. In *Paleoamerican Sites in the Eastern United States (13,450–11,200 B.P.),* edited by E. M. Seibert, pp. 20–54. United States Department of the Interior, National Park Service, National Historic Landmarks Program, Washington, D.C.

2004 Paleoindian Occupations in the Southeastern United States. In *New Perspectives on the First Americans,* edited by B. T. Lepper and R. Bonnichsen, pp. 119–28. Center for the Study of the First Americans, Texas A&M University, College Station.

Anderson, D. G., and M. K. Faught
1998 The Distribution of Fluted Paleoindian Projectile Points: Update 1998. *Archaeology of Eastern North America* 26:163–87.

2000 Palaeoindian Artifact Distributions: Evidence and Implications. *Antiquity* 74:507–13.

Anderson, D. G., and J. C. Gillam
2000 Paleoindian Colonization of the Americas: Implications from an Examination of Physiography, Demography, and Artifact Distribution. *American Antiquity* 65(1):43–66.

2001 Paleoindian Interaction and Mating Networks: Reply to Moore and Moseley. *American Antiquity* 66(3):530–35.

Anderson, D. G., and K. E. Sassaman (editors)
1996 *The Paleoindian and Early Archaic Southeast.* University of Alabama Press, Tuscaloosa.

Anderson, D. G., and J. Schuldenrein (editors)
1985 *Prehistoric Human Ecology along the Upper Savannah River: Excavations at the Rucker's Bottom, Abbeville, and Bullard Site Groups.* Russell Papers. Interagency Archaeological Services, National Park Service, Atlanta.

Anderson, D. G., C. E. Cantley, and A. L. Novick
1982 *The Mattassee Lake Sites: Archaeological Investigations along the Lower Santee River in the Coastal Plain of South Carolina.* Special Publication. Interagency Archaeological Services, National Park Service, Atlanta.

Anderson, D. G., R. J. Ledbetter, and L. D. O'Steen
1990 *Paleoindian Period Archaeology of Georgia.* Georgia Archaeological Research Design Paper 6. Laboratory of Archaeology Series Report 28. University of Georgia, Athens.

Banks, L. D.
1990 *From Mountain Peaks to Alligator Stomachs: A Review of Lithic Sources in the Trans-Mississippi South, the Southern Plains, and Adjacent Southwest.* Oklahoma Anthropological Society Memoir 4. Tulsa.

Binford, L. R.
1980 Willow Smoke and Dogs' Tails: Hunter–Gatherer Settlement Systems and Archaeological Site Formation. *American Antiquity* 45:4–20.

Breitburg, E., J. Broster, A. L. Reesman, and R. G. Strearns
1996 The Coats-Hines Site: Tennessee's First Paleoindian Mastodon Association. *Current Research in the Pleistocene* 13:6–8.

Brookes, S. O.
1979 *The Hester Site, An Early Archaic Site in Monroe County, Mississippi: A Preliminary Report.* Mississippi Department of Archives and History, Jackson.

Broster, J. B.
1982 Paleo-Indian Habitation at the Pierce Site (40Cs24); Chester County, Tennessee. *Tennessee Anthropologist* 7:93–104.

Broster, J. B., and G. L. Barker
1992 Second Report of Investigations at the Johnson Site (40Dv400): The 1991 Field Season. *Tennessee Anthropologist* 17(2):120–30.

Broster, J. B., and M. R. Norton
1990 Lithic Analysis and Paleo-Indian Utilization of the Twelkemeier Site (40Hs173). *Tennessee Anthropologist* 15: 115–31.

1993 The Carson-Conn-Short Site (40Bn190): An Extensive Clovis Habitation in Benton County, Tennessee. *Current Research in the Pleistocene* 10:3–5.

1996 Recent Paleoindian Research in Tennessee. In *The Paleoindian and Early Archaic Southeast,* edited by D. G. Anderson and K. E. Sassaman, pp. 288–97. University of Alabama Press, Tuscaloosa.

Broster, J. B., D. P. Johnson, and M. R. Norton
1991 The Johnson Site: A Dated Clovis-Cumberland Occupation in Tennessee. *Current Research in the Pleistocene* 8:8–10.

Broster, J. B., M. R. Norton, D. J. Stanford, C. V. Haynes, Jr., and M. A. Jodry
1994 Eastern Clovis Adaptations in the Tennessee River Valley. *Current Research in the Pleistocene* 11:12–14.

1996 Stratified Fluted Point Deposits in the Western Valley of Tennessee. In *Proceedings of the 14th Annual Mid-South Archaeological Conference,* edited by R. Walling, C. Wharey, and C. Stanley, pp. 1–11. Panamerican Consultants, Inc. Special Publications 1. Tuscaloosa, Alabama.

Brown, I. W.
1994 Recent Trends in the Archaeology of the Southeastern United States. *Journal of Archaeological Research* 2:45–111.

Butzer, K. W.
1991 An Old World Perspective on Possible Wisconsinan Settlement

of the Americas. In *The First Americans: Search and Research,* edited by T. D. Dillehay and D. J. Meltzer, pp. 137–56. CRC Press, Inc., Boca Raton, Florida.

Cable, J. S.
1982 Organizational Variability in Piedmont HunterGatherer Lithic Assemblages. In *The Haw River Sites: Archaeological Investigation at Two Stratified Sites in the North Carolina Piedmont,* assembled by S. R. Claggett and J. S. Cable, pp. 637–88. Report R–2386. Commonwealth Associates, Inc., Jackson, Michigan.

1996 Haw River Revisited: Implications for Modeling Late Glacial and Early Holocene Hunter–Gatherer Settlement Systems in the Southeast. In *The Paleoindian and Early Archaic Southeast,* edited by D. G. Anderson and K. E. Sassaman, pp. 107–48. University of Alabama Press, Tuscaloosa.

Charles, T.
1981 Dwindling Resources: An Overture to the Future of South Carolina's Archaeological Resources. South Carolina Institute of Archaeology and Anthropology, University of South Carolina, *The Notebook* 13:1–85.

Christen, J. A., and C. D.Litton
1995 A Bayesian Approach to Wiggle-Matching. *Journal of Archaeological Science* 22:719–25.

Claggett, S. R., and J. S. Cable (editors)
1982 *The Haw River Sites: Archaeological Investigations at Two Stratified Sites in the North Carolina Piedmont.* Report R–2386. Commonwealth Associates, Inc., Jackson, Michigan.

Clausen, C. J., A. D. Cohen, C. Emeliani, J. A. Holman, and J. J. Stipp
1979 Little Salt Spring, Florida: A Unique Underwater Site. *Science* 203:609–14.

Coe, J. L.
1964 *The Formative Cultures of the Carolina Piedmont.* Transactions of the American Philosophical Society 54(5). Philadelphia.

Daniel, I. R., Jr.
1998 *Hardaway Revisited: Early Archaic Settlement in the Southeast.* University of Alabama Press, Tuscaloosa.

2001 Stone Raw Material Availability and Early Archaic Settlement in the Southeastern United States. *American Antiquity* 66:237–65.

Daniel, I. R., Jr., and J. R. Butler
1991 Rhyolite Sources in the Carolina Slate Belt, Central North Carolina. *Current Research in the Pleistocene* 8:64–66.

Daniel, I. R., Jr., and M. Wisenbaker
1987 *Harney Flats: A Florida Paleoindian Site.* Baywood Publishing Company, New York.

DeJarnette, D. L., E. Kurjack, and J. Cambron
1962 Excavations at the Stanfield-Worley Bluff Shelter. *Journal of Alabama Archaeology* 8(1-2):1–124.

Dincauze, D. F.
1993a Fluted Points in the Eastern Forests. In *From Kostenki to Clovis: Upper Paleolithic Paleo-Indian Adaptations,* edited by O. Soffer and N. D. Praslov, pp. 279–92. Plenum Press, New York.

1993b Pioneering in the Pleistocene: Large Paleoindian Sites in the Northwest. In *Archaeology of Eastern North America: Papers in Honor of Stephen Williams,* edited by J. B. Stoltman, pp. 43–60. Mississippi Department of Archives and History, Archaeological Report No. 25, Jackson.

Doran, G. H., D. N. Dickel, W. E. Ballinger, Jr., O. F. Agee, P. J. Laipis, and W. W. Hauswirth
1988 Anatomical, Cellular and Molecular Analysis of 8,000-yr-old Human Brain Tissue from the Windover Archaeological Site. *Nature* 323:803–806.

Dragoo, D. W.
1973 Wells Creek: An Early Man Site in Stewart County, Tennessee. *Archaeology of Eastern North America* 1:1–56.

Driskell, B. N.
1994 Stratigraphy and Chronology at Dust Cave. *Journal of Alabama Archaeology* 40:18–33.

1996 Stratified Late Pleistocene and Early Holocene Deposits at Dust Cave, Northwestern Alabama. In *The Paleoindian and Early Archaic Southeast,* edited by D. G. Anderson and K. E. Sassaman, pp. 222–37. University of Alabama Press, Tuscaloosa.

Dunbar, J. S.
1991 Resource Orientation of Clovis and Suwannee Age Paleoindian Sites in Florida. In *Clovis: Origins and Adaptations,* edited by R. Bonnichsen and K. L. Turnmire, pp. 185–213. Center for the Study of the First Americans, Oregon State University, Corvallis.

Dunbar, J. S., and B. I. Waller
1983 A Distribution of the Clovis/Suwannee Paleoindian Sites of Florida: A Geographic Approach. *Florida Anthropologist* 36:18–30.

Dunbar, J. S., S. D. Webb, and M. Faught
1988 Page-Ladson (8Je591): An Underwater Paleo-Indian Site in Northwestern Florida. *Florida Anthropologist* 41:442–52.

Eighmy, J. J. L., and R. S. Sternberg (editors)
1990 *Archaeomagnetic Dating.* The University of Arizona Press, Tucson

Ellis, C., A. C. Goodyear, D. F. Morse, and K. B. Tankersley
1998 Archaeology of the Pleistocene–Holocene Transition in Eastern North America. *Quaternary International* 49/50:151–66.

Engelbrecht, W., and C. Seyfort
1994 Paleoindian Watercraft: Evidence and Implications. *North American Archaeologist* 15:221–34.

Faught, M. K.
1996 *Clovis Origins and Underwater Prehistoric Archaeology in Northwestern Florida.* Ph.D. dissertation, Department of Anthropology, University of Arizona, Tucson.

2004 The Underwater Archaeology of Paleolandscapes, Apalachee Bay, Florida. *American Antiquity* 69:275–89.

Faught, M. K., and J. F. Donoghue
1997 Marine Inundated Archaeological Sites and Paleofluvial Systems: Examples from a Karst Controlled Continental Shelf Setting in the Apalachee Bay, Northeastern Gulf of Mexico. *Geoarchaeology* 12 (5):417–58.

Faught, M. K., J. S. Dunbar, and S. D. Webb
1992 New Evidence for Paleoindians on the Continental Shelf of Northwestern Florida. *Current Research in the Pleistocene* 9:11–12.

Faught, M. K., D. G. Anderson, and A. Gisiger
1994 North American Paleoindian Database: An Update. *Current Research in the Pleistocene* 11:32–35.

Ferring, C. R.
1995 The Late Quaternary Geology and Archaeology of the Aubrey Clovis Site, Texas. In *Ancient Peoples and Landscapes,* edited by E. Johnson, pp. 273–82. Museum of Texas Tech University, Lubbock.

Fiedel, S. J.
1999 Older than We Thought: Implications of Corrected Radiocarbon Dates for Paleo-Indians. *American Antiquity* 64:99–115.

2000 The Peopling of the New World: Present Evidence, New Theories, and Future Directions. *Journal of Archaeological Research* 8:39–103.

2004 Clovis Age in Calendar Years: 13,500–13,000 CALYBP. In *New Perspectives on the First Americans,* edited by B. T. Lepper and R. Bonnichsen, pp. 73–80. Center for the Study of the First Americans, Texas A&M University, College Station.

Frink, D. S.
1992 The Chemical Variability of Carbonized Organic Matter through Time. *Archaeology of Eastern North America* 20:67–79.

1994 The Oxidizable Carbon Ratio (OCR): A Proposed Solution to Some of the Problems Encountered with Radiocarbon Data. *North American Archaeologist* 15:17–29.

1999 The Scientific Basis of Oxidizable Carbon Ratio (OCR) Dating. *SAA Bulletin* 17(5):32–36.

Gagliano, S. M., and H. F. Gregory
1965 *A Preliminary Survey of Paleo-Indian Points from Louisiana.* Louisiana Studies Institute, Northwestern State College, Louisiana Studies 4(1). Natchitoches.

Gardner, W. M. (editor)
1974 *The Flint Run Paleoindian Complex: A Preliminary Report 1971 through 1973 Seasons.* Occasional Paper No. 1. Archaeology Laboratory, Catholic University of America. Washington, D.C.

Gardner, W. M.
1983 Stop Me If You've Heard This One Before: The Flint Run Paleoindian Complex Revisited. *Archaeology of Eastern North America* 11:49–59.

1989 An Examination of Cultural Change in the Late Pleistocene and Early Holocene (circa 9200 to 6800 B.C.). In *Paleoindian Research in Virginia: A Synthesis,* edited by J. M. Wittkofski and T. R. Reinhart, pp. 5–51. Special Publication 19. Archeological Society of Virginia, Richmond.

Gillam, J. C.
1996 A View of Paleoindian Settlement from Crowley's Ridge. *Plains Anthropologist* 157:273–86.

1999 Paleoindian Settlement in Northeastern Arkansas. In *Arkansas Archaeology: Papers in Honor of Dan and Phyllis Morse,* edited by R. C. Mainfort, Jr. and M. D. Jeter, pp. 99–118. University of Arkansas Press, Fayetteville.

Goldman-Finn, N., and B. N. Driskell
1994 Introduction to Archaeological Research at Dust Cave. *Journal of Alabama Archaeology* 40:1–16.

Goodyear, A. C., III
1974 *The Brand Site: A Techno-functional Study of a Dalton Site in Northeast Arkansas.* Arkansas Archaeological Survey, Research Series 7. Fayetteville.

1979 *A Hypothesis for the Use of Cryptocrystalline Raw Materials Among Paleoindian Groups of North America.* Research Manuscript Series 156. South Carolina Institute of Archaeology and Anthropology, University of South Carolina, Columbia.

1982 The Chronological Position of the Dalton Horizon in the Southeastern United States. *American Antiquity* 47:382–95.

1989 A Hypothesis for the Use of Cryptocrystalline Raw Materials among Paleoindian Groups of North America. In *Eastern Paleoindian Lithic Resource Use,* edited by C. Ellis and J. C. Lothrop, pp. 1–9. Westview Press, Boulder, Colorado.

1992 Archaeological and Pedological Investigations at Smiths Lake Creek (38AL135), Allendale County, South Carolina. *Current Research in the Pleistocene* 9:18–20.

1999a The Early Holocene Occupation of the Southeastern United States: A Geoarchaeological Summary. In *Ice Age Peoples of North America,* edited by R. Bonnichsen and K. L. Turnmire, pp. 432–81. Center for the Study of the First Americans, Oregon State University, Corvallis.

1999b Results of the 1999 Allendale Paleoindian Expedition. *Legacy* 4(1-3):8–13. Newsletter of the South Carolina Institute of Archaeology and Anthropology, University of South Carolina, Columbia.

2000 The Topper Site 2000. Results of the 2000 Allendale Paleoindian Expedition. *Legacy* 5(2):18–25. Newsletter of the South Carolina Institute of Archaeology and Anthropology, University of South Carolina, Columbia.

2004 Evidence of Pre-Clovis Sites in the Eastern United States. In *Paleoamerican Origins: Beyond Clovis,* pp. 103–12. Center for the Study of the First Americans, Texas A&M University, College Station.

Goodyear, A. C., III, and T. Charles
1984 *An Archaeological Survey of Chert Quarries in Western Allendale County, South Carolina.* Research Manuscript Series 195. South Carolina Institute of Archaeology and Anthropology, University of South Carolina, Columbia.

Goodyear, A. C., III, S. B. Upchurch, T. Charles, and A. B. Albright
1985 Chert Sources and Paleoindian Lithic Processing in Allendale County, South Carolina. *Current Research in the Pleistocene* 2:47–49.

Goodyear, A. C., III, J. L. Michie, and T. Charles
1990 *The Earliest South Carolinians: The Paleoindian Occupation of South Carolina.* Occasional Papers 2. Archaeological Society of South Carolina, Inc. Columbia.

Goslar, T., and W. Madry
1998 Using the Bayesian Method to Study the Precision of Dating by Wiggle-Matching. *Radiocarbon* 40(1-2).

Gramly, R. M.
1993 *The Richey Clovis Cache: Earliest Americans Along the Columbia River.* Persimmon Press, Buffalo, New York.

Hassan, F. A.
1981 *Demographic Archaeology.* Academic Press, New York.

Hranicky, W. J.
2004 McCary Fluted Point Survey: Points 1000 to 1008—A Continuing Study of Virginia Paleoindian Technology. *Quarterly Bulletin, Archeological Society of Virginia* 59:25–52.

Hubbert, C. M.
1989 Paleo-Indian Settlement in the Middle Tennessee Valley: Ruminations from the Quad Paleo-Indian Locale. *Tennessee Anthropologist* 14(2):148–64.

Jodry, M.
1999 Alternate Routes beyond the Horizon: Certitude, Suspended Disbelief, and Surprise. Paper presented at the Clovis and Beyond Conference, Santa Fe, New Mexico.

Johnson, M. F.
1997 Additional Research at Cactus Hill: Preliminary Description of Northern Virginia Chapter–ASV's 1993 and 1995 Excavations. In *Archaeological Investigations of Site 44SX202, Cactus Hill, Sussex County, Virginia,* edited by J. M. McAvoy and L. D. McAvoy, Appendix G. Virginia Department of Historic Resources, Research Report Series No. 8. Richmond.

Johnson, M. F., and J. E. Pearsall
1993 The Dr. Ben C. McCary Virginia Fluted Point Survey: Nos. 880–885, 891–920. *Quarterly Bulletin of the Archeological Society of Virginia* 48:45–63.

1995 The Dr. Ben C. McCary Virginia Fluted Point Survey: Nos. 921–941. *Quarterly Bulletin of the Archeological Society of Virginia* 50(1):17–31.

1996 The Dr. Ben C. McCary Virginia Fluted Point Survey: Nos.

942–951. *Quarterly Bulletin of the Archeological Society of Virginia* 51(4):178–85.

Jones, B. C., and L. D. Tesar
2000 The Wakulla Springs Lodge site (8Wa329): A Preliminary Report on a Stratified Paleoindian through Archaic site, Wakulla County, Florida. *The Florida Anthropologist* 53(2-3):98–116.

Kelly, A. R.
1938 A Preliminary Report on Archaeological Explorations at Macon, Georgia. *Bureau of American Ethnology Bulletin* 119:1–69.

Kelly, R. L.
1983 Hunter–Gatherer Mobility Strategies. *Journal of Anthropological Research* 39:277–306.

1995 *The Foraging Spectrum.* Smithsonian Institution Press, Washington, D.C.

Kelly, R. L., and L. C. Todd
1988 Coming into the Country: Early Paleoindian Hunting and Mobility. *American Antiquity* 53:231–44.

Killick, D. J., A. J. T. Jull, and G. S. Burr
1999 A Failure to Discriminate: Querying Oxidizable Carbon Ratio (OCR) Dating. *SAA Bulletin* 17(5):32–36.

Lahren, L., and R. Bonnichsen
1974 Bone Foreshafts from a Clovis Burial in Southwestern Montana. *Science* 186:147–50.

Ledbetter, R. J., S. A. Kowalewski, and L. D. O'Steen
1981 Chert of Southern Oconee County, Georgia. *Early Georgia* 9:1–13.

Lepper, B. T., and D. J. Meltzer
1991 Late Pleistocene Human Occupation of the Eastern United States. In *Clovis: Origins and Adaptations,* edited by R. Bonnichsen and K. L. Turnmire, pp. 175–184. Center for the Study of the First Americans, Oregon State University, Corvallis.

Lopinot, N. H., J. H. Ray, and M. D. Conner
1998 *The 1997 Excavations at the Big Eddy Site (23CE426) in Southwest Missouri.* Center for Archaeological Research, Southwest Missouri State University Special Publication No. 2, Springfield.

2000 *The 1999 Excavations at the Big Eddy Site (23CE426).* Center for Archaeological Research, Southeast Missouri State University Special Publication No. 3, Springfield.

Marshall, E.
2001 Pre-Clovis Sites Fight for Acceptance. *Science* 291:1730–1732.

Mason, R. J.
1962 The Paleo-Indian Tradition in Eastern North America. *Current Anthropology* 3:227–83.

McAvoy, J. M.
1992 *Nottoway River Survey: Part I: Clovis Settlement Patterns: The 30 Year Study of a Late Ice Age Hunting Culture on the Southern Interior Coastal Plain of Virginia.* Archeological Society of Virginia Special Publication Number 28. Dietz Press, Richmond, Virginia.

McAvoy, J. M., and L. D. McAvoy
1997 *Archaeological Investigations of Site 44SX202, Cactus Hill, Sussex County, Virginia.* Virginia Department of Historic Resources, Research Report Series No. 8. Richmond, Virginia.

McAvoy, J. M., J. C. Baker, J. K. Feathers, R. L. Hodges, L. J. McWeeney, and T. R. Whyte
2000 Summary of Research at the Cactus Hill Archaeological Site, 40SX202, Sussex County, Virginia: Report to the National Geographic Society in Compliance with Stipulations of Grant #6345-98. National Geographic Society, Washington, D.C.

McCary, B. C.
1951 A Workshop Site of Early Man in Dinwiddie County, Virginia. *American Antiquity* 17:9–17.

1975 The Williamson Paleo-Indian Site, Dinwiddie County, Virginia. *The Chesopiean* 13:48–131. Reprinted in Peck 1985

1984 *Survey of Virginia Fluted Points.* Archaeological Society of Virginia Special Publication No. 12.

1991 *Survey of Virginia Fluted Points.* Archeological Society of Virginia Special Publication 12. 2nd ed., rev. (originally published 1984).

McDonald, J. N.
2000 An Outline of the Pre-Clovis Archeology of SV-2, Saltville, Virginia, with Special Attention to a Bone Tool Dated 14,510 yr CALYBP *Jeffersonia: Contributions from the Virginia Museum of Natural History* 9:1–59.

McGahey, S. O.
1981 The Coldwater and Related Late Paleo-Indian Projectile Points. *Mississippi Archaeology* 16:39–52.

1987 Paleo-Indian Lithic Material: Implications of Distributions in Mississippi. *Mississippi Archaeology* 22:1–13.

1993 Mississippi Paleoindian, Early Archaic Survey. *Mississippi Archaeology* 28:1–44.

1996 Paleoindian and Early Archaic Data from Mississippi. In *The Paleoindian and Early Archaic Southeast,* edited by D. G. Anderson and K. E. Sassaman, pp. 354–84. University of Alabama Press, Tuscaloosa.

Meltzer, D. J.
1984 *Late Pleistocene Human Adaptations in Eastern North America.* Ph.D. dissertation, Department of Anthropology, University of Washington, Seattle.

1988 Late Pleistocene Human Adaptations in Eastern North America. *Journal of World Prehistory* 2:1–53.

Meltzer, D. J., and B. D. Smith
1986 Paleo-Indian and Early Archaic Subsistence Strategies in Eastern North America. In *Foraging, Collecting, and Harvesting: Archaic Period Subsistence and Settlement in the Eastern Woodlands,* edited by S. Neusius, pp. 1–30. Center for Archaeological Investigations, Southern Illinois University, Carbondale.

Michie, J. L.
1996 The Taylor Site: An Early Occupation in Central South Carolina. In *The Paleoindian and Early Archaic Southeast,* edited by D. G. Anderson and K. E. Sassaman, pp. 238–69. University of Alabama Press, Tuscaloosa.

Moore, J. H., and M. E. Moseley
2001 How Many Frogs Does It Take to Leap around the Americas? Comments on Anderson and Gillam. *American Antiquity* 66:526–29.

Morrow, J.
1996 *The Organization of Early Paleoindian Lithic Technology in the Confluence Region of the Mississippi, Illinois, and Missouri Rivers.* Ph.D. Dissertation, Department of Anthropology, Washington University, St. Louis.

Morse, D. F.
1975 Paleoindian in the Land of Opportunity: Preliminary Report on the Excavations at the Sloan Site (3GE94). In *The Cache River Archaeological Project: An Experiment in Contract Archaeology,* assembled by M. B. Schiffer and J. H. House, pp. 93–113. Arkansas Archaeological Survey, Research Series 8. Fayetteville.

1977 Dalton Settlement Systems: Reply to Schiffer (2). *Plains Anthropologist* 22:149–58.

Morse, D. F. (editor)
1997 *Sloan: A Paleoindian Dalton Cemetery in Arkansas.* Smithsonian Institution Press, Washington, D.C.

Morse, D. F., and A. C. Goodyear, III
1973 The Significance of the Dalton Adze in Northeast Arkansas. *Plains Anthropologist* 18:316–22.

Morse, D. F., D. G. Anderson, and A. C. Goodyear, III
1996 The Pleistocene-Holocene Transition in the Eastern United States. In *Humans at the End of the Ice Age: The Archaeology of the Pleistocene-Holocene Transition,* edited by L. G. Strauss, B. V. Eriksen, J. M. Erlandson, and D. R. Yesner, pp. 319–38. Plenum Press, New York.

Nami, H. G., M. R. Norton, D. J. Stanford, and J. B. Broster
1996 Comments on Eastern Clovis Lithic Technology at the Carson-Conn-Short Site (40BN190), Tennessee River Valley. *Current Research in the Pleistocene* 13:62–64.

Neill, W. T.
1958 A Stratified Early Site at Silver Springs, Florida. *The Florida Anthropologist* 11:32–52.

Peck, R. M. (editor)
1985 *The Williamson Site, Dinwiddie County, Virginia.* Privately Printed by Rodney Peck, Harrisburg, North Carolina.

Rolingson, M. A.
1964 *Paleoindian Culture in Kentucky: A Study Based on Projectile Points.* Studies in Archaeology 2. University of Kentucky, Lexington.

Rolingson, M. A., and D. W. Schwartz
1966 *Late Paleo-Indian and Early Archaic Manifestations in Western Kentucky.* Studies in Archaeology 3. University of Kentucky, Lexington.

Sanders, T. N.
1988 The Adams Site: A Paleoindian Manufacturing and Habitation Site in Christian County, Kentucky. In *Paleoindian and Archaic Research in Kentucky,* edited by C. Hockensmith, D. Pollack, and T. Sanders, pp. 1–24. Kentucky Heritage Council Press, Frankfort.

1990 *Adams: The Manufacturing of Flaked Stone Tools at a Paleoindian Site in Western Kentucky.* Persimmon Press, Buffalo, New York.

Shott, M. J.
1996 An Exegesis of the Curation Concept. *Journal of Anthropological Research* 52:259–280.

2002 Sample Bias in the Distribution and Abundance of Midwestern Fluted Bifaces. *Midcontinental Journal of Archaeology* 27:89–123.

Stahle, D. W., E. R. Cook, and J. W. C. White
1985 Tree-Ring Dating of Baldcypress and the Potential for Millennia-long Chronologies in the Southeast. *American Antiquity* 50:796–802.

Steele, J., J. Adams, and T. Sluckin
1998 Modeling Paleoindian Dispersals. *World Archaeology* 30(2):286–305.

Stuiver, M., P. J. Reimer, E. Bard, J. W. Beck, G. S. Burr, K. A. Hughen, B. Kromer, G. McCormac, J. van der Plicht, and M. Spurk
1998 INTCAL98 Radiocarbon Age Calibration, 24,000–0 CALYBP. *Radiocarbon* 40(3):1041–83.

Tankersley, K. B.
1990 Paleoindian Period. In *The Archaeology of Kentucky: Past Accomplishments and Future Directions,* edited by D. Pollack, pp. 73–142. Kentucky Heritage Council, Frankfort.

1991 A Geoarchaeological Investigation of Distribution and Exchange in the Raw Material Economies of Clovis Groups in Eastern North America. In *Raw Material Economies among Prehistoric Hunter–Gatherers,* edited by A. Montet-White and S. Holen, pp. 285–303. University of Kansas Publications in Anthropology 19, Lawrence.

1994 Was Clovis a Colonizing Population in Eastern North America? In *The First Discovery of America,* edited by W. S. Dancey, pp. 95–116. Ohio Archaeological Council, Columbus.

1998 Variation in Early Paleoindian Economies of Late Pleistocene Eastern North America. *American Antiquity* 63:7–20.

Taylor, R. E., C. V. Haynes, and M. Stuiver
1996 Clovis and Folsom Age Estimates: Stratigraphic Context and Radiocarbon Calibration. *Antiquity* 70(269):515–25.

Upchurch, S. B.
1984 Petrology of Selected Lithic Materials from the South Carolina Coastal Plain. In *An Archaeological Survey of Chert Quarries in Western Allendale County, South Carolina,* edited by A. C. Goodyear, III, and T. Charles, pp. 125–60. Research Manuscript Series 195. South Carolina Institute of Archaeology and Anthropology, University of South Carolina, Columbia.

Walker, R. B.
2000 *Subsistence Strategies from Late Paleoindian and Archaic Contexts at Dust Cave, Alabama.* University of Alabama, Office of Archaeological Services, Report of Investigations. Tuscaloosa, Alabama.

Walthall, J. A.
1998 Rockshelters and Hunter-Gatherer Adaptation to the Pleistocene/Holocene Transition. *American Antiquity* 63:223–38.

Walthall, J. A., and B. Koldehoff
1998 Hunter–gatherer Interaction and Alliance Formation: Dalton and the Cult of the Long Blade. *Plains Anthropologist* 43:257–73.

Webb, C. H., J. L. Shiner, and E. W. Roberts
1971 The John Pearce Site (16CD56): A San Patrice Site in Caddo Parish, Louisiana. *Bulletin of the Texas Archaeological Society* 42:1–49.

Webb, W. S.
1951 *The Parrish Village Site.* Reports in Anthropology 7(6). University of Kentucky, Lexington.

Williams, S., and J. B. Stoltman
1965 An Outline of Southeastern United States Prehistory with Particular Emphasis on the Paleoindian Era. In *The Quaternary of the United States,* edited by H. E. Wright and D. G. Frey, pp. 669–83. Princeton University Press, Princeton, New Jersey.

Wittkofski, J. M., and T. R. Reinhart (editors)
1989 *Paleoindian Research in Virginia: A Synthesis.* Special Publication 19. Archeological Society of Virginia, Richmond.

Comparing Clovis and the Western European Upper Paleolithic: What Are the Rules of Evidence?

Michael B. Collins

Abstract

Prehistorians have long debated whether similarities between Clovis of North America and various upper-Paleolithic archaeological manifestations of Western Europe during the late Pleistocene are the result of convergence or of historical relatedness. Debate at present is spirited and centered on the specific issue of similarly produced chipped-stone bifaces of Clovis (North America) and of Solutrean (southwestern Europe) affiliations. Clearly this and many other shared traits beg to be explained, but the effort must effectively address two serious impediments: data limitations and the lack of clear rules of evidence for such inquiries. Little confidence can be placed in the kinds of subjective assessments and emphatic assertions that have always characterized this field of research.

Introduction

Similarities between Clovis material culture and that of several of the upper-Paleolithic manifestations of Europe have often been noted and discussed. Paleolithic prehistorians sometimes refer to North American Paleoindian artifacts as a far-flung expression of the upper Paleolithic (e.g., Bordes 1968:213–19), and Americanists occasionally comment on resemblances between aspects of upper-Paleolithic and North American Paleoindian artifacts (e.g., Boldurian and Cotter 1999:95–97; Haynes 1982, 1987; Wilmsen and Roberts 1978:99–100, 180). Recent suggestions by Bruce Bradley and Dennis Stanford that there may be a connection between Clovis or pre-Clovis in North America and the upper-Paleolithic Solutrean of France and Spain (Stanford and Bradley 2002, and as reported by Begley and Murr 1999:53–55; Dorfman 2000:70; Parfit 2000:60–61; Preston 1997:76–78; Sellet 1998) elicited a firmly stated rebuttal from Solutrean expert Lawrence Straus (2000). In this commentary, I venture into this fray (with a liberal measure of trepidation) to express my views on the subject.

Basically, from my point of view the issue is considerably more complex than either the Bradley/Stanford or the Straus position indicates; therefore, neither position comes close to a resolution. Importantly, a major impediment to resolution is lack of evidentiary principles in our discipline. I do not in any form or fashion presume to resolve the issue with these comments, but I do hope to set the stage for a more objective and comprehensive formulation of the question and examination of the evidence (see Sellet 1998 for a brief but similar comment).

To summarize the two positions briefly, Bradley and Stanford (2002; and quoted in Begley and Murr 1999; Dorfman 2000; Parfit 2000; Preston 1997) suggest that lithic technological similarities between Clovis and Solutrean possibly are great enough to indicate historical derivation of Clovis from Solutrean; Straus (2000:224; also quoted in Parfit 2000:61 and in Preston 1997:77) counters that the similarities are probably convergent, since Solutrean is removed from Clovis by 5,000 km of Atlantic Ocean and 5,000 radiocarbon years. The particular shared traits at the heart of this debate are the form and technological details of Clovis and Solutrean bifaces.

As Bradley and Stanford observe, bifacial lanceolate points and large bifaces produced using a highly skilled and intentional technique of overshot flaking occur in both Clovis and Solutrean to the virtual exclusion of other bifacial industries the world over. Straus argues that, on the contrary, overshot flaking is not unique to Solutrean and Clovis and that, even if it were, it would not be an indicator of relatedness.

Having studied the Solutrean lithics, especially debitage, from Laugerie Haute as part of my dissertation research [Collins 1974],

TARL, PRC5, R7500, The University of Texas at Austin, Austin, TX 78576; e-mail: m.b.collins@mail.utexas.edu invested considerable effort in looking at Clovis lithics, and listened to formal and informal verbal presentations of their views by Stanford and by Bradley, I concur with Bradley and Stanford's basic observation that Solutrean and Clovis bifaces were manufactured using remarkably similar knapping techniques and that those techniques are not characteristic of other bifacial industries of which I am aware.

At that most fundamental level of archaeological analysis, when we invoke the venerable comparative approach, we expect to see similarities and differences between the manifestations being compared. Two or more archaeological cultures are considered "related" if they have enough traits in common, especially if those traits are not likely to be the result of techno-functional responses to similar needs. Historically, such assertions of relatedness have been quite subjective and usually immune to empirical ratification or rejection. Intuitively, more shared traits are generally taken to indicate greater relatedness than are fewer common traits. I know of no satisfactory basis for objectively and empirically judging the degree of relatedness indicated by the similarities itemized below.

A number of traits are, in my opinion, shared between Clovis and the upper Paleolithic of western Europe, and these vary from general to specific, from technological to ideological. My purpose is not to argue that this array of shared traits is indicative of historical relatedness, but to ask, What is the best explanation for these similarities, and how do we objectively approach the issue? Among possible explanations are the usual three alternatives: (1) some degree of direct historical relatedness, (2) indirect historical relatedness ("stimulus diffusion"), or (3) convergence ("independent invention"). These are not new revelations; a number of scholars have made similar comparisons over the years. Examples include Müller-Beck (1966) and Haynes (1987); both considered it likely that Clovis derived from early upper-Paleolithic cultures that spread eastward across Eurasia from eastern Europe. Jelinek (1971) reviewed similarities between Paleoindian and upper-Paleolithic (Magdalenian and especially Solutrean) bone and stone artifacts but concluded that the resemblances were the result of convergence. This history of conflicting interpretations of the same evidence is symptomatic of the failure of prehistorians to establish rigorous criteria for distinguishing historical relatedness from convergence.

Shared Traits of Clovis and the Western European Upper Paleolithic

For purposes of this discussion, I adopt a very simplistic notion of "Clovis" as an archaeological horizon that was widespread across North America between ca. 11,200 and 10,900 RCYBP (Collins 1999a and 1999b; Frison 1991; Haynes 1992, 1993; Stanford 1991). The primary defining artifact characteristic is a large lanceolate fluted projectile point, the "Clovis point." Other common material traits are large bifaces, tools on flakes, two forms of blade cores, blades, tools on blades, percussion-flaked mammoth bone, and various forms of osseous rod objects. Uncommon traits include engraved stones, a perforated bone baton or shaft wrench, a barbed harpoon tip of ivory, chipped-stone adzes, and chipped-stone crescents (Boldurian and Cotter 1999; Collins 1998a, 1999a, 1999b; Collins and Brown 2000; Frison 1991; Frison and Bradley 1999; Hannus 1989; Haynes and Hemmings 1968; Hemmings et al. n.d.; Hester 1972; Lahren and Bonnichsen 1974; Mehringer 1988; Stanford 1991; Webb and Hemmings n.d.; Wilke et al. 1991). Isolated finds of diagnostic Clovis points are the most common kind of "site" throughout the Clovis geographic distribution (e.g., AENA 1982; Dincauze 1993; Hester et al. 1982; Hofman and Wyckoff 1991; MacDonald 1971; Mason 1962; Meltzer and Bever 1995; Ranere and Cook 1991; Titmus and Woods 1991a). In eastern North America most substantive sites seem to be occupations, whereas in western North America kill sites and caches are better known. Oasis localities such as Blackwater Draw manifest multiple site functions. Several important Clovis campsites have been discovered and investigated (e.g., Aubrey [Ferring 1989, 1990, 2001], Carson Conn Short [Broster and Norton 1993; Nami et al. 1996], Gault [Collins 1998a, 1999a; Collins and Brown 2000; Collins et al. 1992; Collins et al. 1991] and Thunderbird [Gardner 1974, 1976, 1977, 1983]), of which only the Aubrey site is fully reported. Caches of Clovis stone, bone, and ivory objects occur with some frequency (Collins 1999a, 1999b; Frison 1991; Green 1963; Stanford and Jodry 1988; Wilke et al. 1991; Woods and Titmus 1985), and a burial of apparent Clovis affiliation is inferred for the Anzick (Montana) site (Lahren and Bonnichsen 1974; Wilke et al. 1991). I think it likely that we have a relatively meager sample of the Clovis archaeological record and, consequently, a very incomplete understanding of Clovis lifeways.

The foregoing is not a satisfactory view of Clovis for most purposes, since it ignores the implications of regional variations, many questions of subsistence, and other important details (Meltzer 1993), but it serves in the present attempt to address a broad, general question.

Equally simplistic versions of the upper-Paleolithic archaeological cultures of western Europe are invoked for the same reason. Specifically, I consider traits ascribed to Aurignacian, Gravettian, Solutrean, and Magdalenian primarily as they are expressed in southwestern Europe (France, Spain, Portugal, and Italy) (Bahn and Vertut 1997; Bocquet-Apel and Demars 2000; Bordes 1968; de Lumley [ed.] 1976; Gamble 1986; Laville et al. 1980; Marshack 1969; 1991; Smith 1966a, 1966b; de Sonneville-Bordes 1960, 1963). Each of these "cultures" (or "civilizations," as they are often called) lasted for several millennia and evolved over time in decided contrast to the brief duration of Clovis. In this abbreviated synopsis I do not distinguish the subperiods or regional variants of these manifestations, nor am I concerned with questions of the lifeways represented by these archaeological patterns.

Aurignacian is a widespread manifestation of long duration in Europe and the Near East between about 33,000 and 26,000 years ago. Its characteristic stone artifacts are blades and tools on blades and, less commonly, tools on flakes. These are often produced with steep continuous stepped retouch ("Aurignacian retouch"). Typical tool forms are blades, blade segments, end scrapers on blades (often carinated), blades with extensive bilateral retouch (many with concave edges called "strangulated" blades), and burins (commonly single-faceted). A minor element is small retouched blades. Carinated scrapers as well as other retouched forms also are found on flakes. Bone "points" occur either with split bases or with points at both ends. Simple parietal art is attributable to the Aurignacian (Bahn and Vertut 1997; Bordes 1968; de Lumley [ed.] 1976).

Between ca. 27,000 and 19,000 years ago (and slightly overlapping with the late Aurignacian) is the Gravettian, found widely in Europe (Spain, France, Italy, Belgium, central Europe, Ukraine, and the Russian plain). It is distinguished by the Gravette point (long blades with one edge steeply retouched), burins on truncated blades, complex burins on blades (type Noailles), uncommon stemmed points, and unifacially retouched flakes. Bone points include beveled-base and bi-conical forms. Parietal as well as portable art (including plaquettes) are found in the Gravettian (Bahn and Vertut 1997; de Lumley [ed.] 1976; Ruspoli 1986). In France, the Gravettian is often included as a late phase of the Perigordian.

Next in the upper-Paleolithic sequence is the Solutrean, lasting approximately from 20,000 to 16,000 years ago in central and southern France, Spain, and Portugal. This is one of the most distinctive of the Paleolithic cultures because of its magnificent bifacially flaked "laurel leaves," "willow leaves," and lanceolate points. Other lithic artifacts include unifacial points on blades, single-shouldered points on blades, endscrapers, and burins. Bone objects are beveled-base points, perforated staffs (batons or shaft wrenches) and, notably, eyed needles. Parietal art includes painting as well as sculpture and bas relief forms; portable art objects, including plaquettes, are present in the late Solutrean (Bahn and Vertut 1997; Bordes 1968; de Lumley 1976; Smith 1966a, 1966b).

Magdalenian, the final culture in the upper-Paleolithic sequence, is best known for its spectacular parietal art and portable art. It is distinctive also for its tools of stone and bone. Burins (often dihedral) are common. There are also endscrapers on blades, borers, backed blades, backed bladelets, and denticulated blades. Burins occur on flakes, as do multiple graver tips. Bone objects are often decorated and include beveled-base as well as bi-conical points, spear throwers, eyed needles, batons, and barbed harpoons. The Magdalenian is dated between 16,000 and 10,000 years ago and has its greatest expression in France, Spain, and parts of Italy and Switzerland (Bahn and Vertut 1997; Bordes 1968; de Lumley [ed.]. 1976; Marshack 1969, 1991; Ruspoli 1986).

In spite of a long-held view that the Clovis archaeological manifestation represents a specialized big-game hunting way of life, a closer look at the evidence suggests that, in fact, the groups responsible for what we call Clovis were generalized foragers with the ability to kill large game (including mammoths) and perhaps a propensity to scavenge carcasses of large animals (Ferring 2001:239–243; Meltzer 1993). In this alternative view, Clovis people were hunter-gatherers who hunted large game along with smaller animals, collected such animals as turtles and tortoises, and almost certainly exploited a variety of edible plants; they were generalists, not specialists. The diversity of habitats occupied by Clovis is a prime indication of a generalized rather than a specialized adaptation.

Interpretations of the upper Paleolithic also have matured from a previous narrow perception of a lifeway wholly occupied with big-game hunting (e.g., "les chasseurs," Sonneville-Bordes 1960) to recognizing a series of complex adaptations utilizing diverse resources in a variety of habitats (e.g., Gamble 1986).

Chipped Stone

Clovis chipped-stone objects consist primarily of bifaces, modified flakes, unmodified blades, tools on blades, and blade cores. Among bifaces are fluted projectile points, preforms of projectile points, and large bifacial cores (Bradley 1982; Collins 1999a, 1999b; Frison 1991; Stanford 1991; Wilke et. al 1991). In their form and technology of manufacture, large Clovis bifaces share attributes with Solutrean bifaces. Both Clovis and Solutrean knappers heat-treated bifaces in early reduction stages on occasion but not invariably (Bordes 1969; Collins 1973; Nami et al. 1996; Wilke et al. 1991). Large thin Clovis points, preforms, and bifaces found at the sites of East Wenatchee, Simon, Drake, Fenn, and others exhibit a significant proportion of flake scars extending completely or almost completely across the width of the piece (Frison and Bradley 1999; Gramly 1993; Stanford and Jodry 1988; Woods and Titmus 1985). Similar flake scars are seen on Solutrean laurel-leaf and willow-leaf bifaces (Clottes 1976; Sacchi 1976; Smith 1966a, 1966b). Overshot flakes, noteworthy in both Solutrean and Clovis debitage for their numbers and for the skillful knapping that produced them, did not result from knapping errors that produced the occasional overshot flake (and often ruined the biface) in most bifacial industries (Collins 1974); Clovis and Solutrean overshot flakes intentionally removed a minimal amount of the far edge of the biface being flaked. Typical Clovis points retain remnants of overshot flake scars (e.g., the "type specimen" from Blackwater Draw [Boldurian and Cotter 1999: Figure 25]). Lanceolate bifacial Solutrean projectile points, often with concave bases, are similar to Clovis points in general (Clottes 1976; Smith 1966a, 1966b) but lack the ground basal and lateral edges and the flutes typical of Clovis points. Smith [1963] reports two fluted Solutrean bifacial points from Laugerie Haute; however, he considers them accidental and therefore insignificant in regard to the origin of Clovis fluting.

Clovis blades and blade cores are similar to those of the Aurignacian, Solutrean, and Magdalenian, although bladelets (or microblades) are almost completely lacking in Clovis assemblages (Collins 1999a, 1999b). Tools on blades are comparatively rare in Clovis, but endscrapers, gravers, lateral unifaces, and denticulated blades occur and resemble similar artifacts in the upper Paleolithic, especially Magdalenian. Burins on blades are conspicuous by their near absence in Clovis.

Osseous Rods

Clovis sites from Florida to Washington have yielded elongate, cylindrical, and conical objects of bone or ivory (Boldurian and Cotter 1999; Dunbar 1991; Gramly 1993; Hester 1972; Lahren and Bonnichsen 1974). Many of these are pointed at one end with either beveled or squared bases, quite similar to those of Solutrean and Magdalenian affiliation (Bordes 1968:159–66; de Lumley [ed.] 1976). Bi-beveled bone rods from the East Wenatchee site are similar in design (though larger in size) to those in the Solutrean. At least two of the bi-beveled bone rods from East Wenatchee have long series of tick marks along one edge. Examined closely, each tick mark in the series on one specimen is seen to consist of two marks of unequal size, similar to engraving documented by Marshack (1991:147–68) on carved bones of Magdalenian age. Simple marks in the series on the second East Wenatchee piece resemble engravings on many bone rods in Solutrean and Magdalenian assemblages.

A single perforated baton or shaft wrench from the Murray Springs Clovis site has counterparts in both the Solutrean and the Magdalenian. A fragmentary barbed harpoon point made of ivory from Glory Hole, Florida, possibly of Clovis affiliation (Hemmings et al. 1999; Webb and Hemmings n.d.), is quite similar to Magdalenian examples (e.g., from Grotte de la Salpêtrière [Escalon de Fonton and Bazile 1976: Figure 5–17, 20]).

Although none has yet been reported, it is highly likely that Clovis bone technology included the eyed needle (since it is present in Folsom and would seem to be essential in Clovis). If, in fact, there were Clovis eyed bone needles, they would be similar to Solutrean and Magdalenian eyed bone needles.

Petroglyphs?

Three petroglyphs and a pictograph on the Colorado Plateau in Utah appear to be depictions of mastodons or mammoths (Castleton 1984, 1987). Paleontological evidence establishes the presence of mammoth and mastodon during the late Glacial (15,000–10,000 RCYBP) on the Plateau (Agenbroad and Mead 1989; Graham and Lundelius 1994). At least four sites in the general region, three in Arizona and one in Colorado, document the association of Clovis artifacts with mammoth remains: Lehner (Haury et al. 1959), Murray Springs (Hemmings 1970), Naco (Haury 1953), and Dent (Haynes 1966). Therefore, it is at least theoretically possible that one or more of the petroglyphs at Manila (Castleton 1984:56), near Moab (Castleton 1984:194), and Indian Creek (Castleton 1987:199) represent the work of Clovis-age people. Petroglyphs of large animals, chiefly horses and bovids, at open-air sites in Portugal, Spain, and France have been attributed on stylistic grounds to the upper Paleolithic, from late Solutrean to Magdalenian (Bahn and Vertut 1997:128–133). At issue in both the European and North American cases is whether or not the art is truly of late-Pleistocene age. This is well discussed by Bahn and Vertut in reference to the European sites, and it applies equally to the Utah sites. The painted mammoth in Ferron Canyon (Castleton 1984:214) is dubious, and on geological grounds I am skeptical that sandstone rock surfaces containing the Utah petroglyphs would survive for 11,000 radiocarbon years. However, I find the content of both the European and the American examples intriguing and recognize the possibility that under ideal conditions rock outcrops and their surfaces might remain stable for what is comparatively brief geologic time. In sum, until the ages and authorships of these petroglyphs are better established, it will remain an open question whether or not open-air rock art depiction of large herbivores is a trait shared between the continents during the late Pleistocene.

Ocher

It is commonly acknowledged that beyond its frequent use as a pigment, ocher was evidently used as a ritual material. Upper-Paleolithic graves contain ocher (Quechon 1976) as does the one known Clovis grave (Wilke et al. 1991). But ocher also had technological applications as an ingredient in adhesives in the Magdalenian (Allain and Rigaud 1989) and possibly in Clovis, given its common occurrence on the haft areas of points (Titmus and Woods 1991b:198). It is also possible that Clovis knappers employed ocher in polishing the lateral margins of projectile points (Titmus and Woods 1991b). It is not certain whether the quantities of ocher indicated in certain Clovis caches (Frison and Bradley 1999:78; Wilke et al. 1991) served ritual or functional purposes.

Plaquettes

Small engraved and sometimes painted stones ("plaquettes") are a common form of portable art in the Magdalenian and a less common one in the Solutrean and Gravettian (Bahn and Vertut 1997:41–45, 63, 84–103). Many of these depict animals, but irregular and geometric motifs are also present in modest numbers. Several small engraved stones of Clovis age have been found at the Gault site (Collins et al. 1991; Collins et al. 1992); one each has been found in early-Paleoindian contexts at Wilson-Leonard (Collins 1998b:151) and at Blackwater Draw (Hester 1972:Figure 93g). These predominantly exhibit geometric designs, but at least one and possibly two possess representations of animals. Among Magdalenian specimens, those of Italy (Marshack 1969) bear the greatest similarities to those of Clovis affiliation with multiple sets of crossed and parallel lines.

Discussion

Were one to accept uncritically all the similarities just discussed, it would be tempting to describe Clovis as a generalized upper-Paleolithic-like cultural manifestation in the Western Hemisphere with technological and expressive traits shared with the Aurignacian, Gravettian, Solutrean, and especially the Magdalenian of western Europe. The technological similarities apply to lithic and osseous tools and details of their manufacture and treatment. Heat treatment, overshot flaking, blade making, and the manufacture of numerous stone tools have common expressions. Osseous artifacts, too, are similar in several forms, even down to the scoring on the beveled surfaces of beveled-base "points" and tick marks along margins of bone rods. Ocher served ornamental as well as technological functions in both Clovis and European cultures of the late Pleistocene. Open-air parietal as well as portable plaquette art forms are traits in common. Woven together and augmented with carefully selected illustrations, these several similarities would be the basis of a plausible argument for some form of historical relatedness between Clovis and the western European upper Paleolithic. Prehistorians have certainly argued passionately for cases of relatedness on skimpier evidence.

However, all the upper-Paleolithic cultures of western Europe are geographically far removed from the Clovis culture area of North America, and all but the Magdalenian are significantly earlier in time. Furthermore, upper-Paleolithic cultures differ in many significant ways from Clovis and comparisons between Clovis material culture and various western European upper-Paleolithic manifestations are debatable to a greater or lesser extent. In fact, debate is precisely what is needed, since these comparisons are one facet of the important question of how the Americas were first peopled. The problem with such a debate is that our discipline has not firmly established the rules of evidence in questions of historical relatedness versus convergence and we are missing much of the relevant evidence.

Clarke (1978:328–33) articulated the view that similar adaptive demands could independently produce parallel technological responses from historically unrelated peoples. The result would be two or more "technocomplexes" sharing a number of traits. How similar do similar adaptive demands have to be? Are the adaptive challenges of Clovis comparable to those of any of the upper-Paleolithic cultures of Western Europe? Are they comparable enough to bring about any, some, or all of the convergences discussed above? What are the criteria that prehistorians should use to address these questions objectively and systematically? No archaeologist's answer to these questions would go unchallenged.

If not convergence, then what of historical connections? Time and space are critical elements to be considered in any attempt to trace historical connections. From western Europe, people or ideas could reach North America on a westerly course across the North Atlantic or an easterly course across Eurasia and then either the North Pacific or Beringia. Neither of these avenues, at our present state of knowledge, enjoys much empirical support. There is an elapsed time of some 10,000 years between the late Aurignacian and Clovis, far too great for a simple process of historical succession; however, Clovis is partly contemporaneous with the late Magdalenian. And as a result of northerly expansion during the late Magdalenian (Bocquet-Apel and Demars 2000), the transatlantic distance between it and Clovis is considerably reduced.

The real issue here is not the specific question of a historical connection between the western European upper Paleolithic and Clovis; it is the much broader question of the origins of the founding populations in the Americas and their cultures. Solutrean and Clovis are just two small pieces that may or may not be contiguous parts of the same puzzle. The peopling of the Americas is a complex problem whose solution depends on evidence from much of the world over the last 20,000 years or so. That evidence must be sufficient to resolve questions of human biological, linguistic, material cultural/technological, and environmental history.

At the moment, evidence of two kinds is missing. First, much of the archaeological map for the interval from 20,000 to 11,000 RCYBP is blank (notably parts of Eurasia, the continental shelf and coastal margins), and it is not known if the gaps are real or represent sampling deficiencies. Second, even for those parts of the world for which archaeological assemblages are known, critical information is lacking. We rarely know the biological characteristics of the people responsible for the assemblages; environmental contexts are often unclear; and all too often published reports of assemblages are only suitable for making superficial comparisons. As the Bradley/Stanford proposal underscores, the intrigue is not in the final form of Solutrean and Clovis bifaces, it is in the compelling similarities of their manufacture. Most reports on lithic assemblages fail to present information on cores, debitage, aborted fragments, and other elements that reveal details of knapping behavior.

Countless comparisons that probe beyond superficial similarities of archaeological assemblages are waiting to be made. At the same time, human biologists, linguists, and paleogeographers have much work to do. Empirically supported answers to the question of the peopling of the Americas will begin to emerge after scholars on both sides of the Atlantic and both sides of the Pacific have thoroughly debated the bulk of the meaningful comparisons now known and that will emerge in the future.

We can view the Solutrean-Clovis debate as a prototype. It has merit in that it goes beyond superficiality, but even at that, it only begins to probe sufficiently into the behavior that produced the artifacts being compared. Lacking so far is any consideration of such topics as hafting technology, functions of the compared tools, and other avenues leading to greater understanding of the cultural contexts of these artifacts. It has merit in that it forces reconsideration of the long-held premise that the Americas had to be reached by people on foot (e.g., Jennings 1978:1–4; and succinctly stated by Coles and Higgs [1969:419], "A land bridge is regarded as being the first necessity"). But it also has ponderous gaps of time and space. The mistake would be to dismiss the possibility without thoroughly debating those gaps; perhaps they are not as fatal to the proposal as they first appear. Consider, for example, that as Straus (2000:224) notes, Clovis is removed from Solutrean by 5,000 km of Atlantic Ocean. From the eastern fringe of the upper-Paleolithic (eastern Gravettian) populations in the central Russian plain (Soffer 1993) it is more than 10,000 km to the Clovis realm, via Beringia, on a route that passes through what is today one of the two coldest places on earth. The mean low temperatures in a large area of Siberia between longitudes 135° and 150° and latitudes 60° and 75° are as cold as Antarctica, and, in fact, are colder in one small locality (Fuller and Sadao 1992). It is certainly debatable whether this was a significantly less challenging land route to the Americas than was a maritime one around the North Atlantic ice fringe.

Before debating the merits of any given possibility of historical relatedness, effort must be directed toward improving the rules of evidence for such debates; we also must bring forth and debate what may at first seem to be improbable scenarios. In doing so, renewed scrutiny of a number of old premises is long overdue; foremost among these is our tendency to underestimate the capabilities of humans in the past, a failing that I refer to as "chronoracism." It is impossible to objectively debate, for example, a maritime route for the first American colonizers if it is presumed that late-Pleistocene humans did not have the capacity to construct and navigate watercraft. The specific outcome of any particular debates is not as important as the confidence that we can place in that outcome.

References Cited

AENA
1982 A Compilation of Fluted Points of Eastern North America by Count and Distribution: An AENA project. *Archaeology of Eastern North America* 10:27–45.

Agenbroad, L. D., and J. I. Mead
1989 Quaternary Geochronology and Distribution of *Mammuthus* on the Colorado Plateau. *Geology* 17:861–64.

Allain, J. and A. Rigaud
1989 Colles et mastics au Magdalenien. In Actes du Colloque International de Nemours 1987, sous la direction de M. Olive et Y. Taborin, pp. 221–23. *Memoirs du Musée de Préhistoire d'Ile de France No. 2.*

Bahn, P. G. and J. Vertut
1997 *Journey through the Ice Age.* University of California Press. Berkeley.

Begley, S., and A. Murr
1999 The First Americans. *Newsweek,* 26 April: 50–57.

Bocquet-Apel, J.-P., and P.-Y. Demars
2000 Population Kinetics in the Upper Palaeolithic in Western Europe. *Journal of Archaeological Science* 27:551–70.

Boldurian, A. T., and J. L. Cotter
1999 *Clovis Revisited: New Perspectives on Paleoindian Adaptations from Blackwater Draw, New Mexico.* The University Museum, University of Pennsylvania. Philadelphia.

Bordes, F.
1968 *Old Stone Age.* McGraw-Hill, New York
1969 Traitement Thermique du Silex au Solutreen. *Compte Rendu des Séances Mensuelles de la Société Préhistorique Française* 7:197.

Bradley, B. A.
1982 Flaked Stone Technology and Typology. In *The Agate Basin Site,* edited by G. C. Frison and D. J. Stanford, pp. 181–208. Academic Press, New York.

Broster, J. B. and M. R. Norton
1993 The Carson-Conn-Short Site (40BN190): An Extensive Clovis Habitation in Benton County, Tennessee. *Current Research in the Pleistocene* 10:3–4.

Castleton, K. B.
1984 *Petroglyphs and Pictographs of Utah, Volume One: The East and Northeast.* Utah Museum of Natural History, Salt Lake City.
1987 *Petroglyphs and Pictographs of Utah, Volume Two: The South, Central, West and Northwest.* Utah Museum of Natural History, Salt Lake City.

Clarke, D. L.
1978 *Analytical Archaeology.* Methuen, London.

Clottes, J.
1976 Les civilizations du Paléolithique supérieur dans les Pyrenées. In *La Préhistoire Française,* sous la direction de H. de Lumley, pp. 1214–31. Editions du Centre National de la Recherche Scientifique. Paris.

Coles, J. M. and E. S. Higgs
1969 *The Archaeology of Early Man.* Frederick A. Praeger, New York.

Collins, M. B.
1973 Additional Notes on Thermal Treatment of Chert in the Solutrean of Laugerie Haute Ouest. *Proceedings of the Prehistoric Society* 39:461–66.
1974 A Functional Analysis of Lithic Technology among Prehistoric Hunter-Gatherers of Southwestern France and Western Texas. Unpublished Ph.D. dissertation, The University of Arizona. Tucson.
1998a Interpreting the Clovis Artifacts from the Gault Site. *TARL Research Notes* 6(1):5–11.
1998b Early Paleoindian Components. In *Wilson Leonard, An 11,000-year Archeological Record of Hunter-Gatherers in Central Texas,* edited by M. B. Collins, pp. 123–159. Studies in Archeology 31, Texas Archeological Research Laboratory, The University of Texas at Austin and Archeology Studies Program, Report 10, Texas Department of Transportation, Environmental Affairs Division, Austin.
1999a *Clovis Blade Technology.* University of Texas Press, Austin.
1999b Clovis and Folsom Lithic Technology on and near the Southern Plains: Similar Ends, Different Means. In *Folsom Lithic Technology,* edited by D. S. Amick. Archaeological Series 12. International Monographs in Prehistory, Ann Arbor.

Collins, M. B. and K. M. Brown
2000 The Gault Gisement. *Current Archeology in Texas* 2(1) 8–11.

Collins, M. B., T. R. Hester, and P. J. Headrick
1992 Engraved Cobbles from the Gault Site, Central Texas. *Current Research in the Pleistocene* 9:3–4.

Collins, M. B., T. R. Hester, D. Olmstead, and P. J. Headrick
1991 Engraved Cobbles from Early Archaeological Contexts in Central Texas. *Current Research in the Pleistocene* 8:13–15.

Dincauze, D. F.
1993 Fluted Points in the Eastern Forests. In *From Kostenki to Clovis: Upper Paleolithic-Paleo-Indian Adaptations,* edited by O. Soffer and N. D. Praslov, pp. 279–92. Plenum, New York.

Dorfman, A.
2000 New Ways to the New World. *Time,* April 17: 70.

Dunbar, J. S.
1991 Resource Orientation of Clovis and Suwannee Age Paleoindian Sites in Florida. In *Clovis: Origins and Adaptations,* edited by R. Bonnichsen and K. L. Turnmire, pp. 185–213. Center for the Study of the First Americans, Oregon State University, Corvallis.

Escalon de Fonton, M., and G. Bazile
1976 Les civilisations du Paléolithique supérieur en Provence littorale. In *La Préhistoire Française,* pp. 1145–56. Sous la direction de H. de Lumley, editions du Centre National de la Recherche Scientifique, Paris.

Ferring, C. R.
1989 The Aubrey Clovis Site: A Paleo-Indian Locality in the Upper Trinity River Basin, Texas. *Current Research in the Pleistocene* 6:9–11.
1990 The 1989 Investigations at the Aubrey Clovis Site, Texas. *Current Research in the Pleistocene* 7:10–12.
2001 *The Archaeology and Paleoecology of the Aubrey Clovis Site (41DN479), Denton County, Texas.* Center for Environmental Archaeology, Department of Geography, University of North Texas, Denton.

Frison, G. C.
1991 The Clovis Cultural Compex: New Data from Caches of Flaked

Stone and Worked Bone Artifacts. In *Raw Material Economies among Prehistoric Hunter-Gatherers,* edited by A. Montet-White and S. Holen, pp. 321–33. University of Kansas, Publications in Anthropology 19. Lawrence.

Frison, G., and B. Bradley
1999 *The Fenn Cache: Clovis Weapons and Tools.* One Horse Land & Cattle Company. Santa Fe, New Mexico.

Fuller, R. B., and S. Sadao
1992 *Fuller Projection Dymaxion Air-Ocean Map of the World, Mean Low Temperatures.* Buckminster Fuller Institute, Los Angeles.

Gamble, C.
1986 *The Palaeolithic Settlement of Europe.* Cambridge University Press, Cambridge.

Gardner, W. M.
1974 *The Flint Run Complex: Pattern and Process during the Paleoindian to Early Archaic.* Occasional Publication No. 1:5–47. Archaeology Laboratory, Catholic University of America, Washington, D.C.

1976 Paleoindian to Early Archaic: Continuity and Change in Eastern North America during the Late Pleistocene and Early Holocene. In *Habitats Humaines Antérieurs a l'Holocene en Amerique,* edited by J. B. Griffin, pp. 19–46. Pretirage, Colloque XVII, IXe Congres, Union Internationale des Sciences Préhistoriques et Protohistoriques. Nice, France.

1977 The Flint Run PaleoIndian Complex and Its Implications for Eastern North American Prehistory. *Annals, New York Academy of Sciences* 288:257–63.

1983 Stop Me If You've Heard This One Before: The Flint Run Paleoindian Complex Revisited. *Archaeology of Eastern North America* 11:49–64.

Graham, R. W., and E. L. Lundelius, Jr.
1994 FAUNMAP: A Database Documenting Late Quaternary Distributions of Mammal Species in the United States. *Scientific Papers* XXV(2), Illinois State Museum, Springfield.

Gramly, R. M.
1993 *The Richey Clovis Cache.* Persimmon Press, Buffalo, N.Y.

Green, F. E.
1963 The Clovis Blades: An Important Addition to the Llano Complex. *American Antiquity* 29:145–65.

Hannus, L. A.
1989 Flaked Mammoth Bone from the Lange/Ferguson Site, White River Badlands Area, South Dakota. In *Bone Modification,* edited by R. Bonnichsen and M. H. Sorg, pp. 395–412. Center for the Study of the First Americans, University of Maine, Orono.

Haury, E. W.
1953 Artifacts with Mammoth Remains, Naco, Arizona, I: Discovery of the Naco Mammoth and the Associated Projectile Points. *American Antiquity* 19:1–14.

Haury, E. W., E. B. Sayles, and W. W. Wasley
1959 The Lehner Mammoth Site, Southeastern Arizona. *American Antiquity* 25:2–30.

Haynes, C. V., Jr.,
1966 Elephant Hunting in North America. *Scientific American* 214 (6):104–12.

1982 Were Clovis Progenitors in Beringia? In *Paleoecology of Beringia,* edited by D. M. Hopkins, J. V. Matthews, Jr., C. W. Schweger, and S. B. Young, pp. 383–98. Academic Press, New York.

1987 Clovis Origins Update. *The Kiva* 52(2):83–93.

1992 Contributions of Radiocarbon Dating to the Geochronology of the Peopling of the New World. In *Radiocarbon after Four Decades,* edited by R. E. Taylor, A. Long, and R. S. Kra, pp. 355–74. Springer-Verlag, New York.

1993 Clovis-Folsom Geochronology and Climatic Change. In *From Kostenki to Clovis,* edited by O. Soffer and N. D. Praslov, pp. 219–36. Plenum Press, New York.

Haynes, C. V., Jr. and E. T. Hemmings
1968 Mammoth-Bone Shaft Wrench from Murray Springs, Arizona. *Science* 159:186–87.

Hemmings, C. A., J. S. Dunbar, and S. D. Webb
1999 Florida's Early Paleoindian Bone and Ivory Tools. In New Directions in First American Studies: Extended Abstracts from the Clovis and Beyond Conference, Santa Fe.

Hemmings, E. T.
1970 Early Man in the San Pedro Valley, Arizona. Unpublished Ph.D. dissertation, University of Arizona, Tucson.

Hester, J. J.
1972 Blackwater Draw Locality No. 1: A Stratified Early Man Site in Eastern New Mexico. *Fort Burgwin Research Center Publication* 8. Rancho de Taos, New Mexico.

Hofman, J. L., and D. G. Wyckoff
1991 Clovis Occupation in Oklahoma. *Current Research in the Pleistocene* 8:29–32.

Jelinek, A. J.
1971 Early Man in the New World: A Technological Perspective. *Arctic Anthropology* 8(2):15–21.

Jennings, J. D.
1978 Origins in *Ancient Native Americans,* edited by J. D. Jennings, pp. 1–41. Freeman & Co., San Francisco.

Lahren, L., and R. Bonnichsen
1974 Bone Foreshafts from a Clovis Burial in Southwestern Montana. *Science* 186:147–50.

Laville, H., J.-P. Rigaud, and J. Sackett
1980 *Rock Shelters of the Perigord.* Academic Press, New York.

Lumley, H. de [direction]
1976 *La Préhistoire Française.* Editions du Centre National de la Recherche Scientifique. Paris.

MacDonald, G. F.
1971 A Review of Research on Paleo-Indian in Eastern North America, 1960–1970. *Arctic Anthropology* 8(2):32–41.

Marshack, A.
1969 Polesini: A Reexamination of the Engraved Upper Palaeolithic Mobiliary Materials of Italy by a New Methodology. *Rivista di Scienze Preistoriche* 24:219–81.

1991 *The Roots of Civilization* (Second Edition). Moyer Bell, New York.

Mason, R.
1962 The Paleo-Indian Tradition in Eastern North America. *Current Anthropology* 3:227–83.

Mehringer, P. J.
1988 Weapons Cache of Ancient Americans. *National Geographic* 174:500–503.

Meltzer, D. J.
1993 Is There a Clovis Adaptation? In *From Kostenki to Clovis,* edited by O. Soffer and N. D. Praslov, pp. 293–310. Plenum Press, New York.

Meltzer, D. J., and M. R. Bever
1995 Paleoindians of Texas: An Update on the Texas Clovis Fluted Point Survey. *Bulletin of the Texas Archeological Society* 66:47–80.

Müller-Beck, H.
1966 Paleohunters in America: Origins and Diffusion. *Science* 152:1191–1210.

Nami, H. G. M. R. Norton, D. Stanford, and J. B. Broster
1996 Comments on Eastern Clovis Lithic Technology at the Carson Conn Short Site (40bn190), Tennessee River Valley. *Current Research in the Pleistocene* 13:62–64.

Parfit, M.
2000 Hunt for the First Americans. *National Geographic* 198(6):40–67.

Preston, D.
1997 The Lost Man. *The New Yorker.* June 16:70–81.

Quechon, G.
1976 Les sepultures des Hommes du Paléolithique supérieur. In *La Prehistoire Française,* sous la direction de H. de Lumley, pp. 728–733. Editions du Centre National de la Recherche Scientifique, Paris.

Ranere, A. J., and R. G. Cooke
1991 Paleoindian Occupation in the Central American Tropics. In *Clovis: Origins and Adaptations,* edited by R. Bonnichsen and K. L. Turnmire, pp. 237–53. Center for the Study of the First Americans. Oregon State University, Corvallis.

Ruspoli, M.
1986 *The Cave of Lascaux: The Final Photographs.* Harry N. Abrams, New York.

Sacchi, D.
1976 Les civilizations du Paléolithique supérieur en Languedoc occidental (Bassin de l'Aude) et en Roussillon. In *La Préhistoire Française,* sous la direction de H. de Lumley, pp. 1174–88. Editions du Centre National de la Recherche Scientifique, Paris.

Sellet, F.
1998 The French Connection: Investigating a Possible Clovis-Solutrean Link. *Current Research in the Pleistocene* 15:67–68.

Smith, P. E. L.
1963 A Fluted Point from the Old World. *American Antiquity* 28:397–99.

1966a *Le Solutreen en France.* Delmas, Bordeaux.

1966b The Solutrean Culture. *Scientific American* 211:86–94.

Soffer, O.
1993 Upper Paleolithic Adaptations in Central and Eastern Europe and Man-Mammoth Interactions. In *From Kostenki to Clovis,* edited by O. Soffer and N. D. Praslov, pp. 31–49. Plenum, New York.

Sonneville-Bordes, D. de
1960 *Le Paléolithique Supérieur en Perigord.* Delmas, Paris.

1963 Upper Paleolithic Cultures in western Europe. *Science* 142:347–55.

Stanford, D.
1991 Clovis Origins and Adaptations: An Introductory Perspective. In *Clovis: Origins and Adaptations,* edited by R. Bonnichsen and K. L. Turnmire, pp. 1–13. Center for the Study of the First Americans, Oregon State University, Corvallis.

Stanford, D., and B. Bradley
2002 Ocean Trails and Prairie Paths? Thoughts about Clovis Origins, pp. 255–271. In *The First Americans.* Memoirs of the California Academy of Sciences, Number 27, San Francisco.

Stanford, D., and M. A. Jodry
1988 The Drake Clovis Cache. *Current Research in the Pleistocene* 5:21–22.

Straus, L. G.
2000 Solutrean Settlement of North America? A Review of Reality. *American Antiquity* 65:219–226.

Titmus, G. L., and J. C. Woods
1991a Fluted Points from the Snake River Plain. In *Clovis: Origins and Adaptations,* edited by R. Bonnichsen and K. L. Turnmire, pp. 119–131. Center for the Study of the First Americans, Oregon State University. Corvallis.

1991b A Closer Look at Margin "Grinding" on Folsom and Clovis Points. *Journal of California and Great Basin Anthropology* 13:194–203.

Webb, S. D., and C. A. Hemmings
n.d. Ivory and Bone Tools from Late Pleistocene Deposits in the Aucilla and Wacissa Rivers, North-Central Florida. In Wetlands Archaeology Conference Proceedings, edited by Barbara Purdy. Oxbow Press, Oxford.

Wilke, P. J., J. J. Flenniken, and T. L. Ozbun
1991 Clovis Technology at the Anzick Site, Montana. *Journal of California and Great Basin Anthropology* 13 (2):242–72.

Wilmsen, E. N., and F. H. H. Roberts
1978 *Lindenmeier, 1934–1974, Concluding Report on Investigations.* Smithsonian Contributions to Anthropology 24. Smithsonian Institution Press, Washington, D.C.

Woods, J. C., and G. L. Titmus
1985 A Review of the Simon Clovis Collection. *Idaho Archaeologist* 8(1):3–8.

More Bits and Pieces: A New Look at Lahontan Chronology and Human Occupation[1]

Amy J. Dansie[1] and Wm. Jerry Jerrems[2]

> Standing on the salt encrusted floor of the old lake and looking toward terraced hills, it does not take much imagination to picture the body of water that must have been here, even though it has taken tremendous imagination to work out the history of the lake that carved those terraces.
>
> Donald K. Grayson (1993:88–89)

Abstract

Many scientists spent years in the vast deserts of the western Great Basin seeking the dateable bits and the geological pieces documenting ancient Lake Lahontan. We review the Lake Lahontan literature and the Nevada State Museum collections, extracting dates addressing lake levels and human occupation between 14,000 and 8000 years ago. Several new bone amino acid AMS dates fill in critical time gaps. These "bits and pieces" are then organized into a new lake and extinction model, correlated with Vance Haynes's Clovis Drought model and dating of the IACP (Intra-Allerød Cold Period) and the Younger Dryas as outlined by Stuart Fiedel. Our model postulates that late-Pleistocene lakes dropped to near desiccation in the Bølling, then rose slightly during the IACP, and rose much more in the early Younger Dryas after a possible drought 11,000 RCYBP. Several new dates support Tom Stafford's findings that the main extinction event dates to the IACP, 11,300 to 11,100 RCYBP. Two possibly cultural horse mandibles from Fishbone Cave, Winnemucca Lake, date 11,350 ± 40 and 11,210 ± 50 RCYBP. New dates were also obtained on short-faced bear (13,820 ± 50 RCYBP) and shrub ox (11,950 ± 50 RCYBP). However, evidence for late survival of some extinct Great Basin megafauna, and possibly Clovis culture, includes two artifacts of mammoth bone and ivory directly dated to 10,340 ± 40 and 10,360 ± 50 RCYBP.

Introduction

The earliest human occupations in the Great Basin cluster around extinct water features inactive since the end of the Pleistocene. Two of these were vast pluvial lakes spread over much the northern Great Basin, Lake Bonneville in the east, and Lake Lahontan in the west. As these giants rose and fell with changing climate, so did hundreds of smaller lakes in the isolated valleys of the Great Basin. Most lakes changed from deep, wave-scoured high shorelines to treacherous mud flats and dry playas in cycles. Rivers raged and trickled in turn. Marshes and meadows followed receding lake edges to their death on the valley floors, only to be flooded again through time. Only Pyramid Lake never died, even when it dropped below the historic low stand during critical periods of Paleoamerican occupation. The nature of the early human landscape in this unique region has eluded the efforts of many dedicated scientists because the timing and extent of the lake level changes were unknown, and for years contradictory proxy data frustrated all efforts to resolve the environmental sequence.

Lake Lahontan was a vast, mountain-rimmed freshwater Pleistocene lake in the northwestern Great Basin, with a large central island (Figure 1). The lake was about 900 ft (274 m) deep at Pyramid Lake, inundating 8,665 square miles (22,442 km^2), when it reached its high stand at an elevation of 4,390 ft (1,338 m) above sea level at 13,070 RCYBP (Adams and Wesnousky 1998;

[1] Revised written version of exhibit presented at the Clovis and Beyond Conference, Santa Fe. 1999

[2] The senior author retired in April 2000 from a 30-year career as Nevada State Museum anthropologist, zooarchaeologist, and anthropology curation manager. Through intimate experience with the actual specimens and original field documents, and years of field work with Jonathan Davis, this study of Lake Lahontan prehistory was built on a bachelor's degree in anthropology from University of Nevada, Reno. The junior author earned a master's in soil science, on the advice of Jonathan Davis, and is currently enrolled in geoarchaeology graduate work at the University of Nevada, Reno.

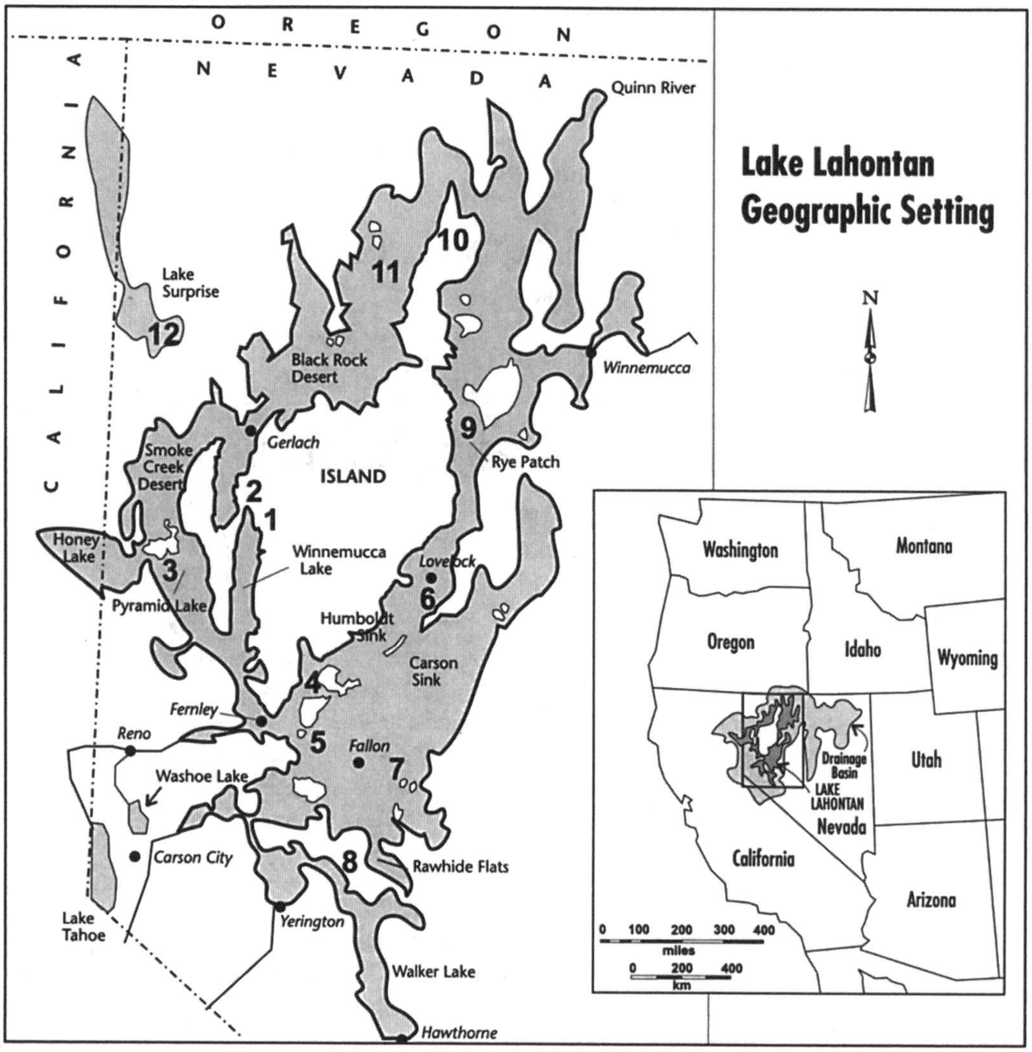

1 **Guano Mt.**
Fishbone Cave
Crypt Cave

2 **Falcon Hill**
Kramer Cave
Shrub ox
Coleman site

3 **Wizards Beach**
Camels and horse
Wizards Beach Man
Ivory point
Fishing cordage

4 **Brady's Hot Spring**
Dansie site
Harvey Clovis site
Jessup embayment to NE
Crew and Frock sites ESE

5 **Sadmat Locality**
Pebble Mound site

6 **Humboldt Sink**
Leonard Rockshelter
Heizer's 3950-foot lithic sites

7 **Grimes Point**
Spirit and Hidden Caves
Grimes Burial Shelter
Petroglyphs

8 **Walker River Canyon**
McGee Point
Bone beds

9 **Rye Patch Reservoir**
Old Humboldt site
Farmdalian fossil site

10 **Handprint Cave**
Above high shoreline

11 **Black Rock Desert, Quinn River arm**
Wallmann Mammoth
Wallmann Bison
Clovis farthest south (lowest)

12 **Duck Flat**
Giant short-faced bear in tufa dome, Lake Surprise

Figure 1. Lake Lahontan high stand, 13,000 rcybp, 4390 ft (1338 m) ASL. Sites and localities discussed in text.

Grayson 1993:92). In this land of interior drainage, six separate rivers (Figure 2C) flow toward a central series of basins, each of which is unique in elevation and morphology (Figure 2A). The basins are connected at sills of widely varying elevation (Figure 2A,B), resulting in complicated hydrology on a vast scale. Lahontan chronology is vital to interpreting archaeological and paleontological contexts in the western Great Basin, owing to the rapidly changing environments throughout the terminal Pleistocene and early Holocene (TP/EH) (Figure 4). Such changes affected human subsistence and settlement decisions and, equally important, the geological preservation of the evidence. We focus on the period of early human occupation, trying to understand the landscapes of northwestern Nevada from 14,000 to 8,000 RCYBP.

We will present a brief history of Lake Lahontan chronology modeling (Figure 3), and offer a new model (Figure 5) incorporating new dates and recent improvements in understanding the dynamic environments facing the first Americans (Haynes et al. 1999). In addition to documenting new archaeological and paleontological dates, we intend this report to be a tool for correlating the previous Lahontan literature, each study providing important parts of the larger puzzle despite dating problems of the past. Also, during this research we were able to correct some significant chronology errors that have spread through the Great Basin archaeological literature and to explain the issues surrounding them.

This report builds on the work of the late Jonathan O. Davis, whose 1982 article, "Bits and Pieces: The Last 35,000 Years in the Lahontan Area," was the last detailed overview attempting to correlate Lahontan geology and archaeology before reliable dates were available. Don Grayson (1993) provides a valuable review of the current scientific issues surrounding this report. We address the chronological gaps he identified so clearly, which were the same gaps Davis battled.

Continental Paleohydrological Models

Rapid environmental change at the Pleistocene-Holocene transition is becoming increasingly well documented, and its impact on humans was undoubtedly significant (Beck and Jones 1997; Huckleberry et al. 2001; Madsen 1999; Straus 1996). The method of continental climatic correlation promises to help us understand the ecological circumstances of early human cultures and

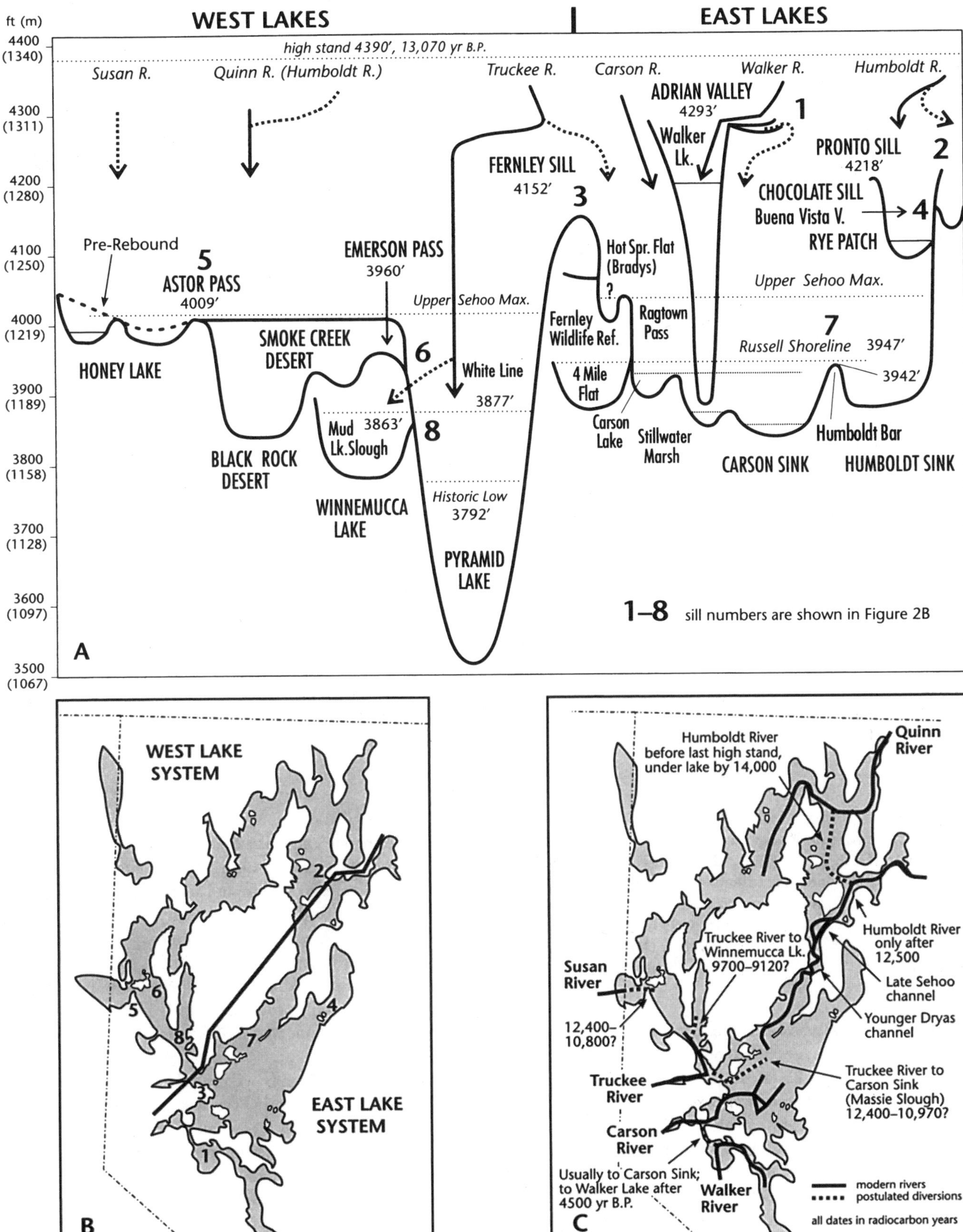

Figure 2. Lake Lahontan. **A,** sills, passes, and sub-basins; **B,** sill locations; **C,** modern rivers and postulated diversions.

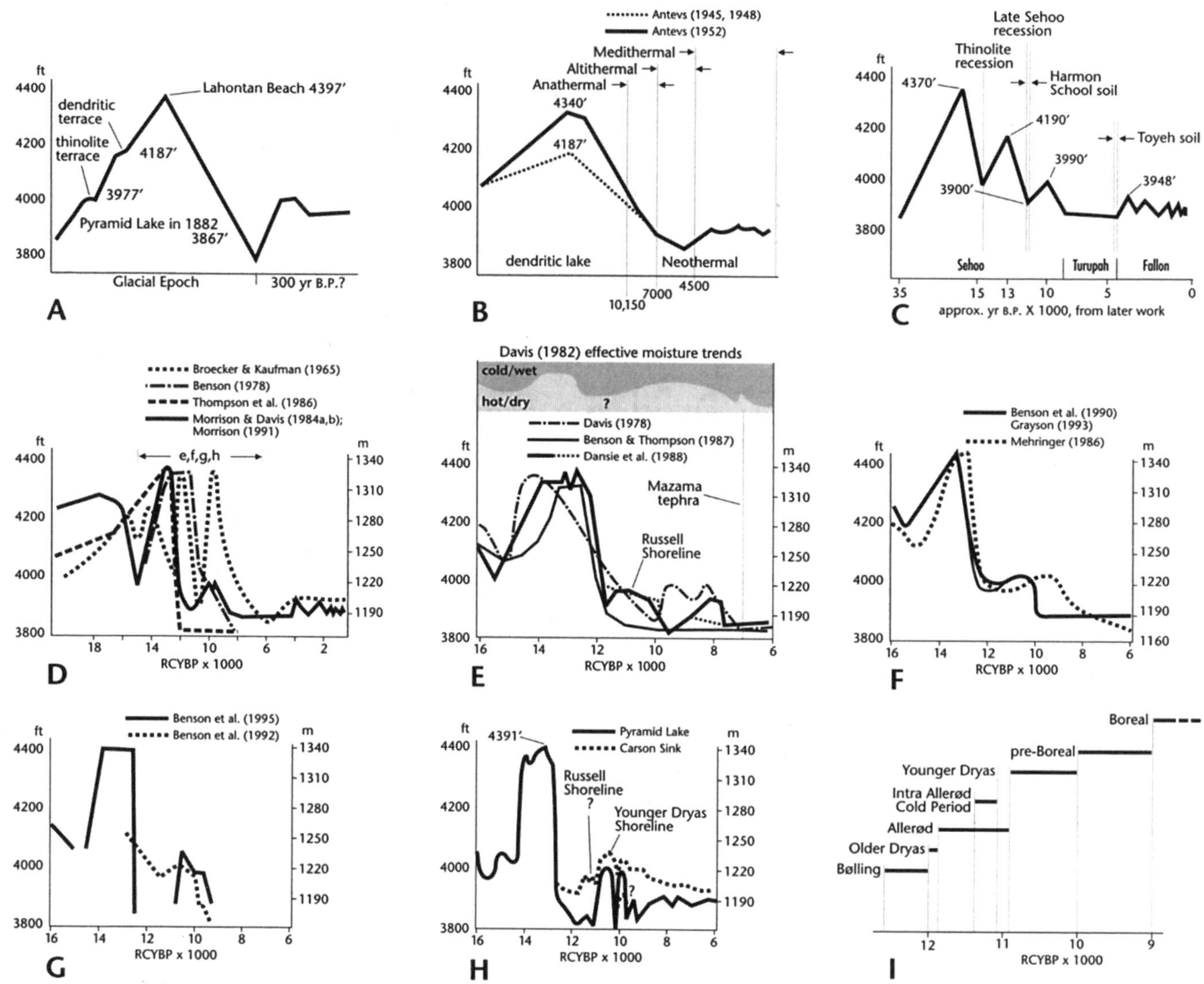

Figure 3. Comparison of Lake Lahontan chronology models. **A,** Russell (1885); **B,** Antevs (1945, 1948, 1952); **C,** Morrison (1964, Fig. 33); **D,** conflicting Lahontan chronology models; **E,** Davis (1978), Benson and Thompson (1987), and Dansie et al. (1988); **F,** Benson et al. (1990), Grayson (1993), and Mehringer (1986); **G,** Benson et al. (1995 and 1992); **H,** Dansie and Jerrems (1999); **I,** scale of time period critical to early human occupation, 12,600–8000 RCYBP.

adaptations in the New World. The Great Basin is theoretically ideal for studies of climate change because all the precipitation is captured in the closed basins, thus giving us a measure of the prevailing effective moisture over a large portion of the continent (the vast drainage basin of Lake Lahontan is shown on Figure 1). Lake recessions and transgressions during the early human occupation period in the Great Basin are particularly important because of growing evidence of a severe drought during terminal Clovis times (Haynes 1991; Haynes et al. 1999). Was there a Clovis Drought in the Great Basin? Many previous models show deeper lake levels during this time period, which would preclude early human occupation in the lowest elevations.

Climate change and fluctuations in lake level before Clovis times also may have affected humans, so we cover the last 14,000 years in lake history, geography and archaeology. Climate change also affects preservation of the evidence; some of the driest archaeological and paleontological assemblages in the world are found in the Lahontan shoreline tufa caves created by the lake. Lahontan lacustrine sediments also preserve bone exceptionally well. Our funding sources focused on bone amino acid dates, so we sought bone to test the research questions addressed here.

Problems with radiocarbon dating (Fiedel 1999; Straus 1996) made correlating Great Basin lakes with phases of global climate change difficult, as Benson et al. (1997) and Madsen (2000b) have encountered. We structure our discussion around the radiocarbon-dated climate phases presented by Fiedel (1999) and present all our results in uncalibrated radiocarbon years (RCYBP). If the correlation with calendar years is ever resolved, these data can be stretched to fit the new calibrations. As we note below, this issue may be pivotal in some Lahontan dating efforts, and we indicate where Fiedel's radiocarbon "plateau" may affect the

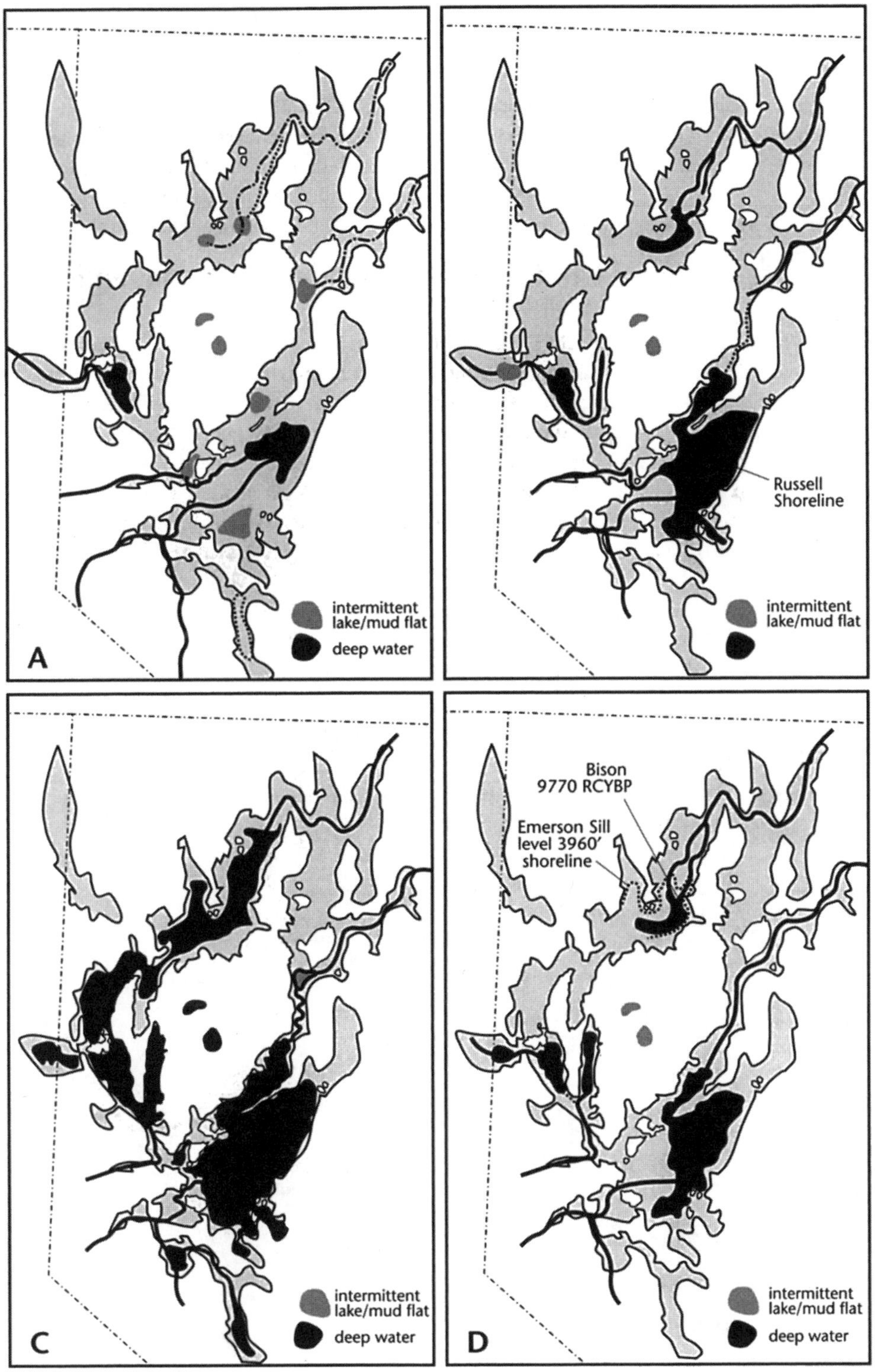

Figure 4. Approximate lake levels at selected time periods. **A,** Bølling/Allerød (ca. 12,300–11,450 RCYBP) and Clovis Drought (ca. 11,080–10,970 RCYBP); **B,** IACP rivers and lakes (11,360–11,100 RCYBP; ca. 3,950 ft in Carson Sink); **C,** maximum Younger Dryas lakes (ca. 10,700 RCYBP; 4,018 ft west, 4,050 ft east; late (upper) Sehoo); **D,** late pre-Boreal lakes (9500–9000 RCYBP; ca. 3,925 ft in Carson Sink).

model in Figure 5B. The use of older radiocarbon assays in new models must also deal with the issue of $\delta^{13}C$, adjustments which were added in the early 1990s to new dates and to some of the republished older ones (Benson 1993:172). We do not address this complex issue directly; instead, we cite the source and number of each date for future evaluation. All our recently obtained dates are corrected for $\delta^{13}C$, but we cannot vouch for all sources.

We combine Haynes's geoarchaeological interpretations (Haynes et al.1999) with Fiedel's (1999) radiocarbon/calendar correlations for the following discussion. Briefly, this model states that the last full glacial conditions terminated abruptly at the onset of the Bølling/Allerød climate phase at 12,600 RCYBP, when temperatures rose sharply and "catastrophic ablation" of the great ice sheets began (Fiedel 1999). According to Haynes, severe drought marked the Allerød; a brief interval called the Intra Allerød Cold Period (IACP), 11,360 to 11,100 RCYBP, ameliorated the drought somewhat (Figure 5). The terminal Allerød, 11,000 years ago, was marked by even greater drying, the "Clovis Drought." Clovis people confronted the dry Allerød climate conditions, when people and mammoths dug wells in desiccated spring ponds in the Plains and Southwest both before and after the IACP (Haynes et al. 1999). Most extinctions, except mammoths, appear to date to the IACP.

At 10,970 RCYBP there was a sudden return to glacial climate. This was marked by a catastrophic deep freeze, followed immediately by increased moisture and continued cold climate for the next 800 radiocarbon years (1,300 calendar years): the Younger Dryas (Fiedel 1999). The initial cold snap probably froze the last remaining surface water over vast areas, and Haynes reports that no mammoths or Clovis artifacts are found above (after) this event (Haynes et al. 1999; Taylor et al. 1996). Immediately above the last mammoth bones in the dried spring ponds is a black mat representing algal growth in the recharged springs (Haynes 1984). This pattern has been observed by Haynes (1991) in many parts of the continent, although Holliday (2000) has another interpretation. Wildly varying climate during the Younger Dryas may have produced mixed signals in the geological record (Madsen 1999, 2000a). Just as suddenly as it started, the Younger Dryas abruptly ended at 10,100 RCYBP (Fiedel 1999).

If there was a "Clovis Drought" in western Nevada, was there also a "black mat" episode reflecting sudden onset of increased moisture? Such extreme climate reversals should have left their mark in the Lahontan Basin, but is there any evidence?

The Quest for a Lahontan Chronological Model

Israel C. Russell (1885) described ancient Lake Lahontan in exquisite detail, including a complex series of well-preserved geomorphic features such as beach terraces, gravel bars, deltas, calcium carbonate deposits (tufa), and lake sediments. The basic stratigraphic and paleontological sequence of the Pleistocene was known in general outline by Russell's time. He saw a correlation between the massive lake and the "Glacial epoch," but was unaware of any way to estimate its age (Russell 1885:270). Lacking

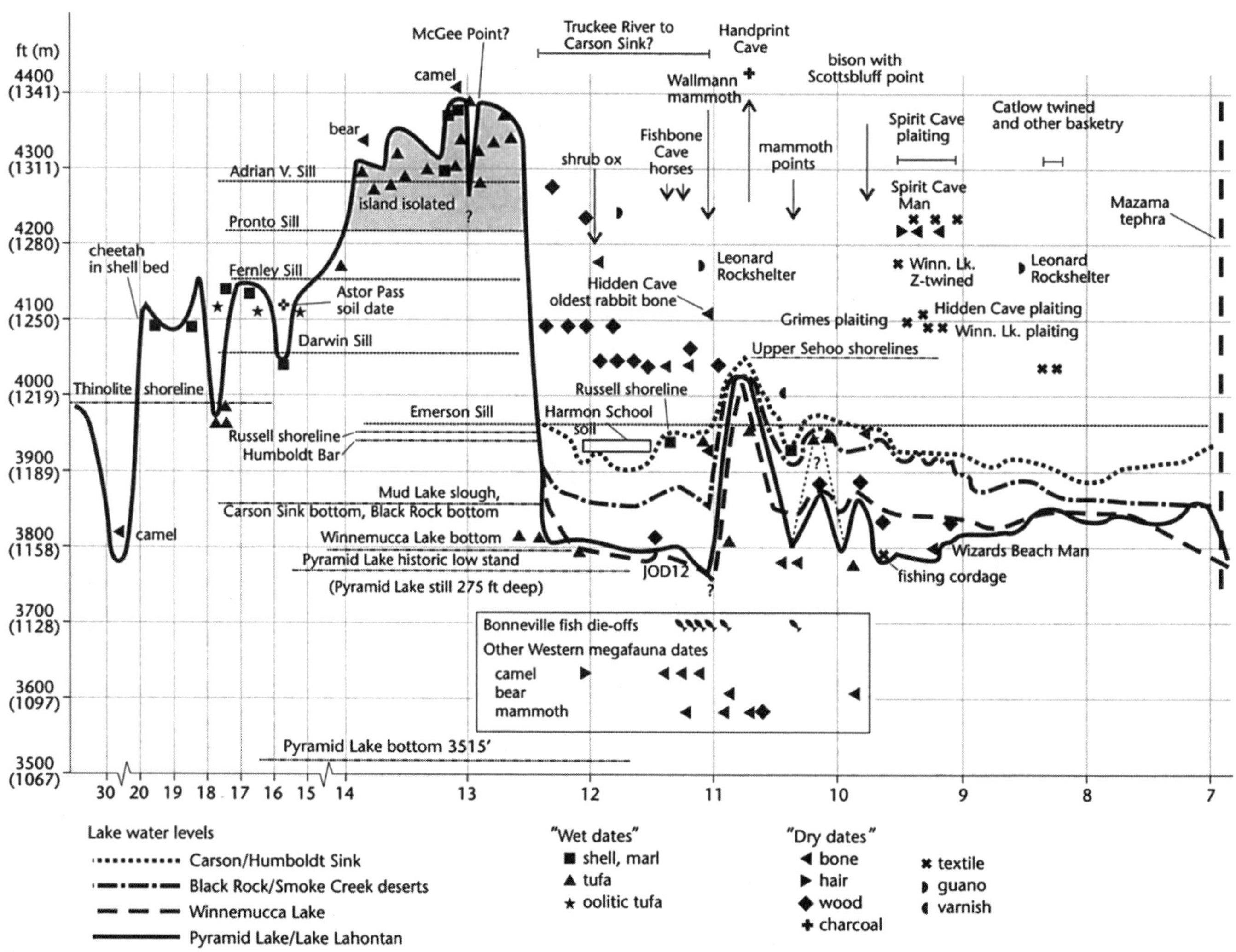

Figure 5A. Late-Pleistocene/early-Holocene Lake Lahontan chronology and archaeology, lake levels and dated specimens.

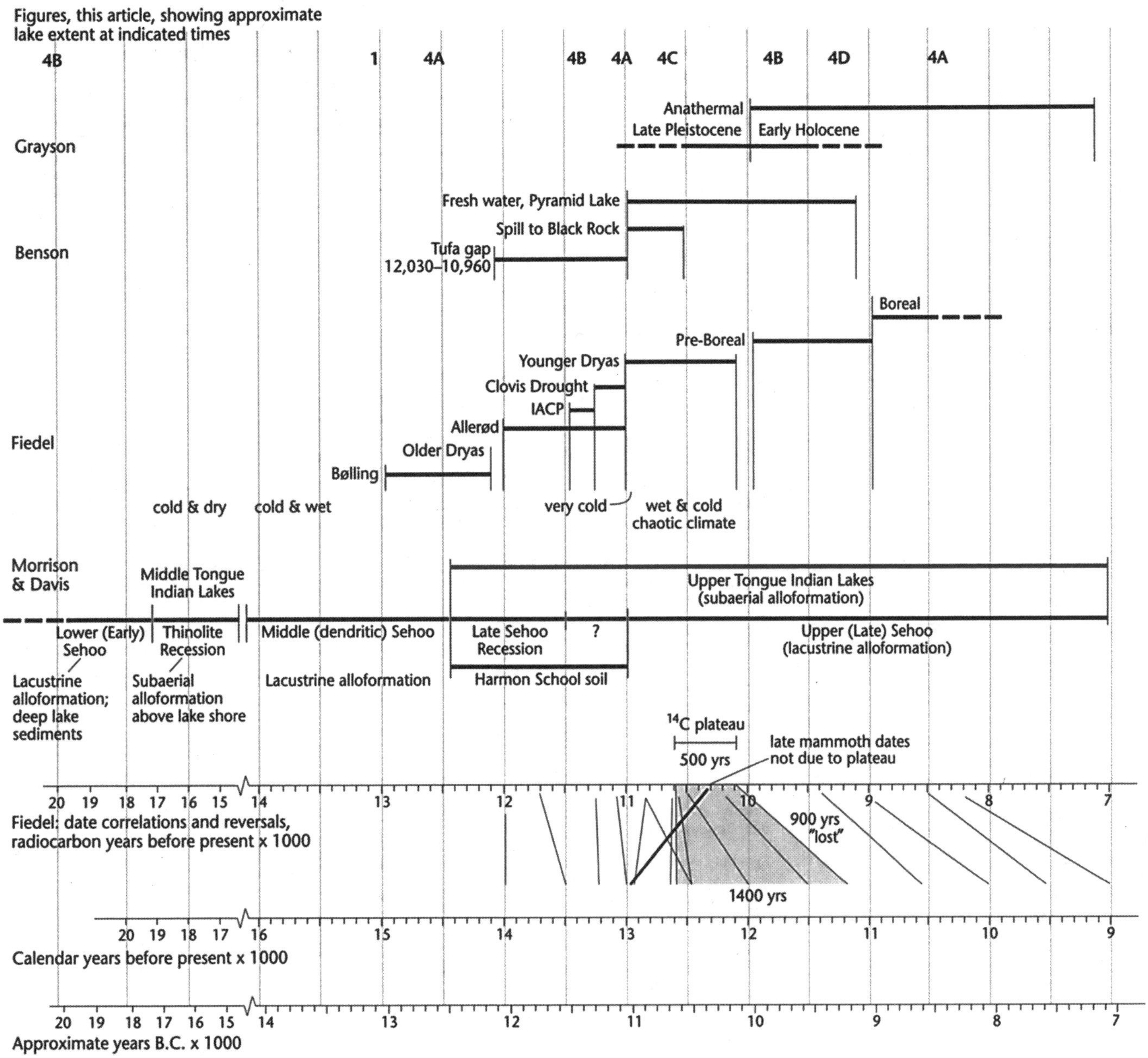

Figure 5B. Late-Pleistocene/early-Holocene Lake Lahontan chronology and archaeology, climate models correlated with geological models.

a means of chronological control, Russell estimated that the final recession of this now-dry lake happened 300 years before his studies (Figure 3A), based on the extraordinary preservation of the lake features.

Beginning in the 1920s, Ernst Antevs (1925, 1948, 1953, 1955) provided a New World climate chronology, based on Old World lake varves that served as the primary basis for archaeological and geochronological interpretations in North America until well after the advent of radiocarbon dating (Grayson 1993; Wormington 1957). Antevs (1945, 1948, 1952) was also one of the first to develop a Lake Lahontan model based on his field work and his climate model (Figure 3B) and on some of Roger Morrison's data with different interpretations (Morrison 1964a). Antevs defined the postglacial Neothermal as a sequence of major temperature regimes: the Anathermal, a time of rising temperature 10,150 to 7000 [varv]; the Altithermal or "long Drought," a time of maximum temperature, 7000 to 4500 [varv]; and the Medithermal, moderate temperature from 4500 [varv] to the present (Antevs 1953). He adjusted the timing of the Altithermal to 7500 to 4000 [varv] in Antevs 1955. Except for these varve-based dates of Antevs, all dates cited herein are in uncalibrated radiocarbon years before present (RCYBP). All radiocarbon dates mentioned in the text are listed on Table 1.

Heizer (1951) was the first Great Basin archaeologist to explore the timing of Lake Lahontan's demise in relation to human

Table 1. Lahontan dates grouped by climate period. Gaps between radiocarbon dates are shown to emphasize correlations between dates, global climate changes (Fiedel 1999; Haynes 1991) and local environmental consequences. All dates are uncalibrated ^{14}C (RCYBP). ■Indicates new date, Dansie and Jerrems 1999, 2000, here. ▲Indicates important re-date. ~~1234~~ Indicates date no longer used. Elevations represent lake levels or estimates; ▼indicates site elevation higher than lake.

Climate period	Date	Gap (yrs)	Sample/climate	Elev. (ft)	Lahontan status/cultural context	Reference
Oldest Dryas (cold)	■13,830 ± 50		Short-faced bear in tufa dome, Duck Flat, NV, NW of Gerlach, AMS amino acid date	n/a	Rising to high stand of Lake Surprise, in high shallow bay, consistent with other dates	Nevada State Museum (NSM) collection, Selsor pers. comm. 2000; Dansie and Jerrems 2000
	13,070 ± 60	750	*Camelops* bone, top of lagoon sands	4390	High stand, Jessup Embayment	Adams and Wesnousky 1998
	12,840 ± 70	230	Pyramid Lake tufa	4360	High tufa date	Benson et al. 1995
Bølling Onset	12,600	240	Ice sheets begin catastrophic ablation			Fiedel 1999
	12,550 ± 80	50	Pyramid Lake tufa	4360	High tufa date	Benson et al. 1995
	12,490 ± 80	60	Pyramid Lake tufa	4346	Last high date, 44 ft below maximum. Rain and snow stop.	Benson et al. 1995
Lahontan Crash	ca. 12,500–12,400		590 ft below highest lake level		Drops 546 ft in 120 yrs	
	12,370 ± 110	120	Pyramid Lake tufa, very low	3800	First low tufa date	Benson et al 1995
	~~12,700 ± 300~~		Pyramid Lake tufa Diaphragm Cave, very low lake elevation indicated	3820	Once wildly inconsistent, now overlaps earliest low Benson date	Broecker and Orr 1958:1026
	ca. 12,300		Harmon School soil; weathering begins on lake clay in Carson Sink?	3920	ca. 1000 yrs of weathering suggested by soil profile	Morrison 1964; Dansie and Jerrems field observations
	12,350 ± 180 12,240 ± 180 12,130 ± 180	20 110 110	Crypt Cave juniper Packrat nests	▼4098	First "dry" dates, confirm crash, no rise after; 292 ft below high stand	Benson and Thompson 1987a
	▲~~12,280 ± 520~~		Fishbone Cave horse #1, re-dated	▼4040	1070 years earlier than re-date	Stafford et al. 1987; Thompson et al. 1986, NSM collections; see new date 11,210 below
Older Dryas begins	12,080	50			Lakes still low	Fiedel 1999
	12,070 ± 210 12,060 ± 260	20 10	Guano Cave juniper and rat dung, packrat nest	4037	Confirm continued dry at moderately low levels	Thompson et al. 1986
	12,040 ± 80	20	Pyramid Lake tufa	3805	Dolomite, not rejected, low	Benson et al. 1995
	12,030 ± 110	10	Pyramid Lake tufa; tufa gap begins	3805	Last tufa before tufa gap, low	Benson et al. 1995
Allerød begins	11,970	60			Lakes still low, tufa gap starts 60 yrs before Allerød begins	Fiedel 1999
	■11,950 ± 50	20	Falcon Hill shrub ox mandible, AMS amino acid date, non-cultural	▼4185	Extinct megafauna survived the initial crash by 450 yrs, but note ca. 600-yr gap in bone dates following this	Dansie and Jerrems 1999; NSM collections
	11,555 ± 500	395	Fishbone Cave, wood, bottom of Level 4 first date run	▼4040	Human-modified firewood? 195 yrs earlier than horse jaw #2; standard deviations overlap	Orr 1956; NSM collections
	11,490 ± 120	65	Pyramid Lake JOD-12 wood, top of lacustrine, end of deep lake deposition	3796	Pyramid Lake very low, below ca. 3810 ft for 880 yrs? 540 yrs after last tufa date, 880 yrs after first low date	Davis 1978; Valastro 1978; misprinted in Lagoni 1985
	~~11,380 ± 120~~		Winnemucca Lake tufa	3957	16-ft rise in 30 yrs? Date not used in Benson et al. 1995, rejected (lakes still low)	Benson et al 1992:23
IACP begins	11,360	130	Colder, less evaporation		Also slightly wetter?	Fiedel 1999
	■11,350 ± 40	10	Fishbone Cave Level 4 horse #2 mandible, UCR AMS amino acid date	▼4040	Human occupation? Some extinct fauna survive; onager-like equid. Agate biface associated.	*Densie and Jerrems 1999; NSM Collections
	11,300 ± 130	50	Kent Ranch deep water marl, Stillwater, Carson Sink	▼3875 3947	First lake rise? Truckee River diversions to Carson Sink? Supporting evidence for Russell Shoreline lake at 3947 ft	Elston et al 1988:372; Currey 1988
	ca. 11,300		Tufa on Harmon School soil begins to form	3920	1070 yrs consistent with degree of weathering	Jerrems field observation; Morrison 1964a
	11,270 ± 170	30	Falcon Hill rat dung	4254	Rats and their food survive	Benson and Thompson 1987a
	11,220 ± 110	50	Huntington mammoth, Utah		Died high in Wasatch Mt. bog with bear gnawing marks	Bonneville area; Madsen 2000

Table 1 (cont'd).

Climate period	Date	Gap (yrs)	Sample/climate	Elev. (ft)	Lahontan status/cultural context	Reference
	▲■11,210 ± 50	10	Fishbone Cave Level 4 horse #1 mandible, UCR AMS amino acid date	▼4040	130 yrs younger than other horse instead of 920 yrs older. Cultural? Agate biface.	New re-date Selsor pers. comm. Nov. 2000, this paper; NSM collections
Clovis begins	11,200	10	13,200 calendar yr		All but mammoths and *Arctodus* bear extinct?	Fiedel 1999; Haynes 1991
	11,200 ± 250	0	Fishbone Cave wood Level 4, average of 4 new runs	▼4040	Official L-245, possible human associated: firewood/flooring?	Broecker and Orr 1958; typo in Orr 1974 table: 11,250 ± 250, repeated by Hattori 1982,"mat"
	11,199 ± 570	1	Leonard Rockshelter bat guano on last lake sands	▼4177 3920–3940?	Lake marsh elevation estimated, marsh in upper valley = Russell Shoreline?	Heizer 1951
	11,120 ± 120	79	Hidden Cave *Lepus* collagen	▼4103	Oldest of new "dry" dates showing lake below cave, also first sign of life near cave	Mullen 1997 in Rhode et al. 2000
	11,100 ± 110	20	Carson Sink Upsal; Hogback tufa	3947	Russell Shoreline	Elston et al. 1988:372; see Benson et al. 1992:23; Rhode et al. 2000
IACP ends	11,100	0	Pyramid Lake tufa gap continues		Only Carson Sink has evidence of lake rise during IACP	Fiedel 1999
	11,080 ± 300	20	Wallmann mammoth whole bone date. New AMS 10,630 ± 70 protein preservation: 0.03%.	▼3968 lake below 3950	Huge bull in drying slough on Black Rock desert. Cultural? May be much older, not reliable dates	Stout 1986; Dansie et al.1988; Selsor pers. comm. 2000; enamel date pending at UCR; NSM collections
Clovis drought	11,000–10,970	80	Severe drought at end of long drying trend		Mammoths becoming extinct?	Haynes 1991; Haynes et al. 1999
Catastrophic cold snap?	10,970	0	Freezes last water holes, final extinction? 7–15° C drop in 10 yr.		Hot springs and Pyramid Lake may have been refugia.	Haynes 1991; Fiedel 1999
Younger Dryas begins	10,970	0	Wetter, cooler		Chaotic climate change	Fiedel 1999
	10,960 ± 80	10	Pyramid Lake tufa gap ends	3811	First Pyramid Lake tufa after gap, lake rising again	Benson et al. 1995
	10,900 ± 300	60	Fishbone Cave wood "L-245" first run	4040	Cultural firewood?	Orr 1974, NSM collections
	10,870 ± 75	30	Huntington giant short-faced bear *Arctodus* found with mammoth		350 yrs after mammoth death, but feeding on it: *Cultural?* Late survival of extinct bear.	Lake Bonneville area; Madsen 2000
Clovis ends	10,800	70	12,900 calendar yr			Fiedel 1999
	10,740 ± 70	60	Handprint Cave, fine small stemmed point, charcoal (above high stand)	▼4416 4000?	Oldest cultural date in Lahontan basin; similar points above 4000 ft. Bovid hair association.	Bryan 1988; Gruhn and Bryan 1988; Wallmann pers. comm.; Bryan and Tuohy 1999; NSM collections
	Younger Dryas? "by" 10,500		Retransgression of Lahontan lakes after crash	4051 east, 4018 west	Breached bar to Rawhide Flats but did not fill it. 104 ft higher than Russell Shoreline in east lake system	Adams and Wesnousky 1998; Rhode et al. 2001 (east); Benson et al. 1995:28 (west)
^{14}C Plateau starts	10,600	140	12,600 calendar yr			Fiedel 1999
	10,460 ± 180	140	Pyramid Lake organic matter below varnish, Anaho Island	4002	in wave-cut feature, spill to Smoke Creek/Black Rock	Benson et al. 1992:23
	10,450 ± 80	10	Winnemucca Lake tufa	3952	Lakes moderate	Benson et al. 1995
	10,380 ± 80	70	Humboldt bar *Anodonta*	3929	Lakes drop to marshes	Benson et al. 1992:23
	■10,360 ± 50	20	Pyramid Lake ivory Clovis-style rod, oldest dated artifact.	3800	First positive cultural mammoth in Nevada. Pyramid Lake very low?	Dansie and Jerrems 1999; NSM Ting collection
	■10,340 ± 40	20	Mammoth rib harpoon, 2 parts of 1 leister? Dates mammoth death	3800	Pyramid Lake, same age, location (same animal?) as ivory rod	Dansie and Jerrems 1999 NSM Ting collection
Younger Dryas ends	10,100	240	11,200 calendar yr; 900 calendar yrs missing in ^{14}C dates		Global transitional climate; no Lahontan Lake data for last 240 yrs of Younger Dryas	Fiedel 1999
"Anathermal" begins	10,100	0				Antevs 1948
	~~10,080 ± 80~~	20	Winnemucca Lake tufa	3952	370 yrs after last tufa date, lake rise: tufa date problem?	Benson et al. 1992:23; not used in Benson et al. 1995
	~~9990 ± 80~~	90	Winnemucca Lake tufa	3952	Last lakes around 3950? Or tufa date problem?	Benson et al. 1995

Table 1 (cont'd).

Climate period	Date	Gap (yrs)	Sample/climate	Elev. (ft)	Lahontan status/cultural context	Reference
Pre-Boreal	10,000	-10	Holocene drying begins			Fiedel 1999
	9970 ± 140	30	Pyramid Lake delta wood	3863	Conflict: lakes drop ca. 90 ft? Or rise 63 ft from 3800 at 10,340?	Prokopovich 1983:188
	9960 ± 50		Mineral Hill *Arctodus* or very large *Ursus*?		Eastern Nevada, latest extinct bear date?	Hockett and Dillingham 2000
	9920 ± ?	50	Tufa on pelican wing, Pyramid Lake	3790	Lakes drop 73 ft	NSM collections, Benson pers. comm. to D. R. Tuohy
	~~9860~~ ± ~~80~~	60	Tufa, Winnemucca Lake	3952	Last tufa date, too young?	Benson et al. 1995
	9780 ± 135	80	Pyramid Lake delta wood	3863	Mud Lake slough level, spill to Winnemucca Lake	Prokopovich 1983:188; Benson et al. 1992:23
	■9770 ± 50	10	Wallmann bison, Black Rock Desert, AMS amino acid date	3950	Scottsbluff point associated, meadows on Quinn River playa?	Dansie and Jerrems 1999; Dansie et al. 1988. NSM collections
	9720 ± 100	50	Pyramid Lake wood, delta	3835	25 ft below Mud Lake slough sill	Born 1972:36
	9660 ± 170	60	Sagebrush fishing line, Pyramid Lake	3785	Oldest textile, very low Wizards Beach area	Tuohy 1988; NSM collections
	9540 ± 120	80	Z-twined basketry, Shinners A, Falcon Hill	▼4185 3830	Oldest basketry date in Nevada, water level (?) marsh optimum	Hattori 1982; NSM collections
	9470 ± 60	70	Grimes burial shelter plain weave tule and hemp	▼4080 3925	Oldest Spirit Cave–type matting; all suggests marsh at 3925 ft	Dansie 1997; Fowler et al. 2000; NSM collections
	9415 ± 25	55	Spirit Cave man average of 7 AMS amino acid dates	▼4239 3925	Plain weave tule and hemp tule matting, twined tule mat, hair, bone.	Tuohy and Dansie 1997; Fowler et al. 2000; NSM collections
	9329 ± 50	86	Hidden Cave entrance	▼4103 3925	Plain weave tule and hemp matting	Fowler et al. 2000 NSM collections
	9300 ± 70 9270 ± 60	29 30	Spirit Cave Burial #1 matting and bone AMS	▼4239 3925	Plain weave tule and hemp matting	Tuohy and Dansie 1997; Fowler et al. 2000; NSM collections
	9225 ± 60	45	Wizards Beach man, 2 UCR bone dates average	▼3795 3790?	Pyramid Lake near historic low level. (Beta Analytic date 9500)	Tuohy and Dansie 1997; Kirner et al. 1997; NSM collections
	9220 ± 50	5	Chimney Cave, Guano Mt.	▼4098 3830?	Plain weave tule and hemp matting	Fowler et al. 2000; NSM collections
	9140 ± 60	80	Crypt Cave, Guano Mt.	▼4098 3830?	Plain weave tule and hemp matting	Fowler et al. 2000; NSM collections
	9120 ± 60	10	Crypt Cave, Guano Mt.	▼4098 3830?	Plain weave tule and hemp matting	Tuohy and Dansie 1997; Fowler et al. 2000; NSM collections
	9040 ± 50	80	Spirit Cave cremation bag youngest in this set. One individual in three bags.	▼4239 3925?	Plain weave tule and hemp matting, twined hemp bags. 430-yr spread for plain weave	Tuohy and Dansie 1997; Owsley and Jantz 1999:82; Fowler et al. 2000; NSM collections
Pre-Boreal ends, Boreal begins	9000	40				Fiedel 1999
	8800 ± 90	200	Pyramid Lake delta wood	3831	Pyramid still low	Born 1972:36
	8660 ± 300	140	Leonard Rockshelter bat guano	▼4177	Marsh in the Humboldt Sink	Heizer 1956
	8380 ± 120	280	Z-twined matting, Shinners A, Falcon Hill	▼4185 3830?	660-yr cultural gap	Hattori 1982:14; NSM collections
	▲■8370 ± 50	10	Catlow twined, Fishbone Cave, 1999 re-date of feature	▼4040 3830?	Associated with burial #2, sagebrush twining, netting and pelican skin; see next date	Dansie and Jerrems 1999; NSM collections
	8220 ± 50	150	Fishbone Cave twined sagebrush matting/sandal	▼4040 3830?	Re-date of "oldest textile," burial #2	K. Selsor pers. comm. 2000; Fowler and Hattori 2000
	7980 ± 610	400	Nicolarsen site tufa dome, Winnemucca Lake textile	▼3960 3830?	400-yr cultural gap; textiles, burials, atlatl associated	Hattori 1982:14; NSM and Berkeley collections
	▲7830 ± 350	150	Fishbone Cave netting, burial #2	▼4040 3830?	Published Orr date, lab notes show 8200; see 8370 above	Orr 1974; NSM collections

occupation, starting with Antevs's chronology and applying some of the first radiocarbon dates ever run by Libby. Heizer (1951:95) correlated the oldest cultural materials (Humboldt culture in the caves, Granite Point culture on the shorelines) with Anathermal lakes of 3,950 ft (1,204 m) in the Humboldt Sink, noting that the lakes had never been as high since (Figure 3B). Heizer (1951) also noted that "dry dates" resting on lake deposits in both the Lahontan and Bonneville basins (11,199 ± 570 RCYBP, C-599, at Leonard Rockshelter, and 11,151 ± 570 RCYBP at Danger Cave), were important correlations documenting the pluvial lake recessions. Heizer said little about the earlier Danger Cave sheep dung date of 11,453 ± 600 RCYBP. The 1950s Great Basin pluvial lake model with which archaeologists were most familiar, was of a gradually declining high stand through 7000 RCYBP (Figure 3B) (Antevs 1952; Heizer 1951). The related assumption was that lake levels around 3,950 ft (1,204 m) in the Humboldt sink must have been Anathermal in age (after 10,150 RCYBP according to Antevs), and that human occupation of lower elevations could not be older than this time in the Lahontan Basin. These new radiocarbon dates were beginning to push the possibility of human occupations on the lower shore lines back to 11,500–11,200 RCYBP, and Heizer (1951:95) spoke of possibly finding "Pre-Anathermal man" in the Lahontan Basin. An excellent discussion of this early radiocarbon-dating era and the archaeological problems in Great Basin lake correlation can be found in Jennings (1957:94–98).

Tufa and Stratigraphic Chronologies

Early efforts to establish a radiocarbon-dated Lahontan chronology began with Phil Orr's "Operation Tufa" in 1954 (Orr 1956a, b). His explorations of tufa caves at Winnemucca Lake, Nevada, revealed compelling evidence of Pleistocene human occupation associated with extinct horse bones and camel teeth in Level 4 of Fishbone Cave. Orr (1956a; Broecker et al. 1956) reported radiocarbon dates of 11,555 ± 500, 10,900 ± 300, and 11,200 ± 250 RCYBP (L-245) on wood from Level 4 of Fishbone Cave. Orr was aware that his first dry dates were similar to the Leonard Rockshelter and Danger Cave dates noted above, all around 11,200 RCYBP. He did not like the 11,555 ± 500 date, and he re-dated the wood until he was satisfied with the L-245 published date of 11,200 ± 250 RCYBP. Orr believed the rise and fall of Lake Lahontan could be dated by radiocarbon assays on the tufa deposits that mark the various lake levels on the hillsides around the ancient lake. However, Orr placed more faith in the tufa dates than in the dates on wood and other organic materials. This faith resulted in published reports with contradictory chronological indicators (Orr 1956a), including changing the reported site elevations[3] (Wormington 1957) and averaging widely disparate dates to support his model that Lake Lahontan was still at a higher elevation than Fishbone Cave at 11,500 RCYBP (thus rejecting his earliest dry-wood date). According to Orr, the lake then conveniently dropped below the cave for a human occupation at 11,200 RCYBP, then rose again to a high stand around 9700 RCYBP, even though the older dry materials had not been inundated. This contradiction was emphasized later with even more conflicting tufa-based data showing a late high stand in Broecker and Kaufman (1965) (Figure 3D, dotted line).

At the same time, Roger Morrison (1952; 1964a, b) conducted an extensive geological survey of the Carson Sink. Morrison inferred a terminal high stand date of 12,500 RCYBP from continental climatic correlations (Figure 3C) (Morrison 1964a: Plate 10; Morrison and Frye 1965). He called this high stand the "middle member" of the Sehoo formation, which he ultimately concluded was the highest lake stand of the Sehoo formation (Figure 3D) (Morrison 1991; Morrison and Davis 1984a, b; Reheis and Morrison 1997). Immediately following the last high stand, Morrison clearly defined a Late Sehoo recession to "near desiccation" (down to 3,920 ft/1,195 m) around 11,500 RCYBP (exactly the opposite of Orr's claim). This desiccation interval is evidenced by the formation of the Harmon School Soil (Morrison 1964a: 19, 75, 114, 127), marked at the type locality by a subaerially weathered zone of deep lake clay with pedogenesis (soil development), confirmed by Jerrems. Following this dry interval, Morrison (1964a) defined a moderate lake rise to 3,990 ft (1,216 m), which he named the upper member of the Sehoo formation (Late Sehoo). At the Harmon School locality, the upper Sehoo is marked by lithoid tufa slabs resting on a disconformity on top of the Harmon School soil, which are buried by lacustrine silt and sand on which the later Toyeh soil was developed. This final return of Lake Lahontan, the upper Sehoo, was schematically graphed by Morrison between 10,500 and 7500 RCYBP, with a peak around 10,000 RCYBP. This lake was far below the highest lake levels, but higher than any lakes since that time (Figure 3C, D), meeting the same definition, almost, of the Anathermal lakes mentioned above. Forty feet separates the lake shores mentioned by Heizer (1951) and Morrison (1964a), with a significant difference in geographic extent (Figure 4B, D).

Morrison compared his soil pedogenic and stratigraphic model with Antevs and Russell's lake models (Morrison 1964b: Figure 38), shown here in Figure 3A–C. The postulated 9700 RCYBP high stand of Broecker and Orr (1956, 1958) appeared absurd, despite the apparent quality of the scientific data and analysis. Similar problems of conflicting information were found in the Lake Bonneville area of Utah. This combination of conflicting tufa and dry organic dates contradicting stratigraphic geological reconstructions led Marie Wormington (1957:196) to conclude:

> In view of the stratigraphic evidence, it is difficult to believe that these dates can be correct. If they are, all previous interpretations of the terminal history of Lakes Lahontan and Bonneville must be wrong. For archaeologists the geological situation is one of complete confusion. There is little that they can do but bide their time and await clarification from the geologists.

This confusing situation has remained in effect for decades, leaving archaeologists with no chronological environmental structure for early human occupation studies in the Great Basin. Despite the sincere effort of many dedicated scientists, archaeologists have bided their time for over 40 years awaiting "clarification from the geologists."

Building on Broecker and Orr's (1956, 1958) tufa chronology, Broecker and Kaufman (1965) documented two maximum Lahontan lake episodes at 11,700 and 9700 RCYBP (Figure 3D, dotted line), completely contradicting Morrison's model. In 1978, Jonathan O. Davis (1978a) approached Lahontan chronology from a new perspective that built on Morrison's geological and pedogenic data, and added tephrochronology (volcanic ash stratig-

[3] Wormington (1957:195) states: "The exact height of Fishbone Cave is not known. There is a discrepancy of 105 feet between the estimated elevation given on page 3 and that given on page 10 of Orr's (1956b) report. These are respectively, 4120 and 4225 feet altitude." [The real elevation is 4,048 ft.]

raphy) to the chronometric arsenal. Davis emphasized the blatant discrepancies in these previous lake models and noted that Steven Born's (1972) evidence of a low stand at Pyramid Lake, based on a wood date around 9700 RCYBP, threw greater doubt on the Broecker and Kaufman model.

Davis's (1978a, 1983) revised stratigraphic chronology, based on tephrochronology, extended the time span of Morrison's Sehoo formation, the last lake cycle, starting ca. 35,000 years ago, and extended the pre-Sehoo lake cycles significantly farther into the past. Like Morrison, Davis (1978a) showed the last deep lake declining significantly around 12,500 years ago from a high stand ca. 14,000 RCYBP. He documented a wood date of 11,490 ± 130 RCYBP (TX-2563) from a very low elevation near the Pyramid Lake delta (3,796 ft/1,157 m), on the top of deep lake (lacustrine) clay (Davis 1978a; Valastro et al. 1978). Davis mentioned this date at least four times (1978a: 40, 60, 62, 80), but never again cited this evidence explicitly. The date was lost in a misleading listing by Lagoni (1985:64), who erroneously reported this date 144 ft higher (3,940 ft/1,201 m) as a lacustrine clay wood date, citing Valastro et al. 1978. It was then cited for years as evidence of a higher lake stand during this critical period. The Davis (1978a) 11,490 RCYBP wood date is actually the first radiocarbon-dated evidence of Morrison's "late Sehoo recession" (time of Harmon School Soil) between the middle Sehoo (the last high stand) and the upper (late) Sehoo (the last moderate lakes), even though Davis did not include this date in his lake model curve (Figure 3E, dot-dash line).

Both Morrison and Davis emphasized the occurrence of archaeological materials along the "upper Sehoo" shorelines (around 3,990–4,000 ft/1,216–1,219 m); and Tuohy (1968, 1970, 1977, 1984, 1985) commented on the apparent significance of moderate elevation lake shores and early Great Basin human occupation, as did Grosscup (1956). Tuohy emphasized the lower elevation of Clovis points compared with Stemmed Points. Richard Shutler (Shutler 1960, 1961; Shutler and Shutler 1959) also examined the correlation of Lahontan lake levels and early human occupation and documented Clovis points on the lake bed of Washoe Lake, near Reno. None of Orr's archaeological observations or tufa dates could be correlated with these open lithic sites, and the archaeological record remained more consistent with the undated stratigraphic and geomorphic reconstruction of lake history than with the tufa-based chronologies. Figures 3D and 3E demonstrate the stratigraphic and chronological confusion which confronted archaeologists through the 1970s and 1980s.

A new generation of tufa dates, using careful selection criteria aimed at eliminating contamination errors, was reported by Larry Benson (1978, 1981). He showed the fall from the high stand at ca. 11,000 RCYBP (Figure 3D, dot-dash line), without a subsequent transgression to the high stand at 9700 RCYBP as indicated by Broecker and Orr (1958). Benson (1978, 1981) displayed no hint of the upper Sehoo following the major recession from the last high stand. Such a lake model would mean, under the existing archaeological model, that Clovis sites should be only higher than any later Paleoindian cultures in the Lahontan Basin, which they clearly were not.

Davis's original assessment of Lake Lahontan's demise (Davis 1978a) showed a gradual recession from a ca. 14,000 RCYBP high stand (Figure 3E, dot-dash). This interpretation changed when Davis (1982:60) stated that "Lake Lahontan fell rapidly starting about 12,500 RCYBP" after a 13,000 RCYBP high stand, a full 1,500 years earlier than shown by Benson (1981). The Davis (1982) climatic model was built on a continental climate correlation, when the Two Creekan interval was thought to be a time of "sudden lessening" of effective moisture between 12,500 and 11,000 RCYBP. This period is now called the Bølling/Allerød. Figure 3D, dark line, shows the Morrison and Davis (1984a, b) geological and continental climate chronology applied to Lake Lahontan, rejecting the confusing tufa dates available at the time.

This stage of Lahontan model building was a time of confusion and frustration. In his Hidden Cave chapter, Davis (1985:80) was clear about his source of chronological information at the time of writing: "Morrison's major work in the Fallon area was before the advent of radiocarbon dating, so that the estimates of absolute age in the discussion which follow are derived from later work (Benson 1978; Davis 1978)." The terminal Lahontan evidence from Hidden Cave lacked any dates from before the last high stand at ca. 13,000 until 6900 RCYBP, the Mazama tephra (a 6,100-year gap). Davis said of the critical Lahontan event (Davis 1985:101): "STRATUM X: This is clean pebble gravel, deposited by waves as Lake Lahontan receded from Hidden Cave for the last time. The age of this event is uncertain, but the best approximation from Benson's data is about 10,000 RCYBP." Pyramid Lake delta dates reported by Born (1972) suggested the lake had dropped by 9720 RCYBP, and Davis combined the delta dates with Benson's (1978) tufa dates to estimate the 10,000 RCYBP minimal age for the last Sehoo recession below Hidden Cave. Why the discrepancies?

The Davis 1985 chapter was actually an unrevised draft written in 1981 (Thomas, editorial footnote to Davis 1985:80). Through an unfortunate accident in the timing of major shifts in Lahontan chronology, Thomas (1985) caused a significant and misleading error to be ensconced in one of the best Lahontan Basin prehistories ever published. Thomas (1985:361) said "Lake levels rose once again by 15,000 RCYBP, *covering the cave until about 7500 RCYBP* [emphasis added]," a grossly misleading statement. What Davis (1985:103) actually said was: "The cave was briefly above water again about 15,000 RCYBP, then deeply drowned until about 10,000 RCYBP. From 10,000 until about 7500 RCYBP, the interior was moist," not under water. This in itself is also a misleading statement, since the current estimate of the age of the final recession past Hidden Cave is 12,400 RCYBP.

Davis's 1982 "Bits and Pieces" had been published when the Hidden Cave report went to press, but even though, like Morrison, he knew something was wrong with the first Benson tufa dates, Davis could not find any *dated* evidence to contradict them. The irony is that Davis did have the date; he alludes to it so obliquely in 1982 that most Lahontan students have missed it: "Benson (1978) showed the 13,000 to 11,500 high stand had occurred, but that the 9500 year high stand had not. *Valastro et al. (1978) present a few noncarbonate radiocarbon dates that Davis (1978a) integrated with tephrostratigraphy.*" (Davis 1982:57–58;

emphasis added). Davis was alluding to the 11,490 RCYBP date on wood he collected from the top of the last deep-lake lacustrine sediments at Pyramid Lake, but doesn't even claim it as his own sample. Because of the lack of a Thinolite Unit at low levels of Pyramid Lake, Davis (1978a) concluded that this sample technically lay on top of the lower member of the Sehoo formation, which was ca. 17,000–16,000 RCYBP years ago. The wood actually marked the top of the middle Sehoo, the 11,490 RCYBP date fitting the Morrison sequence very well. Davis (1982), following Morrison and Frye (1965), places the Harmon School Soil after the last high stand and before 11,000 (when it seems to have become wetter after a drying trend); one would think he would link the wood date with this dry interval. The problem was that it implied a low lake exactly when all the other dates indicated a high stand, the same gross incongruity Marie Wormington lamented 25 years before.

However, even though Davis's wood date, from the lowest accessible elevation in the Lahontan Basin, fit the geology and emerging continental climate sequence quite well, the hope for accurate tufa dating won again: "But if Broecker's tufa dates cannot be used in constructing a lacustrine chronology, we are left with too few dates in too large an area through too long a time to define details" (Davis 1982:58). All Davis could do at the time was mention his conflicting evidence obscurely, implying that the dates might support a different interpretation. To emphasize it in 1982 would have meant rejecting Benson's new dates as well as the Broecker and Kaufman (1965) dates on the basis of a single date that defied all the tufa and shell dates. In fact, it continued to defy all other models until Benson himself rejected all the contaminated tufa dates, leaving the low wood date right in the middle of a tufa gap at Pyramid Lake (Benson et al. 1995).

As discussed below, just after the Hidden Cave report was published new studies firmly established the Lahontan decline at least to Hidden Cave levels by 12,400 RCYBP, so this is not news to Great Basin specialists. However, having heard Hidden Cave tour guides mention the 7500 RCYBP lake date, we hope to clarify this issue for a broader audience. This correction also clarifies the setting for the sparse late-Pleistocene fauna in Hidden Cave (Grayson 1985), giving them plenty of time to accumulate in the cave after the last lake stand and before extinction. Dates reported by Mullen (1997) on the small terrestrial mammals from Hidden Cave confirm this new view, with dates ranging from 11,120 to 10,100 RCYBP in the early levels (Rhode et al. 2000).

Dates on the Last Recession from the High Stand

The first clear radiocarbon evidence of a rapid decline of the last high stand appears in Thompson et al. (1986; Thompson et al. 1987) (Figure 3D, dashed line), pushing back the time of the last high stand to before 12,500 RCYBP. Evidence of lake recessions includes what we call "dry dates," based on terrestrial specimens that either show no evidence of subsequent lake water wetting or were clearly above water when deposited (Figure 5, upper right, lists date types). The new Thompson, et al. (1986) model of rapid lake recession was built on a combination of dry dates on packrat middens, deltaic wood, and carefully selected tufa samples. This combination of chronological data paved the way for ongoing refinements of Lake Lahontan chronology that continue to this day. One of the pioneering efforts in this important research was the improved dating of bone by Thomas Stafford, which includes the first date on the Fishbone Cave horse bones, 12,280 ± 580 RCYBP (Stafford et al. 1987; Thompson et al. 1987). The most dramatic new insight of this research was firm evidence of a rapid fall of Lake Lahontan of over 300 ft (91 m) in 200 years. Benson and Thompson (1987a, b) confirmed this rapid lake decline, including evidence of levels approaching the historic low stand at Pyramid Lake shortly after the high stand (Figure 3E, thin line). But what happened after that? Although important progress had been made, the lake sequence and the consequent environmental changes affecting early humans were still unclear.

A promising breakthrough seemed to be the newly named Russell Shoreline ca. 3,947 ft (1,203 m) dated to between 11,300 ± 130 (Beta 24291, marl) and 11,100 ± 110 (Beta 24290, tufa) (Currey 1988, 1990; Elston et al. 1988). These new dates were difficult to fit into prevailing models (Figure 3E, Dansie et al. 1988), and still are (discussed below under IACP lake evidence).

Except for these dates, which seemed a bit early and a bit low, there was still no dated evidence of an upper Sehoo lake that correlated with the geology and archaeology, although Benson et al. (1990) hint at a possible resurgence. Two years later Benson et al. (1992) reported the first evidence of water levels correlated with spill from Pyramid Lake to the Smoke Creek Desert over the Emerson Pass sill (3,960 ft/1,207 m), suggesting the upper Sehoo lake rise (Figure 2A, B sill #6; Figure 3G, dotted line). However, the Benson et al. 1992 model completely obscured the rapid fall so thoroughly documented in 1986 and 1987 (Benson and Thompson 1987a; Thompson et al. 1986) (Figure 3D dashed line; Figure 3E thin line). Furthermore, Benson et al. (1992) showed no recession below Emerson sill immediately following the high stand crash (Figure 3G dotted line), even though it was suggested in 1987 (Figure 3E thin line). It was still unclear what kind of landscape early human occupants might have faced or where their dateable sites might be found. The last large overviews of Great Basin archaeology and paleoenvironments cited Benson et al. 1990 or similar models as shown in Figure 3F (Grayson 1993; Mehringer 1986).

A major change in Lahontan chronology occurred when Benson et al. (1995) documented a gap in tufa dates between 12,030 and 10,960 RCYBP, representing Pyramid Lake levels below the historic low stand (Figure 3G heavy line, Figure 5). The deltaic wood date indicating a significant recession at 11,490 RCYBP (Davis 1978a) is an important non-tufa organic date that supports Benson's tufa gap evidence for a low stand (below ca. 3,800 ft/1,158 m) during the Allerød (Figure 5). With this new development, the once wildly contradictory Broecker and Orr (1958:1026) 12,700 ± 300 RCYBP date on the low-elevation tufa diaphragm in Diaphragm Cave (3,820 ft/1,165m) now overlaps the Benson et al. (1995) first low date of 12,370 ± 110 (Table 1). These data also support Morrison's (1964a, 1991) definition of the late Sehoo recession and provide an approximate date for the beginning of the Harmon School soil-forming interval at ca. 12,300 RCYBP.

The tufa gap was followed by evidence of a return to "relatively fresh water" after 10,960 RCYBP in the Pyramid Lake and Winnemucca Lake basins. Water flowed into the Smoke Creek/Black Rock Deserts by ca. 10,500 RCYBP (Benson et al. 1995:24, 28), filling all these connected basins to 4,019 ft/1,225 m (Benson et al. 1995, 1996). This documented resurgence of Lake Lahontan correlates well with Morrison's upper Sehoo (Morrison 1964a, 1991) and appears to support the basic paleohydrological model of Haynes et al. (1999). The later part of the tufa gap may correlate with the Clovis Drought, and the subsequent rise of the upper Sehoo lakes may correlate with the colder and wetter Younger Dryas (Benson et al. 1995; Fiedel 1999; Haynes 1991; Haynes et al. 1999). The freshwater lakes appear to survive at moderate levels until ca. 9600 RCYBP (Benson et al. 1995), but the Winnemucca Lake tufa dates of 9990 and 9860 RCYBP showing levels at 3,953 ft (1,205 m) are doubtful to us, as shown on Figure 5 and Table 1.

In general outline, these newer models based on radiocarbon dates are consistent with Morrison's original geological models both in time estimate and degree of hydrological change, and they fit Haynes's Clovis Drought and Younger Dryas model extremely well. This growing convergence of once widely disparate hydrological models (Figure 3A–G) is encouraging, but what about the people?

New Bits and Pieces in Lahontan Chronology

Dansie and Jerrems began to assemble all available materials related to Lahontan chronology in 1999. Utilizing approximately 200 radiocarbon dates, we constructed a preliminary graph of lake elevations through time. The discoveries outlined above led us to believe that we had made some important progress toward a culturally relevant chronological model filling in the "Early Man" void. However, other than three cases there were virtually no cultural or paleontological dates for the terminal Pleistocene in the Lahontan Basin. Gruhn and Bryan's (1988) significant charcoal date of 10,700 ± 70 RCYBP (Beta-21885) was associated with the exquisite small Stemmed Point (Figure 6B) and agate endscraper from Handprint Cave above the Black Rock Desert. Bryan and Tuohy 1999 discuss this find and mention the association of this date with bovid hair as well. The other early date, Orr's Fishbone Cave "date" of 11,200 (or 11,555 or 10,200) RCYBP did not date the textiles, despite the confusing literature to the contrary. The 11,080 RCYBP date on the Wallmann mammoth (Stout 1986) was dubious, both for cultural association and date reliability (all discussed below). The 12,280 RCYBP date on the Fishbone Cave horse, which seemed to support the growing belief that megafauna may have become extinct before humans entered the western Great Basin, cast great doubt on the cultural indicators.

This dearth of dates prompted us to search the old collections for specimens that would date more than one component of our model. Taking advantage of ongoing research at Erv Taylor's UCR Radiocarbon Laboratory (Kirner et al. 1996, 1997; Taylor et al. 1995), we also sought bone and tooth pairs that would contribute to the new quest for reliable enamel dates. Dates on bone and hair pairs produced significant scientific discoveries in the last six years, and the superb bone preservation in some Lake Lahontan contexts is bound to yield more surprises. For example, early human remains and related textile assemblages in the Nevada State Museum—Spirit Cave Man and Wizards Beach Man—were recently dated between 9470 and 9040 RCYBP (Kirner et al. 1996, 1997; Tuohy and Dansie 1997) (Table 1; Figure 1, #3 and #7).

In presenting our results, we combine the Haynes/Fiedel model applied to the Lahontan Basin with the results of our dating project; we also include some new geological data generated by other researchers, which is not widely known. The results, a selection from over 300 assembled dates, are listed with annotations in Table 1 and shown on the Lahontan chronology chart, Figure 5.

Oldest Dryas, ca. 17,000–12,600 RCYBP

Lake Lahontan dropped from fairly high levels before 17,000 RCYBP to near 4,000 ft (1,220 m) around 15,000 RCYBP during the Thinolite Recession (Morrison 1964a, 1991). The lake rose to above 4,300 ft (1,311 m), the last high stand, around 14,000 RCYBP, isolating the island (Figures 1 and 5). Deep lake sediments of clay and silt in valley bottoms, vast deltas, high shore tufa deposits, and beach bars are the main evidence of the middle Sehoo formation.

The McGee Point

No overview of Lahontan geoarchaeology would be complete without mention of the McGee point (Figure 6A), an obsidian biface found in direct association with mammoth or mastodont bones; it was buried 25 ft below the top of the "upper lacustral clays," and seems well within the middle Sehoo lake sediments (McGee 1889; Russell 1885:247). The site location in the Walker River canyon is shown in Figure 1, #8. Russell reported extensive vertebrate bone beds in the same geological stratum. Based on Russell's geological description, the McGee point would possibly date to a brief drop in the high stand before 13,300, suggested by some of Benson's tufa dates (Benson 1981; Benson et al. 1995). It is apparently related to the Stemmed Point tradition, consistent with Bryan's hypothesis (Bryan 1988; Bryan and Tuohy 1999) of pre-Clovis Stemmed Points in the far West. It is also reminiscent of Cressman's (1960) report of artifacts found below the 13,000 RCYBP terminal flood deposits farther north, as mentioned in Bryan and Tuohy (1999). However, we also have an alternative hypothesis of a later time, discussed below.

High Stand Megafauna Dates

Two dates on extinct megafauna address the timing of the last high stand and regional correlation of pluvial lakes. A *Camelops* carpal (foot) bone at the Jessup embayment firmly dates the last and maximum high stand of Lake Lahontan to 13,070 ± 60 RCYBP (NSRL-3014) (Adams and Wesnousky 1998). It is from

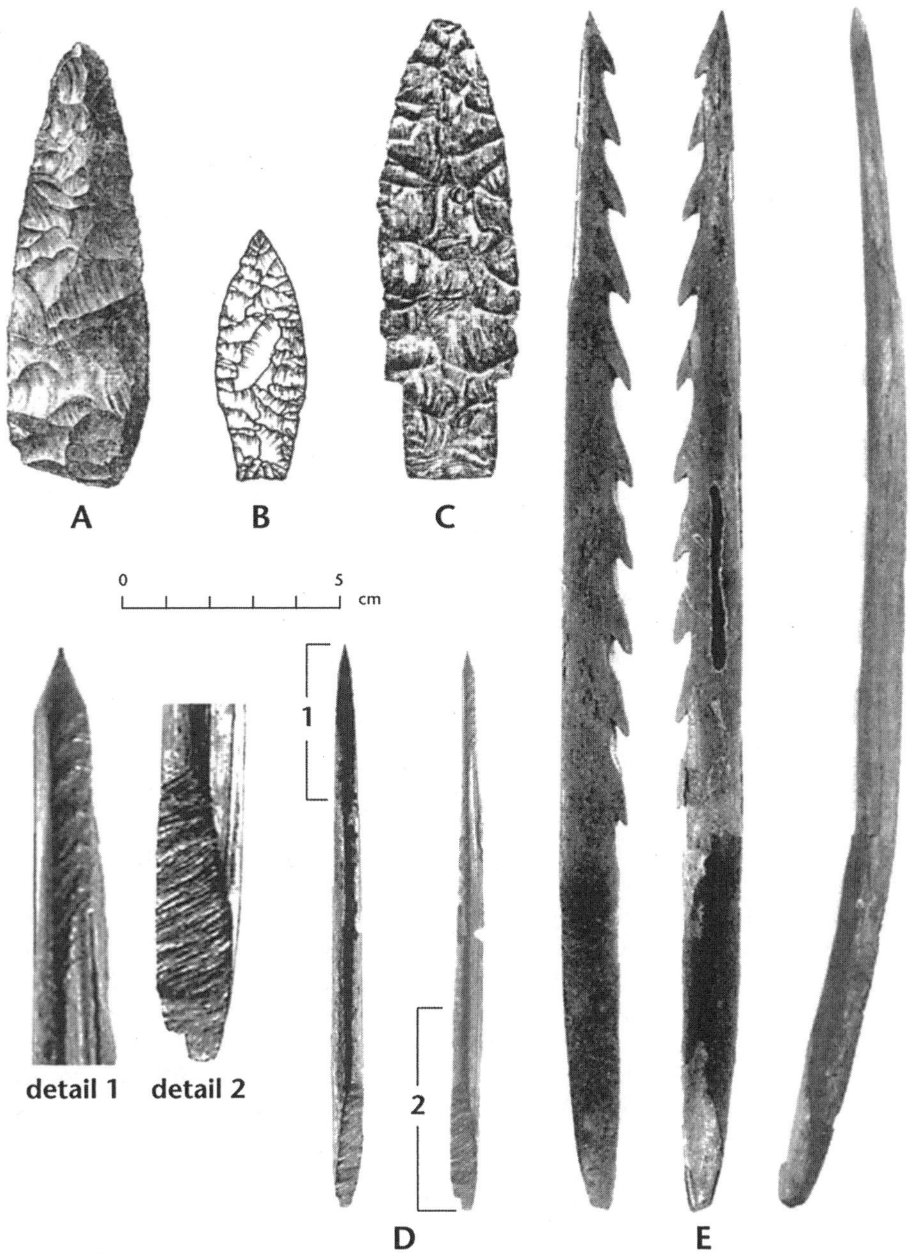

Figure 6. Some artifacts discussed in the text. **A,** McGee point, Walker River Canyon (Russell 1885: Fig. 33:247); **B,** small agate stemmed point, Handprint Cave, 10,740 ± 70 RCYBP (Gruhn and Bryan 1988, from drawing by Maggie Brown, in Bryan 1988, Figure 1:55); **C,** Scottsbluff point with Wallmann bison, 9770 ± 50 RCYBP, Black Rock Desert (drawing by Steve Wallmann; Dansie et al. 1988:176, Fig.9); **D,** mammoth ivory point, 10,360 ± 50 yr RCYBP, Pyramid Lake (3,800 ft); **E,** mammoth rib leister/harpoon, 10,340 ± 50 RCYBP, Pyramid Lake (3,800 ft). **C** and **D** were scanned from original specimen by A. Dansie.

the top of lagoonal sands under a perched back beach playette, at the very highest Sehoo lake shore (4,393 ft/1,339 m). The lake fluctuated before and after this date just below this high stand for around 1,500 years (Figure 5) (Adams and Wesnousky 1998; Benson 1978, 1981; Benson and Thompson 1987a), then began its recession shortly after the camel's death.

Our date on the remains of a short-faced bear (*Arctodus*) encased in tufa from Duck Flat northwest of Gerlach, Nevada (Figure 1, #12), also dates the Pleistocene Lake Surprise rise toward the last high stand at 13,820 ± 50 RCYBP (UCR-3919/CAMS-69401, Dansie and Jerrems 2000). This location at ca. 4,720 ft (1,439 m) was a shallow bay at the last high stand not far from Lake Lahontan; and the 800 years between these two bone dates is consistent with the amount of tufa deposition on the bear skeleton. The excellent preservation of the bone and the intact anatomical position (D. J. Gray, field notes, NSM) is consistent with the death of the bear at the edge of rising lake water. Both of these dates, although technically "dry" dates, are also "wet" dates, providing exceptional lake evidence. These dates also directly addresses the question of synchronicity of the northern Great Basin lake stands with the Lahontan system (Nials 1999:10), at least for lakes at the same latitude.

Bølling 12,600–12,080 RCYBP

The Lahontan Crash, ca. 12,500–12,400 RCYBP

Lake Lahontan apparently remained at relatively high levels for the first 100 years of the Bølling. Benson et al. (1995) documented the last high Pyramid Lake tufa date (4,360 ft/1,329 m) at 12,490 ± 80 RCYBP, the first low date at 12,370 ± 110 RCYBP (3,806 ft/1,160m) and the last low tufa date (3,806 ft/1,160 m) at 12,030 ± 110 RCYBP. We summarized above the dry date evidence that the lake receded rapidly after 12,500 RCYBP (Table 1). These data suggest the lake dropped 546 ft (166 m) in 120 years, at a rate of 4.5 ft (1.4 m) per year: the Lahontan Crash (Benson and Thompson 1987a; Dansie et al. 1988:189; Davis 1982; Thompson et al. 1986). The close agreement of the evidence of the Lahontan Crash with the onset of the dry Bølling at 12,600 RCYBP (Fiedel 1999) is compelling evidence of continental climatic correlation.

Older Dryas 12,080–11,970 RCYBP

We have no Great Basin data to report from this short cold period. However, the next dated specimen, which follows the end of the Older Dryas by only 20 radiocarbon years, may apply to this climate interval.

Allerød 11,970–10,970 RCYBP

Megafauna Survive the Crash

Our date on the Falcon Hill shrub ox (*Euceratherium*) maxilla, dated at 11,950 ± 50 RCYBP (UCR-3780/ CAMS-59155), shows a moderately late survival of this rare extinct "sheep-ox." While not directly related to Lahontan lake levels, it does demonstrate that some megafauna survived the initial onset and duration of the Bølling and the Older Dryas; its presence is synchronous with the onset of the Allerød 11,970 RCYBP (Fiedel 1999). This shrub ox was assumed to predate human occupation (Hattori 1982; Heizer and Baumhoff 1970; Shutler 1961), and there is no evidence of human association.

Harmon School Soil Formation

Good evidence exists for continued low lake levels throughout the Bølling/Allerød. Davis's (1978a) wood date on top of the last deep lake sediments of 11,490 RCYBP demonstrates that Pyramid Lake had dropped to low levels at least by the middle Allerød, if not long before in the early Bølling (Benson et al. 1995; Broecker and Orr 1958). Our assembled data indicate that this late Bolling/early Allerød interval is the only candidate for the time of the Harmon School soil formation in the Carson Sink.

Intra-Allerød Cold Period (IACP) 11,360–11,100 RCYBP

Fishbone Cave Horses

Orr identified a "horse splint bone awl" from Level 3 in Fishbone Cave, which is clearly an artifact (Orr 1956a, Figure 10), and claimed a large agate biface and uniface (Orr 1956a, Figure 11) were directly associated with the marrow-fractured mandibles and long bones of two horses from Level 4. The small horses (a younger and an older individual) are actually onagers, similar to the wild hemiones of the Asian steppes. Orr (1956a) also claimed these horse bones and artifacts were associated with a partial human burial including textiles. We chose direct dating of the critical specimens from this unique setting, since the field and laboratory documents are adequate for independent stratigraphic analysis and new dating methods are minimally destructive.

Stafford (et al. 1987) dated one of two Fishbone Cave horse mandibles (#1) at 12,280 ± 520 RCYBP (AA-759) as part of the Lahontan chronology studies of Benson and Thompson (1987a:73). We thought it was vital to determine the age of the other mandible (#2) because they both exhibit strong evidence of cultural modification. Fishbone Cave horse mandible (#2) dated 11,350 ± 40 RCYBP (UCR-3783/CAMS-59288, Dansie and Jerrems 1999), a date 930 years later than the other horse mandible (#1). Because of this wide discrepancy, we re-dated the Fishbone Cave horse mandible #1; and the new amino acid date of 11,210 ± 50 RCYBP (UCR-3920/CAMS-69408; Dansie and Jerrems 2000) shows this horse to be 140 years younger, instead of 930 years older than mandible #2. We cannot account for the discrepancy in the bone dates; however, we accept both UCR dates, since they are more recently assayed (see Stafford 1994; Stafford et al. 1987; Taylor et al. 1992; Taylor et al. 1995).

These Fishbone Cave bone dates fall within the Intra-Allerød Cold Period (IACP), 11,360–11,100 RCYBP. The large accumulation of juniper twigs and broken and twisted sagebrush in the Level 4 deposits of Fishbone Cave is suspected to be cultural firewood, owing to the large, fairly uniform size of the wood fragments. Even with all the dating using old methods, there is still a large quantity of Level 4 wood that could be dated. Shredded bark bedding (Orr 1956a) may also be represented in the bulk samples. The wood date of 11,200 ± 250 RCYBP (L-245), a laboratory average of a series of separate runs, was on this material, collected by Broecker himself from Level 4 to test a new radiometric gas method (Broecker and Orr 1958; NSM files). This official L-245 date compares favorably with the two new horse bone dates, especially horse #1, only 10 years older. The other two Fishbone Cave wood dates, submitted by Orr from Level 4, 11,555 ± 500 RCYBP (no lab number) and 10,900 ± 300 RCYBP, ("L-245"), were separate assays, not averages as some literature implies. Orr's field diagrams and his Figure 4 (1956a:17) document stratigraphic correlation of one horse mandible on the dated "juniper bark." All three of these wood dates are consistent with the new horse mandible dates and may represent multiple occupations of the cave during the IACP.

Orr (1956a:Figure 11) was clear that the large agate biface and uniface were the only artifacts that were positively recovered from Level 4; in fact, he clearly describes finding the "large chipped knife" lying directly on the broken diaphragm[4] (Level 5), 1 ft from a horse bone (Orr field notes, June 25, 1954). Orr (1956a) states: "These two are the only chipped stone objects definitely established as belonging to Level 4. . . . [They] show

[4] A tufa diaphragm is a flat tufa sheet about 3 to 6 inches thick, sloping down to the rear of the cave, that formed wall to wall in some lakeside caves. It appears to be formed like flowstone (travertine) near the lake surface on wave-built sand berms choking the mouth of the cave (Broecker and Orr 1958). When the sand washes away, a closed cavity exists above it until the diaphragm is broken into flat "paving stones," which fall to the floor of the cave.

definite indications of patination and rechipping, with a thin deposit of lake tufa on one indicating that they were already 'antiques' some 11,000 years ago." We consider these artifacts to be consistent with the numerous cuts and impact fractures on the horse bones, and the taphonomic pattern suggests field butchering and marrow snacking by the hunters (analysis in preparation by authors). Blood residue analysis of the two artifacts failed to produce any residue traces (M. Newman pers. comm. 2000).

The small, defensible cave, with an original opening only 42 inches high, may have been used during the IACP for field butchering and protection from the elements. The virtual absence of carnivore-gnawing on the impact-fractured horse bones is one of the taphonomic indicators supporting (but not proving) human modification (G. Haynes pers. comm. 1999). Furthermore, numerous bone impact flakes were found by the authors throughout the bulk wood sample from Level 4, demonstrating the bones were broken in the cave. No herbivore dung was found in the cave (Orr field notes, June 25, 1954), which would have indicated possible hoofed animal trampling should one have been able to squeeze into the cave. Finally, no serious rock fall occurred after the wood and bark layer was deposited on the fallen diaphragm, precluding natural fracturing of the bones within the cave. All these lines of evidence suggest human transport and modification rather than animal habitation in Fishbone Cave; consequently the two agate artifacts are probably associated with the horse bones.

IACP Lake Evidence

The Russell Shoreline, representing a lake rise to 3,949 ft (1,203 m) at 11,300 ± 130 RCYBP (Beta-24291, Elston et al. 1988), was defined based on a deep-water marl date at Stillwater Slough. The Upsal Hogback tufa date of 11,100 ± 110 RCYBP (Beta-24290) was also cited as evidence for the last moderate lake (Currey 1990; Elston et al.1988:378). This is the only evidence of a lake rise between the Lahontan Crash and the Younger Dryas. Decreased evaporation during the cold IACP may also have been accompanied by increased available moisture; and there are local reasons (possible river diversions) why the Carson Sink may have filled slightly while other basins remained virtually dry. Because the Carson Sink and Humboldt Sink are connected at 3,940 ft (1,200 m, the natural height of the Humboldt Bar), if the Carson Sink rose to 3,949 ft (1,203 m, the Russell Shoreline), so would the Humboldt Sink (Figure 2 A and B, sill #7).

We note that Currey (1990:201) later cites the date for the Russell Shoreline as 11,100 RCYBP in his text, and 11,000–10,000 RCYBP in his abstract, correlating the Russell Shoreline with the Younger Dryas–age Gilbert Shoreline of Lake Bonneville. Grayson (1993:95) and Madsen (1999, 2000b) discuss this problem and conclude that the Russell Shoreline is probably contemporaneous with the Bonneville basin's Gilbert Shoreline of Younger Dryas age. However, Madsen uses the date 11,200 RCYBP for the beginning of the Younger Dryas (Madsen 1999:77). An IACP age of the Russell Shoreline in the Carson/Humboldt sinks would correlate with the Leonard Rockshelter basal bat guano date (11,199 RCYBP, Heizer 1951) for this lake stand ca. 3,940 ft (1,200 m). At this level a vast marsh would form in the upper Humboldt Sink for the bats to thrive on. The age of the Russell Shoreline remains unresolved, but taking all the dates at face value, including Fiedel's (1999), we plot it as IACP in age.

Clovis Drought 11,000 RCYBP

The Wallmann mammoth was a huge male that died in the drying Jackson Creek slough on the Quinn River arm of the Black Rock Desert. A copy of the skeleton is on display at the Nevada State Museum, depicting its death scene. This nearly complete mammoth skeleton has always eluded confident chronological assessment (Dansie et al. 1988), although the virtually non-collagenous bone apatite date of 11,080 ± 300 RCYBP (GX-11067, Stout 1986) would seem to be the perfect confirmation of a mammoth on the drying lake bed of the Black Rock Desert during the Clovis Drought. Attempts to re-date the Wallmann mammoth revealed less than 0.03% amino acid preservation, casting doubt on the original date (K. Selsor pers. comm. 2000). The Wallmann mammoth could be Farmdalian in age, ca. 25,000 RCYBP (Dansie et al. 1988). Hopefully, the pending enamel dating project will yield a more reliable result (Jerrems 2000; R. E. Taylor pers. comm. 2000).

An agate uniface and possible manuport, found within the slough hole with the Wallmann mammoth skeleton, remain problematical (Dansie et al. 1988). Five other, smaller mammoths discovered a few miles away at the Delong site (Livingston 1992) are geologically estimated to have died between 11,000 and 12,000 years ago, during a time of ecological stress (Livingston 1999). Whether or not the Wallmann or Delong mammoths are evidence of the Clovis Drought cannot be determined at this time, but not for lack of trying. We found no direct evidence of the Clovis Drought in the Lahontan Basin; but we record the prediction of this drought on our final chronology chart (Figure 5) as a drop to near desiccation in all basins except Pyramid Lake, which probably dropped below the historic low stand.

Younger Dryas 10,970–10,100 RCYBP

Lake Evidence

Ken Adams (Adams and Wesnousky 1998; Rhode et al. 2000) reports shorelines of 4,052 ft (1,235 m) in the eastern Lahontan system (Carson and Humboldt sinks) and 4,000 ft (1,219 m) in the western system (Pyramid and Winnemucca lakes and Black Rock Desert, Figure 2B). His data indicate these shorelines clearly postdate the Lahontan crash and are probably correlated with the Younger Dryas. Indirect evidence of this return of the moderately high lakes, Morrison's upper Sehoo, is found in the archaeological distributions on the Black Rock Desert. A distinctive small Stemmed Point from Handprint Cave (Figure 6B), at an elevation of ca. 4,416 ft (1,346 m), was associated with bovid hair and a charcoal date of 10,700 ± 70 RCYBP (Beta-21885, Bryan 1988; Gruhn and Bryan 1988). Steve Wallmann (S. Wallmann pers. comm. 1989) reports that this same point type is found exclusively above an elevation of 4,000 ft (1,219 m) on the Quinn River arm of the Black Rock Desert. This indirect evidence suggests a date for the Younger Dryas lake maximum of ca. 10,700 RCYBP.

Benson et al. (1995) reported a tufa date of 10,450 ± 80 RCYBP (CAMS-4432) from Winnemucca Lake that also supports a lake stand of at least 3,952 ft (1,205 m). At Pyramid Lake they saw evidence of an even higher stand that suggests the entire Black Rock Desert had filled and all the western lakes (Figure 2B) had reached an elevation of 4,018 ft (1,224 m), roughly the same level Adams reconstructed from independently analyzed shoreline data. Benson, unable to find any tufa in the Pyramid Lake basin to match the wave-cut cliffs marking this lake rise, resorted to dating this event with rock varnish. We believe Benson's observation of the wave-cut tufa and rock face may be good evidence of this Younger Dryas lake rise. Although varnish dates have yet to be verified, the oldest Pyramid Lake varnish date, 10,460 ± 180 (TO-167), is consistent with other data. Benson's (Benson et al. 1992) lake model includes the varnish results (Figure 3G, dotted line), and we show one of them in Figure 5 to demonstrate the elevation of the wave-cut feature.

Archaeological Evidence: Ivory Point and Mammoth Rib Harpoon from Pyramid Lake

Of particular interest is an assemblage of fishing equipment from the historic low stand of Pyramid Lake (Ting 1967, 1968; Tuohy 1988, 1990). Previous dates on fishing cordage from this locality showed cultural use of Wizards Beach at 9,660 ± 170 RCYBP (GX-13744, Tuohy 1988). This fits nicely with Benson's new data (Benson et al. 1995) indicating the end of relatively fresh water in the Pyramid Lake basin at 9600 RCYBP. Low lake levels at Pyramid Lake do not necessarily mean the lake is unproductive; the lake trout are known to survive such recessions (Coffin and Cowan 1995; La Rivers 1994:293).

Several bone artifacts show possible affinities with Clovis technology, including single-beveled, crosshatched bases and short tapered, conical tips. We decided to test the Clovis Drought model by dating these artifacts. We hypothesized that they would date from one of two possible periods on either side of the upper Sehoo: (1) between 11,200 and 10,960 RCYBP, the "Clovis Drought"; or (2) between 9700 and 9220 RCYBP, based on the Wizards Beach dates from the low stand.

Surprisingly, the black ivory point (Figure 6D) from 3,800 ft (1,158 m), near the Needles on the northwest shore of Pyramid Lake, dates to 10,360 ± 50 RCYBP (UCR-3796/CAMS-59286, Dansie and Jerrems 1999). The serrated bone harpoon or leister (Figure 6E), also from 3,800 ft (1,158 m), made of mammoth rib, produced a statistically identical date of 10,340 ± 40 RCYBP (UCR-3795/CAMS 59285, Dansie and Jerrems 1999). The dates of these artifacts suggest they could even be from the same individual mammoth.

These dates, quite late for mammoth, might indicate late survival through the Clovis Drought in the Pyramid Lake area. They are also 440 years after Clovis (Fiedel 1999) and may be another example of late mammoth survival in and near the Great Basin. Owl Cave (a.k.a. the Wasden site) in Idaho produced evidence of mammoth associated with large Folsom-like points. The mammoth bone collagen was dated 10,920 ± 150, and the charcoal under it dated 10,640 ± 85 RCYBP (Bryan and Tuohy 1999; Miller and Dort 1978). Taylor et al. (1996) commented on this late survival of mammoth in the far West.

The two mammoth bone and ivory implements were likely designed for taking the giant Pyramid Lake cutthroat trout, which were over 41 lbs historically (Sigler and Sigler 1987:114) and at least 60 lbs in the past (Wheeler 1967:100). John C. Fremont (1845) described trout 2 to 4 ft long at Pyramid Lake in 1844. Such potential leads to speculation that the last mammoth hunters may have turned to this terminal Lahontan fishery for survival. If the identification as fishing implements is correct, and the setting certainly supports it, these are the oldest directly dated fishing implements in the New World. With a deeply crosshatched single-beveled base, the serrated leister/harpoon (Tuohy 1990) may be a previously unrecognized Clovis related artifact type. Similar "points" found in Florida date to late Clovis times (Hemmings, pers. comm. 2001; Hemmings et al. 2002 in press; Webb and Hemmings 2000; Dunbar et al. this volume).

Of related interest we note that a date of 10,380 ± 80 RCYBP (Beta-29024, Benson et al. 1992) on *Anodonta* shell from the Humboldt bar at 3,929 ft (1,198 m), virtually the same date as our ivory point and bone leister from Pyramid Lake, supports their interpretation as evidence of lakes lowering from the maximum Younger Dryas lake stand to a shallow freshwater habitat (Schneider 2000). Peter Mehringer (Mehringer pers. comm. 1999) agrees with this timing of a mid–Younger Dryas lake recession in the northern Great Basin.

The age of the ivory point and mammoth rib harpoon dates the death of the animal, not necessarily the date of deposition. The artifacts presumably could have been made and used until deposited later, obscuring climatic implications of a low stand during the middle Younger Dryas. A few other suspect tufa dates suggest that the Lahontan Basin lakes were somewhat higher during this period. These artifacts could, however, reflect a low lake level at the time of death, which would be consistent with Benson's evidence that spill to the Black Rock Desert may have ceased at nearly the same time (Benson et al. 1995), as shown on Figure 5B. Morrison has plotted a "bi-lobed" upper Sehoo for decades; this middle Younger Dryas recession would probably correlate with the drop between two peaks (Figure 3D, heavy line), although our evidence suggests a much greater recession at Pyramid Lake at this time (Figure 5) than Morrison shows for the Carson sink. We found no non-tufa dates to indicate where the lakes were in the late Younger Dryas.

Younger Dryas/Pre-Boreal Transition 10,100–10,000 RCYBP

Benson reports tufa dates of 10,080 ± 60 RCYBP (LDGO-1743L) and 9,990 ± 80 RCYBP (CAMS-4431) for lake levels around 3,952 ft (1,205 m), slightly below the Emerson sill but 90 ft above Mud Lake slough (Benson et al. 1992, 1995). We found no other dates in the Lahontan Basin from this 100-year interval between the Younger Dryas and the Pre-Boreal; therefore we consider these tufa dates suspect.

Pre-Boreal 10,000–9,000 RCYBP (Neothermal Onset)

Lake Evidence

There are several early pre-Boreal environmental dates that tell somewhat conflicting stories regarding lake levels (details in Table 1). Five Pyramid Lake dates show the lake below 3,860 ft (1,177 m). Two are on deltaic wood: 9970 ± 140 RCYBP (I-8193) and 9780 ± 135 RCYBP (I-8194, Prokopovich 1983). Another date of 9920 RCYBP is from tufa on pelican wing bone found at 3,790 ft (1,155 m) (Benson pers. comm. to Donald R. Tuohy). One of the most important dates in this set is on fishing cordage from the historic low stand of Pyramid Lake, 3,785 ft (1,154 m), dated 9660 ± 170 RCYBP (GX-13744, Tuohy 1988). Further evidence of a pre-Boreal low stand at Pyramid Lake is the date of 9225 ± 25 RCYBP on the skeleton of Wizards Beach Man (UCR-3445/CAMS-26369 and 26370, Kirner et al. 1996; Tuohy and Dansie 1997). Only one tufa date shows a Winnemucca Lake level 92 ft higher within this same time period, 9860 ± 80 RCYBP (CAMS-4430), at 3,952 ft (1,205 m), nearly up to Emerson sill (Figure 2, sill #6) (Benson et al. 1995). While all these dates might be accurate lake level indicators reflecting rapid fluctuations, we believe the wood, bone, and textile dates are stronger evidence for lower lake levels. We discuss this further below.

Wallmann Bison

The Wallmann bison (*Bison bison*) associated with a Scottsbluff point (Figure 6C) from the Black Rock Desert (Dansie et al. 1988), was dated 9770 ± 50 RCYBP (UCR-3782/CAMS-59156, Dansie and Jerrems 1999) (Figure 1, map #11). This date is early within the Cody complex time range (Stanford 1999:321; Wallmann and Amick 1991), and about 4,000 years early for small bison (McDonald 1981). The Old Humboldt site (Dansie 1984, 1987) produced further evidence that small bison evolved early in the Great Basin, perhaps during the 2,500-year interval (14,000–12,500 RCYBP) when the large island (Figure 1) was isolated by the last high lake stand. Most of the northwestern Great Basin terrain was either steep mountains or under water at this time. The island, large enough for significant grasslands, held two internal basins where meadows or marshes may have formed. If only for a few hundred years, when the rain stopped 12,600 years ago, the grasslands would have been overgrazed long before the island was connected to the mainland as the water receded. The idea that a smaller form of bison could evolve and survive in such an island setting is not unreasonable.

This bison date of 9770 RCYBP also marks a time when there was no lake on the Black Rock basin above 3,950 ft (1,204 m), which is below Emerson sill, thus supporting the low lake evidence cited above. Meadows and grasslands probably grew between the Quinn River and Jackson Slough channels at this time. Bird bone and staining from a "brown mat" above these partially weathered bison bones also strongly suggest that a later marsh developed over the bison kill site (Dansie et al. 1988). While this marsh could represent the deltaic edge of a vast lake in the Black Rock and Smoke Creek Deserts, it could also be a local phenomenon created by the earlier lake stand stabilized at the Emerson sill level. A temporary beach bar could have impounded water above the shoreline, creating a local marsh on the Quinn River.

Cluster of Pre-Boreal Archaeological Dates

Developing from the dating project at Spirit Cave, a large body of cultural data indirectly suggests a series of shallow lake marshes in the Lahontan Basin. Table 1 lists the 19 pre-Boreal dates used in our lake model. This large cluster of dates spans the entire 1,000 years of the pre-Boreal, with no more than a 91-year gap between them. We discuss these further below, in the pre-Boreal Marsh section.

Early Boreal 9000–7980 RCYBP

Two environmental dates, on deltaic wood at Pyramid Lake (8800 ± 90 RCYBP WIS-374, Born 1972) and bat guano from Leonard Rockshelter (8660 ± 300 RCYBP L-281, Heizer 1956), reveal moderately low water levels that were probably still generally productive at the onset of the Boreal. However, the last pre-Boreal cultural date in the Lahontan Basin (9040 RCYBP, previous paragraph) is followed by a 660-year cultural date gap that ends with a suite of textile dates between 8380 and 8200 RCYBP (Dansie and Jerrems 1999; Fowler and Hattori 2000). Another 400-year cultural date gap is followed by a Winnemucca Lake textile date at 7980 ± 600 RCYBP (Hester 1974:2), ending our time frame coverage (Table 1). Madsen (1999) discussed evidence that significant climate change occurred at 8000 RCYBP. We cover only our new dates here; note, however, that there are only seven cultural radiocarbon dates in our database from 7900 to 6000 RCYBP, each separated by several hundred years. This suggests sparse human occupation after 8000 RCYBP. We thus agree with Madsen, but we have no new evidence of climate change during this period in the Lahontan Basin.

Correcting Old Errors at Fishbone Cave

One of our research priorities was to clarify cultural association in early sites. We dated what was assumed to be the "oldest radiocarbon dated textile in western North America" (Adovasio 1986:199, Figure 6 d; Hyland and Adovasio 2000:144). This twined sagebrush matting fragment from Level 3 (Orr 1956a: Figure 8) was mistakenly assigned an associated date of 11,200 ± 250 RCYBP by Orr (L-245, Broecker and Orr 1958; Broecker et al. 1956; Orr 1974). The small piece of twined sagebrush bark textile was associated with remnants of Burial 2 in Fishbone Cave (P3e/321), but not with the shredded bark and wood of Level 4 discussed above. In the Dirty Shame Rockshelter publication, Andrews et al. (1986:216) also mention the "oldest basketry in the world" and state the date as 9300 ± 25 [sic] B.C. (thereby incorporating Orr's 1974 typo of 11,250 ± 250 RCYBP and adding a typo of their own). Andrews et al. (1986:216) cited Orr (1974) and Rozaire

(1974) in discussing the oldest sandals, stating that "this specimen" (implying they were the same specimen) derived from Level 4 of Fishbone Cave was associated with the 9300 B.C. date. However, Orr (1974) does not show a sandal from Level 4, only the delicate fragment of sage bark twining. The sandal shown in Rozaire 1974 is not the same specimen as the twined sagebrush matting, despite the similarity in shape; rather, it appears to be a tule sandal from Level 3 (P3e/291) based on our direct comparison of the specimen, catalogs, and the photograph. Willig (1991:97) further entrenched these assumptions about the "oldest textile" as a fact when she stated that "Fishbone Cave, Nevada, produced a date of 11,200 ± 250 RCYBP (L-245) on a fragment of twined juniper bark found in association with a human burial, basketry fragments, and two large chert knives (Adovasio 1986; Orr 1956a, 1974)." Willig also mistakenly listed the L-245 date as a shell date on her Table 1. Considering the accumulation of reporting errors and ambiguities by Orr and assumptions made by later scholars, the evidence for the "oldest textile in the world" seemed worth sorting out.

The "oldest textile" is preserved by a lump of burial sediment that retains clear impressions of Catlow Twined basketry (Orr 1956a: Figure 8). Our date on one of three small fragments of Catlow Twined basketry (Orr 1956a: Figure 7) directly associated with the lump yielded a date of 8370 ± 50 RCYBP (UCR-3779/CAMS-59287, Dansie and Jerrems 1999), confirming our assessment that Phil Orr never dated the twined sagebrush matting as he implied but never actually stated. This is the oldest Catlow Twined basketry currently known in the Great Basin; and indirectly supports the old date on netting associated with this feature, published as 7830 ± 350 RCYBP (L-289kk, Broecker and Kulp 1957; Broecker and Orr 1958; Orr 1956a, Figure 6; Orr 1974). Orr, in trying to make the burial old enough to belong with the wood dates from Level 4, caused great confusion to anyone trying to work with just some of his documents. The maximum net age (7830 + 350 = 8180) was listed as 8200 RCYBP in Orr's lab notes and site records, even with the ± 350 added to it in one place.

Catherine Fowler and Eugene Hattori recently had the twined sagebrush sample ("oldest textile") directly dated to 8220 ± 50 RCYBP (Fowler and Hattori 2000; K. Selsor pers. comm. 2000). Although 150 years later, this date is generally consistent with our date on the associated Catlow Twined fragment, taking long-term curation of this finely woven textile into account. The new date on this feature clearly separates the earliest textile and burial components in Fishbone Cave from the extinct horse bones by 3,000 years. These textiles, netting, and a prepared pelican skin (Orr 1956a: Figure 9) are probably associated with the partial human burial (#2), but there is enough variation in the three dates to doubt the contemporaneity of all the components, even though they are all probably over 8,000 years old.

Another piece of Orr's model, the Fishbone Cave "horse splint bone awl" with ligaments attached, was assumed, again, to be evidence of extinct fauna utilization (Orr 1956a). We extracted a small sample, dated to 3090 ± 50 RCYBP (UCR-3789/CAMS-59283, Dansie and Jerrems 1999). This date confirms our refutation of its original identification by Orr, who even specified that it was the left fourth metapodial of an extinct horse. It is not even close in form or structure to the several onager splint bones from the cave available for comparisons. The thickness, density, and form of the highly modified bone are consistent with a bighorn sheep metapodial awl, common in Great Basin Archaic assemblages.

This bone awl is another lesson in the dangers of assuming. In view of the fact that many archaeologists have published on the assumed horse bone awl even after close examination (such as Dansie et al. 1988; Hattori 1982), we must caution that the shrub ox and the short-faced bear must also be reanalyzed. The horses also have yet to be evaluated within the current Pleistocene equid taxonomy, a daunting task.

Discussion

Jonathan Davis said, "There is no 'Lahontan Chronology'; there is a complex family of chronologies, which are fewer when water is high, and more numerous when it is low" (Davis 1982:54). The Basin profile chart (Figure 2A) shows the complex relationships of the Lahontan family. These new bone, ivory, and textile dates, combined with guano, shell, wood, and tufa dates, suggest that we must scrutinize differences between sub-basins and not assume that all will rise and fall at the same rate at the same time (Figure 5B). Figure 4 shows approximate lake levels at different times, and Figure 2C shows possible river diversions during the Bølling/Allerød and Younger Dryas periods. As Davis emphasized, there is an important distinction between environmental change and climate change, and the Lahontan Basin has several examples where this distinction is critical to archaeological interpretations.

River Diversions and Early Human Occupation

Truckee River to Carson Sink

The recent work of Adams (et al. 1999) shows that isostatic rebound may have significantly affected river diversions in ways not previously considered in late Lahontan chronology[5]. When the lake was deep, the land was pushed downward by the weight of the water; after the water receded, the land rebounded over time. After the last high stand, starting 12,500 RCYBP, the Truckee River may have first flowed east onto the newly exposed, nearly flat delta plain at Fernley (Figure 2B, west side of #3), ultimately spilling south to the Carson Sink (Figure 2C) over Darwin sill (Figure 2, east side of #3). Such a diversion may have created a wet environment in the Fernley sink, especially during the IACP, 11,360–11,100 RCYBP. This diversion may also have maintained a wetter Carson sink sub-basin in an otherwise dry climate, especially with the Walker River diverted at Adrian Valley into the Carson River (Figure 2A, B, sill #1; Figure 2C).

A moderately deep lake in the Carson Sink during the IACP

[5]The hypothesized Farmdalian diversion, during the Wizards Beach Recession, was much older, ca. 25,400 years ago (Dansie et al. 1988).

could explain the notable absence of any early point types at the lower elevations and their rarity in the Humboldt Sink (Kelly 1983); maybe both lake beds were under water most of this time (Rhode et al. 2000). This postulated IACP lake might also explain the otherwise anomalous dates (11,300 to 11,100 RCYBP) on the 3,947-ft (1,203-m) Russell Shoreline (Elston et al. 1988; Rhode et al. 2000). These Russell Shoreline dates are well within the tufa gap at Pyramid Lake (Benson et al. 1995) and thus might appear contradictory.

We hypothesize that the Harmon School soil interval represents lake levels below 3,920 feet (1,195 m), only 60 ft deep, between 12,300 and 11,300 RCYBP in the vast Carson Sink. The water rose in the early IACP, the waves washed away the top of the Harmon School soil (creating an unconformity), and then tufa began to form in relatively shallow water. The lake continued to rise another 27 ft to the Russell shoreline, depositing shoreline tufa at ca. 3,953 ft (1,205 m) and moderately deep water marl at Stillwater Marsh (Elston et al. 1988; Rhode et al. 2000).

During this same time interval (and in much earlier recessions), Truckee River diversion into the Carson Sink basin may have been responsible for carving the Massie Slough channel on the north side of the Carson River delta, grading down to the edge of the lowest playa about 3,920 ft (1,195 m). Such a major river diversion in a time of general drought may someday help us understand the Paleoamerican sites along the beach bars above this slough, including the Sadmat site (Figure 1, #5) and large pebble mound sites (Dansie 1981; Irwin-Williams, Dansie and Davis 1986; Irwin-Williams, Dansie, Davis, and Jacobsen 1986; Irwin-Williams et al. 1990; Tuohy 1981). Three sites (Dansie, Crew, and Frock), characterized by heavily patinated, thick bipointed bifaces and advanced side-struck technology, are located on desert pavement shorelines above 4,200 ft (1,280 m). Each is adjacent to a lookout hill in a straight line linking the two areas that would benefit most from the IACP Truckee River diversion through Massie Slough, when megafauna moved from the island to the Carson and Fernley Sinks (Figure 1, #4). All have intact back beach playettes, which still enhance and catch even minor rainfall, much like the pebble mounds. Even with the river diverted, the Bølling and early Allerød were apparently relatively hostile in the Lahontan Basin, with intermittent Truckee River flows reaching the Carson Sink but not Pyramid Lake.

Pre-rebound Susan River flow may explain how Pyramid Lake could have been kept alive if the Truckee River were diverted to the Carson Sink (Figure 2). Before isostatic rebound, the Susan River may have flowed through Honey Lake Valley and over Astor Pass directly onto the northwest shore of Pyramid Lake at Wizards Beach. Adams et al. (1999) document the tectonic evidence and suggest this event may have occurred during the terminal Pleistocene, after 12,500 RCYBP. Dansie and Davis (Dansie et al. 1988) describe other evidence of this stream flow (rusty channel gravel and sand) on Wizards Beach.

Truckee River Diversion to Winnemucca Lake

Another potentially controversial river diversion issue was first addressed by Russell (1885). He documented a bifurcation of the Truckee River at the Mud Lake Slough, whereby half the river flowed to Winnemucca Lake and half to Pyramid Lake, when Pyramid Lake was lower than the slough threshold. He also described a complete diversion of the Truckee River into Winnemucca Lake caused by a debris dam which lasted only one season. Though a significant diversion or bifurcation might be difficult to accept (Hattori 1982), we believe the archaeological evidence supports such an event during the pre-Boreal (10,000–9000 RCYBP). Cultural dates between 9540 and 9120 RCYBP at Winnemucca Lake strongly suggest the presence of a marsh at the north end at the same time Pyramid Lake was near its historic low stand (9920, 9660 and 9220 RCYBP), 70 ft lower than the slough (Table 1). There are enough springs along the west side of Winnemucca Lake to attract some occasional use during times when the lake was dry, but the springs are far removed from the Guano Mountain caves on the northeast shore. The lowest point of Winnemucca Lake, the first to pond, is toward the north end (Russell 1885). Bifurcation or diversion may have occurred several times during the Pre-Boreal and later (Table 1; Figure 5).

Pre-Boreal Marshes

Additional evidence of pre-Boreal marshes exists in the Carson Sink. The Spirit Cave–type plain weave matting made of tule and hemp (Fowler et al. 2000; Tuohy and Dansie 1997), found at three Grimes Point sites (Table 1, Figure 1 map #7), and the small marsh fish found in the intestines of Spirit Cave Man (Eiselt 1997), may be evidence of a marsh level of about 3,925 ft (1,146 m) on the flats below the cave. Such a water level would be sufficiently shallow that marshes could develop in a vast area of undulating terrain, free of the destructive wave action of deeper lakes. Morrison (1964a) described a terminal Sehoo Carson River delta at 3,925 ft (1,146 m), representing a relatively brief lake stand lower than, and following, the upper Sehoo lakes. The 450-year span of the plain weave matting in the Grimes Point area (9450 to 9000 RCYBP, Fowler et al. 2000; Tuohy and Dansie 1997) would fit closely Morrison's delta and brief 3,925-ft (1,146 m) still-stand description. The continued shrinking of the marshes and lakes after 9000 RCYBP is indicated by a variety of data. Later lake rises to and above these levels remain controversial and are beyond our scope (Rhode et al. 2000).

Correlations

It seems that most of the megafauna did not survive the IACP (11,360–11,100 RCYBP, Fiedel 1999; Haynes 1984; Stafford pers. comm. 1999), and our results support Stafford's findings very closely. None of the five recently assayed dates on camels and horses in the Great Basin are younger than the middle IACP. The eastern Nevada Sunshine Well locality *Camelops* date to 11,330 ± 60, 11,340 ± 50, and 11,910 ± 50 RCYBP (Beck and Jones 2001 in prep; Dansie et al. 2001 in prep; Huckleberry et al. 2001; K. Selsor pers. comm. 2001). The Smith Creek Cave camelid hair dates to 12,060 ± 450 RCYBP (Bryan 1988:67), and the Jessup embayment Lahontan high-stand camel dates to 13,070 ± 60 (Adams and Wesnousky 1998). Our two horse (onager) dates are 11,350 ± 50 and 11,210 ± 50. The only shrub ox date is 11,950 ± 50. Our date on the extinct bear, *Arctodus*, is 13,820 ± 50 RCYBP.

In contrast, late survival of two species of Great Basin megafauna is one of the biggest surprises. The Huntington mammoth from Utah, dated 11,220 ± 110 RCYBP (AA-4936, Madsen 2000a), another IACP megafauna date, is within the date range for mammoths elsewhere (Haynes et al. 1999). However, the giant short-faced bear (*Arctodus simus*) found with it, at 10,870 ± 75 RCYBP (AA-6974, Madsen 2000a), dates to several hundred years after the main extinction event in North America (Grayson 1991; Haynes et al. 1999). Hockett and Dillingham (2000) report even later survival of possible *Arctodus* or very large *Ursus* from Mineral Hill, eastern Nevada, with two dates: an amino acid date by Stafford of 9960 ± 50 RCYBP and a Beta Analytic date of 9710 ± 40 RCYBP. Our two dates on the mammoth ivory point and mammoth rib harpoon average 10,350 RCYBP, and there is evidence of late mammoth dated 10,920 RCYBP and 10,600 RCYBP (Bryan and Tuohy 1999) in Idaho, bridging the "post-extinction" time span to our latest dates.

These data suggest Great Basin megafauna die-offs in the early IACP and mid–Younger Dryas. An increased frequency of megafauna radiocarbon dates can be interpreted as reflecting a higher frequency of deaths, greatly increasing the statistical probability of preservation and discovery. The megafauna die-offs are supported by dates on fish die-offs in the Bonneville basin (Broughton et al. 2000; Madsen 2000b). The fish dates may date each of the three postulated times of ecological stress, the IACP (11,270 ± 135, 11,263 ± 83, and 11,181 ± 85), terminal Allerød/Clovis Drought (11,065 ± 105 and 10,910 ± 60), and mid–Younger Dryas (fish gone by 10,350 ± 80), the same time our data show lakes declining. Late survival of mammoth and short-faced bear in the Great Basin may actually be the exceptions that complement the Clovis Drought extinction model. If a sudden polar freeze caused the final extinctions in most places, the hot springs so abundant in the Great Basin, and Pyramid Lake, over 200 feet deep at the lowest known level, may have provided just enough refuge to allow survival into the Younger Dryas of the most aggressive herbivores and omnivores: the mammoth and the giant short-faced bear.

One gap in our model is the absence of any sign of life on the last lake sediments in the dry caves after the Lahontan Crash, between ca. 12,400 and 11,500. In all but one (Falcon Hill) of the dry tufa caves in our sample (Table 1), most "first dry" organic dates on the cave bottoms cluster between 11,300 and 11,200 RCYBP (Leonard Rockshelter, Danger Cave, Fishbone Cave, Hidden Cave), with a few around 11,500 showing some terrestrial life. Note the 400-year date gap between the Falcon Hill shrub ox and the first two wood dates on Table 1. While it may mean the caves were under water longer than our model allows, the packrat and tufa gap dates suggest it must be something else.

Could it be that virtually nothing happened near the caves for almost a thousand years? Only a small drift of windblown sand marks this interval in Fishbone Cave. All kinds of life reappear at 11,350 RCYBP, with an abrupt end at 11,100 RCYBP—the exact span of the IACP. We conclude the IACP was initially wetter, bringing life, including humans, back to the lake beds and rivers from their upland refugia, but turned very cold and dry toward the end, killing off all the new life within a few hundred years. By the time of the Clovis Drought, only a few mammoths and large bears survived, along with the hardy small bison. A wide range of elevations and springs helped them survive these climate changes in the Great Basin.

The earliest Stemmed Point may be the McGee point, perhaps geologically dated between 14,000 and 12,500 (Figure 5B). This interpretation would be consistent with Bryan's dates at Smith Creek Cave, which indicate the Stemmed Point tradition may date to at least 12,000 RCYBP in the Great Basin (Bryan 1988; Bryan and Tuohy 1999) and possibly earlier. Although we agree that Stemmed Points may be older than Clovis in the West, there is a chance the McGee point is not the proof it seems. If the Younger Dryas were a time of near catastrophic rainfall, perhaps the 25 ft of lacustrine silt deposits burying the McGee point are from a time when the Walker Lake basin filled locally from the Walker River and surrounding mountains. Silt from the high-stand lake beds upstream in Mason Valley was sufficiently abundant to quickly bury a late-Allerød or early–Younger Dryas site in the river canyon when the lake rose above the site, ca. 10,800 RCYBP. Severely jumbled and missing sediment beds in this region (Benson 1988) and giant meander scars on the Humboldt River (Adams et al. 1999; Davis 1978b, 1978c, 1990) remind us to be cautious in assuming slow uniformitarianism in all cases, especially with the rapid and severe Younger Dryas climate changes.

We also agree with Bryan and Tuohy (1999) that some Clovis artifacts in Nevada are early Clovis Drought–era phenomena about 11,100 years old, and that some may be terminal or Western Clovis, several hundred years after the last Clovis in the Southwest and Plains and contemporaneous with *later* Stemmed Points and the Younger Dryas. This would account for the divergent elevations of some Clovis and Stemmed Points: the Harvey site, Washoe Lake bed, Black Rock lowest elevations (Shutler and Shutler 1959; Tuohy 1968, 1977, 1985), and the Dietz site (Willig 1984, 1988) have divergent elevations. The co-occurrence of Western Clovis with other stemmed-point assemblages are found at the Old Humboldt site (Rusco and Davis 1987) and many other sites in the Great Basin (Beck and Jones 1997).

We see a growing body of evidence that the Stemmed Point tradition may have ancient roots in the Great Basin. The possibly mid-high-stand McGee point suggests that high-stand human occupations may yet be found. If so, pre-Clovis archaeological sites may be found in unique settings of the Lahontan Basin from the highest to the lowest elevations. The relatively late occurrence of mammoth bone and ivory fishing implements, made in a distinctive manner previously recognized as Clovis style, suggests a new vision of the possible fate of Clovis in America at the last refuge of the mammoth, Pyramid Lake. Our examination of the Ting collection (Ting 1968) revealed that the megafauna bone points became more plentiful as the lake dropped to its historic low stand in 1967, leading to the inescapable hypothesis that if there was a Clovis Drought in the Lahontan Basin, there should be mammoths, Clovis artifacts, and early stemmed-point artifacts below the historic low stand of 3,796 ft (1,157 m) at Pyramid Lake, very well preserved, some dating to 11,000 and some to 10,350 RCYBP, along with megafauna fossils dating to 25,500–35,000 RCYBP in equally good condition (Dansie et al. 1988). Testing this hypothesis would be politically incorrect at this time.

Fiedel (1999) discusses a radiocarbon plateau between 10,600 and 10,100 RCYBP, where 900 calendar years are "lost." The seeming short interval of 630 radiocarbon years for the Upper Sehoo lakes may actually be more like 1,200 calendar years, long enough for the Younger Dryas lakes to rise and fall. This is also enough time for the development of the complex variety of Stemmed Points around the 3,950- to 4,040-ft shorelines throughout the Lahontan Basin, and equivalent streamside or lakeshore sites all across the Great Basin (Beck and Jones 1997). The correlation of calendar years and radiocarbon dates discussed in Fiedel (1999) is shown on the bottom of Figure 5. The thin diagonal lines are samples with two dates, radiocarbon and calendar years. Invoking the ^{14}C plateau does not bring our late ivory and mammoth bone points close enough to 11,000 RCYBP (dark line) to discount their unexpected late date. These dates are much farther away from the calendar calibrations than any in Fiedel, in the opposite direction.

Summary and Conclusions

We summarize our new dates and their direct implications in our published abstract (Dansie and Jerrems 2004), but here we summarize some new perspectives inspired by the data presented. The terminal Pleistocene as a time of drastic and sudden changes—from full glacial to severe drought, then a sudden return to near-glacial weather—is a new vision of the past for many of us, one accentuated by the timing and magnitude of Lake Lahontan water-level changes between 13,000 and 8000 years ago (Figure 5). The Bølling/Allerød onset was not dry or hot enough to kill off all the megafauna or their food; they survived until the last two or three hundred years of this drought. Lake Lahontan almost died during this time, with the Harmon School soil forming on the lowest shorelines.

Many megafauna survived the early Allerød, only to die during the IACP. Some megafauna possibly survived even longer in the Lahontan Basin. Here the latest horse dates (11,350 and 11,210 RCYBP) are from the early IACP, but the last mammoth dates are mid–Younger Dryas, 10,350 RCYBP. The last dates for Great Basin short-faced bear or Pleistocene-size *Ursus* may be even later, ca. 9960 RCYBP.

We model the Russell Shoreline as a moderate lake in the Carson and Humboldt Sinks during the IACP. The higher lakes generally above 3,960 ft (1,207 m) in both the west and east lake systems are all probably late Pleistocene, Younger Dryas in age, 10,970–10,100 RCYBP, not Anathermal in age as originally modeled by Antevs (Figure 3B). The deepest lakes, around 4,000 ft and above (1,219 m), lasted from about 10,800 to 10,400 RCYBP, ending halfway through this colder, wetter Younger Dryas period. It appears that all these upper Sehoo lakes and any earlier IACP lakes followed a major desiccation interval instead of a gradually shrinking Ice Age lake; Roger Morrison was right about most of it all along (Figure 3D, G, H).

The separation of low-elevation classic Clovis points from the higher Stemmed Points along late Sehoo lake shores is probably one Lahontan version of the 10,970 RCYBP "Black Mat" (Haynes 1991). We presented additional evidence of rapid hydrologic change, marking this transition across the Lahontan Basin. Older archaeological sites might be deeply buried in Younger Dryas sediments with only a few calendar years separating them from later sites, or they might have been washed away by waves and river floods. However, the probable existence of earlier Stemmed Points and later Clovis artifacts makes a challenging task even more complex. We had hoped to clarify the predictable times and elevations available to human occupation around Lake Lahontan. While it may be easier than before, as one would expect, in many ways it is even more complicated now. The variety of lithic and textile cultural traits spanning the terminal Pleistocene and early Holocene, combined with the variation in physical anthropology and evidence of interpersonal violence before 9000 RCYBP (Owsley and Jantz 1999), hints of a dynamic human landscape in a time of massive environmental change.

The diversion of the Truckee River into the Carson Sink, a new idea in the archaeological models of the Lahontan Basin, might explain many otherwise enigmatic bits and pieces. The Susan River flowing through Honey Lake Valley to Pyramid Lake before isostatic rebound may also explain the concentration of Paleoamerican artifacts from the northwest corner of Pyramid Lake all the way to eastern Honey Lake Valley, with virtually no early materials at the modern delta end of the lake.

Radiocarbon dates can be equivocal for various reasons. Some, especially tufa dates, have led to serious misunderstanding of the ancient earth history we study so earnestly. But some that do not seem to fit are actually vital clues that our models are simply "wrong." Surely this model will be revised with new data, but preliminary as it is, it provides a chronological and geomorphic structure that can be tested in the field. It was not just the timing of the water levels that influenced human occupations, it was also tectonic effects on rivers through the terminal Pleistocene that created habitats no longer possible after rebound. Despite our difficulty with Phil Orr's records and methods, we want to say he did his best to save the only dry Lahontan cave with extinct fauna and human artifacts in direct association, and just in the nick of time. Likewise, none of the other errors we tried to correct reflect negatively on the value of the scientist's contributions, nor on our considerable respect for their efforts.[6]

But we still wonder, why is Fishbone Cave the only dry Lahontan cave with possible evidence of a human presence at

[6] Orr's enthusiasm to penetrate Lake Lahontan's mysteries saved the unique treasures of Fishbone Cave. Guano miners would have destroyed this fragile, exceptional site in an afternoon; and we thank Orr for saving these vital bits and pieces of Lahontan prehistory. Orr excavated Fishbone Cave when he did because guano miners had beaten him to Guano and Crypt Caves, his second field season (Orr 1952, 1956a). There may never be another dry Lahontan cave, but its value would lie in being undisturbed; looting would destroy the value of the artifacts.

11,200–11,300 RCYBP, during the cold IACP, or anytime before the extinctions? In this land of superb dry cave preservation all around Lake Lahontan's shores, with abundant perishables from just a few centuries later, why haven't other caves yielded similar evidence of possible early human occupation? One thing is clear, early people in the Lahontan Basin rarely used the dry caves. It remains a mystery why, with their abundant perishable assemblages, the lithic assemblages in the dry cave collections do not match the abundant stemmed-point assemblages around the lakeshores and streams in the terminal Pleistocene.

Bits and Pieces Tell a Story

So what do these bits and pieces tell us, in plain English? In essence, the turbulent times may have proceeded as in the following scenario, knowing full well there are serious gaps and leaps of imagination here, but all based on the data reported here. There might also be a not-so-hidden warning about rapid climate change as the natural moral of the story.

Despite the 200 hundred years of new moisture and food, horses and camels and scores of Pleistocene animals of all sizes became extinct by 11,100 RCYPB. The giant short-faced bears continued to hunt down or out-compete every last survivor of the main extinction that could not outrun them, and gnawed on frozen carcasses for hundreds of years, going up the mountains as they warmed again. A few mammoths survived in pockets, able to defend themselves from the bears. Humans had many fewer predators, but then, they also had much less food and water. These people were tough.

Acknowledgments

The authors wish to thank the Trustees of the Calhoun Foundation (formerly the Truman Orr Foundation) for their years of generous support in the Nevada State Museum's ongoing curation, analysis and reporting of the Winnemucca Lake dry cave collections. We also appreciate the vision of the Nevada State Legislature in their generous support of the radiocarbon dating program that made this work possible. Also, we wish to thank R. Erv Taylor of the University of California, Riverside, and his expert staff at the radiocarbon laboratory, Donna Kirner, Karen Selsor and Debbie George, for the years of generous assistance in dating these ancient treasures of the past (all the UCR/CAMS amino acid dates reported herein). Finally, we especially want to thank the Nevada State Museum Department of Exhibits for their remarkable dedication and excellence in producing the "poster" exhibit shown at the Clovis and Beyond conference in Santa Fe, the University of Nevada Library, and the Nevada State Museum.

We dedicate this report, and the exhibit to which it relates, to the late Jonathan O. Davis and to Donald R. Tuohy, without whom it would not have been possible.

References Cited

Adams, K. D., and S. G. Wesnousky
1998 Shoreline Processes and the Age of the Lake Lahontan Highstand in the Jessup Embayment, Nevada. *Geological Society of America Bulletin* 110(10):1318–32.

Adams, K. D., S. G. Wesnousky, and B. G. Bills
1999 Isostatic Rebound, Active Faulting, and Potential Geomorphic Effects in the Lake Lahontan Basin, Nevada and California. *Geological Society of America Bulletin*, 111(12):1739–56.

Adovasio, J. M.
1986 Prehistoric Basketry. In *Handbook of North American Indians*, Volume 11: *Great Basin*, edited by W. L. d'Azevedo, pp. 194–205. Smithsonian Institution Press, Washington, D.C.

Andrews, R. L., J. M. Adovasio, and R. C. Carlisle
1986 Sandals. In Perishable Industries from Dirty Shame Rockshelter, Malheur County, Oregon. *University of Oregon Anthropological Papers* 34.

Antevs, E.
1925 Quaternary Climates: On the Pleistocene History of the Great Basin. *Carnegie Institution of Washington, Publications* 352:510–14. Washington, D.C.
1945 Correlations of Wisconsin Glacial Maxima: *American Journal of Science*, V.243-A, Daly Volume, p. 1–39.
1948 Climatic Changes and Pre-White Man. In The Great Basin with Emphasis on Glacial and Post Glacial Times. *Biological Series* 10(7), *University of Utah Bulletin* 38(20) pp. 168–91. Salt Lake City.
1952 Cenozoic Climates of the Great Basin: *Geologische Rundschau*, v. 40, no. 1, p. 94–108.
1953 Geochronology of the Deglacial and Neothermal Ages. *Journal of Geology.* 61(3):195–230.
1955 Geologic-climatic Dating in the West. *American Antiquity* 20(4):317–35.

Beck, C., and G. T. Jones
1997 The Terminal Pleistocene/Early Holocene Archaeology of the Great Basin. *Journal of World Prehistory* 11(2):161–236.
2001 in prep. *The Paleoarchaic Archaeology of Eastern Nevada: Part 1. The Sunshine Locality*, edited by C. Beck and G. T. Jones.

Benson, L. V.
1978 Fluctuation in the Level of Pluvial Lake Lahontan during the Last 40,000 Years. *Quaternary Research.* 9(3):300–18.
1981 Paleoclimatic Significance of Lake-level Fluctuations in the Lahontan Basin. *Quaternary Research.* 16:390–403.
1988 Preliminary Paleolimnologic Data for the Walker Lake Subbasin, California and Nevada. *U.S. Geological Survey Water Resources Investigations Report* 87-4258.
1993 Factors Affecting ^{14}C ages of Lacutrine Carbonates: Timing and Duration of the Last Highstand Lake in the Lahontan Basin. *Quaternary Research* 39:163–74.

Benson, L. V., J. Burdett, S. Lund, M. Kashgarian, and S. Mensing
1997 Nearly Synchronous Climate Change in the Northern Hemisphere during the Last Glacial Termination. *Nature* 388:263–65.

Benson, L. V., D. R. Currey, R. I. Dorn, K. R. Lajoie, C. G. Oviatt, S. W. Robinson, G. I. Smith, and S. Stine
1990 Chronology of Expansion and Contraction of Four Great Basin Lake Systems during the Past 35,000 Years. In P. A. Meyers and L. V. Nenson (eds.), Paleolakes and Paleooceans. *Paleogeography, Paleoclimatology, Paleoecology* 78:241–86.

Benson, L. V., D. R. Currey, and S. Hostetler
1992. Lake-Size Variations in the Lahontan and Bonneville Basins between 13,000 and 9000 ^{14}C yr B.P. *Paleogeography, Paleoclimatology, Paleoecology* 117:1–30.

Benson, L.V., M. Kashgarian, and M. Rubin
1995 Carbonate Deposition, Pyramid Lake Subbasin, Nevada: 2. Lake Levels and Polar Jet Stream Positions Reconstructed from Radiocarbon Ages and Elevations of Carbonates (Tufas) Deposited in the Lahontan Basin. *Palaeogeography, Palaeoclimatology, Palaeoecology* 117:1–30.

Benson, L. V., and R. S. Thompson
1987a Lake-level Variation in the Lahontan Basin for the Past 50,000 Years. *Quaternary Research.* 28:69–85.
1987b The Physical Record of Lakes in the Great Basin. Chapter 11

In W. F. Ruddiman and H. E. Wright Jr. (eds.), North America and Adjacent Oceans during the Last Deglaciation, Vol. K-3:241–60, *The Geology of North America*, Geol. Soc. Am., Boulder, Colo.

Benson, L., L. White, and R. Rye
1996 Carbonate Deposition, Pyramid Lake Subbasin, Nevada: 4. Comparison of the Stable Isotope Values of Carbonate Deposits (Tufas) and the Lahontan Lake-level Record. *Palaeogeography, Palaeoclimatology, Palaeoecology* 122:45–76.

Born, S. M.
1972 Late Quaternary Historic, Deltaic Sedimentation, and Mudlump Formation at Pyramid Lake, Nevada. Reno: University of Nevada Desert Research Institute, Center for Water Resources Research.

Broecker, W. S., and A. Kaufman
1965 Radiocarbon Chronology of Lake Lahontan and Lake Bonneville II, Great Basin. *Geological Society of America Bulletin* 76(5):537–66.

Broecker, W. S., and P. C. Orr
1956 Late Wisconsin History of Lake Lahontan. *Geological Society of America Bulletin* 67(12, pt. 2):1675–76. (Abstract.)

1958 Radiocarbon Chronology of Lake Lahontan and Lake Bonneville. *Bulletin of the Geological Society of America* Bulletin 69(8):1009–32.

Broecker, W. S., and J. L. Kulp
1957 Lamont Natural Radiocarbon Measurements IV. *Science* 126:1324–34.

Broecker, W. S., J. L. Kulp, and C. S. Tucek
1956 Lamont Natural Radiocarbon Measurements III. *Science* 124:154–65.

Broughton, J. M., D. B. Madsen, and J. Quade
2000 Fish Remains from Homestead Cave and Lake Levels of the Past 13,000 Years in the Bonneville Basin. *Quaternary Research* 53:392–401.

Bryan, A. L.
1979 Smith Creek Cave. In The Archaeology of Smith Creek Canyon, Eastern Nevada. Edited by D. R. Tuohy and D. Rendall, pp. 165–251. *Nevada State Museum Anthropological Papers No. 17,* Carson City.

1988 The Relationship of the Stemmed Point and Fluted Point Traditions in the Great Basin. In Early Human Occupation in Far Western North America: The Clovis-Archaic Interface. Edited by J. A. Willig, C. M. Aikens, and J. L. Fagan, pp. 53–74. *Nevada State Museum Anthropological Papers No. 21,* Carson City.

Bryan, A. L., and D. R. Tuohy
1999 Prehistory of the Great Basin/Snake River Plain to About 8500 Years Ago. In *Ice Age Peoples of North America*, edited by R. Bonnichsen and K. L. Turnmire, pp. 249–80. Center for the Study of the First Americans, Oregon State University Press, Corvallis.

Coffin, P. D., and W. F. Cowan
1995 *Recovery Plan for the Lahontan Cutthroat Trout.* U.S. Fish and Wildlife Service, Region 1. Portland, Oregon.

Cressman, L. S. and collaborators
1960 Cultural Sequences at the Dalles, Oregon. *Transactions of the American Philosophical Society*, n.s. 60(10).

Currey, D. R.
1988 Isochronism of Final Pleistocene Shallow Lakes in the Great Salt Lake and Carson Desert Regions of the Great Basin. *AMQUA Program with Abstracts*, 10:117.

1990 Quaternary Paleolakes in the Evolution of Semidesert Basins, with Special Emphasis on Lake Bonneville and the Great Basin, U.S.A. *Palaeogeography, Palaeoclimatology, Palaeoecology* 76:189–214.

Dansie, A. J.
1981 Pebble Mound Complexes in Northwestern Nevada. *Nevada Archaeologist* 3(1):16–28, Carson City.

1984 Analysis of Faunal Remains from 26Pe670, 1983 Excavations. In 1983 Excavations at Archaeological Site 26Pe670 Rye Patch Reservoir, Nevada, edited by J. O. Davis. *Desert Research Institute, University of Nevada System Technical Report* No. 38, Reno.

1987 Archaeofauna from 26Pe670. In *Studies in Archaeology, Geology and Paleontology at Rye Patch Reservoir, Pershing County, Nevada.* By M. K. Rusco and J.O. Davis, pp. 69–73. Nevada State Museum Anthropological Papers 20.

Dansie, A. J., J. O. Davis, and T. W. Stafford, Jr.
1988 The Wizard's Beach Recession: Farmdalian (25,500 yr B.P.) Vertebrate Fossils Co-occur with Early Holocene Artifacts. In Early Human Occupation in Far Western North America: The Clovis-Archaic Interface, edited by J. A. Willig, C. M. Aikens, and J. L. Fagan. *Nevada State Museum Anthropological Papers,*21:153–200.

Dansie, A. J., and J. Jerrems
1999 Lahontan Chronology and Early Human Occupation in the Western Great Basin: A New Look at Old Collections. Poster presented at the Center for the Study of the First Americans Clovis and Beyond Conference, Santa Fe.

2000 Lahontan Chronology and Paleogeography: A Dynamic Landscape for Early Human Occupation. Paper presented at the 27th Great Basin Anthropological Conference, Ogden, Utah.

2004 Lahontan Chronology and Early Human Occupation in the Western Great Basin: A New Look at Old Collections. In *New Perspectives on the First Americans,* edited by B. T. Lepper and R. Bonnichsen, 55–63. Center for the Study of the First Americans, Texas A&M University, College Station.

Dansie, A. J., D. R. Tuohy, and P. Hutchinson
2001 (ms in prep) History of Field Work at Sunshine Well, Long Valley, Nevada, by Amateurs and Nevada State Museum Staff, 1971 to 1993. In *The Paleoarchaic Archaeology of Eastern Nevada: Part 1. The Sunshine Locality,* edited by C. Beck and G. T. Jones.

Davis, J. O.
1978a Quaternary Tephrochronology of the Lake Lahontan Area, Nevada and California. *Nevada Archaeological Survey Research Papers* 7, Reno.

1978b Late Sehoo Discharge of the Humboldt River: Stratigraphic Archaeology at the North Valmy Power Plant, Humboldt County, Nevada (abstract). *Geological Society of America Abstracts with Program*, 10(7):386.

1978c Great Meander Scars on the Humboldt River, Nevada: Chronology and Paleohydrologic Significance. Abstracts of Papers Presented at the Sixteenth Great Basin Anthropological Conference, Oct. 20–21, Reno.

1982 Bits and Pieces: The Last 35,000 years in the Lahontan area. In Man and Environment in the Great Basin, edited by D. B. Madsen and J. F. O'Connell, pp. 53–75. *Society for American Archaeology Papers* 2.

1985 Chapter 7. Sediments and Geological Setting of Hidden Cave, pp. 80–103. In The Archaeology of Hidden Cave, Nevada. By D. H. Thomas. *American Museum of Natural History, Anthropological Papers* Vol. 61: Part 1.

1990 Giant Meanders on Humboldt River near Rye Patch Nevada Due to Catastrophic Flooding: *Geological Society of America Abstracts with Program*, 22(7):A309.

Eiselt, B. S.
1997 Fish Remains from the Spirit Cave Paleofecal Material: 9,400 Year Old Evidence for Great Basin Utilization of Small Fishes. *Nevada Historical Society Quarterly* 40(1):117–39.

Elston, R .G., K. L. Katzer, and D. R. Currey
1988 Chronological Summary, pp. 367–78. In Preliminary Investigation in *Stillwater Marsh:Human Prehistory and Geochronology,* Intermountain Research Reports, edited by C. Raven and R. Elston. Report submitted to U.S. Fish and Wildlife Service, Portland.

Fiedel, S. J.
1999 Older than We Thought: Implications of Corrected Dates for Paleoindians. *American Antiquity* 64(1):95–115.

Fowler, C. S., and E. M. Hattori
2000 Early Textile Complexes in the Western Great Basin. Pg. 45 in *Program and Abstracts*, 27th Great Basin Anthropological Conference, October 5–7, Ogden, Utah.

Fowler, C. S., E. M. Hattori, and A. J. Dansie
2000 Ancient Matting from Spirit Cave, Nevada: Technical Implications. Pp. 119–39 In P. B. Drooker and L. D. Webster, editors, *Beyond Cloth and Cordage: Archaeological Textile Research in the Americas.* University of Utah Press, Salt Lake City.

Fremont, J. C.
1845 *Report of the Exploring Expedition to the Rocky Mountains in the Year 1842 and to Oregon and North California in the years 1843–44.* U.S. Senate, 28th Cong. 2nd Session, Exec. Do. 174. Gales and Seton, Printers, Washington, D.C.

Grayson, D. K.
1985 Chapter 10. The Paleontology of Hidden Cave: Birds and Mammals, pp. 125–61. In The Archaeology of Hidden Cave, Nevada. By David H. Thomas. *American Museum of Natural History, Anthropological Papers* Vol. 61: Part 1.

1991 Late Pleistocene Mammalian Extinctions in North America: Taxonomy, Chronology and Explanations. *Journal of World Prehistory* 5:193–232.

1993 *The Desert's Past: A Natural Prehistory of the Great Basin.* Smithsonian Institution Press, Washington, D.C.

Grosscup G. L.
1956 The Archaeology of the Carson Sink Area. *University of California Survey Reports*, No. 33. Berkeley.

Gruhn, R., and A. Bryan
1988 The 1987 Archeological Fieldwork at Handprint Cave, Nevada. *Nevada Archaeologist* 6(2):1–13.

Hattori, E. M.
1982 The Archaeology of Falcon Hill, Winnemucca Lake, Washoe County, Nevada. *Nevada State Museum Anthropological Papers* 18. Carson City.

Haynes, C. V., Jr.
1984 Stratigraphy and Late Pleistocene Extinction in the United States. In *Quaternary Extinctions: A Prehistoric Revolution*, edited by P. S. Martin, and R. G. Klein. University of Arizona Press.

1991 Geoarchaeological and Paleohydrological Evidence for a Clovis-Age Drought in North America and Its Bearing on Extinction. *Quaternary Research* 35:438–50.

Haynes, C. V., Jr., D. J. Stanford, M. Jodry, J. Dickenson, J. L. Montgomery, P. H. Shelley, I. Rovner, and G. Agogino
1999 A Clovis Well at the Type Site 11,500 B.C.: The Oldest Prehistoric Well in America. *Geoarchaeology* 14(5):455–70.

Hemmings, C.A., J. S. Dunbar, and S. D. Webb
2002 (in press) Florida's Early Paleoindian Bone and Ivory Tools. *Clovis and Beyond Abstracts.*

Heizer, R. F.
1951 Preliminary Report on the Leonard Rockshelter Site, Pershing County, Nevada. *American Antiquity* 17(2)89–98.

1956 Recent Cave Explorations in the Lower Humboldt Valley, Nevada. *Reports of the University of California Archaeological Survey.* Papers on California Archaeology: 37–43, pp. 50–57.

Heizer, R. F., and M. A. Baumhoff
1970 Big Game Hunters in the Great Basin: A Critical Review of the Evidence. In Papers on the Anthropology of the Western Great Basin. *Contributions of the University of California Archaeological Research Facility,* No. 7, pp. 1–12. Berkeley.

Hester, T. R.
1974 Archaeological Materials from Site NV-Wa-197, Western Nevada: Atlatl and Animal Skin Pouches. *University of California Archaeological Research Facility Contributions*, No. 21, pp. 1–36.

Hockett, B., and E. Dillingham
2000 Late Pleistocene to Middle Holocene Faunal Changes at Mineral Hill and Pintwater Caves: Implications for Climatic Changes in the Central and Southern Great Basin. Program and Abstracts, 27th Great Basin Anthropological Conference, Ogden.

Holliday, V.
2000 Folsom Drought and Episodic Drying on the Southern High Plains from 10,900–10,200 ^{14}C yr B.P. *Quaternary Research* 53:1–12.

Huckleberry, G., C. Beck, G. T. Jones, A. Holmes, M. Cannon, S. Livingston, and J. M. Broughton.
2001 Terminal Pleistocene/Early Holocene Environmental Change at the Sunshine Locality, North-Central Nevada, U.S.A. *Quaternary Research* 55:303–12.

Hyland, D. C., and J. M. Adovasio
2000 The Mexican Connection: A Study of Sociotechnical Change in Perishable Manufacture and Food Production in Prehistoric New Mexico, pp. 141–159. In P. B. Drooker and L. D. Webster, editors, *Beyond Cloth and Cordage: Archaeological Textile Research in the Americas.* University of Utah Press, Salt Lake City.

Irwin-Williams, C., A. Dansie, and J. O. Davis
1986b Prehistoric Water Harvesting. Paper presented at the 20th Great Basin Anthropological Conference, Las Vegas, Nevada.

Irwin-Williams, C., A. Dansie, J. O. Davis, and R. Jacobsen
1986a Reconstructing Prehistoric Water Harvesting Techniques in the Northern Great Basin. Paper presented at the Western Social Science Association conference (by Dansie), Reno, Nevada and at the Society for American Archaeology conference in New Orleans (by Irwin-Williams).

Irwin-Williams,C.C., C. B. Osmond, A. J. Dansie and L. F. Pitelka
1990 Man and Plants in the Great Basin, pp. 1–16 in Plant Biology of the Basin and Range, edited by C. B. Osmond, L. F. Pitelka and G. M. Hidy. *Ecological Studies* Volume 80, Springer-Verlag, Berlin.

Jennings, J. D.
1957 Danger Cave. *University of Utah Anthropological Papers* 27. Salt Lake City.

Jerrems, J.
2000 Bone Chronology Roundup for Paleoarchaic (Indian) Issues in the Great Basin. Paper presented at the 27th Great Basin Anthropological Conference, Ogden, Utah.

Kelly, R.
1983 *An Examination of Amateur Collections from the Carson Sink, Nevada.* Technical Report No. 10, Nevada Bureau of Land Management, Reno.

Kirner, D. L., R. Burky, K. Selsor, D. George, R. E. Taylor, and J. R. Southon
1997 Dating the Spirit Cave Mummy: The Value of Reexamination. *Nevada Historical Society Quarterly,* Vol. 40(1):54–56 Spring 1997, Reno.

Kirner, D. L., R. Burky, R. E. Taylor, A. Dansie, and D. Tuohy
1996 AMS Radiocarbon Dates from Spirit Cave, Nevada: Implications for Early Holocene Occupation of the Western Great Basin. Paper presented at the Society for American Archaeology, 61st Annual Meeting, April 10–14. 1996, New Orleans.

Lagoni, J. R.
1985 The Geomorphology of the Truckee River Delta, at Pyramid Lake, Nevada, and Implications on the Decline of Lake Lahontan, Master of Science Thesis in Geology. University of Nevada, Reno.

La Rivers, I.
1994 Fishes and Fisheries of Nevada. University of Nevada Press. (Originally published as La Rivers 1962, Nevada State Fish and Game Commission.)

Livingston, S. D.
1992 The DeLong Mammoth Locality, Black Rock Desert, Nevada. *Current Research in the Pleistocene* 8:94–97.

1999 "Black Rock Desert Mammoths." Lecture presented at the Nevada State Museum Humphreys Lecture series, Carson City.

Madsen, D. B.
1999 Environmental Change during the Pleistocene-Holocene Transition and Its Possible Impact on Human Populations, pp. 75–82. In *Models for the Millennium: Great Basin Anthropology Today*, edited by C. Beck, University of Utah Press, Salt Lake City.

2000a A High-Elevation Allerød-Younger Dryas Megafauna from the West-Central Rocky Mountains. In Intermountain Archaeology, D. B. Madsen, and M. D. Metcalf, eds. *University of Utah Anthropological Papers* No. 122.

2000b Late Quaternary Paleoecology in the Bonneville Basin. Bulletin 130, Utah Geological Society, Department of Natural Resources.

McDonald, J. N.
1981 *North American Bison: Their Classification and Evolution.* University of California Press, Berkeley.

McGee, W. J.
1889 An Obsidian Implement from Pleistocene Deposits in Nevada. *American Anthropologist,* Vol. II, October.

Mehringer, P. J., Jr.
1986 Prehistoric Environments. In *Handbook of North American Indians*, Volume 11:*Great Basin,* edited by W. L. d'Azevedo, pp. 31–50. Smithsonian Institution Press, Washington, D.C.

Miller, S. J., and W. Dort, Jr.
1978 Early Man at Owl Cave: Current Investigation at the Wasden Site, Eastern Snake River Plain, Idaho. In *Early Man in America from a Circum-Pacific Perspective*, edited by A. L. Bryan, pp. 129–39. Occasional Papers No. 1, Department of Anthropology, University of Alberta, Edmonton.

Morrison, R. B.
1952 Late Quaternary Climatic History of the Northern Great Basin [abs.]: *Geological Society of America Bulletin,* v. 63, no. 12, pt. 2, p. 1367.

1964a Lake Lahontan: Geology ofSouthern Carson Desert, Nevada *U.S. Geological Survey Professional Paper* 401, 156 p.

1964b Soil Stratigraphy; Principles, Applications to Differentiation and Correlation of Quaternary Deposits and Landforms, and Applications to Soil Sciences. Ph.D. dissertation, University of Nevada, Reno, 178 pp.

1991 Quaternary Stratigraphic, Hydrologic, and Climate History of the Great Basin, with Emphasis on Lakes Lahontan, Bonneville, and Tecopa. In *The Geology of North America,* Vol. K-2, Quaternary Nonglacial Geology: Coterminous U.S. The Geological Society of America, 1991.

Morrison, R. B., and J. O. Davis
1984a Quaternary Stratigraphy and Archeology of the Lake Lahontan Area; A Re-assessment, In J. Lintz, ed., Western geological excursions, v. 1, Reno, Mackay School of Mines, Geological Society of America 1984 annual meeting field trip 13 Guidebook, p. 252–81.

1984b Supplementary guidebook for GSA field trip 13, Quaternary Stratigraphy and Archeology of the Lake Lahontan Area; A Re-assessment. Reno, Desert Research Institute, University of Nebraska, Social Science Center Technical Report 41, 50 p.

Morrison, R. B., and J. C. Frye
1965 Correlation of the Middle and Late Quaternary Successions of the Lake Lahontan, Lake Bonneville, Rocky Mountain (Wasatch Range), Southern Great Plains and Eastern Midwest Areas. *Nevada Bureau of Mines,* Report 9. Mackay School of Mines, University of Nevada, Vernon E. Scheid, Director.

Mullen, C. O.
1997 Mammalian Response to Pleistocene/Holocene Invironmental Change in the Great Basin: The Jackrabbit's Tale. Ph.D. dissertation: Ecology, Evolution and Conservation Biology, University of Nevada, Reno.

Nials, F. L.
1999 Geomorphic Systems and Stratigraphy in Internally-drained Watersheds of the Northern Great Basin: Implications for Archaeological Studies. University of Nevada, Reno, Sundance Archaeological Research Fund, Technical Paper No. 5.

Orr, P. C.
1952 Preliminary Excavation of Pershing County Caves. *Nevada State Museum, Dept. of Archaeology Bulletin* 1:1–21. Carson City.

1956a Pleistocene Man in Fishbone Cave Pershing County, Nevada. *Nevada State Museum, Dept. of Archaeology Bulletin* 2:1–20. Carson City.

1956b Operation Tufa, manuscript on file, Nevada State Museum.

1969 A New Radiocarbon Dated Cat Skull from Crypt Cave, Nevada. *Santa Barbara Museum of Natural History Dept. of Geology* Bulletin 2:1–8.

1974 Notes on the Archaeology of the Winnemucca Caves, 1952–58. *Nevada State Museum Anthropological Papers* 16:47–59, Carson City.

Prokopovich, N. P.
1983 Alteration of Alluvium by Natural Gas in the Pyramid Lake Area, Nevada. *Bulletin, Association of Engineering Geologists.* 20(2):185–96.

Reheis, M. C., and R. Morrison
1997 High, Old Pluvial Lakes of Western Nevada. *BYU Geology Studies 1997*, VOL. 42, Part 1.

Rhode, D., K. D. Adams, and R. G. Elston
2000 Geoarchaeology and Holocene Landscape History of the Carson Desert, Western Nevada, pp. 45–74 *Great Basin and Sierra Nevada: GSA Field Guide 2.* D. R. Lageson, editor. Geological Society of America.

Rozaire, C.
1974 Analysis of Woven Materials from Seven Caves in the Lake Winnemucca Area, Pershing County, Nevada. *Nevada State Museum Anthropological Papers Number 16*, Collected Papers No. 4. Carson City.

Rusco, M. K., and J. O. Davis
1987 Studies in Archaeology, Geology, and Paleontology at Rye Patch Reservoir, Pershing County, Nevada. *Nevada State Museum Anthropological Papers* 20. Carson City.

Russell, I. C.
1885 Geological History of Lake Lahontan, a Quaternary Lake of Northwestern Nevada. Monograph of the U.S. Geological Survey 11.

Schneider, J.
2000 The Paleoenvironmental Significance of the Presence of *Anodonta sp.*, a Fresh Water Mussel, in Archaeological Sites in the Great Basin. Paper presented at 27th Great Basin Anthropological Conference, Ogden, UT.

Shutler, R., Jr.
1960 Archaeological Excavations and Survey in Washoe and Churchill Counties, Nevada. Manuscript on file, Nevada State Museum. Carson City.
1961 Correlation of Beach Terraces with Climatic Cycles of Pluvial Lake Lahontan, Nevada. *Annals of the New York Academy of Sciences* 95:513–20. New York.

Shutler, M. E., and D. Shutler, Jr.
1959 Clovis-Like Points from Nevada. *The Masterkey* 33(1):30–32. Los Angeles.

Sigler, W. F., and J. W. Sigler
1987 *Fishes of the Great Basin: A Natural History.* University of Nevada Press, Reno.

Stafford, T. W.
1994 Accelerator C-14 Dating of Human Fossil Skeletons: Assessing and Results on New World Specimens. In R. Bonnichsen and D. G. Steele, editors, *Method and Theory for Investigating the Peopling of the Americas.* Center for the Study of the First Americans, Oregon State University Press, Corvallis.

Stafford, T. W., Jr., A. J. T. Jull, K. Brendel, R. C. Duhamel, and D. Donahue
1987 Study of Bone Radiocarbon Dating Accuracy at the University of Arizona, NSF Accelerator Facility for Radioisotope Analysis. *Radiocarbon,* Vol. 29, No. 1, pp. 24–44.

Stanford, D.
1999 Paleoindian Archaeology and Late Pleistocene Environments in the Plains and Southwestern United States. In *Ice Age Peoples of North America,* R. Bonnichsen and K. L. Turnmire, editors, pp. 281–339. Center for the Study of the First Americans, Oregon State University, Corvallis.

Stout, B.
1986 Discovery and C14 Dating of the Black Rock Desert Mammoth. *Nevada Archaeologist* 5(2):21–23.

Straus, L. G.
1996 The World at the End of the Last Ice Age. Chapter 1. In Straus, Eriksen, Erlandson and Yesner, editors, *Humans at the End of the Ice Age.* Plenum Press, New York.

Taylor, R. E., P. E. Hare, C. A. Prior, D. L. Kirner, L. Wan, and R. B. Burky
1995 Radiocarbon Dating of Biochemically Characterized Hair. *Radiocarbon* 37:1–11.

Taylor, R. E., C. V. Haynes, Jr., and M. Stuiver
1996 Clovis and Folsom Age Estimates: Stratigraphic Contexts and Radiocarbon Calibration. *Antiquity* 70:515–525.

Taylor, R. E, A. Long, and R. S. Kra, eds.
1992 *Radiocarbon after Four Decades: An Interdisciplinary Perspective.* Springer-Verlag, New York.

Thomas, D. H.
1985 The Archaeology of Hidden Cave, Nevada. Vol. 61: Part 1. *Anthropological Papers of the American Museum of Natural History,* New York.

Thompson, R. S., L. Benson, and E. Hattori
1986 A Revised Chronology for the Last Pleistocene Lake Cycle in the Central Lahontan Basin. *Quaternary Research* 25:1–9.

Thompson, R .S., E. M. Hattori, and D. R. Tuohy
1987 Paleoenvironmental and Archaeological Implications of Early Holocene–Late Pleistocene Cave Deposits from Winnemucca Lake, Nevada. *Nevada Archaeologist* 6(1):34–38.

Ting, P. C.
1967 A Pyramid Lake Surface Artifact Assemblage Located at or near the 3800 Foot Elevation. *Nevada Archaeological Survey Reporter* 1(8):4–11. Reno.
1968 Bone Points from Pyramid Lake. *Nevada Archaeological Survey Reporter* 2(3):4–13. Reno.

Tuohy, D. R.
1968 Some Early Lithic Sites in Western Nevada. In "Early Man in Western North America," Symposium of the Southwestern Anthropological Association, San Diego, 1968, chaired by C. I. Williams. *Eastern New Mexico Contributions in Anthropology* 1(4). Portales.
1970 The Coleman Locality: A Basalt Quarry and Workshop near Falcon Hill, Nevada. *Nevada State Museum Anthropological Papers* 15:143–205.
1977 More on Clovis Points from Washoe Lake, Nevada. *Nevada Archaeological Survey Reporter* 10(4):1–7.
1981 A Brief History of the Discovery and Exploration of Pebble Mounds, Boulder Cairns, and Other Rock Features at the Sadmat Site, Churchill County, Nevada. *The Nevada Archaeologist* 3(1):4–15.
1984 Implications of Obsidian Hydration Readings and Source Determination for 28 Presumed "Early Man" Points from Nevada. In Obsidian Studies in the Great Basin, R. E. Hughes, editor. *Contributions of the University of California Archaeological Research Facility* No. 45:193–221.
1985 Notes on the Great Basin Distribution of Clovis Fluted and Folsom Projectile Points. *Nevada Archaeologist* 5(1):15–18.
1988 Artifacts from the Northwestern Pyramid Lake Shoreline. Ch. 11 In J. A. Willig, C. M. Aikens, and J. L. Fagan, editors, Early Human Occupation in Far Western North America: The Clovis-Archaic Interface. *Nevada State Museum Anthropological Papers* 21, pp. 201–16. Carson City.
1990 Pyramid Lake Fishing: The Archaeological Record, pp. 121–158 in Wetland Adaptations in the Great Basin. J. C. Janetski and D. B. Madsen eds. *Museum of Peoples and Cultures Occasional Papers* 1. Brigham Young University.

Tuohy, D. R., and A. Dansie
1997 New Information Regarding Early Holocene Manifestations in the Western Great Basin. In *Nevada Historical Society Quarterly,* Vol. 40(1):24–53 Spring 1997, Reno.

Valastro, S. Jr., E. M. Davis, and J.G. Varela
1978 University of Texas at Austin Radiocarbon Dates XII. *Radiocarbon,* 20(2):245–73.

Wallmann, S., and D. S. Amick
1991 Cody Complex Occupation in the Black Rock Desert, Nevada. *Current Research in the Pleistocene* Vol. 8.

Webb, S. D., and C. A. Hemmings
2000 Ivory and Bone Tools from Lake Pleistocene Deposits in the Aucilla River, North-Central Florida. In *International Wetlands Conference Number 4 Monograph.* Edited by B. Purdy.

Wheeler, S. S.
1967 *The Desert Lake: The Story of Nevada's Pyramid Lake*. Caxton Printers, Ltd. Caldwell, Idaho.

Willig, J. A.
1984 Geoarchaeological Research at the Dietz Site and the Question of Clovis Lake Marsh Adaptation in the Northern Great Basin. In Geoarchaeology in the Northwest: Recent Applications and Contributions edited by J. A. Willig. *Tebiwa, The Journal of Idaho Museum of Natural History* 21:56–69, Pocatello.

1988 Paleo-Archaic Adaptations and Lakeside Settlement Patterns in the Northern Alkali Basin, Oregon. In Early Human Occupation in Far Western North America: The Clovis-Archaic Interface, edited by J. A. Willig, C. M. Aikens, and J. L. Fagan, pp. 417–82. *Nevada State Museum Anthropological Papers No. 21,* Carson City.

1991 Clovis Technology and Adaptation in Far Western North America: Regional Pattern and Environmental Context, pp. 91–118. In *Clovis Origins and Adaptations*, R. Bonnichsen and K. L. Turnmire (eds) Center for the Study of the First Americans, Oregon State University Press, Corvallis.

Wormington, H. M.
1957 *Ancient Man in North America*. Popular Series No. 4, Denver Museum of Natural History, Denver.

The Ryan/Harley Site 8Je1004: A Suwannee Point Site in the Wacissa River, North Florida

James S. Dunbar,[1] C. Andrew Hemmings,[2] Pamela K. Vojnovski,[3] S. David Webb,[4] and William M. Stanton[5]

Abstract

The Ryan-Harley site (8JE1004) is the first middle-Paleoindian site in the southeast U.S. to yield both diagnostic lithic artifacts, namely the Suwannee point, and faunal remains including extinct late-Pleistocene species. Bone specimens from the site were too mineralized to radiocarbon date, but the frequency, variety, and preservation of excavated tools and faunal material represents a middle-Paleoindian campsite. The site is remarkable because it yielded at least three and possibly as many as seven late-Pleistocene species (i.e., *Tapirus veroensis, Equus sp., Ondontra zibethicus, Mammut americanum, Paramylodon harlani, Holmesina septentrionalis*) that became regionally extinct before the beginning of the Holocene at ca. 10,000 RCYBP. In the desert Southwest Pleistocene megafauna became extinct prior to the Allerød/Younger Dryas boundary ca. 11,000 RCYBP. Evidence from the Ryan-Harley site indicates that Pleistocene megafauna in the southeastern U.S. survived beyond that boundary during Younger Dryas until ca. 10,500 RCYBP. The faunal remains also contain several extant species including medium-sized and smaller mammals, especially deer; reptiles, especially turtles and alligators; and fish.

[1]Senior Archaeologist, Florida Bureau of Archaeological Research, C.A.R.L. Archaeological Program, 500 South Bronough Street, R. A. Gray BLDG., Tallahassee, FL 32399-0250; e-mail: jsdunbar@dos.state.fl.us

[2]Florida Museum of Natural History, 222 Dickinson Hall, University of Florida, Gainesville, FL 32611-7800; e-mail: hemmings@grove.ufl.edu

[3]Identa-Bonze Archaeological Consultant, 3080 Black Fox Drive, Loganville, GA 30052; e-mail: flwerpower1066@yahoo.com

[4]Distinguished Research Curator and Professor of Zoology and Geology, Florida Museum of Natural History, 222 Dickinson Hall, University of Florida, Gainesville, FL 32611-7800; e-mail: webb@flmnh.ufl.edu

[5]Archaeologist, Bureau of Natural and Cultural Resources, 578-A South Appleyard Drive, Tallahassee, FL 32304; e-mail: william.stanton@dep.state.fl.us

Because the Ryan-Harley site artifacts share a number of Clovis or Clovis-like traits, we believe that the Suwannee toolkit evolved from Clovis, its predecessor.

Introduction

Investigations of the Ryan-Harley site in the Wacissa River have revealed a mid-Paleoindian site. The Ryan-Harley site is believed to represent a post-Clovis, Suwannee-point campsite located a little deeper than 1 m below the present low-water stage of the Wacissa River. The site is unique because bone as well as stone has been preserved. Attempts to radiocarbon-date the site have been futile because associated charcoal has not been found and organic silts have replaced the collagen content of the bone samples submitted for evaluation. Two radiocarbon dates on oxidized roots provide evidence that the site's Holocene inundation took place about 4,500 years ago. Subsequent to Holocene inundation, roots that penetrated the Suwannee stratum were preserved under totally saturated conditions and represent a post-depositional invasion into the Suwannee level.

The purpose of this chapter is to discuss the significance of the faunal remains and artifact assemblage from the Ryan-Harley site. Even though the area of test excavations was confined to a relatively small area of 7 m^2 threatened by erosion and the sample of cultural materials is small, we believe it is of sufficient size and importance to be reported. Primary consideration is given to the in situ specimens from the Suwannee component, but displaced specimens of obvious Paleoindian age are also mentioned. Specimens recovered from displaced contexts are treated as secondary to our primary findings. The site's stratigraphy will be discussed elsewhere.

As far as we know, this is the first Suwannee-point site discovered in the southeastern United States having bone preservation. Ryan and Harley Means discovered the site while exploring the Wacissa River bottom on successive site visits (Means and Means in press). Ryan is a wildlife biologist with the Coastal Plains Institute, and Harley is a geologist with the Florida Geological Survey. Their ecologist father, Dr. Bruce Means, subsequently accompanied the brothers on a surface-collecting inspection of the site. We are indebted to them for recognizing and reporting their discovery to the Florida Bureau of Archaeological Research. The site was initially investigated by the BAR in 1998 and was subsequently the subject of salvage excavations carried out by the Aucilla River Research Project of the Florida Museum of Natural History. Throughout the investigations, archaeologists with the Conservation and Recreation Lands (CARL) survey of the Bureau of Archaeological Research were involved, with Melissa Memory and Jonathan Lammers providing much of the critical field assistance. The Wacissa River tract is a state-owned CARL property managed by the Florida Fish and Wildlife Conservation Commission. Funding for the fieldwork came from grants awarded by the National Geographic Society and the Division of Historical Resources, Florida Department of State.

The small town of Wacissa takes its name from the river and is located near the river's headwater just below the Cody Scarp in Jefferson County (Figure 1). In the Big Bend area of the Florida Panhandle southwards onto the peninsula, coastal rivers like the Wacissa have small drainage areas and low gradients. The Wacissa is unusual because its existence is dominated by artesian discharge from the Floridan aquifer system. The Wacissa originates in several headsprings and then courses southerly through flat swampy terrain before its junction with the lower Aucilla River in two primary as well as other smaller confluences. The Wacissa River basin is shallow and broad compared with the deep, often underground or entrenched surface channels of the Aucilla River. The Wacissa is noted for having a varied channel system including sections dominated by one or two wide, shallow channels as well as sections dominated by innumerable small braided channels. The erosive ability of the Wacissa is not as destructive as many other river systems because its drainage area is small, its channel is short (less than 24 km, or 15 miles), and its gradient is almost zero (between its headwater and sea level the difference is less than 6 m). This is an important distinction since the concept of a braided-channel system usually brings to mind desert-like arroyo gullies and the potential for aggressive erosion. This is not the case in the Wacissa River, which occupies a swampy basin 1.5 to 5 km wide without stream terraces (Yon 1966). Nevertheless, erosion takes place in focused ("point source") areas where, for one reason or another, the flow dynamics have changed and down-cutting occurs in newly rerouted channels or in blowouts of existing channels. The Ryan-Harley site appears to represent a rerouted channel cutting through new ground.

The Ryan-Harley site is located in a braided channel of the Wacissa River south of the Goose Pasture boat landing. The site is hidden in a bewildering network of low-energy braided channels in the middle of the Wacissa's heavily vegetated swamp forest. The Page-Ladson site (8Je591) in the Aucilla River is located about 1½ km south of the Ryan-Harley site. Page-Ladson has also produced waisted Suwannee points, but no evidence suggestive of campsite activity.

Field Investigations

Three brief underwater archaeological surveys took place at the Ryan-Harley site in June and August of 1998. As a result of those surveys, a 2-week salvage effort was undertaken in May 1999 in an attempt to plot and document in situ materials being threatened by erosion. The results of this survey indicate that an in situ component of the site had been eroding into the river along a 20-m section of channel. The first surveys of the site demonstrated that Paleoindian (Suwannee point), early-Archaic (Bolen and Kirk serrated point), and Woodland (Deptford ceramic) artifacts occurred within the limits of the site. It is important to note that the dominant artifact concentration belongs to the Suwannee component (Figure 2). The Suwannee component has no apparent contamination from Bolen or younger cultural materials. This paper will discuss the Suwannee component and its significance.

In situ specimens were either excavated by hand and piece plotted or recovered from the floating artifact screens. The most significant excavation took place in a 7-m^2 area of channel bot-

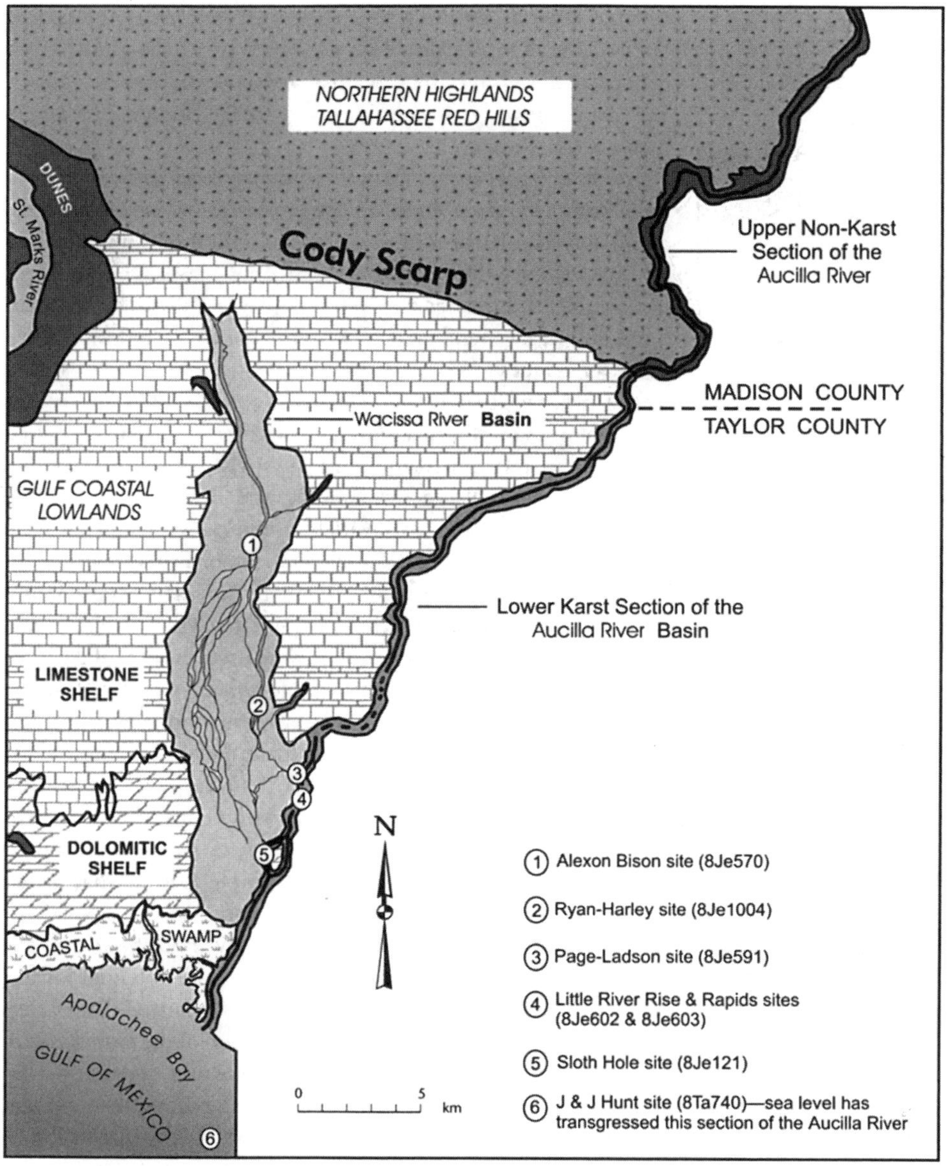

Figure 1. Physiographic map of southern Jefferson County, north Florida, depicting the karst sections of the Aucilla and Wacissa rivers (adapted from Yon 1996; Fernald and Purdum 1998).

tom. In these 7 test units, about 2.5 m^2 of the Suwannee component was found to be intact and totally buried, about 1.5 m^2 had been exposed and was actively eroding, and about 3 m^2 had already eroded below the Suwannee-point level (Figure 2). Controlled surface collections produced a sample of displaced artifacts and fossils, many of which are suspected to have originated from the Suwannee level.

The sediment column in the site area is about 2.5 m thick and rests on the Oligocene-age Suwannee Limestone formation. The Suwannee-point level is 10–15 cm thick and occupies the approximate middle of the section from about 1.05 m to 1.20 m in the stratigraphic column.

A thicker sediment column lies several meters west of the Suwannee component and appears to represent the center of a late-Pleistocene paleo-channel that once flowed through the area at an oblique angle to the modern channel. The paleo-channel has been completely filled with channel cut followed by channel-fill deposits. Pleistocene channel-lag deposits represent the channel-cut levels; the channel-fill levels are represented by still water–deposited, fine-grained botanical material. The deepest channel deposit includes freshwater shell lag lying below sand lag. Above, fine-grained Holocene peat completely fills the old channel. From the center of the old channel eastward, the shell lag interfaces with and has a facies contact with up-slope silt deposits. This up-slope silt level forms the surface upon which sandy loam and the Suwannee artifacts accumulated. It is not the intent of this chapter to discuss a comprehensive stratigraphic reconstruction of the site; in fact, there is much to be learned about the site's relationship to the Pleistocene wetland environment that once existed during Suwannee times. However, it is important to note that the paleo-channel likely represents an interval of pre-Suwannee-age high-water tables. It is our assumption that Suwannee point makers represent a middle-Paleoindian manifestation at the Ryan-Harley site. They appear to have used the site when it is unclear if water flowed through the Wacissa basin. Nevertheless,

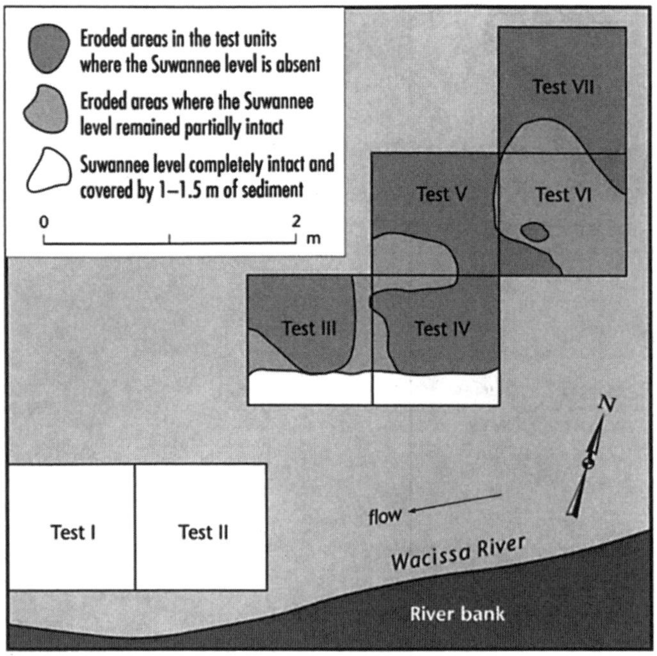

Figure 2. Ryan-Harley site (8Je1004) test units.

the Wacissa basin appears to have been a wetland with standing surface water.

Faunal Analysis

The faunal analysis focuses on a sample of faunal bone excavated from test units at the Ryan Harley site in 1999. A total of 368 bone specimens were recovered from the Suwannee level. The sample is extremely varied, containing taxa from all five major classes of vertebrates including mammalian species now extinct or that no longer range in Florida.

The Spanish explorer d'Escalante Fontaneda, writing of his experiences in Florida in the 1540s and 1550s, stated that the Indians in southwest Florida ate alligators, snakes, turtles, and "many more disgusting reptiles which, if we were to continue enumerating, we should never be through" (Escalante Fontaneda et. al 1944). They also ate opossum, eels, fish, fox, deer, and oysters, as well as "bread" made from a root that grew at the bottom of lakes and rivers. Fontaneda's description reflects the importance of aquatic environments in obtaining food. The reliance of certain cultural groups on wetland resources can be traced into prehistoric context in Florida but it has never before been traced to a Paleoindian context. At Dust Cave site in Alabama, the late-Paleoindian component dating no older than ca. 10,500 RCYBP included wetland species (Goldman-Finn and Driskell 1994; Walker 1998).

Methodology

The faunal sample from the site was collected from 6 of the 7 1-by-1-m test units that recovered in situ deposits (Figure 2). The material from these units was water screened through 1/8-inch (3.2 mm) hardware cloth at the site. All the samples, recovered from a single stratum, represent a Suwannee-point middle-Paleoindian occupation. Therefore, for quantitative purposes, the samples from all six units were combined to form a total sample for the site. The remains were identified using the faunal comparative collections of BAR in Tallahassee and the Vertebrate Paleontology Range of the Florida Museum of Natural History in Gainesville, Florida.

So that faunal assemblages may be compared with those from other sites, standard methods of quantification were employed, including fragment count, bone weight, and minimum numbers of individuals (MNI). However, numerous biases are inherent in the sample, and the figures given should be read with caution and used only as a general guide. The sample is small; in hindsight, a finer screen mesh or column samples would likely have increased the recovery of small faunal remains as well as small artifacts.

Minimum numbers of individuals were computed using paired elements in relation to individual size (Klein and Cruz-Uribe 1984; Reitz and Wing 1999; Wing and Brown 1979). When paired elements were not available, non-paired elements were used (for example, basioccipital elements of fish species) or, in some cases, a single fragment. Owing to the fragmentary nature of the sample, whole elements were not always available, and in such cases fragmentary specimens were used. A total of 31 individuals were identified in the sample, which is well below the optimal number of 200 individuals recommended by Wing and Brown (1979:118–121) to constitute an "adequate" sample. As stated before, this sample is smaller than desired, and priority should be given to obtaining a larger sample for future analysis.

Another bias affecting the sample is screen size. As the smallest screen mesh used was 1/8 inch, the sample is somewhat biased against smaller species, including small fish. Additionally, factors affecting bone degradation may significantly affect the assemblage (Nicholson 1996). Studies have shown that the potential for fish utilization in riverine environments is great (Garson 1980; Limp and Reidhead 1979). The varied composition of the faunal assemblage recovered from the site was dominated by wetland species, including some fish, and suggests that wetland resources were an important part of the fauna exploited by the site's inhabitants. As the technology to exploit wetlands was available (Peterson et. al 1984), fish may have played a more important role than is currently shown in the totals. Further testing at the site with a smaller mesh size may help support this hypothesis.

For the purpose of this analysis, specimens that could only be identified to the Class level were divided into size categories. Large mammals were determined to be deer size and larger. Medium-size mammals were determined to be anything larger than a rabbit but smaller than a deer. Small mammals were determined to be rabbit size and smaller. Large birds were determined to be anything goose size and larger. Medium birds were determined to be duck size, and small birds were determined to be perching bird size.

Amphibians

Amphibians contributed only six fragments weighing 1.5 g to the total in-place assemblage. These species are not considered a primary food source, and their presence may be purely commensal.

Anura, the frogs and toads, are found throughout the state in a variety of habitats, although most are found in wetlands. One frog or toad bone fragment weighing 0.1 g and representing a single individual was recovered from the in situ deposits.

Amphiuma spp. Amphiumas live in the shallows of lakes, ponds, streams and ditches. Nocturnal in nature, they feed on crayfish, frogs, small snakes, aquatic insects, worms, fish and their eggs, and other aquatic organisms. These creatures have strong, powerful jaws and will bite if handled carelessly. Amphiumas nest at the water's edge or in shallow water beneath a log, in matted vegetation, or other place of concealment (Bartlett and Bartlett 1999). Three fragments of this genus collectively weighing 1.2 g and representing a single individual were recovered from the *in situ* deposits.

Siren lacertina The Greater Siren is common in ponds, lakes, slow rivers, canals, drainage ditches, and other such areas in Florida. Adults may be found in open water and amid the snags, hydrilla, and other vegetation. Hatchlings and juveniles may often be found in mats of open vegetation. This large salamander forages actively under the cover of darkness, but may also be active on cloudy days and periods of low barometric pressure. They bury deep into the mud, creating a moisture-retaining, multilayered slime cocoon (Bartlett and Bartlett 1999). They become dormant when their ponds become dry. One fragment of this species weighing 0.1 g and representing a single individual was recovered from the in situ deposits.

Birds

A total of 18 fragments of bird bone collectively weighing 6.1 g and representing at least three individuals were recovered from the in situ deposits. None of these were identified beyond the class level. Eight fragments represent medium to large birds, three represent medium-sized birds, and seven represent small birds. Based on these identifications, it is difficult to determine the habitat of the species represented or their possible migratory habits.

Fish

Amia calva Bowfin inhabit quiet, usually clear waters of low-gradient, sluggish streams, swamps, and oxbow lakes with vegetation. The species spawns in early spring in shallow waters. The male of the species will clear vegetation and excavate a shallow nest, where he protects the eggs until they hatch, and then guards the young for several weeks more (Boschung et. al 1998). Six bowfin fragments collectively weighing 2.5 g and representing two individuals were recovered from the in situ deposits at the site.

Lepisosteus spp. Gars typically inhabit quiet, weedy, often stagnant backwater. The longnose gar (*Lepisosteus osseus*), the most widespread of the genus today, prefers backwaters, large creeks, and lakes and can be observed lying motionless near the surface of quiet waters (Boschung et. al 1998). Three gar fragments collectively weighing 0.4 g and representing a single individual were recovered from the in situ deposits.

Micropterus salmoides Largemouth bass prefer quiet, clear to slightly turbid streams, ponds, lakes, and reservoirs, often with vegetation (Boschung et. al 1998). Five fragments of this species, weighing 4.3 g and representing a single individual, were recovered from the in situ deposits.

Lepomis microlophus The redear sunfish also prefers clear, quiet pools of warm streams, ponds, lakes, and reservoirs with vegetation or other cover (Boschung et. al 1998). One fragment of redear sunfish weighing 1.8 g and representing a single individual was recovered from the in situ deposits.

Ictaluridae Bullhead catfishes prefer pools and backwaters of sluggish streams, ponds, and lakes, usually in areas with heavy vegetation (Boschung et. al 1998). Two catfish fragments weighing 2.6 g and representing two individuals were recovered from the in situ deposits.

Mammals

Ondatra zibethicus The muskrat, the most aquatic of the microtines, has partially webbed feet. Its scaly, hairless tail is laterally flattened and is used as a rudder (Kurtén and Anderson 1980:266). This animal spends most of its time in marshes, lakes, or along stream banks, where it builds marsh houses or burrows in banks with underwater entrances. It feeds on aquatic plants, molluscs, and aquatic vertebrates (Kurtén and Anderson 1980). Five muskrat fragments weighing 7.9 g and representing two individuals were recovered from the in situ deposits.

Sylvilagus spp. The range of these rabbits is based on the species. *Sylvilagus* species inhabit marshes and wetlands, but may inhabit dryer areas as well. Three rabbit fragments weighing 5.9 g and representing at least a single individual were recovered from the in situ deposits.

Mustela vison Mink only inhabit areas along streams and lakes, and their presence is a good indicator of nearby permanent water. Mink are amphibious and den along stream banks. They are excellent swimmers, and their diet includes a range of animals, including crayfish, fish, frogs, birds, muskrats, and other riparian species (Kurtèn and Anderson 1980). Minks tend to be solitary and nocturnal, which would explain their low numbers (a single specimen) at the site. Kurtén and Anderson (ibid.) note that mink are uncommon in Pleistocene assemblages. One mink mandible weighing 1.0 g and representing a single individual was recovered from the in situ deposits.

Odocoileus virginianus The white-tailed deer inhabits woodlands, forest edges, and stream borders (Kurtén and Anderson 1980) and is a ruminant browser. The species was and still is common throughout Florida, and its widespread occurrence on Paleoindian sites testifies to its importance as both a food and raw-material source (Dunbar and Vojnovski in press). White-tailed deer was recovered from all test units and is the single largest mammal contributor. A total of 34 fragments of this species, collectively weighing 206.2 g and representing at least two individuals, were recovered from the in situ deposits. In addition, 36 fragments of probable white-tailed deer weighing 74.3 g were also recovered.

Procyon lotor Raccoons are found in wooded areas along streams and lakes, with the highest densities reported in mixed forest-wetland environments. The late-Pleistocene/early-Holocene specimens were generally larger than those living in the same areas today (Kurtèn and Anderson 1980). Raccoons are omnivorous and feed on a wide variety of plant and animal species. Much of their food is obtained from the water. They are nocturnal in habit. Four fragments of raccoon bone, collectively weighing 9.1 g and representing two individuals, were recovered from the in situ deposits.

Tapirus veroensis Modern tapirs are browsers that feed on soft vegetable tissues. They are semiaquatic and live in moist forests (Kurtèn and Anderson 1980). Numerous finds of *Tapirus* have been found in Florida. Six tapir tooth fragments weighing 4.9 g and probably representing a single individual were recovered from the in situ deposits.

Equus sp. Because the identification of the Ryan-Harley horse was based on a single tooth and was not identified to the species, a discussion of the genus is difficult. One horse tooth weighing 15.2 g was recovered from the in situ deposits.

Mammut or *Mammuthus* See discussion below.

Holmesina septentrionalis See discussion below.

Megalonyx jeffersoni See discussion below.

Reptiles

Alligator mississippiensis American alligators are found primarily in freshwater but occasionally inhabit estuarine or saltwater habitats. They are restricted to the close vicinity of a permanent water source (Bartlett 1999). Thirteen fragments of alligator bone collectively weighing 87.7 g and representing an unknown number of individuals were recovered from the in situ deposits. In addition, two fragments of probable alligator bone weighing 5.3 g were also recovered.

Trionyx ferox The Florida soft-shelled turtle prefers nonflowing waters and is common to abundant in ponds, lakes, swamps, marshes, cypress heads, and other similar habitats. The species basks extensively on banks, exposed snags, mats of floating vegetation, and tangles of subsurface sticks (Bartlett 1999). The species is known to wander occasionally from its normal habitation area. Two fragments of this species, collectively weighing 10.2 g and probably representing a single individual, were recovered from the in situ deposits.

Trachemys scripta Two subspecies (*scripta* and *elegans*) exist for *T. scripta*. However, the specimens recovered from the site most likely belong to the subspecies *scripta*, the yellow-bellied slider. This species is now common in the northern one-third of Florida, including the Panhandle. It is common to abundant in usually heavily vegetated ponds, lakes, canals, ditches, and other similar bodies of water (Bartlett 1999). A total of 26 fragments of this species, collectively weighing 127.3 g and representing at least one individual, were recovered from in situ deposits.

Kinosternidae Mud and musk turtles most often occur in shallow freshwater, but some are present in brackish water. They inhabit ponds, lakes, swamps, marshes, cypress heads, flooded flatwoods, river edges, and estuaries (Bartlett 1999). Nineteen fragments of this family collectively weighing 11.0 g and representing at least two individuals were recovered from the in situ deposits from the site.

Gopherus polyphemus Gopher tortoises inhabit sandy, open-scrub habitats, turkey oak–longleaf pine associations, sandy, vegetated coastal dunes, and other well-drained habitats that contain ample low herbaceous plants which are consumed by the species (Bartlett 1999). Gopher tortoises dwell in colonies, but are slow to reproduce and have high hatchling mortality. A single fragment of gopher tortoise carapace weighing 1.3 g was recovered from the in situ deposits at the site. This represents the only exclusively upland species in the assemblage. Local populations of this species could easily be depopulated, given its low reproduction rate and high hatchling mortality. It is doubtful whether this species could be a sustaining resource.

Agkistrodon piscivorus Cottonmouth moccasins occur in marshes, lakes, ponds, and streams with wooded shores, sluggish waterways, low areas near water, and some coastal areas (Wright and Wright 1957). One cottonmouth vertebra weighing 0.4 g and representing a single individual was recovered from the in situ deposits.

Geochelone crassiscutata See discussion below.

Results

The in situ deposits at the Ryan-Harley site produced 368 vertebrate fauna fragments weighing a total of 871 g. Fourteen taxa were identified to the species level, four to the genus level, seven to the family level, and four to the order level, representing five vertebrate classes (Table 1). By weight, the largest faunal contributors to the assemblage are reptiles (48.53 percent), followed closely by mammals (44.07 percent). Unidentified vertebrates were next (4.35 percent), followed by bony fish (2.15 percent), birds (0.70 percent), and amphibians (0.17 percent). When the percentages by weight for wetland versus mixed versus upland environments are examined, wetland resources account for 51.72 percent of the total weight; mixed resources account for 48.13 percent; upland resources, represented by a single gopher tortoise, account for 0.15 percent of the total.

The overall percentage of wetland resources included within the faunal assemblage may indicate that the site's inhabitants utilized most extensively the catchment areas closest to the campsite. In addition to species that normally inhabit wetlands, such as fish, turtles, alligators, snakes, and amphibians, species that normally range on higher ground often venture into wetland areas for water or to forage and may be captured on these occasions.

The environmental picture provided by the in situ faunal assemblage is that of a shallow, low-energy freshwater stream or perhaps pond as a permanent nearby water source. It is likely that the Wacissa River basin was, at least seasonally, a series of freshwater ponds during climatic conditions that were becoming dryer. All terrestrial mammals identified in the sample are known to frequent wetland habitats (horse, raccoon, muskrat, rabbit, mink, tapir, deer). The vast majority of reptiles, with the exception of the gopher tortoise, required an aquatic habitat. Fish species present prefer quiet or stagnant waters as habitat.

The horizontal distribution of bone specimens throughout

the test units includes *Odocoileus virginianus* cervical vertebrae recovered in articulated position. However, most of the larger bones were fragmentary, including all the long bone. About 60 percent of the *Odocoileus virginianus* (17) and probable *Odocoileus virginianus* (37) long bones displayed green fractures (Figure 4). For specimens of large mammal long bone other than *Odocoileus virginianus* (10), 50 percent show green fractures. For medium-sized mammal long bone specimens (7), about 30 percent display green fractures. Of two *Alligator mississippiensis* long bone fragments, one displayed a green fracture. Considered together, 54.7 percent of the long bone specimens display green fractures. Thus, of 71 fragmentary specimens of large and medium-sized mammal long bone, 54.7 percent display green fractures, suggesting marrow was extracted at the site. The faunal assemblage includes mammals, birds, reptiles, amphibians, and fish, most of them wetland species.

We are unaware of any method that might determine if the turtle shell from the site was cracked opened culturally or fell apart naturally. Given the associated concentration of lithic debitage and tools and the high percentage of green-fractured long bone, we believe much of the faunal remains represents midden refuse. However, we also recognize that a smaller, unquantifiable percentage of animal bone may represent scavengers or other commensal animals that coincidentally died at the site.

Scanning electron microscopy (SEM) on modified bone specimens from the site has thus far proven to be of little value. Two specimens excavated from context had what appeared to be rodent gnaw marks. However, closer inspection under SEM showed that the fine detail of the original modification had been obscured by acidic reduction of the outer surface. Rodent gnawing is still the likely origin of the marks since they occur in paired sets. An important point to make here is the imperfect preservation of the bone material. In addition, organic silts have replaced the bone collagen. Likely sources for this modification are the site's former subaerial exposure during the late Pleistocene and the late-Holocene preservation of organic roots and rootlets in the sediment column after inundation.

Late-Pleistocene fauna remains of extinct species recovered from nearby displaced context include *Hesperotestudo crassiscutata*, *Mammut americanum*, *Paramylodon harlani*, *Holmesina septentrionalis*, and additional specimens of *Ondatra zibethicus*, *Tapirus veroensis*, and *Equus* sp. In addition, the remains of *Canis familiaris*, which may be Holocene in age, is interesting because of its possible Paleoindian affinity. Some of the armor scutes of *Holmesina septentrionalis* and the Proboscidean ivory fragment were particularly convincing as having recently been displaced from stratigraphic context, judging by the location of their recovery adjacent to uneroded deposits and similarities in color and preservation with the in situ material.

Another interesting aspect of the Proboscidean ivory is that it is an artifact. It is a small fragment of a beveled and roughened ivory shaft (Figure 3C) similar to the type of specimens being documented by Hemmings from the Sloth Hole site (8Je121). Like other out-of-context recoveries, the ivory artifact was not included in the faunal analysis but is worth mentioning because of its possible affiliation with the mid-Paleoindian Suwannee component.

Some of the fauna remains recovered from displaced context are also important. Myliobatidae (eagle rays), *Mugil* spp. (mullet), and *Arius felis* (hardhead catfish) are migratory saltwater, brackish, and freshwater species that are found today in the near-coastal Aucilla and Wacissa rivers (Hoese and Moore 1998). During the mid-Paleoindian timeframe the coast would have been too far inland and the passage obstructed by too many discontinuous karst surface channels for such a migration. In addition, some of the freshwater turtles from the displaced fauna material represent *Pseudemys* spp., which is adapted to flowing river habitats. Thus far, the turtle specimens from the Suwannee component are suggestive of still or sluggish water habitats. In other words, there appears to be a qualitative difference between the fauna recovered from the Suwannee level and specimens recov-

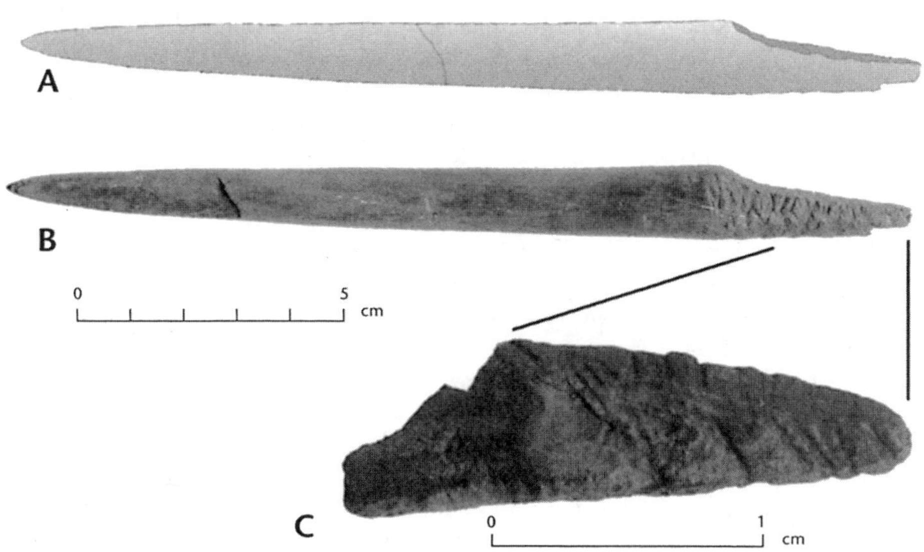

Figure 3. Ivory shaft of mastodon tusk. **A**, digital model of a carved ivory shaft shown in **B**; **B**, ivory shaft (foreshaft) from the Page/Ladson site; **C**, proximal ivory shaft fragment from the Ryan-Harley site. Photo courtesy of Florida Bureau of Archaeological Research.

ered from displaced context. The fauna from displaced context reflect a greater diversity of species, additional habitats and, in one instance, a habitat range that did not exist for three marine species during the late Pleistocene.

Lithic Analysis

A total of 193 lithic artifacts were recovered from in situ context with the fauna specimens. The artifacts include a Suwannee preform with flute or flutelike basal thinning, several bifaces, a variety of uniface tools, and debitage that includes one overshot flake. A concentration of heat-exploded lithic fragments indicates a

Figure 4. Green-fractured mammal long bone; in situ recoveries from the Ryan-Harley site (8Je1004). Photo courtesy of Florida Bureau of Archaeological Research.

fire pit or hearth may lie undisturbed beneath the eastern riverbank. Debitage from the Suwannee component is consistent with late-stage tool production and tool maintenance. Thus far, there is no evidence of quarrying or initial-stage reduction at this site.

Artifacts recovered from displaced context include three waisted Suwannee points, a Greenbriar-like Bolen point, a Kirk Serrated point, Deptford ceramic sherds, and bone pins. The ever-present scatter of modern refuse was also present despite the site's remote location in the middle of the Wacissa River's swamp forest. The staining and patination of lithic artifacts is interesting. The Suwannee and Greenbriar-like Bolen points were similarly patinated. However, the Kirk Serrated was stained dark brown. The Kirk Serrated fell from the wall of a stratigraphic section being profiled, apparently dislodged from the lower portion of the channel fill, which had the consistency of somewhat compressed and degraded peat. The Kirk point is stained the same color as the channel fill deposits. The Bolen was recovered from the lag sand below organic channel fill. The Kirk and Bolen points came from the paleo-channel. The test units excavated in the Suwannee level are 8–10 m west of the paleo-channel. The Suwannee-point level is also elevated 1 m or more above the paleo-channel and appears to have been originally deposited on the old riverbank. Many of the known and suspected Suwannee artifacts were collected directly adjacent to the eroding Suwannee component, not in the paleo-channel.

Trace element analysis of chert samples to determine quarry source location has proven to be useless in Florida. Chert identification by rock character and fossil inclusions has, thus far, proven to be the best means to evaluate prehistoric lithic resource utilization (Upchurch et. al 1982).

Coastal plains chert of Suwannee Limestone formation is available locally in three forms: silicified cap rock, concretionary or cannonball chert, and silicified flowstone. A second type of chert originating from the Miocene St. Marks Formation is available several kilometers west of the Wacissa basin. The Suwannee preform is interesting because the raw material is reminiscent of St. Marks–formation chert event, although the process of silicification obliterated any telltale microfossils.

Contrary to the hypothesis that Paleoindians preferred homogenous microcrystalline chert over large-grained or heterogeneous chert (Goodyear 1983), not all the lithic specimens recovered from the Ryan-Harley site conform to that standard. In fact, one scraper and a utilized flake were manufactured from large-grained chert. The occurrence of large-grained chert might reflect limited supplies of locally available lithic resources; alternatively, it might reflect a purposeful use for certain kinds of tools or tool uses.

Suwannee Limestone chert from local sources differs qualitatively and quantitatively. First, the most readily available but dubious quality is silicified cap rock. It outcrops in numerous places throughout the Wacissa and Aucilla river basins and occurs in irregular shapes and sizes. Although sometimes vitreous, it is most often vuggy or riddled with quartz or calcite inclusions. Concretionary or Cannonball chert, which occurs where the Suwannee limestone is dolomitic, is found in spherical and oblong sausage shapes and is usually not vuggy but may have calcite inclusions. Its primary characteristic is that it differs in grain size from one concretionary band to the next. A typical sphere of cannonball chert may have a dozen or more concretionary bands, some fine grained and some large grained. Finally, flowstone chert is dense, translucent, and semi-glassy. Flowstone chert is found with and often incorporated into outcrops of Suwannee Limestone. With few exceptions, almost all the known outcrops of Cannonball and flowstone chert were inundated by Holocene water table rise. Inundated quarries of Cannonball and flowstone chert are known and have been recorded in the Aucilla-Wacissa river basins (for example, Fossil Hole 8Je1997 in the Aucilla River).

Based on high count or microfossils, some artifacts, including an endscraper, were manufactured from a type of Suwannee Limestone chert that appears to have been imported from quarry sources east of the Aucilla-Wacissa basins. The chert has copious amounts of foraminifera incorporated into its lithology. Of the chert outcrops inspected in the Aucilla-Wacissa basin, fossil inclusions are less abundant. Suwannee chert sources known to have abundant foraminifera occur from the Steinhatchee River eastward, a distance of 80 km or more away. Most of the lithic artifacts from the Ryan-Harley site appear to be made of local silicified caprock and Cannonball chert.

Lithic Tools Recovered in Context

Biface Tools

Five bifaces were recovered from in situ context.

Suwannee preform base with flutelike basal thinning (Figures 5B and 5C) This diagnostic specimen appears to have broken in production. The preform is slightly recurvate, similar to the Suwannee points recovered from displaced context. The preform also fits comfortably within the range of variation that Ripley Bullen (1975) declared as representative for the type. It is also quite possible that completing the point would have resulted in additional waisting. The flutelike thinning on the obverse side (Figure 5C), 31.0 mm long, was done in an earlier stage of preform manufacture, as is evident by the removal of subsequent overlapping lateral and basal reduction flakes. The reverse side has two basal thinning flakes 16.3 mm and 16.8 mm long. The specimen is not laterally or basally ground, and a few flake platforms remain intact along the laterals.

Figure 5. Ryan-Harley site (8Je1004) preforms from in situ context. **A**, distal end of preform; **B**, fluted side; **C**, outline of flute. Photo courtesy of Florida Bureau of Archaeological Research.

Preform (Figure 5A) This specimen, which is missing its base, is plano-convex in cross section and excurvate. It appears to have snapped in manufacture owing to excessive hinge fracturing on its convex face. The preform may have originally been recurvate; however, with the base missing it is impossible to determine if it was excurvate or recurvate before breaking.

Preform tip fragments Two tip fragments snapped in manufacture were recovered. One is biconvex in cross section and the other plano-convex.

Bifacial blocky chunk This specimen has a nondescript irregular form.

Other Tools and Lithics

Uniface ovoid scrapers (Figure 7) Three medium-sized (ca. 7–15 mm long) uniface ovoid scrapers were recovered in situ. One complete and two broken ovoid scrapers were recovered in context. Both broken specimens apparently snapped during use and were utilized after breaking. The complete specimen appears to be a combination side-, endscraper.

Uniface endscraper on a blade-flake (Figure 9) This particularly interesting uniface tool has heavy use-wear along the distal end on the uniface side. In addition, the dorsal ridge of the blade flake midsection is worn smooth, a feature that does not occur proximally or distally. There is no evidence of hafting abrasion along the lateral edges of the blade to suggest hafting. Thus it seems either that some kind of unidentified use wore down the blade ridge or that it was hafted and lateral hafting wear did not occur despite heavy wear to the ventral end and dorsal ridge. The endscraper was made on a percussion-struck blade-flake.

Uniface small conical or turtleback scrapers (Figure 8B) One complete and two broken conical scrapers ca. 40–50 mm in diameter were recovered. One of the broken specimens had exploded when exposed to fire. The other specimen displayed a break, presumably incurred during use. The complete specimen has a number of denticulate points, some of which were worn by use.

Uniface scraper on flakes One scraper recovered was manufactured from coarse-grained chert.

Thumbnail scraper One specimen recovered was manufactured on a small flake.

Utilized flakes One coarse-grained and five microcrystalline

Figure 6. Overshot flake from Ryan-Harley site (8Je1004).

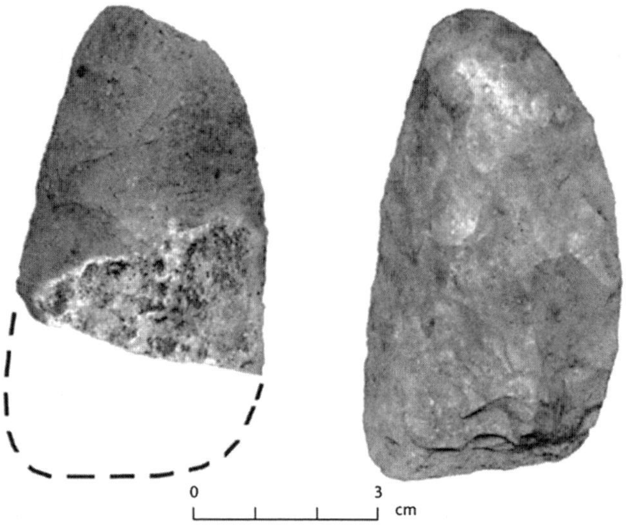

Figure 7. Ovoid scrapers from the Ryan-Harley site (8Je1004). Photo courtesy of Florida Bureau of Archaeological Research.

flakes show utilization. The coarse-grained specimen appears to be made from Cannonball chert.

Overshot flake (Figure 6) One distal end of an overshot flake struck from a large biface was recovered. Like the endscraper, it was made from Suwannee Limestone chert that likely originated east of the Aucilla-Wacissa river basin.

Bipolar flake or core One specimen recovered has a bulb of percussion on each end, one shattering from the obverse and the other from the reverse side.

Debitage with striking platforms A total of 61 flakes and flake fragments had surviving striking platforms, 8 with ground platforms as well as unprepared platforms. The remaining debitage had back-flaked, crushed, or partially missing platforms.

Debitage without platforms A total of 110 flake fragments were recovered in situ.

Heat-exploded specimens Of the 193 lithic specimens recovered, 20 specimens concentrated near the eastern riverbank displayed pot-lid fractures from heat explosion. The heat-exploded specimens include the conical scraper, debitage, and other detritus.

Lithic Tools from Displaced Context Believed to Belong to the Suwannee Component

Biface Tools

Waisted Suwannee point (Figure 10A–10C) Three of the recurvate points were surface collected. Specimen I has one small flutelike basal thinning flake 8 mm long. Specimen B was found in two pieces that cross-mended. Specimen B also has two overshot flakes on its obverse distal end and a third overshot-like flake removed from the reverse side from the right lateral edge at the base. Specimen C is neither fluted nor has overshot flaking.

Point tips Three point tips are suspected to represent the distal ends of Suwannee points.

Preform Two preform fragments are suspected to represent parts of broken Suwannee points.

Other Tools and Lithics

Abrader One extremely vuggy silicified abrader stone shows heavy wear on one surface. As much as a coarse rock can be, this specimen is almost polished from use-wear.

Uniface ovoid scrapers Four ovoid scrapers similar to those recovered in context came from displaced context.

Uniface small conical or turtleback-like scrapers Three specimens were recovered in situ.

Horse-hoof scraper One specimen was recovered in situ.

Miscellaneous scrapers Seven scrapers of random form were recovered in situ.

Sidescraper on thin blade-flake (Figure 11) This specimen is almost identical to specimens recovered from the Sloth Hole site (8Je121), a Clovis site in the Aucilla River.

Figure 8. Uniface tools from the Ryan-Harley site (8Je1004). **A**, scraper; **B**, heat-exploded turtleback scraper. Photo courtesy of Florida Bureau of Archaeological Research.

Results

The lithic toolkit from the Ryan-Harley site is as similar to Clovis as it is dissimilar to Dalton or Bolen lithic tool assemblages. Distinguishing the Suwannee tool kit has not generally been straightforward because some of the most notable sites have not presented stratigraphic separation of Suwannee and Bolen materials. For example, the Harney Flats site (8Hi507), a Paleoindian base camp in Hillsborough County, Florida, produced Suwannee and Bolen implements from the same or slightly separate stratigraphic levels. Lithic tools from the site included oval and conical scrapers, endscrapers, tools on blades or blade-like flakes, Suwannee and Bolen points, triangular-shaped hafted spokeshaves, adzes, microliths, and assorted other lithic tools (Daniel and Wisenbaker 1987).

Suwannee projectile points from the Ryan-Harley site are of the waisted or recurvate variety (Dunbar and Hemmings 2001). The uniface tool kit includes a variety of well-crafted mostly ovoid scrapers, snub-nosed endscrapers, and various flake and blade-flake tools. Conspicuously absent from the Ryan-Harley Suwannee tool kit are the notched and triangular assortment of hafted uniface tools as well as bifacial Dalton-like adzes and

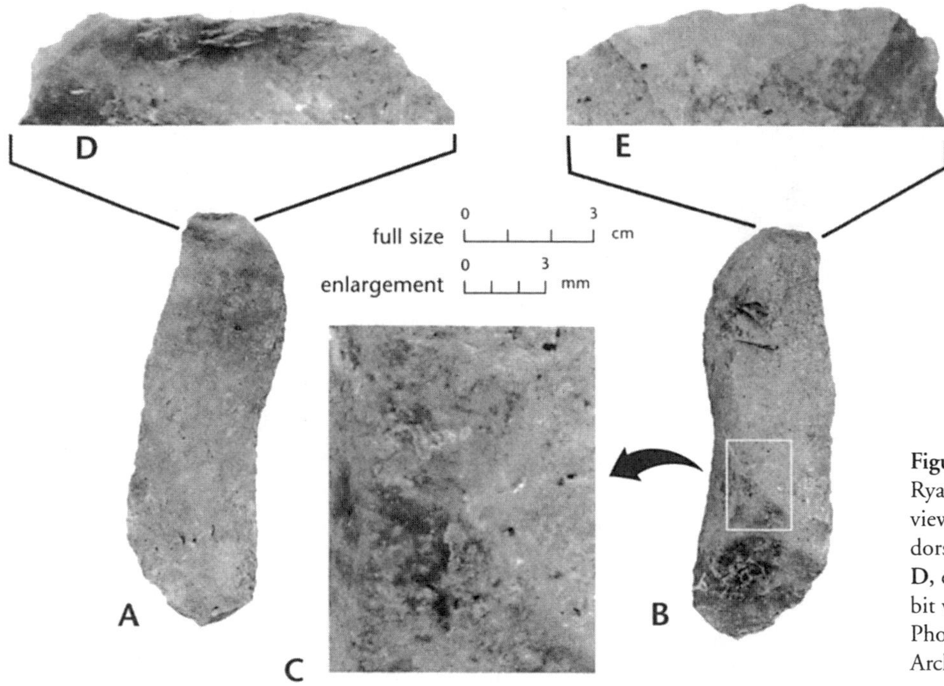

Figure 9. Endscraper on blade-flake from Ryan-Harley site (8Je1004). **A,** ventral view; **B,** dorsal view; **C,** enlargement of dorsal ridge showing blade crest wear; **D,** enlargement showing ventral working bit wear; **E,** enlargement of dorsal bit. Photo courtesy of Florida Bureau of Archaeological Research.

microliths that are part of the Bolen (Purdy 1981) and Dalton (Goodyear 1974) assemblages. Thus the Suwannee-age toolkit is distinctively unlike documented Bolen assemblages (Milanich 1994) because it lacks the microliths, Dalton-like and Aucilla adzes, and Hendrix and Edgefield scrapers as well as the triangular-shaped hafted endscrapers, spokeshaves, and Waller knives.

The Waisted Suwannee points (Figures 10, 13, 14, and 15) is both similar, yet distinct compared with the recurvate or waisted-Clovis type (Figure 12) (Dunbar and Hemmings 2003; Goodyear 1983; Perino 1985; Prufer 1962). The Clovis-waisted type is a fluted, overshot-flaked point similar to the more typical parallel-sided to slightly recurvate Clovis; but like the Suwannee, it is recurvate. The average ratio of maximum width to hafting width of the waist is 1.2:1 for both Suwannee (n = 9) and Clovis (n = 7) waisted with a variance of ± 0.1. Specimens used in this determination include all Suwannee-waisted points from the Ryan-Harley and Page-Ladson sites as well as a few others and all the Clovis-waisted points from the Sloth Hole site as well as a few others.

Bolen points are opposite-beveled, side-notched or corner-notched points that are often manufactured as smaller points compared with the lanceolate Suwannee form. Bolen points share few features in common with the Suwannee type. While it appears to be uncommon, some waisted Suwannee points are opposite-beveled, which is an edge feature common on Bolen points (Daniel and Wisenbaker 1987). None of the Ryan-Harley specimens are opposite-beveled.

Figure 10. Suwannee points from displaced context at Ryan-Harley site (8Je1004). Photo courtesy of Florida Bureau of Archaeological Research.

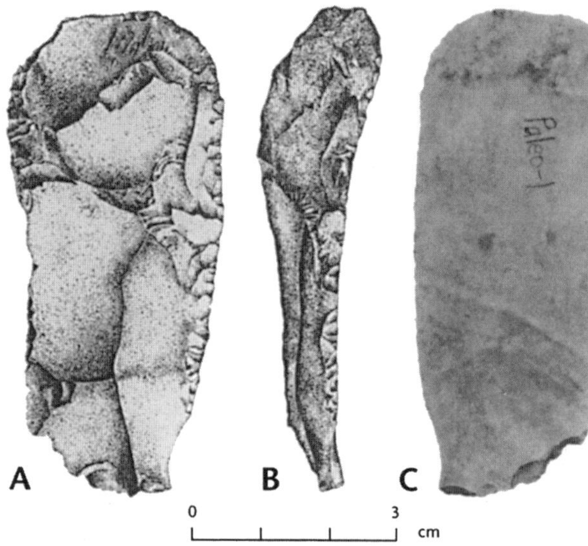

Figure 11. Uniface tool on blade-flake from Ryan-Harley site (8Je1004). **A**, dorsal view; **B**, side view; **C**, ventral view. Photo courtesy of Florida Bureau of Archaeological Research.

In contrast, most features of the Suwannee point assemblage from the Ryan-Harley site resemble the Clovis assemblage in several respects. These include the presence of occasional fluting, overshot flaking (Figure 13), variation in size (Figure 14), and a similar variety of uniface tools. Based on the materials recovered from displaced context, this would also include ivory shafts or foreshafts. The lithic tool assemblage from Ryan-Harley suggests that the waisted Suwannee assemblage represents the mid-Paleoindian evolved form of Clovis-waisted. If this proves to be true, it also means that the Paleoindian tradition in the southeast United States had an uninterrupted continuity of in-place development and evolution, as we might well expect.

Discussion and Conclusions

The apparent significance in the Southeast is that Pleistocene megafauna survived beyond the Allerød boundary into the early cold phase of the Younger Dryas. The data suggest that Suwannee point makers, whom we believe to be post-Clovis offspring, relied not only on megamammals but on various other species as well. If there is a notion that Suwannee point–making peoples in the southeastern United States were singly devoted to hunting big game or upland mammals, it is not supported by the evidence from the Ryan-Harley site. Wetland species including tapir, turtles, alligators, fish, birds, and amphibians also appear to have been sources of nutrition. Many of these species also suggest the Suwannee toolkit was more extensive than we know, albeit archaeologically perishable.

The most significant aspect of the Ryan-Harley site is the faunal remains. The faunal assemblage at Ryan-Harley suggests at least some continuity in dietary procurement compared with Dust Cave in northern Alabama. Dust Cave produced a Holocene fauna assemblage dominated by migratory birds and aquatic species in late-Paleoindian context. Notably absent were any Pleistocene megafauna remains. The late-Paleoindian occupation at Dust Cave is estimated to date between ca. 10,500 RCYBP and 10,000 RCYBP (Goldman-Finn and Driskell 1994; Walker 1998). The Ryan-Harley site is believed to represent a middle-Paleoindian site dating somewhere between ca. 11,000 RCYBP and 10,500 RCYBP, perhaps as late as ca. 10,300 RCYBP. Considering the small sample size, it seems remarkable to have such a variety of species present including extinct late-Pleistocene forms.

The Ryan-Harley site yields evidence indicating Pleistocene megafauna survived in the Southeast in post-Clovis times, which should mean during the Younger Dryas. To propose otherwise would place the Suwannee culture older than has otherwise been recognized (Anderson et. al 1996). To assume a pre–Younger Dryas age for the Suwannee culture would mean it was a Clovis contemporary or its predecessor (Stanford 1991). At the Horn Rock Shelter 2 (41Bq46) in central Texas a waisted Suwannee or Suwannee-like point was recovered at and below Folsom levels of the site (Redder 198540-42 and Figure 2G), suggesting a middle- to early-Paleoindian context. The Horn Rock Shelter is outside the southeastern U.S. and, therefore, outside the typical distribution; thus the Horn Rock Shelter may or may not be a temporal indicator.

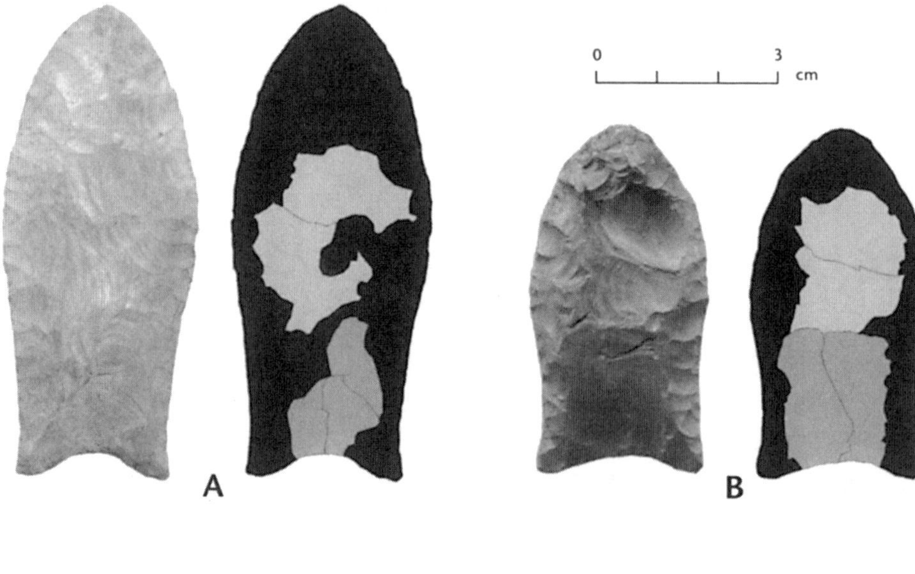

Figure 12. Clovis waisted points. **A**, specimen collected as an isolated find about 1 km south of the Ryan-Harley site (8Je1004); **B**, specimen from the Sloth Hole site (Je121). Photo courtesy of Florida Bureau of Archaeological Research.

Figure 13. Suwannee-waisted points with Clovis-like features. All are isolated finds from the Santa Fe River. Photo courtesy of Florida Bureau of Archaeological Research.

Even though there have not been any Southeastern Suwannee point sites dated by radiometric means and the stratigraphic position of the Suwannee tool kit has not been placed in relation to other Paleoindian tool assemblages (Simpson and Clovis, for example), most researchers place the Suwannee tool assemblage as middle Paleoindian (Anderson et al. 1996; Ellis et al. 1998; Goodyear 1999). We also believe that the Suwannee point tool assemblage represents a middle-Paleoindian context, but also recognize the chronology remains to be demonstrated.

The occurrence of numerous Clovis-like traits suggests the Ryan-Harley site may represent an older Suwannee-point site compared with sites such as Harney Flats that produced an opposite-beveled waisted Suwannee point along with opposite-beveled Bolen points at or very nearly at the same stratigraphic elevation (Daniel and Wisenbaker 1987).

In an unpublished letter dating to or just after WWII and again in the premier issue of the *Florida Anthropologist*, Clarence Simpson (Simpson 1948; Simpson n.d.) reported finding partially articulated mastodon with lanceolate points and ivory rods in the Ichetucknee River, north Florida. The only lithic point

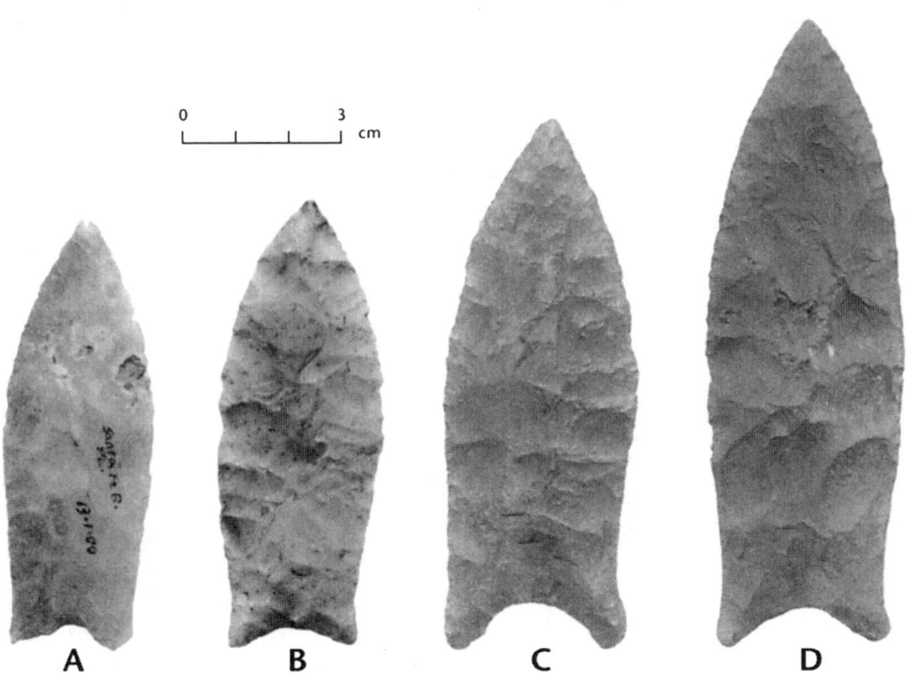

Figure 14. Size variation in Suwannee-waisted points. **A**, isolated find from Santa Fe River; **B**, from the Page/Ladson site (8 Je591); **C**, **D**, isolated finds from the Santa Fe River.

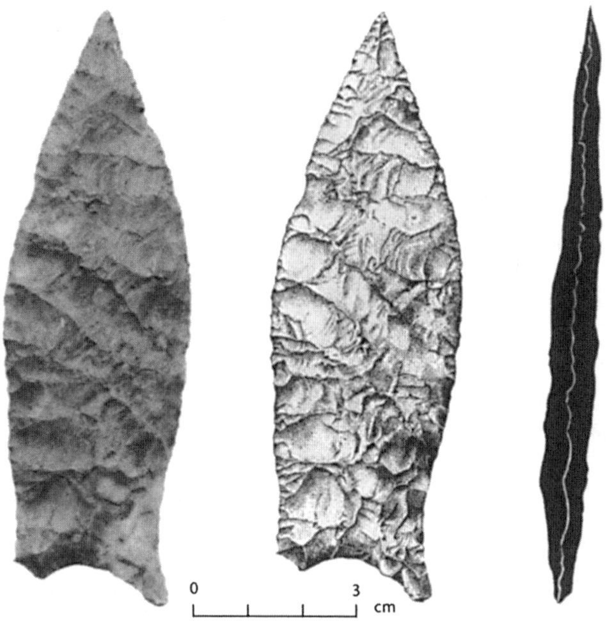

Figure 15. Suwannee-waisted point type from the Ryan-Harley site (8Je1004). Photo courtesy of Florida Bureau of Archaeological Research.

depicted (Simpson 1948; Simpson n.d.) is a waisted Suwannee. The point here is that another Suwannee site has been previously documented and includes extinct Pleistocene megafauna in association with a lithic scraper as well as a waisted Suwannee point and carved ivory artifacts from nearby but displaced context.

Wilfred T. Neill (1958) tested the first widely accepted Paleoindian site in Florida, the stratified Silver Springs site (8Mr92). From the lowest cultural level of the site he recovered a Clovis waisted point in situ (Neill 1958, Figure 3a) along with a probable preform (Neill 1958, Figure 3d). An additional Clovis waisted point was recovered out of context (Neill 1958, Figure 3b), along with a single example of a classic Clovis type (Neill 1958, Figure 3c). The Sloth Hole site (8JE121) has produced five waisted Clovis points as well as one classic Clovis. The Sloth Hole Clovis points occur with the largest presence of carved ivory shafts or foreshafts from a single site in North America (Hemmings 1999).

Of the Florida Clovis waisted points inspected, their characteristic shape, fluting, and overshot flaking represent diagnostic features. As previously mentioned, their maximum width to waist width ratio is the same as the sample of waisted Suwannee points measured. Although much less common, one of the waisted Suwannee points from displaced context at the Ryan-Harley site has overshot flaking and the preform recovered in context has a flute or flutelike basal thinning. The primary difference is that the Clovis waisted points have late-stage flutes. In addition, an overshot flake fragment was recovered in situ along with a tool assemblage that more closely resembles Clovis than any other known stone tool tradition in Florida.

Thus the initial surface collections from the Ryan-Harley site include Suwannee points and a variety of uniface tools. Also from displaced context the remains of *Hesperotestudo crassiscutata*, *Mammut americanum*, *Paramylodon harlani*, *Holmesina septentrionalis*, and additional specimens of *Ondatra zibethicus*, *Tapirus veroensis*, and *Equus* sp. were collected along with numerous smaller vertebrates. Test excavation led to the recovery of additional Suwannee tools, including a fluted preform, and associated fauna including Pleistocene horse, tapir, and muskrat. Within an approximate area of 4 m^2, 193 stone artifacts and 368 faunal remains were recovered. A total of 561 specimens were collected in context from a natural level 10–15 cm thick. Altogether that is an average of about 140 specimens per square meter. The stone artifacts represent 34.4 percent of the materials present, and faunal bone represents 65.6 percent.

Stone tool patina, the tight vertical distribution of site remains, the occurrence of articulated elements, the high percentage of green-fractured long bone, and the occurrence of an assemblage of small, medium, and large animal bone indicates a culturally derived accumulation. The artifact suite suggests lithic tool manufacture and maintenance as well as food preparation and consumption and refuse discard. Concentrations of diagnostic artifacts, lithic debitage, and green-fractured bone (mostly deer size) indicate that this site was utilized as a camp or that at least some kind of prolonged activity took place requiring a diversity of tools and fauna exploitation. Recent geoarchaeological analysis of the Ryan-Harley site sediment column determined there is no evidence of post-depositional reworking of un-eroded sections of the site (Balsillie et al. in press). Although undated, the Suwannee point level of the Ryan-Harley site represents an undisturbed Paleoindian campsite.

Five samples of bone, four from extant species and one from an extinct form, were submitted for radiometric dating. However, all the specimens were so mineralized they could not be dated. Thus the Suwannee component has not been successfully dated even though two radiometric dates from oxidized roots recovered in and just above the artifact-producing level yielded dates of 4,420 ± 60 RCYBP (Beta 132151) and 4,590 ± 70 RCYBP (Beta 123575). We believe these dates represent the approximate time of mid-Holocene inundation of this section of the Wacissa River basin. After that, further oxidation of Holocene plant materials stopped. The Ryan-Harley site appears to be a naturally inundated middle-Paleoindian campsite represented by midden refuse. If so, this is the first site of its kind to be documented.

One of the unexpected surprises that would never have occurred in an upland site was ca. 4,500 years of root and rootlet that had invaded the soil and remained preserved within its stratigraphic column. The peat-like appearance of the Suwannee level is from post-depositional plant growth. Remarkable as it might seem, Holocene root invasion did not appreciably disturb the cultural remains, at least in the 10- to 15-cm-thick sandy loam of the Suwannee-point level.

Based on the site's location adjacent to a paleo-channel and the faunal remains recovered, the Ryan-Harley site appears to represent an occupation that took place when the Wacissa basin experienced slow or intermittent flow or still-water pond conditions. When occupied, the surrounding terrain continued to support wetland species and the turtle species suggestive of still-water conditions.

The Ryan-Harley site has expanded our understanding of Paleoindian subsistence patterns on a local and regional level and will continue to provide the opportunity to compare these data with continent-wide hypotheses regarding Paleoindian subsistence strategies. The faunal remains from the Ryan-Harley site suggest a varied diet consisting of some of the last late-Pleistocene megafauna along with other mammals, birds, reptiles, amphibians, and fish. These remains reflect a strong reliance on mammals, especially deer, and also the exploitation of reptiles, especially turtles and alligators, and fish. The fauna assemblage is rich in wetland as well as mixed wetland and upland species.

This and other site investigations by the Aucilla River Research Project provide substantial evidence that sites in the Aucilla and Wacissa River basins have suffered minimal erosion since the late Pleistocene. In other words, they contain some of the best-preserved late-Pleistocene sites in the southeastern United States because they now lie well preserved in underwater or wetland settings. The importance of the Ryan-Harley site is not only that it is the first Suwannee site with bone preservation but also that it lies within the broad expanse of the Wacissa River basin, suggesting there will be others like it. While the Ryan-Harley site may not resolve questions about Southeastern Clovis and megafauna interactions, it does promise to shed light on the cultural transition away from big-game availability toward a reliance on Holocene species. The transition from Paleoindian to early-Archaic game procurement in the Southeast is important because there appears to have been no interruption of in-place cultural development. Presently, the Ryan-Harley site offers a view of the cultural adaptations of necessity as Suwannee people and their economy began to adapt to the decline of late Pleistocene megafauna and towards a time of early-Holocene environmental stress and scarce surface water.

References Cited

Anderson, D. G., L. O'Steen, and K. E. Sassaman
1996 Environmental and Chronological Considerations. In *The Paleoindian and Early Archaic Southeast*, edited by D. G. Anderson and K. E. Sassaman, , pp. 3–15. University of Alabama Press, Tuscaloosa, AL.

Bartlett, R. D., and P. P. Bartlett
1999 *A Field Guide to Florida Reptiles and Amphibians*. Gulf Publishing Company, Houston, TX.

Boschung, H. T., J. D. Williams, D. W. Gottshall, D. K. Caldwell, and M. C. Caldwell
1998 *National Audubon Society Field Guide to North American Fishes, Whales and Dolphins*. Alfred A. Knopf, New York, NY.

Bullen, R. P.
1975 *A Guide to the Identification of Florida Projectile Points*. Revised edition. Kendall Books, Gainesville, FL.

Daniel, R. I., and M. Wisenbaker
1987 *Harney Flats: A Florida Paleo-Indian Site*. R. W. Moeller, editor. Baywood Publishing Company, New York, NY.

Dunbar, J. S., and C. A. Hemmings
2001 Florida Paleoindian Points and Knives. New Perspectives in First Americans Studies, Center for the Study of the First Americans, Texas A&M University, College Station.

d'Escalante Fontaneda, Hernando, D. O. True, and B. Smith
1944 *Memoir of D d'Escalente Fontaneda Respecting Florida Written in Spain, about the Year 1575*. University of Miami and the Historical Association of Southern Florida Miscellaneous Publications, No. I, Miami, FL.

Garson, A. G.
1980 Comment upon the Economic Potential of Fish Utilization in Riverine Environments and Potential Archaeological Biases. *American Antiquity* 45(3):562–67.

Goldman-Finn, N. S., and B. N. Driskell
1994 *Journal of Alabama Archaeology*. Preliminary Archaeological Papers on Dust Cave, Northwest Alabama 40, No. 1 & 2. The Alabama Archaeological Society, Moundville, AL.

Goodyear, A. C.
1974 *The Brand Site: A Techno-functional Study of a Dalton Site in Northeast Arkansas*. Publications on Archeology: Research Series; No. 7 Arkansas Archeological Survey, Fayetteville, AK.

1983 A Hypothesis for the Use of Crytocrystalline Raw Materials among Paleoindian Groups of North America. In *Eastern Paleoindian Lithic Resource Use.*, pp. 1–XXi. C. J. Ellis and J. C. Lothrop ed. Westview Press, Boulder, San Francisco & London.

Hemmings, C. A.
1999 *The Paleoindian and Early Archaic Tools of Sloth Hole (8Je121): An Inundated Site in the Lower Aucilla River, Jefferson County, Florida*. Master's thesis, Department of Anthropology, University of Florida, Gainesville, FL.

Hoese, H. D., and R. H. Moore
1998 *Fishes of the Gulf of Mexico, Second Edition*. Texas A&M University Press, College Station, TX.

Klein, R. G., and K. Cruz-Uribe
1984 *The Analysis of Animal Bones from Archeological Sites. Prehistoric Archeology and Ecology*. University of Chicago Press, Chicago, IL.

Kurtèn, B., and E. Anderson
1980 *Pleistocene Mammals of North America*. Columbia University Press, New York, NY.

Limp, W. F., and V. A. Reidhead
1979 An Economic Evaluation of the Potential of Fish Utilization in Riverine Environments. *American Antiquity* 44(1):70–78.

Milanich, J. T.
1994 *Archaeology of Precolumbian Florida*. University Press of Florida, Gainesville, FL.

Neill, W. T.
1958 A Stratified Early Site at Silver Springs, Florida. *Florida Anthropologist* 11:38–48.

Nicholson, R. A.
1996 Bone Degradation, Burial Medium and Species Representation: Debunking the Myths, and Experiment-based Approach. *Journal of Archaeological Science* 23:513–33.

Perino, G.
1985 *Selected Preforms, Points, and Knives of the North*. Gregory Perino, Idabel, OK.

Peterson, J. B., N. D. Hamilton, J. M. Adovasio, and A. L. McPherron
1984 Netting Technology and the Antiquity of Fish Exploitation in Eastern North America. *Midcontinental Journal of Archaeology* 9(2):199–225.

Prufer, O. H.
1962 Fluted Points and Ohio Prehistory. *The Explorer* 4(3):14–21.

Purdy, B. A.
1981 *Florida's Prehistoric Stone Technology*. University Presses of Florida, Gainesville, FL.

Reitz, E. J., and E. S. Wing
1999 *Zooarchaeology*. Cambridge University Press, Cambridge, England.

Simpson, J. C.
1948 Folsom-like Points from Florida. *Florida Anthropologist* 1(1):11–15.
n.d. Yuma Type Points from Florida. In *Misc. Document Files*. Florida Museum of Natural History, Department of Anthropology, Gainesville, FL.

Upchurch, S. B., R. N. Strom, and M. G. Nuckels
1982 *Methods of Provenance Determination of Florida Cherts*. Department of Geology, University of South Florida, Tampa, FL.

Walker, R. B.
1998 *The Late Paleoindian through Middle Archaic Faunal Evidence from Dust Cave, Alabama*. Doctoral dissertation, University of Tennessee. Knoxville, TN.

Wing, E. S., and A. B. Brown
1979 *Paleonutrition: Method and Theory in Prehistoric Foodways*. Studies in Archaeology, Academic Press, New York, NY.

Wright, A. H., and A. A. Wright
1957 *Handbook of Snakes of the United States and Canada*. Handbooks of American Natural History, Comstock, Ithaca, N.Y.

Yon, J. W.
1966 *Geology of Jefferson County, Florida*. Florida Geological Survey Bulletin No. 48, Tallahassee, FL.

Rapid Clovis Colonization of the Americas: Chronological Evidence and Archaeological Analogues

Stuart J. Fiedel

Abstract

Paleoindians of the Clovis culture colonized the Americas within 600 years or less (ca. 13,500–12,900 CALYBP). The archaeologically attested movement of Thule whale hunters from Alaska to Greenland between A.D. 900 and 1050 is an analogous case of rapid long-distance expansion. Climate change and inter-societal competition may have spurred migration in both cases. Thule Inuit replaced precursors of the Dorset culture throughout the Arctic; if there were pre-Clovis people in the Americas, they were replaced by Clovis in North America, and by makers of Clovis-derived fishtail points in South America.

The Paleoindian colonization of America was almost incredibly rapid. The earliest reliable dates yet obtained for a Clovis assemblage come from the Aubrey site in Texas (Ferring 1995): about 11,600 RCYBP, or ca. 13,500 CALYBP (an unpublished date for the Anzick burial, presented at the *Clovis and Beyond* conference, is ca. 11,550 RCYBP [Sellars 1999]).[1] Clovis was everywhere in North America by the time of the Younger Dryas onset at about 12,900 CALYBP (or 11,200–10,700 RCYBP) (Fiedel 1999a). Soon after initial expansion, the Clovis complex underwent regional diversification, as Folsom developed in the West and mid-Paleo forms like Debert and Cumberland/Barnes appeared in the East. Dates of ca. 10,700–11,000 RCYBP from Cerro La China and Cueva del Medio show that Paleoindian hunters making fishtail points had reached the southern tip of South America by about 12,900 CALYBP (Flegenheimer and Zarate 1997; Nami 1994, 1996). The stylistic and technological derivation of these points from the Clovis tradition is clear (Lynch 1983; Morrow and Morrow 1999; Nami 2000; Pearson 1998), and indicates that these people were Clovis descendants.

So, it apparently took only some 300 to 600 years for Clovis hunters to fill North America; and in the same time span, an offshoot population trekked some 4,000 miles to Tierra del Fuego. Some archaeologists, such as Whitley and Dorn (1993), have expressed incredulity about this remarkable event. They argue that there is no known analogue for such rapid population growth and movement by hunter-gatherers; even farmer-pastoralists who invaded and occupied new territories, such as the Bantu, moved more slowly. But the diagnostic artifacts and their associated dates cannot be dismissed; either Paleoindians did expand very rapidly, or else a peculiar suite of fluted points, spurred endscrapers, gravers, etc., was borrowed by preexisting populations inhabiting a wide range of diverse habitats across the continent. Such wholesale borrowing seems improbable (Storck 1991). Only a few sites—Monte Verde, Chile (Dillehay 1997), Meadowcroft, Pennsylvania (Adovasio et al. 1990), and Cactus Hill, Virginia (McAvoy and McAvoy 1997)—offer dubious hints of a pre-Clovis population. If precursors really existed in these areas, they seem to have been entirely replaced by Clovis.

In fact, there are several well-known cases of comparably rapid hunter-gatherer expansion and migration; for example, the Numic migration into the Great Basin and adjacent regions (between ca. A.D. 1000 and 1500) (Madsen and Rhode 1994), and the migration of Athapaskans from the Yukon to the Southwest during the same period (Opler 1983). Unfortunately, archaeologists cannot agree on the timing and process of Numic migration; and the Athapaskans left virtually no archaeological record dur-

Senior Archeologist, Louis Berger Group, 2300 N St. NW, Washington, D.C. 20037; e-mail: sfiedel@louisberger.com

[1]Stafford (pers. comm.) doubts the accuracy of the 11,550 RCYBP Anzick date, and the new, precise ^{14}C dates indicate an age of ca. 11,040 RCYBP (Morrow and Fiedel in preparation). Clovis dates earlier than 11,200 RCYBP should be regarded skeptically.

ing their trek to the south. I will focus instead on a migration that is archaeologically well documented, such that there is no disagreement concerning either the dates or the cultural processes involved. The Thule culture, ancestral to recent Inuit or Eskimos, developed on the west coast of Alaska and spread by a swift coastal migration from northern Alaska to Greenland and Labrador (McGhee 1984). Thule people traversed this distance of ca. 1,500 to 2,000 miles in less than 150 years, between A.D. 900 and 1050. In the process, they replaced a long-resident precursor population of the Dorset culture, which had occupied the Arctic coast for 1,500 years. Whether there was face-to-face competitive contact between Dorset and Thule, or whether Dorset collapsed suddenly due to environmental change and Thule moved into the vacuum, remains uncertain.

Thule people covered more than 1,500 miles in 100 to 150 years, sloughing off sufficient populations along the migration route to fill the intervening territory and to create a visible archaeological record. By the most direct route, Clovis ancestors would have covered about 1,300 miles getting from Alberta to New Mexico. They could have done this in less than 100 years. In fact, they previously may have trekked through the ice-free corridor in a matter of months, leaving no obvious traces of their passage. At a reasonable pace of 15 miles per day, it would have taken 80 days to walk the 1,200 miles through the corridor.

Both the Paleoindian and Thule migrations occurred against a backdrop of climatic change. It would be simplistic, however, to attribute migration in each case to either amelioration or deterioration of climate. The first pulse of Thule movement along the coast occurred during a warm interval, the Medieval Climatic Optimum; but a second phase of expansion into more interior environments occurred after A.D. 1200, as the climate cooled (McGhee 1984:373–4). The few dated late-Pleistocene sites known in far Northeast Asia—Ushki and Berelekh (Dikov 1996; Mochanov and Fedoseeva 1996)—suggest that initial settlement of that region by ancestors of the Paleoindians may have occurred around 14,000 RCYBP (17,000 CALYBP), after the coldest part of the last glaciation. The earliest sites in central Alaska, such as Broken Mammoth (Holmes 1996) and Swan Point (Holmes et al. 1996), have produced radiocarbon dates of ca. 11,700 RCYBP (13,700 CALYBP), which may imply that initial human occupation of eastern Beringia followed the abrupt Bølling warming at 12,600 RCYBP (14,700 CALYBP). Alaskan temperatures were warmer than present during the Bølling-Allerød interstadial (Elias 2000). At 11,600 RCYBP, there is evidence of a sudden onset of aridity in central Alaska, which caused lakes to dry up (Abbott et al. 2000). Droughts may have reduced the lacustrine resources on which Clovis ancestors depended (such as waterfowl, which accounted for 40 percent of the meat in the Nenana occupation at the Broken Mammoth site [Yesner 1998]). This economic stress may have been the impetus that spurred emigration southward through the widening ice-free corridor. Clovis expansion within North America, south of the ice sheets, also occurred at a time of climatic stress, during the Intra-Allerød Cold Period (ca. 13,250–13,100 CALYBP) and the ensuing 200-year warm period that created a drought in the Southwest (Haynes 1991).

Social factors may also have played a role in inducing migration. Mason (1998) suggests that the west Alaskan Birnirk culture, ancestral to Thule, had become marginalized in competition with Old Bering Sea, Ipiutak, and Punuk groups over access to whaling sites and other vital resources in the Bering Sea region. This social pressure may have caused their movement to the Point Barrow area around A.D. 800–900. However, it is likely that occasional unpredicted collapses of local resources at Barrow were a more immediate cause of Thule migration (Mason 1998). Clovis ancestors in Northeast Asia may have experienced similar social pressure. Radiocarbon dates now show that pottery was spreading through the Amur Basin at 13,000–12,000 RCYBP (Kuzmin et al. 1997). Judging from the New World record, Clovis ancestors accepted neither ceramics nor an earlier innovation, microblade technology. Neither Clovis people nor their possible Alaskan ancestors of the Nenana complex used microblades[2]. However, the Denali culture, which entered central Alaska by 10,700 RCYBP or earlier (West 1996), was an offshoot of the microblade-based Dyuktai culture of Siberia. The evident cultural diversity of terminal-Pleistocene Northeast Asia and Alaska raises the possibility of inter-societal conflicts that could have precipitated the eastward and subsequent southward migration of marginal groups.

Presumably, both in the Thule and Clovis cases, the colonizing population was thinly scattered across the landscape. Historic Eskimo densities were sometimes on the order of one person per 100–200 square miles (Arima 1984:460; Balikci 1984:429). The Cree and Athapaskans of the subarctic forests, who lived by hunting caribou, fishing, and trapping fur-bearing mammals, maintained densities of about one person per 80 square miles (200 km^2) (Helm 1968; Rogers and Leacock 1981:173). In view of recent reconstructions of Paleoindian environment and economy, a subarctic analogy seems most appropriate, particularly for eastern North America (Custer and Stewart 1990). At a density similar to that of recent subarctic hunters, it would have taken about 125,000 Paleoindians to fill North and South America below the ice front. Assuming doubling per generation (every 20–25 years), it would take 160–200 years for an initial population of 500 to reach this size. In radiocarbon time, this process would unfold within a one-sigma time frame; it would appear effectively instantaneous.

Both empirical observation of Australian bands (Birdsell 1968) and computer simulation of hunter-gatherer societies (Wobst 1976) suggest that the minimal effective size of human mating

[2]The Nenana component at Swan Point seems to include a microlithic element (Holmes 1998). However, stratigraphic/temporal separation of macroblade and later microlithic assemblages is clear at several other Nenana sites, and Goebel (pers. comm.) suspects that post-depositional mixing occurred at Swan Point. On technological grounds, neither Nenana nor Clovis ancestry can be traced *directly* to any one of the microblade-dominated cultures of Late Glacial Northeast Asia. Ushki 1, Level 7, while not stylistically similar to Clovis, does show that cultures relying on bifaces and blade tools and lacking wedge-shaped microblade cores were present in western Beringia (Kamchatka) prior to 10,800 RCYBP (Dikov 1996).

networks (and dialect tribes) is about 500 individuals. Based on these numbers, we can speculatively model a multitiered Paleoindian social landscape, with micro-bands of ca. 25 people each occupying 5,200-km^2 territories, aggregating seasonally in macro-bands of ca. 100 to 150. Four to six macro-bands might form a network of about 500 people for exchange of spouses and trade items. This social entity would encompass a territory of about 100,000 km^2; envisioned as an approximate circle, this area would have a radius of about 180 km. A comparable social entity may be recognizable in the archaeological record of terminal-Pleistocene Northwest Europe, where three stylistically defined ethnic groups each occupied an area of ca. 100,000 km^2 (Price 1991:199). In the continental U.S., there might have been 100 such territories, with a total population of about 50,000, around 12,500 CALYBP. In a similar vein, Anderson (1995) hypothesized the existence of 15 middle-Paleoindian macro-bands east of the Mississippi, each in a territory of ca. 20,000–80,000 km^2. This social landscape would represent a late, stabilized situation; social organization during initial expansion into unbounded territory could have been quite different.

The Thule-Clovis analogy can be pushed further. Thule specialized in whale hunting; Clovis specialized in mammoth hunting. Thule assemblages contain distinctive carved harpoon heads of the Sicco type; Clovis points also appear emblematic of an ethnic identity (Haynes 1991:315). Sicco harpoon heads, carried by the initial Thule migrants all the way to Greenland, were replaced by the Thule 3 type within less than 200 years (McGhee 1984). Similarly, the classic Clovis form gave way to regional stylistic variants (Folsom, Debert, Barnes, Fell fish-tail) about 400 years after the initial pulse of expansion. Hunting giant mammals is a dangerous enterprise, and success brings commensurate social standing. Historic Alaskan Eskimo whaling captains or *umialiks* enjoyed high status (Sheehan 1985). There are hints in the archaeological record, such as the elaborate burial of the Anzick child (Lahren and Bonnichsen 1974), that there may have been more status differentiation in Clovis society than we usually assume. Among their other perquisites, Alaskan Eskimo chiefs and successful hunters often had two or more wives (Ray 1984:286). If Clovis hunters were also polygynous, this might help to account for the high birth rate that must be assumed in explaining the rapid colonization process. Such a practice would have exacerbated the scarcity of potential wives that has been speculatively suggested as a motivation for Paleoindian journeys across great distances (MacDonald 1998).

There may have been significant differences between the Clovis and Thule migrations. The Inuit ancestors had several types of watercraft that allowed them to move much faster than people on foot; and for terrestrial travel, they may already have developed dogsleds (Harp 1983:141). Thus, their movement through coastal waters could have been much more rapid than the overland trek that Clovis people undertook. However, there is no reason to preclude the use of skin-covered boats, like the Mandan bull boats, by Paleoindians (Engelbrecht and Seyfert 1994; Jodry 1999). Use of hides to cover pole frames is assumed to have been a common method of house construction in upper Paleolithic Eurasia. The same technique would have sufficed for constructing simple boats. While such craft might not have been adequate for coastal voyages, they would have been useful for exploring major rivers such as the Missouri. The use of dog-drawn travois cannot be precluded, either; since the New World dog seems, on the evidence of both skeletal morphology (Olsen and Olsen 1977) and DNA (Leonard et al. 2002; Vila et al. 1997), to be descended from Old World wolves, Paleoindian ancestors probably had domesticated dogs when they left Asia. Possible evidence of Paleoindian dogs has been reported from Tagua Tagua, Chile (Palma 1969).

It has been said (Meltzer 1995:26) that Thule people simply expanded their range into a familiar environment, for which they already possessed sophisticated technological and social adaptations; in contrast, Paleoindians had to adjust repeatedly to unfamiliar terrain and climate as they moved southward. However, this contrast may be overstated. The Arctic coastal regions, in fact, do not constitute a uniform environment; and Thule material culture and social organization changed rapidly as the migrants encountered new local conditions. Whale hunting techniques used in northern Alaska had to be altered in the eastern Arctic, and this resulted in a new settlement pattern (McGhee 1984:371). Localized adaptations, sometimes entailing abandonment of whaling and a shift to caribou, fish, and seal, emerged during the second phase of Thule expansion, before ca. A.D. 1200–1300 (McGhee 1984:371). These significant changes in basic subsistence patterns had been adopted less than 300 years after the Thule migration started. If Paleoindians were similarly flexible and opportunistic, we should not be surprised to find evidence of major subsistence reorientation only a few centuries after initial colonization; e.g., the marine-focused sites on the southern coast of Peru, recently dated to ca. 11,000–10,700 RCYBP (Sandweiss et al. 1998, 2000). In the case of Clovis, the migrants moved through late-Pleistocene environments that may have been less sharply differentiated than those of the Holocene, particularly with respect to faunal distributions. Paleontologists often note the association in fossil assemblages of animals that are not sympatric today (e.g., FAUNMAP Working Group 1996). More importantly, the well-documented Paleoindian focus on hunting of large migratory animals such as mammoths, bison, and horses would have permitted, indeed compelled, wide-ranging movement across environmental boundaries (Kelly and Todd 1988). Besides, if they stuck to the familiar mountain-edge environment as they moved south, Paleoindians could have gotten at least as far as New Mexico without having to change their lifeways; the Athapaskans did the same thing around A.D. 1400.

Finally, we have to consider the most important ostensible difference between Thule and Clovis. Thule folk were intruding into an environment that had already been occupied for more than 2,500 years; the resident Dorset population had been there for 1,500 years. Although Thule people evidently collected and sometimes imitated Dorset artifacts, there is no clear indication that they actually encountered living inhabitants in Dorset settlements; nor is there linguistic or cultural evidence of the absorption of Dorset people into Thule culture (Park 1993). Whether Thule expansion or environmental catastrophe was responsible, the Dorset people seem to have disappeared from the landscape abruptly and totally. In contrast to this Arctic situation, most of us have assumed, until recently, that Clovis people or their immediate ancestors entered a truly New World, with no humans in it. If Clovis was first, the Thule migration might be a less appropriate analogue than the expansion of the Arctic Small Tool Tradition into the same region centuries earlier (ca. 4200–3600 RCYBP), when the Arctic was an uninhabited *terra incognita* (Ellis 1998; Fiedel 1992:151; Maxwell 1976). That earlier migration was also very rapid; ASTT people may have settled in northern Labrador as early as 4000 RCYBP (Fitzhugh 1976).

The Thule-Clovis analogy is more appropriate if we accept the exiguous and debatable evidence (Fiedel 1999b, 2000a) from Monte Verde, Meadowcroft, and Cactus Hill, that Clovis people encountered precursor and/or contemporaneous but ethnically distinct populations. In that case, the hypothesized precursors become analogous to the Dorset culture. Perhaps pre-Clovis people, unable to cope with terminal-Pleistocene climatic oscillations, had died out by the time Clovis arrived (if one takes the reported dates at face value, a 5,000-year hiatus separates the "pre-Clovis" level at Cactus Hill from the Clovis level (ca. 15,000 RCYBP (18,000 CALYBP) vs. 10,900 RCYBP (12,900 CALYBP). Or Clovis people and their precursors competed for the same resources and habitats, and the intruders succeeded in displacing (or absorbing) the residents. If Clovis bands systematically tracked their predecessors, the upshot might have been a sort of leapfrogging from one prime habitat to the next that allowed them to fill the continent more rapidly than they could have done if they had explored virgin territory. If there was a precursor population already in place when Clovis expansion occurred, or if competing non-Clovis (e.g., Western Stemmed [Bryan 1991]) people were moving simultaneously across the land, the fluted points might be seen in a different light. Fluting may have assured secure fastening of the point to the spear or foreshaft; it may have taken on magical significance (Bradley 1993) as a crucial prelude to hunting success; but perhaps it also served an important social function, as a stylistic marker by which Clovis folk distinguished themselves from ethnically distinct competitors.

On the other hand, if Clovis people ventured into a landscape devoid of other humans, their rapid expansion may have resulted primarily from the quest for selected habitats. They may have been looking for lakes. Clovis migration appears to coincide with a widespread drought in the Southwest (probably corresponding to the warm spike evident in Greenland ice cores at the end of the Allerød, ca. 13,100 CALYBP). Potable water may have been a major determinant of their settlement pattern (Haynes et al. n.d.). Lakes would also have provided a suitable habitat for fish and waterfowl (Fiedel 2000b) and might also have been magnets for thirsty megafauna. Paleoindians could have followed the obvious tracks of mammoth herds from one such oasis to the next (G. Haynes 1991, 2002:210), perhaps moving several hundred miles with each shift of their base camp to a new hunting ground. Thus, there may be a similarity in the island-hopping of the early Polynesian Lapita people (ca. 3500 CALYBP) (Bellwood 1987), the leapfrog migration of Linearbandkeramik people between patches of arable soil in Central Europe (ca. 7000 CALYBP) (Fiedel and Anthony 1979, 2003), and Clovis migration between dispersed mammoth refugia, which would explain the seemingly explosive character of each of these population movements.

References Cited

Abbott, M. B., B. P. Finney, M. E. Edwards, and K. R. Kelts
2000 Lake-Level Reconstruction and Paleohydrology of Birch Lake, Central Alaska, Based on Seismic Reflection Curves and Core Transects. *Quaternary Research* 53(2):154–66.

Adovasio, J. M., J. Donahue, and R. Stuckenrath
1990 The Meadowcroft Rockshelter Radiocarbon Chronology 1975–1990. *American Antiquity* 55(2):348–54.

Anderson, D. G.
1995 Paleoindian Interaction Networks in the Eastern Woodlands. In *Native American Interaction: Multiscalar Analyses and Interpretations in the Eastern Woodlands*, edited by M. S. Nassaney and K. E. Sassaman, pp 1–26. University of Tennessee Press, Knoxville.

Arima, E. Y.
1984 Caribou Eskimo. In *Arctic, Handbook of North American Indians*, vol. 5, edited by D. Damas, pp. 397–414. Smithsonian Institution Press, Washington, D.C.

Balikci, A.
1984 Netsilik. In *Arctic, Handbook of North American Indians*, vol. 5, edited by D. Damas, pp. 415–30. Smithsonian Institution Press, Washington, D.C.

Bellwood, P.
1987 *The Polynesians: The Prehistory of an Island People.* Thames and Hudson, London.

Birdsell, J. B.
1968 Some Predictions for the Pleistocene Based on Equilibrium Systems among Recent Hunter-Gatherers. In *Man the Hunter*, edited by R. B. Lee and I. De Vore, pp. 229–40. Aldine, Chicago.

Bradley, B. A.
1993 Paleoindian Flaked Stone Technology in the North American High Plains. In *From Kostenki to Clovis: Upper Paleolithic-Paleoindian Adaptations*, edited by O. Soffer and N. D. Praslov, pp. 251–62. Plenum Press, New York.

Bryan, A. L.
1991 The Fluted-Point Tradition in the Americas—One of Several Adaptations to Late Pleistocene American Environments. In *Clovis: Origins and Adaptations*, edited by R. Bonnichsen and K. L. Turnmire, pp. 15–34. Center for the Study of the First Americans, Oregon State University, Corvallis.

Custer, J. F., and R. M. Stewart
1990 Environment, Analogy, and Early Paleoindian Economies in Northeastern North America. In *Research in Economic Anthropology*, edited by K. B. Tankersley and B. L. Isaac, pp. 303–22. Supplement No. 5, JAI Press, Greenwich, Connecticut.

Dikov, N. N.
1996 The Ushki sites, Kamchatka Peninsula. In *American Beginnings: The Prehistory and Palaeoecology of Beringia*, edited by F. H. West, pp. 244–50. University of Chicago Press, Chicago.

Dillehay, T. D.
1997 *Monte Verde, a Late Pleistocene Settlement in Chile, Volume 2, The Archaeological Context and Interpretation*. Smithsonian Institution Press, Washington, D.C.

Elias, S. A.
2000 Late Pleistocene Climates of Beringia, Based on Analysis of Fossil Beetles. *Quaternary Research* 53(2):229–35.

Ellis, C.
1998 The Fluted Point Tradition and the Arctic Small Tool Tradition: What's the Connection? Paper presented at the 31st Annual Chacmool Conference, University of Calgary, Alberta, November 1998.

Engelbrecht, W. and C. Seyfert
1994 Paleoindian Watercraft: Evidence and Implications. *North American Archaeologist* 15:221–34.

FAUNMAP Working Group
1996 Spatial Response of Mammals to Late Quaternary Environmental Fluctuations. *Science* 272:1601–06.

Ferring, C. R.
1995 The Late Quaternary Geology and Archaeology of the Aubrey Clovis Site, Texas: A Preliminary Report. In *Ancient Peoples and Landscapes*, edited by E. Johnson, pp. 273–81. Museum of Texas Tech University, Lubbock.

Fiedel, S. J.
1992 *Prehistory of the Americas*, second edition. Cambridge University Press, Cambridge.
1999a Older than We Thought: Implications of Corrected Dates for Paleoindians. *American Antiquity* 64 (1):95–116.
1999b Artifact Provenience at Monte Verde: Confusion and Contradictions. *Scientific American Discovering Archaeology* 1(6): Special Report 1–12.
2000a The Peopling of the New World: Present Evidence, New Theories, and Future Directions. *Journal of Archaeological Research* 8(1):39–103.
2000b Quacks in the Ice: Waterfowl, Paleoindians, and the Discovery of America. Paper presented at 65th annual meeting of Society for American Archaeology, Philadelphia, April 6.

Fiedel, S. J., and D. Anthony
1979 The Diffusion of the Early European Neolithic: A Reconsideration. Paper delivered at the Second Eastern European Archaeology (Hleb i Vino) Conference. University of Pennsylvania, Philadelphia.
2003 Deerslayers, Pathfinders and Icemen: Origins of the European Neolithic as Seen from the Frontier. In *Colonization of Unfamiliar Landscapes, the Archaeology of Adaptation,* edited by M. Rockman and J. Steele, pp. 144–68. Routledge, London and New York.

Fitzhugh, W.
1976 Paleoeskimo Occupations of the Labrador Coast. In Eastern Arctic Prehistory: Paleoeskimo Problems. *Memoirs of the Society for American Archaeology* 31, edited by M.S. Maxwell, pp. 103–18.

Flegenheimer, N. and M. Zárate
1997 Considerations on Radiocarbon and Calibrated Dates from Cerro La China and Cerro El Sombrero, Argentina. *Current Research in the Pleistocene* 14:27–28.

Harp. E., Jr.
1983 Pioneer Cultures of the Sub-Arctic and the Arctic. In *Ancient North Americans*, edited by J. D. Jennings, pp. 115–47. W. H. Freeman, San Francisco.

Haynes, C. V., Jr.
1991 Geoarchaeological and Paleohydrological Evidence for a Clovis-Age Drought in North America and its Bearing on Extinction. *Quaternary Research* 35:438–50.

Haynes, C. V., Jr., D. J. Stanford, M. Jodry, J. Dickenson, J. L. Montgomery, P. H. Shelley, I. Rovner, and G. A. Agogino
1999 A Clovis Well at the Type Site 11,500 BC: The Oldest Prehistoric Well in America. *Geoarchaeology* 14(5):455–70.

Haynes, G.
1991 *Mammoths, Mastodonts, and Elephants*. Cambridge University Press, Cambridge.
2002 *The Early Settlement of North America: The Clovis Era.* Cambridge University Press, Cambridge.

Helm, J.
1968 The Nature of Dogrib Socioterritorial Groups. In *Man the Hunter*, edited by R. B. Lee and Irven DeVore, pp. 118–25. Aldine, Chicago.

Holmes, C. E.
1996 Broken Mammoth. In *American Beginnings: The Prehistory and Palaeoecology of Beringia*, edited by F. H. West, pp. 312–17. University of Chicago Press, Chicago.
1998 New Data Pertaining to Swan Point, the Oldest Microblade Site Known in Alaska. *Current Research in the Pleistocene* 15:21–22.

Holmes, C. E.,R. Vander Hoek, and T. E. Dilley
1996 Swan Point. In *American Beginnings: The Prehistory and Palaeoecology of Beringia*, edited by F. H. West, pp. 319–323. University of Chicago Press, Chicago.

Jodry, M.
1999 Alternate Routes beyond the Horizon: Certitude, Suspended Disbelief, and Surprise. Paper presented at the Clovis and Beyond conference, Santa Fe, October 30.

Kelly, R. L., and L. C. Todd
1988 Coming into the Country: Early Paleoindian Hunting and Mobility. *American Antiquity* 53(2):231–44.

Kuzmin, Y. V., A. J. T. Jull, Z. S. Lapshina, and V. E. Medvedev
1997 Radiocarbon AMS Dating of the Ancient Sites with Earliest Pottery from the Russian Far East. *Nuclear Instruments and Methods in Physics Research B* 123:496–97.

Lahren, L., and R. Bonnichsen
1974 Bone Foreshafts from a Clovis Burial in Southwestern Montana. *Science* 186:147–50.

Leonard, J. A., R. K. Wayne, J. Wheeler, R. Valadez, S. Guillen, and C. Vila
2002 Ancient DNA Evidence for Old World Origin of New World Dogs. *Science* 298:1613–16.

Lynch, T. F.
1984 The Paleo-Indians. In *Ancient South Americans*, edited by J. D. Jennings, pp. 87–138. W. H. Freeman, San Francisco.

MacDonald, D. H.
1998 Subsistence, Sex and Cultural Transmission in Folsom Culture. *Journal of Anthropological Archaeology* 17: 217–39.

Madsen, D. B., and D. Rhode, editors
1994 *Across the West: Human Population Movement and the Expansion of the Numa*. University of Utah Press, Salt Lake City.

Mason, O. K.
1998 The Contest between the Ipiutak, Old Bering Sea, and Birnirk Polities and the Origin of Whaling during the First Millennium A.D. along Bering Strait. *Journal of Anthropological Archaeology* 17:240–325.

Maxwell, M. S. (editor)
1976 *Eastern Arctic Prehistory: Paleoeskimo Problems. Memoirs of the Society for American Archaeology* 31.

McAvoy, J. M., and L. D. McAvoy
1997 *Archaeological Investigations of Site 44SX202, Cactus Hill, Sussex County, Virginia.* Virginia Department of Historic Resources, Research Report Series No. 8, Richmond.

McGhee, R.
1984 Thule Prehistory of Canada. In *Arctic, Handbook of North American Indians,* vol. 5, edited by D. Damas, pp. 369–76. Smithsonian Institution Press, Washington, D.C.

Meltzer, D. J.
1995 Clocking the First Americans. *Annual Review of Anthropology* 24:21–45.

Mochanov, Y. A., and S. A. Fedoseeva
1996 Berelekh, Allakhovsk Region. In *American Beginnings: The Prehistory and Palaeoecology of Beringia,* edited by F. H. West, pp. 218–21. University of Chicago Press, Chicago.

Morrow, J. E., and T. A. Morrow
1999 Geographic Variation in Fluted Projectile Points: A Hemispheric Perspective. *American Antiquity* 64(2):215–31.

Nami, H. G.
1994 Reseña Sobre los Avances de la Arqueología Finipleistocénica del Extremo Sur de Sudamérica. *Revista Chungara* 26(2):145–63.
1996 New Assessments of Early Human Occupations in the Southern Cone. In *Prehistoric Mongoloid Dispersals,* edited by T. Akazawa and E. J. E. Szathmáry, pp. 256–69. Oxford University Press, Oxford.
2000 Technological Comments on Some Paleoindian Lithic Artifacts from Ilaló, Ecuador. *Current Research in the Pleistocene* 17:104–108.

Olsen, S. J., and J. W. Olsen
1977 The Chinese Wolf Ancestor of New World Dogs. *Science* 197(4303):533–35.

Opler, M. E.
1983 The Apachean Culture Pattern and Its Origins. In *Southwest, Handbook of North American Indians,* vol. 10, edited by A. Ortiz, pp. 368–92. Smithsonian Institution Press, Washington, D.C.

Palma, J.
1969 El Sitio de Tagua-Tagua en el Ambito Paleo-Americano. *Actos del V Congreso Nacional de Arqueologia,* Museo Arqueológico de La Serena, Chile, pp. 315–25.

Park, R. W.
1993 The Dorset-Thule Succession in Arctic North America: Assessing Claims for Culture Contact. *American Antiquity* 58(2):203–34.

Pearson, G. A.
1998 Pan-American Paleoindian Dispersals as Seen through the Lithic Reduction Strategies and Tool Manufacturing Techniques at the Guardiria Site, Turrialba Valley, Costa Rica. Paper presented at 63rd annual meeting of Society for American Archaeology, Seattle, March 26.

Price, T. D.
1991 The View from Europe: Concepts and Questions about Terminal Pleistocene Societies. In *The First Americans: Search and Research,* edited by T. D. Dillehay and D. J. Meltzer, pp. 185–208. CRC Press, Boca Raton.

Ray, D. J.
1984 Bering Strait Eskimo. In *Arctic, Handbook of North American Indians,* vol. 5, edited by D. Damas, pp. 285–303. Smithsonian Institution Press, Washington, D.C.

Rogers, E. S., and E. Leacock
1981 Montagnais-Naskapi. In *Subarctic, vol. 6, Handbook of North American Indians,* edited by J. Helm, pp. 169–89. Smithsonian Institution, Washington D.C.

Sandweiss, D. H., H. McInnis, R. L. Burger, A. Cano, B. Ojeda, R. Paredes, M. del Carmen Sandweiss, and M. R. Glascock
1998 Quebrada Jaguay: Early South American Maritime Adaptations. *Science* 281(5384):1830–32.

Sandweiss, D. H., B. Tanner, D. Sanger, F. Andrus and D. Piperno
2000 Paleoindian-Age Domestic Structure at a Peruvian Fishing Site. Paper presented at 65th annual meeting of Society for American Archaeology, Philadelphia, April 6.

Sellars, V. S.
1999 The Accuracy of AMS ^{14}C Dates on Human Bone and Associated Vertebrate Remains—Determining Chronologies for Late Pleistocene Human Occupation of the New World. Poster presented at "Clovis and Beyond" conference, Santa Fe.

Sheehan, G.
1985 Whaling as an Organizing Focus in Northwestern Alaskan Eskimo Societies. *In Prehistoric Hunter-Gatherers: The Emergence of Cultural Complexity,* edited by T. D. Price and J. A. Brown, pp. 123–54. Academic Press, San Diego.

Storck, P. J.
1991 Imperialists without a State: The Cultural Dynamics of Early Paleoindian Colonization as Seen from the Great Lakes Region. In *Clovis: Origins and Adaptations,* edited by R. Bonnichsen and K. L. Turnmire, pp. 153–62. Center for the Study of the First Americans, Oregon State University, Corvallis.

Vila, C., P. Savolainen, J. E. Maldonado, I. R. Amorim, J. E. Rice, R. L. Honeycutt, K. A. Crandall, J. Lundeberg, and R. K. Wayne
1997 Multiple and Ancient Origins of the Domestic Dog. *Science* 276(5319):1687–89.

West, F. H., editor
1996 *American Beginnings: The Prehistory and Palaeoecology of Beringia.* University of Chicago Press, Chicago.

Whitley, D. S., and R. I. Dorn
1993 New Perspectives on the Clovis vs. Pre-Clovis Controversy. *American Antiquity* 58(4):626–47.

Wobst, M.
1976 Locational Relationships in Palaeolithic Society. *Journal of Human Evolution* 5:49–58.

Yesner, D. R.
1998 Origins and Development of Maritime Adaptations in the Northwest Pacific Region of North America: A Zooarchaeological Perspective. *Arctic Anthropology 35* (1):204–22.

Evidence for Pre-Clovis Sites in the Eastern United States

Albert C. Goodyear

Truth is the daughter of time.
–John Trapp, 17th-century Puritan writer

Abstract

Over the past 25 years, a number of archaeological sites in eastern North America have manifested evidence of human occupations dating earlier than 11,500 RCYBP. These sites include Meadowcroft Rockshelter, Penn.; Cactus Hill, Va.; Saltville-2, Va.; and Topper, S.C. Except for Topper, pre-Clovis (earlier than 11,500 RCYBP) remains were encountered incidental to conducting normal research. Topper was intentionally tested for the possibility of pre-Clovis remains. This paper is a brief review of the geological contexts, dating, and artifactual evidence of these sites. Collectively, they indicate a late-glacial time range of approximately 12,000 to 16,000 RCYBP. Technologically, Meadowcroft Rockshelter and Cactus Hill are the most similar, with evidence of bifacial points, unifaces, and prismatic blades. Although probable bone and ivory artifacts are present, lithics at Saltville- 2 are expedient and minimal, making comparisons difficult. Topper, which is a chert quarry, is distinctive in that it has no evidence of bifaces and is dominated by small flake tools with an emphasis on burin-like tools. More geoarchaeological fieldwork is needed to target landforms that possess sediments deposited from 18,000 to 12,000 RCYBP. The Southeast may be a good place to prospect for these sites, given its milder climate during late-glacial times.

The purpose of this paper is to present briefly archaeological evidence of human occupation of the eastern United States prior to 11,500 RCYBP, the conventional pre-Clovis temporal boundary. Four sites are reviewed for their artifacts, geological contexts, and dating. While questions, reservations, and rejections exist within the profession concerning some or all of these sites, nevertheless it is believed that they sufficiently meet these criteria and thus constitute evidence.

Director of Allendale Paleoindian Expedition, S.C. Institute of Archaeology and Anthropology, University of South Carolina, 1321 Pendleton St., Columbia, SC 29208-0071; e-mail: goodyear@sc.edu

Meadowcroft Rockshelter, Pennsylvania

The most long-standing case for archaeological evidence of pre-Clovis (earlier than 11,500 RCYBP) humans in the eastern United States is the famous Meadowcroft Rockshelter, located about 48 km southwest of Pittsburgh, Pennsylvania (Figure 1). Meadowcroft Rockshelter is within the Cross Creek drainage, a tributary of the Ohio River, located about 12 km away. The site is situated in a sandstone rockshelter with an area of about 65 m^2 under the overhang. Altogether, 11 distinct strata have been defined, with Stratum I (culturally sterile shale) at the bottom followed by artifact-bearing strata II–XI. Cultural material was recovered within a 4.6-m-deep deposit formed by attrition, roof fall, and sheet wash accumulation. Within this 4.6-m deposit, a remarkably complete string of 52 ^{14}C dates in nearly perfect chronological order has been obtained, spanning nearly 20 millennia (Adovasio et al. 1998).

Multidisciplinary excavations at the rockshelter, directed by James M. Adovasio in cooperation with other scientists and sponsored by the University of Pittsburgh, started in 1973 and continued variously through the 1970s and 1980s; the last major excavation, sponsored by Mercyhurst College with which Adovasio is associated, was conducted in 1994–1995. Some 75 publications have been produced for the site, ranging from articles to monographs (Adovasio et al. 1999).

Evidence of pre-Clovis occupation lies within the lower and middle portions of the lithostratigraphic unit called IIA. Enough lithic artifacts were recovered to define the Miller complex. This complex consists of thin bifaces, including one lanceolate point, the Miller Lanceolate (Figure 2); small prismatic blades; retouched flake tools and blades (Figure 3); and debitage related to late-stage core and biface reduction and tool kit maintenance. Overall, the stone tools and debitage constitute a highly curated tool kit. It is significant that the lithic raw materials are representative of subsequent fluted-point tool kits, including exotic cherts

Figure 1. Sites in the eastern United States with evidence of pre-11,500 RCYBP occupations.

such as found in Flint Ridge, Ohio, Kanawha cherts of West Virginia, Pennsylvania jaspers, and local Monongahela chert (Adovasio et al. 1999). The Miller complex is further defined by surveys done in the Cross Creek watershed, where other lanceolate points, small prismatic blades, and small polyhedral blade cores have been recovered. According to Adovasio et al. (1999:418), this complex has a Eurasiatic and Siberian appearance. These authors also note that small blades and polyhedral cores are absent from subsequent Paleoindian fluted-point assemblages in this region, reinforcing the technological distinctiveness of the Miller complex.

The six oldest radiocarbon dates with artifacts unquestionably associated range in age from 12,800 ± 870 RCYBP to 16,175 ± 975 RCYBP (Adovasio et al. 1999:Figure 1). Given the large standard deviations of some of the dates, a conservative estimate would be from 12,000 to 15,000 RCYBP (Adovasio et al. 1990:352). These dates were obtained on charcoal from features such as fire pits, fire floors, and other charcoal concentrations. One ^{14}C date of 19,600 ± 2400 RCYBP was obtained on carbonized simple-plaited basketry. Taking a conservative view, at two standard deviations, this date would come forward in time to 14,800 RCYBP, in line with the other pre-Clovis dates above.

Because of its antiquity and non-Clovis lithic assemblage, Meadowcroft had a dramatic impact on New World archaeology, one that has been surrounded by controversy ever since. Criticisms have focused on the lack of Pleistocene flora and fauna, the apparent absence of the Pleistocene-Holocene transition in the stratigraphy, and possible contamination of the ^{14}C dates. The presence in the pre-Clovis levels of bona fide artifacts as well as hearths and other human-related fire features has not been contested. With better paleoenvironmental and geoarchaeological data available since the 1970s for the late-glacial environments of eastern North America, criticism of the plant and animal remains and stratigraphy has abated (Adovasio et al. 1998). The most persistent questions address possible contamination of the ^{14}C dates by particulate and soluble humates from coal. Despite detailed microscopic searches, no evidence of particulate contamination has been detected (Adovasio and Carlisle 1988:713; Adovasio et al. 1990:351). As for soluble humates contaminating the lower levels, a recent soil micromorphology study by Goldberg and Arpin (1999:340) concludes:

> We see no evidence of groundwater saturation of any strata nor do we see evidence of any other mechanisms by which particulate or non-particulate contamination could have been introduced into the sediments in general and into the charcoal samples in particular.

After 25 years since its first publication, Meadowcroft Rockshelter appears to be what the investigators originally claimed. In recent years other sites have joined a growing group of sites that also manifest evidence of people being in North America before Clovis.

SV-2, Saltville, Virginia

Saltville Valley is a well-known late-Quaternary fossil locality situated in the Valley and Ridge Province of the Appalachian Mountains of southwest Virginia (Figure 1). Paleontological and ar-

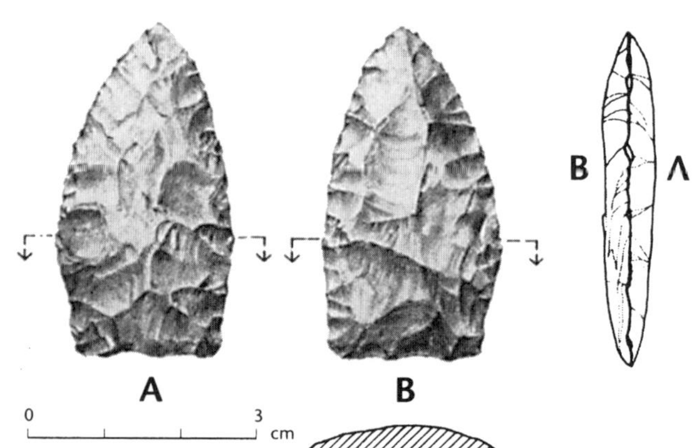

Figure 2. Miller Lanceolate projectile point from Meadowcroft Rockshelter (36WH297). By S. Patricia, courtesy of J. M. Adovasio, Mercyhurst Archaeological Institute, Erie, Pennsylvania.

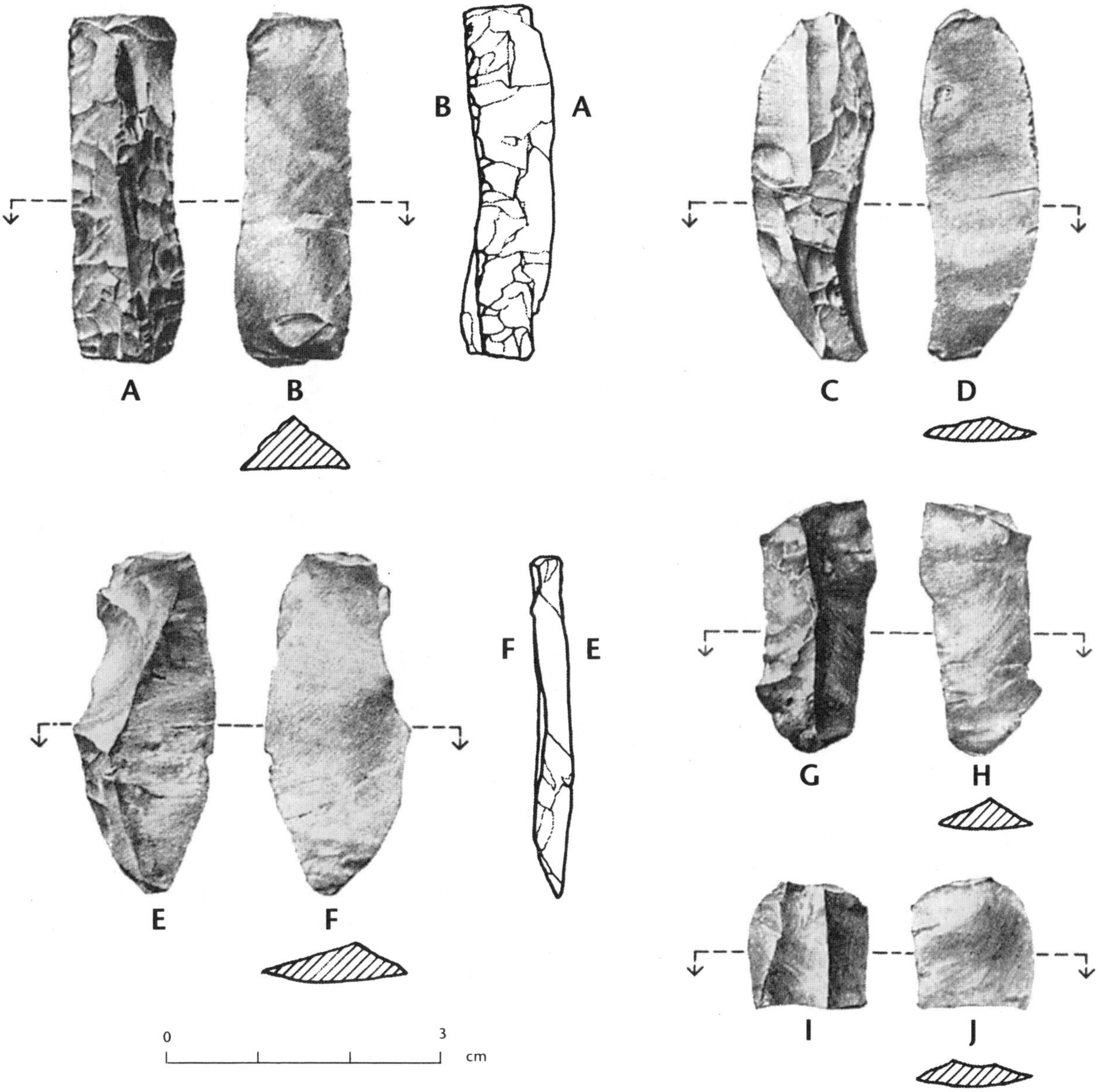

Figure 3. Examples of small prismatic blades from Meadowcroft Rockshelter (36WH297). By S. Patricia, courtesy of J. M. Adovasio, Mercyhurst Archaeological Institute, Erie, Pennsylvania.

chaeological excavations have been conducted there over the past 20 years by paleogeographer Jerry McDonald, a research associate with the Smithsonian Institution and the Virginia Museum of Natural History. The valley has an unusual history: about 13,500 RCYBP the river became a lake and the valley bottom was sealed with mud, yielding excellent preservation of late-glacial plant and animal remains. Excavations have revealed a detailed, finely resolved biostratigraphic record well supported by ^{14}C dates back to 15,000 RCYBP (McDonald 1995, 2000; McDonald and Kay 1999:196; Wisner 1996).

The critical area of interest for archaeology at SV-2 (44SM37) lies within an approximately 1-m-thick alluvial, colluvial, and lacustrine deposit spanning some 5,000 radiocarbon years, which overlies a pedestal of resistant bedrock. In the lower part of this deposit, at least three discrete radiocarbon-dated horizons exist, ranging in age from 14,510 to 13,000 RCYBP, each bearing archaeological evidence of human occupation.

The oldest horizon, which extends over most of the bedrock pedestal, is integrated by skeletal and dental remains of what may be a single proboscidean individual. This horizon is dated by a single AMS bone collagen date of 14,510 ± 80 RCYBP on what is considered to be a bone tool (McDonald and Kay 1999:196). The bone is a tibia, probably of musk ox (*Bootherium bombifrons*). This specimen is intriguing because of its percussion shaping and microwear patterns (Figure 4). Based on his independent microwear study, Marvin Kay (pers. comm. 1999) is convinced the "tibia was intentionally fractured, was hand-held, and was used as a mattock hide beaming tool." Other objects and unusual spatial concentrations of bone were also found in the lowest horizon in a distribution that implies human agency.

The unusual condition and distribution of proboscidean bone suggests to McDonald that a single mastodon was butchered and burned (McDonald 2000). Also found were two hand-size sandstone objects that could have served as hand axes or wedges, together with what appears to be an artificially grooved piece of bone broken after the groove was made. A partially worn chert flake (Figure 5A) and a thin chunk of chert with flaking and possible wear damage from use as a wedge (M. Kay, pers. comm. 1999) were also associated with the mastodon layer.

A middle horizon bearing evidence of human occupation consists of two feature-like arrangements situated on or within a stratigraphically discrete fluvial sand layer. The first is an unlikely assortment of large pebbles and cobbles of lithics that would be suitable for use as tools or the production of stone tools. These materials were found in a small depression about 15 cm deep and 70 cm in diameter. This depression is thought to be natural, since other similar depressions were observed set into the surface of the sand layer. This shallow basin was rapidly filled by fluvial puddling, with at least five stages evident. All objects thought to be related to humans lie in or partly within the upper stratum of the fill. In addition to fish skull bones and a piece of charcoal, 12 pieces of microdebitage were found, 3 of which were morphologically similar to flakes of bifacial retouch. Wood twigs found in the upper stratum of this depression were radiocarbon-dated at 13,950 ± 70 RCYBP (McDonald 2000). The second featurelike manifestation consists of seven concretions and one prismatic column of weathered bedrock found in an unusual orientation in a maroon mud-filled depression within the sand lense. Five of the concretions and the section of weathered bedrock are elongate; all are oriented vertically with their upper ends at the same depth. Moreover, the five concretions were formed of gray mud, which strongly indicates that they were intrusive into the maroon mud. McDonald (2000) is of the opinion that these geologically anomalous objects indicate intentional human placement.

The latest horizon is represented by a single unusual middenlike concentration of bivalves at least 2 m in one dimension, which was found within a rill eroded into the colluvium. The middenlike mass contains over 200 shells of clam, over 500 pieces of small vertebrate bones and teeth, charcoal, and 125 pieces of microdebitage, including a small number of chert flakes. Some of the flakes appear to be biface reduction flakes (Figure 5C). Based on stratigraphy, McDonald (2000) interprets the date of the faunal concentration at 13,500 to 13,000 RCYBP; it was apparently deposited during a low-water stage that occurred soon after the lake formed. The faunal contents suggest people were harvesting burrowing clams and collecting fish and amphibians in ponds during times of lowered lake waters.

The compelling aspect of the SV-2 pre-Clovis evidence resides in the two unusually modified bone objects and the chert flakes. These bone objects and the unusual spatial array of reduced proboscidean bones have not been observed in the Saltville Valley paleontological excavations except on the bedrock pedestal. The chert material is exotic to the Saltville Valley. Microdebitage has not been found in river channel sediments, nor are there naturally occurring tool-quality chert cobbles from which the chert flakes could have originated. SV- 2 possesses extraordinary geological stratigraphy, with temporally discrete depositional units of high integrity and resolution. The corresponding ^{14}C dates also bear witness to the integrity of the stratigraphy.

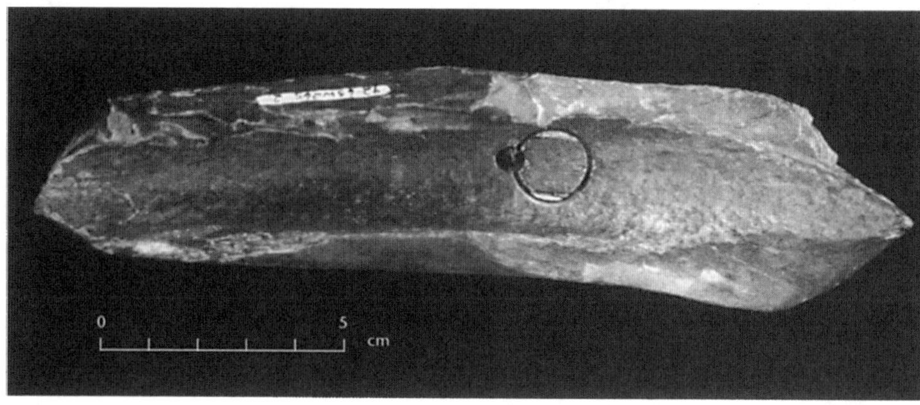

Figure 4. Tibia tool (VMNH 721) from lowest pre-Clovis horizon at SV-2 (44SM37). Photo courtesy of Jerry McDonald.

Cactus Hill, Virginia

Cactus Hill (44SX202) is a stratified multicomponent site in a sand dune approximately 1.8 m thick overlooking the Nottoway River in the interior coastal plain of Virginia (Figure 1). The site

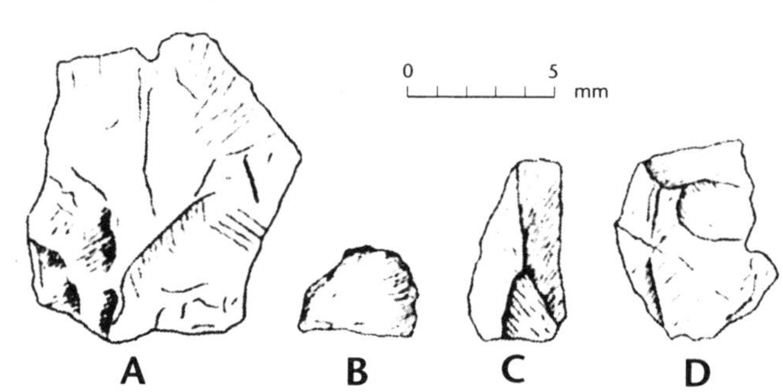

Figure 5. Microflakes of probable human origin from pre-Clovis horizons at SV-2 (44SM37). Courtesy of Jerry McDonald.

has been excavated concurrently by two groups, one led by Joseph McAvoy of the Nottoway River Survey, the other by Michael Johnson, an archaeologist with Fairfax County, Virginia, working with the Archaeological Society of Virginia. The primary publication on the site is authored by Joseph and Lynn McAvoy (McAvoy and McAvoy 1997), and Johnson published a report in the same year (Johnson 1997).

Based on several hundred square meters of excavation, a well-documented and -dated archaeological sequence has been established, beginning with Clovis and ending with the 18th century. This sequence is based on numerous projectile points and other diagnostic stone tools in correct stratigraphic order, plus ^{14}C-dated features. This conventional sequence is contained within the upper 1 m of the site. At about a meter below surface, a bona fide Clovis occupation has been recovered, including several fluted points, typical Paleoindian unifaces, and prismatic blades. One Clovis-age ^{14}C date of 10,920 ± 250 RCYBP on hard Southern pine charcoal has been obtained from a hearth. The lithic raw materials associated with the Clovis horizon consist of quartzite from the Nottoway River and familiar Paleoindian cherts, such as that from the famous Williamson site located nearby.

In the 1993 excavations, about 7 cm below a Clovis surface that produced two fluted points, a hearth-like concentration of charcoal was found that contained seven quartzite flakes and three quartzite prismatic blades (Figure 6) (McAvoy and McAvoy 1997:103). White-pine charcoal from this concentration yielded a ^{14}C date of 15,070 ± 70 RCYBP. In three other areas of the site, prismatic blades of quartzite were also found underneath what are suspected to be Clovis surfaces. During the 1996 field season, which was undertaken to determine whether additional sub-Clovis artifacts could be found, six more lithic clusters were found consisting of quartzite prismatic blades. Another hearth-like concentration of charcoal from one of these clusters produced a ^{14}C date of 16,670 ± 730 RCYBP (McAvoy 1997). Including all the quartzite artifact clusters found below the Clovis level, McAvoy (1997) believes two separate sub-Clovis lithic horizons are present; the lower horizon consists of large prismatic blades, followed by a zone of smaller blades. Two thin, basally thinned trianguloid-to-lanceolate bifaces that were also found may be associated with the upper smaller blade clusters (Figure 6, A and B).

Questions about the integrity of the intricate but shallow stratigraphy led to a follow-up field and laboratory study in 1997–1998 (McAvoy et al. 2000). Sedimentological analysis showed that all the sands are eolian in origin; the dune is a "border source" type known to have formed during full- and late-glacial periods along the Atlantic coastal plain (cf. Ivester et al. 2001). Additional charcoal samples were sought in the dune below the artifact-bearing levels to determine whether ancient natural charcoal sources may have contaminated the site. Even though flotation was used, very little charcoal adequate for radiocarbon dating was found in the lower portion of the dune. One sample, collected about 2 cm below a hearth, dated 16,940 ± 50 RCYBP, very close in age to the previous date of 16,670 ± 730 RCYBP from charcoal also collected from a hearth. About 70 cm below this hearth, near the base of the dune, a single piece of conifer charcoal dated 19,700 ± 130 RCYBP. This charcoal, which is well below the lowest artifact zone, is thought to be from a forest fire.

In sum, Cactus Hill has a number of archaeological and stratigraphic characteristics that lend credibility to a pre-11,500 RCYBP human occupation. A well-documented Clovis occupation is present, including a Clovis-age ^{14}C date. At 10–15 cm below this zone in several places in the site, a distinctive quartzite industry is found, typified by prismatic blades, polyhedral cores, and trianguloid-to-lanceolate bifaces. Three ^{14}C dates from late-glacial plant remains recovered from hearths range from 15,000 to 17,000 RCYBP. Radiocarbon dating of known arboreals, sedimentology, and optically stimulated luminescence (OSL) dating of quartz sands all show an in-place stratigraphic sequence (McAvoy et al. 2000).

Topper Site, South Carolina

The Topper site, 38AL23, is a stratified multicomponent site situated on a Pleistocene terrace of the Savannah River in Allendale County, South Carolina (Figure 1). The site complex includes a terrestrial chert outcrop and quarry (38AL139) exposed on the hillside and in an adjacent creek bed, which is mantled by sands (Goodyear 1999a; Goodyear and Charles 1984). The artifact-bearing sediments on the terrace below are represented by two different geologic deposits. The upper 1.0 to 1.4 m of sands, which are colluvial (slopewash) in origin, contain an essentially complete Holocene archaeological sequence beginning with Clovis and ending with the 18th century. Below that, continuing to about 2.2 m below surface, are Pleistocene-age sands that are alluvial in origin. These dual-deposit sands more than 2 m thick lie unconformably over a scoured gray silty clay Pleistocene terrace, which is about 2 m thick. The pre-Clovis lithic assemblage lies on this Pleistocene terrace and within the overlying alluvial sands (Figure 7, 2d). The pre-Clovis-age levels have been excavated yearly since 1998, the year of discovery. As of the 2002 season, a total area of 146 m^2 of the pre-Clovis occupation had been excavated.

The upper colluvial artifact-bearing sands are dated by diagnostic prehistoric ceramics and projectile points. Time-sensitive artifacts are in order by depth, indicating a gentle burial of the Holocene record with minimal human and natural disturbances. Features with charcoal associated are rare. Small dispersed charcoal fragments present in the upper meter have been translocated down the profile through bioturbation and are thus unreliable for ^{14}C dating. Clovis has been recognized at the site by the presence of typical transversely flaked "rowboat"-shaped bifaces (cf. Morrow 1996), with end thinning or early-stage fluting (Goodyear 2000:Fig. 2). The base of one Clovis point has been excavated in its proper position as well as a possible second one (Goodyear and Steffy 2003). OSL dating at the base of the colluvium produced a date of 13,500 ± 1000 CALYBP (Forman 2002, 2003), close to the accepted date of 13,000 CALYBP of Clovis (Taylor et al. 1996).

Proceeding downward (Figure 7), the contact of the colluvium (3b1) and the alluvial sand (2d) below has been dated by OSL at 15,200 ± 1500 CALYBP (Forman 2002, 2003). Below

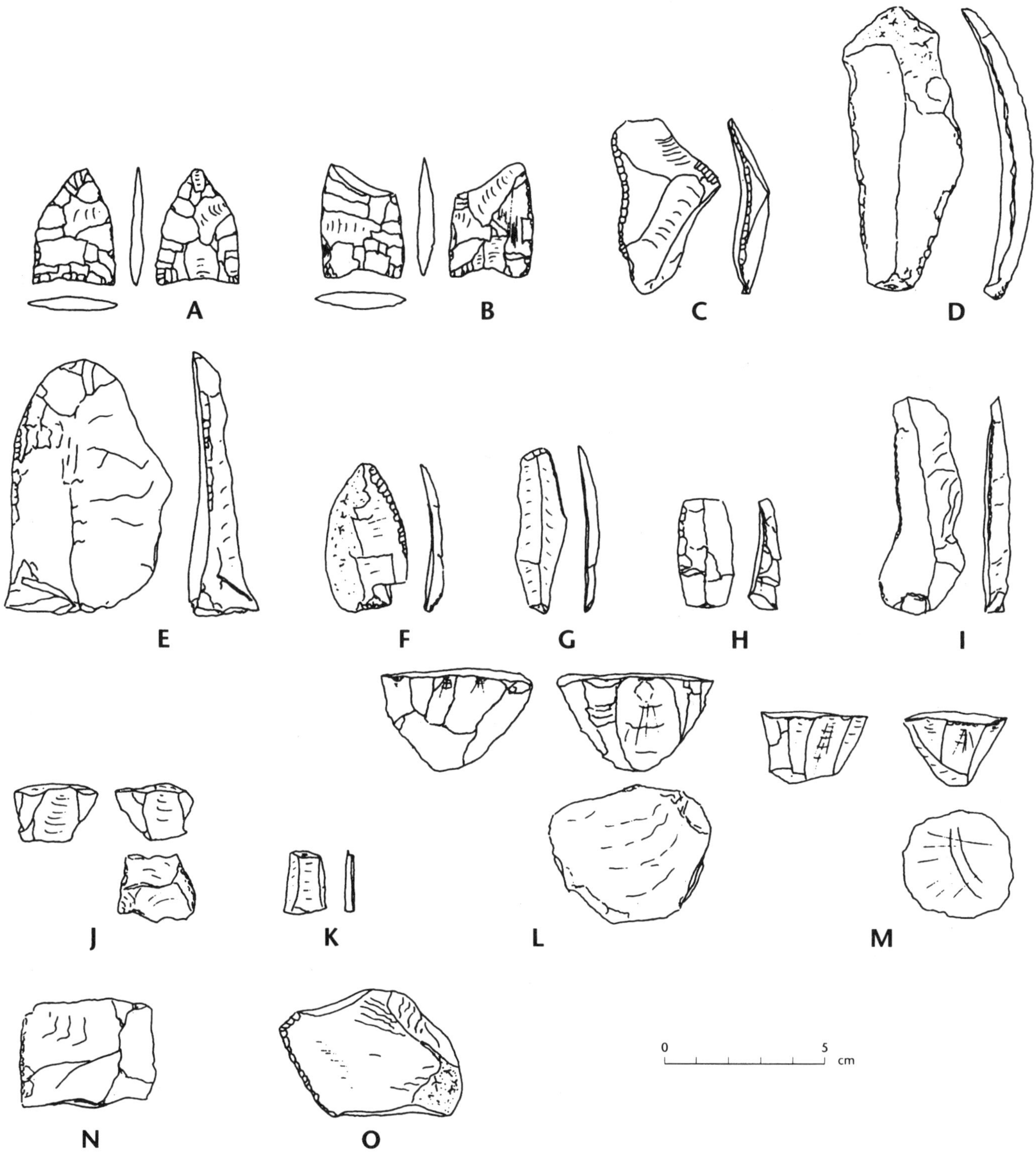

Figure 6. Various projectile points, prismatic blades, blade cores, and utilized flakes from pre-Clovis levels from Cactus Hill (44SX202). Courtesy of Joseph McAvoy.

this, in the upslope area of the site, lies a paleosol (3a) formed in colluvium that separates Clovis from the Pleistocene alluvial sands below. Project soil morphologist John Foss (pers. comm.) estimates it would take 2000 to 4000 years for this soil to develop (cf. Foss and Segovia 1984). These data indicate that the pre-Clovis artifact-bearing alluvial sands below are at least 16,000 to 20,000 years old. Since no organic materials suitable for radiocarbon dating have been observed in the Pleistocene sediments, we cannot fix a maximum time constraint for the pre-Clovis occupation. Underneath the Pleistocene terrace, infinite ^{14}C dates in excess of 50,000 yr B.P. have been obtained on plant remains (Figure 7:2a).

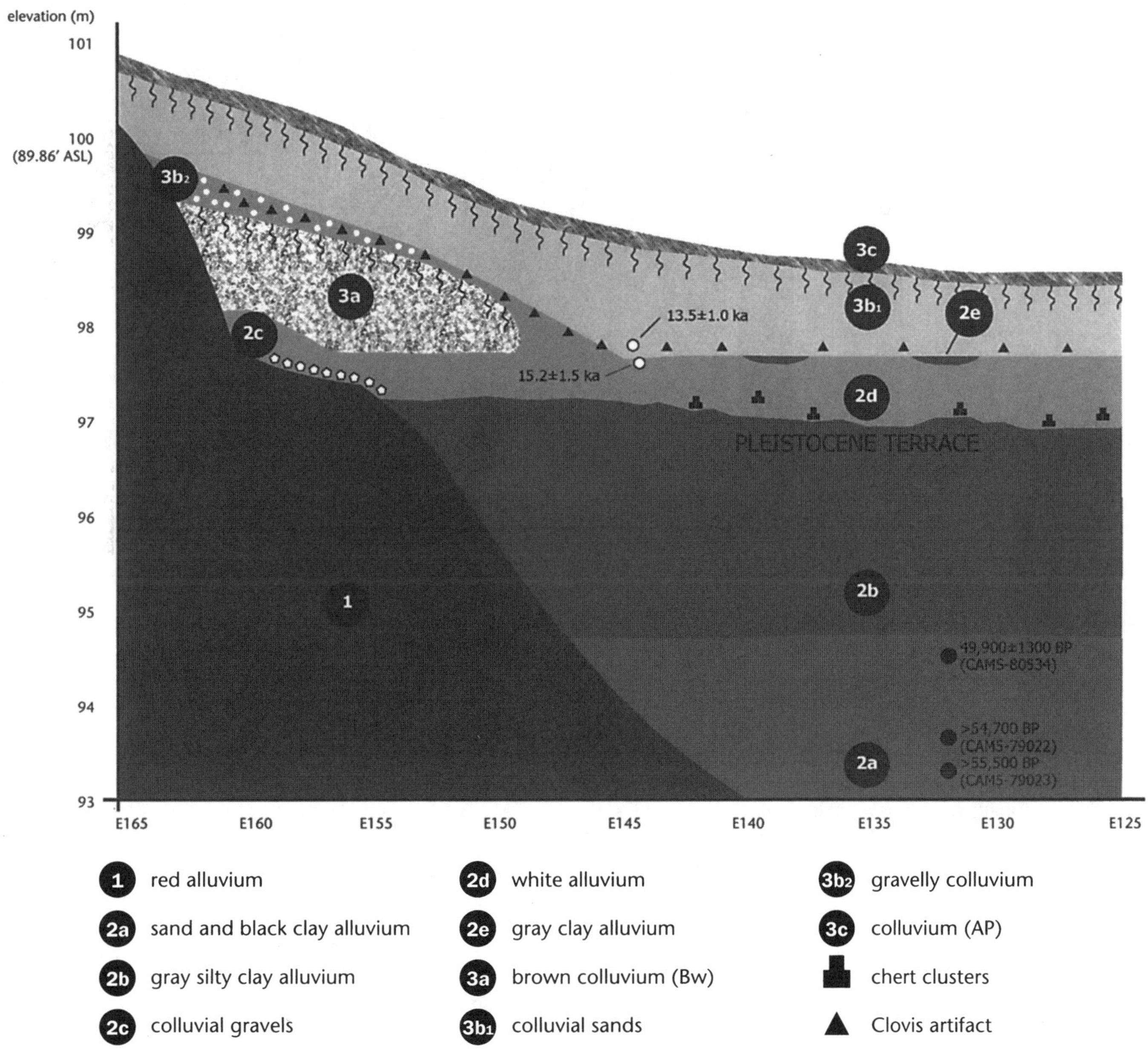

Figure 7. Stratigraphic profile of the Topper site showing relationships of geological units and archaeological components. (Based on Michael R. Waters et al. 2003.)

The unusual chert artifacts (Figure 8) said to be pre-Clovis occur stratigraphically primarily in the lower half of the Pleistocene sands (Figure 7, 2d). In one area of the site these lithics are associated with what appear to be loci where chert cobbles were processed. These concentrations (Figure 9) are found on the terrace, away from the natural source of the chert cobbles at the base of the hill. They appear as clusters of broken chert, some lying in discrete piles on common surfaces (Figure 9). They are neither fluvially formed nor deformed by water action. Because they have no negative bulbs, the broken chert cobbles are interpreted as cores made by smashing, probably one cobble against another, since no hammerstones large enough to break these cobbles have been recovered thus far. Small quartz cobbles (less than 8 cm) are present on the hillside associated with older alluvial deposits. These were also gathered and used as hammerstones and cores for quartz flake tools. The fractured chert cobbles have marked force lines in multiple directions with hard terminations often resulting in hinges. Some have been retouched bifacially and unifacially on their margins for use as cobble tools such as choppers and spokeshaves. Numerous small flakes (less than 2 cm) recovered among these chert clusters, with striking platforms and bulbs of force, are a by-product of retouching cobbles for tools as well as uniface production on flakes.

Flakes created by smashing the cobbles were modified into a variety of small tools, many of which may even be described as microlithic (Figure 8). A few steeply chipped unifaces have been found, some retouched on all edges (Figure 8I), others on the end or side of a flake. Some flakes have been unifacially retouched

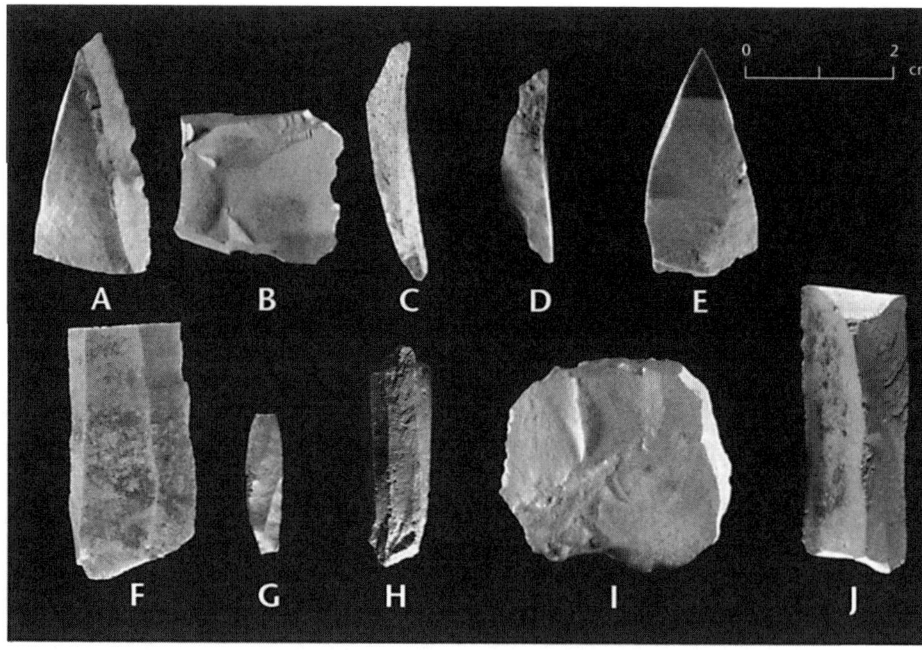

Figure 8. Lithic artifacts from the pre-Clovis occupation at the Topper site; **A,B,E,** bend-break tools; **C,D,** bend-break spalls; **F,G,** blades; **H,** possible microblade core; **I,** scraper; **J,** blade-like tool. (SCIAA-USC photo by Daryl P. Miller.)

to form spokeshaves. No bifaces or the distinctive reduction flakes from biface manufacture have been recovered. Numerous utilized flakes are present as well as flakes that have been modified by the bend-break method of fracture. The bend-break pieces, numbering in the hundreds, appear to have been created for burin or chisel use (Figure 8A,B,E). Bend-break manufacturing yields trihedral and quadrahedral spalls that mimic burin spalls (Figure 8C,D). These spalls do not have striking platforms or bulbs of force created by burinating a flake. Furthermore, although the bend-break pieces morphologically resemble classic burins (e.g., Figure 8A,E), no burins have been recovered with the diagnostic negative bulb. Such burin-like implements are often classified as expedient or psuedo-burins in Holocene-age assemblages in North America (cf. Crabtree 1973:49). However, bend-break tools on flakes are known for Clovis (Ferring 2001:155–156; McAvoy and McAvoy 2003:195) and for Folsom, especially the radial-break type (Root et al. 1999:146–151). Topper is unusual, however, in that bend-break artifacts constitute a primary component of the technology.

Blades and even bladelets, although few in number, have been found at Topper. Figure 8F is a midsection of a well-made prismatic blade with three prior detachments. A few small blades have been found, notably Figure 8G, which is a whole blade with a bulb of force and three parallel blade scars. Figure 8H may be a microblade core, although what would be the platform end is missing and the chert is very weathered. Figure 8I, a scraper chipped around its circumference, gives the strong impression of being a core tablet from rejuvenating a small blade core. A detailed analysis will be necessary to segregate what may be small blades from the numerous bend break spalls.

The chert artifacts from both the pre-Clovis and Clovis levels at Topper are heavily weathered, a characteristic widely noted for Tertiary-age South Atlantic coastal plain cherts of marine origin (Clark and Purdy 1979; Hurst and Kelly 1961; Upchurch 1984). Chemical weathering can be so severe that in some cases superficial landmarks such as force lines and eraillures are eroded off flake surfaces, thereby also removing microwear traces. A preliminary examination of 50 of the less severely weathered pre-Clovis chert artifacts from Topper found only six with possible use striae, with one having microplating (Kay 2002). Despite weathering, however, some pieces show macroscopic nibbling and edge damage. A comprehensive, experimentally based study of use wear will be needed to determine the full analytical potential of microwear at Topper.

In sum, unusual microlithic artifacts associated with a smash-core technology have been found in a stratified deposit underneath Clovis-age sediments and Clovis-related lithic technology. OSL dating alone indicates these artifacts are in excess of 15,000 calendar years old; the paleosol lying between the Clovis materials and the Pleistocene alluvium would add 2000 to 4000 years to that age. This technology is remarkable for its non-bifacial character and its emphasis on microlithic-sized artifacts, specifically bend-break burin-like tools. Although this evidence is clearly associated with chert quarrying, it seems clear that more than just lithic extraction was taking place there. The emphasis on microlithic tools, as well as core choppers, suggests some type of craft activity. The presence of microliths may imply the manufacture of organic artifacts from bone, antler, wood, and ivory.

It is also significant that the artifacts are associated with Ice Age Savannah River alluvium, deposited when the river was able to flood at the Topper site elevation. Sometime prior to the arrival of Clovis people, the Savannah River bed was scoured and cut down to near its present elevation. Scouring made available for the first time river cobble chert, which was heavily exploited

Figure 9. Example of pre-Clovis chert clusters from the 2002 block excavation at the Topper site (SCIAA-USC photo).

by Clovis people and all succeeding prehistoric populations during the Holocene.

Conclusions

Four sites located in the eastern U.S. have been briefly reviewed for evidence of human occupations prior to 11,500 RCYBP. In each case, credible evidence of artifacts in deposits that predate that age has been recovered. Given the presence of bifacial projectiles at Meadowcroft and Cactus Hill and perhaps Saltville (SV-2) based on debitage, it is not difficult to see Clovis emerging from these technologies. Even the lithic raw materials at Meadowcroft are typically Paleoindian. The Topper site, based on evidence recovered thus far, makes a strong break from these sites. Although Topper is unmistakably a quarry site, there are too many small tools requiring hafting for effective use to interpret it simply as that. The microlithic-size flake tools, many of them apparently expedient, are in a real sense non-Clovis. Located at 33° N latitude, Topper is also the southernmost of these sites. It lies on the late-glacial ecotone, between the temperate and climatically stable Southeastern coastal plain, with its near modern flora, and the glacial-influenced mixed conifers and northern hardwoods of the North (Delcourt and Delcourt 1981).

The acceptance of Monte Verde in Chile as a pre-12,000 RCYBP site and other sites of similar age in North America should cause an acceleration in field investigations of Pleistocene landscapes. Since full- and late-glacial rivers were flowing at higher elevations, Pleistocene river banks on T2 terraces should be targeted for testing. In the Southeast, many alluvial terraces underneath Clovis and other Paleoindian manifestations experienced scouring, which would obliterate evidence of pre-Clovis occupations (cf. Goodyear 1999b). It seems clear that for open, nonrockshelter sites, the context will be essentially geological. The work of interdisciplinary teams is essential to prove conclusively the age and reality of such sites.

Acknowledgments

I would like to thank the organizers and sponsors of the Clovis and Beyond conference for the invitation to present an earlier version of this paper. James Adovasio, Jerry McDonald, Marvin Kay, Joe and Lynn McAvoy, Mike Johnson, and Mike Waters were most helpful in supplying information, reports, and graphics. Daryl P. Miller is thanked for help with Topper artifact photography. Tony Pickering helped create Figure 7. Jim Chandler aided the technical editing. The members of the Allendale Paleoindian Expedition are recognized for their assistance with the Topper site excavations. Clariant Corporation, owners of the Topper site, are acknowledged for their generous support of our research there. Last, the editor Rob Bonnichsen is thanked for his patience as well.

References Cited

Adovasio, J. M., and R. C. Carlisle
1988 The Meadowcroft Rockshelter. *Science* 239:713–14.

Adovasio, J. M, J. Donahue, and R. Stuckenrath
1990 The Meadowcroft Rockshelter Radiocarbon Chronology 1975–1990. *American Antiquity* 55:348–54.

Adovasio, J. M., D. Pedler, J. Donahue, and R. Stuckenrath
1998 Two Decades of Debate on Meadowcroft Rockshelter. *North American Archaeologist* 19:317–41.
1999 No Vestige of a Beginning nor Prospect for an End: Two Decades of Debate on Meadowcroft Rockshelter. In *Ice Age Peoples of North America: Environments, Origins, and Adaptations of the First Americans*, edited by R. Bonnichsen and K. L. Turmire, pp. 416–31. Oregon State University Press, Corvallis.

Clark, D. E., and B. A. Purdy
1979 Electron Microprobe Analysis of Weathered Florida Chert. *American Antiquity* 44 (3):517–24.

Crabtree, D. E.
1973 The Obtuse Angle as a Functional Edge. *Tebiwa,* the Journal of the Idaho State University Museum, Pocatello, Idaho.

Delcourt, P. A., and H. R. Delcourt
1981 Vegetation Maps for Eastern North America: 40,000 Yr. B.P. to the Present. In *Geobotany II,* edited by R. C. Romans, pp. 123–65. Plenum, New York.

Ferring, C. R.
2001 *The Archaeology and Paleoecology of the Aubrey Clovis Site (41DN479) Denton County, Texas.* Center for Environmental Archaeology, Department of Geography, University of North Texas, Denton.

Forman, S.
2002 OSL Dating and Topper. Paper presented at the Allendale-Topper Conference, Columbia, South Carolina

2003 Luminescence Dating of Quaternary Sediments. *The Mammoth Trumpet* 18 (3):10–13. Center for the Study of the First Americans, Texas A&M University, College Station.

Foss, J. E., and A. V. Segovia
1984 Rates of Soil Formation. In *Groundwater As a Geomorphic Agent,* edited by R. G. LaFleur, pp. 1–17. Allen and Unwin, Boston.

Goldberg, P., and T. L. Arpin
1999 Micromorphological Analysis of Sediments from Meadowcroft Rockshelter: Implications for Radiocarbon Dating. *Journal of Field Archaeology* 26 (3):325–42.

Goodyear, A. C.
1999a Results of the 1999 Allendale Paleoindian Expedition. *Legacy,* 4 (1-3):8–13. Newsletter of the South Carolina Institute of Archaeology and Anthropology, University of South Carolina.

1999b The Early Holocene Occupation of the Southeastern United States: A Geoarchaeology Summary. In *Ice Age Peoples of North America,* edited by R. Bonnichsen and K. L. Turnmire, pp. 432–81. Oregon State University Press, Corvallis.

2000 The Topper Site 2000: Results of the 2000 Allendale Paleoindian Expedition. *Legacy* 5 (2):18–25. Newsletter of the South Carolina Institute of Archaeology and Anthropology, University of South Carolina, Columbia.

Goodyear, A. C., and T. Charles
1984 *An Archaeological Survey of Chert Quarries in Western Allendale County, South Carolina.* Research Manuscript Series 195. Institute of Archaeology and Anthropology, University of South Carolina, Columbia.

Goodyear, A. C., and K. Steffy
2003 Evidence for a Clovis Occupation at the Topper Site, 38AL23, Allendale County, South Carolina. *Current Research in the Pleistocene* 20:23–25.

Hurst, V. J., and A. R. Kelly
1961 Patination of Cultural Flints. *Science* 28:251–56.

Ivester, A. H., D. S. Leigh, and D. I. Godfrey-Smith
2001 Chronology of Inland Eolian Dunes on the Coastal Plain of Georgia, USA. *Quaternary Research* 55:293–302.

Johnson, M. F.
1997 Additional Research at Cactus Hill: Preliminary Description of Northern Virginia Chapter—ASV's 1993 and 1995 Excavations. In *Archaeological Investigations of Site 44SX202, Cactus Hill, Sussex County, Virginia,* edited by J. M McAvoy and L. D. McAvoy, Appendix G. Virginia Department of Historic Resources, Research Report Series No. 8, Richmond.

Kay, M.
2002 Stone Tool Microscopy. Paper presented at the Allendale-Topper Conference, Columbia, South Carolina.

McAvoy, J. M.
1997 Addendum: Excavation of the Cactus Hill Site, 44SX202, Areas A–B, Spring 1996: Summary Report of Activities and Findings. In *Archaeological Investigations of Site 44SX202, Cactus Hill, Sussex County, Virginia,* edited by J. M. McAvoy and L. D. McAvoy. Virginia Department of Historic Resources, Research Report Series No. 8, Richmond.

McAvoy, J. M., and L. D. McAvoy, editors
1997 *Archaeological Investigations of Site 44SX202, Cactus Hill, Sussex County, Virginia.* Virginia Department of Historic Resources, Research Report Series No. 8., Richmond.

McAvoy, J. M., and L. D. McAvoy
2003 *The Williamson Clovis Site, 44DW1, Dinwiddie County, Virginia:* An Analysis of Research Potential in Threatened Areas. Virginia Department of Historic Resources, Research Report Series No. 13, Richmond.

McAvoy, J. M., J. C. Baker, J. K. Feathers, R. L. Hodges, L. J. McWeeney, and T. R. Whyte
2000 Summary of Research at the Cactus Hill Archaeological Site, 44SX202, Sussex County, Virginia: Report to the National Geographic Society in Compliance with Stipulations of Grant 6345-98.

McDonald, J. N.
1995 Evidence of Paleo-Indians at Saltville, Virginia, 13,950 Years Ago. Unpublished manuscript in possession of author.

2000 An Outline of the Pre-Clovis Archaeology of SV-2, Saltville, Virginia, with Special Attention to a Bone Tool Dated 14,510 yr B.P. *Jeffersoniana* 9:1–59. Contributions from the Virginia Museum of Natural History, Martinsville.

McDonald, J., and M. Kay
1999 Pre-Clovis Archaeology at SV-2, Saltville, Virginia. Abstracts of the Society for American Archaeology 64th Annual Meeting, p. 196.

Morrow, J. E.
1996 The Organization of Early Paleoindian Lithic Technology in the Confluence Region of the Mississippi, Illinois and Missouri Rivers. Doctoral dissertation on file with the Department of Anthropology, Washington University. St. Louis.

Root, M. J., J. D. William, M. Kay, and L. K. Shifrin
1999 Folsom Ultrathin Biface and Radial Break Tools in the Knife River Flint Quarry Area. In *Folsom Lithic Technology, Explorations in Structure and Variation,* edited by D. S. Amick, pp. 144–68. International Monographs in Prehistory, Archaeological Series 12, Ann Arbor, Michigan.

Taylor, R. E., C. V. Haynes, Jr., and M. Stuiver
1996 Clovis and Folsom Age Estimates: Stratigraphic Context and Radiocarbon Calibration. *Antiquity* 70 (269):515–25.

Upchurch, S. B.
1984 Appendix A, Petrology of Selected Lithic Materials from the South Carolina Coastal Plain. In *An Archaeological Survey of Chert Quarries in Western Allendale County, South Carolina,* by A. C. Goodyear and T. Charles. University of South Carolina, Institute of Archaeology and Anthropology, Research Manuscript Series 195, Columbia, South Carolina.

Waters, M. R., T. W. Stafford, S. Forman, and J. E. Foss
2003 Geoarchaeological Investigations at the Topper Site, South Carolina. Manuscript in preparation.

Wisner, G.
1996 Saltville Site Has Evidence of 14,000-Year-Old Feasts. *The Mammoth Trumpet* 1, (4), pp. 1; 18–20. Center for the Study of the First Americans, Corvallis, OR.

Clovis, Pre-Clovis, Climate Change, and Extinction

C. Vance Haynes, Jr.

Abstract

The route for initial peopling of the Americas is not known in spite of recent opinions to the contrary. A coastal route and the ice-free corridor route are still working hypotheses that need further testing. Clovis progenitors are also hypothetical. However, prevailing biases suggest a coastal route to the exclusion of an ice-free corridor route, and Clovis originating from an unproven pre-Clovis population in the Americas. Presented here are data to suggest that an ice-free passage during the Allerød warm period may have extended from the unglaciated Yukon Plateau to the Pelly Valley to the Liard Valley to the interior Plains of Canada in time for people of the Nenana complex in Central Alaska between 11,800 and 11,500 RCYBP to move to the Great Plains and become progenitors of the Clovis complex by 11,500 RCYBP. Whereas Clovis expansion throughout North America coincides with extinction of Rancholabrean fauna at about 10,890 RCYBP, "black mat" stratigraphy indicates the event was too sudden to have been caused by either human predators alone or climate change alone. Instead a combination of predation, drought, and the Younger Dryas deep-freeze may have been the cause of the Rancholabrean termination.

Introduction

Several recent events in American archaeology have served to reconstitute interpretations regarding the peopling of America. None of them necessarily preclude previous interpretations because all are hypothetical. For example, the ice-free corridor route is not a "dead" issue as some would have us believe. Conversely an active ice-free corridor does not preclude a coastal migration route. And even a transatlantic route has recently been revived by Stanford and Bradley (Hall 2000) and challenged by Straus (2000).

Furthermore, if Monte Verde is a valid 12,500 RCYBP[1] archaeological site (Dillehay 1989, 1997), this fact does not mean that sites previously rejected for lack of compelling evidence have any more validity now than they did under earlier scrutiny. Nor does it mean that equivocal pre-Clovis sites are now less so.

It has been said that the "Clovis-first" model has caused some archaeologists not to explore pre-Clovis sediments, but what archaeologist worth his or her salt would refrain from excavating below a Clovis level because of a preconception of no archaeology being there? The concept is certainly unscientific. Most archaeologists always look below the lowest occupation with the hope of finding even earlier evidence of human presence if the sediments are of Pleistocene age. Some of us have spent the major part of a lifetime looking for evidence of Clovis progenitors or pre-Clovis occupants of America, and have provided for this possibility in chronological classifications (Haynes 1969).

No archaeologist should preclude the possibility of pre-Clovis occupation of the Americas. The Clovis technocomplex has to have had progenitors. The question is, Were the most immediate progenitors in Beringia, North America, or even western Europe? The differences in opinions are based upon different interpretations of the available evidence, and no reasonable interpretation should be excluded from the realm of possibilities. Anyone who is convinced of having all the answers has left the realm of science and entered that of faith. For objective scientific inquiry, we need to keep in mind T. C. Chamberlin's (1965) concept of multiple working hypotheses.

A Plea for Multiple Hypotheses

Scientists, in trying to understand and explain aspects of nature and the universe, often have to be inconclusive and deal with uncertainties by constructing hypotheses. Webster's New Colle-

Regents Professor Emeritus, Departments of Anthropology and Geosciences, University of Arizona, Tucson, AZ 85721.

[1]All ages are stated as uncalibrated radiocarbon years before present (i.e., 1950).

giate Dictionary (1979) defines hypothesis as "a tentative assumption made in order to draw out and test its logical or empirical consequences." However, Chamberlin has warned that a single or preferred hypothesis can become a ruling theory. He states:

> The moment one has offered an original explanation for a phenomenon which seems satisfactory, that moment affection for his intellectual child springs into existence; and as the explanation grows into a definite theory, his parental affections cluster about his intellectual offspring, and it grows more and more dear to him, so that, while he holds it seemingly tentative, it is still lovingly tentative, and not impartially tentative. So soon as this parental affection takes possession of the mind, there is a rapid passage to the adoption of the theory. There is an unconscious selection and magnifying of the phenomena that fall into harmony with the theory and support it, and an unconscious neglect of those that fail of coincidence. The mind lingers with pleasure upon the facts that fall happily into the embrace of the theory, and feels a natural coldness toward those that seem refractory. Instinctively there is a special searching-out of phenomena that support it, for the mind is led by its desires. There springs up, also, an unconscious pressing of the theory to make it fit the facts, and a pressing of the facts to make them fit the theory. When these biasing tendencies set in, the mind rapidly degenerates into the partiality of paternalism (p. 755).

To compensate for one's biases the solution offered by Chamberlin is to pursue multiple working hypotheses:

> The effort is to bring up into view every rational explanation of new phenomena, and to develop every tenable hypothesis respecting their cause and history. The investigator thus becomes the parent of a family of hypotheses: and, by his parental relation to all, he is forbidden to fasten his affections unduly upon any one (p. 756).

The last statement is easier said than done. However, these eventually boil down to two or three hypotheses that are constantly being tested by new data and interpretations that often may not lead to a definitive answer or solution without more investigations producing more pertinent facts. Collective opinions may swing back and forth; but the hope is that some hypotheses will be disproved and the correct hypothesis will eventually prevail as, for example, Darwinian evolution and Wegener's continental drift. Platt (1964) recommends using "Strong Inference" when a way is seen to eliminate one of two competing hypotheses. However, new facts can lead to new hypotheses to add to the hopper of multiple working hypotheses. Old, once abandoned hypotheses may be reconstituted on the bases of new evidence. In this way we are forever attempting to approximate reality.

In the future the scientific investigation of all potential pre-Clovis sites must include on-site evaluation of the evidence by experienced and objective geologists and archaeologists within a reasonable time of its discovery. The standards today should be no less than they were at Folsom in 1927. It is apparent, as always with equivocal evidence, that interpretations of the data become more or less subjective depending on the bias of the interpreter. Therefore, the best recourse is to fall back on Chamberlain's (1965) principle of multiple working hypotheses until future research sheds more light on the problem.

Recently, attorney Mark Tumeo (1999) contrasts the approaches of lawyers versus engineers and scientists. A significant difference between lawyers and scientists is that lawyers have to arrive at a decision pragmatically in order to argue the case. Winning the argument is what it takes to win the case, not presenting multiple hypotheses. However, unlike jurisprudence, in science it is truth we seek, not winning the argument.

Having stated these ideas about scientific objectivity, I proceed by presenting a somewhat biased interpretation of the migration of Clovis progenitors, the Clovis occupation of North America, and the pre-Clovis situation in America. It is simply my interpretation based upon what I consider the best available evidence. However, alternative interpretations are possible and are discussed, in keeping with the Method of Multiple Working Hypotheses. The peopling of America is viewed in the light of North American climate change and its effect on the geologic record. It is assumed that Clovis progenitors, though primarily hunters, were opportunistic generalists exploring a rapidly changing environment (Kelly and Todd 1988; Hoffecker, et al. 1993), initially with essentially no competition from other humans.

Clovis Progenitors

Who are some candidates for Clovis progenitors? Probably not Monte Verdeans (Dillehay 1989, 1997). Aside from the equivocal aspects pointed out by Fiedel (1999a), the Monte Verde artifact assemblage is too unlike the Clovis artifact assemblage to be the precursor, though it is not an impossibility. Monte Verde is not the breakthrough that the Folsom discovery was in 1927 (Haynes 1999). The Nenana complex of Alaska is a much better candidate, as Goebel et al. (1991), Hamilton and Goebel (1999), and Yesner (1998) have claimed. There are significant differences; but considering the environmental differences between central Alaska and the Great Plains with its abundant megafauna, the corridor filter, and the time factor, significant changes are to be expected.

As Fiedel (2000) indicated, the timing is reasonable. The earliest Nenana radiocarbon date so far is 11,800 RCYBP at Walker Road (Goebel et al. 1996), and the earliest Clovis age is 11,590 ± 90 RCYBP at the Aubrey site in north central Texas (Ferring, 1995) and 11,550 ± 60 RCYBP for the Anzick Clovis burial (Stafford et al. 1991 and pers. comm. 1998). Is 250 years (more or less depending on calibration and actual ages) enough time for Clovis progenitors to go 5,000 km? That is 15 to 20 generations at an average rate of 20 km per year. As discussed later, this is probably a gross underestimate of their rate of travel. In fact, Surovell (2000) has devised a mathematical model for Clovis mobility and expansion that suggests it is easier (cheaper) to raise children with more frequent camp moves than with staying put for extended periods.

Deglaciation and Global Climate Change

There can be little doubt that global climate is the major control or driving force behind deglaciation. Recent research on the geochronology of climate change of the northern hemisphere in relation to American archaeology has been reviewed by Fiedel (1999b). Comparing the record of deglaciation and the limited archaeological record to the ice-core records and those of paleolakes (Patterson and Hammer 1987; Stuiver et al. 1995), some interesting correlations exist that may not be completely fortuitous (Figure 1). The rapid and uninterrupted retreat of the

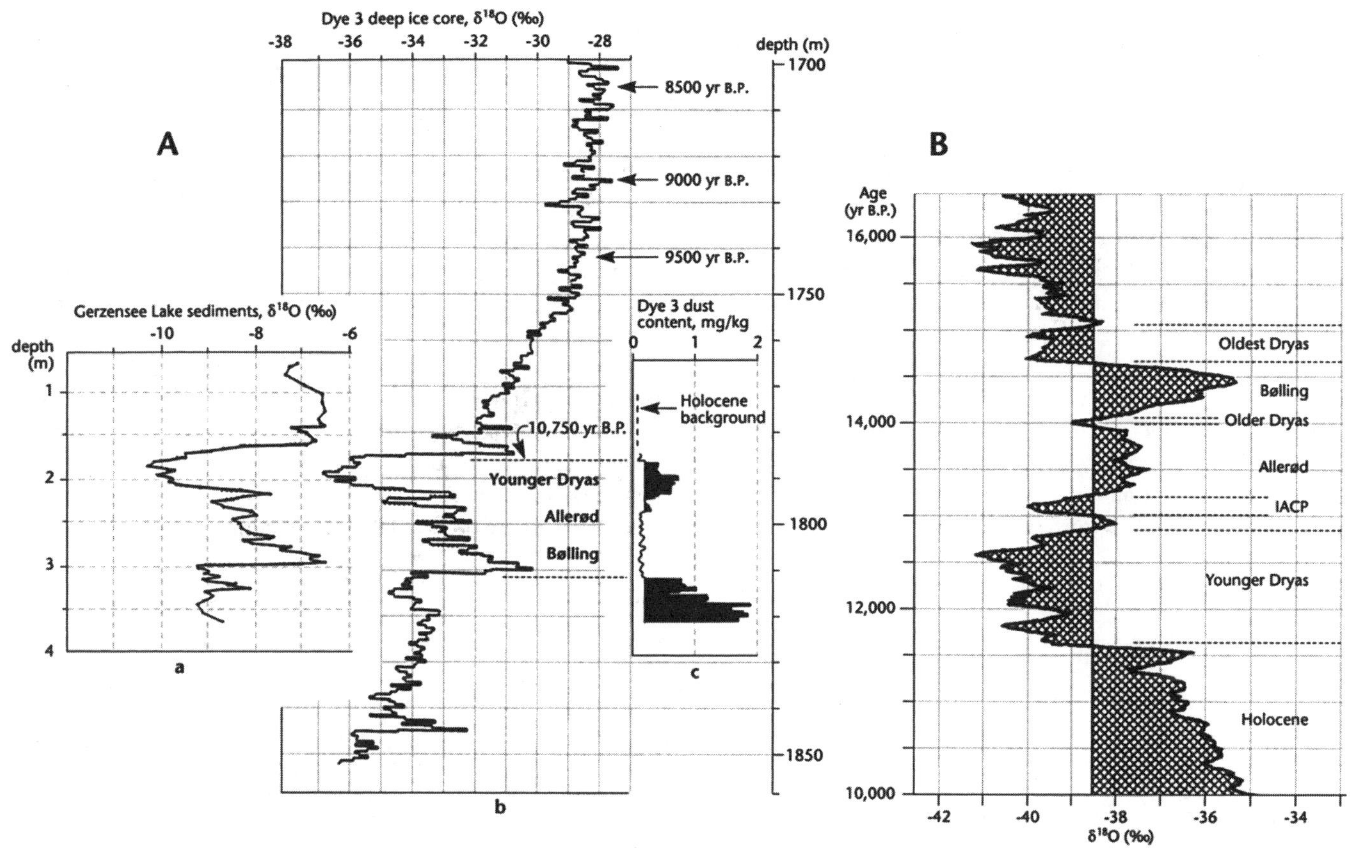

Figure 1. A, Comparison of $\delta^{18}O$ in Lake Gerzensee (**a,** ages in calibrated ^{14}C years), $\delta^{18}O$ in the Dye 3 Greenland ice core (**b,** ages in real years); and the dust concentration in the Dye 3 ice core; and the Bølling, Allerød, and Younger Dryas pollen zones. (From Patterson and Hammer 1987, Fig. 6. b.) **B,** Correlation of the oxygen isotope paleoclimate record of Greenland Ice Sheet Project 2 (GISP2) ice core from 16,500 to 10,000 calibrated radiocarbon years. (From Stuiver et al. 1995, Fig. 11).

Laurentide ice between 13,000 and 12,000 RCYBP is probably in response to the warm Bølling climatic period preceded and ended by brief cold periods, Oldest and Older Dryas, which appear to have had little or no effect in retarding deglaciation. From about 12,000 to 10,800 RCYBP, the fluctuating warm Allerød period with an Inter-Allerød Cold Period (IACP) furthered deglaciation. In North America, the Two Creeks interstade (TCI) more or less coincides with the Allerød of Europe. Kaiser (1994) has shown by dendrochronology that the TCI lasted at least 252 years from 12,050 to 11,750 RCYBP.

After the inception of Lake Agassiz about 11,700 RCYBP, it attained an area of approximately 134,000 km^2; and drained to the Mississippi system before rapidly draining to the Atlantic Ocean via the Great Lakes and the Saint Lawrence Valley about 10,900 yr RCYBP (Leverington et al. 2000). West of Lake Agassiz between 12,000 and 10,700 RCYBP, a dynamic complex of 18 rapidly changing meltwater spillways existed across the northern Great Plains, the result of catastrophic floods draining 13 large ice-dammed lakes (Kehew and Teller 1994).

Deglaciation began in the Great Lakes region with the retreat of Port Huron ice; and reached a maximum retreat at the height of the TCI, just before the Two Creeks forest bed drowned around 11,800 RCYBP, the earliest time Nenanans may have started down the ice-free corridor. The forest was drowned by a rising Lake Michigan as ice of the Two Rivers lobe readvanced and blocked the straits of Mackinac and the Cheboygan outlets (Fullerton 1980; Hansel et al. 1985). Post-Clovis fluted-point makers appear to have occupied the Great Lakes areas soon after the Two Rivers advance and during the time Lake Algonquin existed (Deller and Ellis 1988; Gramly 1988; Storck 1988), essentially during the Younger Dryas cold period, a rapid return to glacial conditions that lasted about 1000 years.

The Ice-free Corridor

When did an ice-free corridor become passable to humans during deglaciation? The potential significance of an ice-free corridor to the peopling of the New World was proposed by Johnston (1933) without benefit of chronological control. It was revived when renewed radiocarbon dating of the Two Creeks forest bed (Broecker and Ferrand 1963) indicated that the major glacial retreat of the Two Creeks interstade was earlier than the radiocarbon ages of Clovis sites (Haynes 1964, 1980). Wilson and Burns (1999) provide an objective update on this controversial issue. Some believed there was no closure in late glacial time (Reeves 1973); but this view is seriously challenged by recent evidence indicating that the last Laurentide glaciation was the most extensive of the Wisconsinan, and coalesced with the Cordilleran ice sheet as well as with piedmont and valley glaciers of the eastern Cordillera (Jackson et al. 1997; Young et al. 1994). Although most researchers find compelling evidence for closure for at least a brief period (Rutter 1980, 1984), recent evidence indicates closure or conditions precluding large mammals for about 9,000 years (Burns 1996).

The best graphic assessment of corridor deglaciation is that of Dyke and Prest (1987), who have updated the chronology of deglaciation proposed by Prest in 1969. In a series of maps of deglaciation in 1,000-year intervals they show a narrow, at least 25-km-wide separation between the Laurentide ice margin and glaciers of the eastern Cordillera by 13,000 RCYBP (Figure 2) between Fort Liard and the Athabasca River about 200 km northwest of Edmonton. The model of Marshall and Clarke (1999) shows an even wider neck by 13,000 RCYBP. At this stage the corridor probably would not have been an attractive passage for either people or megafauna (Mandryk 1996). Farther south the ice had already significantly retreated, forming an ever-widening gap as if emerging from the neck of a funnel. The plotting of more than 50 fluted-point locations (Haynes 1980; Wilson and Burns 1999) and the location of a Clovis cylindrical proboscidean-bone rod or point (Wilmeth 1968) on the Dyke and Prest maps shows that only a few points fall in ice-free areas of the 13,000 RCYBP sheet (Figure 2). These locations suggest that all these fluted points postdate 13,000 RCYBP.

Between 13,000 and 12,000 RCYBP, Dyke and Prest (1987) state that "net retreat of the Laurentide margin . . . was more rapid than previous rates everywhere. . . ." "Uninterrupted retreat is inferred for most areas" during this time, which corresponds to the Bølling warm interval (Stuiver et al. 1995). By 12,000 RCYBP the narrowest part of the ice-free corridor, in the Fort Liard area, was at least 80 km wide (Figure 3). The maps of Morlan and others (1999) show a similar gap but with more detail of Cordilleran deglaciation. All but a few of the fluted-point locations fall in ice-free areas. Therefore 12,000 RCYBP is likely the earliest period

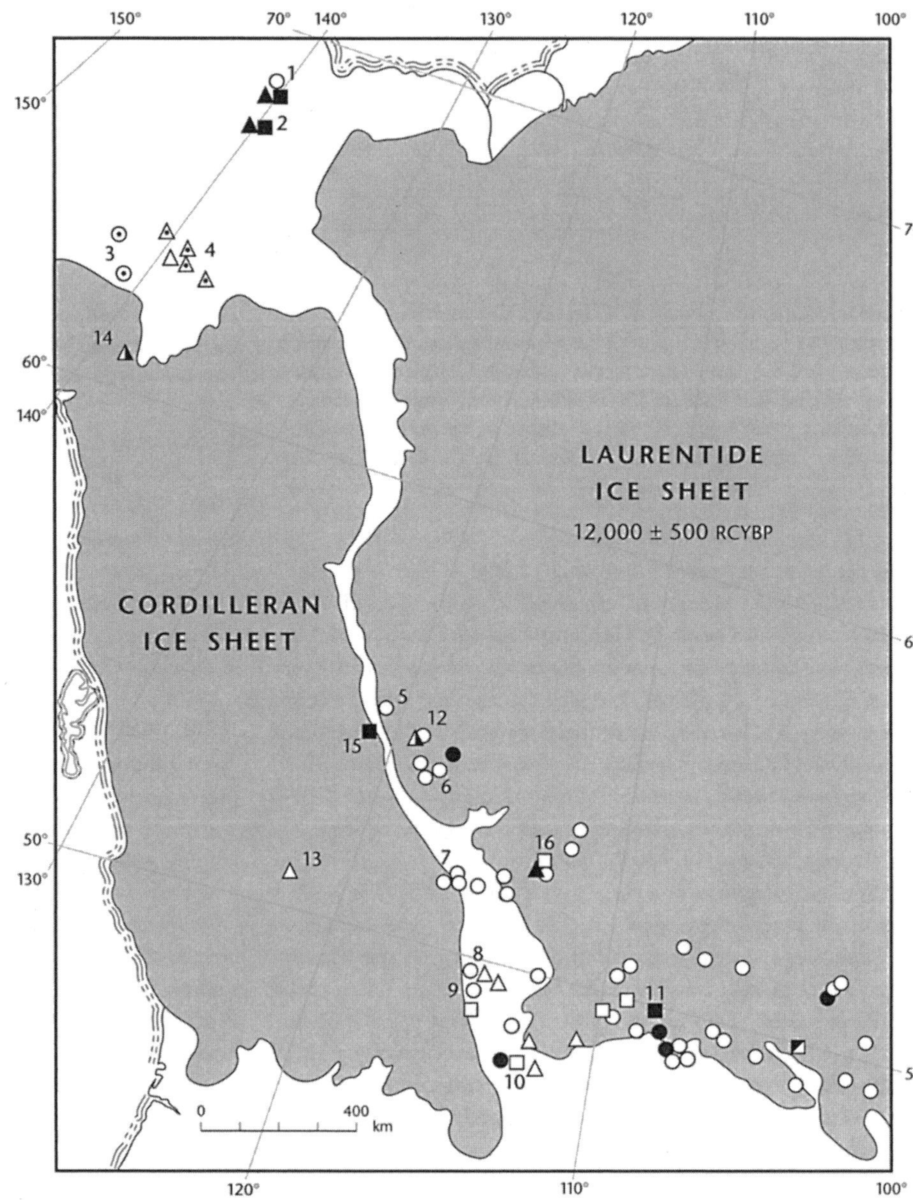

Figure 2. Generalized map of ice-free corridor at 13,000 RCYBP after Dyke and Prest (1987).

- ● Clovis point
- ○ other fluted points
- ⊙ Nenana sites

radiocarbon-dated bones:
- ▲ large mammal >11,500 RCYBP
- △ large mammal 11,500–11,000 RCYBP
- arctic ground squirrel bone or nest >11,000 RCYBP

radiocarbon-dated wood or organic matter:
- minimum age for deglaciation
- ■ mammoth bone ≥11,500 RCYBP
- □ mammoth bone < 11,500 RCYBP
- Clovis mammoth-bone point or rod

1 Old Crow basin
2 Bluefish Caves
3 Tanana River Nenana sites
4 Dawson-Klondike area
5 Charlie Lake Cave
6 Birch Hills
7 fluted-point concentration in upper Pembina Valley
8 Vermillion Lakes site
9 Sibbald Flat
10 St. Mary Lake mammoth tracks
11 Kyle mammoth
12 Boone Lake
13 Lochnore Creek, Fraser River Valley
14 Kluane Lake, Yukon Territory
15 Portage Mountain Dam mammoth tusk
16 Clover Bar gravel pit, east Edmonton, Alberta

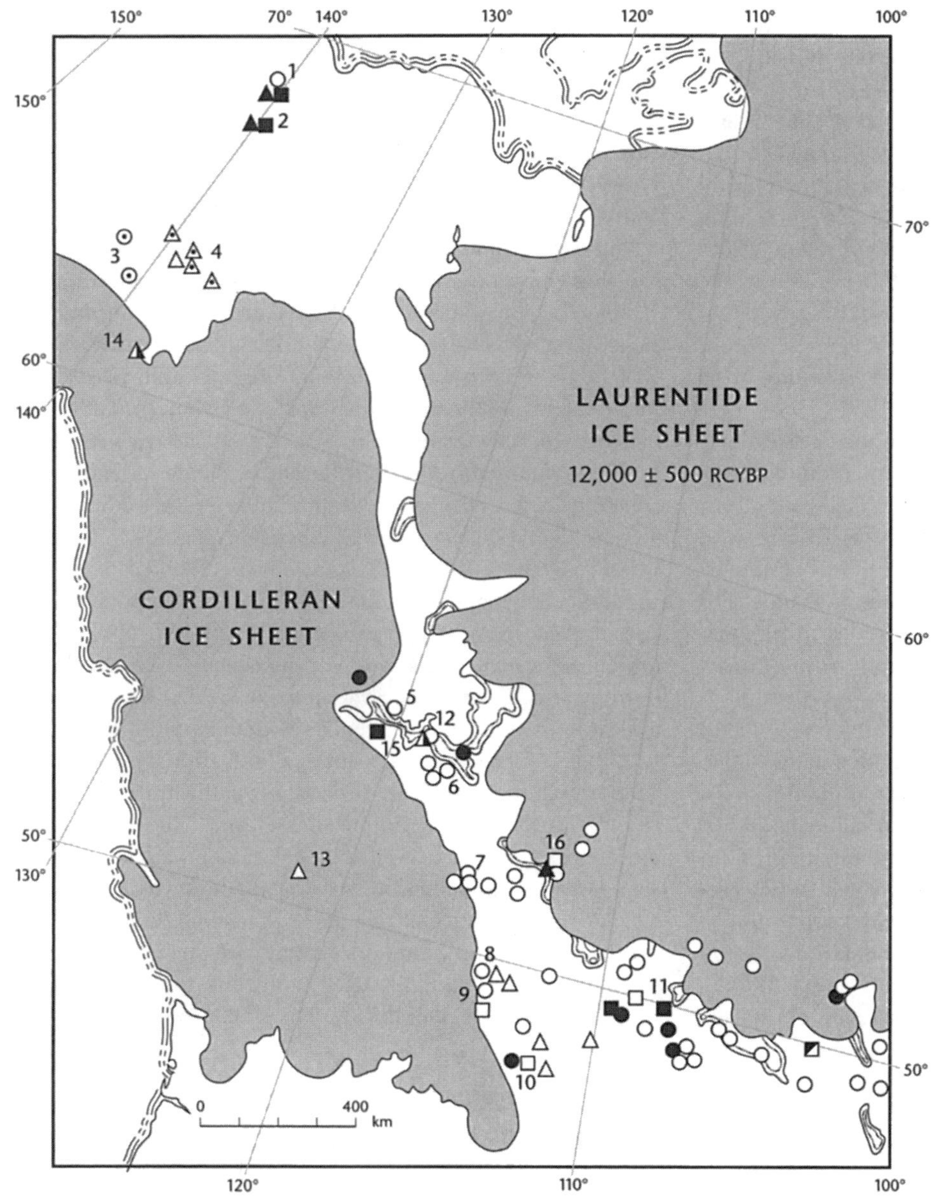

Figure 3. Generalized map of the ice-free corridor at 12,000 RCYBP, modified after Morlan et al. (1999) and Jackson and Duk-Rodkin (1996).

- ● Clovis point
- ○ other fluted points
- ⊙ Nenana sites

radiocarbon-dated bones:
- ▲ large mammal >11,500 RCYBP
- △ large mammal 11,500–11,000 RCYBP
- arctic ground squirrel bone or nest >11,000 RCYBP

radiocarbon-dated wood or organic matter:
- ▲ minimum age for deglaciation
- ■ mammoth bone ≥11,500 RCYBP
- □ mammoth bone < 11,500 RCYBP
- Clovis mammoth-bone point or rod

1. Old Crow basin
2. Bluefish Caves
3. Tanana River Nenana sites
4. Dawson-Klondike area
5. Charlie Lake Cave
6. Birch Hills
7. fluted-point concentration in upper Pembina Valley
8. Vermillion Lakes site
9. Sibbald Flat
10. St. Mary Lake mammoth tracks
11. Kyle mammoth
12. Boone Lake
13. Lochnore Creek, Fraser River Valley
14. Kluane Lake, Yukon Territory
15. Portage Mountain Dam mammoth tusk
16. Clover Bar gravel pit, east Edmonton, Alberta

during which the ice-free corridor could have been transgressed, provided that environmental conditions such as muskeg, runoff, catastrophic floods, and wind chill factors would have allowed it. A recently opened corridor would have been a formidable environment for people to penetrate, one characterized by proglacial lakes, catastrophic floods, and ever-changing drainages (Teller and Kehew 1994) as well as stagnant ice and hummocky ground moraine (Mandryk 1996); but eventually muskeg, lake, and floodwater would have given way to herb- and shrub-dominated vegetation, steppe grassland, and riparian areas (MacDonald and McLeod 1996) as drainages became integrated. With the eastern part of the corridor being a land of proglacial lakes (Teller 1995), waterfowl would have been abundant during summer months (Fiedel n.d.). Evidence of the first occupants of the corridor necessarily must be extremely rare, but might be found under eolian sediments blanketing uplands and the earliest terraces formed during degradation and integration of the drainages; but as Wilson and Burns (1999) state, some evidence could be deeply buried. The Bow River between Calgary and the Front Range aggraded as a braided channel as paraglacial deposits were washed from the uplands before and after 11,300 ± 290 RCYBP (RL-757), a radiocarbon age on a Pleistocene mammal bone in the upper part of the 10-m terrace (T-4) gravels (Jackson and MacDonald 1982). Most of the rivers flowing from the Rocky Mountains to the prairies probably have similar deposits and geochronologies (Wilson 1996; Wilson and Churcher 1984).

On their Web site, Morlan and others (1999) show the distribution of radiocarbon ages on megafauna and ground squirrel nests in the corridor and elsewhere in millennial intervals. Between 12,500 and 11,500 RCYBP the only radiocarbon date within the corridor is on bison bone from the Clover Bar gravel pit on the North Saskatchewan River in east Edmonton (**16,** Figures 2, 3, and 4) dated 11,620 ± 170 RCYBP (AECV-1203c). Not in their data bank is a mammoth tusk from glacial gravels of the Peace River at the Portage Mountain Dam about 50 km east of Fort Saint John (**15,** Figures 2, 3, and 4) that initially provided an age of 11,600 ± 1000 RCYBP (I-2244A) (Rutter 1977). However, it has been redated to 25,800 ± 320 RCYBP (GSC-2859) by the Geological Survey of Canada. At Boone Lake (**12,** Figures 2, 3, and 4) a fragment of a poplar tree branch was recovered in a core dated 11,700 ± 269 RCYBP (SFU-223), providing a minimum age for deglaciation of the Saddle Hills area (White et al. 1985). The Kyle mammoth, dated 12,035 ± 299 RCYBP (S-246), may be an archaeological site (Wilson and Burns 1999); it is located in southwestern Saskatchewan (**11,** Figures 2, 3, and 4) in proglacial lake sediments. The locality is well south of the neck of the corridor.

Burns (1996), after evaluating the radiocarbon ages of vertebrate bones in the corridor, proposes that a mammoth steppe environment, as per Guthrie (1982), was established there at least by 11,600 RCYBP. He further proposes that the lag time between deglaciation and faunal recolonization probably was on the order of hundreds of years rather than thousands. Wilson (1996) described the Alberta part of the corridor as having open poplar woodlands with major patches of grassland, with *B. bison antiquus,* as far north as Peace River at least by 11,500 yr B.P.

From the upper Yukon drainage to Medicine Hat, Alberta, Morlan and others (1999) show eight localities of mammal bones dated between 11,500 and 11,000 RCYBP. Considering that at least some of the bone ages could be minimum values because of bone's notorious reputation for younger contamination, it appears that the corridor was open and passable to people and game animals at least by 11,600 RCYBP; and, therefore, at the time of occupation of central Alaska by Nenana Paleoamericans 11,800 to 11,300 RCYBP (Hamilton and Goebel 1999; Yesner 1998).

Between 12,000 and 11,000 RCYBP, essentially during the Allerød warm period, there appear to be no significant barriers to ingress and egress via the ice-free corridor south of the Mackenzie mountains (Figure 4). Recently deglaciated areas, initially wetlands, soon became open prairies veined with riparian habitats (White et al. 1985), even though some areas of stagnant ice may have remained (Mandryk 1996). By 11,000 RCYBP all the fluted-point locations are well within the ice-free corridor where the narrowest point is now over 200 km wide. To the southeast ca. 120 km the area of Charlie Lake Cave (**5,** Figure 4), with its in situ fluted point, appears to have been deglaciated by 11,000 RCYBP, and became a prairie woodland soon thereafter when post-Clovis fluted-point makers utilized the cave (Fladmark et al. 1988). Fluted-point concentrations occur in the Birch Hills area (**6,** Figure 4) and in the upper Pembina Valley (**7,** Figure 4). Farther south is the Vermillion Lakes stratified site (**8,** Figure 4) with an unidentified occupation dated 10,770 ± 175 RCYBP (SFD-348) on the lowest charcoal sample (Fedje et al. 1995). The earliest occupations of the site (Groups 1 and 2, ca. 10,800–9600 RCYBP) occur in eolian black organic silts separated by sterile eolian silts and colluvial gravels and debris flows that accumulated, as the radiocarbon dates indicate, during the Younger Dryas cold period.

Recent geological work in northwestern Canada (Jackson and Duk-Rodkin 1996) indicates that although a Peel Valley route to the Mackenzie Valley opened soon after 13,000 RCYBP, passage southward along the east side of the Mackenzie mountains (Figure 4) was blocked by glacial Lake Mackenzie until ca. 11,000 RCYBP (Lemmen et al. 1995), therefore too late for Clovis progenitors to pass that way on dry land (Mandryk, in press). Jackson and Duk-Rodkin's map (1996, Figure 14.2) for the period 12,000 to 11,000 RCYBP shows a potential route through the Yukon Plateau west of the Mackenzie mountains via the Tintina Trench and the Pelly River to the upper Liard River. Passage depends on this route being deglaciated by between 11,800 to 11,600 RCYBP. Jackson and Duk-Rodkin (1996) describe the northern Cordillera as a complex of stagnating ice and ice-dammed lakes before 13,000 RCYBP. The maps of Morlan and others (1999) show the area between the St. Elias Mountains and the Mackenzie Range deglaciated by 12,000 RCYBP. Arctic ground squirrel nests in the Klondike District (**4,** Figure 3) have yielded radiocarbon ages of 13,350 ± 265 RCYBP, 12,200 ± 100 RCYBP, and 11,170 ± 205 RCYBP, indicating arctic steppe conditions during Bølling and Allerød times. Not in their data bank is a radiocarbon date of 12,500 ± 200 RCYBP as a minimum age for deglaciation of the Kluane Lake area, Yukon Territory (**14,** Figure 3) by glaciers from the St. Elias Range (Denton and Stuiver 1967). If an ungulate bone date of 11,365 ± 1000 RCYBP (GX-2612) for the Lochnore Creek area (**13,** Figure 3) is valid, it indicates significant deglaciation of the upper Fraser Valley of British Columbia during Allerød time. These radiocarbon age data suggest that Cordilleran deglaciation had progressed at least as far as Morlan and others (1999) show on their map for 11,500 to 10,500 RCYBP (Figure 4). Therefore, it is possible, if not likely, that people could have traversed the Pelly and upper Liard Valleys 11,800 to 11,600 RCYBP. This would bring them into the Interior Plains at Nelson Forks (**17,** Figure 4).

Stratigraphic Manifestations of Paleoclimatic Events at Paleoindian Sites

Over much of the United States, particularly in non-glacial valleys of low-order streams, there is a remarkably consistent sequence of stratigraphic events that can be interpreted in relation to global climate change that influenced the peopling of America (Haynes 1984). These events probably had a causal effect on extinction of the Pleistocene megafauna, a termination that was geologically instantaneous.

These data are based on geological and archaeological investigations of at least 50 sites or localities in the U.S. (Table 1) conducted over the past 50 years (Figure 5). Also, from these data it is possible, in many cases, to reconstruct the Clovis-age landscape and see the stratigraphic abruptness of the termination of the Pleistocene megafauna (Haynes 1998).

The geochronology at these sites reveals a consistent pattern of geological and hydrological processes that can be reasonably interpreted in terms of Northern Hemisphere climate change as presently understood from pollen records and oxygen isotope variations in Greenland ice cores (Haynes 1998, n.d.). The stratigraphy at many of these sites can be reduced to a generalized se-

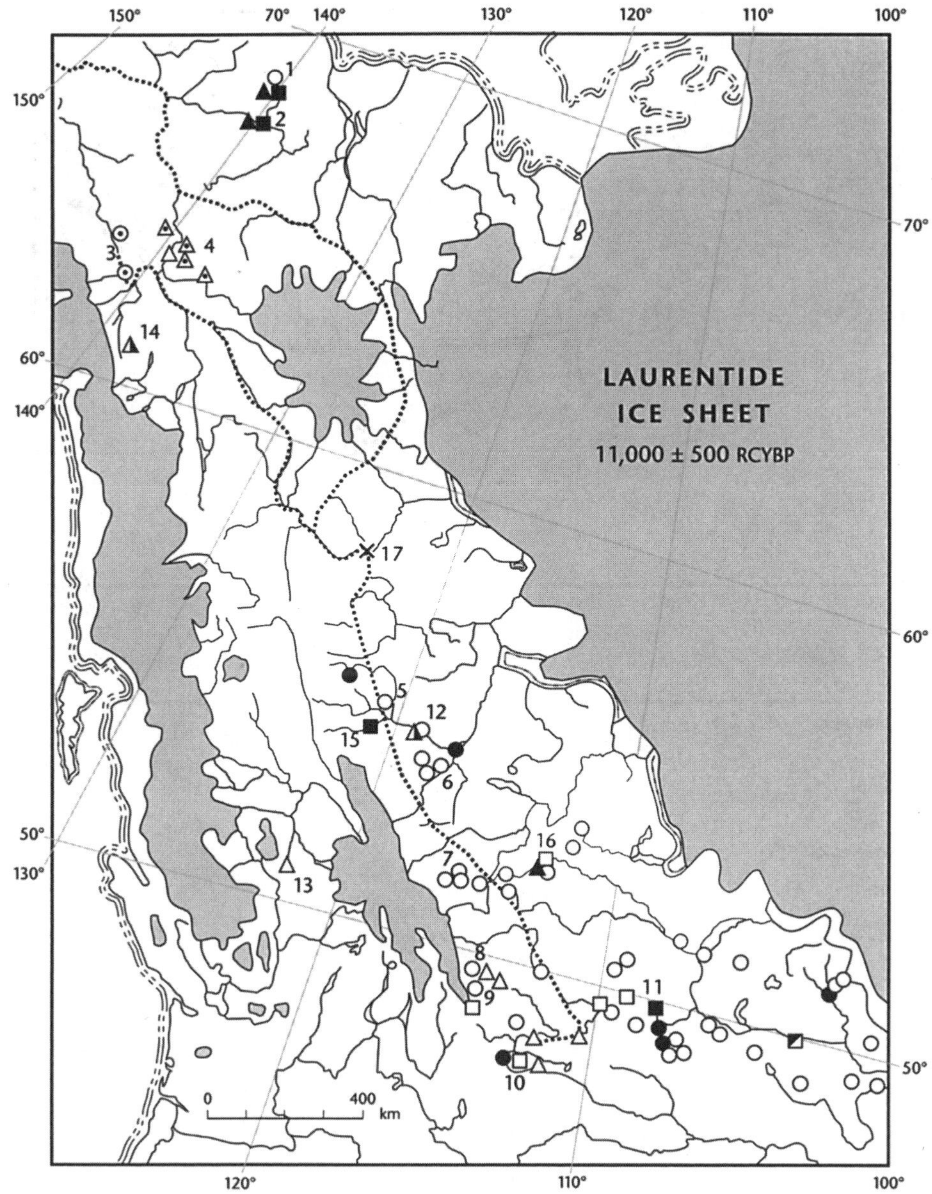

Figure 4. Generalized map of ice-free regions at 11,000 RCYBP modified after Morlan et al. (1999) and Jackson and Duk-Rodkin (1996). Dotted lines are potential routes of Clovis progenitors.

quence, shown in Figure 6. Not all sequences fit this model, but enough do to be convincing of a direct response to climate change as a cause.

Stratum α usually consists of late-Pleistocene lake beds, spring deposits, or alluvial terrace deposits in excess of 13,000 RCYBP associated with a significant complement of Rancholabrean fauna and no archaeological evidence. The upper part represents shrinking lakes, desiccating wetlands, reduced spring heads, and stream degradation concomitant with deglaciation. By 13,000 RCYBP water tables had fallen to their lowest levels in at least 15,000 years and streams were grossly underfit for their full-glacial age valleys. During the Bølling-Allerød warm periods, springs, lakes, and some streams dried up as emergent water tables fell below the ground surface during the Clovis drought (Haynes 1991a), seen as two-fold at some sites. Uplands became stable during degradation, but many valley sides were eroded to form the Z_1 contact.

Stratum β_1 represents terminal-Pleistocene channel sands at many localities, and Clovis artifacts occur in the top 10 cm at some sites. Clovis people interacted with the last of the Rancholabrean megafauna at spring heads or along spring-fed streams (Stratum β_1) during the Allerød warm period. Termination of the megafauna occurred at the end of the Allerød, at the same time that Clovis hunters appear to have been most active, and immediately before a sudden return to the glacial climate of the Younger Dryas.

Stratum β_{2a}, the stratigraphic manifestation of the Younger Dryas cold period, includes gray, black, or brown organic pond or lake sediments (i.e., diatomite or peaty clay), wet-meadow soils (aquolls) (Reider 1990), or black organic mats (Haynes 1991a) reflecting a rise of water tables to either emergence as ponds or saturation of lowland surfaces. The mollic Younger Dryas deposits are devoid of in situ Clovis artifacts and megafauna

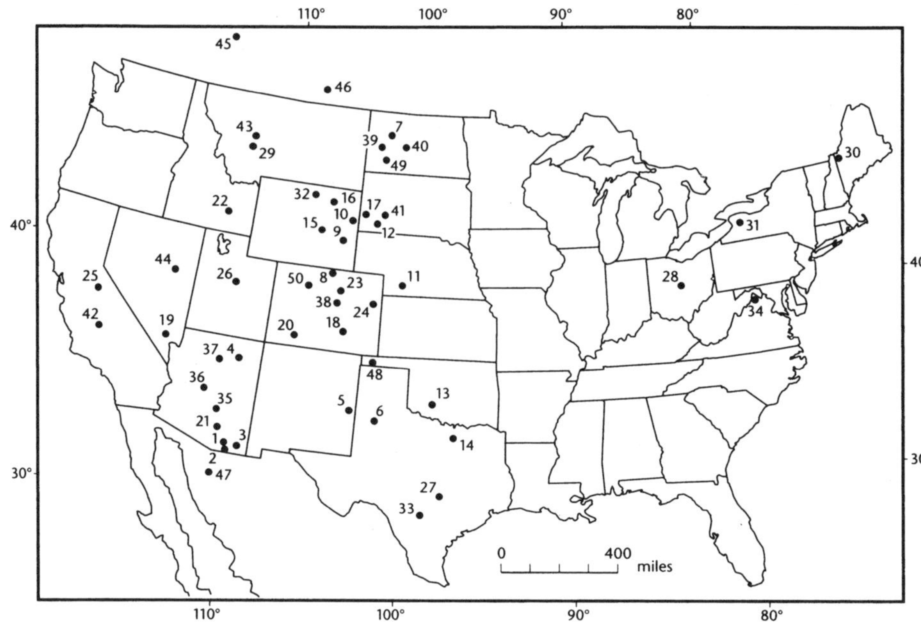

Figure 5. Map of the conterminous United States showing sites and locations with "black mats" of Younger Dryas age. Sites are identified in Table 1.

Table 1. Localities with Younger Dryas–age dark organic deposits shown in Figure 5.

1	Murray Springs and Escapule Clovis sites and the BLM mammoth locality, Ariz.	**26**	Huntington Canyon mammoth, Utah
2	Lehner Clovis site, Ariz.	**27**	Wilson Leonard site, Texas
3	Naco, Leikem, and Navarette Clovis sites, Ariz.	**28**	Burning Tree mastodon site, Ohio
4	Red Peak Valley, Blackmesa, Ariz.	**29**	Indian Creek site, Mont.
5	Blackwater Draw Clovis site, N.M.	**30**	Vail site, Maine
6	Lubbock Lake site, Texas	**31**	Arc and Hiscock sites, N.Y.
7	Knife River flint quarries and Folsom sites in the Leonard paleosol, N.D.	**32**	Sisters Hill site, Wyo.
8	Lindenmeier Folsom site, Colo.	**33**	Richard Beene site, Texas
9	Hell Gap site, Wyo.	**34**	Thunderbird Clovis site, Va.
10	Agate Basin and Sheamen sites, Wyo.	**35**	Chapo Ranch locality, Davidson Canyon, Ariz.
11	Medicine Creek Basin, Neb. and Brady soil of the region	**36**	M&M mastodon locality, Ariz.
12	Lange/Ferguson Clovis site, S.D.	**37**	Kaibito Plateau, Ariz.
13	Domebo Clovis site, Okla.	**38**	Lamb Spring site, Colo.
14	Aubrey Clovis site, Texas	**39**	Theodore Roosevelt National Park, N.D.
15	Henrick mammoth site, Wyo.	**40**	Flaming Arrow site, N.D.
16	Carter/Kerr-McGee site, Wyo.	**41**	Badlands National Park, S.D.
17	Jim Pitts site, S.D.	**42**	Tulare Lake sites, Calif.
18	Trout Creek Folsom site, Colo.	**43**	McHaffie site, Mont.
19	Southern Nevada "black mats"	**44**	Sunshine site, Nev.
20	Black Mountain Folsom site, Colo.	**45**	Vermillion Lake site, Alta.
21	Elgin mammoth locality, Ariz.	**46**	Niska site area, Sask.
22	Simon Clovis site, Idaho	**47**	El Bajillo site, Sonora
23	Frazer and Jurgens sites, Colo.	**48**	Nall site, Okla.
24	Jones-Miller site, Colo.	**49**	J. B. Spring, East Rainy Butte, N.D.
25	Clarks Flat site, Calif.	**50**	Barger Gulch, Wyo.

other than bison. Instead they contain Folsom, Goshen-Plainview, Agate Basin, and Hell Gap artifacts at many sites in the Great Plains. The Z_2 contact of β_{2a} with β_1 is conformable in some areas and erosional in others, and merges with the Z_1 contact at the edges of the β_1 channel to mark the Z_{1-2} surface. It forms the topography of the Clovis-age landscape.

Stratum β_{2b} represents eolian silt and/or fine sand that was thickened in low areas by slope wash and redeposited as overbank alluvium in some areas. A brief episode of erosion is apparent at some sites by the Z_3 contact, representing a brief hiatus that occurred sometime between 10,000 and 9000 RCYBP (9150–9400 and 8600 RCYBP). The transition from Paleoindian to Archaic lifestyles occurred during β_{2b} deposition, which was followed by a brief period of stability before a major episode of erosion and deflation occurred that removed β_2 in many areas, particularly along trunk streams, during the Altithermal warm, dry period.

Specific Localities with Younger Dryas–Age "Black Mats"

The evidence of this sequence is provided by the geochronology of more than 50 sites or localities (Table 1 and Figure 5), only a few of which are presented here. Proceeding more or less from north to south, the following numbers refer to localities of Table 1 and Figure 5.

7. The Knife River Flint region of North Dakota contains the Leonard paleosol 13,000–7000 RCYBP (Figure 7) as the stratigraphic host for Folsom and Goshen-Plainview

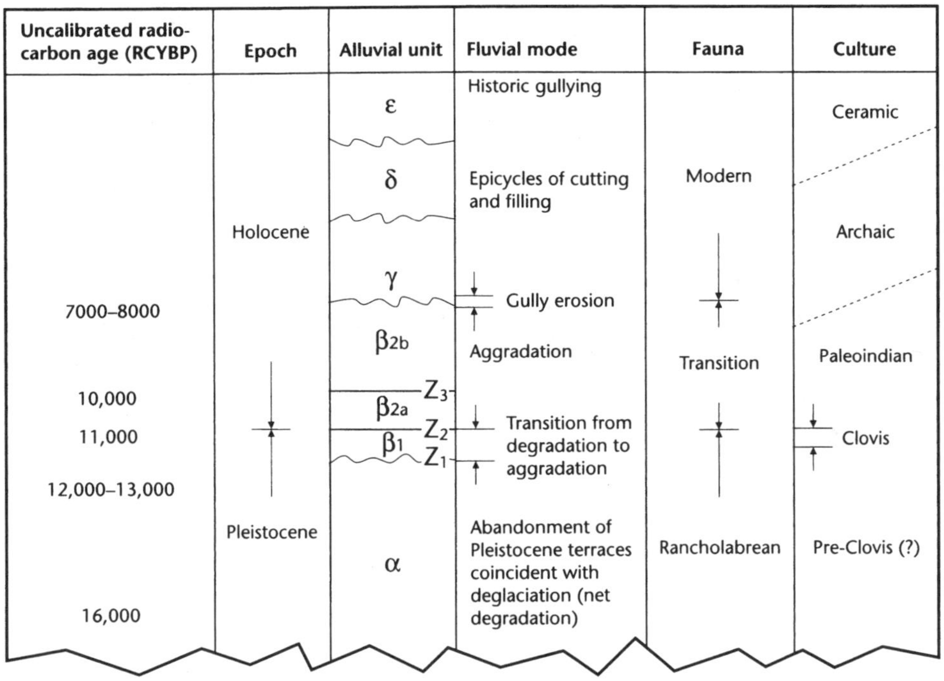

Figure 6. Generalized stratigraphic sequence for the Pleistocene-Holocene transition modified after Haynes 1984.

archaeology (Kuehn 1996; Root et al. 1996; Van Nest 1995). Contamination by Cenozoic lignite and by younger soils appears to have affected the radiocarbon ages of what is probably a Younger Dryas–age mollic paleosol.

31. At the Arc site in New York near the southern edge of Lake Erie, fluted points, bifaces, and a hollow-base bone point occurred on a glacial till and were buried by a black organic peat deposit (Figure 8) with a Younger Dryas age of 10,370 RCYBP. The occupational surface is an erosional contact that truncates lacustrine deposits of Lake Tonawanda, indicating desiccation of the lake by the time Paleoindians first occupied the site (Tankersley et al. 1997). At the nearby Hiscock site, Clovis artifacts appear to be associated with a period of low water and a feature that could be a waterhole dug by mastodons (Laub 1994; Laub and G. Haynes 1998). At both Arc and Hiscock the correlation with the Clovis drought in the West and the Allerød warm period of northern Europe seems reasonable.

16. The Carter/Kerr-McGee site of northeast Wyoming contains a Folsom component a in black argiaquoll with a date

STANLEY A. AHLER

Figure 7. The Leonard paleosol (β_{2a}) exposed at Lake Ilo, in the Knife River flint area of North Dakota (7, Figure 5).

K. B. TANKERSLEY

Figure 8. At the Arc site on the southern edge of Lake Erie (**31,** Figure 5), black peat of Younger Dryas age overlies Clovis artifacts resting on glacial till.

of 10,400 ± 300 RCYBP (Frison 1991; Reider 1990) and stratigraphically above a black level containing a Goshen-Plainview projectile point.

10. At the Agate Basin site in east central Wyoming, several black organic bands separated by colluvial fine sand representing Younger Dryas time host an archaeological sequence with Folsom overlain by Agate Basin and Hell Gap technocomplexes (Frison and Stanford 1982).

8. At the Lindenmeier site in northern Colorado, the Folsom levels are in a black aquoll with three ^{14}C ages averaging 10,600 ± 60 RCYBP and truncated by the Z_3 contact about 9600 RCYBP (Haynes, Beukens, et al. 1992). The paleoaquoll is of Younger Dryas age.

11. Recent investigations of May (1992) in the Medicine Creek Basin of Nebraska have shown buried soil YY of Schultz and others (1951) to be between 10,700 and 10,000 RCYBP, essentially Younger Dryas age. The Cody complex archaeology of Davis (1962) occurs on top of this mollic paleosol (Conyers 2000).

5. At the Clovis-type site in Blackwater Draw, New Mexico, Stratum β_1 is represented by four substrata (B_1, B_2, B_3, and C) of spring-laid sand separated by brief erosional intervals believed to represent desiccation and drought between 13,000 and 11,000 RCYBP (Haynes 1991a), during the Bølling-Allerød periods. The fluctuating spring discharge during this time may reflect the climatic fluctuations of the Bølling-Allerød periods such as the Inter Allerød Cold Period. Clovis folk were there during the Allerød period and before the deposition of the Folsom-age diatomite (Stratum D) in a Younger Dryas–age pond that dried to a wet-meadow shortly after 9800 RCYBP.

Green (1992) has always claimed that the earliest Clovis presence coincides with the deposition of the gray sand (Stratum B). It could, therefore, be as old as 12,000 RCYBP. I have suggested that the two or three Clovis artifacts found therein could have sunk into unit B when it was fluidized by ascending spring water (Haynes, Saunders, et al. 1992). Inspection of two of the projectile points in 1997 revealed that the Clovis point is highly polished, typical of spring action; but the other point is unpolished obsidian and almond-shaped, not unlike some Chindadn points from central Alaska. It may be as old as Green proposed or it could be intrusive. Unfortunately, neither find is adequately documented.

6. Stratigraphy of the Lubbock Lake site, Texas, like Blackwater Draw, contains a Folsom level in a Younger Dryas–age diatomite overlying fluvial sand with Clovis-age radiocarbon dates (Holliday 1985; Stafford 1981).

14. At the Aubrey site in Texas, the E/C contact (Z_{1-2} contact) is the surface occupied by the Clovis people, with two activity areas dated 11,560 RCYBP. Dark gray organic clay of Stratum E (Stratum b_{2a}) has two radiocarbon ages of 10,390 ± 80 and 10,940 ± 80 RCYBP (Ferring 1995). It appears to be the result of a rise in the water table during Younger Dryas time.

19. Numerous black mats in southern Nevada occur in association with spring deposits, and most are of Younger Dryas age (Quade et al. 1998).

The upper San Pedro Valley of Arizona (**1–3,** Figure 5) has four Clovis sites, two more probable Clovis sites, and skeletons of at least eight young mammoths of Clovis age but without artifacts (Haynes 1998). All are directly under black mats of Younger Dryas age (Figure 9) (Jull et al. 1999).

1. At Murray Springs, a bison kill and a mammoth kill by a

Figure 9. The black organic mat (β_{2a}) at the Murray Springs Clovis site, Curry Draw, Arizona (**1,** Fig. 5) covers the Clovis occupation surface (Z_{1-2}). Extinction of the Pleistocene megafauna occurred on this contact, and only *Bison b. antiquus* survived elsewhere.

dry streambed (β_1) occurred 10,940 RCYBP upon an erosional surface (Z_{1-2}) covered by a black organic mat (β_{2a}) (Figure 10) with a white marl facies, both of which formed during Younger Dryas time between 10,800 and 9700 RCYBP (Jull et al. 1999; Taylor et al. 1996).

On the left bank of Horsethief Draw 4 km southeast of Murray Springs, a young mammoth that may have escaped the hunters died with two Clovis points in its chest at the Escapule site and was buried by the black mat (β_2) (Hemmings and Haynes 1969).

2. At the Lehner site (Haury et al. 1959) 16 km south of Murray Springs, the remains of 12 mammoths and 13 Clovis points are dated 10,980 RCYBP (Taylor et al. 1996); and are associated with Mammoth Kill Creek (β_1), located directly under the "black mat" (β_{2a}) that formed 10,800 to 9700 RCYBP (Haynes and Haury 1982).

3. Located 18 km to the east on the left bank of Greenbush Draw, the Naco mammoth skeleton with eight Clovis points in direct association (Haury et al. 1955) is probably one that got away from the mammoth hunt at Lehner. It lies directly under pond facies of the black mat (β_{2a}).

Upstream, a tenth of a kilometer at the Navarrete site (Huckell n.d.) are the remains of another mammoth with

Figure 10. At the Murray Springs Clovis site (**1,** Figure 5) the black mat and its marl facies (white) cover the channel sand (β_1, contact Z_2) and the Clovis-age surface (Z_{1-2}) away from the channel. The upper contact of β_{2a} with β_{2b} has been made irregular and diffuse through soil formation (pedoturbation).

probable association with Clovis artifacts under dark gray mud facies of the Younger Dryas–age black mat (β_{2a}).

A kilometer farther upstream, the Leikem mammoth site (Johnson and Haynes n.d.), in the right bank, may be another mammoth with a single Clovis point in it that got away from a Clovis attack. It lies on channel sand (β_1) and directly under dark gray mud facies of the black mat; that is, on the Z_{1-2} contact.

Skeletons of young mammoths on the Z_{1-2} contact but without artifacts (Haynes et al. n.d.) include the Schaldack site with two skeletons, the Donnet mammoth with a healed broken rib, the BLM mammoth (Figure 11) and two more downstream, all directly under the black mat (β_{2a}). These are all within 5 km of Murray Springs (**1,** Figure 5). There are many other occurrences of the Younger Dryas–age black mat in the Upper San Pedro Valley. Only those known to have faunal associations have been mentioned here.

Throughout the United States, no extinct megafaunal remains other than bison, and no in situ Clovis artifacts, have ever been found within the Younger Dryas deposits or black mats (Figure 11); and no Folsom artifacts have ever been found in situ in underlying strata or in conclusive association with mammoth. The Pleistocene termination of horse, donkey, dire wolf, lion,

Figure 11. At the BLM (V005) mammoth locality (**1,** Figure 5), without artifact association, the mammoth skeleton lay upon the Clovis-age surface (Z_{1-2}) and was buried by the black mat (β_{2a}). The mammoth tooth, in contact with the black mat, represents one of the last Rancholabrean faunal elements and the β_{2a}/α contact (Z_{1-2}) is a prime candidate for the Pleistocene-Holocene boundary in the San Pedro Valley of Arizona.

camel, mammoth, and others appears to have been geologically instantaneous.

Causes of Pleistocene Extinction

The occurrence of so many Clovis sites with stratigraphic evidence of drought toward the end of the Allerød warm period, and the termination of the Pleistocene megafauna at the onset of the Younger Dryas return to glacially cold conditions, make possible several explanations for the cause of the extinction. Martin's (1967) overkill hypothesis makes humans the sole cause. Lundelius and Graham (1999) invoke climate. MacPhee and Marx (1999) suggest a hyperdisease did in the fauna. I do not believe any one of these alone was responsible. Instead, I see the incredible coincidence of drought, rise of the Clovis population, and the return to a glacial climate as Younger Dryas began as indicating multiple causes of extinction. Animals under stress concentrated at dwindling water sources only to be done in by Clovis hunters, but many relatively young tender mammoths in the San Pedro Valley of Arizona died without Clovis impact. Did a long-lasting Younger Dryas deep freeze deny water to them? It is another hypothesis to add to the hopper. Nowhere is the abruptness of the Rancholabrean termination better exemplified than in the San Pedro Valley of Arizona, where skeletal remains, abundant on the Z_{1-2} contact, are absent from the black mat (Figure 11). This absence can be tested wherever Stratum β_{2a} occurs.

Pre-Clovis Strata and Pre-Clovis Sites

At many stratified Clovis and Folsom sites, as well as at thousands of geological localities throughout North America, there are late-Pleistocene pre-Clovis-age strata (α and early β_1), commonly with associated Rancholabrean fauna, yet without any evidence of human presence. However, the three best potential exceptions are the Meadowcroft Rockshelter in Pennsylvania, the Schaefer and Hebior mammoth localities in Wisconsin, and the Cactus Hill site in Virginia.

The age of Stratum IIa at Meadowcroft still remains unresolved (Adovasio et al. 1990; Haynes 1991b) in spite of reports to the contrary (Adovasio et al. 1999; Goldberg and Arpin 1999). Repeated pleas for the radiocarbon analysis of nuts and seeds that could resolve the questions of age and potential contamination have gone unheeded. The advantage of radiocarbon-dating seeds is that contamination by mobile organics would not be a problem. These would be completely removed during standard chemical pretreatment, whereas the "charcoal" from hearths in lower Stratum IIa is so soluble that most of it goes into solution during pretreatment, raising a question of its true nature and origin (Haynes 1991b). Further, the soluble organics are approximately twice the age of the organic residue. So what is the source of these humic acids and how did they get into the "charcoal" if not by ground water?

In keeping with the seed-dating approach are two recent radiocarbon analyses of hawthorn pits from the Shawnee-Minisink Clovis site of ca. 10,900 RCYBP (Dent 1999 and pers. comm.) that substantiate the earlier, less precise radiocarbon ages (McNett 1985). Comparable results from seeds from lower Stratum IIa at Meadowcroft could validate previous radiocarbon ages. Other results could either demonstrate intrusion or indicate an early-Holocene occupation. Furthermore, any humic acids extracted from seeds from Stratum IIa should also be dated. Regardless of the outcome, the excavators at Meadowcroft could then be commended for applying the scientific method much as did Wendorf in regard to Wadi Kubbaniya in Egypt, where 18,000 RCYBP barley proved to be very late and therefore intrusive (Desruisseaux 1983).

In fact, the Miller Lanceolate projectile point from Meadowcroft Stratum IIa is not unlike some of the Chesrow projectile points from Wisconsin (Overstreet 1993), which Overstreet (1996) now believes may be pre-Clovis on the basis of recent dating of mammoth bones in southeastern Wisconsin. Bone XAD gelatin indicates ages between 12,310 ± 60 RCYBP and 12,520 ± 50 RCYBP, in close agreement with two spruce wood dates of 12,220 ± 80 RCYBP and 12,480 ± 130 RCYBP (Overstreet and Stafford 1997). Both the Schaefer mammoth and the Hebior mammoth are "intimately" associated with chipped-stone tools (Overstreet 1996; Overstreet et al. 1993, 1995). The radiocarbon ages for the Schaefer and Hebior mammoth sites suggest human occupation of the western side of the Michigan basin between the inner and outer Port Huron glacial advances and during the high stand of the Glenwood II phase of glacial Lake Michigan (Overstreet 1993, Figure 6). The sites are located about 190 km south of the Two Creek forest locality; if valid, they would have been occupied before the Two Creek forest bed was flooded and likely, therefore, before an ice-free corridor became passable. Unfortunately, the stone artifacts are not diagnostic (no projectile points have been found). If they are a part of the Chesrow complex as proposed by Overstreet (1996), they appear to be something other than Clovis—perhaps proto-Clovis? On the other hand, on typological grounds Chesrow projectile points appear to be early Archaic or late Paleoindian rather than early Paleoindian. Here again, skeptical inquiry and on-site evaluation are needed to have a Folsom-like breakthrough.

The Cactus Hill site in Virginia contains a Clovis level in eolian dunes adjacent to the Nottoway River. It also contains a flake and blade level 10–15 cm below the Clovis level (Johnson 1998; McAvoy 1997). Both Joe McAvoy and Mike Johnson, though working independently, have welcomed site visitation and skeptical inquiry at Cactus Hill, and are testing multiple hypotheses. The "pre-Clovis" artifacts are mostly made of quartzite that is readily available in local river gravels, whereas the Clovis level contains artifacts of several nonlocal lithic sources. Radiocarbon ages on dispersed charcoal flecks and organic matter range from about 16,000 to 9250 RCYBP, indicating that redeposition of older charcoal may be a problem rather than "down drift" of younger charcoal as claimed (McAvoy 1997). The 9250 RCYBP charcoal came from *under* a large tool, and associated decayed root fragments produced a postmodern radiocarbon age. The pre-Clovis level may actually be an earlier Clovis level in which the two short ("pentagonal") points of exotic lithic material are simply severely reworked Clovis points discarded because the makers considered them no longer serviceable. The points are unusually thin, with all evidence of fluting, if there ever was any, completely removed. Perhaps the almost exclusive use of local quartzite is because the makers were pioneers who had yet to discover other chert sources such as Cattail Creek at the Williamson Clovis site about 16 km to the west. By the time of the second occupation, years or decades later, other and better lithic sources had been found.

Maybe the "pentagonal" points are proto-Clovis artifacts made by Clovis progenitors. In fact, they are similar to some forms of Chindadn-Nenana points from central Alaska (Cook 1996), and Yesner (1998) has made a strong case for the Nenana sites with Chindadn points being sites of the pioneering occupants 11,800 RCYBP. Mammoths appear to have become extinct in Alaska before 12,000 RCYBP. The Nenana people may have known of them only from skeletal remains and perhaps oral tradition. If they eventually reached the interior plains of Canada a few decades later, the sight of living mammoths would have been an awesome experience. Upon finding that puny Chindadn points were not very effective in dispatching mammoths, they soon developed the Clovis point (Haynes 1987; Pearson 1997). Perhaps Cactus Hill has evidence of a pioneering band of Nenana folk and Clovis progenitors. If so, their arrival there would probably be no earlier than about 11,700 RCYBP to allow a century to get from central Alaska to the coastal plain of Virginia. On the other hand, progenitors in Alaska could be older than the oldest Nenana ^{14}C date presently known; for example, Bluefish Caves (**2,** Figure 1), where mammoth bone is dated 12,845 ± 250 RCYBP and musk ox bone is dated 14,370 ± 130 RCYBP (Cinq-Mars and Morlan 1999).

The Exploration of North America 11,700–11,600 RCYBP

Since the models of Haynes (1966) and Martin (1967; Mosimann and Martin 1975), periodically there have been several new models of the colonization of the New World (e.g., Hassan 1981; Meltzer 1995). Hoffecker and others (1993) suggest that the vegetation of Beringia from 14,000 to 12,000 RCYBP was a complex mosaic of arctic tundra and steppic river valleys, outwash plains, and south-facing slopes. They consider that people were not present in North America before 12,000 RCYBP and that Beringia was colonized at the beginning of the Allerød in response to environmental change including the reappearance of trees that would provide firewood. Their view that an ice-free corridor was open at least by 14,000 RCYBP was before the more recent evidence mentioned earlier. Steele et al. (1995) consider the ice-free corridor, open between 13,000 and 12,000 RCYBP, to be the most likely route to the mid-continent and beyond. Whittington and Dyke (1984), experimenting with the Mosimann and Martin (1975) model, eliminated the wave front concept and concluded that extinction could occur in as little as 250 years. Kelly (1996) emphasized the fact that there are no ethnographic analogs for the peopling of America.

The most recent model is that of Anderson and Gillam (2000), who considered the least-cost pathway to be northward along the Arctic coast and then southward down the Mackenzie valley (which, as stated earlier, Lemmen and others (1995) consider blocked by glacial Lake Mackenzie until ca. 11,000 RCYBP) by people who had no concept of what lay ahead. They considered founding population sizes of 25, 50, and 175 individuals, doubling every 100, 134, and 268 years; group fissioning occurred when the population reached twice the average group size. They considered group range values, before fissioning, of 25-, 50-, 100-, 200-, and 400-km-diameter areas. Starting with an initial population of 25 doubling every 268 years, when group (band) size reaches or exceeds 50 and a range diameter of 25 km, they fill the New World in 4,556 years with at least 3 million people. However, they conclude that the time would be much less when areas of ice, water, deserts, etc. are avoided. Their most liberal time to populate the hemisphere is 600 years. As Anderson and Gillam (2000) conclude, the modeling possibilities are boundless.

In the model that follows, I consider that part of the founding population of interior Alaska are pioneers who seasonally explore new ground by following river valleys southeastward, observing active deglaciation and taking advantage of new discoveries. The routes taken are derived from maps showing detailed relief data (Trident Press International 1998).

The Nenana occupation of Central Alaska began at least by 11,800 RCYBP. Both the Broken Mammoth and Swan Point sites (**3,** Figure 3) occur in the Tanana River Valley, which leads headward to the divide with the upper Yukon River basin. Arctic ground squirrel nests in the Klondike District (**4,** Figure 3) with radiocarbon ages between 13,350 and 11,170 RCYBP indicate arctic steppe conditions in the region Nenana people would cross in moving up the Yukon Valley or the Pelly Valley into the deglaciating Yukon Plateau. Over several generations, Nenana people may have explored the Yukon Plateau as far south as the upper Liard Valley; and witnessed annual widening of the corridor between the Mackenzie and St. Elias ranges under the influence of relatively warm, dry summer climate (MacDonald and McLeod 1996) during Allerød warming. Each summer they may have explored farther to the south while subsisting on edible plants, big game, fish, and waterfowl along lakes and rivers draining from stagnating ice and retreating valley glaciers. Base camps may have shifted farther south every year or so. At the same time, conditions on the Interior Plains continued to improve as Laurentide ice retreated. Lakes and rivers could be traversed by boat and on ice in winter, and dogs may have been used as pack animals (Henderson 1994). A hypothetical scenario as to how this movement might have taken place follows.

One summer may have taken a dozen Nenana people from the upper Tanana Valley to the divide with the White River Valley, then via the Yukon Valley to the Pelly River in the Tintina Trench and a winter camp near Faro. Next spring the Pelly Valley would lead them around the northeast flanks of the Pelly Mountains and over a divide near Finlayson Lake to the upper Frances River. The Frances Valley would lead them to the Liard Plains near Watson Lake, from where, with boats if a proglacial lake existed, they would follow the Liard Valley to Nelson Forks. As they exited the mountains, they would have their first view of the vast Interior Plains of

Canada, extending to the east as far as the eye could see. Laurentide ice would be at least 200 km to the east.[2]

At this point they might decide to return to the Yukon Plateau to inform others of their discovery. The return journey could be made in half the time of the outward journey because of travel over familiar ground and no side trips to explore tributaries. The next summer they returned to Nelson Forks with a much larger group, perhaps 50 people, the leader having talked several other bands into joining what amounted to a tribe for exploration southward.

Early on, the question had to arise as to where the southward exploration might lead them. It is very likely that there were people whose curiosity and quest for knowledge of the beyond would urge them ever southward. Eventually a band leader would decide to go as far south as possible and not return to the base camp as winter approached. This decision could have taken them from a winter camp in the Nelson Forks area of British Columbia to perhaps Lake Peace in Alberta by the end of summer. While camped on the north side, exploratory excursions south of the lake could be made by crossing on the winter ice. By spring time they would be ready to continue southward.

Up to this point, living mammoths may have been unknown to Nenana folk. The former existence of mammoths in central Alaska was known by the finding of skeletal remains that provided bone and ivory for tool making (Yesner 1998). Folklore may also have provided word of the former existence of these creatures. The first sign of living mammoths may have been the discovery of spoor, dung, and tracks south of Lake Peace, which would have created great excitement and a strong incentive to move onward.

As the band moved southeastward, paralleling the foothills of the Canadian Rockies, they probably avoided the Laurentide ice and recently deglaciated areas. Even so they had to cross many drainages carrying snowmelt northeastward out onto recently deglaciated plains, with many lakes and abandoned channels formed as drainage integration responded to isostatic adjustment and outburst floods as ice dams gave way. Some band members may have become victims of such catastrophes; but in spite of the geologic hazards, the abundance of deer, elk, ducks, and geese made for good hunting. While dried meat and pemmican were basic staples, they probably consumed fresh meat and fruits as the opportunities arose.

The first sighting of living mammoths may have been between Dawson Creek and Grand Prairie in Alberta, where poplar trees were growing around Boone Lake by 11,700 RCYBP (White et al. 1985). This sight created intense jubilation among the hunters. Hunting was delayed as they observed the herd in order to develop a hunting strategy. The first attack was a miserable failure, as they succeeded only in wounding a young female mammoth and infuriating the matriarch. Their atlatl spears were obviously too small and light to penetrate mammoth hide effectively. The more they attacked the mammoths, the more wary the herd became. This failure led to heavier spears tipped with larger stone points; but in order to facilitate hafting in a longitudinally notched foreshaft, someone decided to remove longitudinal flakes from both sides of the nearly completed biface. With this new armament and better understanding of mammoth behavior, hunters were eventually able to bring down a young adult female mammoth, perhaps near Red Deer, Alberta.

In the meantime, as small herds of bison were encountered, a new strategy had to be perfected. While not overly concerned at first, bison, like mammoths, soon became wary of this new predator on the prairie. Whereas mammoths provided bone and ivory as well as meat, the hides were too thick and heavy for portable tepee construction. Bison, on the other hand, provided larger hides than deer or elk, requiring fewer of them for tepees. However, they did require thinning by scraping to make them more flexible for easy handling and lighter for transport by dog travois. Deer and elk hides remained the preferred material for garments.

After wintering in the Red Deer area, perhaps one group decided to return to the Tanana Valley to tell others of their discoveries. The two groups agreed to meet at the Red Deer camp in the third summer. The main group moved farther southeast, hunting, fishing, and gathering whatever plant foods were edible. From the Cypress Hills they moved to a new winter camp near Manyberries, Alberta. The following summer they followed the Milk River into Montana, and from near Havre they moved up the Big Sandy Valley to the upper Missouri. Following the White Cliffs, they turned south when they reached the mouth of the Judith River; and continued south to Judith Gap, from where they took a southwesterly course. The Potter Creek Valley led them around the north end of the Crazy Mountains and into the Shields valley, where they made their sixth winter camp near Wilsall, Montana. The tribal leader's infant son, having drowned, was buried with great ceremony at the base of a small rock cliff and with many offerings of well-worked stone and bone at what we know today as the Anzick site (Jones and Bonnichsen 1994; Lahren and Bonnichsen 1974).

By this scenario it took them six years to travel 5,700 km from the upper Tenana Valley to the Anzick site, estimated by multiplying valley lengths on maps by a factor of 1.5 to estimate actual kilometers traveled. Assuming summer travel is from 1 May to 30 September, there are 153 days each year for six years for a rate of 6.2 km/day. If they traveled at this rate during the winters instead of making winter camps, the journey would have taken 2.6 years. Even if they traveled at half this rate, the trip could have been made in 6 to 12 years.

Hundreds to perhaps as many as 2,000 campsites, depending on summer-only or all-year travel, should have been left behind between the Tanana Valley and the Shields Valley of Montana. Discovery of the physical evidence of such a journey by the first group will be a very rare serendipitous event, but so are all Paleoamerican site discoveries.

A similar scenario can be invoked for the spread of Clovis

[2] The maps of Lemmen et al. (1995, Fig. 4e and d) show Laurentide ice in contact with interfluves along the Rocky Mountain front as late as 11,000 RCYBP with proglacial Lake Liard extending partway up the Liard Valley. This interpretation appears to be incompatible with data presented here on Figure 4, i.e., Charlie Lake Cave and Boone Lake would be under Laurentide ice until after 11,000 RCYBP.

people across the United States by doubling the straight-line map distances to account for meandering kilometers and exploration, and assuming a rate of only 2 km/day while exploring new ground. Eventually bands would have broken away from the main group to explore in new directions, with agreement to rendezvous every year or so. From Anzick via North Platte, Nebraska, to the Aubry Clovis site near Denton, Texas, is about 1,900 km; and would take about 5.2 years to go twice the distance. Anzick to Clovis, New Mexico, a distance of 1,550 km, would take about 4.2 years, Anzick to East Wenatchee about 2 years, Anzick to Kimmswick, Missouri, about 5 years, and about 3 more years to reach the Williamson site in Virginia. One band may have reached the Simonsen site in southern Idaho in less than 2 years and gone on to Borax Lake, California, and the Pacific coast in about 2 years. Near East Wenatchee, Washington, a band came under an ash fall of the Glacier Peak eruption about 11,200 yr B.P. (Mehringer and Foit 1990). In order to appease the gods as day turned into night, the shaman ordered offerings of the most magnificent Clovis artifacts ever known (Gramly 1993).

As North America became better known, rates of movement would likely have at least doubled when going from one known point to another via familiar territory. Movements were undoubtedly planned to visit occasionally known lithic source areas, known habitats of megafauna, and patches of desirable vegetation at the proper time of year. As bands proliferated, tribal rendezvous would likely have been annual events to exchange information, trade items, and genes. Band populations probably increased significantly under favorable conditions, but declined during times of drought. As the net decline of water tables during Allerød warming caused spring discharge and traditional watering places to dry up, megafauna under stress became especially vulnerable to hunters when concentrated around water holes. When water tables reached their lowest level, the climate suddenly reversed and became glacially cold during the Younger Dryas. Prolonged winter deep freezes may have further denied access to water for the megafauna. All but bison were terminated at the onset of Younger Dryas. Reduced numbers of bison survived in favorable areas because of their greater numbers. Human population may also have declined because Folsom evidence does not appear at the base of the immediately overlying stratum (β_{2a}) at most Folsom sites. Instead, as at the Blackwater Draw Clovis site, it is a few centimeters above the basal contact (Z_{1-2}) of the host stratum; that is, the diatomaceous earth (β_{2a}) of Younger Dryas age at Blackwater Draw.

Conclusions

The hypothetical scenarios presented here indicate that the peopling of North America south of the glacial margins could have occurred in much less time than the multiple centuries traditionally allowed for it. Five circumstances that support, but do not necessarily prove, the conclusion that a band of hunter-gatherers in central Alaska at least by 11,700 RCYBP could have moved through an ice-free route and into the interior plains of Canada by 11,600 RCYBP are: (1) Allerød global warming 11,800 to 11,000 RCYBP, (2) Nenana folk in the upper Tanana Valley in Alaska at the same time, (3) the appearance of mammoth steppe biome over an ice-free route at least by 11,600 RCYBP, (4) the TCI lasting at least 252 years between 12,050 and 11,750 RCYBP, and (5) Clovis folk in Montana and central Texas by 11,550 RCYBP and throughout the conterminous United States on the Z_{1-2} surface over the next five centuries. During the early Allerød warming peak at 11,700 RCYBP, the journey to Montana could have been accomplished in six summer seasons, or less if they continued southward during the winters. Imagine the excitement if Clovis folk or their successors met other people, perhaps in the Great Basin, people who had arrived along the West Coast via a coastal route (Mandryk and others, in press; Surovell, in press).

Acknowledgments

A very preliminary version of this paper was presented at the Clovis and Beyond Conference, Santa Fe, New Mexico, October 1998. I thank Robson Bonnichsen and Forest Fenn for organizing that excellent and successful gathering of professional and avocational archaeologists and geoarchaeologists. Support for my research has been provided by the National Geographic Society, the National Science Foundation (grants DBS-9119834 and SBR-9725044), and the Regents Professor research fund of the University of Arizona. Various versions of this paper have been read by Bruce Huckell, Richard Morlan, Stuart Fiedel, and Todd Surovell. Their comments and suggestions are appreciated as are those of four anomalous reviewers. Expert word processing was provided by Barbara Fregoso and technical editing was provided by Carol Gifford, both of the Department of Anthropology, University of Arizona. Jim Abbott of Tucson, Arizona, created the computer graphics.

References Cited

Adovasio, J. M., J. Donahue, and R. Stuckenrath
1990 The Meadowcroft Rockshelter Radiocarbon Chronology 1975–1990. *American Antiquity* 55:348–54.

Adovasio, J. M., D. Pedler, J. Donahue, and R. Stuckenrath
1999 No Vestige of a Beginning nor Prospect for an End: Two Decades of Debate on Meadowcroft Rockshelter. In *Ice Age Peoples of North America,* edited by R. Bonnichsen and K. L. Turnmire, pp. 416–31. Center for the Study of the First Americans, Oregon State University, Corvallis.

Anderson, D. G., and J. C. Gillam
2000 Paleoindian Colonization of the Americas: Implications from an Examination of Physiography, Demography, and Artifact Distribution. *American Antiquity* 65(1):43–66.

Broecker, W. S., and W. R. Ferrand
1963 Radiocarbon Age of the Two Creek Forest Bed, Wisconsin. *Geological Society of America Bulletin* 74(6):795–802.

Burns, J. A.
1996 Vertebrate Paleontology and the Alleged Ice-Free Corridor: The Meat of the Matter. *Quaternary International* 32:107–12.

Chamberlin, T. C.
1965 The Method of Multiple Working Hypotheses. *Science* 148: 754–59.

Cinq-Mars, J., and R. E. Morlan
1999 Bluefish Caves and Old Crow Basin: A New Report. In *Ice Age Peoples of North America,* edited by R. Bonnichsen and K. L. Turnmire, pp. 200–12. Center for the Study of the First Americans, Oregon State University, Corvallis.

Conyers, L. B.
2000 Subsurface Mapping of a Buried Paleoindian Living Surface, Lime Creek Site, Nebraska, USA. *Geoarchaeology: An International Journal* 15(8):799–817.

Cook, J. P.
1996 Chapter 6—Healy Lake. In *American Beginnings,* edited by F. H. West, pp. 223–27. University of Chicago Press, Chicago.

Davis, E. M.
1962 Archaeology of the Lime Creek Site in Southwestern Nebraska. University of Nebraska State Museum, Special Publication No. 3.

Deller, D. B., and C. J. Ellis
1988 Early Palaeo-Indian Complexes in Southwestern Ontario. In *Late Pleistocene and Early Holocene Paleoecology and Archeology of the Eastern Great Lakes Region,* edited by R. S. Laub, N. G. Miller, and D. W. Steadman. *Bulletin of the Buffalo Society of Natural Sciences* 33:251–63.

Dent, R.
1999 Shawnee Minisink: New Dates on the Paleoindian Component. Abstracts of the 64th Annual Meeting of the Society for American Archaeology, Chicago.

Denton, G. H., and M. Stuiver
1967 Late Pleistocene Glacial Stratigraphy and Chronology, Northeastern St. Elias Mountains, Yukon Territory, Canada. *Geological Society of America Bulletin* 78(4):485–510.

Desruisseaux, P.
1983 The Archaeological Revolution That Wasn't. *The Chronicle of Higher Education* 16 November 1983.

Dillehay, T. D. (editor)
1989 Monte Verde, a Late Pleistocene Settlement in Chile Vol. 1: Paleo-environment and Site Context. Smithsonian Institution Press, Washington and London.
1997 Monte Verde, a Late Pleistocene Settlement in Chile Vol. II: The Archaeological Context and Interpretation. Smithsonian Institution Press, Washington and London.

Dyke, A. S., and V. K. Prest
1987 Late Wisconsin and Holocene History of the Laurentide Ice Sheet. *Géographic Physique et Quaternaire* 61(2):237–63.

Fedje, D. W., J. M. White, M. C. Wilson, D. E. Nelson, J. S. Vogel, and J. E. Southon
1995 Vermillion Lakes Site: Adaptations and Environments in the Canadian Rockies during the Latest Pleistocene and Early Holocene. *American Antiquity* 60(1):81–108.

Ferring, C. R.
1995 The Late Quaternary Geology and Archaeology of the Aubrey Clovis Site, Texas: A Preliminary Report. In *Ancient Peoples and Landscapes,* edited by E. Johnson, pp. 273–81. Museum of Texas Tech University, Lubbock.

Fiedel, S. J.
1999a Artifact Provenience at Monte Verde; Confusion and Contradictions. *Discovering Archaeology* 1(6):1–23 (Special Report).
1999b Older Than We Thought: Implications of Corrected Dates for Paleoindians. *American Antiquity* 64(1):95–115.
2000 The Peopling of the New World: Present Evidence, New Theories, and Future Directions. *Journal of Archaeological Research* 8(1):39–103.
2000 Quacks in the Ice: Waterfowl, Paleoindians, and the Discovery of America. Society for American Archeology Abstracts of the 65th Annual Meeting, Philadelphia, p. 124.

Fladmark, K. R., J. C. Driver, and D. Alexander
1988 The Paleoindian Component at Charlie Lake Cave (HbRf39), British Columbia. *American Antiquity* 53(2):371–84.

Frison, G. C.
1991 *Prehistoric Hunter of the High Plains,* 2nd edition. Academic Press, San Diego.

Frison, G. C., and J. D. Stanford
1982 *The Agate Basin Site: A Record of the Paleoindian Occupation of the Northwestern High Plains.* Academic Press, New York.

Fullerton, D. S.
1980 Preliminary Correlation of Post-Erie Interstadial Events (16,000–10,000 Radiocarbon Years before Present), Central and Eastern Great Lakes Region, and Hudson, Champlain, and St. Lawrence Lowlands, United States and Canada. U. S. Geological Survey Professional Paper 1089.

Goebel, T., R. Powers, and N. Bigelow
1991 The Nenana Complex of Alaska and Clovis Origins. In *Clovis Origins and Adaptations,* edited by R. Bonnichsen and K. L. Turnmire, pp. 49–79. Center for the Study of the First Americans, Oregon State University, Corvallis.

Goebel, T, W. R. Powers, N. H. Bigelow, and A. S. Higgs
1996 Chapter 7—Walker Road. In *American Beginnings,* edited by F. H. West, pp. 356–63. University of Chicago Press, Chicago.

Goldberg, P., and T. L. Arpin
1999 Micro Morphological Analysis of Sediments from Meadowcroft Rockshelter, Pennsylvania: Implications for Radiocarbon Dating. *Journal of Field Archaeology* 26(3):325–44.

Gramly, R. M.
1988 Palaeo-Indian Sites South of Lake Ontario, Western and Central New York State. In *Late Pleistocene and Early Holocene Paleoecology and Archeology of the Eastern Great Lakes Region,* edited by R. S. Laub, N. G. Miller, and D. W. Steadman, pp. 265–80. *Bulletin of the Buffalo Society of Natural Sciences* 33.
1993 *The Richey Clovis Cache.* Persimmon Press, Buffalo.

Green, F. E.
1992 Comments on the Report of a Worked Mammoth Tusk from the Clovis Site. *American Antiquity* 57(2):331–37.

Guthrie, R. D.
1982 Mammals of the Mammoth Steppe as Paleoenvironmental Indicators. In *Paleoecology of Beringia,* edited by D. M. Hopkins, J. V. Matthews, Jr., C. E. Schweger, and S. Young, pp. 307–29. Academic Press, New York.

Hall, D. A.
2000 The North Atlantic Hypothesis. *Mammoth Trumpet* 15(2):1–7.

Hamilton, T. D., and T. Goebel
1999 Late Pleistocene Peopling of Alaska. In *Ice Age Peoples of North America,* edited by R. Bonnichsen and K. L. Turnmire, pp. 156–99. Center for the Study of the First Americans, Oregon State University, Corvallis.

Hansel, A. K., D. M. Michelson, A. F. Schneider, and C. E. Larson
1985 Late Wisconsinan and Holocene History of the Lake Michigan Basin. In *Quaternary Evolution of the Great Lakes,* edited by P. F. Karrow and P. E. Calkin, pp. 39–53. Geological Association of Canada, Special Paper 30.

Hassan, F. A.
1981 *Demographic Archaeology.* Academic Press, London.

Haury, E. W., E. Antevs, E. and J. F. Lance
1953 Artifacts with Mammoths Remains, Naco, Arizona. *American Antiquity.* 19(1):1–4.

Haury, E. W., E. B. Sayles, and W. W. Wasley
1959 The Lehner Mammoth Site, Southeastern Arizona. *American Antiquity.* 25:2–30.

Haynes, C. V., Jr.
1964 Fluted Projectile Points: Their Age and Dispersion. *Science* 145 (3639):1408–13.
1966 Elephant-Hunting in North America. *Scientific American* 214(6):104–12.

1969 The Earliest Americans. *Science* 166:709–15.

1980 The Clovis Culture. In *Ice-Free Corridor and the Peopling of the New World,* edited by N. W. Rutter and C. E. Schwager, pp. 115–21. *Canadian Journal of Anthropology* 1.

1984 Stratigraphy and Late Pleistocene Extinction in the United States. In *Quaternary Extinctions: A Prehistoric Revolution,* edited by P. S. Martin and R. G. Klein, pp. 345–53. University of Arizona Press, Tucson.

1987 Clovis Origin Update. *Kiva* 52(2):83–92.

1991a Geoarchaeological and Paleohydrological Evidence for a Clovis Age Drought in North America and Its Bearing on Extinction. *Quaternary Research* 35(3):438–50.

1991b More on Meadowcroft Radiocarbon Chronology. *The Review of Archaeology* 12(1):8–14.

1998 Geochronology of the Stratigraphic Manifestations of Paleoclimatic Events at Paleoindian Sites. Abstracts of the 63rd Annual Meeting of the Society for American Archaeology, Seattle.

1999 Monte Verde and the Pre-Clovis Situation in America. *Scientific American Discovering Archaeology* 1(6):17–19 (Special Report).

n.d. Younger Dryas black mats and other stratigraphic manifestations of climate change and extinction during the Pleistocene-Holocene transition in North America. Unpublished manuscript.

n.d. The Rancholabrean Termination: Sudden Extinction in the San Pedro Valley, Arizona, 11,000 B.C. Unpublished manuscript, 22 pp. Paper presented at the Proceedings of the International Conference on Mammoth Site Studies, University of Kansas, 1998.

Haynes, C. V., Jr., and E. W. Haury
1982 Archaeological Investigations at the Lehner Site, Arizona, 1974–1975. *National Geographic Society Research Reports* 14:325–34.

Haynes, C. V., Jr., R. P. Beukens, A. J. T. Jull, and O. K. Davis
1992 New Radiocarbon Dates from Some Old Folsom Sites: Accelerator Technology. In *Ice Age Hunters of the Rockies,* edited by D. J. Stanford and J. S. Day, pp. 83–100. University of Colorado Press, Niwot.

Haynes, C. V., Jr., J. J. Saunders, D. Stanford, and G. A. Agogino
1992 Reply to F. E. Green's Comments on the Clovis Site. *American Antiquity* 57(2):338–44.

Hemmings, E. T., and C. V. Haynes, Jr.
1969 The Escapule Mammoth and Associated Projectile Points, San Pedro Valley, Arizona. *Journal Arizona Academy of Science.* 5:184–88.

Henderson, N.
1994 Replicating Dog Travois Travel on the Northern Plains. *Plains Anthropologist* 39(148):145–59.

Hoffecker, J. F., W. R. Powers, and T. Goebel
1993 The Colonization of Beringia and the Peopling of the New World. *Science* 259:46–53.

Holliday, V. T.
1985 Archaeological Geology of the Lubbock Lake Site, Southern High Plains of Texas. *Geological Society of America Bulletin* 96(12):1483–92.

Huckell, B. B.
n.d. The Navarrete Site: A Third Clovis Locality on Greenbush Draw. Unpublished manuscript for *Clovis Occupation of the San Pedro Valley, Arizona,* edited by C. V. Haynes, Jr. and B. B. Huckell.

Jackson, L. E., Jr., and A. Duk-Rodkin
1996 Quaternary Geology of the Ice-free Corridor: Glacial Controls on the Peopling of the New World. In *Prehistoric Mongoloid Dispersals,* edited by T. Akazawa and E. J. E. Szathmary, pp. 214–27. Oxford University Press, New York.

Jackson, L. E., Jr., and G. M. MacDonald
1982 Paraglacial Origin for Terraced River Sediments in Bow Valley, Alberta. *Canadian Journal of Earth Sciences* 19(12):2219–31.

Jackson, L. E., Jr., F. M. Phillips, K. Shimamura, and F. C. Little
1997 Cosmogenic ^{36}Cl Dating of the Foothills Erratics Train, Alberta, Canada. *Geology* 25(3):195–98.

Johnson, A. E., and C. V. Haynes, Jr.
n.d. The Leikum Mammoth Site near Naco Arizona: A Possible Clovis Site. Unpublished manuscript for *Clovis Occupation of the San Pedro Valley, Arizona,* edited by C. V. Haynes, Jr. and B. B. Huckell.

Johnson, M. F.
1998 The Cactus Hill Site (44SX202) and Its Implications for the Peopling of the Southeast. Paper presented at the 63rd Annual Meeting of the Society for American Archaeology, Seattle.

Johnston, W. A.
1933 Quaternary Geology of North America in Relation to the Migration of Man. In *The American Aborigines: Their Origin and Antiquity,* edited by D. Jenness, pp. 9–45. University of Toronto Press, Toronto.

Jones, S., and R. Bonnichsen
1994 The Anzick Clovis Burial. *Current Research in the Pleistocene* 11:42–44.

Jull, A. J. T., C. V. Haynes, Jr., D. J. Donahue, G. S. Burr, and J. W. Beck
1999 Radiocarbon Ages at Murray Springs, Arizona and the Influence of Climate Change on Clovis Man. *Revue d'Archéometrie,* Actes du colloque "C14 Archéologie," 1998, pp. 339–43.

Kaiser, K. F.
1994 Two Creeks Interstade Dated through Dendrochronology and AMS. *Quaternary Research* 42:288–98.

Kehew, A. E., and J. T. Teller
1994 History of Late Glacial Runoff along the Southwestern Margin of the Laurentide Ice Sheet. *Quaternary Science Reviews* 13:859–77.

Kelly, R. L.
1996 Ethnographic Analogy and Migration to the Western Hemisphere. In *Prehistoric Mongoloid Dispersions,* edited by T. Akazwa and E. J. E. Szathmary, pp. 228–40. Oxford University Press, New York.

Kelly, R. L., and L. C. Todd
1988 Coming into the Country: Early Paleoindian Hunting and Mobility. *American Antiquity* 53:231–44.

Kuehn, D. D.
1996 The Aggie Brown Member of the Oahe Formation: A Late Pleistocene/Early Holocene Marker Horizon in Western North Dakota. *Current Research in the Pleistocene* 13:121–23.

Lahren, L., and R. Bonnichsen
1974 Bone Foreshafts from a Clovis Burial in Southwestern Montana. *Science* 186:147–50.

Laub, R. S.
1994 The Pleistocene/Holocene Transition in Western New York State: Fruits of Interdisciplinary Studies at the Hiscock Site. In *Great Lakes Archaeology and Paleoecology: Exploring Interdisciplinary Initiatives in the Nineties,* edited by R. I. McDonald, pp. 155–67. The Quaternary Sciences Institute, University of Waterloo, Waterloo, Ontario.

Laub, R. S., and G. Haynes
1998 Fluted Points, Mastodons, and Evidence of Late-Pleistocene Drought at the Hiscock Site, Western New York State. *Current Research in the Pleistocene* 15:32–34.

Lemmon, D. S., A. Duk-Rodkin, and J. N. Bednarski
1995 Late Glacial Drainage Systems along the Northwestern Margin of the Laurentide Ice Sheet. *Quaternary Science Review* 13:805–828.

Leverington, D. W., J. D. Mann, and J. T. Teller
2000 Changes in the Bathymetry and Volume of Glacial Lake Agassiz Between 11,000 and 9300 ^{14}C Yr B.P. *Quaternary Research* 54(-):174–181.

Lundelius, E. L., Jr., and R. Graham
1999 The Weather Changed: Shifting Climate Dissolved Ancient Animal Alliances. *Discovering Archaeology* 1(5):48–50, 53.

MacDonald, G. M., and T. K. McLeod
1996 The Holocene Closing of the "Ice-free"Corridor: A Biogeographical Perspective. *Quaternary International* 32:87–95.

MacPhee, R. D. E., and P. A. Marx
1999 Mammoths and Microbes: Hyperdisease Attacked the New World. *Discovering Archaeology* 1(5):54–56, 59.

Mandryk, C. A. S.
1996 Late Wisconsinan Deglaciation of Alberta: Processes and Paleogeography. *Quaternary International* 32:79–85.

in press The Ice-Free Corridor (or not?): An Inland Route by Any Other Name Is Not So Sweet nor Adequately Considered. Chacmool Proceedings, Calgary, Alberta.

Mandryk, C. A. S., H. Josenhans, D. F. Fedje, and R. W. Matthews
2000 Late Quaternary Paleoenvironments of Northwestern North America: Implications for Inland Versus Coastal Migration Routes. *Quaternary Science Reviews* 20:301–314.

Marshall, S. J., and G. K. C. Clarke
1999 Modeling North American Freshwater Runoff through the Last Glacial Cycle. *Quaternary Research* 52:300–315.

Martin, P. S.
1967 Prehistoric Overkills. In *Pleistocene Extinctions: The Search for a Cause,* edited by P. S. Martin and H. E. Wright, Jr., pp. 75–120. Yale University Press, New Haven.

May, D. W.
1992 Landforms, Alluvial Stratigraphy, and Late-Quaternary Environments in the Medicine Creek Basin, Southwestern Nebraska. Report for the Department of Anthropology, University of Colorado, Boulder, and the U. S. Department of the Interior, Bureau of Reclamation, Grand Island Office, Nebraska.

McAvoy, J. M.
1997 *Archaeological Investigations of Site 44SX202, Cactus Hill, Sussex County, Virginia.* Virginia Department of Historic Resources.

McNett, C. W., Jr. (editor)
1985 *Shawnee Minisink: A Stratified Paleoindian Archaic Site in the Upper Delaware Valley of Pennsylvania.* Academic Press, Orlando.

Mehringer, P. J., Jr., and F. F. Foit, Jr.
1990 Volcanic Ash Dating of the Clovis Cache at East Wenatchee, Washington. *National Geographic Research* 6(4):495–503.

Meltzer, D. J.
1995 Clocking the First Americans. *Annual Reviews of Anthropology* 24:21–45.

Morlan, R. E., A. S. Dyke, and R. N. McNeely
1999 Mapping Ancient History. Web page, http://www.geoserv.org

Mosimann, J. E., and P. S. Martin
1975 Simulating Overkill by Paleoindians. *American Scientist* 63: 305–313.

Overstreet, D. F.
1993 Chesrow: A Paleo-Indian Complex in the Southern Lake Michigan Basin. *Case Studies in Great Lakes Archaeology No. 2.* Great Lakes Archaeology Press, Milwaukee.

1996 Still More on Cultural Contexts of Mammoth and Mastodon in the Southwestern Lake Michigan Basin. *Current Research in the Pleistocene* 13:36–38.

Overstreet, D. F., D. J. Joyce, K. F. Hallim, and D. Wasion
1993 Cultural Contexts of Mammoth and Mastodon in the Southwestern Lake Michigan Basin. *Current Research in the Pleistocene* 10:75–77.

Overstreet, D. F., D. J. Joyce, and D. Wasion
1995 More on Cultural Context of Mammoth and Mastodon in the Southwestern Lake Michigan Basin. *Current Research in the Pleistocene* 12:40–42.

Overstreet, D. F., and T. W. Stafford, Jr.
1997 Additions to a Revised Chronology for Cultural and Non-Cultural Mammoth and Mastodon Fossils in the Southwest Lake Michigan Basin. *Current Research in the Pleistocene* 14:70–71.

Patterson, W. S. B., and C. U. Hammer
1987 Ice Core and Other Glacial Data. In *North America and Adjacent Oceans During the Last Deglaciation,* edited by W. F. Ruddiman and H. E. Wright, Jr., pp. 91–109. Geological Society of America, The Geology of North America, Boulder, Colorado, V. K-3.

Pearson, G. A.
1997 Non-Mongoloid Pleistocene Expansions: Old and New Ideas on the Origins of the First Americans. Paper presented at the 24th Annual Meeting of the Alaska Anthropological Association, Whitehorse, Yukon.

Platt, J. R.
1964 Strong Inference. *Science* 146:347–353.

Prest, V. K.
1969 Retreat of Wisconsin and Recent Ice in North America. Geological Survey of Canada, map 1257A.

Quade, J., R. M. Forester, W. L. Pratt, and C. Carter
1998 Black Mats, Spring-fed Streams, and Late-Glacial-Age Recharge in the Southern Great Basin. *Quaternary Research* 49(2):129–148.

Reeves, B. O. K.
1973 The Nature and Age of the Contact Between the Laurentide and Cordillera Ice Sheets in the Western Interior of North America. *Arctic and Alpine Research* 5:1–16.

Reider, R. G.
1990 Late Pleistocene and Holocene Pedogenic and Environmental Trends at Archaeological Sites in Plains and Mountain Areas of Colorado and Wyoming. In *Archaeological Geology of North America,* edited by N. P. Lasca and J. Donahue, pp. 335–360. Geological Society of America, Centennial Special Vol. 4, Boulder.

Root, M. J., D. MacDonald, and A. M. Emerson
1996 The Bobtail Wolf Site: An Update on Recent Excavations. *Current Research in the Pleistocene* 13:38–40.

Rutter, N. W.
1977 Multiple Glaciation in the Area of Williston Lake, British Columbia. *Geological Survey of Canada Bulletin* 273:1–31.

1980 Late Pleistocene History of the Western Canadian Ice-Free Corridor. *Canadian Journal of Anthropology* 1:1–8.

1984 Pleistocene History of the Western Canadian Ice-free Corridor. In *Quaternary Stratigraphy of Canada,* edited by R. J. Fulton, pp. 49–56. Geological Survey of Canada Paper 84-10.

Schultz, L. B., G. C. Lueninghoener, and W. D. Frankforter
1951 A Graphic Résumé of the Pleistocene of Nebraska. *Bulletin of the University of Nebraska State Museum* 3:1–41.

Stafford, T., Jr.
1981 Alluvial Geology and Archaeological Potential of the Texas Southern High Plains. *American Antiquity* 46(3):548–65.

Stafford, T. W., Jr., P. E. Hare, L. A. Currie, A. J. T. Jull, and D. Donahue
1991 Accelerator Radiocarbon Dating at the Molecular Level. *Journal of Archaeological Sciences* 18:35–72.

Steele, J., C. Gamble, and T. Sluckin
in press Estimating the Velocity of Paleoindian Expansion into South America. In *People as Agents of Environmental Change,* edited by R. Nicholson and T. O'Connor. Oxbow Press, Oxford.

Storck, P. L.
1988 The Early Palaeo-Indian Occupation of Ontario: Colonization or Diffusion? In *Late Pleistocene and Early Holocene Paleoecology and Archeology of the Eastern Great Lakes Region,* edited by R. S. Laub, N. G. Miller, and D. W. Steadman, pp. 243–250. *Bulletin of the Buffalo Society of Natural Sciences* 33.

Straus, L. G.
2000 Solutrean Settlement of North America? A Review of Reality. *American Antiquity* 65(2):219–26.

Stuiver, M., P. M. Grootes, and T. F. Braziunas
1995 The GISP 2 $\delta^{18}O$ Climate Record of the Past 16,500 years and the Role of the Sun, Ocean, and Volcanoes. *Quaternary Research* 44:341–54.

Surovell, T. A.
2000 Early Paleoindian Women, Children, Mobility, and Fertility. *American Antiquity* 65(3):493–508.
in press Can a Coastal Migration Explain Monte Verde? Submitted to *American Antiquity.*

Tankersley, K. B., S. Vanderlaan, J. D. Holland, and S. Bland
1997 Geochronology of the Arc Site: A Paleoindian Habitation in the Great Lakes Region. *Archaeology of Eastern North America* 25:31–44.

Taylor, R. E., C. V. Haynes, Jr., and M. Stuiver
1996 Clovis and Folsom Age Estimates: Stratigraphic Context and Radiocarbon Calibration. *Antiquity* 70(269):515–25.

Teller, J. T.
1995 History and Drainage of Large Ice-dammed Lakes along the Laurentide Ice Sheet. *Quaternary International* 28:83–92.

Teller, J. T., and A. E. Kehew
1994 Introduction to the Late Glacial History of Large Proglacial Lakes and Meltwater Runoff along the Laurentide Ice Sheet. *Quaternary Science Reviews* 13:795–99.

Trident Press International
1998 *The 21st Century World Atlas.* Naples, Florida.

Tumeo, M. A.
1999 Understanding between Scientists and Attorneys for Better Service to Business and Community. *The Federal Lawyer* 46(7):19–20.

Van Nest, J.
1995 Chapter 4—Geology of the Alkali Creek site. In *Alkali Creek: A Stratified Record of Prehistoric Flint Mining in North Dakota,* edited by M. D. Metcalf and S. A. Ahler, pp. 53–109. Metcalf Archaeological Consultants, Inc, Eagle, Colorado.

Webster's New Collegiate Dictionary
1979 Riverside Press, Cambridge, MA.

White, J. M., R. W. Mathews, and W. H. Mathews
1985 Late Pleistocene Chronology and Environment of the "Ice-Free Corridor" of Northwestern Alberta. *Quaternary Research* 24:173–86.

Whittington, S. L., and B. Dyke
1984 Simulating Overkill: Experiments With the Mosimann and Martin Model. In *Quaternary Extinctions,* edited by P. S. Martin and R. G. Klein, pp. 451–65. University of Arizona Press, Tucson.

Wilmeth, R.
1968 A Fossilized Bone Artifact from Southern Alberta. *American Antiquity* 33:100–101.

Wilson, M. C.
1996 Late Quaternary Vertebrates and the Opening of the Ice-Free Corridor, with Special Reference to the Genus *Bison. Quaternary International* 32:97–105.

Wilson, M.C., and J. A. Burns
1999 Searching for the Earliest Canadians: Wide Corridors, Narrow Doorways, Small Windows. In *Ice Age Peoples of North America,* edited by R. Bonnichsen and K. L. Turnmire, pp. 213–48. Center for the Study of the First Americans, Oregon State University, Corvallis.

Wilson, M. C., and C. S. Churcher
1984 The Late Pleistocene Bighill Creek Formation and its Equivalents in Alberta: Correlative Potential and Vertebrate Paleofauna. In *Correlations of Quaternary Chronologies,* edited by W. C. Mahaney, pp. 159–75. Geo Books, Norwish, UK.

Yesner, D. R.
1998 Human Colonization of Eastern Beringia and the Question of Mammoth Hunting. Paper presented at the International Conference on Mammoth Site Studies, University of Kansas, Lawrence.

Young, R. R., Burns, J. A., Smith, D. G., Arnold, L. D., and R. B. Rains
1994 A Single, Late Wisconsin, Laurentide Glaciation, Edmonton Area and Southwestern Alberta. *Geology* 22:683–86.

Envisioning Water Transport Technology in Late-Pleistocene America

Margaret A. Jodry

Abstract

Human use of watercraft dates back at least thirty thousand years and some researchers propose that this technology enabled the peopling of Australia closer to fifty thousand years ago. Due to the vagaries of preservation across millennia, direct evidence of Paleolithic rafts or boats is lacking despite other data substantiating their use. Archeologists infer the existence of Paleolithic watercraft from indirect evidence such as the transport of stone across bodies of water that separate locations where raw stone was acquired from places where it was used and discarded (i.e., obsidian gathered from islands inaccessible from the mainland without the use of boats). This article takes as its starting position the likelihood that watercraft technologies were among transportation and mobility options employed during early explorations and settlements of the Americas. While early colonization models increasingly envision possible maritime travel, they are relatively mute on possible uses for watercraft as aids in inland exploration, mobility, and settlement. This paper draws ideas from examinations of ethnographic and archaeological records to provide a springboard for expanding mobility models beyond their present tendency to view pedestrian travel as the dominant, if not nearly exclusive, mode of inland movement in late-Pleistocene America.

> By the end of the Pleistocene, man lived in many places that he could have reached only by crossing water. To find it necessary to postulate "land bridges" in accounting for his dispersal among the islands or across straits and wider bodies of water is to ignore the fact that for many inhabitants of the world's islands, seacoasts, and river banks, water has not separated places; it has joined them (Edwards 1965:v).

National Museum of Natural History, 304, MRC 112, Smithsonian Institution, Washington D. C. 20560; e-mail: Jodry.Pegi@nmnh.si.edu

Introduction

Technological skills of watercraft construction and navigation confer advantages with regard to moving people and goods; maintaining and promoting social networks across distances; exploring new country and resources; and extending one's reach while fishing, water fowling, sea mammal hunting, and collecting plants, shells, and rocks. This paper examines the proposition that watercraft technology was an integral part of many late-Pleistocene lifeways in the Americas, not only along coasts but *inland* as well (Engelbrecht and Seyfert 1994).

Recurrent movement from place to place is a reality for most hunter-gatherers, and studies of mobility represent productive avenues for investigating human adaptation and expansion at the end of the Pleistocene. Patterns observed in the directions and distances over which objects are carried, particularly those made of materials that can be traced to points of origin, provide indispensable clues to the movements of prehistoric peoples. To gain greater insight concerning patterns of landscape learning and mobility, we need to accurately model *how* travel was accomplished. In other words, what physical and social constraints and/or opportunities shaped human mobility at various times and places. Important in this regard are the possible roles played by transportation aids. To travel by foot, relying exclusively on human load bearing, is one proposition; to employ watercraft or dog-assisted carrying capacity is quite another.

Models of inland mobility and range expansion typically concern themselves with pedestrian travel. Ideas about how Paleoindian groups entered and expanded within the New World have traditionally focused on terrestrial/pedestrian transport. Recently, the probability of boat travel along the Pacific Coast has gained wider acceptance. Less consideration has been given to the question of how people migrating along the coast may have explored inland. This paper considers variation in ethnographic water transport technologies in the Americas to gain inspiration for current studies of late-Pleistocene human adaptations.

Archaeological evidence indicates watercraft technology developed early and remained an option exercised frequently, though differently, by prehistoric peoples thereafter where suitable construction materials and waterways permitted. Human remains from Santa Rosa Island off the California coast predict the use of watercraft in this area by 10,970 RCYBP, temporally coincident with Clovis culture. Maritime models of human colonization of the Americas increasingly consider the role of boats as transportation aids. In contrast, views of human backpacking and pedestrian travel continue to dominate predictive models of inland colonization and mobility, to the near exclusion of alternative possibilities. It is somewhat perverse that several inland colonization models for North America predict that people dispersed along rich biotic zones associated with river drainages, yet typically omit from discussion the potentially important role of water transport in this process. Comparative review of aboriginal transportation in the Americas indicates that "boats of some kind were found everywhere except where streams and other natural bodies of water were absent or too small to navigate" (Driver and Massey 1957:287). In response to a perceived need to more broadly address watercraft technology in models of late-Pleistocene colonization and adaptation in the Americas, I present a synopsis of archaeological evidence for early maritime travel and early use of watercraft in coastal, riverine, and lacustrine adaptations. The paper begins with a global focus, then returns to the Americas.

The earliest compelling evidence for human use of watercraft is the colonization of Australia by Paleolithic seafarers more than 40,000 years ago (Bowler and Magee 2000; Gamble 1994; Jones 1979; Mulvaney and Kamminga 1999; Figure 3). Even during periods of low sea level, an ocean trench nearly 2 km deep and 80 km wide separated the Sunda peninsula of Southeast Asia (Borneo, Java, and Sumatra) from the paleocontinent of Sahul (Australia and New Guinea). Sea crossings of up to 100 km were required "where the mariners could not have seen land in any direction" (Mulvaney and Kamminga 1999:111). By 28,740 yr B.P. people were living in rockshelters on Buka Island in the northern Solomon Islands (Wickler and Spriggs 1988). This colonization required a sea voyage of no less than 180 km from New Ireland, off the northeast coast of Sahul (Mulvaney and Kamminga 1999). The types of watercraft used are unknown, but rafts of bamboo or palm logs have been suggested.

By 30,000 years ago ancient people were navigating the open seas between the Japanese mainland and the Izu Islands (Oda 1990). Paleolithic travel in the Japanese archipelago is visible archaeologically, due to the presence on the Japanese mainland of obsidian originating on the outer islands of Kozushima and Onbasejima (Suzuki 1973, 1974). Despite a late-Pleistocene sea level lowering of 100–140 m, a wide strait made it impossible to obtain Kozushima obsidian without water transport such as dugout canoes or rafts (Aikens and Higuchi 1982; Oda, 1990:64). By 9000–8000 yr B.P. people of the Jomon culture possessed very highly developed watercraft technology, which they used to procure sea mammals and deep-water fish and to transport obsidian boulders weighing up to 19 kg (Oda, 1990:67).

Obsidian transport in the Aegean Sea began no later than 12,800 yr B.P., based on the identification of volcanic glass from the Cycladic island of Melos in Franchthi Cave 150 km away in Greece (Renfew and Aspinall 1990). It is important to note that the initial indications of early watercraft travel, primarily reported beginning in the 1970s, consist of indirect evidence for movement of people and lithic materials. Rare remains of prehistoric boats were discovered later; the oldest examples known to this author were reported in the 1990s.

Direct evidence of early boats includes the remains of two wooden pirogues excavated from Seine River sediments in the Paris Basin of France, radiocarbon dated to 8059 ± 53 RCYBP (ARC. 1197; 9245–8710 CALYBP) and 7991 ± 53 RCYBP (ARC. 1196; 9040–8620 CALYBP) (Bonnin 1995, 1996). The oldest watercraft excavated in Africa is a fully preserved dugout canoe from Dufuna in northeast Nigeria, found buried beneath 5 m of water-laid sediments. Radiocarbon dates for this dugout are 7670 ± 110 RCYBP (KI-3587) and 7264 ± 55 RCYBP (KN-4683) (Breunig 1996). Intriguing evidence of early boats from Scandinavia was excavated from the Huseby Klev site in western Sweden. Several pieces of resin dated between 9100 and 8600 yr B.P. exhibit impressions of wood slats and cordage preserved in what are thought to represent resinous coatings of boat sealant and caulking (Nordquist 1994, 1995). During the early occupation, the Huseby Klev site was located on an island in the outer Swedish archipelago, a location that further corroborates the use of watercraft.

These examples, and others from the Americas discussed below, demonstrate that the archaeological record of Pleistocene and early-Holocene watercraft technology is not silent. It *is*, however, decidedly soft-spoken, if not heavily reliant on sign language. As noted by McGrail (1998:2), who writes from the dual perspectives of seaman and archaeologist,

> The direct evidence for all forms of early water transport is undoubtedly poor and in some cases nonexistent, as has been recognized by Clark (1952:292), Cunliffe (1978:290) and Hassal (Wilde, 1978:79) when considering the Bronze Age, the Iron Age, and the Roman Period in N.W. Europe. On the other hand many archeologists have failed to evaluate and draw deductions from what evidence there is, and with some notable exceptions—for example Grahame Clark in his *Prehistoric Europe, the economic basis* (1952:84–90, 282–92—discussion of the maritime aspects of prehistoric and medieval Europe has been left to non-archeologists, often with an unsatisfactory outcome.

If deducing evidence of water transport for medieval times is a challenge, the narrow windows resulting from inadequate preservation and postglacial coastline inundation further daunt efforts to identify such evidence from the late Pleistocene. Regardless, the physical challenges are compounded by an intellectual predilection for land-oriented reconstruction of postglacial travel in the Americas.

New World coastal colonization models featuring maritime watercraft technology (Fladmark 1979; Heusser 1960) are being revisited with renewed focus (Dixon 1999, 2001; Erlandson and Moss 1996; Fedje and Josenhans 2000; Mandryk et al. 2001; Stanford and Bradley 2002). Meanwhile, models of inland colonization and adaptation are less inclusive (certainly less explicit) regarding the significant potential of water transport to facilitate range exploration and exploitation. Paradoxically, models that specifically highlight riverine corridors for early colonization and settlement (i.e., Anderson 1990, 1996; Anderson and Gillam 2000; Dincauze 1993) do not directly address the role of inland watercraft technology within these early adaptations. A single reference to watercraft by Anderson and Gillam (2000:49) suggests that some use of this technology "may have been needed to make a Northwest Coastal entry route attractive" to Paleoindians. Explicit discussion of inland water travel serves to expand understanding of long-distance social networks (Keenlyside 1985:84, 1991:170–72; Mayer-Oakes 1984:236; Wright 1994:52), early hunting and fishing strategies (Engelbrecht and Seyfert 1994:226), material transport options (Engelbrecht and Seyfert 1994:227; Frison and Bradley 1999:83), and logistical challenges involved in crossing the Mississippi River and its major tributaries.

To sharpen research focus regarding the potential contributions of water transport technology to Paleoindian adaptations, the following section presents ethnographic information regarding broad categories and uses of aboriginal watercraft. Next, selected examples of demonstrated and postulated uses of watercraft by late-Pleistocene and early-Holocene peoples in the Americas are discussed. The paper concludes that water transport technology has implications beyond late-Pleistocene models of coastal entry and maritime adaptation. The technology's potential contribution to Paleoindian mobility, social interaction, subsistence practices, and settlement patterns should be considered wherever navigable waterways and watercraft construction materials existed. The need to address inland watercraft travel explicitly when constructing models of human behavior during the late Pleistocene is clear.

Aboriginal Watercraft

Over 70 percent of the earth is covered by water. With the exception of the most arid environments, it is nearly impossible to travel significant distances without encountering a body of water that might present itself as a travel obstacle—or perhaps as a potential travel route. Variation in necessity, habitat, and climate coupled with human ingenuity and communication resulted in the development of many forms of aboriginal watercraft (Casson 1994; Edwards 1965; Herron 1998; Heyerdahl 1957, 1979; Hornell 1946; McGrail 1998). Rafts, reed boats, bark and dugout canoes, and skin boats provided native peoples with expedient and long-term, seasonal and year-round means of navigating New World waterways. I propose that Paleoindian counterparts to many of these craft are to be expected in the Americas, and that other forms unknown ethnographically should not be ruled out.

Floats, Rafts, and Reed Boats

There is general agreement that initial watercraft consisted of such simple flotation devices as inflated skins and buoyant wood (Hornell 1946; McGrail 1998). A fisherman in the Chilean Province of Coquimbo used a hide float as transport while fishing in 1944 (Edwards 1965:17) and another existed in use until 1975 (T. Dillehay pers. comm. 2003). Spanish chronicles report that the native peoples of Ica and Arica on the Chilean coast used rafts supported by sealskin floats to travel to islands in the west. Their watercraft consisted of two inflated pontoons made of male seal or sea lion skins, which were made waterproof by coats of seal fat overlain with additional layers of clay mixed with grease. The pontoons were tied together, and covered with a deck of wooden cross pieces (Edwards 1965:17–19).

> The hide floats were well adapted for launching and landing in the surf and on rocky shores. They offered little resistance to moving water, and floated like corks in rough seas that would be hazardous for larger and heavier craft (Edwards 1965:103).

Dillehay reports the use of such in the 1920s and 1930s.

Watercraft incorporating hide floats "may still be seen in everyday use from Africa and the Near East through southern Asia, and north through Tibet, China, and Manchuria to the Amur River" (Edwards 1965:90). They range from single skins used as swimming floats to large rafts with numerous skins lashed beneath a wooden framework. Log rafts were widely used on inland waterways wherever there were buoyant timber species. McGrail (1998:44) notes that rafts,

> have been used for inter-island traffic, in for example Oceania (Hornell, 1946:64–80), and in the coastal waters of S. America (Edwards 1965:64–80), Formosa (Taiwan) and India (Hornell, 1946:61–68), and off other countries between latitudes of 40°N and 40°S where sea temperatures are suitable for flow-through water transport.

Rafts

Rafts served quite often as an expedient or temporary form of watercraft throughout most of North America, both for seafaring and river- and lake-faring. In the interior of northern California, where streams were too small for canoe navigation, the Cahto people traveled on rafts made of five or six logs tied together (Myers 1978:246). Eastern Miwok fashioned rafts to cross streams in California's Sierra Mountains by lashing together a pair of logs. They used tule boats more widely in other portions of their range (Levy 1978:406). The Western Pomo of northern California regularly visited offshore islands to hunt seals and sea lions and to collect mussels. They made these ocean crossings "on rafts of driftwood bound together with vegetal fibers" (Bean and Theodoratus 1978:289). Lillooet of southwest British Columbia used rafts in addition to canoes.

Log rafts were used in still water. Single sheets of cottonwood, spruce, or red cedar (or in the Bridge River area, paper birch) were used for manufacturing bark canoes. Dugout canoes carved from red cedar, cottonwood, or Sitka spruce were still in use among the Lillooet in the 1970s. These watercraft were maneuvered up the rivers by use of canoe poles and in the Lillooet Lake by paddles (Bouchard and Kennedy 1998:181).

On the Northern Plains, Blackfoot people improvised rafts using travois and lodgepoles, and also by making a float of skin lodge coverings (Ewers 1955). Swimming adults pushed or towed these expedient craft, which could carry baggage and children. These rafts differed from bullboats described below.

Rafts were used frequently in the Southwest as an expedient means for crossing waterways (Driver and Massey 1957:288,

292). The people of San Felipe Pueblo in New Mexico traversed the Rio Grande by rafts in the late 1600s to reach farmlands, which extended up and downstream on either side of the river (Strong 1979:393). Cane rafts were employed by some groups in southeastern North America (Swanton 1946). In South America rafts were used extensively as ferries on rivers in Colombia and Ecuador (Edwards 1965:61).

Among the most sophisticated use of rafts were pre-Columbian sailing vessels on the northwest coast of South America. These aboriginal trading balsas, equipped with native cotton sails and paddles, were nearly the size of 16th-century Spanish ships. Bartholomé Ruiz, a pilot on reconnaissance for Pizarro in 1526, was one of the first Europeans to encounter a native seagoing balsa in the waters off the coast of northern Ecuador (chronicled in the *Relacion Samaano-Xeréz,* cited in Edwards 1965:67–68; Smith and Haslett 1999; Szaszdi 1978). These rafts could carry more than 25 tons and keep both cargo and 20 people dry. Balsa logs were tied together to form a floating platform or hull, which was overlain by a deck of cane. The use of more than one mast, and triangular fore-and-aft-rigged sails are reconstructed from historical records (Edwards 1965; Smith and Haslett 1999). A recently built replica, *La Mantena-Huancavilva,* voyaged for several weeks off Ecuador and Colombia before traveling to Costa Rica. During this trip, Smith and Haslett (1999) made the following observations:

> In the log-hull construction lies the tremendous security and stability of this vessel; shipped water, rather than swamping the vessel, slips between the logs, which remain buoyant. . . . The lack of a rudder, indicates that the vessel was steered by some other mechanism. This was undoubtedly the *guare* . . . or planks . . . inserted through gaps between the hull logs, forming vanes or skags below the waterline. The resistance of these planks to water beneath the vessel change the sailing balance, and thereby allow steering (Smith and Haslett 1999:17). . . . Modern yachts and sailboats can manage to sail up to 45 degrees into the wind. *La Mantena–Huancavilva* managed to sail within 60 degrees of the wind, which is comparable to the caravels used by early Spanish and Portuguese explorers in the 16th Century (Smith and Haslett 1999:26). . . . Clearly, the exploration and continued use of a maritime route on the Pacific coast of Western South America and Central America would have to be carried out by nothing less than fully-capable sailing craft. Those who assume that the northward-flowing Humboldt current in this region affords a simple drift route to the North do not appreciate the complexities of sailing vessels on this coastline (Smith and Haslett 1999:27).

Native sailing vessels are known also from Egypt by approximately 5500 yr B.P. Sails were made of woven reeds, papyrus, and linen with lines of palm, papyrus, or grass fibers (Casson 1994:21). Grass mat sails, apparently of aboriginal origin, are reported for *umiaks* in Alaska (Driver and Massey 1957:292).

Reed Boats

From prehistoric times, bundles of buoyant reeds harvested in marshy habitats were used to make raft-like boats (Edwards 1965:1). Both temporary as well as longer-lived watercraft were made to facilitate travel and subsistence activities on inland lakes and marshes and along the Pacific coast. Paiute in the western Great Basin traditionally used tule- and cattail-constructed watercraft on lakes and marshes to gather waterfowl eggs and to drive flightless molting ducks onshore for capture.

> The men made boat-like rafts of dry pithy tules tied in bundles, which they maneuvered in and out of the clumps of reeds in search of eggs, not caring whether they were fresh or partially incubated (Wheat 1967:9). . . . Back in the marshes the ducklings were getting fat and noisy and the old ducks were molting their flight feathers. Men on their tule rafts drove the helpless ducks out of the water onto land, where, half running and half airborne, they tried to escape the swiftest of the Indian runners in a hockey-like chase (Wheat 1967:11).

These boats were made of two cigar-like bundles of tules, tied with ropes of twisted cattails and bound together at the stern and prow. A gunwale made of cattail leaves prevented eggs, game, and weapons from sliding off the boat. An example illustrated by Wheat (1967:46) is about 2.8 m long and light enough to be easily lifted with one hand (see also Hunn et al. 1998:533). A boat of this size could carry one person, but was often used to carry weapons and game while the hunter waded or swam, pushing the boat ahead. Larger boats were made to carry two or more people. Poling rather than paddling propelled them. A single cutting tool was needed in their construction. In earlier times, Paiute used sickles made of deer scapula or "thin sheets of slate with fluted edges" to harvest and trim the reeds (Wheat 1967:41). After several hours of use, the boats absorbed water and became heavy, but did not sink. Drying onshore restored their former buoyancy. Tule balsas lasted no more than a single season and could be abandoned without an afterthought, to be built anew the following year. Annual construction ensured that the technology was observed and practiced regularly, contributing to the preservation of this technology to modern times.

Native peoples of California navigated tule boats around San Francisco and Monterey Bays with double-bladed paddles. These watercraft were over 3 m long and held up to 4 people. They were used to gain access to offshore islands and seabird rookeries (Serra 1955). Men paddled out to collect eggs of cormorant and other species, and to club seal and sea lion pups. On the Carquinez Straits they used tule boats to bring in enormous catches of salmon (Font 1933). Among the rushes and cattails of coastal marshes, tule balsas were used to harvest eggs, ducks, and geese. The Coast Miwok of northern California caught bay fish in a seine strung between two reed boats (Kelly 1978:416). Yokuts people living near the upper end of the San Joaquin Valley used reed boats as their sole watercraft on rivers and lakes. Their "average-sized balsa comfortably held six persons and their possessions" (Wallace 1978:452).

Throughout the Plateau region of western North America, tule boats were used on lakes and streams in addition to bark and dugout canoes and hide-covered bullboats (Kootenai, Hewes 1998:630; Coeur d'Alene, Palmer 1998:318; and Spokane, Ross 1998:275).

> Canoes and rafts were used primarily to gain access and transport great quantities of roots from root fields south of the Spokane River. Wood and bundled tule rafts, anchored with large weir weights,

were used by men when diving for freshwater mussels (Ross 1998:275).

Reed boats are no longer used across most of their former North American range, but in parts of South America they persist (Edwards 1965:1). José de Acosta observed the following on the Peruvian coast in 1509:

> [The Indians] make floats of sedges or reeds, well tied, which they call balsas. They carry them on their shoulders to the shore, toss them energetically into the water, climb aboard, and thus like horsemen they enter the sea. . . . They carry on these bundles their nets and lines . . . going out to sea to fish in the Callao de Lima, for . . . each one sits on his little balsa like a . . . jockey, cutting through the waves of the sea, which is very rough where they fish. . . . On reaching the shore, they set their boats on their shoulders, and later take them apart and put the bundles on the beach so they will drain and dry (edited from citation in Edwards 1965:2).

In recent times such boats are used to gather reeds and to hunt birds in the shallows and marshes close to shore on Ecuadorian highland lakes. Native people use similar watercraft in highland Peru to fish, hunt birds, gather eggs, and catch frogs (Edwards 1965:9). These rafts are more expedient than Peruvian coastal boats or the Aymará balsas of Lake Titicaca. Reed boats are known from west central Argentina in areas characterized by intermittent swamps, saline lakes, and shallow streams, and from Mexico at Lake Chapala, Lake Tlaxcala, and in Nayarit (Edwards 1965:85).

A striking use of reed boats is depicted on an Egyptian tomb in Saqqara dated about 4400 yr B.P. Hunters are illustrated standing on reed boats spearing hippopotami, which apparently were recovered by means of attached lines (Casson 1994:16). In the Old World, reed boats range "discontinuously from eastern Asia to southern Africa, and from Europe to Australia and Tasmania" (Edwards 1965:87). They represent an early and widespread form of watercraft whose temporal range may well extend to Paleoindian times in the Great Basin, Plateau, Pacific coast, South American highlands, Mexico, and elsewhere, where buoyant plant materials such as tule, cattails, or totora reeds were formerly available.

Bark and Dugout Canoes

Bark canoes were a dominant form of native watercraft in the Subarctic, Maritime Provinces, and Great Lakes (Driver and Massey 1957:289). Aboriginal distribution in South America was coeval with the fiords, inlets, and channels characterizing the Chilean archipelago (Edwards 1965:30). Dugout canoes were widely distributed along North America's Northwest coast and in the Plateau, Eastern Woodlands, Northeast, and Gulf Coasts, as well as in Central America and tropical South America (Driver and Massey 1957:290). Bark and dugout canoes were used in southwest and southeast Alaska, much of the Plateau, along the upper and central reaches of the Mississippi River system, the southwestern Great Lakes, and in the Middle Atlantic region. The introduction of metal tools reduced the labor of making dugouts and led, historically, to their increased manufacture and distribution in North and South America. Bark canoes and reed boats were the dominant earlier forms in many areas (Driver and Massey 1957; Edwards 1965).

Bark Canoes

> The Indian bark canoes were most efficient watercraft for use in forest travel; they were capable of being propelled easily with a single-bladed paddle. This allowed the paddler, unlike the oarsman, to face the direction of travel, a necessity in obstructed or shoal waters and in fast-moving streams. The canoes, being light, could be carried overland for long distances, even where trails were rough or nonexistent. Yet they could carry heavy loads in shallow water and could be repaired in the forest without special tools. . . . Bark canoes were designed for various conditions: some for use in rapid streams, some for quiet waters, some for the open waters of lakes, some for use along the coast. Most were intended for portage in overland transportation as well. They were built in a variety of sizes, from small one-man hunting and fishing canoes to canoes large enough to carry a ton of cargo and a crew, or a war-party, or one or more families moving to new habitations. Some canoes were designed so they could be used, turned bottom up, for shelter ashore (Adney and Chapelle 1983:3).

Constructing birchbark canoes was a task that often involved the cooperative efforts of males and females over the course of a couple of weeks. Some manufacturing activities produced potentially recognizable archaeological signatures, including small diameter postholes in the outline of a canoe and concentrations of large rocks discarded subsequent to their expedient use as construction weights (illustrations in Adney and Chapelle 1983:40–49). Manufacturing other bark canoes types left a more ephemeral record (i.e., flat-bow bark canoe construction illustrated in Brunton 1998:224).

At every stage in canoe construction, from harvesting the proper trees to employing appropriate techniques for splitting and bending different types of wood, technology entailed detailed knowledge of wood characteristics. While paper birch (*Betula papyrifera*) was preferred, many barks were used including spruce (*Picea*), elm (*Ulmus*), chestnut (*Castenea dentata*), hickory (*Carya*), basswood (*Tilia*) and cottonwood (*Popylus*) (Adney and Chapelle 1983:15). Koyukon people in the Alaskan forest strip birchbark in the spring and early summer, a practice which does not kill the tree (Nelson 1983:53).

Long, durable black spruce roots (*Picea mariana*) often were used to lace the bark pieces together. These flexible roots grow very close to the ground surface, where they are easily harvested with a sharp stick or by hand. They can be obtained up to 20 ft long with a maximum diameter "no larger than a lead pencil" (Adney and Chapelle 1983:16). White cedar (*Thuja occidentalis* L.) is well suited for the frame because it splits cleanly and readily when dry and well seasoned, and fallen timber can be used. Alternatively, if black spruce or hard maple (*Acer saccharum* Marsh. or *Acer migrum* Michx.) is employed, the wood is split while green. Coating the seams with tree resin mixtures (white or black spruce) helps ensure a waterproof craft. Ritual and prayer are integral aspects of harvesting materials and constructing canoes (Nelson 1983).

> A prayer was said to the spirit of the great cedar tree before felling it, that it might fall in the desired direction. A prayer was said be-

> fore the bark of the standing tree was stripped off, explaining what it was to be used for. It was a courteous apology as well as a prayer of thankfulness and recognition of so useful a material (Stewart 1982:162).

Woodworking tools traditionally used in canoe construction include a stone axe or adz and stone knives; wood, horn, and stone wedges; wooden mauls to drive wedges, sharp pointed sticks, bone awls, hot water, and stone weights characterized by at least one flat side (Adney and Chapelle 1983:17–20; Osgood 1940:96–104, 359–78). Favorable features of a building site included shade (to prevent the bark from drying too quickly), smooth ground firm enough to hold upright stakes, and proximity to water where the boat would be launched.

> Probably the commonest place is along the river bank where the shavings can be thrown into the water so no women of menstrual age will step on them . . . sometimes Indians make a canoe while away from a village in a spring camp (Osgood 1940:368, Ingalik, southwest Alaska).

Birchbark hunting canoes are considered an essential tool for every Ingalik Athapascan male of adolescent age or older, who lives along the lower Yukon of central Alaska (Osgood 1940:368). On spring hunting trips, when game has been killed, two canoes may be lashed together and a platform built between them to transport the meat. Ingalik canoes also are used for visiting fish traps and nets and for hauling the fish harvest. In summer, Ingalik people "seldom go anywhere except in their canoes" and use them to travel from one village to another "on rivers, lakes, sloughs, or wherever there is enough unobstructed water over which a canoe can pass" (Osgood 1940:370).

A possible early archaeological record of the use of sewn bark canoes for fishing and fowling comes from the site of Star Carr, Yorkshire. A boat paddle excavated from this lakeshore site is dated approximately 9500 yr B.P. (Clark 1954:12). The landscape that surrounded Star Carr was characterized by birch forests; wood suitable for dugout canoes was lacking.

Dugout Canoes

Some wood canoes were fashioned from hollow logs (Walker 1998:Figure 2), but many were deliberately "dug out" as the name of the watercraft implies.

> The method of manufacture of the dugout canoe differed little from area to area. It consisted mainly of alternate burning and adzing away of the charred wood. The process took weeks, or even months if interrupted by other activities; certainly longer than the making of a bark canoe (Driver and Massey 1957:290). . . . Winthrop (1636) described the Indians as going into the woods and stripping a large, suitable tree of its bark so that it would die slowly over the following year . . . the base of the tree might be burned and /or cut . . . until the tree fell over. . . . A groove would be cut on the upper surface of the tree and a series of fires lit. The Narragansett would return periodically to scrape the charred wood from the upper burned surface and to rekindle a fire. The entire process would take about two weeks, although during a single day no more than two to four hours were actually involved in physical labor (Brose and Greber 1982:248–49).

Controlled burns and scraping implements (stone adz, shell) were primary aboriginal tools for dugout manufacture. An experienced primitive technologist who recently made a dugout using reconstructed aboriginal methods notes the following,

> [An adz is] admirably designed for cutting off high spots, as opposed to gouging out trenches where fire is more efficient. This balance between fire and adz became the rule of progress. After the char was scraped and the burned surfaces cleaned with a wet handful of river grass, it took only a few minutes of chopping to even the floor for the next burn. . . . About twenty individual burns were required, varying from one to four hours each. There had been over thirty hours of "active" burning time (i.e. not counting time for kindling to burn down, scraping char, etc.). The 600 lb. log I started with now weighed less than 150 lbs. It was 12 feet long, 19 inches in beam, and 11 inches deep (outside dimensions). Thickness varied from 1 to 1½ inches amidships, two and three inches at the ends (Dina 1996:50–54).

Klamath and Modoc peoples of southern Oregon and northern California used fire and elkhorn wedges driven by knot-headed mauls to hollow out ponderosa pine canoes (Stern 1998:452). Algonquian people in North Carolina felled trees by "well-controlled burning of their base; then they were further worked with fire and scraping tools of shell or stone axes" (Feest 1978:275). Their dugout canoes could carry 20 people. The woodworking tool kit of the Thompson peoples in British Columbia included ground-stone adzes and axes, hammerstones and mauls, antler wedges, stone knives and drills, and beavertooth crooked knives (Wyatt 1998:193).

A class of Paleoindian stone tools, which have been identified as woodworking implements used to make dugouts, is Dalton adzes (Morse 1971; Morse and Goodyear 1973). The working end of these robust bifaces is convex in transverse section and unifacially beveled in longitudinal section. Their lateral edges are heavily ground, as is the poll opposite the bit. A large number were made from flat, elongated pebbles or cobbles (Goodyear 1974:41). Dalton adzes were first recognized in the waterway-rich terrain of northeastern Arkansas, where they occur in campsites, caches, and burials (Morse and Goodyear 1973). They appear to date between 10,500 and 9900 yr B.P. (Driskell 1996; Goodyear 1982, 1999; Morse 1997).

Use-wear analysis indicates that Dalton adzes were used on charred and dry wood (Gaetner 1994). Of 15 adzes studied, 11 had charcoal residue on their working ends, which appeared as "amorphous, angular black inclusions wedged into or beneath the well-developed charred-wood polish" (Yerkes and Gaetner 1997:63). Parallel longitudinal striations on the dorsal and ventral surfaces resulted from abrasion against charred wood. Hafting wear suggests that these tools were bound in leather and set in seasoned-wood or antler handles.

Dalton settlement patterns are strong evidence of routine navigation of the central Mississippi Valley waterways (Walthall and Koldehoff 1998), a point discussed later in the paper. Bradley (1997:57) suggests that Dalton projectile points and knives represent an in situ development directly out of Clovis technology. Perhaps Dalton adzes represent further development in woodworking technology from Clovis antecedents. Collins (1999:50) draws attention to "thicker bifaces, less refined than point preforms, [that] occur in Clovis assemblages. The function of these is not known, and their numbers seem to be relatively low." He cites an example from the Yellow Hawk site near the headwaters of the Brazos River (Mallouf 1989:Figure 10a), and another from the Clovis-type site at Blackwater Draw (Hester 1972:Figure 90k). I suggest that such bifaces might represent woodworking tools suitable for constructing dugout canoes. Several possible candidates for woodworking tools are robust bifaces illustrated from the Thunderbird site in Virginia (Boyer 1974:Figures 60, 67–72) and a large bifacial "scraper plane" illustrated from the Adams site, Kentucky (Sanders 1990:Figure 38). Directed examination and use-wear study of these and other Clovis assemblages for evidence of woodworking tools akin to (and precursors for?) Dalton adzes may be instructive.

Ethnographic Uses of Dugout and Birchbark Canoes

An area of North America where canoe travel dominated all other forms of transportation (including pedestrian) is the Northwest Coast (Driver and Massey 1957:290). Watercraft technology provided an important means of procuring and transporting the harvest from a great natural abundance of fish, shellfish, sea mammals, and waterfowl.

The Northern Coast Salish made a variety of canoes for hunting and fishing, including a small canoe used for lake travel and beaver hunting. They had different styles of paddles for use in rough weather or for night hunting (Kennedy and Bouchard 1990:447). Watercraft gave the Nootkans of Vancouver Island access to offshore halibut banks and the means to harpoon seals, porpoises, sea lions, and whales on water. They used canoes to hunt migratory waterfowl on dark nights by holding a mat before a fire, which cast a shadow that drew birds, which could then be netted (Arima and Dewhirst 1990:394). Canoes were also used to dredge dentalia shells from deep water with long-handled rakes. These shells were exchanged prehistorically with groups from the Plateau, and thus canoes played an important role in these social relations (Galm 1994, and references therein).

In the Plateau region, canoes were used for a wide range of subsistence activities including night fishing, often with torches (Hewes 1998:625, 630). Trident leisters thrust vertically into the water from canoes were used to catch trout and whitefish from lakes and rivers (Hewes 1998:628, 632). In Nez Perce territory, canoes facilitated dip-netting of lamprey eels (Hewes 1998:626). Nez Perce hunters also drove "deer and elk into larger rivers where they could be easily dispatched from canoes" (Walker 1998:421). Along the Columbia River and on islands within it, native people "moored canoes to shore and used them as platforms from which to take spring chinook with bident harpoons, gaff hooks, and dip net . . . many fished throughout the year, often from dugout canoes, using seines or leisters" (Stern 1998:397). Klamath women used dugouts to collect the seeds of yellow pond-lily (*Nuphar polysepalum*), which formed a substantial portion of their diet (Hunn et al. 1998:530).

In northeastern North America, the distribution of birchbark (Northern) and dugout canoes (Southern) overlapped in southern New England, where a dugout could be made of pine, oak, or chestnut in 10 to 12 days by one person using stone tools (Salwen 1978:164). Coastal natives of New England speared lobsters and crabs from canoes and harpooned sturgeons, which were "attracted to the surface by torches at night" (Snow 1978:138).

The Beothuk of Newfoundland used well-designed birchbark canoes with a light wooden frame to kill caribou at deer fences along rivers and to venture to islands 57 km offshore to bring back eggs from bird rookeries (Reynolds 1978:102–3; Pastore 1992:18). Other subarctic groups also used canoes to spear caribou as they crossed rivers and lakes during fall migrations (Rogers and Smith 1981:132). Cree hunted from small canoes and used larger traveling canoes to move families and possessions (Honigmann 1981:221). Chipewyan used bark canoes primarily to hunt caribou and to cross rivers, rather than for long-distance movement, because their "travel routes tended to run across, rather than with, the drainage systems" (Smith 1981:280). Summer travel among the Slavey living in northwest Canada "was primarily by water in shallow-draft, one- or two-man canoes sheathed with spruce bark or birchbark" (Asch 1981:341). Small lakes and waterways crossing low-lying plains characterize their territory.

Some canoes were intended to serve as expedient watercraft:

> In forest travel it was not always possible or practical to portage a canoe for a long distance simply to make a short water passage somewhere along the route. War parties and hunters, therefore, often found it necessary to build a temporary canoe, one that could be utilized for a limited water passage and then abandoned. Since such a limited use did not warrant expenditure of much time or labor on construction, the canoe was prepared quickly from readily available material and in order to meet these requirements many Indian tribes developed canoe forms and building techniques somewhat different from the more elaborate construction using birch or spruce bark. (Adney and Chapelle 1983:212).

Looking to South America, use of watercraft technology was critical to the indigenous peoples in the Chilean archipelago, where dense vegetation and rugged terrain made land travel difficult. Alacaluf traveled primarily by water, with land journeys limited to short hunting trips. Bark canoes supported their maritime procurement of shellfish, birds, sea lion, and fish. A traditional canoe collected in 1904 was "made of beech bark, sewed with baleen strips" (Bird 1946:66–67). It measured nearly 4 m long, with a maximum width of 67 cm and a depth inside the hull of 50 cm.

Yahgan living in the mountainous island archipelago of Tierra del Fuego occasionally hunted whales in the open sea and hunted seals from both land and canoe (Cooper 1946a:84). Their subsistence activities included the harvest of mussel, fish, porpoise, birds, bird eggs, limpets, crabs, and sea urchins. Cormorants were taken at night by torchlight with clubs and a gorge hook. Yahgan traveled almost exclusively by water, using a gondola-shaped bark canoe as their sole form of watercraft (Lothrop 1932:253–54). It was manufactured of three strips of beech bark harvested in the spring when sap was running. Traditionally, Yaghan used a bone chisel or a mussel shell knife to cut the bark, a bone barking tool to strip it, and a small chisel to smooth it. The bark pieces were sewn together with baleen or "with the stringy seams of wild celery grass" and caulked with moss mixed with mud (Cooper 1946a:88). The canoes leaked much and were bailed. Construction and repair of Yahgan canoes traditionally were allocated to males. Females managed and maneuvered the craft (including during sea mammal hunts), maintained the fire set in the middle of the boat, and moored the canoe offshore in kelp beds (Cooper 1946a:96).

Owing to their great durability compared with reed and skin boats, dugout canoes dominate the archaeological record of preserved remains of early watercraft. Dugouts provide a *minimum* date for watercraft use in particular areas, but this should not be taken to represent the *earliest* use of boats in a given region. Data from Florida exemplify this observation.

The remains of a dugout canoe from DeLeon Springs, Florida, represent the oldest watercraft from North America known to this author. Newsom and Purdy (1990:167) report its radiocarbon date as 5120 RCYBP (cf. 7000 yr B.P. reported in Walthall and Koldehoff 1998:261). In 1990 there were nearly 200 canoes reported from Florida, the largest group of prehistoric and historic watercraft in the world (Newsom and Purdy 1990:164). In 2000 an additional 93 dugout canoes were discovered at a single site located east of Gainesville in north-central Florida. A drought lowered the water level in Newnan's Lake (historically known as Lake Pithlachocco), where dugouts of Southern hard pine (>77 percent) and cypress (4 percent) were uncovered along a 2.4-km stretch of the northern shoreline (Memory et al. 2001). After 53 of the boats were excavated and radiocarbon-dated, one returned an anomalously young date, while 52 range in age between 500 and 5300 calibrated radiocarbon years before present (Memory et al. 2001). Seventy percent of the Newnan's Lake dugouts date between 3000 and 5000 yr B.P. Newsom and Purdy (1990:176) explain that most prehistoric dugouts recovered in Florida were preserved in peat deposits that developed after 6000 yr B.P.; therefore, the probability of finding watercraft in Florida older than 6000 yr B.P. is slim.

Early dugouts elsewhere in the world tend to date around 8000 yr B.P., perhaps reflecting a greater number of dugout canoes in use around that time (population increase?) or more favorable conditions for their preservation. When coastal areas were inundated by postglacial sea level rise, a large sample of watercraft older than 8000 yr B.P. lay out of archaeological reach until underwater recovery methods improved in recent times. This long-standing sampling bias, which favors boat preservation post–8000 yr B.P., contributes to the impression that dugout canoes were not widely used prior to that time. Skin, reed, and bark boats likely represent even earlier forms, but these watercraft are considerably more perishable than dugout canoes. Nonetheless, just as a single site in Florida (Newnan's Lake) surprised everyone recently with its incredibly rich record of canoes, so too might archaeologists yet document preserved remains of late-Pleistocene watercraft. Envisioning this as a possibility is an important first step toward increasing the chances of making *and recognizing* such a discovery.

Skin Boats

> The Eskimos used several kinds of boats for hunting and other subsistence activities. Before the advent of trade with the outside, the principal crafts were the well-known kayak, a single seat skin-covered canoe with a cockpit where the hunter sat dressed in waterproof clothing made from walrus intestine; and the umiak, an open, flat-bottomed skin canoe built of walrus hide stretched and sewed over a wooden lattice frame (Hughes 1984:250).

Skin watercraft of various kinds were used in the Arctic, in the North American maritime worlds of the Northwest and Northeast; and in the Subarctic, Plains, and Southeast. Skin boats also were used along South America's west coast and in the Pampas of Argentina. The hides of sea mammals, guanaco, moose, caribou, bison, elk, and other animals provided outer coverings or floats for these craft.

Umiaks

Umiaks are large open skin boats constructed of driftwood and covered with split female walrus hides or seal skins. Traditionally, they served in the arctic as hunting boats for whales, walruses, and seals, and as transportation and shelter for families and possessions while traveling. They have a carrying capacity of up to 4 tons and formerly were propelled by oars and sails (Hughes 1984:251, 273). An umiak is suited to hunting in the ice floes; "because of its lightness, it could quickly be pulled up out of the water and its tough skin sides were not easily punctured by the sharp edges of the ice" (Hughes 1984:215). Historically, every individual or family did not own one because they were relatively costly to manufacture and to maintain. A family or group that had access to such an economic asset was more prosperous than others without such access (Ricklis 1982:139). d'Anglure (1984:484) cites a ratio of 10 umiaks for 200 people for the Canadian Arctic at Aufaluk (Haven 1773) and Labrador (Taylor 1974).

Adney and Chapelle (1983:181) argue that the umiak was more widely used prior to European contact than existing records indicate. It was needed for family migration by sea and enabled people to cross other large bodies of water. Umiaks are known from Kodiak Island through the Aleutians, along the north and east coasts of Alaska to the Mackenzie River, and along portions of the Siberian coast. Umiaks were a technological necessity that enabled the rapid dispersal of Thule Eskimo into the eastern Arctic about 1000 yr B.P. (McGhee 1984:371). The earliest cold-adapted maritime peoples who traveled the northernmost coasts of the Americas likely used similar types of watercraft.

On the north Alaskan coast, water transport is vital. Traditionally such transport was mainly by umiak, which constituted the key whaling vessel (Spencer 1984:329). Umiak examples still in use in the 1970s ranged in length from 5 to nearly 7 m. The frame was made "of whatever materials were conveniently at hand, such as wood obtained from driftwood and bone" and was covered in walrus or seal skin (six bearded seal skins being required for an average-size boat) (Spencer 1984:329). Walrus skins were split and carefully stitched by women to create the essential watertight covers. Greenland umiaks had a heavier frame and were used primarily to move family and household effects between hunting grounds (d'Anglure 1984:486, Figure 11).

In the Canadian Arctic "the combined use of umiak, kayaks, and dogs allowed bands of 20 to 30 people to move in summer along the coasts and the large watercourses with weapons and baggage" (d'Anglure 1984:484).

> They could travel several hundred miles under sail or oars to exploit the elongated zones of their territory, along the coasts or up the rivers, or to search outside their territory for rare materials such as wood for construction, steatite, or walrus ivory. . . . At the entrance to Wakeham Bay Low (1901) met an umiak with masts rowed by two women and steered by a man; it contained three other women, some children, a dog team and sled, tents, and baggage and was escorted by eight men and youths in kayaks. In Ungava Bay Low encountered another umiak accompanied by seven kayaks, some thirty people, the dogs towing an umiak, going to caribou hunting territory some 50 miles up the Riviére Arnaud (Low 1901). Dogs pulled an umiak like a sled, when the current was strong (d'Anglure 1984:485).

Domestic dogs are reported from the Folsom component at the Agate Basin site in Wyoming at nearly 10,600 yr B.P. (Frison and Stanford 1982), from the Jones-Miller site in Colorado at 10,200 yr B.P. (Stanford 1999), from burials in the Midwest at 8500 yr B.P. (Morey and Waint 1992), and from the Zhokhov site in the Siberian Arctic at 8000 yr B.P. (Pitul'ko and Kasparov 1996:14).

Kayaks

If umiaks are analogous to modern-day sports utility vehicles such as Ford Explorers, kayaks are Triumph convertibles. Used widely in the Arctic and parts of the Subarctic, kayaks are highly maneuverable decked hunting canoes carefully built to meet local conditions (Adney and Chapelle 1983:191).

> The basic requirements in nearly all kayaks are the same; to paddle rapidly and easily, to work against strong wind and tide or heavy head sea, to be maneuverable, and to be light enough to be readily lifted from the water and carried. The low freeboard required makes decking a necessity. In general, the kayak is designed to carry one paddler. . . . Where it is desireable to portage the kayak over ice or land for a great distance the boat is very light and is capable of being carried like a large basket, by inserting one arm under the decking at the manhole or cockpit, but where such a requirement is not an important factor, the kayaks are often large and heavy. In the majority of types, the degree of seaworthiness obtained is very great. (Adney and Chapelle 1983:192).

Aleut kayaks (*baidarkas*) were of shallow draft and long in proportion to beam; they could travel up to 7 miles an hour and move against a current (Lantis 1984:171–72).

> In the foggy Aleutian climate, men had to navigate by the currents, which they learned by instruction and experience, at any given time and place taking account also of wind direction and tide. In an emergency they could travel at night, seeing almost nothing but feeling the direction of the water beneath them and the wind above them (Lantis1984:173).

Kayaks gave Aleuts access to the marine foods on which they depended including sea mammals, fish, sea urchins, birds, and bird eggs. They used kayaks to surround and harpoon sea otters and to fish for halibut and cod with hooks tied to lines made of seaweeds nearly 150 fathoms long (Lantis 1984:175). The prehistoric people living at the Anangula site on the eastern Aleutian Island of Ananiuliak 7200–8700 years ago sustained themselves on a maritime diet (McCartney and Veltre 1996:448). Burned bones of whales and large birds dominate the site's small faunal assemblage, and fishing line weights were recovered (McCartney and Veltre 1996:446, 448). Boats were required to reach Anangula from nearby Umnak Island (a source of obsidian used at the site) and to cross several inter-island passes between there and the Alaska peninsula (Black 1974, 1975, 1976). Boats likely aided in the final evacuation of the site, which was forced by catastrophic ash fall from a volcanic eruption (Black 1975; McCartney and Turner 1966). Navigating the notoriously treacherous Aleutian waters by nearly 8700 yr B.P. presupposes an even earlier (and likely extended) period of technological development of the necessary skills and knowledge base.

Aleutian kayaks were typically covered with sea lion skin, Alaskan kayaks were most often covered with bearded seal skin, and in the central Arctic caribou skins were sometimes used (Adney and Chapelle 1983:194). The latter had seams rubbed with fish oil and ocher. Caribou and Netsilik Eskimo used such kayaks to hunt caribou at water crossings.

> Thousands or even tens of thousands of caribou can concentrate against water barriers posed by the major lake and river system and funnel into crossing places at lake narrows or slow river sections where the Caribou Eskimo waited to lance . . . the swimming herds from their kayaks (cf. Arima 1975:147–53). Such hunting was very productive, and surplus meat was preserved by sun-drying in thin slices laid on flat rock or gravel (Arima 1984:448).

An example of this hunting strategy, observed among the Innu (Naskapi) at a narrows on Mistinibi Lake in Northern Labrador in 1906, underscores the productivity of such a hunt (Loring 1997). Four men and as many boys,

> had speared no less than twelve or fifteen hundred deer [caribou] in a few weeks. From three to five hundred carcasses, skinned and washed out, were hauled up on the beach, drying hard and black in the sun and the cool September wind (Cabot 1912:235–244, as cited in Loring 1997:195).

In a relatively brief intensive hunt a small Innu group was able to slay an impressive number of caribou using favorable geography, knowledge of prey behavior, and technological aids including lance and boats. The subsequent butchering, meat and fat preservation, and skin preparation were considerable tasks organized by females. The three households involved comprised 20 to 25 people (Loring 1997). When groups of this size among the neighboring Inuit Eskimo divided to travel in smaller family units, kayaks and dogs facilitated transport.

> With a kayak and two to four dogs a family of four, an adult couple and two children, could travel along the interior lakes and rivers for one to three months in the summer and early autumn hunting caribou and fishing for lake trout. The dogs carried bags of provisions and utensils, and even sometimes an infant, and dragged the tent poles. On her back the woman carried perhaps a child and a bundle of skins for the tent or for bedding. The man in the kayak transported everything that would fit inside or on top. . . . When a hunting region was reached, the man went on foot in search of game, carrying only a caribou skin, his weapons, and some dried food. The skin served as protection if he had to sleep out, and as a boat if

he had to cross a watercourse—stuffing it with brush, tying the edges together, and using it as a float while paddling with a caribou scapula (d'Anglure 1984:485).

Bullboats

Among expedient aboriginal watercraft, different forms of skin boats were particularly widespread. Of these, bullboats are well known from the North American Plains where they were used to ferry people and belongings across and along streams. Bullboats consisted of bowl-shaped basketwork frames (often made of willows or other saplings found along streams) covered with bison or other hides (Adney and Chapelle 1983:220; Stewart 2001:339). The bottom of the vessel was flat, and typically a single bison hide was used to make a boat somewhat greater than a meter in diameter.

On major waterways of the North American Plains, such as the Missouri River, bullboats were left on the river bank for regular use, "elsewhere they were improvised from tipi covers, parfleches, or any available hides, as the occasion demanded" (Driver and Massey 1957:289). They were paddled, poled and pulled by swimmers.

> Probably all the Plains Indians living near streams once used the bull-boat, but existing records show only the Mandan, Omaha, Kansas, Hidatsa, and Assiniboin to have used it. The Blackfoot (Siksika) and Dakota are said to have used some kind of a skin boat in which their tepee poles were employed as a temporary frame, but nothing is recorded of their form (Adney and Chapelle 1983:220).

Wolf-chief, a Hidatsa who lived along the Missouri, related that he used a bull-boat and a noose to snag logs floating down the river "as it was nearer to bring wood from the Missouri than to go to the timber for it" (Wilson 1924:209). Hidatsa women sometimes carried these craft to get firewood, the boat bottom resting on their necks and held by a pack strap. Wood was cut to fit, then loaded in the bull-boat and ferried home on the river (Wilson 1924:209–10). Dogs carried bullboats on travois during hunting trips.

> Once I went hunting with my husband. We took a dog and carried a boat on a travois. . . . We came back in the bull-boat carrying a doe and two fawns, and the dog, besides ourselves. We bound our travois to the boat in such a way that the skin saddle of the travois was out of the water, but the basket and lower ends of the poles dragged in the current. . . . It was not unusual for my husband when going upstream to hunt deer to take one of our dogs with a bull-boat on his travois. He would then float downstream in the boat with his game as I have described (Buffalo-Bird-Woman, cited in Wilson 1924:210–11).

After a Hidatsa hunting trip to the Yellowstone River, bullboats transported people, meat, and hides a distance of nearly 200 river miles down the Missouri (Lehmer 2001). This scenario of returning home by means of a downstream float trip in a temporary skin boat is repeated in ethnographic literature from a variety of places. Driver and Massey (1957:289) note the use of bullboats in the Great Lakes and Southeast as well as in Argentina, Central Asia, Mesopotamia, Ireland, Wales, and the Mediterranean.

In the Northeast, the Beothuk in Newfoundland used temporary canoes fashioned with caribou skins on inland waterways. Penobscots in Maine made moosehide canoes after a hunt to carry home game. This watercraft was dismantled later and the hide reused for other purposes (Snow 1978:139). Beginning in the late 1800s, Indians of the Mackenzie Mountains in Canada built mooseskin boats in headwater areas and used them to descend rivers carrying loads of dried meat (Gillespie 1981:332, 334). The size and shape of these mooseskin boats resembled *umiaks* rather than bullboats. Earlier, spruce-bark canoes had been used to transport dried meat. Hare of northwestern Canada occasionally made moosehide canoes for summer transport on lakes and rivers (Savishinsky and Hara 1981:318). Bullboats were traditionally known among Kootenai of the North American Plateau (Hewes 1998:630) and Tehuelche of Patagonia and the Pampas, who used guanaco bullboats to ferry themselves across rivers (Bourne 1853:133–34; Moreno 1879:256–57 as cited in Cooper 1946b:146).

Discussion

Boats are ubiquitous in their current distribution throughout the world wherever suitable waterways are present. Traditionally they were constructed to suit local needs and conditions from a variety of plant and animal materials known to have been available in many areas during the early postglacial period. Watercraft manufacture was compatible with Paleoindian technologies for working animal skins, wood, and other plant materials. Collectively, ethnographic and archaeological records indicate a time depth of no less, and probably considerably more, than 8000 to 10,000 years for the nearly worldwide distribution of watercraft technology.

Surely early peoples in the Americas are neither considered too technologically unsophisticated, nor too disinterested to devise transportation aids (including watercraft), which allowed them greater range in mobility, subsistence, and social interaction. In light of the significant advantages provided by water transport technology, it appears judicious to seek a better understanding of its probable roles within regional Paleoindian adaptations during the Pleistocene-Holocene transition.

While living in an archaeological camp in Alaska on the banks of a Yukon River tributary, I became acutely aware of a key point regarding waterways and information flow. Although the number of people camped within 50 km or more was relatively small, word-of-mouth knowledge of one another's whereabouts and activities was extensive. People observed, and were observed, as boats passed one another's camps. Local etiquette dictated that people stop to share coffee *while exchanging news*. Strangers stopped, as did friends and acquaintances. Though the human population along that particular stretch of the Porcupine River was dispersed in the summer season, travel along the waterway insured that people had contact with one another, either directly or through down-the-line dissemination of information. Riverine travel linked people together. *Sparse populations during early Paleoindian times may have been able to remain in closer contact over greater distances when water travel was part of their way of life.*

Waterway Travel in Late-Pleistocene/ Early-Holocene America

In this section, I turn to the archaeological record of North America, and to a limited extent South America, to explore current thinking regarding late-Pleistocene and early-Holocene peoples and water transport technologies. The discussion is organized geographically and moves from east to west. It is selective rather than comprehensive in scope.

North America

Southeast

Much of southeastern North America is prime country for watercraft travel. The region includes abundant riverine waterways, Atlantic and Gulf coastlines, a southern latitude, and "a good portion of this area is and was subtropical in climate" (Goodyear 1999:433). The Southeast has produced a greater number and diversity of fluted points than elsewhere in the Americas, leading researchers to propose a Clovis homeland in this area (Anderson 1990; Mason 1962:235; Stanford 1991).

Anderson (1990, 1996) suggests that New World colonization occurred sometime after 12,000 yr B.P., with movement down a postulated Ice Free Corridor. He proposes that people, upon exiting this corridor, followed the Missouri drainage system southeastward toward the Mississippi River and the Eastern Woodlands. "Each river would have offered a favorable artery to the south and eastward, assuming movement was along and not across channels" (Anderson 1990:185). From the Mississippi, people gained access to three main tributaries that likely served as settlement and communication corridors: the Ohio, Cumberland, and Tennessee Rivers (Anderson 1990:187; Williams and Stoltman 1965). Collectively, these three waterways produced the largest concentrations of Paleoindian projectile points known from the Eastern Woodlands (Anderson 1990:185–87, 1996:34). Anderson suggests that people settled in these areas for an estimated span of four to twelve generations prior to further dispersal.

As populations grew, group fissioning is thought to have prompted further exploration, expansion, and regional differentiation by 11,000 yr B.P. "Thus colonization of the region was rapid, and long-distance movement was probably commonplace" (Anderson 1990:189). Though Anderson's model does not explicitly address the use of watercraft, the presence or absence of water transport technology is a significant factor in this inland colonization scenario. Watercraft would enable people to travel quickly and to arrive in the Southeast quite early as Anderson predicts. Moreover, water transport would enable small founding populations to maintain social contact and mating networks across greater distances, perhaps accelerating the timing of further dispersal into adjacent areas.

There is compelling evidence of significant use of watercraft by Dalton people approximately 10,500 to 9900 yr B.P. along the central Mississippi Valley and its major tributaries. Dalton is a cultural horizon distinguished by a regionally variable Paleoindian projectile point/knife style in eastern North America. Dalton points are thought to derive directly from Clovis points in the Arkansas-Missouri area (Bradley 1997; Ellis et al. 1998; Goodyear 1982). Walthall and Koldehoff (1998) argue that highly mobile bands of Dalton people established alliance networks characterized by intense interaction between 10,500 and 10,000 yr B.P. They suggest that along a 700-km segment of the central Mississippi River and its tributaries, neighboring Dalton groups were in close and sustained contact in a biotically rich environment "where band-packing likely occurred early and where evidence of regional stylistic zonation is first recognized" (Walthall and Koldehoff 1998:269). Dalton social networks are evidenced archaeologically by the distribution (from bluff margin chert sources to widespread Dalton sites) of large, distinctive, well-crafted stone bifaces known as Sloan points, made of Crescent Burlington chert (Walthall and Koldehoff 1998:261–266).

Such networks served as buffering mechanisms for stabilizing social relationships in order to mitigate the effects of resource risk and interband discord. A symbolic representation of this social interaction was the Dalton long blade, referred to by archaeologists as the Sloan point, a ritual token that circulated among males who were likely members of related cult lodges or similar totemic descent groups (Walthall and Koldehoff 1998:269).

Highly prized Burlington chert from the Crescent Quarries in east central Missouri could be strategically accessed and transported by water. The Crescent Hills are bounded to the north by the Missouri River and are bisected by the Meramec River just above its confluence with the Mississippi. Average blocks of Crescent Burlington toolstone weighing 13–14 kg and measuring 30–50 cm in length was used to manufacture Sloan points, ranging in length from 11 to 38 cm (Walthall and Koldehoff 1998:264). The distribution of these bifaces, extending along the central Mississippi River, the lower Missouri and Illinois river drainages, and the upper St. Francis River and adjacent tributaries, suggests widespread riverine travel and social interaction (Walthall and Koldehoff 1998:Figure 1).

Dalton sites extend along both banks of the early-Holocene course of the central Mississippi River from above its confluence with the Missouri and Illinois rivers to northeast Arkansas (Walthall and Koldehoff 1998:259). "Repeated crossings of the deeply entrenched early Holocene Mississippi River and its floodplain in either direction without water transport is unlikely" (Walthall and Koldehoff 1998:262). Regular movement of Dalton people and toolstone across the Mississippi is inferred from the distribution of Crescent Burlington chert at Nochta and other Dalton sites in the Missouri uplands (Higgins 1990).

Historically, the Michigamea used dugout canoes in this same territory to move along the Mississippi River and tributary streams, where they had "villages in the American Bottom and seasonal encampments downstream in northeastern Arkansas" (Walthall and Koldehoff 1998:262). Dalton woodworking adzes may suggest that Dalton people traveled and exploited the rich waterways of the central Mississippi Valley using similar watercraft.

Clovis people may have procured lithic raw materials during

waterway travel as well. Two extensive Clovis lithic workshops are situated on the banks of major Southeastern waterways, where good-quality raw materials were obtained in quantity, the Carson-Conn-Short site located along a Pleistocene channel of the Tennessee River in northwest Tennessee (Broster and Norton 1993, 1996) and the Adams site situated along the Little River tributary of the Cumberland drainage in southwest Kentucky (Sanders 1990). As previously mentioned, use-wear study seeking evidence of Clovis woodworking tools akin to (and precursors of?) Dalton adzes is needed. A large bifacial "scraper plane" from the Adams site (Sanders 1990:Figure 38) and several bifaces from the Thunderbird site discussed further below are particularly intriguing.

Middle Atlantic

The Shawnee Minisink site in southeastern Pennsylvania provides a rare window into Clovis riverine subsistence choices, including fishing (McNett 1985a). Moreover, its setting suggests the possibility of watercraft use. During the late Pleistocene, this favorable campsite occupied "a fairly open field bordering a wood's edge" on a south-facing terrace near the confluence of Brodhead Creek and the Delaware River (Dent and Kauffman 1985:75). The upper Delaware constitutes a broad waterway flowing between the Pocono Mountains to the west and the Kittatinny Ridge to the east, which range upwards to 425 and 455 m elevation, respectively.

Shawnee Minisink is situated 5 km upstream of the Delaware Water Gap, a spectacular craggy portal in the steep, rocky Appalachian Mountains through which the Delaware River provides a ready, easily negotiated transport corridor more than 2 km wide. Use of watercraft in this location would have greatly facilitated mobility and communication along this major river, which flows to the Atlantic Ocean. In historic times, the Lenni Lenape (Delaware) traveled the upper Delaware River and adjoining streams and lakes using small portable birchbark canoes and dugouts (Goddard 1978:217). Presently the Delaware Water Gap near the Shawnee Minisink site is a major recreational area devoted to canoeing, fishing, and hiking.

Flotation of Clovis hearth sediments yielded calcined, fragmented fish vertebrae (species unidentifiable) and carbonized seeds of grape (*Vitis* sp.), blackberry (*Rubus* sp.) and hawthorn plums (*Crataegus* sp.). An uncalibrated AMS radiocarbon date of 10,940 ± 90 (Beta-101935) was determined for a charred hawthorn plum seed from this campfire (R. Dent pers. comm. 1998). A late-summer/early-fall occupation is postulated based on the harvest season of hawthorn plums and other seeds including *Chenopodium* sp. and *Acalypha virginica* (Dent and Kauffman 1985:73). Species identification of wood charcoal from the Paleoindian component indicates the former presence of birch (*Betula* sp.), willow (*Salix* sp.), oak (*Quercus* sp.), and white pine (*Pinus strobus*), materials suitable for watercraft construction.

The Clovis lithic assemblage, made predominately of locally available black chert, indicates limited tool manufacture but extensive tool use and rejuvenation (Marshall 1985:195). A single Clovis point (made of Onondaga chert) was recovered, and a diverse array of hafted scrapers dominates the tool assemblage. Use-wear analysis suggests that many of these scrapers were used to work bone and wood (Evans 1985). The tools "occur in two major clusters, each containing several different kinds of end-scrapers, leading to the conclusion that a rather complex manufacturing process was underway" near hearths (McNett 1985b:322). Evans (1985:252) suggests that,

> During the Paleoindian occupation it would appear that the Shawnee Minisink site was used chiefly for a specific task that involved small, short-edged, steep-angled endscrapers . . . many of which were hafted in various ways . . . that the endscrapers were used at the site is indicated both by traces of wear and by the frequent breakage of these tools at the haft . . . the majority of the unifacially utilized tool edges in the Paleoindian component indicate use on hard materials, such as bone or the hardwoods that would have been available in the river valley. . . . Few of the endscrapers showed the polish or extensive rounding . . . that has been associated with hide working.

Fish scaling was not supported by use-wear evidence; "more likely would have been the use of the scrapers to produce wood or bone fishing paraphernalia (hooks, spear tips, etc), if indeed a riverine orientation for tool use at Shawnee Minisink is supportable" (Evans 1985:252). In the 17th century the Delaware fished for shad, striped bass, sturgeon, eels, and other species using "large stone-weighted seines, smaller nets set on poles, fish traps, weirs, hooks, and bows and arrows. Fish and shellfish were commonly sun-dried on racks and preserved" (Goddard 1978:217). Whether watercraft maintenance or manufacture was included among the Paleoindian activities undertaken at Shawnee Minisink is unknown, as is the nature of Clovis fishing technology. McNett (1985c:84) estimates that greater than 50 percent of the Paleoindian component remains deeply buried at the site. The distribution of phosphorus that was revealed in auger testing indicates "the heaviest occupation occurred to the northeast of the portion excavated."

Another significant Paleoindian campsite situated beside a major stream in the middle Atlantic is the Thunderbird site in northern Virginia (Gardner 1974, 1979). Post mold patterns preserved in the Clovis level suggest the former presence of a structure measuring 8 by 3 m, built near a spring-fed creek that drained to the South Fork of the Shenandoah River (Gardner 1974:21). Here for nearly 3000 years Paleoindian people selectively camped along tributaries of the Shenandoah River, where high-grade jasper was available.

During the terminal Pleistocene the middle Shenandoah Valley was characterized by a broad floodplain containing numerous backwater swamps and sloughs through which the river alternately flowed as a braided or downcutting stream (Gardner 1974:9). Dating to at least 10,500 yr B.P., these sloughs represent seasonally important habitats for birds, animals, and a variety of floral species. Boats may have been as useful to Clovis people accessing the resources in these swamps and sloughs, as watercraft evidently were for Dalton people exploiting similar resources in the central Mississippi Valley. Suitable woods and animal skins for watercraft construction were available in the Middle Atlantic from late-Pleistocene times.

Clovis and later Paleoindians obtained jasper in the limestone uplands on the east side of the South Fork of the Shenandoah River, which they transported to the Thunderbird site located on the opposite bank (1974:22). When the principal investigator, Bill Gardner, traveled between the Thunderbird site and associated jasper quarries, he often ferried across by boat. Early peoples may have done likewise, depending on the changing volume and configuration of the river during 3,000 years of Paleoindian occupation in the area.

Paleoindians brought a significant amount of toolstone across the Shenandoah River from the upland jasper quarries. The Thunderbird lithic assemblage includes more than 520 complete and broken bifaces that were excavated and surface collected in 1971 and 1972. Boyer (1974:115–17) illustrates and reports summary measurements for 11 preliminary classes of large bifaces from Thunderbird. Of interest to this paper are four biface groups for which "wear" was reported on the distal and lateral edges (Classes H, I, J and K), and a fifth group described as "crude chopper tools" (Class G). Table 1 lists Boyer's descriptive statistics and observations of edge modification for these five groups of large bifaces. Further analysis of the Thunderbird collection might shed light on the presence of woodworking tools suitable for constructing dugouts or other watercraft.

A rich Clovis record is documented in the middle Atlantic for the upland drainage divide separating the Susquehanna and Delaware river systems, above their respective discharges into the Atlantic Ocean. This north-south divide (known as the Delmarva Peninsula) currently forms the eastern margin of the Chesapeake Bay on the eastern shores of Maryland and Virginia. Lowery (2001) reports over 350 Clovis points from two principal, yet relatively short, tributaries (Choptank and Chester) of the Susquehanna River.

Many of the upland Clovis sites on the Delmarva Peninsula are associated with the former heads of spring-fed creeks, such as the Paw Paw Cove site, where Clovis point manufacture occurred beside a hearth (Lowery 1989, 1992). A high incidence of complete fluted points in private surface collections suggests terrestrial hunting losses. Local outcrops of glacial gravels were utilized as raw material sources. The benefits of watercraft in this upland setting, where Clovis sites are associated with first-order tributaries, is expected to have been negligible; however, the relatively dense distribution of fluted points in the uplands bordering the lower Susquehanna River system gives some indication of the rich record that likely lined the banks of this major waterway and the estuarine environments at its mouth. Here the advantages of watercraft in subsistence and other activities would have been considerable; however, these are the very settings inundated by the rising waters of the Atlantic Ocean and the formation of the Chesapeake Bay. Dredging and oyster harvesting activities in the Chesapeake Bay have yielded Clovis points, documenting their common occurrence along the currently drowned Susquehanna River channel.

The possibility of identifying underwater Clovis sites preserved in low-energy settings in leeside bays of the Chesapeake near the former banks of the Susquehanna River exists and is being pursued (D. Lowery, pers. comm. 2001). A Younger Dryas–age loess provides a protective mantle of variable thickness over these Clovis deposits, as is the case at the Shawnee Minisink site (Lowery 2001).

Atlantic shorelines and major river systems that provided open water corridors through terminal-Pleistocene and early-Holocene forested environments are features common to the middle Atlantic and the Southeast regions. The importance of these corridors as foci for early-Paleoindian colonization and interaction is well recognized; however, the relative advantages and disadvantages with regard to energy costs of mobility for pedestrian travel compared with watercraft transport along these corridors during early- and later-Paleoindian times requires study.

The distribution of lithic raw materials in the eastern United States indicates that Paleoindian people routinely transported themselves across the many broad rivers characterizing the region. Unless we presume that swimming or pedestrian travel over ice constituted the primary seasonal options for crossing rivers, it is time to contemplate the ramifications of watercraft alternatives.

> There is . . . strong evidence for progressive movement of exceptional or exotic stones carried west to east and south to north, for distances of hundreds of kilometers. . . . Some assemblages, especially in the north, have evidence of stone sources exploited from several directions, implying elaborate travels or networks of communication and trade. While it is necessary to be aware that stone can be moved by other than human agents, it is significant that most of the movements take the stone upriver and upglacier from its sources and across major drainage divides. . . . In the context of this observation, the impressive distances involved gain significance. (Dincauze 1993:286.)

Table 1. Attributes of apparently utilized bifaces from the Thunderbird site. From Boyer (1974:115–17).

	Class G	Class H	Class I	Class J	Class K
Length range (mm)	58–69	48–64	51–67	45–54	54–63
Width range (mm)	46–61	38–49	26–36	27–37	35–42
Thickness range (mm)	24–35	9–16	7–13	10–18	17–23
Mean weight (g)	108	33	19	18	44
Use-wear observations	Moderate to heavy wear on distal end and immediate edges points to use as crude chopper tools.	Light to moderate wear on distal and lateral edges.	Light to moderate wear on one or sometimes both lateral edges and distal end.	Light to moderate wear extending to all edges.	Moderate to heavy wear on lateral edges and distal end.
Description	Made on short thick flakes.	Short flakes w/prominent platform and convex distal end.	Made from a sturdy flake and are distinctly beveled.	Very similar to Class I, beveled, but not as well made.	Similar to Classes I and J, though larger and cruder.

Saint Lawrence River, Champlain Sea, and the Far Northeast
The St. Lawrence River not only connected the interior of the continent to the Atlantic Ocean, but with its network of tributaries also provided extensive highways for prehistoric peoples (Wright 1994:66).

Between 11,000 and 10,000 years ago, Paleoindian populations colonized the recently deglaciated far Northeast, where environments remained relatively open longer than was the case in the eastern Great Lakes or southern New England (Abel and Fuerst 1999). This section discusses archaeological sites associated with early deglaciated landscapes surrounding the St. Lawrence waterway.

Champlain Sea
During the last glacial period, the lower Saint Lawrence and Champlain valleys in New England were depressed below sea level by the weight of the Laurentide Ice Sheet. With retreat of glacial ice from the mouth of the Saint Lawrence by 12,500 yr B.P., marine transgression created the Champlain Sea, an inland brackish water environment extending westward from the Atlantic Ocean approximately 1800 km toward present-day Lake Ontario (Abel and Fuerst 1999; Davis and Jacobson 1985). By approximately 10,200 yr B.P., isostatic rebound in combination with further retreat of Laurentide Ice subsequently drained the Champlain Sea and reintroduced dry land to Saint Lawrence and Champlain lowlands.

During its geologically brief existence, cold-water marine mammals colonized the Champlain Sea. Late-Pleistocene marine fauna included beluga whales (*Delphinapterus leucas*), bowhead whales (*Balaenoptera physalus*), harbor porpoise (*Phoecoena phocena*), and 20 species of seals. Represented among the latter were ringed seals (*Phoca hispida*) and bearded seals (*Erignathus barbatus*), species frequently hunted by humans (Harington 1977). Based on a demonstrated association of over 30 fluted points with a former shoreline of the Champlain Sea in present-day northern Vermont, Loring (1980:28) proposes that early peoples may have honed their skills hunting marine mammals such as migratory beluga whales and seals in near-shore environments of this ancient sea.

Recently, Loring reexamined one of the Champlain Sea–associated fluted points illustrated in his 1980 publication (Loring 1980:Figure 3b). This nearly complete specimen is missing a basal ear and exhibits an impacted tip. It is made of an exceptionally fine grained blue-gray quartzite, which was originally thought to be an odd variant of local Cheshire quartzite. Based on subsequent fieldwork in Labrador during which Loring gained extensive familiarity with Ramah chert, he now considers it most probable that the fluted point in question is made of Ramah chert (S. Loring pers. comm. May 2001, Arctic Studies Center, Smithsonian Institution). Recent visual comparison with examples of Cheshire quartzite further confirmed for Loring that Ramah chert is the likely source of the raw material. Subsequent laser microprobe analyses have determined that the mineralogical signatures of the fluted point and Ramah chert compare favorably, but both differ from the mineralogical signatures of Cheshire quartzite samples (S. Loring pers. comm. June 2003).

Ramah chert sources are unknown south of Hamilton Inlet, northern Labrador, which is greater than 1300 km (straight-line distance) northeast of the fluted point find in Vermont (Fitzhugh 1972:43). Given glacial conditions, pedestrian travel from Hamilton Inlet to the eastern shores of the Champlain Sea is deemed highly unlikely; and coastal transport by watercraft is implied. Until approximately 10,000 yr B.P. Hamilton Inlet was separated from the mouth of the Champlain Sea by a glacial tongue extending to the Atlantic Ocean. Transport between the two regions required seagoing watercraft along a "treacherous and storm-ridden coast" (Fitzhugh 1972:43). People of the Moorehead phase of the Maritime Archaic were fully involved in trade and transport of Ramah chert bifaces by 7500 yr B.P. (Fitzhugh 1975, 1978; McGhee and Tuck 1975). The recovery of a toggling harpoon head and a walrus tusk in the L'Anse Amour burial mound near the Strait of Belle Isle, Labrador, suggests that Maritime Archaic people

> had extended their maritime adaptation to ice-edge or open water hunting of marine mammals as large and dangerous as the walrus. The presence of ground stone adzes and gouges early in the local tradition suggests a woodworking industry capable of constructing watercraft useful for inshore fishing and limited open-water hunting. (McGhee and Tuck 1975:119–20).

An earlier developmental phase logically precedes the fully maritime adaptation known to exist by 7500 yr B.P. The Vermont fluted point, if truly made of Ramah chert, may suggest that the archaeological record of maritime adaptation and Atlantic travel in the far Northeast extends back to the terminal Pleistocene.

Canadian Maritime Provinces and Saint Lawrence Waterway
The Canadian Maritimes were among the final deglaciated landscapes in North America to be colonized by people who used fluted points. While radiocarbon-dated Paleoindian components are at a premium in the far Northeast, the Debert site in central Nova Scotia establishes the presence of fluted points in the region by 10,600 yr B.P. (MacDonald 1968). Most lithic materials at Debert were obtained along the Bay of Fundy shoreline, approximately 50 km distant.

Keenlyside (1991:166) speculates that Paleoindians entered the Maritime Provinces along the Atlantic coast, then traveled major river systems to the interior. The coast in the vicinity of the Gulf of Saint Lawrence was free of glacial ice by approximately 13,500 yr B.P. (Dumais 2000: 82). "Seasonal rounds must have brought various groups living along the Atlantic seaboard and major waterways into frequent contact with one another" (Keenlyside 1991:170). These groups are thought to have developed "a significant reliance on marine resources, in particular herding sea mammals, such as seal and walrus" (Keenlyside 1991:163).

> The unique early postglacial environment of the Canadian Maritimes provided an ideal setting for the evolution of a primarily land-based people to one with a greater reliance on coastal and marine resources. This expansion of the resource base necessitated the development of new technologies to exploit these new environments. The presence of a sophisticated sea-mammal hunting technology as early as 8000 years ago on the southern Labrador coast (McGhee and Tuck

> 1976) implies a considerably earlier developmental stage. I suspect that this may have taken place among the late Paleoindian peoples of the southern Gulf. Resource-rich bays, such as St. Peters, Basin Head, Savage Harbour, and others found along Prince Edward Island's north shore, appear to have been particularly attractive during this developmental period. Envisaged is a gradual increase in the reliance on marine resources, perhaps combined with improved water transportation and the development of a more sophisticated marine hunting technology (Keenlyside 1991:172).

Triangular, strongly barbed projectile points from Prince Edward Island are thought to represent late-Paleoindian atlatl darts or end blades for thrusting spears used to hunt sea mammals.

> The barbed points are closely reminiscent of Northwest Coast sea-mammal hunting implements, such as those employed historically by the West Coast Makah people. In the East Coast context, these specimens would have been particularly well suited for use on marine mammals, such as seal or walrus (Keenlyside 1991:170).

At the time of the earliest documented occupation of the Gulf of St. Lawrence about 9500 yr B.P., waterborne mobility is evident and people appear to have been exploiting both maritime and interior riverine resources (Wright 1994:52).

Most of the Gaspé Peninsula of Quebec (on the southern shore of the Saint Lawrence Waterway) became ice-free around 10,500 to 10,000 years ago, with localized ice remaining until nearly 9000 yr B.P. (Dumais 2000:84). North-south valleys opening on the Saint Lawrence formed estuaries, which Dumais (2000:85) suggests were selectively exploited by early inhabitants of the Gaspé coast. A tundra landscape characterized the area until nearly 9300 yr B.P. Cold climate and the proximity of the Laurentide Ice Sheet (lying just inland from the north shore of the Saint Lawrence) delayed the development of a closed forest until approximately 7800 to 7200 yr B.P.

Dumais (2000:99–103, Figure 18) proposes a migration of Plano people from the northeastern Plains west of the Great Lakes to the Gaspé Peninsula approximately 9500 to 9000 yr B.P. Makers of Plano projectile points (termed Saint Anne points in the far Northeast) occupied a series of habitation and quarry sites along the lower Saint Lawrence Valley (i.e., Rimouski, La Matre, and Mitis), which are associated with postglacial Goldthwait Sea estuaries. Another site, Varney Farm, overlooking the Nezinscot River in western Maine, is AMS radiocarbon-dated between 8380 and 8700 yr B.P. (Cox and Petersen 1997; Petersen et al. 2000). Although faunal material was not recovered, immunological analysis of protein residue from tools from the Canadian sites produced positive reactions to deer, elk, bear, rabbit, rat, canid, felid, fish, and sea lion antiserum (Dumais 2000:106). The latter is judged to be an indication of pinneped hunting (Chalifoux 1999:77).

Local coastal exposures of chert and siltstone were exploited by Plano flintknappers at Gaspé Peninsula sites. Wright (1994:52) accepts the discovery of Gaspé Peninsula chert at a site located on an island of Lake Saint Francis 800 km upstream as indirect evidence of watercraft travel by Plano peoples (see also Dumais 2000:102).

> At the earliest known period of human occupation of the St. Lawrence River there was considerable mobility, either as a result of extensive seasonal rounds or trade . . . we are now permitted to infer that these early hunters not only possessed some form of efficient watercraft but exploited maritime resources as well as interior riverine resources. With regard to the question of whether this evidence indicates trade or extensive seasonal rounds it should be noted that environmental conditions at this time . . . would have favored a very free ranging form of settlement pattern along the southern margin of the ablating continental glacier. The possession of watercraft also would have markedly increased mobility (Wright 1994:52).

The clear correlation between these coastal Plano settlements and former estuaries that existed when the Goldthwait sea inundated the downstream portions of valleys that opened on the St. Lawrence is concordant with the idea of early maritime adaptation. Atlantic salmon and seals are expected to have been particularly abundant in these settings during the spring-summer seasons.

Emphasis on Caribou Hunting in Models of Northeastern Paleoadaptations

Ethnographic analogs based on arctic and subarctic peoples (such as the Naskapi/Innu and others) have been employed to model Northeastern Paleoindian adaptations as those of specialized caribou hunters (Fitting 1966; MacDonald 1968; Spiess 1979; Spiess et al. 1985; Storck 1988). Loring (1997) and Levine (1997) caution that a complementary suite of subsistence strategies buffered caribou hunting in traditional arctic and subarctic lifeways, including fishing and migratory waterfowl and sea mammal hunting.

Loring introduces an interesting point with possible ramifications for early-Paleoindian adaptations beyond those he proposes. He (1997:211) suggests that the heavy reliance on caribou hunting by the Innu was a relatively recent historical development.

> Archaeological research in Labrador provides evidence that the "classic" Naskapi/Innu caribou hunting specialization is a late (post-contact) phenomenon that comes about partially in response to Inuit and European usurpation of the coast and the loss of access to maritime resources. During the preceding late Prehistoric Period in Labrador, caribou apparently served as an important resource option but access to maritime resources remained a critical feature of the Indian economy. The "classic" Naskapi/Innu caribou hunting stereotype appears to be a unique historical event in Labrador.

This forms the basis of his argument that the Innu situation provides an inappropriate analog for investigating early-Paleoindian economies. I wonder if widespread coastal inundation due to postglacial sea level rise may have temporarily, yet significantly, disrupted coastal subsistence practices of some early-Paleoindian groups, prompting them to compensate for temporarily lost maritime resources by concentrating on caribou in similar fashion to that documented historically for the Innu. Although the causes of lost maritime resources varied in the two instances, response to disrupted access to these significant resources may have prompted a similar choice—a temporary subsistence shift toward greater reliance on caribou.

Distribution of Munsungun Lake Chert: Watercraft-aided Procurement and Transport?

Pollock et al. (1999) have characterized seven categories of chert from the Munsungun Lake outcrops in northern Maine, which were exploited by makers of fluted points since the recession of the late-Wisconsin Ice Sheet. Fluted points made of Munsungun cherts are identified at sites in Maine, Massachusetts, New Hampshire, and New Jersey, often in association with drainages (Pollack et al. 1999:271). Given the lacustrine-dominated environs of the Munsungun chert quarries and their setting near the headwaters of two major drainage systems, the probability that watercraft were used to access the area and to transport chert from it must be given serious consideration.

> Munsungun Lake is centrally located, and is easily accessible through the Penobscot and Aroostook River drainage systems. Nomadic human groups could easily reach, and would naturally pass through the area using various waterways (Pollock et al. 1999:271).

As strongly as anywhere in North America, the use of watercraft to procure and transport lithic material may be indicated in the case of the Munsungun Lake chert outcrops.

Eastern Great Lakes

> Most water trails were seasonal, that is, easier in the high water of spring and early summer than in fall or winter. Although water routes were preferred by most travelers, some land routes were relatively easy and generally took no more than twice the time of a water route between the same endpoints. . . . The most significant attribute of the geography of the divide between the upper Mississippi and the Great Lakes drainage is that instead of mountains at the heads of the streams there are numerous "short" . . . easily traversed portages by which one could carry a canoe and its freight, or during high water in the spring, pull, push, or even paddle a canoe across the divide (Little 1987:57).

Early peoples, particularly during the Parkhill phase (Ellis and Deller 1997:13), frequently chose campsites along the shorelines of active glacial lakes (Deller and Ellis 1992a; Jackson 1990; Jackson et al. 2000; Roosa 1977; Storck 1984, 1997). Many of these locations are in accord with subsistence activities oriented toward fishing, waterfowling, and intercept hunting of caribou. Traditionally, the latter activity was facilitated by watercraft.

The Fisher site in the Huron Basin of southern Ontario was situated at the base of a peninsula on the western margin of a large shallow lagoon. It "would potentially have been a favorable habitat for fish, particularly during spawning" as well as a good location to hunt large mammals (Storck 1997:278). Makers of fluted points had immediate access to the lake, a downwind location, and good visibility of the surrounding countryside (Storck 1984:287). During the occupation(s), the site was associated either with the Main stage of Lake Algonquin dated about 11,200–10,400 yr B.P. (Karrow et al. 1975) or Lake Ardtrea dated approximately 10,500 yr B.P. (Jackson et al. 2000).

Use-wear analyses of tools from the site indicate that fish processing and woodworking took place on a low terrace near the water's edge (Area C) (Tomenchuk 1997:143). Strong evidence suggests that "most of the economically significant fishes in the drainage system of present-day Lake Huron probably inhabited glacial Lake Algonquin by 11,800 yr B.P." when water temperatures are estimated to have been comparable to that of Lake Superior today (Tomenchuk 1997:122–23). Use-wear traces developed on experimental tools used to remove fish spines were judged a close match for utilized flakes identified as fish processing tools. Perciformes (i.e., wall-eye, perch, sauger) may have been the family of fish exploited. Tomenchuk (1997:126) proposes that four individuals shared in woodworking activities in the fish processing area. Whether this woodworking was related to the manufacture of fishing gear such as weirs, drying equipment, or boats is unknown. Use-wear analysis of tools from a nearby low ridge suggests large mammal butchering, hide processing, and bone, wood, and antler working (Tomenchuk 1997:135).

Nearly 38 km to the southeast, the Banting site provides evidence of Paleoindians camped on an island in Lake Algonquin near the mouth of a small bay. About 11,200–10,400 years ago, this island was accessible only by crossing water or ice.

> In the winter, caribou movements on the frozen lake, such as those in the arctic today . . . would have been easily monitored . . . during warm seasons of the year, islands, such as the one on which the Banting site is located, may have provided forage and refuge for caribou and opportunities to flush the animals into the lake where they could be killed in boats (Storck 1984:288).

Other sites associated with the southern shores of glacial Lake Algonquin include Hussey, which occupied the crest of a peninsula projecting into the lake, and a group of early-Paleoindian localities 50 km to the east near the Gainey site of Udora (Storck 1988; Storck and Spiess 1994). A cache of informal woodworking tools thought to have been used to process jack pine or black spruce roots was excavated at the latter (Storck and Tomenchuk 1990). Nelson (1983:50) notes that "good basket roots are found only in specific places–along certain lakeshores, for example"

A final site, Parkhill, is located near the southeast end of present-day Lake Huron. During the Paleoindian occupation (about 10,500 yr B.P.), Parkhill was situated on the axis of a sharp bend in a strandline overlooking Lake Algonquin. This location may have been adjacent to a corridor of caribou movements along the shoreline (Deller and Ellis 1992a; Storck 1984).

Interregional communication across more than 700 km separating the eastern Great Lakes from southern New England is evident during the Parkhill Phase (approximately 10,500 yr B.P.) by the location of the Neponset site near Boston (Carty and Spiess 1992). Barnes projectile points exhibiting full-facial fluting, a preform fluted from the tip, gravers, radial break tools, and miniature points characterize this assemblage. Carty and Spiess (1992:19) note that while the Neponset site appears unrelated to other sites in the New England–Maritime region, it closely resembles Parkhill assemblages to the west.

I suggest that a combination of watercraft-enhanced mobility, direct social interaction, and down-the-line communication may account for interregional stylistic projectile point similarities and striking assemblage correspondences between Parkhill (Great Lakes) and Folsom (Rocky Mountain/Plains) groups, and to some extent Cumberland (Southeast) peoples (see Deller and Ellis 1992b). Whether climatic influences and biotic change associated with the Younger Dryas interval contributed to this apparent dispersal of technological approach bears future investigation. The previously discussed movement of late-Paleoindian Plano people from the northeastern Plains to the Saint Lawrence

Valley between approximately 9400 and 8400 yr B.P. also points toward watercraft-supported mobility.

Eastern Periphery of Great Plains

The transport of lithic materials from source areas to distant sites has provided archaeologists working in Australia, Japan, Greece, and the Northwest and California coasts of North America with keys to population movements that imply probable use of watercraft by early peoples. This approach is applied here to the fluted-point record along the eastern periphery of the Plains to examine the possibility of watercraft travel on and across rivers. Comprehensive coverage is not attempted; rather the aim is to suggest a fruitful avenue for future research.

High-quality Knife River flint has been identified from the Bostrom Clovis site in Illinois (Koldehoff 1983:227) and sites in nearby St. Louis County, Missouri (Perino 1967:102), over 1,500 km from the source area in southwestern North Dakota (Tankersley 1991:Table 17.1). Transport to these sites implies crossing the Missouri River. To reach the Bostrom site, people traversed the Mississippi River as well.

Two Clovis points and a Folsom preform of Knife River flint are reported from the Loess Hills of southwestern Iowa, which are situated on the opposite side of the Missouri River from the lithic source area (Billeck 1998:406). The Loess Hills sites are located 900 km downstream from the source, near the confluence of the Missouri and Platte rivers, strongly suggesting the possibility that Folsom people were traveling on (not exclusively along the banks of) these water routes. Perhaps the farthest occurrence is a possible Knife River Clovis point visually identified (Gramly 1988:Table 3) from the Lamb site in Genesee County, New York, a distance of nearly 2,000 km involving multiple river crossings.

A Folsom preform base and a Folsom point (recycled and corner-notched by later people) of South Bend chert from the Missouri River near its confluence with the Platte drainage in Nebraska were recovered in the Loess Hills of southwestern Iowa (Billeck 1998:Table 2; Morrow and Morrow 1999:Table 1). The South Bend chert source is relatively close as a bird flies (about 8–10 km), but travel from source to site suggests Folsom people were crossing the Missouri River. A preform base of Florence B chert from the northern Flint Hills of Nebraska was found on the opposite side of the Missouri River in southwestern Iowa, at least 120 km from the closest source area (Morrow and Morrow 1999:75). A complete Folsom point from south central Iowa is tentatively identified as being made of Smoky Hill jasper (Niobrara) from a source area some 470 km to the southwest on the Nebraska-Kansas border (Morrow and Morrow 1999:74–75). If this material identification is correct, it represents another occurrence of Folsom quarrying on the west side of the Missouri River prior to transport and hunting on the east side of this major tributary of the Mississippi River.

The distribution of Clovis points made of Hixton silicified sandstone from southwestern Wisconsin implies that makers of fluted points crossed the Ohio River into present-day central Kentucky. Likewise, Clovis points made of Hopkinsville chert from southwestern Kentucky are found on the north side of the Ohio River in Indiana and Ohio (Tankersley 1991:Figure 17.4).

Holen (2002) reports the presence in a Clovis collection from Nebraska of Edwards Plateau chert from Texas, indicating transport across multiple intervening rivers. Additional crossings of the Arkansas River by Clovis people are indicated by the presence of Alibates chert from the Texas Panhandle, as well as Edwards Plateau chert in the Drake Clovis Cache in northern Colorado (Stanford and Jodry 1988). Folsom people transported Alibates northward across the Arkansas River and carried Black Forest silicified wood from the Plains east of Colorado Springs southward across the same drainage (Jodry 1999; Meltzer et al. 2002). The use of skin boats by Great Plains Paleoindians is therefore considered likely along major rivers.

Further west, obsidian from the Malad source in south-central Idaho is represented in the Fenn Clovis Cache from the area of southwest Wyoming, northwest Utah, and southeast Idaho (Frison 1991:41; Hughes 1994:369). Also represented in the cache are Utah jasper and Green River chert from sources located near the Green River. Frison and Bradley (1999:83) note that these materials may have been transported, in part, on this waterway.

Comparative empirical data regarding the relative frequencies of lithic materials transported from source areas to sites across waterways of different size and character, during successive Paleoindian time periods, is needed to investigate the early use and development of watercraft technology in the Americas. Traditionally in Paleoindian studies, raw material distribution data are investigated and reported primarily with reference to pedestrian mobility patterns and social boundaries. The aim of this paper (and an earlier version, Jodry 1999) is to encourage expansion of our research questions to address explicitly the question of water transport technology.

Great Basin

Nearly all the fluted-point sites in the Great Basin are located along the margins of late-Pleistocene/early-Holocene lakes and marshes (Grayson 1993:238). While direct evidence of subsistence activities (and transport) is poorly represented, association with highly productive wetland habitats seems clear. The archaeological record of fishing in the Great Basin begins in earnest approximately 9600 yr B.P. (Tuohy 1990:124), but earlier suggestions exist.

The oldest known fishing line from the lower Truckee River Basin in western Nevada is radiocarbon-dated to 9660 ± 170 RCYBP (GX-13744) (Tuohy 1988:225). This three-ply fishing line from Wizards Beach is made of sagebrush bark. Follett (1982:188) reports 9600 yr B.P. as the earliest date on fish remains from 10 archaeological sites associated with Winnemucca Lake in Washoe County, Nevada. During historic times, fishing constituted a nearly year-long subsistence option. "This was due in part to the winter and spring timing of the two cutthroat runs, the late spring kuyui run, and the spring runs of the lesser food fishes, such as the Tahoe suckers, redsides, speckled dace and the tui chub" (Tuohy 1990:122). Farther west, in the Fort Rock Basin of south-central Oregon, the earliest fish bones recovered from Connley Caves lie immediately above a level dated to 9800 ± 250 RCYBP (Gak-1743) (Greenspan 1990:213).

Dansie and Jerrems (1999) document a barbed leister or harpoon made of mammoth rib bone from Pyramid Lake dated to 10,340 ± 50 RCYBP (UCR-3795/CAMS 59285), and a mammoth ivory point dated to 10,360 ± 50 RCYBP (UCR-3796/CAMS 59286). They propose that these tools may have been designed for procuring Pyramid Lake cutthroat trout, which ranged in excess of 27 kg (60 lbs.). While near-shore pedestrian fishing may be represented, the possible use of tule boats on these lakes during the late-Pleistocene/early-Holocene deserves serious consideration. Identification of two species of small fish as well as what appears to be ground bulrush seeds in the intestines of Spirit Cave Man documents these marsh resources as dietary items 9400 years ago (Eiselt 1997; Napton 1997). Tule mats dated between 9200 and 9400 yr B.P. (Dansie 1997; Tuohy and Dansie 1997) indicate a sophisticated fiber technology based on woven bulrush, hemp, and other marsh plants. As discussed previously, harvesting of such plants as well as waterfowl, eggs, and fish was facilitated historically in the Great Basin and in comparable environments throughout the world by the use of water transport technology, notably boats constructed of tule bundles.

Pacific Coast

Pacific shorelines in far northern (Aleutians to the Northwest Coast) and far southern (Tierra del Fuego to southern Chile) America are maritime zones of dense fiords, island archipelagos, and high forested mountains where indigenous people traveled more by boat than by other means. New World models featuring maritime peoples are being tested at Pacific coastal sites, where major progress has been made during the last decade in reconstructing environmental and cultural parameters that characterized early adaptive patterns of humans in the New World.

Southeast Alaska and the Northwest Coast

Early maritime adaptations in North America have been investigated at excavated sites in the Alexander Archipelago of southeastern Alaska, including Hidden Falls (Davis 1989, 1996), Ground Hog Bay 2 (Ackerman 1968; 1996b; Ackerman et al. 1979) and 49-PET-408 (Dixon et al. 1997). The early occupations at the former sites date between approximately 9,500 and 10,000 yr B.P., respectively (Ackerman 1996b:429). Boats were required to reach them and would have been needed for maritime hunters and fishers to use offshore resources (Ackerman 1996b:429).

The bones of a young adult male who lived ca. 9200 yr B.P. (Dixon et al. 1997) were recovered from a cave (49-PET-408) on Prince of Wales Island (Dixon 1999:117). Isotopic analysis indicates that his diet consisted primarily of marine foods such as ringed seal, sea otter, and fish (Dixon 2001:286). Since Prince of Wales was not connected to the mainland at this time and some of the materials from which associated stone tools were made do not originate on the island, use of watercraft is implied. "These people were experienced at navigating year-round open water along rugged forested coastlines characterized by fjords, islands, rocky headlands, and calving glaciers" (Dixon 1999:178). By 8,200 yr B.P., the preserved remains of bottom fish, rock fish, and marine mammals recovered from Locality 1 at the Chuck Lake site on Heceta Island further confirm a maritime economy supported by water transport technology (Ackerman et al. 1985; Dixon 1999; Okada et al. 1989).

Early peoples who employed a microblade technology were transporting obsidian among islands (Sitka and Sumez) and the mainland (southeast Alaska and northern British Columbia) by 9500 to 9000 yr B.P. (Ackerman 1992, 1996a; Carlson 1994; Davis 1989; Erlandson et al. 1991). Rising sea level inundated many sites older than 9500 yr B.P., challenging efforts to document even earlier occurrences of maritime travel as evidenced by obsidian transport along the coast. Shorelines dating between 9400 and 12,500 yr B.P. are submerged to depths as great as 150 m (Josenhans et al. 1997). Focused efforts to locate submerged archaeological sites associated with early postglacial landscapes have produced encouraging results on the continental shelf of British Columbia. Fedje and Josenhans (2000:100) have identified productive intertidal shellfish beds containing species used by coastal people, as well as a basalt flake associated with a paleoriver terrace inundated by sea level rise approximately 10,000 yr B.P. A coastal landscape "viable for human occupation and movement before 12,200 yr B.P." is indicated (Fedje and Josenhans 2000:102).

Southern California

The Southern California coast has a well-documented record of early maritime adaptation (Erlandson 1994). Paleoindians settled the California Channel Islands (separated from the mainland throughout the Pleistocene) no later than 10,400 yr B.P. (Erlandson et al. 1996; Erlandson et al. 1999) and perhaps closer to 11,000 yr B.P. (Johnson et al. 2000). This is the earliest definitive evidence of the use of seafaring watercraft in North America.

Human remains from the Arlington Springs site on Santa Rosa Island recently yielded an AMS radiocarbon determination on bone collagen of 10,970 ± 80 yr B.P. (CAMS-16810) (Johnson et al. 2000). The presence of this individual on an offshore island demonstrates the existence of maritime-adapted people capable of navigating seaworthy boats. A deeply buried shell midden (CA-SRI-6) exposed in a sea cliff at the mouth of Arlington Canyon on this island produced remains from a deeply buried residential campsite dated to 9300 yr B.P. Cisco chert artifacts from the midden are thought to originate on San Miguel Island nearly 15 km to the west (Erlandson et al. 1997). Black abalone dominates the dietary assemblage, which includes other shellfish, fish, birds, and a single sea mammal (species uncertain). The 17 types of shellfish recovered indicate the former presence of an open-coast rocky-shore habitat, "by far the most productive intertidal habitats found on the Channel Islands throughout prehistory" (Erlandson et al. 1999:259). Most of the fish represented (i.e., sheephead and rockfish) inhabit rocky near-shore or kelp bed environments. Meat from marine sources (primarily shellfish, 85 percent) contributed over 98 percent to the total reconstructed meat yield. During the occupation, the site was located near the edge of a coastal plain some 2 km from the coastal shore.

A 9900-year-old shell midden at Running Springs on San Miguel Island is small and inconspicuous. Erlandson (2000:23) proposes that this type of site "is exactly what the earliest island campsites should look like if they were left behind by small mobile groups of people." Early components from Daisy Cave, also on San Miguel Island, date between 10,600 and 9400 yr B.P. (Erlandson et al. 1996). Extensive early-Holocene midden deposits here yielded shellfish, 18 taxa of marine fish, pinniped and sea bird bones, fish gorges, and sea grass cordage. This important site provides a rare detailed view of early Channel Island adaptation (Torben et al. 2001).

The early-site occupants appeared to have fished primarily in kelp beds or rocky shores in the productive waters adjacent to the site, with technologies that included boats, nets, and hook-and-line fishing (Torben et al. 2001:609).

> It now appears that net fishing and nearshore marine fishing were fundamental and optimal strategies used by some of the earliest Pacific Coast peoples not just to survive, but to thrive (Torben et al. 2001:610).

Torbin et al. (2001:609) argue that many of the technological foundations of maritime societies, including seaworthy boats, nets, fishhooks, and fish spears, were "in place by the earliest well-documented occupations of the Pacific Coast of the Americas (12,000 to 10,000 years ago)."

Southern Peru

Recent excavations in south coastal Peru at the site of Quebrada Jaguay 280 suggest that terminal-Pleistocene–age foragers may have practiced transhumance, moving between the Andean highlands and the coast (Sandweiss et al. 1998). The identification of obsidian from the adjacent Andean highlands in the terminal-Pleistocene levels suggests that early peoples traveled to the mountains or maintained social networks with those living there. Approximately 11,100 to 10,000 yr B.P. these people exploited shellfish, crabs, and fish while living near the coast. The dominance of a single species (*Sciaenae*, or drums) of small fish of a restricted length range implies a netfishing procurement strategy. A few specimens of knotted cordage from early-Holocene deposits at the site may represent parts of fishnets (Sandweiss et al. 1998:1830).

At the nearby site of Quebrada Tacahuay, dated between 10,770 and 10,530 yr B.P., obsidian flakes from the highlands 130 km to the east were again recovered at a coastal campsite where people hunted cormorants and other seabirds and fished for anchovies (Keefer et al. 1998). The upwelling Humboldt Current along the steep continental shelf off Peruvian and Chilean shores creates "one of the richest marine environments in world" (Dillehay 2000:154). A record of prehistoric exploitation of these resources from very early times is abundantly evident.

Conclusions

Paleolithic use of watercraft from Australia to the Japanese archipelago between 40,000 and 30,000 yr B.P. implies that fully modern humans developed seafaring technology many millennia prior to the time people are generally predicted to have arrived in the Americas. Nevertheless, New World models of Paleoindian mobility and inland colonization are overwhelmingly oriented toward pedestrian travel, which fails to take into account a global archaeological record indicative of early incorporation of water transport technology into subsistence and settlement practices.

Early evidence of use of boats in North America is derived from the occupation of California's Northern Channel Islands between 10,400 and 10,970 ± 80 RCYBP (CAMS-16810). The latter is an uncalibrated radiocarbon date obtained on human bone from the Arlington Springs site on Santa Rosa Island (Johnson et al. 2000). The presence of humans in the New World is securely dated to 12,500 yr B.P. at the Monte Verde site in south central Chile (Dillehay and Pino 1997) and 12,890 ± 90 RCYBP (AA20125) at Piedra Museo (El Puesto rockshelter) in Argentina (Miotti and Salemme 1998, cited in Borrero 1999:329). If coastal-entry models proposing that some peoples initially arrived in the Americas aboard sea craft are confirmed (Dixon 2001; Easton 1992; Fladmark 1979; Gruhn 1988; Heusser 1960; Mandryk et al. 2001; Stanford and Bradley 2002), then water transport technology in the New World may prove to be at least two millennia older than currently demonstrated.

Distribution of lithic raw materials, particularly on islands, points toward early watercraft use. Carlson (1994:318, 348) reports obsidian trade in all regions of the Northwest by 9000 yr B.P., and between islands and the mainland of British Columbia as early as 9500 yr B.P. Galm (1994) describes exchange of obsidian and marine shell across the adjacent Interior Plateau by 10,000 yr B.P. Communication among groups and the transport of lithic materials in northwest North America took advantage of water transport along the coast and the main tributaries of the Columbia and Fraser river systems. Evidence of long-distance lithic transport from quarry to campsites along the Saint Lawrence River by 8500 yr B.P. provides compelling suggestions of early watercraft use, as does the distribution of Munsungun Lake cherts in the Northeast.

Exploitation of islands off the California and British Columbia coasts (Erlandson et al. 1999), identification of submerged coastal sites (Faught 1988; Fedje and Josenhans 2000; Jackson et al. 2000; Stright et al. 1999), and selective occupation of lacustrine and riverine shorelines by early peoples in the Northeast, Southeast, Great Lakes, and Great Basin leave little doubt that marine and lacustrine/marsh habitats were significant targets of subsistence exploitation in early postglacial times. Comparative ethnography emphasizes the nearly ubiquitous, worldwide distribution of water transport technology in these settings. Clark cautions that while "comparative ethnography can prompt the right questions, only archaeology in conjunction with the various natural sciences . . . can give the right answers" (1952:357). The goal of this paper is to prompt questions concerning the potential significance of water transport technology for early human dispersal, landscape learning, and adaptation in the Americas, not only in coastal settings but also in the interior of the American continents. We are unlikely to arrive at a fuller understanding of Paleoindian mobility and dispersal if persistent views regarding Paleoindians *afoot* continue to obscure potential insights offered by testing models of Paleoindians *afloat and afoot*.

The former presence of such quintessentially perishable items of material culture as Thule Eskimo snowhouses are archaeologically visible owing to *recognition of the tools used to make them* (snow knives and snow probes, McGhee 1984:372) and the *recognition of the possibility* that such houses formerly existed. So too might Paleoindian watercraft be identified archaeologically by recognizing tools used in their manufacture. Additional clues lie in the cultural transport of lithic raw materials across major rivers, the possible association of campsites with portage points along waterways, faunal indicators of deep-sea maritime exploitation, and demonstrated use of islands in the Northeast during warm seasons. Greater recognition that water transport technology existed in the Americas during the late Pleistocene is likely to result in more realistic models of Paleoindian mobility, subsistence, and settlement, and a greater likelihood that models inclusive of watercraft technology will be rigorously tested.

Coastal-entry models featuring watercraft did not benefit from widespread serious consideration until geological, paleoenvironmental, and archaeological data undermined the perceived viability of the Ice-Free Corridor paradigm for terrestrial colonization, while concurrently demonstrating that the northwest coast of North America was not a prohibitively inhospitable landscape during the terminal Pleistocene. There are many more researchers currently willing to embrace the notion of seafaring Paleoindians than existed 10 years ago. Hopefully, within five years there will be more researchers testing models of inland waterway travel and communication than exist presently.

Acknowledgments

I thank Rob Bonnichsen and Dennis Stanford for the opportunity to participate in the Clovis and Beyond symposium, which provided the impetus for this paper. Their comments, those of Tom Dillehay, two anonymous reviewers, and a technical editor are much appreciated. My interest in the topic stems from a 1994 journal article in *North American Archaeologist* by Engelbrecht and Seyfert, whom I also thank. Any errors or omissions remain my own.

References Cited

Abel, T. J., and D. N. Fuerst
1999 Prehistory of the St. Lawrence Headwaters Region. *Archaeology of Eastern North America* 27:1–53.

Ackerman, R. E.
1968 Archeology of the Glacier Bay Region, Southeast Alaska. Department of Anthropology Report of Investigations 44. Washington State University, Pullman.

1992 Earliest Stone Industries on the North Pacific Coast of North America. *Arctic Anthropology* 29 (2):18–27.

1996a Early Maritime Culture Complexes of the Northern Northwest Coast. In *Early Human Occupation of British Columbia*, edited by R. L. Carlson and L. Dalla-Bona, pp. 123–32. University of British Columbia Press, Vancouver.

1996b Ground Hog Bay, Site 2. In *American Beginnings: The Prehistory and Palaeoecology of Beringia*, edited by F. H. West, pp. 424–49. University of Chicago Press, Chicago.

Ackerman, R. E., T. D. Hamilton, and R. Stuckenrath
1979 Early Cultural Complexes on the Northern Northwest Coast. *Canadian Journal of Archaeology* 3:195–209.

Ackerman, R. E., K. C. Reid, J. D. Gallison, and E. Roe
1985 Archeology of Heceta Island: A Survey of 16 Timber Harvest Units in the Tongass National Forest, Southeastern Alaska. Project Reports 3, Center for Northwest Anthropology. Washington State University, Pullman.

Adney, E. T., and H. I. Chapelle
1983 *The Bark Canoe and Skin Boats of North America*. Smithsonian Institution Press, Washington, D.C.

Aikens, C. M., and T. Higuchi
1982 *Prehistory of Japan*. Academic Press, New York.

Anderson, D. G.
1990 The Paleoindian Colonization of Eastern North America, A View from the Southeastern United States. In *Early Paleoindian Economies of Eastern North America*, edited by K. B. Tankersley and B. L. Isaac, pp. 163–216. Research in Economic Anthropology Supplement 5. JAI Press Inc., Greenwich, Connecticut.

1996 Models of Paleoindian and Early Archaic Settlement in the Lower Southeast. In *Paleoindian and Early Archaic Southeast*, edited by D. G. Anderson and K. E. Sassaman, pp. 16–28. University of Alabama Press, Tuscaloosa.

Anderson, D. G., and J. C. Gillam
2000 Paleoindian Colonization of the Americas: Implications from an Examination of Physiography, Demography, and Artifact Distribution. *American Antiquity* 65(1):43–66.

Arima, E. Y.
1975 A Contextual Study of the Caribou Eskimo Kayak. National Museum of Man. Mercury Series, Ethnology Service Paper 25, Ottawa.

1984 Caribou Eskimo. In *Arctic*, edited by D. Dumas, pp. 447–62. *Handbook of North American Indians,* Volume 5. Smithsonian Institution, Washington, D.C.

Arima, E., and J. Dewhirst
1990 Nootkans of Vancouver Island. In *Northwest Coast*, edited by W. Suttles, pp. 391–411. *Handbook of North American Indians*, Volume 7. Smithsonian Institution, Washington, D.C.

Asch, M. I.
1981 Slavey. In *Subarctic*, edited by J. Helm, pp. 338–49. *Handbook of North American Indians*, Volume 6. Smithsonian Institution, Washington, D.C.

Bean, L. J., and D. Theodoratus
1978 Western Pomo and Northeastern Pomo. In *California*, edited by R. F. Heizer, pp. 289–305. *Handbook of North American Indians*, Volume 8. Smithsonian Institution, Washington, D.C.

Billeck, W. T.
1998 Fluted Point Distribution in the Loess Hills of Southwestern Iowa. *Plains Anthropologist* 42:401–09.

Bird, J.
1946 The Alacaluf. In *The Marginal Tribes*, edited by J. H. Steward, pp. 55–79. *Handbook of South American Indians*, volume 1, United States Printing Office, Washington, D.C.

Black, R. F.
1974 Late-Quaternary Sea Level Changes, Umnak Island, Aleutians: Their Effects on Ancient Aleuts and Their Causes. *Quaternary Research* 4:264–81.

1975 Late-Quaternary Geomorphic Processes: Effects on the Ancient Aleuts of Umnak Island in the Aleutians. *Arctic* 28:159–69.

1976 Geology of Umnak Island, Eastern Aleutians Islands, as Related to the Aleuts. *Arctic and Alpine Research* 8:7–35.

Bonnin, P.
1995 Prospection Subaquatique Nandy, Le-Coudray-Montceaux. *Bilan Scientifique de la région Ile-de-France, 1994.* Ministère de la culture, Direction du patrimoine, sous-direction de l'archéologie, Paris.

1996 Prospection Subaquatique Nandy, Le-Coudray-Montceaux. *Bilan Scientifique de la région Ile-de-France, 1995.* Ministère de la culture, Direction du patrimoine, sous-direction de l'archéologie, Paris.

Borrero, L. A.
1999 The Prehistoric Exploration and Colonization of Fuego-Patagonia. *Journal of World Prehistory* 13(3):321–55.

Bouchard, R. T., and D. Kennedy
1998 Lillooet. In *Plateau,* edited by D. E. Walker, Jr., pp. 174–90. *Handbook of North American Indians,* Volume 12. Smithsonian Institution, Washington, D.C.

Bourne, B. F.
1853 *The Captive in Patagonia or Life Among the Giants.* Boston.

Bowler, J. M., and J. W. Magee
2000 Redating Australia's Oldest Human Remains: A Sceptic's View. *Journal of Human Evolution* 38:719–26.

Boyer, W. P., Jr.
1974 A Preliminary Analysis of the Thunderbird Site Bifaces. In *The Flint Run Paleo-Indian Complex: A Preliminary Report 1971–73 Seasons,* edited by W. M. Gardner, pp. 114–18. Occasional Publication Number 1, Archeology Laboratory, Department of Anthropology, The University of America, Washington, D.C.

Bradley, B. A.
1997 Sloan Site Biface and Projectile Point Technology. In *Sloan: A Paleoindian Dalton Cemetery in Arkansas,* edited by D. F. Morse, pp. 53–57. Smithsonian Institution Press, Washington, D.C.

Breunig, P.
1996 The 8000-Year-Old Dugout Canoe from Dufuna (NE Nigeria). In *Aspects of African Archaeology,* edited by G. Pwiti and R. Soper, pp. 461–68. Papers from the 10th Congress of the PanAfrican Association for Prehistory and Related Studies. University of Zimbabwe Publications, Harare.

Brose, D. S., and I. Greber
1982 The Ringler Archaic Dugout from Savannah Lake, Ashland County, Ohio: With Speculations on Trade and Transmission in the Prehistory of the Eastern United States. *Midcontinental Journal of Archaeology* 7:245–82.

Broster, J. B., and M. R. Norton
1992 The Carson-Conn-Short Site (40BN190): An Extensive Clovis Habitation in Benton County, Tennessee. *Current Research in the Pleistocene* 8:8–10.

1996 Recent Paleoindian Research in Tennessee. In *The Paleoindian and Early Archaic Southeast,* edited by D. G. Anderson and K. E. Sassaman, pp. 288–97. The University of Alabama Press, Tuscaloosa.

Brunton, B. B.
1998 Kootenai. In *Plateau,* edited by D. E. Walker, Jr., pp. 223–37. Handbook of North American Indians, Volume 12. Smithsonian Institution, Washington, D.C.

Cabot, W. B.
1912 *In Northern Labrador.* Richard Badger, Boston.

Carlson, R. L.
1994 Trade and Exchange in Prehistoric British Columbia. In *Prehistoric Exchange Systems of North America,* edited by T. G. Baugh and J. E. Ericson, pp. 307–61. Plenum Press, New York

Carty, F. M., and A. E. Spiess
1992 The Neponset Paleoindian Site in Massachusetts. *Archaeology of Eastern North America* 20:19–37.

Casson, L.
1994 *Ships and Seafaring in Ancient Times.* The University of Texas Press, Austin.

Chalifoux, E.
1999 Late Paleoindian Occupation in a Coastal Environment: A Perspective from La Matre, Gaspé Peninsula, Quebec. *Northeast Anthropology* 57:69–79.

Clark, J. G. D.
1952 *Prehistoric Europe: The Economic Basis.* Methuen, London.

1954 *Excavations at Star Carr:* An Early Mesolithic Site at Seamer Near Scarborough Yorkshire. University Press, Cambridge.

Collins, M. B.
1999 *Clovis Blade Technology.* University of Texas Press, Austin.

Cooper, J. M.
1946a The Yahgan. In *The Marginal Tribes,* edited by J. H. Steward, pp. 127–68. *Handbook of South American Indians,* Volume 1. Bureau of American Ethnology Bulletin 143. Smithsonian Institution, Washington, D.C.

1946b The Patagonian and Pampean Hunters. In *The Marginal Tribes,* edited by J. H. Steward, pp. 127–68. *Handbook of South American Indians,* Volume 1. Bureau of American Ethnology Bulletin 143. Smithsonian Institution, Washington, D.C.

Cox, B. J., and J. B. Petersen
1997 The Varney Farm (36-57): A Late Paleoindian Encampment in Western Maine. *The Maine Archaeological Society Bulletin* 37(2):25–49.

d'Anglure, B. S.
1984 Contemporary Inuit of Quebec. In *Arctic,* edited by David Dumas, pp. 683–866. *Handbook of North American Indians,* Volume 5. Smithsonian Institution, Washington, D.C.

Dansie, A. J.
1997 Notes on Textiles Associated with the Spirit Cave Burials. In a reprint of Cave Burials near Fallon, Nevada, by S. M. Wheeler. *Nevada Historical Society Quarterly* 40(1):17–23.

Dansie, A. J., and J. Jerrems
1999 Lahonton Chronology and Early Human Occupation in the Western Great Basin: A New Look at Old Collections. Poster presented at the Center for the Study of the First Americans "Clovis and Beyond" Conference, Santa Fe.

Davis, R. B., and G. L. Jacobsen, Jr.
1985 Late Glacial and Early Holocene Landscapes in Northern New England and Adjacent Areas of Canada. *Quaternary Research* 23(3):341–368.

Davis, S. D.
1989 The Hidden Falls Site, Baranof Island, Alaska. *Aurora: Alaska Anthropological Association Monograph Series.* Fairbanks.

1996 Hidden Falls. In *American Beginnings: The Prehistory and Palaeoecology of Beringia,* edited by F. H. West, pp. 413–24. University of Chicago Press, Chicago.

Deller, D. B., and C. J. Ellis
1992a The Early Paleo-Indian Parkhill Phase in Southwestern Ontario. *Man in the Northeast* 44:15–54.

1992b *Thedford II: A Paleo-Indian Site in the Ausable River Watershed of Southwestern Ontario*. Memoirs of the Museum of Anthropology 24, University of Michigan, Ann Arbor.

Dent, R. J., and B. E. Kauffman
1985 Aboriginal Subsistence and Site Ecology as Interpreted from Microfloral and Faunal Remains. In *Shawnee Minisink, A Stratified Paleoindian-Archaic Site in the Upper Delaware Valley of Pennsylvania*, edited by C. W. McNett, Jr., pp. 55–79. Academic Press, Orlando.

Dillehay, T. D.
2000 *The Settlement of the Americas, A New Perspective*. Basic Books, New York.

Dillehay, T., and M. Pino
1997 Site Setting and Stratigraphy. In *Monte Verde: A Late Pleistocene Settlement in Chile*, Vol. 2, edited by T. D. Dillehay, pp. 25–40. Smithsonian Institution Press, Washington, D.C.

Dina, J.
1996 In Search of the Dugout Canoe. *Bulletin of Primitive Technology* 12:50–54.

Dincauze, D. F.
1993 Fluted Points in the Eastern Forests. In *From Kostenki to Clovis Upper Paleolithic-Paleo-Indian Adaptations*, edited by O. Soffer and N. D. Praslov, pp. 279–92. Plenum Press, New York.

Dixon, E. J.
1999 *Bones, Boats & Bison: Archeology and the First Colonization of Western North America*. The University of New Mexico Press, Albuquerque.

2001 Human Colonization of the Americas: Timing, Technology and Process. *Quaternary Science Reviews* 20:277–299.

Dixon, E. J., T. H. Heaton, T. E. Fifield, T. D. Hamilton, D. E. Putnam, and F. Grady
1997 Late Quaternary Regional Geoarchaeology of Southeast Alaska Karst: A Progress Report. *Geoarchaeology* 12(6):689–712.

Driskell, B. N.
1996 Stratified Late Pleistocene and Early Holocene Deposits at Dust Cave, Northwest Alabama. In *The Paleoindian and Early Archaic Southeast*, edited by D. G. Anderson and K. E. Sassaman, pp. 315–30. The University of Alabama Press, Tuscaloosa.

Driver, H. E., and W. C. Massey
1957 *Comparative Studies of North American Indians*. Transactions of the American Philosophical Society, new series 42 (2), Philadelphia.

Dumais, P.
2000 The La Martre and Mitis Late Paleoindian Sites: A Reflection on the Peopling of Southeastern Quebec. *Archaeology of Eastern North America* 28:81–112.

Easton, N. A.
1992 Mal de Mer above Tierra Incognita; or, What Ails the Coastal Migration Theory? *Arctic Anthropology* 29(2):28–42.

Edwards, C. R.
1965 *Aboriginal Watercraft on the Pacific Coast of South America*. Ibero-Americana 47. University of California, Berkeley.

Eiselt, B. S.
1997 Fish Remains from the Spirit Cave Paleofecal Material: 9,400 Year Old Evidence for Great Basin Utilization of Small Fishes. *Nevada Historical Quarterly* 40:117–39.

Ellis, C., and D. B. Deller
1997 Variability in the Archaeological Record of Northeastern Early Paleoindians: A View from Southern Ontario. *Archaeology of Eastern North America* 25:1–30.

Ellis, C. J., A. C. Goodyear, D. F. Morse, and K. B. Tankersley
1998 Archaeology of the Pleistocene-Holocene Transition in Eastern North America. In *As the World Warmed: Human Adaptations across the Pleistocene/Holocene Boundary*, edited by B. V. Eriksen and L. G. Strauss, pp. 151–66. *Quaternary International* 49/50.

Engelbrecht, W. E., and C. L. Seyfert
1994 Paleoindian Watercraft: Evidence and Implications. *North American Archaeologist* 15:221–34.

Erlandson, J. M.
1994 *Early Hunter-Gatherers of the California Coast*. Plenum Press, New York.

2000 A Stratified Association of Mammoth Remains and Archaeological Materials at Running Springs, San Miguel Island, California. *Current Research in the Pleistocene* 17:22–24.

Erlandson, J. M., M. Douglas, J. Kenneth, R. J. Behl, and I. Hough
1997 The Cisco Chert Source on San Miguel Island, California. *Journal of California and Great Basin Anthropology* 19:124–30.

Erlandson, J., D. Kenneth, L. Ingram, D. Guthrie, D. Morris, M. Tveskov, G. West, and P. Walker
1996 An Archaeological and Paleontological Chronology for Daisy Cave (CA-SMI-261), San Miguel Island, California. *Radiocarbon* 38:355–73.

Erlandson, J. M., M. L. Moss, and R. E. Hughes
1990 Archaeological Distribution and Trace Element Geochemistry of Volcanic Glass from Obsidian Cove, Sumez Island, Southeast Alaska. *Canadian Journal of Archaeology* 16:89–95.

Erlandson, J. M., and M. L. Moss
1996 The Pleistocene-Holocene Transition along the Pacific Coast of North America. In *Humans at the End of the Ice Age, The Archaeology of the Pleistocene-Holocene Transition,* edited by L. G. Strauss, B. V. Eriksen, J. M. Erlandson, and D. R. Yesner, pp. 278–301. Plenum Press, New York.

Erlandson, J. M., T. C. Rick, R. L. Vellanoweth, and D. J. Kenneth
1999 Maritime Subsistence at a 9300 Year Old Shell Midden on Santa Rosa Island, California. *Journal of Field Archaeology* 26:255–65.

Evans, J.
1985 Shawnee Minisink Site. In *Shawnee Minisink, A Stratified Paleoindian-Archaic Site in the Upper Delaware Valley of Pennsylvania*, edited by C. W. McNett, Jr., pp. 221–59. Academic Press, Orlando.

Ewers, J. C.
1955 *The Horse in Blackfoot Indian Culture*. Bureau of American Ethnology Bulletin 159. Smithsonian Institution, Washington D.C.

Faught, M. K.
1988 Inundated Sites in the Apalachee Bay area of the Eastern Gulf of Mexico. *Florida Anthropologist* 41:185–209.

Fedje, D. W., and T. Christensen
1999 Modeling Paleoshorelines and Locating Early Holocene Coastal Sites in Haida Gwaii. *American Antiquity* 64:635–52.

Fedje, D. W., and H. Josenhans
2000 Drowned Forests and Archaeology on the Continental Shelf of British Columbia, Canada. *Geology* 28:99–102.

Feest, C. F.
1978 North Carolina Algonquians. In *Northeast*, edited by B. G. Trigger, pp. 271–81. *Handbook of North American Indians*, Volume 15. Smithsonian Institution, Washington, D.C.

Fitting, J. E.
1966 The Paleo-Indian Occupation of Holcombe Beach. In *The*

Paleo-Indian Occupations of the Holcombe Beach, edited by J. E. Fitting, J. DeVisscher, and E. J. Wahala, pp. 109–40. Anthropological Papers Number 27, Museum of Anthropology, University of Michigan, Ann Arbor.

Fitzhugh, W. W.
1972 Environmental Archeology and Cultural Systems in Hamilton Inlet, Labrador: A Survey of the Central Labrador Coast From 3000 B.C. to the Present. *Smithsonian Contributions to Anthropology*, Number 16. Washington, D.C.
1974 A Maritime Archaic Sequence from Hamilton Inlet, Labrador. *Arctic Anthropology* 7(2):117–38.
1978 Maritime Archaic Cultures of the Central and Northern Labrador Coast. *Arctic Anthropology* 15(2):61–95.

Fladmark, K. R.
1979 Routes: Alternate Migration Corridors for Early Man in North America. *American Antiquity* 44:55–69.

Follett, W. I.
1982 An Analysis of Fish Remains from Ten Archaeological Sites at Falcon Hill, Washoe County, Nevada, with Notes on Fishing Practices of the Ethnographic Kuyuidtkadt Northern Paiute. In *The Archaeology of Falcon Hill, Winnemucca Lake, Washoe County, Nevada*, edited by E. M. Hattori, pp. 179–203. Nevada State Museum Anthropological Papers Number 18. Carson City.

Font, P.
1933 *Complete Diary, a Chronicle of the Founding of San Francisco*, edited by H. E. Bolton. University of California Press, Berkeley.

Frison, G. C.
1991 *Prehistoric Hunters of the High Plains, second edition*. Academic Press, San Diego.

Frison, G., and B. Bradley
1999 *The Fenn Cache: Clovis Weapons and Tools*. One Horse Land and Cattle Company, Santa Fe.

Frison, G. C., and D. J. Stanford, (editors)
1982 *The Agate Basin Site: A Record of the Paleoindian Occupation of the Northwestern Plains*. Academic Press, New York.

Gaetner, L. M.
1994 Determining the Function of Dalton Adzes from Northeast Arkansas. *Lithic Technology* 19:97–109.

Galm, J. R.
1994 Prehistoric Trade and Exchange in the Interior Plateau of Northwestern North America. In *Prehistoric Exchange Systems of North America*, edited by T. G. Baugh and J. E. Ericson, pp. 275–305. Plenum Press, New York.

Gamble, C.
1994 *Timewalkers: The Prehistory of Global Colonization*. Harvard University Press, Cambridge.

Gardner, W. M. (editor)
1974 The Flint Run Paleo-Indian Complex: A Preliminary Report 1971–73 Seasons. Occasional Publication Number 1, Archeology Laboratory, Department of Anthropology, The Catholic University of America, Washington, D.C.

Gardner, W. M.
1979 The Thunderbird Paleo-Indian Site and the Middle Shenandoah Valley Research Program, 1971–1974. *National Geographic Society Research Reports, 1970 Projects*, pp. 165–75. Washington, D.C.

Gillespie, B. C.
1981 Mountain Indians. In *Subarctic*, edited by June Helm, pp. 326–37. *Handbook of North American Indians*, Volume 6. Smithsonian Institution, Washington, D.C.

Goddard, I.
1978 Delaware. In *Northeast*, edited by B. G. Trigger, pp. 213–39. *Handbook of North American Indians*, Volume 15. Smithsonian Institution, Washington, D.C.

Goodyear, A. C., III
1974 The Brand Site, A Techno-Functional Study of a Dalton Site in Northeast Arkansas. Arkansas Archeological Survey, Research Series Number 7.
1982 The Chronological Position of the Dalton Horizon in the Southeastern United States. *American Antiquity* 47:382–95.
1999 The Early Holocene Occupation of the Southeastern United States: A Geoarchaeological Summary. In *Ice Age Peoples of North America: Environments, Origins, and Adaptations of the First Americans*, edited by R. Bonnichsen and K. L. Turnmire, pp. 432–81. Oregon State University Press, Corvallis.

Gramly, R. M.
1988 Palaeo-Indian Sites South of Lake Ontario, Western and Central New York State. In *Late Pleistocene and Early Holocene Paleoecology and Archaeology of the Eastern Great Lakes Region*, edited by R. S. Laub, N. G. Miller, and D. W. Steadman, pp. 227–40. Bulletin of the Buffalo Society of Natural Sciences, Volume 33.

Grayson, D. K.
1993 *The Desert's Past, A Natural Prehistory of the Great Basin*. Smithsonian Institution Press, Washington, D.C.

Greenspan, R. L.
1990 Prehistoric Fishing in the Northern Great Basin. In *Wetland Adaptations in the Great Basin*, edited by J. C. Janetski and D. B. Madsen, pp. 207–32. Museum of Peoples and Cultures, Occasional Papers Number 1, Brigham Young University, Provo.

Gross, J. I.
1974 Some Patterns of Chipping Activity of the Thunderbird Clovis Component. In *The Flint Run Paleo-Indian Complex: A Preliminary Report 1971–73 Seasons*, edited by W. M. Gardner, pp. 100–105. Occasional Publication Number 1, Archeology Laboratory, Department of Anthropology, The University of America, Washington, D.C.

Gruhn, R.
1988 Linguistic Evidence in Support of the Coastal Route of Earliest Entry into the New World. *Man* 23:77–100.

Hamilton, T. D., and T. Goebel
1999 Late Pleistocene Peopling of Alaska. In *Ice Age Peoples of North America: Environments, Origins, and Adaptations of the First Americans*, edited by R. Bonnichsen and K. L. Turnmire, pp. 156–99. Oregon State University Press, Corvallis.

Harington, C. R.
1977 Marine Mammals in the Champlain Sea and the Great Lakes. *Annals of the New York Academy of Sciences* 288:508–37.

Haven, J.
1773 A Brief Account of the Dwelling Places of the Eskimaux to the North of Nagvack to Hudsons Strait, Their Situation and Subsistence. Archives of the Moravian Church, London.

Herron, R. D.
1998 The Development of Asian Watercraft: From the Prehistoric Era to the Advent of European Colonization. Unpublished Dissertation, Texas A&M University, College Station. UMI Dissertation Services, Number 9830920, Ann Arbor.

Hester, J. J.
1972 *Blackwater Draw Locality No. 1: A Stratified Early Man Site in Eastern New Mexico*. Fort Burgwin Research Center Publication 8. Ranchos de Taos, New Mexico.

Heusser, C. J.
1960 *Late-Pleistocene Environments of North Pacific North America: An Elaboration of Late-Glacial and Postglacial Climatic, Physiographic, and Biotic Changes*. Special Publication 35, American Geographical Society, New York.

Hewes, G. W.
1998 Fishing. In *Plateau*, edited by D. E. Walker, Jr., pp. 620–40. *Handbook of North American Indians*, Volume 12. Smithsonian Institution, Washington, D.C.

Heyerdahl, T.
1957 Guara Navigation: Indigenous Sailing off the Andean Coast. *Southwest Journal of Anthropology* 13:134–43.
1979 *Early Man and the Ocean, A Search for the Beginnings of Navigation and Seaborne Civilizations*. Doubleday & Company, Inc., Garden City, New York.

Higgins, M. J.
1990 The Nochta Site: The Early, Middle and Late Archaic Occupations. American Bottom Archaeology FAI-270 Site Reports 21. University of Illinois Press, Urbana.

Honigmann, J. J.
1981 West Main Cree. In *Subarctic*, edited by J. Helm, pp. 217–30. *Handbook of North American Indians*, Volume 6. Smithsonian Institution, Washington, D.C.

Hornell, J.
1946 *Water Transport: Origins and Early Evolution*. Cambridge University Press, Cambridge.

Hughes, C. G.
1984 Siberian Eskimo. In *Arctic*, edited by D. Damas, pp. 247–61. *Handbook of North American Indians*, Volume 5. Smithsonian Institution, Washington, D.C.

Hughes, R. E.
1994 Mosaic Patterning in Prehistoric California–Great Basin Exchange. In *Prehistoric Exchange Systems of North America*, edited by T. G. Baugh and J. E. Ericson, pp. 363–83. Plenum Press, New York.

Hunn, E. S., N. J. Turner, and D. H. French
1998 Ethnobiology and Subsistence. In *Plateau*, edited by D. E. Walker, Jr., pp. 525–45. *Handbook of North American Indians*, Volume 12. Smithsonian Institution, Washington, D.C.

Jackson, L. J.
1990 Interior Paleoindian Settlement Strategies: A First Approximation for the Lower Great Lakes. In *Early Paleoindian Economies of Eastern North America*, edited by K. B. Tankersley and B. L. Isaac, pp. 95–142. Research in Economic Anthropology Supplement 5. JAI Press Inc., Greenwich, Connecticut.

Jackson, L. J., C. Ellis, A. V. Morgan, and J. H. McAndrews
2000 Glacial Lake Levels and Eastern Great Lakes Palaeo-Indians. *Geoarchaeology* 15:415–40.

Jodry, M. A.
1999 Folsom Technological and Socioeconomic Strategies: Views from Stewart's Cattle Guard and the Upper Rio Grande Basin, Colorado. Ph.D. Dissertation, Department of Anthropology, American University, Washington, D.C. UMI Dissertation Services No. 9997500, Ann Arbor.

Johnson, J. R., T. W. Stafford, Jr., H. O. Aije, and D. P. Morris
2000 Arlington Springs Revisited. *Proceedings of the Fifth Channel Islands Symposium*, edited by D. R. Brown, H. W. Chaney, and K. L. Mitchell, pp. 541–45. Santa Barbara Museum of Natural History, Santa Barbara.

Josenhans, H. W., D. W. Fedje, R. Pienitz, and J. R. Southon
1997 Early Humans and Rapidly Changing Holocene Sea Levels in the Queen Charlotte Islands–Hecate Strait, British Columbia, Canada. *Science* 277:71–74.

Jones, R.
1979 The Fifth Continent: Problems concerning the Human Colonization of Australia. *Annual Review of Anthropology* 8:445–66.

Karrow, P. F., T. W. Anderson, A. H. Clarke, L. D. Delorme, and M. R. Sreenivasa
1975 Stratigraphy, Paleontology, and Age of Lake Algonquin Sediments in Southwestern Ontario, Canada. *Quaternary Research* 5(1):49–85.

Keefer, D. K., S. deFrance, M. Moseley, J. Richardson III, D. Satterlee, and A. Day-Lewis
1998 Early Maritime Economy and El Niño Events at Quebrada Tacahuay, Peru. *Science* 281:1833–35.

Keenlyside, D. L.
1985 Late Palaeo-Indian Evidence from the Southern Gulf of St. Lawrence. *Archaeology of Eastern North America* 13:79–92.
1991 Paleoindian Occupations of the Maritimes. In *Clovis Origins and Adaptations*, edited by R. Bonnichsen and K. L. Turnmire, pp. 163–73. Center for the Study of the First Americans, Corvallis.

Kelly, I.
1978 Coast Miwok. In *California*, edited by R. F. Heizer, pp. 414–25. *Handbook of North American Indians*, Volume 8. Smithsonian Institution, Washington, D.C.

Kennedy, I. D., and R. T. Bouchard
1990 Northern Coast Salish. In *Northwest Coast*, edited by W. Suttles, pp. 441–52. *Handbook of North American Indians*, Volume 7. Smithsonian Institution, Washington, D.C.

Kline, D.
1985 Early History of Archaeological Research at the Shawnee Minisink Site. In *Shawnee Minisink, A Stratified Paleoindian-Archaic Site in the Upper Delaware Valley of Pennsylvania*, edited by C. W. McNett, Jr., pp. 15–19. Academic Press, Orlando.

Koldehoff , B.
1983 Paleoindian Chert Utilization and Site Distribution in Southwestern Illinois. *The Wisconsin Archeologist* 64:210–38.

Lantis, M.
1984 Aleut. In *Arctic*, edited by D. Damas, pp. 161–84. *Handbook of North American Indians*, Volume 5. Smithsonian Institution, Washington, D.C.

Lehmer, D. J.
2001 Plains Village Tradition: Postcontact. In *Plains*, Part 1, edited by R. J. DeMallie, pp. 245–55. *Handbook of North American Indians*, Volume 13. Smithsonian Institution, Washington, D.C.

Levine, M. A.
1997 The Tyranny Continues: Ethnographic Analogy and Eastern Paleo-Indians. In *Caribou and Reindeer Hunters of the Northern Hemisphere*, edited by L. J. Jackson and P. T. Thacker, pp. 221–44. World Archaeology Series 6. Avebury, Aldershot, United Kingdom.

Levy, R.
1978 Eastern Miwok. In *California*, edited by R. F. Heizer, pp. 398–413. *Handbook of North American Indians*, Volume 18. Smithsonian Institution, Washington, D.C.

Little, E. A.
1987 Inland Waterways in the Northeast. *Midcontinental Journal of Archaeology* 12:55–76.

Lothrop, S. K.
1932 Aboriginal Navigation off the West Coast of South America. *Journal of the Royal Anthropological Institute* 62:229–56.

Loring, S.
1980 Paleo-Indian Hunters and the Champlain Sea: A Presumed Association. *Man in the Northeast* 19:15–41.
1994 On the Trail to the Caribou House: Some Reflections on Innu Caribou Hunters in Northern Ntessinan (Labrador). In *Caribou and Reindeer Hunters of the Northern Hemisphere*, edited by L. J. Jackson and P. T. Thacker, pp. 185–220. World Archaeology Series 6. Avebury, Aldershot.

Low, A. P.
1901 Report of an Exploration of Part of the South Shore of Hudson Strait and Ungava Bay. Canada. Geological Survey. Summary Report for 1899, *Annual Report Volume II* (L):1–41. Ottawa.

Lowery, D.
1989 The Paw Paw Cove Paleoindian Site Complex, Talbot County, Maryland. *Archaeology of Eastern North America* 17:143–163.
1992 Excavations at the Paw Paw Cove Paleoindian Site, FY 1992. Manuscript on file at the University of Delaware Center for Archaeological Research, Newark, Delaware.
2001 Archaeological Survey of the Chesapeake Bay Shorelines Associated with Acconack County and Northampton County, Virginia, Volumes 1 & 2. Report Prepared for the Virginia Department of Historic Resources, Portsmouth, Virginia.

MacDonald, G. F.
1968 *Debert: A Palaeo-Indian Site in Central Nova Scotia*. Anthropology Papers of the National Museum of Canada No. 16. Ottawa.

Mallouf, R. J.
1989 A Clovis Quarry Workshop in the Callahan Divide: The Yellow Hawk Site, Taylor County, Texas. *Plains Anthropologist* 34:81–103.

Mandryk, C. A. S., H. Josenhans, D. W. Fedje, and R. W. Mathewes
2001 Late Quaternary Paleoenvironments of Northwestern North America: Implications for Inland versus Coastal Migration Routes. *Quaternary Science Reviews* 20:301–14.

Marshall, S. B.
1985 Paleoindian Artifact Form and Function at Shawnee Minisink. In *Shawnee Minisink, a Stratified Paleoindian-Archaic Site in the Upper Delaware Valley of Pennsylvania*, edited by C. W. McNett, Jr., pp. 165–209. Academic Press, Orlando.

Mason, R. J.
1962 The Paleo-Indian Tradition in Eastern North America. *Current Anthropology* 3:227–83.

Mayer-Oakes, W. J.
1984 Fluted Projectile Points—A North American Shibboleth Viewed in South American Perspective. *Archaeology of Eastern North America* 12:231–247.

McCartney, A. P., and C. G. Turner, III
1966 Stratigraphy of the Anangula Unifacial Core and Blade Site. *Arctic Anthropology* 3:28–40.

McCartney, A. P., and D. W. Veltre
1996 Anangula Core and Blade Site. In *American Beginnings: The Prehistory and Palaeoecology of Beringia*, edited by F. H. West, pp. 443–50. University of Chicago Press, Chicago.

McGhee, R.
1984 Thule Prehistory of Canada. In *Arctic*, edited by D. Dumas, pp. 369–376. Handbook of North American Indians, Volume 5. Smithsonian Institution, Washington, D.C.

McGhee, R., and J. Tuck
1975 *An Archaic Sequence from the Strait of Belle Isle, Labrador*. Archaeological Survey of Canada Paper No. 34. Mercury Series. National Museum of Man, Ottawa.

McGrail, S.
1998 Ancient Boats in North-West Europe. The Archaeology of Water Transport to A.D. 1500. Second Edition. Longman, London.

McNett, C. W., Jr.
1985b The Shawnee Minisink Site: An Overview. In *Shawnee Minisink, A Stratified Paleoindian-Archaic Site in the Upper Delaware Valley of Pennsylvania*, edited by C. W. McNett, Jr., pp. 321–25. Academic Press, Orlando.
1985c Artifact Morphology and Chronology at the Shawnee Minisink Site. In *Shawnee Minisink, A Stratified Paleoindian-Archaic Site in the Upper Delaware Valley of Pennsylvania*, edited by C. W. McNett, Jr., pp. 83–120. Academic Press, Orlando.

McNett, C. W., Jr. (editor)
1985a *Shawnee Minisink, A Stratified Paleoindian-Archaic Site in the Upper Delaware Valley of Pennsylvania*. Academic Press, Orlando.

Meltzer, D. J., L. C. Todd, and V. T. Holliday
2002 The Folsom (Paleoindian) Type Site: Past Investigations, Current Studies. *American Antiquity* 67(1):5–36.

Memory, M., B. Swann, and D. Ruhl
2001 Paddling through Time: The Story from Lake Pithlachocco, the Place of Long Boats. Paper delivered at a 53rd Annual Meeting of the Florida Anthropological Society, San Augustine. Paper on file at the Paleoindian/Paleoecology Program, Department of Anthropology, National Museum of Natural History, Smithsonian Institution.

Miotti, L., and M. Salemme
1998 Hunting and Butchering Events during the Late Pleistocene and Early Holocene in Piedra Museo (Patagonia, southernmost South America). Paper presented at the 8th International Congress of the International Council for Archaeozoology, Victoria.

Morey, D. F., and M. D. Waint
1992 Early Holocene Domestic Dog Burials from the North American Midwest. *Current Anthropology* 33(2):224–29.

Moreno, F. P.
1879 Viaje á la Patagonia austral . . . 1876–1877. Buenos Aires.

Morrow, T. A., and J. E. Morrow
1999 On the Fringe: Folsom Points and Preforms in Iowa. In *Folsom Lithic Technology: Explorations in Structure and Variation*, edited by D. S. Amick, pp. 65–81. International Monographs in Prehistory, Archaeological Series 12. Ann Arbor.

Morse, D. F.
1971 The Hawkins Cache: A Significant Dalton Find in Northeast Arkansas. *The Arkansas Archeologist* 12:9–20.
1997 *Sloan: A Paleoindian Dalton Cemetery in Arkansas*. Smithsonian Institution Press, Washington, D.C.

Morse, D. F., and A. C. Goodyear III
1973 The Significance of the Dalton Adz in Northeast Arkansas. *Plains Anthropologist* 18:316–22.

Mulvaney, J., and J. Kamminga
1999 *Prehistory of Australia*. Smithsonian Institution Press, Washington, D.C.

Myers, J. E.
1978 Cahto. In *California*, edited by R. F. Heizer, pp. 244–48. *Handbook of North American Indians*, Volume 8. Smithsonian Institution, Washington, D.C.

Napton, K. L.
1997 The Spirit Cave Mummy: Coprolite Investigations. *Nevada Historical Society Quarterly* 40(1):97–104.

Nelson, R. K.
1983 *Make Prayers to the Raven, A Koyukon View of the Northern Forest.* University of Chicago Press.

Newsom, L. A., and B. A. Purdy
1990 Florida Canoes: A Maritime Heritage from the Past. *Florida Anthropologist* 43:164–80.

Nordquist, B.
1995 The Mesolithic Settlements of the West Coast of Sweden—with Special Emphasis on Chronology and Topography of Coastal Settlements. In *Man and the Sea in the Mesolithic, Coastal Settlement Above and Below Sea Level,* edited by A. Fischer, pp. 185–96. Oxbow Monograph 53, Oxford.
1994 Huseby Klev–Marine Archaology on Land. *NewsWARP, The Newsletter of the Wetland Archaeology* Research Project. 16:24–27.

Oda, S.
1990 A Review of Archaeological Research in the Izu and Ogasawara Islands. In *Man and Culture in Oceania,* 6:53–79.

Okada, H., A. Okada, Y. Kotani, W. M. Olson, Y. Nishimoto, and S. Okino
1989 Heceta Island, Southeastern Alaska Anthropological Survey in 1987. Department of Behavioral Science, Faculty of Letters, Hokkaido University, Sapporo, Japan.

Osgood, C.
1940 Ingalik Material Culture. *Yale University Publications in Anthropology* 22.

Palmer, G.
1998 Coeur d'Alene. In *Plateau,* edited by D. E. Walker, Jr., pp. 313–26. *Handbook of North American Indians,* Volume 12. Smithsonian Institution, Washington, D.C.

Parkington, J.
1981 The Effects of Environmental Change on the Scheduling of Visits to the Elands Bay Cave, Cape Province, S. A. In *Patterns of the Past,* edited by I. Hodder, G. Isaac, and N. Hammond, pp. 341–59. Cambridge University Press, Cambridge.

Pastore, R. T.
1992 *Shanawdithit's People, The Archaeology of the Beothuks.* Atlantic Archaeology, Ltd., St. John's, Newfoundland.

Perino, G.
1967 Early Projectile Points Found. *Central States Archaeological Journal* 14:103–106.

Petersen, J. P., R. N. Bartone, and B. J. Cox
2000 The Varney Farm Site and the Late Paleoindian Period in Northeastern North America. *Archaeology of Eastern North America* 28:113–40.

Pitul'ko, V. V., and A. K. Kasparov
1996 Ancient Arctic Hunters: Material Culture and Survival Strategy. *Arctic Anthropology* 33 (1):1–36.

Pollock, S. G., N. D. Hamilton, and R. Bonnichsen
1999 Chert from the Munsungun Lake Formation (Maine) in Palaeoamerican Archaeological Sites in Northeastern North America: Recognition of its Occurrence and Distribution. *Journal of Archaeological Science* 26:269–93.

Renfew, C., and A. Aspinall
1990 Aegean Obsidian and Franchthi Cave. In Excavations at Franchthi Cave, Greece. Fascicle 5, *Les Industries lithiques taillées de Franchthi,* Tome II, Les Industries Du Mésolithique Et Du Néolithique Initial, by Catherine Perles with the collaboration of Patrick C. Vaughn, Colin Renfew, and Arnold Aspinall, pp. 257–70. Indiana University Press, Bloomington.

Reynolds, B.
1978 Beothuk. In *Northeast,* edited by B. G. Trigger, pp. 101–108. *Handbook of North American Indians,* Volume 15. Smithsonian Institution, Washington, D.C.

Ricklis, D.
1982 *Northern Nomadic Hunter-Gatherers, A Humanistic Approach.* Academic Press, New York.

Rogers, E. S., and J. G. E. Smith
1981 Environment and Culture in the Shield and Mackenzie Borderlands. In *Subarctic,* edited by J. Helm, pp. 130–45. *Handbook of North American Indians,* Volume 6. Smithsonian Institution, Washington, D.C.

Roosa, W. B.
1977 Great Lakes Paleo-Indian: The Parkhill Site, Ontario. In *Amerinds and Their Paleoenvironments in Northeastern North America,* edited by W. S. Newman and B. Salwen, pp. 349–54. *Annals of the New York Academy of Sciences* 288. New York.

Ross, J. A.
1998 Spokane. In *Plateau,* edited by D. E. Walker, Jr., pp. 271–82. *Handbook of North American Indians,* Volume 12. Smithsonian Institution, Washington, D.C.

Salwen, B.
1978 Indians of Southern New England and Long Island: Early Period. In *Northeast,* edited by B. G. Trigger, pp. 160–76. *Handbook of North American Indians,* Volume 15. Smithsonian Institution, Washington, D.C.

Sanders, T. N.
1990 *Adams: The Manufacturing of Flaked Stone Tools at a Paleoindian Site in Western Kentucky.* Persimmon Press Monographs in Archaeology, New York.

Sandweiss D. H., H. McInnis, R. L. Burger, A. Cano, B. Ojeda, R. Paredes, M. Del Carmen Sandweiss, and M. D. Glascock
1998 Quebrada Jaguay: Early South American Maritime Adaptations. *Science* 281:1830–33.

Savishinsky, J. S., and H. S. Hara
1981 Hare. In *Subarctic,* edited by J. Helm, pp. 314–25. *Handbook of North American Indians,* Volume 6. Smithsonian Institution, Washington, D.C.

Serra, J.
1955 *Writings of Junipero Serra.* Edited by Antonine Tibesar. Academy of American Franciscan History, Washington.

Smith, C. M., and J. F. Haslett
1999 Construction and Sailing Characteristics of a Pre-Columbian Raft Replica. *Bulletin of Primitive Technology* 20:13–31.

Smith, J. G. E.
1981 Chipewyan. In *Subarctic,* edited by J. Helm, pp. 271–84. *Handbook of North American Indians,* Volume 6. Smithsonian Institution, Washington, D.C.

Snow, D. R.
1978 Eastern Abenaki. In *Northeast,* edited by B. G. Trigger, pp. 137–47. *Handbook of North American Indians,* Volume 15. Smithsonian Institution, Washington, D.C.

Spencer, R. F.
1984 North Alaska Coast Eskimo. In *Arctic,* edited by D. Damas, pp. 320–37. *Handbook of North American Indians,* Volume 5. Smithsonian Institution, Washington, D.C.

Spiess, A. E.
1979 *Reindeer and Caribou Hunters: An Archaeological Study*. Academic Press, New York.

Spiess, A. E., M. L. Curran, and J. Grimes
1985 Caribou (*Rangifer tarandus*) Bones from New England Paleoindian Sites. *North American Archaeologist* 6:145–60.

Stanford, D. J.
1991 Clovis Origins and Adaptations: An Introductory Perspective. In *Clovis: Origins and Adaptations*, edited by R. Bonnichsen and K. L. Turnmire, pp. 1–13. Center for the Study of the First Americans, Oregon State University, Corvallis.

1999 Analysis and Interpretation of Hell Gap Hunting Strategies at the Jones-Miller site. In *Le Bison: Gibier Et Moyen De Subsistance Des Hommes Du Paléolithique Aux Paléoindiens Des Grandes Plaines*, edited by J. P. Brugal, F. David, J. G. Enloe, and J. Jaubert, pp. 437–54. Actes du colloque international, Toulouse 1995, Editions APDCA, Antibes, 1999

Stanford, D. J., and B. Bradley
2002 Ocean Trails and Prairie Paths? Thoughts about Clovis Origins. In *The First Americans: The Pleistocene Colonization of the New World*, edited by N. Jablonski, pp. 255–71. Memoirs of the California Academy of Sciences, Number 27. San Francisco.

Stanford, D. J., and M. A. Jodry
1988 The Drake Clovis Cache. *Current Research in the Pleistocene* 5:21–22.

Stern, T.
1998 Cayuse, Umatilla, and Walla Walla. In *Plateau*, edited by D. E. Walker, Jr., pp. 395–419. *Handbook of North American Indians*, Volume 12. Smithsonian Institution, Washington, D.C.

Stewart, F. H.
2001 Hidatsa. In *Plains*, Part 1, edited by R. J. DeMallie, pp. 329–48. *Handbook of North American Indians*, Volume 13. Smithsonian Institution, Washington, D.C.

Stewart, H.
1982 *Indian Fishing, Early Methods on the Northwest Coast*. University of Washington Press, Seattle.

Strong, P. T.
1979 San Felipe Pueblo. In *Southwest*, edited by Alfonso Ortiz, pp. 390–406. *Handbook of North American Indians*, Volume 9. Smithsonian Institution, Washington, D.C.

Storck, P. L.
1984 Glacial Lake Algonquin and Early Paleo-Indian Patterns in Southcentral Ontario. *Archaeology of Eastern North America* 12:286–301.

1988 Recent Excavations at the Udora Site: A Gainey/Clovis Occupation Site in Southern Ontario. *Current Research in the Pleistocene* 5:23–24.

1997 *The Fisher Site, Archaeological, Geological and Paleobotanical Studies at an Early Paleo-Indian Site in Southern Ontario, Canada*. Memoirs of the Museum of Anthropology, University of Michigan, Number 30. Ann Arbor.

Storck, P. L., and A. E. Spiess
1994 The Significance of New Faunal Identifications Attributed to an Early Paleoindian (Gainey Complex) Occupation at the Udora Site, Ontario. *American Antiquity* 59:121–42.

Storck, P. L., and J. Tomenchuk
1990 An Early Paleoindian Cache of Informal Tools at the Udora Site, Ontario. In *Early Paleoindian Economies of Eastern North America*, edited by K. B. Tankersley and B. L. Isaac, pp. 45–93. Research in Economic Anthropology Supplement 5. JAI Press Inc., Greenwich, Connecticut.

Stright, M. J., E. M. Lear, and J. F. Bennett
1999 *Spatial Data Analysis of Artifacts Redeposited by Coastal Erosion: A Case Study of McFaddin Beach, Texas*. OCS Study, MMS 99-0068. U.S. Department of the Interior, Mineral Management Service. Herndon, Virginia.

Swanton, J. R.
1946 *The Indians of the Southeastern United States*. Bureau of American Ethnology, Bulletin 43. Smithsonian Institution, Washington, D.C.

Suzuki, M.
1973 Chronology of Prehistoric Human Activity in Kanto, Japan. Part One: Framework for Reconstructing Prehistoric Human Activity in Obsidian. *Journal of the Faculty of Science, University of Tokyo* 4:241–318.

1973 Chronology of Prehistoric Human Activity in Kanto, Japan. Part Two: Time-space Analysis of Obsidian Transportation. *Journal of the Faculty of Science, University of Tokyo* 4:395–486.

Szaszdi, A.
1978 En Torno a lo Balsa de Salango, Ecuador, Que Capturo Bartolome Ruiz. *Anuario de Estudios Americanos* 35:453–554. Seville.

Tankersley, K. B.
1991 A Geoarchaeological Investigation of Distribution and Exchange in the Raw Material Economies of Clovis Groups in Eastern North America. In *Raw Material Economies Among Prehistoric Hunter-Gatherers*, edited by A. Montet-White and S. Holen, pp. 285–304. University of Kansas Press, Publications in Anthropology, No. 19. Lawrence.

Taylor, J. G.
1974 *Labrador Eskimo Settlements of the Early Contact Period*. National Museum of Man, Publications in Ethnology 9. Ottawa.

Tomenchuk, J.
1997 A Parametric Use-Wear Study of Artifacts from Areas C and C-east. In *The Fisher Site, Archaeological, Geological and Paleobotanical Studies at an Early Paleo-Indian Site in Southern Ontario, Canada*, edited by P. L. Storck, pp. 95–161. Memoirs of the Museum of Anthropology, University of Michigan, Number 30. Ann Arbor.

Torben C. R., J. M. Erlandson, and R. L. Vellanoweth
2001 Paleocoastal Marine Fishing on the Pacific Coast of the Americas: Perspectives from Daisy Cave, California. *American Antiquity* 66(4):595–613.

Tuohy, D. R.
1988 Artifacts from the Northwestern Pyramid Lake Shoreline. In *Early Occupation in Far Western North America: The Clovis-Archaic Interface*, edited by J. A. Willig and J. L. Fagan, pp. 201–16. Nevada State Museum Anthropological Papers Number 21. Carson City.

1990 Pyramid Lake Fishing: The Archaeological Record. In *Wetland Adaptations in the Great Basin*, edited by J. C. Janetski and D. B. Madsen, pp. 121–58. Museum of Peoples and Cultures, Occasional Papers Number 1, Brigham Young University, Provo.

Tuohy, D. R., and A. J. Dansie
1997 New Information regarding Early Holocene Manifestations in the Western Great Basin. *Nevada Historical Society Quarterly* 40(1):24–54.

Walker, D. E., Jr.
1998 Nez Perce. In *Plateau*, edited by D. E. Walker, Jr., pp. 420–38. *Handbook of North American Indians*, Volume 12. Smithsonian Institution, Washington, D.C.

Wallace, W. J.
1978 Southern Valley Yokuts. In *California*, edited by R. F. Heizer,

pp. 448–70. *Handbook of North American Indians*, Volume 8. Smithsonian Institution, Washington, D.C.

Walthall, J. A., and B. Koldehoff
1998 Hunter-Gatherer Interaction and Alliance Formation: Dalton and the Cult of the Long Blade. *Plains Anthropologist* 43(165): 257–73.

Wheat, M. W.
1967 *Survival Arts of the Primitive Paiutes*. University of Nevada Press, Reno.

Wickler, S., and M. Spriggs
1988 Pleistocene Human Occupation of the Solomon Islands, Melanesia. *Antiquity* 72:703–6.

Williams, S., and J. B. Stoltman
1965 An Outline of Southeastern United States Prehistory with Particular Emphasis on the Paleoindian Era. In *The Quaternary of the United States*, edited by H. E. Wright and D. G. Frey, pp. 669–83. Princeton University Press, Princeton.

Wilson, G. L.
1924 Hidatsa Horse and Dog Culture. *Anthropological Papers of the American Museum of Natural History* 15:127–311. New York: American Museum Press.

Winthrop, J., Jr.
1636 An Account of the Narragansett Country . . . Letter from J. Winthrop Jr. to J. Winthrop Sr., April, 1636. Massachusetts Historical Society Collection Fourth Series, Volume 6 [Reprinted 1863]. Boston, Massachusetts.

Wright, J. V.
1994 The Prehistoric Transportation of Goods in the St. Lawrence River Basin. In *Prehistoric Exchange Systems of North America*, edited by T. G. Baugh and J. E. Ericson, pp. 47–71. Plenum Press, New York.

Wyatt, D.
1997 Thompson. In *Plateau*, edited by D. E. Walker, Jr., pp. 191–202. *Handbook of North American Indians*, Volume 12. Smithsonian Institution, Washington, D.C.

Yerkes, R., and L. M. Gaetner
1997 Microwear Analysis of Dalton Artifacts. In *Sloan: A Paleoindian Dalton Cemetery in Arkansas*, edited by D. Morse, pp. 58–71. Smithsonian Institution Press, Washington, D.C.

Yesner, D. R
1985 Human Adaptation at the Pleistocene-Holocene Boundary (circa 13,000 to 8,000 BP) in Eastern Beringia. In *Humans at the End of the Ice Age: The Archaeology of the Pleistocene-Holocene Transition*, edited by L. G. Strauss, B. V. Eriksen, J. M. Erlandson, and D. R. Yesner, pp. 255–76. Plenum Press, New York.

Late-Wisconsinan Mammoth Procurement in the North American Grasslands

Eileen Johnson

Abstract

The extent of the late-Pleistocene North American grasslands and the resources they contained are significant aspects in examining the late-Pleistocene peopling of the landscape, and evidence of early occupations. During this time, climatic regimes across this vast stretch of land generally were cooler with greater precipitation or precipitation coming at a different season than today. Mammoth localities throughout this grasslands region associated with early peoples are limited, and 13 localities were selected that met the research criteria, a protocol whose methodology is based on taphonomy (effects on bone after an animal dies), biotechnology (constraints of bone biology and how humans apply that to modify bone), and morphometrics (measurement of shape and size). Cultural modification to the mammoth bone was categorized as marks on bone (two types of mark were identified) and bone breakage (dynamic impact fracturing). Marks were involved with disarticulating and defleshing; breakage was involved with bone quarrying to produce cores and blanks. A construct was proposed and the database compared with expectations to assess the role of mammoth procurement in late-Wisconsinan lifeways, including carcass scavenging. A mammoth carcass represented meat-related subsistence and raw material economy. Strategies developed by late-Wisconsinan grasslands peoples to obtain these resources came to an end as mammoth become extinct.

Introduction

Two main species of mammoth inhabited the continental U.S. during the late Pleistocene. *Mammuthus columbi* occurred coast to coast within the Southern to Western tier of states (Agenbroad 1984:97; Graham 1986:169; Graham and Lundelius 1994:418–19; Maglio 1973; Todd and Roth 1996:200–201). Although *M. primigenius* (the second species) is abundant in Alaska, it is more restricted in its distribution (primarily the upper Midwest into mid-eastern Canada), and recorded localities are not as numerous (Agenbroad 1984:97; Graham and Lundelius 1994:421, 420–23; McAndrews and Jackson 1988:162). While *M. columbi* localities commonly are associated with early peoples (e.g., Frison and Todd 1986; Hannus 1989; Hemmings 1970; Johnson 1991; Leonhardy 1966; Saunders and Daeschler 1994), such an association for *M. primigenius* is rare (McAndrews and Jackson 1988:161; Overstreet 1996a; Overstreet et al. 1993).

During the late Pleistocene, the North American grasslands stretched from the Basin of Mexico through the Great Plains and eastward (Figure 1). The extent of these grasslands and the resources they contained are significant aspects in examining the late-Pleistocene peopling of the landscape, and evidence for early occupations. Issues of recognition of evidence, the role of human activities in creating and disturbing bone beds, and subsistence strategies are being investigated in various areas of this extensive grasslands, in different landscape settings, climate, and environments. A protocol is being developed whose methodology is based on taphonomy (effects on bone after an animal dies), biotechnology (constraints of bone biology and how humans apply that to modify bone), and morphometrics (measurement of shape and size). Research addressed here is centered on mammoth remains. The time span is a ca. 7,000-year period from ca. 18,000 to 11,000 yr B.P., encompassing 13 localities.

Theoretical Perspective

Little is known about the late-Wisconsinan peoples on the Great Plains, and few attempts have been made to place them within a broader perspective. A framework for interpreting this early archaeological record focuses on people-land relationships, expand-

Museum of Texas Tech University, Box 43191, Lubbock, TX 79409-3191; e-mail: eileen.johnson@ttu.edu

Figure 1. The core North American grasslands.

ing on Johnson (1991). The evolutionary ecology view of culture is through explanatory mechanisms of rational choice and natural selection as causal factors (Smith and Winterhalder 1992:39; Winterhalder and Smith 1992:21). Culture as adaptation is social learning, a flexible strategy that combines heritable variation (genetically inherited patterns of behavior) with individual learning (acquired adaptation) (Richerson and Boyd 1992:70). Cultural changes have their basis in the actions of individuals within ecologically situated choices (Smith and Winterhalder 1992:39; Winterhalder and Goland 1997:126). Primary condition factors for determining adaptations, then, are the nature and distribution of resources (Kaplan and Hill 1992:167).

Optimal foraging strategies (Bettinger 1980, 1991; Kelly 1995; Smith and Winterhalder 1992; Winterhalder and Smith 1981) consider environmental variables such as seasonality, potential hazards, and resources (size, distribution, and predictability). Seasonal movement of people is viewed as a mechanism for balancing maintenance and conservation of a resource with optimal exploitation (Hayden 1981:375). The optimal strategy is the option within the alternatives that yields the highest net return or fitness for the individual (Smith 1991:228). It is the best plan or decision made by the individual under the circumstances.

Decision making (rational choice) is viewed within a system of devising strategies for the most favorable resource procurement. These decisions are made in a parametric context where the variables of the physical environment are independent of each other (Winterhalder and Smith 1992:8–9). Choice (elements over which the hunter-gatherer has intentional control) and constraint (elements over which no control exists) are the factors determining outcome.

Assumptions are made about the environment, the people, and the selection and value of the prey to the hunter-gatherer in the prey-choice model. Prey is assumed to be distributed randomly across the landscape; people assess the costs and benefits for different prey species; natural selection has acted on the prowess of each hunter; and calories are the major objective. Pursuit and handling costs, food value of prey species, and prey abundance are equal independent variables (Winterhalder and Smith 1992:16–17). Diet breadth refers to the total number of resources included in the diet, counting down a ranked list (Kaplan and Hill 1992:171).

Patch-choice models differ in assumptions about resource distributions and net gain; but they follow the principle of maximizing effort and have similar constraints as the prey-choice models. Patch-choice models examine how long a hunter-gatherer remains in a patch in terms of profitability of exploitation. Nutrient differences rather than just calories (different currencies) also play a role in exploitation time in a patch (Kaplan and Hill 1992:180, 184).

Subsistence is examined further in terms of a cost-benefit analysis in exploiting resources. A cost-benefit analysis involves time, energy, and risk management. Risk management considers both the success/failure rate (i.e., the degree of certainty) of procuring resources and the safety of individuals. The success/failure rate is tied to predictability of resources, particularly with seasonal resources that experience wide variability over the years (Bettinger 1980, 1991; Keene 1983; Torrence 1983, 1989; Winterhalder and Smith 1981) and uncertainty, i.e., an individual's lack of information about a particular situation (Cashdan 1990:1–2; Smith 1991:231). Risk management pertains to coping mechanisms and decision making in response to unpredictable variation in economic and environmental conditions and differing levels of information. A trade-off exists between the cost of acquiring information and the value of increased knowledge. Risk can be avoided or buffered, while uncertainty is reduced by acquiring additional information (Kaplan and Hill 1992:187).

Risk-reduction models examine combinations of mean and variance that maximize the probability of survival by avoiding critical subsistence shortfalls (Winterhalder and Goland 1997:134). Environmental and climatic conditions make recurrent episodes of scarcity common for hunter-gatherers. These unpredictable shortfalls are ameliorated through adaptive strategies that mitigate the effects of scarcity. Alternatively, success can bring a temporary glut, particularly with large game animal resources. Hunter-gatherers use food sharing and regional exchange (reciprocal access to resources) as strategies of risk reduction in response to the daily unpredictable returns (Smith 1991:233; Winterhalder and Goland 1997:123). For storage to be an important aspect of hunter-gatherer subsistence, alternations of scarcity and high yield of a resource would need to occur with some regularity (Winterhalder and Goland 1997:137).

Risk models provide the framework to examine the advantages and disadvantages of sharing, and specify additional conditions required in order to perpetuate sharing in a social group (Smith 1991:236–37). The degree of sharing or reciprocal access varies from hunter-gatherer group to group, and from one type of resource to another. The benefits of riskier resources are greater when shared, and that sharing increases the nutritional well-being of the members (Kaplan and Hill 1985; Smith, 1991:237). Costs involved in sharing include those associated with maintenance and enforcement.

Climate and Environmental Background

During the late Wisconsinan (ca. 25,000 to 11,000 yr B.P.), climatic regimes across this vast stretch of land generally were cooler with greater precipitation or precipitation coming at a different season than today. Fluctuations existed, particularly in effective moisture leading to periods of drier conditions. Regional grassland-to-woodland balances were affected by both climate and disturbance factors, with the most important variables being precipitation and fire (Fredlund and Tieszen 1997; Sims 1988; Wright and Bailey 1982). The grasslands composition varied across the extensive territory, generally from south to north but also from east to west as it does today (Coupland 1992a). Bodies of water occurred throughout as rivers, lakes, ponds, marshes, playas, and salinas. Typical Rancholabrean herd herbivores (mammoth, bison, horses, camel) ranged throughout the grasslands extent along with attendant large carnivores (Arroyo-Cabrales et al. 1998; Graham and Lundelius 1987, 1994; Johnson 1987, 1989, 2000).

The late-Pleistocene North America grasslands is divided into the Mexican, Southern, and Northern grasslands. The Mexican grasslands form a corridor from northern Mexico to the Basin of Mexico, bounded on the east by the Sierra Madre Oriental and on the west by the Sierra Madre Occidental ranges. This general region today is the desert grassland province (Coupland 1992a; Schmutz et al. 1992).

In general, the climatic regime was cool and moist up to ca. 18,000 yr B.P. (Last Glacial Maximum), then cool and dry in the Basin of Mexico and cool and moist in northern Mexico until the end of the Wisconsinan (Metcalfe et al. 2000). In northern Mexico, a cool and moist climate was marked by strong summer cooling, reduced summer monsoons, fewer winter frosts, and a substantial increase in winter rainfall (Metcalfe et al. 2000; Van Devender 1986, 1990; Van Devender and Burgess 1985). The grassland had extensive lakes and widespread pinyon-juniper woodland (Bradbury 1997; Metcalfe et al. 1997, 2000). A temperate grassland prevailed from 20,000 to 12,000 yr B.P., with desert insect species replacing grassland ones after 12,000 yr B.P. (Elias and Van Devender 1990).

Within the Basin of Mexico, a mixed pine and oak forest and grassland setting expanded and contracted while lake levels in the Basin and water chemistry fluctuated (Bradbury 1989, 1997; González-Quintero, 1986; Lozano-García and Ortega-Guerrero 1994, 1998; Lozano-García et al. 1993; Metcalfe et al. 2000). Between 18,500 and 14,000 yr B.P., the surrounding grassland was widespread. From 14,000 to 10,000 yr B.P., the grassland continued to be reduced and the woodland expanded (Bradbury 1989, 1997; Caballero-Miranda 1997; Caballero and Ortega-Guerrero 1998; Lozano-García et al. 1993; Metcalfe et al. 2000). At paleo Lake Texcoco, the lake was very shallow by 16,000 yr B.P.; and between 14,500 to 11,600 yr B.P., the lake plain was dry (Bradbury 1989, 1997; Brown 1985; Metcalfe et al. 2000).

Grassland environments were prevalent on the Southern grasslands during the late Wisconsinan, although wooded areas existed at certain times in the eastern portion and as riparian woodlands along waterways (Bryant and Holloway 1985; Hall 1991, 2001; Hall and Valastro 1995; Holliday 1987; Toomey et al. 1993; Wilson 1966). The general climate was an equitable, humid, maritime one, with a reduced annual temperature range; and enhanced winter precipitation, mild winters, and cooler summers than today (Harragan 1983; Porter 1983). Greatly influenced by the Gulf of Mexico, changes in sea-surface temperatures altered air mass trajectories that carried moisture inland (Barry 1983; Friedman 1983; Imbrie et al. 1983). Lake levels in the western section of the region (Southern High Plains) rose and fell in response to elevated groundwater and increased precipitation. Changes in precipitation, temperature, humidity, evaporation, cloud cover, and wind affected the lake levels, causing the fluctuations (Reeves 1965, 1966; Smith and Street-Perrott 1983). River valleys in the eastern section underwent moderate incising and aggradation. Immense changes in stream loads and discharges due to fluctuating glacial meltwaters and climate changes affected the river valleys, many experiencing river metamorphosis where today their streams are too small for the valleys in which they flow (Baker 1983; Epps 1973; Hall 1990; Waters and Nordt 1995).

Vegetation on the Southern High Plains around 18,000 yr B.P. was a treeless *Artemisia* grassland, while the Edwards Plateau was a grassland with deciduous trees and pinyon pines in the dissected valleys and riparian areas (Hall and Valastro 1995). Along the southeastern edge, a savanna environment existed of a grassland interrupted with scattered trees (Bryant 1977; Camper 1991). Open grassland environments existed by 12,500 yr B.P. across the Southern Plains, including parklands along the valleys of the Southern High Plains, and savannas or scrub grasslands elsewhere (Bryant 1977; Johnson 1986, 1987; Lundelius et al. 1983). The grassland between 15,000 to 11,000 yr B.P. in central Texas was characterized by a mixed plant community composed of 50 to 60 percent C_3 cool season grasses and trees, and 45 to 50 percent C_4 warm season grasses (Nordt et al. 1994).

On the Northern grasslands, grassland environments were prevalent during the late Wisconsinan, although riparian woodlands along waterways and coniferous forests occurred adjacent to the ice margin (Fredlund 1995; Fredlund and Jaumann 1987; Rousseau and Kukla 1994; Ruhe 1983). The general climate was a cool-dry, continental to periglacial one, with a reduced annual temperature range, cooler summers than today, and a short growing season (Fredlund 1995; Porter 1983; Rousseau and Kukla 1994). Stream entrenchment was occurring near the end of the late Wisconsinan, with stream valleys underfit today (Arbogast and Johnson 1994; Johnson and Logan 1990; Madole 1991). Greatly influenced by the Laurentide Ice Sheet, changes in the ice sheet and air temperatures altered air mass trajectories that carried dry air and reduced precipitation. Around 18,000 yr B.P., an augmented westerly jet stream and a more winterlike general circulation occurred south of the ice margin (Barry 1983:402). Periods of loess deposits signaling increased aridity occurred between 29,000 and 12,000 years ago, reaching a progressively smaller areal extent and thinning thickness. While less seasonal variation occurred in air circulation, arctic air dominated in winter and an enhanced summer air circulation through Alberta lows created synoptic airflow patterns that allowed the extensive eolian deposition (Barry 1983:402). The climate was drier, colder, and windier than today.

After 24,000 yr B.P., the central upland grassland-sage steppe experienced woodland expansion and contraction, particularly in escarpments and riparian areas (Fredlund 1995:74). Grassland in western South Dakota by 14,000 yr B.P. consisted almost exclusively of cool-season C_3 plants. By 11,000 yr B.P., grassland composition had changed to a mixed C_3-C_4 assemblage (Fredlund and Tieszen 1997:214). Along the eastern edge in southern Wisconsin, a spruce parkland was the characteristic upland vegetation by ca. 13,200 yr B.P. (Fredland et al. 1996; Huber and Rapp 1992; Maher 1982). By 12,000 years ago, the mixed woodlands began declining as the grassland expanded through the Northern Plains (Baker and Waln 1985; Fredlund and Jaumann 1987:174–75).

Mammoth Localities

Mammoth localities throughout the Great Plains associated with early peoples are limited. That limitation is further reduced by restricting the localities to those site collections meeting the research criteria. Thirteen localities have been selected based on the following criteria: 1) presence of mammoth remains; 2) presence of cultural modification to the mammoth bone; 3) collection from buried deposits; 4) available appropriate radiocarbon ages; and 5) confirmation of cultural modification to bone through review of the collection by the author (cf. Johnson 1985, 1989, 1991; Johnson and Shipman 1986). Data from various localities have been published elsewhere (Johnson 1985, 1989, 1991; Johnson and Holliday 1997; Johnson and Shipman 1986; Johnson et al. 1994). The localities are summarized (Table 1), with a longer description for those not published (Arroyo-Cabrales et al. 2000, 2001; Johnson 2000).

A number of assemblages have complex life histories involving various natural modifications, including carnivore activity and trampling. These types of marks are identified on established criteria (Blumenschine and Salvaggio 1991; Capaldo and Blumenschine 1994; Johnson 1985; Lyman 1994; Olsen and Shipman 1988; Shipman 1981a, 1989), classification of carnivore activity, and scanning electron microscopy (SEM) in order to segregate and distinguish them from cultural marks.

The types of cultural modification to the mammoth remains are central to the case being developed here. Modification is categorized as marks on bone and as bone breakage. Cultural marks, initially identified on established criteria (Shipman, 1981a, 1988; Shipman and Rose, 1983, 1984; Lyman, 1994) for most localities, have been analyzed in greater detail on mammoth remains from the Wisconsin localities for additional identification criteria (Johnson 2000; Johnson et al. 2000).

A number of consistencies are noted that indicate a regular or standardized way of producing cultural marks in particular places on the bones. The marks themselves have a series of common features that characterize them, and the marks appear in a regular pattern. Those features, overall pattern, and the internal complexity of some marks within that typical pattern argue that these marks are not random occurrences. Two mark types are identified as caused by human agency. Although not all features occur with each mark, taken as a whole, what is characteristic of the mark types can be defined. Mark type 1 has a V-shaped trough that is straight, short, and narrow with a central line or crack, steep-sided and shallow-sided trough walls, striations in the steep-sided wall, and commonly blunted or tapered termini. Trough structure varies from simple to complex, depending on the number and configuration of strokes. Location placement primarily is on diaphyses along muscle ridges, tuberosities, crests, and edges. Overall, the damage morphology that this type of mark causes is a clean, precise incision or slice into the cortical surface. Mark type 1 is interpreted as an incised or cut mark as the result of a sharp-edged object slicing into the bone (Johnson 2000). It is the classic cut mark identified by Shipman (1981a, 1988; Shipman and Rose 1983, 1984).

Mark type 2 has a rough U-shaped trough; straight, short to long, and relatively broad, even-sided trough walls that have rents along the edges; bone chevrons along the bottom; depressed and crushed bone at the entry terminus; and commonly blunted termini. Trough structure is generally simple with one stroke. The usual location is on articular surfaces and across facets. Overall, the damage morphology is that of a hard pointed to rounded object propelled or forced along the cortical surface (Johnson 2000). Mark type 2 is interpreted as a pushed or pry mark. These two types of marks are involved with carcass processing, specifically butchering activities involving disarticulation and defleshing.

Mammoth bone as a production resource is a significant aspect of carcass utilization. A focused, quick blow causes combined tensions that result in a helical fracture in fresh long bones. Fresh bone broken by a hand-held tool has a circular depressed area (inverted cone) at the impact point caused by localized compression failure. Impact creates ring cracks, crushed bone, and cone flakes (technological flakes from the interior of the impact zone). Fracture fronts radiate from the impact zone until they merge, intersect, are deflected, or their energy is dissipated. Radial diaphyseal segments are produced through fracture and intersecting fracture fronts (Johnson 1985, 1991). Breakage is purposeful to cause segmentation and selected segments used in production endeavors. Bone quarrying uses proboscidean bone to produce cores and blanks in the same way toolstone is quarried. The great thickness of compact bone makes large cortical flakes. Bone cores exhibit prepared platforms, and flakes struck from those cores bear remnants of their prepared platform.

Northern Grasslands

A cluster of four mammoth localities (Figure 2) in southeastern Wisconsin is situated within a glacial landscape formed by the

Table 1. Data on human-modified mammoth bones from late-Pleistocene North American grasslands localities.

Locality	General provenience	Composition (MNI, age, sex)	Human-induced bone modification	Age (RCYBP)	Illustration
Mud Lake	lacustrine deposits	1 subadult, sex unknown	multiple-stroke mark type 1 (incised) and mark type 2 (pushed)	ca. 13,530 on purified collagen	Figure 3A
■Fenske	lacustrine deposits	1 young adult, sex unknown	multiple-stroke mark type 1 (incised)	ca. 13,530 on purified collagen	Figure 3B
Schaefer	lacustrine deposits	1 young adult, male?	single-stroke mark type 1 (incised)	ca. 12,300 on purified collagen	Figure 12a in Johnson 2000:29
Hebior	lacustrine deposits	1 young adult, male?	single-stroke mark type 1 (incised)	ca. 12,300 on purified collagen	Figure 12b in Johnson 2000:29
Lubbock Lake	stratum 1 (terrace)	1 individual, age and sex unknown	none	ca. 11,500–17,500 on organic sediments	NA
Lubbock Lake	substratum 1B (gravel bar)	1 young adult & 2 juveniles; 1 female, 2 sex unknown	single-stroke mark type 1 (incised) and mark type 2 (pushed); helical fracture surfaces, intersecting fracture fronts, initial and rebound impact points; cone flakes, cone flake scars; flake struck from bone core	ca. 11,100 on wood	Figures 5, 6 Figures 17a,b,d, 18 in Johnson 1989:456–457; Figures 5.15a, 5.15b in Johnson 1985:198–199
Sand Creek	lacustrine deposits	1 individual, age and sex unknown	helical fracture surfaces, intersecting fracture fronts, cone flakes	ca. 13,450 on organic sediments	Figure 6
Blackwater Draw, Locality No. 1	spring-laid sand (gray sand); mammoth #2	5 mature adults, sex unknown	single-stroke mark type 1 (incised)	ca. 11,630–11,040 on carbonized seeds	none published
Blackwater Draw, Locality No. 1	spring-laid sand (brown sand wedge)	1 individual, age and sex unknown	suggested helical fracture surfaces, intersecting fracture fronts	ca. 10,800 on organic sediments	Figure 3 in Johnson and Holliday 1997:334
Miami	lacustrine deposits	1 adult, 2 juveniles, 2 age unknown; 3 females, 2 sex unknown	multiple-stroke mark type 1 (incised)	11,400–10,350 on organic sediments	Figure 11 in Johnson 1989:452
Cooperton	fluvial deposits	1 subadult male	helical fracture surfaces, intersecting fracture fronts	ca. 17,575–20,400 on bone and tooth	Plates 3, 6, 7 in Anderson 1975: 151, 159, 161
Domebo	clayey-silt deposits	1 subadult female	none	ca. 11,200 on bone and organic sediments, ca. 11,045 on wood	NA
Duewall-Newberry	fluvial deposits	1 subadult male	single-stroke mark type 1 (incised), helical fracture surfaces, intersecting fracture fronts, initial and rebound impact points, cone flakes, cone flake scars	terminal-Pleistocene stratigraphy	Figures 7–12, 15–16 in Steele and Carlson 1989:418–421, 424–425
Bonfire Shelter	stratum H-1	1 individual, age and sex unknown	helical fracture surfaces, intersecting fracture fronts	ca. 12,400 on diffuse charcoal	none published
Bonfire Shelter	stratum E	1 individual, age and sex unknown	helical fracture surfaces, intersecting fracture fronts	12,400–10,200	none published
Tocuila	lahar deposits	7 individuals; mature adult to juvenile; sex unknown	helical fracture surfaces, intersecting fracture fronts, impact point, chattering, hackle marks, cone flakes; bone core and flake struck from bone core	11,188 on charcoal, seeds, and bone	Figures 8, 9, 10

■ New information now available (Joyce, pers. comm. 2004) indicates a different radiocarbon age and genus for the Fenske remains. The ca. 13,500 RCYBP age came from the partial humerus. Purified collagen from the femur yielded an age of ca. 11,200 RCYBP (Joyce and Blazina-Joyce, 2004). The new age may indicate the presence of two animals two thousand years apart or a curatorial problem, as the Mud Lake and Fenske remains were housed together. While first identified as mastodon and then mammoth, the Fenske femur once again is being identified as mastodon. Neither the new age nor change of genus alters the general pattern discussed for mammoth procurement, but rather broadens the pattern to a proboscidean one.

wasting ice of the Lake Michigan lobe or water plane fluctuations of glacial Lake Chicago; they lie in lowlands between moraine ridges. At Mud Lake and Fenske, a few limb parts were found in the course of farming and construction; the excavated Schaefer and Hebior carcasses were disassembled bone piles at the edge of quiescent ponds (Overstreet 1998; Overstreet and Stafford 1997).

In examining for evidence of natural modification, the role of water in bone pile formation was investigated through both the traditional ANOVA and goodness-of-fit approach of discrete data in 10-degree increments (Johnson 1995; Kreutzer 1986, 1988; Shipman 1981b) and the more rigorous Kolmogorov-Smirnov test based on continuous data. Both tests indicated no significant departure from a random distribution of bone. Water trans-

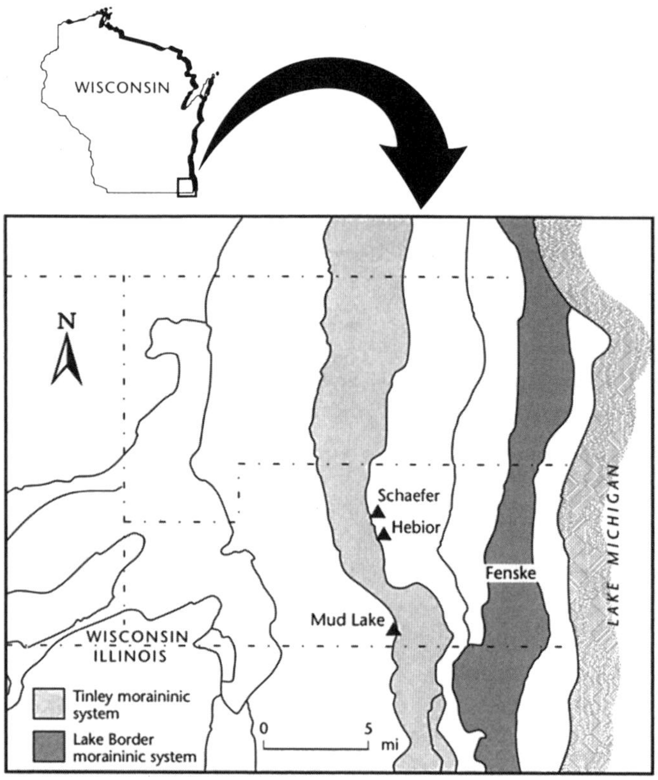

Figure 2. Location of southeastern Wisconsin proboscidean localities.

port did not contribute to forming the bone beds, nor did water movement disturb them. Limited carnivore activity at the Wisconsin localities indicated underutilization of the carcasses (G. Haynes 1982, 1983) that may support a short exposure time. Damage was principally at the cancellous ends of long bones (Johnson 2000).

A small number of elements exhibited a large number of cultural marks (n = 225). Treated as continuous data, the length and width of the marks were plotted. Based on length, the marks were divided into a shorter group and a longer group. The shorter group was divided into two subgroups: short and narrow marks, and short and broader marks. The shorter group consisted of both type 1 and 2 marks; the longer group includes only type 2 marks (pushed). Although the mark types are not mutually exclusive, generally the short and narrow subgroup consisted of type 1 marks (incised), and the short and broader subgroup consisted of type 2 marks. The groupings and fairly tight clusters indicated consistent mark sizes (Johnson 2000; Johnson et al. 2000).

Longer type 2 marks (pushed) are found only on the medial diaphysis of the ulna and anterior carpals of the Mud Lake limb (Figure 3a). Much shorter type 2 marks are located on the proximal end of the ulna (on the anterior cortical surface between the anconeal and olecranon processes) and are paired with or adjacent to similar marks within the articular surface (semilunar notch). The longer type 2 marks are made by the same type of pointed tool as the shorter type 2 marks, forced in a forward-moving motion. The abundance and length of the marks indicate some resistance in accomplishing the task at hand. The type of damage done to the anterior of the carpals is the result of disjointing actions to disassemble the foot to obtain the fatty foot pads (Saunders and Daeschler 1994). The shorter pushed marks on the ulna are also likely the result of disjointing actions.

A number of cut marks on the Fenske and Mud Lake long bones are paired and appear in a chevron pattern, with one set of termini touching, near, or offset from the other (Figure 3b). This pattern is produced by crisscross motion of the tool. Additional parallel and perpendicular variations of this pattern on the same diaphysis indicates various hand/tool motions used to overcome resistance met in removing the muscle bundle. Furthermore, mul-

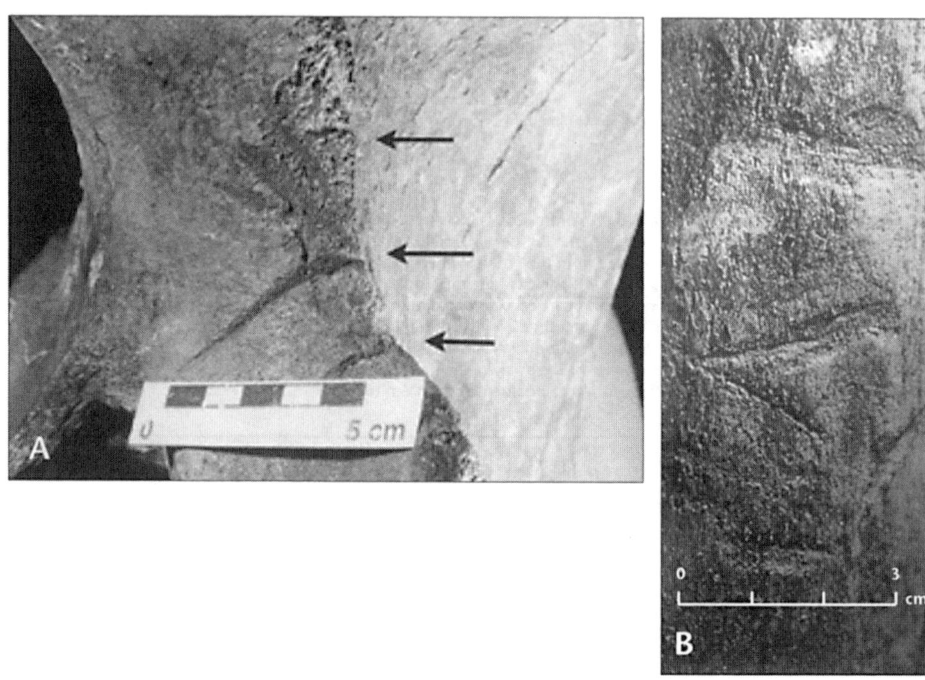

Figure 3. Cultural marks on limb elements. **A,** short mark type 2 on the semilunar notch of the Mud Lake ulna (47Kn246-1); **B,** complex mark type 1 on the diaphysis of the Fenske Lake femus (47Kn240-1).

tiple-stroke marks are found on the Fenske and Mud Lake remains but not on the remains of the other two carcasses. Requiring more than one stroke indicates difficulty in removing tissue or separating joints.

Southern Grasslands

Eight localities in a more central part of the vast grasslands (Figure 4) are situated in a non-glacial landscape formed by deposition and erosion, from aggrading and downcutting to dissolution and eolian action. They are located along streams or at the edge of paleo-lake basins.

Lubbock Lake Landmark (Johnson 1987; Johnson and Holliday 1989) is located in an entrenched meander of Yellowhouse Draw where springs have flowed since at least the late Pleistocene (Figure 4). The cultural and natural history record is contained within well-stratified, well-dated deposits (Holliday 1995, 1997; Holliday et al. 1983, 1985; Johnson 1987). For both late-Pleistocene occupations, identified mark types on bones and the pattern of the cultural marks indicate butchering activities to acquire meat from fresh carcasses of various herd herbivores (Johnson 1987; Johnson and Shipman 1986). Based on the bone breakage pattern (Table 1), people from the latter occupation also used the mammoth carcass as a technological resource (Johnson 1985).

Insufficient materials were recovered from the earlier occupation to examine the role of water in depositing or transporting bone (Shipman 1981b). However, broken edges were sharp, and the bone was neither degraded nor water-modified. For the upper occupation, water did not play a role in site formation, but has disturbed the site through realignment (Johnson 1995; Kruetzer 1988).

The primary types of cultural marks on mammoth remains at Lubbock Lake are the same found in the Wisconsin localities, indicating cutting and prying (defleshing and disarticulating) during butchering. In the earlier of the two occupations, only type 1 marks (incised) have been identified on various elements from extinct bison, camel, and horse; no cultural marks are found on the mammoth remains. In the latter occupation, mammoth remains exhibit both type 1 marks (incised) and type 2 (pushed). These marks, principally on mammoth limbs, are the result of stripping muscle bundles and removing fatty foot pads. The subtle marks (particularly type 1) indicate butchering of fresh carcasses. The remains are from a single event involving a small family group (Johnson 1987).

Helically fractured mammoth long bones, radial diaphyseal segments, debris with intersecting fracture fronts, and cone flakes indicate bone quarrying occurring along a point bar at Lubbock Lake in the latter occupation. A cortical bone flake with a rem-

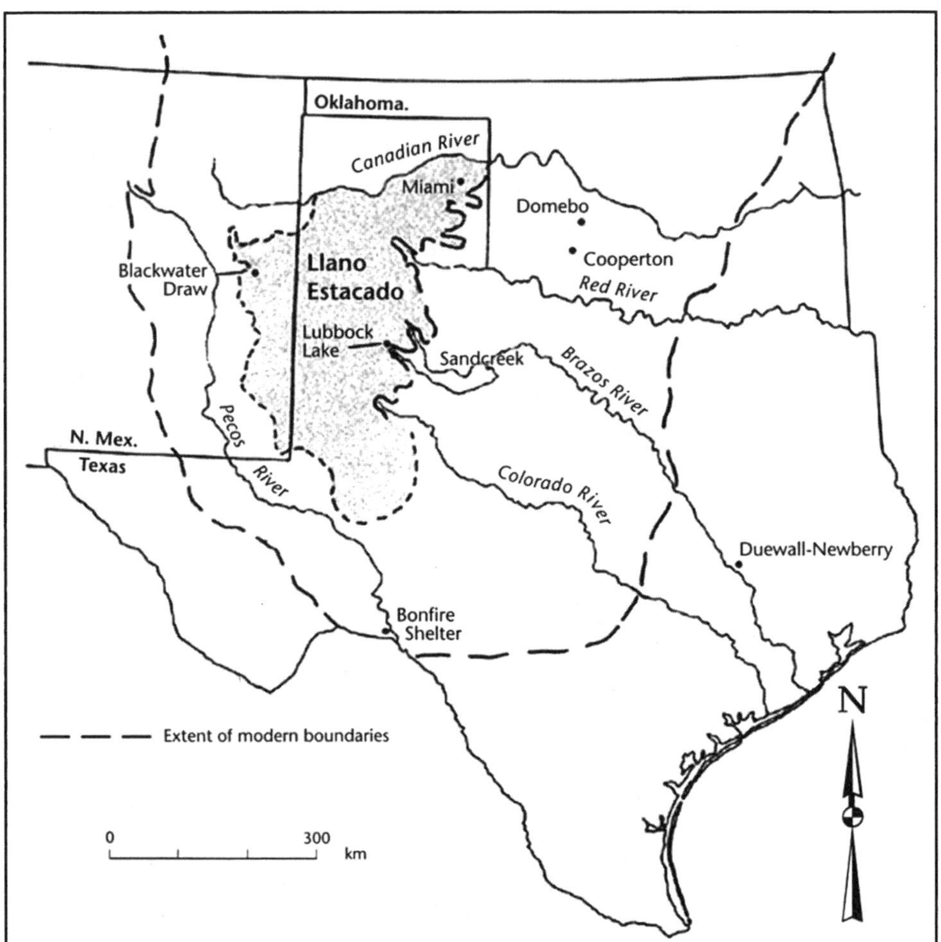

Figure 4. Location of Southern grasslands mammoth localities.

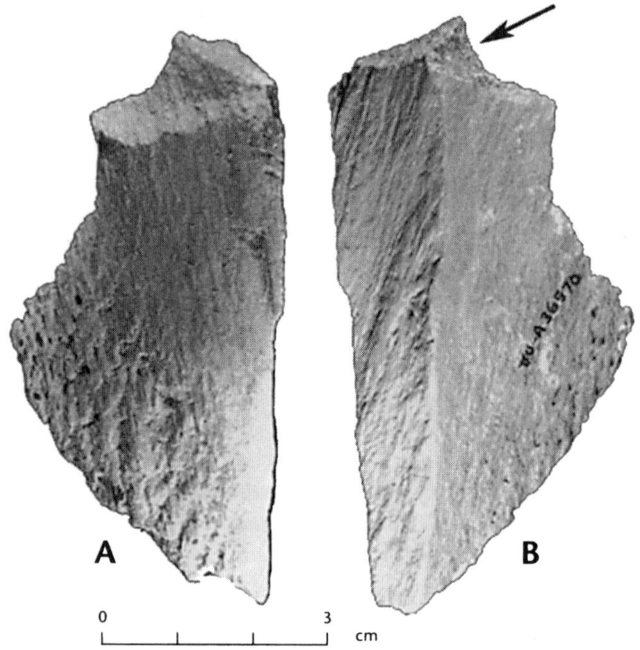

Figure 5. Cortical bone flake (TTU-A36570) produced from thick mammoth cortical bone, from the Area 2 mammoth locality at the Lubbock Lake Landmark. **A,** ventral side; **B,** dorsal side; arrow points to remnant platform.

nant platform (Figure 5) points to core production and use as one of the purposes of bone quarrying (Johnson 1985).

The Sand Creek mammoth (Johnson et al. 1994), located in the heavily dissected landscape along the eastern escarpment of the Southern High Plains (Figure 4), lay along an arroyo in Yellowhouse Canyon about 32 km downstream from Lubbock Lake. The lack of subaerial weathering damage indicates the bones were buried soon after deposition. Not enough material was recovered to examine the role of water in depositing or transporting the bones (Shipman 1981b). However, edges and exposed trabecular surfaces are sharp and distinct, with no evidence of fluvial polishing and rounding. Furthermore, the lake-basin deposits did not indicate a high-energy regime.

Fracturing features and debris (Table 1) are evidence of human influence. The cone flakes are consistent with those found at Lubbock Lake (Figure 6); smaller cone flakes from small ring cracks were found in place overlapping each other. Bone quarrying of the Sand Creek mammoth carcass is for unknown purposes.

Blackwater Draw Locality No. 1 (the Clovis site; Hester 1972; Sellards 1952) is in a paleo-lake basin 2 km north of and draining into Blackwater Draw (Figure 4). Although the stratified sequence of late-Quaternary deposits (Haynes 1975, 1995; Haynes et al. 1992; Holliday 1995, 1997) is continuous across the paleobasin, considerable facies variation exists, particularly on the microstratigraphic level. The most extensive excavation revealed at least five mammoth carcasses on top of the gray sand (Hester 1972; Hughes 1984; Lundelius 1972), along with Clovis points and a variety of lithic tools. The mammoth remains are an accumulation of individual events at the edge of the paleopond (Saunders 1980). Remains from mammoth #2 exhibit type 1 (incised) marks. These marks, located on the diaphysis of a humerus and along diaphyseal segments of ribs (Johnson 1989, 1991), are delicate and uncomplicated, indicating butchering of a fresh carcass.

Remains of a very partial mammoth carcass, recovered in the brown sand wedge (Table 1), represent a later event in the paleobasin than that involving the five mammoth. Investigation of the role of water indicated a bone deposit neither laid down nor disturbed by fluvial action (Johnson and Holliday 1997:339–41). No cultural marks were found on the mammoth remains, although the neural spine of a vertebra exhibited a helical fracture. A humerus and diaphyseal segment were mapped but now are missing from the collection. Their representation and adjacent positions suggested a helical-fractured humerus (the breakage pattern is similar to that of the humerus at Lubbock Lake) and its associated large humeral diaphyseal segment. If correct, then bone quarrying occurred with this carcass. Although the specific purpose is unknown, the large diaphyseal segment would appear similar to one at Lubbock Lake that has been interpreted as a blank from which mammoth bone foreshafts and wrenches would have been made (Johnson, 1985:203). Mammoth bone foreshafts were recovered in another part of the paleobasin (Hester 1972:117).

Figure 6. Cone bone flakes from the zone of impact on shafts of mammoth long bones; arrows point to impact area; Sand Creek (TTU-A1-15393) obverse (**A**) and reverse (**A-1**); Lubbock Lake (TTU-A4140) obverse (**B**) and reverse (**B-1**); on **B-1,** note the flake scars from adjacent cone flake detachments.

Miami (Holliday et al. 1994; Sellards 1938) is located on the high plains surface in an extinct playa basin at the northeastern edge of the Southern High Plains (Figure 4). The carcasses represent a single event involving a small family group (Saunders 1980). All the elements are highly weathered, with deep desiccation cracks and rectilinear fractures common. None of the elements exhibit helical fracturing. Type 1 (incised) marks occur on ribs and a fibula (Table 1). A series of marks on a rib have been noted as possible hack marks (Johnson 1989:452). Upon reconsideration based on the Wisconsin data, these marks are complex, multiple-stroke type 1 marks, indicating some resistance in removing tissue.

Cooperton (Anderson 1975) is located along the edge of a small tributary creek (Figure 4). The hind legs of the disarticulated carcass are missing owing to erosion that exposed the locality. The right scapula and both humeri are helically fractured (Table 1). The humeri are broken in the same manner as that at Lubbock Lake (mid-diaphysis above the supracondyloid crest). Some of the diaphyseal segments could be refitted, while others were not recovered in the excavations. No cultural marks are identified. Some elements exhibit rectilinear fractures, indicating weathering of the bone. Bone quarrying occurred with this carcass for unknown purposes (Anderson 1975:171; Johnson 1991).

Located at the headwaters of a tributary creek to the Washita River (Figure 4), Domebo (Leonhardy 1966) yielded a disarticulated carcass. None of the elements are broken, nor are cultural marks found on accessible bones (most are still encased in plaster jackets). Based on the points and tools recovered and intact bone, the carcass was utilized for meat acquisition. The condition of the carcass at the time of acquisition could not be determined (Johnson 1991).

The disarticulated Duewall-Newberry mammoth (Carlson et al. 1984; Steele and Carlson 1989), located along the Brazos River in central Texas (Figure 4), has type 1 (incised) marks on the ribs (Table 1). These are fine, uncomplicated marks that indicate little resistance removing tissue. The carcass was fresh during butchering. Both humeri and the right femur are helically fractured, exhibiting dynamic loading points, cone flakes, and flake scars (Steele and Carlson 1989:424–25, Figures 15 and 16). The elements are broken in the same manner as those at Lubbock Lake. Large diaphyseal segments and cortical flakes indicate bone was quarried to produce blanks and cores (Johnson 1985, 1991).

Bonfire Shelter (Bement 1986; Dibble and Lorrain 1968) is at the southern end of the Southern Plains, on the Stockton Plateau in a rockshelter along Mile Canyon, a tributary of the Rio Grande (Figure 4). Limited mammoth remains have been recovered around small limestone blocks in two strata. A tibial diaphyseal segment from Stratum H-1 and a long-bone diaphyseal segment from Stratum 1E are helically fractured (Table 1). These elements indicate that limited bone quarrying occurred, probably to produce diaphyseal blanks (Johnson 1985, 1991).

Mexican Grasslands

At the southern end of the grasslands, Tocuila (Arroyo-Cabrales et al. 1999; Corona-M. and Arroyo-Cabrales 1997; Morett et al. 1998a, 1998b) is located in the Basin of Mexico (Figure 7) at what was an edge of Lake Texcoco. The locality is stratified, with the main unit that contains the bones being a mud flow caused by volcanic activity (lahar). A late-Pleistocene channel had been downcut prior to the mammoth-bearing event, and the lahar had flowed through and filled the channel (Arroyo-Cabrales et al. 1999). The disarticulated nature of the remains indicates the mudflow did not kill the mammoth. Whether the mammoth remains had already been laid down in the channel (either as an accumulation or a catastrophic event) and the mud flow covered them, or whether they were carried in by the mud flow is being debated (Arroyo-Cabrales et al. 1999; Morett et al. 1998a, 1998b; Siebe et al. 1999).

Human activity (Table 1) is indicated by dynamic impact fracturing features on long bone segments and fracturing debris (Arroyo-Cabrales et al. 2000, 2001; Johnson et al. 2001). A triangle-shaped femoral radial segment exhibits a helical fracture at the apex, crushing and small flake removal along the opposite edge, and a series of large facets along the cortical surface below the crushed and small flake removal area. This specimen is interpreted as a bone core with a prepared platform and scars from the removal of a number of large cortical flakes (Figure 8). Another specimen has a number of facets on the cortical surface, an area of crushed bone and small flake removal at the top of the faceted area, and an undulating surface on the reverse side with a large diffuse bulge. This specimen is interpreted as a cortical bone flake with remnant platform preparation (Figure 9). The cortical flake conjoins with the central flake scar on the bone core (Figure 10). The Tocuila and Lubbock Lake cortical flakes are morphologically similar and share the same features. The same relationship exists for the large cone flakes from Tocuila, Lubbock Lake, and Sand Creek. Limited evidence of human involvement

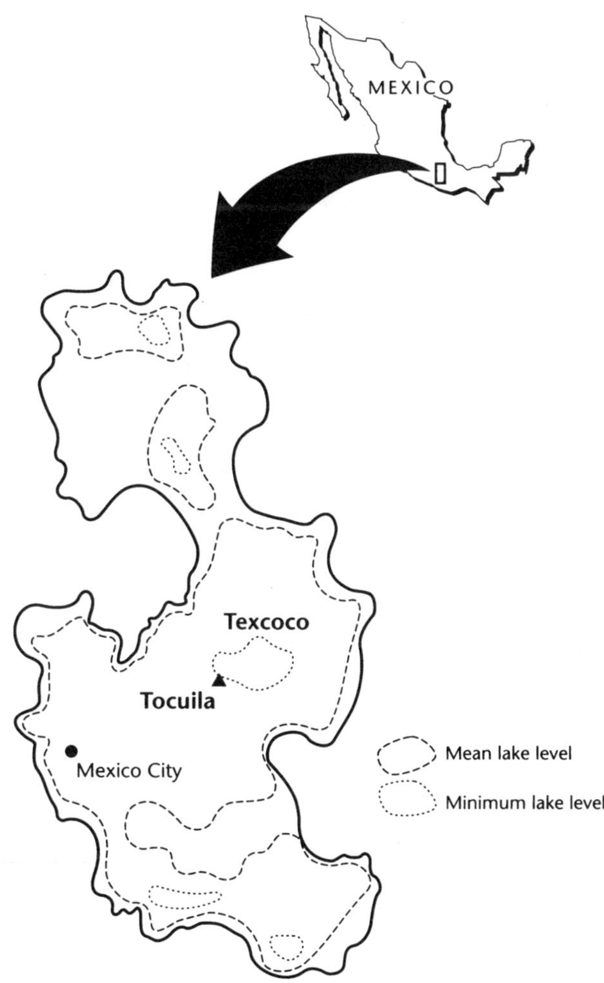

Figure 7. Location of Mexican grasslands mammoth locality in the Basin of Mexico.

with mammoth at Tocuila, principally bone breakage, is interpreted as bone quarrying to produce cores for transport elsewhere (Arroyo-Cabrales et al. 2001; Johnson et al. 2001).

Construct

The current construct builds on that proposed by Johnson (1991:228–29) in examining organization and subsistence of Southern Plains Clovis-age hunter-gatherers; and it examines the role of mammoth in the Plains hunter-gatherer economy. It is based in part on the second and third set of propositions in the earlier construct (subsistence and procurement patterns; and game animal processing, site type recognition, and food sharing). If targeted hunting was practiced, then only a single species would be involved, reflected as single-species kill sites. Targeted hunting would be on a structured basis as a communal hunt. Kill sites could be identified on the basis of complete to nearly complete carcasses of large game animals. If scavenging was practiced, then that activity would appear as a single-species kill site with similar carcass treatment. Distinguishing scavenging from targeted hunting would involve the species (if different from normal target prey) and number of animals represented. Lastly, if the climate was equitable and food supply steady, then food storage was unnecessary and daily food obtained would be pooled and shared among the group.

For this construct, an equitable climate exhibiting less seasonal variation and cooler temperature regimes than today existed across the North American grasslands in a general parkland mosaic of extensive grasslands and wooded areas along streams and forested foothills for late-Wisconsinan times. During this time, fluctuations occurred in lake levels and water chemistry, available moisture, stream volume, and boundaries of grassland vs. woodlands or forest. Grassland composition and character varied by region, with temperate grasslands in the Mexican region, mixed temperate and cool season grasslands in the Southern Plains, and cool season grasslands in the Northern Plains.

Despite fluctuation, the major climatic and environmental parameters remained similar through the several-thousand-year period. The high-diversity environment created through mosaic vegetation and faunal communities provided a wide diet breadth and a stable food supply. Large game animals, primarily herd herbivores, were a patch resource discontinuously distributed across the landscape. Late-Wisconsinan Plains peoples had a broad-based subsistence strategy with a procurement plan of encounter and targeted hunting and scavenging. Procurement was for subsistence and economic purposes, and daily food obtained would be pooled and shared among the group.

A number of propositions are based on this extended statement. If the extension to the late Wisconsinan (particularly since the late glacial maximum) of the Southern Plains Clovis construct is valid, then similar sites and activities would have occurred. If the modulating climate between humid and drier conditions and a rise in annual temperature as much as 2° C were tempered by constants of grasslands habitat, herd herbivores, and available surface water, then subsistence patterns would not be

Figure 8. Bone core (TOC-1-281) from Tocuila produced on thick mammoth cortical bone. **A,** flake scars and crushing from remnant platform preparation; **B,** cortical flake scars.

altered. If earlier late-Wisconsinan Plains peoples practiced a foraging (generalist) strategy as did later Southern Plains Clovis peoples, then social organization would have been an egalitarian band society.

If an annual round governed by a rainy and non-rainy cycle is practiced, then patch versus continuously distributed resources played different roles in the subsistence strategy depending on the time-of-year cycle. If a broad-spectrum subsistence was practiced, then acquiring meat (protein and fat) would be secondary to plants, but significant in terms of nutrition and the social value of sharing food. If targeted hunting was practiced as a secondary method, then a single large game animal species would be the prey target; and that species would represent something beyond basic subsistence needs. If the target prey was a high-risk species, then the number of hunters and hunting strategy would depend on the risk and number of prey animals involved. This dependence would be reflected at sites in terms of the number and population dynamics of prey animals, carcass treatment, resources obtained, and the time of year or place in the annual round. If scavenging was part of the procurement strategy, then that activity would involve high-risk target species or high-ranking prey species. If scavenging was practiced, then a scavenging event may be distinguished from a targeted event by the number and population dynamics of the animals, species represented (if different from targeted prey species), carcass treatment, resources obtained, and the time of year or place in the annual round. If procurement was for purposes other than subsistence, then that purpose would be reflected at sites in terms of carcass treatment, resources obtained, and toolkit.

If food storage was not practiced, if the climate was not equitable, if animal resources were not stable, and if food was not obtained daily, then pooling and sharing of game animal resources with the group would be for purposes other than or in addition to immediate nutritional needs. If food sharing was for other than or in addition to nutritional needs, then that activity would be related to the target prey in terms of the risk, the number of animals, the quantity of resources obtained, the size of the group sharing the food, and the time of year or place in the annual round.

Lastly, if the procurement strategy involved both hunting and scavenging, then the costs and benefits of each should be discerned. Hunting is a high-risk, high-yield activity (the hunt may be unsuccessful, the hunter may be hurt, and the yield depends on the number and size of animals taken); on the other hand, scavenging is a low-risk, variable-yield activity (the carcass is already available with no threat by the animal, and the yield depends on the condition of the carcass and the size of the animal). Mammoth hunting required advanced planning, personnel coordination, search procedures, and handling and transporting efforts. In hunting, either a small number of hunters would ambush or surround a single mammoth, or many hunters would kill first the matriarch and then the milling family members. Mammoth are a high-risk game animal that could be dangerous and unpredictable, particularly in a family group (G. Haynes 1992). Danger would be the principal cost affecting hunting strategy. Risk in this case is not concerned with the unpredictability of where to find game or with information sharing, but with the safety of the individual or group. The benefits for a small hunting party would be a large amount of meat, and the reduction in unpredictability in securing meat over a period of time, as well as the reduced danger in dispatching only one animal. Handling and transporting costs may

Table 2. General age groupings of mammoth procurement localities (all dates approximate RCYBP).

18,000	13,500	12,300	11,000
Cooperton	Mud Lake	Schaefer	Lubbock Lake■
	➤Fenske	Hebior	Blackwater Draw▲
	Sand Creek	Bonfire Shelter◗	Blackwater Draw▼
			Bonfire Shelter◆
			Miami
			Domebo
			Duewall-Newberry
			Tocuila

■gravel bar occupation
▲Mammoth #2, El Llano Dig #1
▼Jelinek mammoth
◗Stratum 1-H
◆Stratum E

➤New information now available (Joyce, pers. comm. 2004) indicates a different radiocarbon age and genus for the Fenske remains. The ca. 13,500 RCYBP age came from the partial humerus. Purified collagen from the femur yielded an age of ca. 11,200 RCYBP (Joyce and Blazina-Joyce, 2004). The new age may indicate the presence of two animals two thousand years apart or a curatorial problem, as the Mud Lake and Fenske remains were housed together. While first identified as mastodon and then mammoth, the Fenske femur once again is being identified as mastodon. Neither the new age nor change of genus alters the general pattern discussed for mammoth procurement, but rather broadens the pattern to a proboscidean one.

Table 3. Summary information on mammoth procurement localities (all dates approximate RCYBP).

Region	Locality	Activity	Age
Northern Grasslands	Mud Lake	meat acqusition by scavenging; stiffened carcass	13,500
	➤Fenske	meat acqusition by scavenging; stiffened carcass	13,500
	Schaefer	meat acqusition by hunting or scavenging; fresh carcass	12,300
	Hebior	meat acqusition by hunting or scavenging; fresh carcass	12,300
Southern Grasslands	Cooperton	bone quarrying through scavenging; fresh bone	17,575–20,400
	●Lubbock Lake	meat acqusition by hunting or scavenging; fresh carcass	17,000–11,500
	Sand Creek	bone quarrying through hunting or scavenging; fresh bone	13,450
	◗Bonfire Shelter	bone quarrying through hunting or scavenging; fresh bone	12,400
	◆Bonfire Shelter	bone quarrying through hunting or scavenging; fresh bone	12,400–10,200
	Duewall-Newberry	meat acquisition and bone quarrying through hunting or scavenging; fresh carcass	12,000–11,000
	Miami	meat acqusition through scavenging; stiffened carcass	11,415–10,345
	▲Blackwater Draw	meat acqusition by scavenging; fresh carcass	11,630–11,040
	Domebo	meat acqusition by hunting or scavenging; fresh carcass	11,200
	■Lubbock Lake	meat acquisition and bone quarrying through hunting or scavenging; fresh carcass	11,100
	▼Blackwater Draw	bone quarrying through hunting or scavenging; fresh carcass	10,800
Mexican Grasslands	Tocuila	bone quarrying through hunting or scavenging; fresh carcass	11,188

◗Stratum 1-H
◆Stratum E
●terrace occupation
■gravel bar occupation
▲Mammoth #2, El Llano Dig #1
▼Jelinek mammoth

➤New information now available (Joyce, pers. comm. 2004) indicates a different radiocarbon age and genus for the Fenske remains. The ca. 13,500 RCYBP age came from the partial humerus. Purified collagen from the femur yielded an age of ca. 11,200 RCYBP (Joyce and Blazina-Joyce, 2004). The new age may indicate the presence of two animals two thousand years apart or a curatorial problem, as the Mud Lake and Fenske remains were housed together. While first identified as mastodon and then mammoth, the Fenske femur once again is being identified as mastodon. Neither the new age nor change of genus alters the general pattern discussed for mammoth procurement, but rather broadens the pattern to a proboscidean one.

be high because of the large amount of meat and the need to preserve it until consumed. In communal hunting where several animals are killed, the cost would be greater because of the larger group sharing the meat supply, the social ramifications in forming an aggregate group for the hunt, increased coordination efforts and handling and transportation costs, and the greater number of people exposed to danger. These costs would be balanced by the massive amount of meat obtained in a multiple kill, the opportunity to share information and establish reciprocal obligations, and other social activities.

Scavenging is a different approach to procurement. Although it still requires handling and transporting efforts, the activity can be done by one person with minimal preparation and coordination. Time spent could be opportunistic rather than an organized search. Locating a suitable carcass to scavenge, however, is highly unpredictable. People are not the only beings that scavenge carcasses. Scavenging puts people in direct competition with large carnivores, both as primary predators and scavengers. Competition from large carnivores is a high risk factor that may have affected the scavenging activities of people. Avoiding direct competition by partitioning activities would minimize this danger. Offsetting these risks were the benefits of the potential amount of meat, availability of carcasses, and elimination of certain danger in hunting live mammoth.

Discussion

The database totals 16 entries that span the ca. 7,000- to 8,000-year period from the late glacial maximum to terminal Pleistocene (Table 1). This time range assumes that available radiocarbon ages for the localities are valid, but those from Cooperton (Anderson 1975) are most suspect based on the problem of free carbonates in the Cooperton soil profile and the pretreatment used (Hass and Banewicz 1980; Hassan et al. 1977). However, comparable dates from similar sites in the Central Plains (Holen 1994, 1996, 1999, 2004; May and Holen 1993) underscore the plausibility that people procured mammoth at this early time on the North American grasslands. Nevertheless, the overall record is spotty at best, and a ca. 4,500-year gap exists between these earliest sites and later ones. The database sites cluster around four time periods (Table 2). Within this database, not all activities associated with mammoth procurement are represented at sites earlier than the latest period (Table 3).

Table 4. Age and gender of mammoth from single-carcass localities.

Locality	Skeletal age	Gender
Mud Lake	<19– <33	not determined
➤Fenske	25–45	not determined
Schaefer	<28	male?
Hebior	>25–<28	male?
Cooperton	subadult	male
Domebo	subadult	female
Duewall-Newberry	subadult	male
Sand Creek	not determined	not determined
Blackwater Draw▲	not determined	not determined
Blackwater Draw▼	not determined	not determined

▲Mammoth #2, El Llano Dig #1

▼Jelinek Mammoth

➤New information now available (Joyce, pers. comm. 2004) indicates a different radiocarbon age and genus for the Fenske remains. The ca. 13,500 RCYBP age came from the partial humerus. Purified collagen from the femur yielded an age of ca. 11,200 RCYBP (Joyce and Blazina-Joyce, 2004). The new age may indicate the presence of two animals two thousand years apart or a curatorial problem, as the Mud Lake and Fenske remains were housed together. While first identified as mastodon and then mammoth, the Fenske femur once again is being identified as mastodon. Neither the new age nor change of genus alters the general pattern discussed for mammoth procurement, but rather broadens the pattern to a proboscidean one.

Figure 9. Cortical bone flake (TOC-1-534) from Tocuila, exhibiting arises from prior flake removal, feather termination, and remnant platform preparation (proximal end).

In comparing the current construct with the database, five areas are examined in terms of a hunter-gatherer foraging strategy: 1) basis for hunting; 2) specialized site; 3) carcass treatment; 4) storage feasibility; and 5) climatic regime. Highly diverse environments within the North American grasslands made available a wide variety of animal and plant resources. Although not all resources were utilized, accumulated data across the grasslands indicate a wide diet breadth, with mammoth being one of several large game animals. Ten localities in the database involve an individual mammoth. Lubbock Lake and Miami appear to be targeted sites where small family units were cropped. What the Tocuila mammoth represent is unclear, as information is not yet available on the age, gender, and carcass condition of those mammoth.

The question is whether mammoth were hunted when encountered or only when targeted. If multiple mammoth within an appropriate age structure represent the targeted hunting of a small herd, then a single mammoth may be an opportunistic hunt. Female African elephants are rarely found outside the herd group, but when males reach sexual maturity (mid to late teens), they frequently become solitary (Douglas-Hamilton and

Figure 10. Refitted cortical bone flake from Tocuila with parent bone core.

Douglas-Hamilton 1975; Eltringham 1982; G. Haynes 1991). Assuming the behavior of mammoth was analogous to that of African elephants, a single mammoth hunted on encounter would be a sexually mature male. Of the few carcasses of individual mammoth whose sex has been determined, both males and females are represented (Table 4). The majority are male, however, suggesting that encounter hunting of solitary male mammoth may have occurred.

Evidence of mammoth butchering is based on cultural modification to the bones. Marks are accidental occurrences, by-products of carcass processing. While their absence on an element does not necessarily mean that the bone was not processed, their presence, type, location, complexity, and intensity provide clues to the processing strategy and tasks involved. For the mammoth localities in the database, these marks appear primarily on a highly select category of elements, and then only on particular bones. Certain parts of the body were targeted, primarily the limbs and occasionally the axial skeleton (mainly the ribs). Mammoth carcasses could be turned using cordage (G. Haynes 1991:182). However, in rough terrain or boggy conditions, the difficulty of turning a carcass probably would preclude that effort. The more efficient strategy would be to remove as much as possible through the available side (Frison 1986:131).

The incised and pushed marks indicate two major butchering activities, those of cutting actions in defleshing, and prying actions in disarticulating. These activities were focused on the limbs, which provided at least five major resources: large muscle masses (meat), tendons, fatty foot pads, marrow, and bone as a raw material. However, just as not all the carcass was used, not all the limb resources were utilized. Muscle bundles were stripped and fatty foot pads retrieved; tendons were most likely harvested, although no direct evidence has been found; marrow was not sought. Obtaining raw material concentrated primarily on the upper limbs. Very directed activities took place within limited task parameters. While mammoth hunting may have been a targeted activity, a locality also may be considered a type of specialized site where specific resources were extracted from certain parts of the carcass.

Bone quarrying is another specialized activity that requires fresh mammoth bone. The earliest evidence of mammoth procurement appears to be for the purpose of quarrying bone. No evidence of meat acquisition activities is seen at Cooperton; or at La Sena in the Northern Plains at the same time (Holen 1994, 1999; Holen and May 1993). Mammoth procurement for bone quarrying is not a subsistence activity, but rather a technological one whose aim is to secure raw material in transportable, usable form. Mammoth procurement for this purpose appears to have been both an activity undertaken in concert with butchering, and an independent activity.

Bone quarrying is oriented towards the production of cores and tool blanks. Although evidence of bone quarrying in the Northern Plains is absent in this database, the activity is indicated at other Northern Plains localities (Hannus 1989, 1997; Holen 1994, 1996, 1999, 2004) that span the same 7,000-year period and appear to represent individual mammoth. Mammoth bone quarrying is a North American grasslands-wide technological activity practiced over a great period of time. It depended on the cortical thickness of mammoth limb bones, and was dropped from the grasslands hunter-gatherer technological repertoire when mammoth became extinct.

Fresh bone, however, does not imply a kill. These resources remain viable for a period of time after death. Skeletal muscles and tendons are the last tissues to decompose (Gill-King 1997:97–98). Bone begins to lose moisture at the time of death. The rate of moisture loss, the length of time since death, exposure time, climatic regime, and environmental conditions influence bone fracturing ease and fracture response (Johnson 1985). Bone and other tissues from a recently dead mammoth may be as suitable as from a freshly killed one.

The difference in carcass treatment is striking, and the localities where butchery occurs fall in two distinct groups (Table 3). Modern elephant fasciae are not readily torn by hand when separating muscle bundles; a cutting tool is necessary. However, in a fresh carcass (pliable tissues with a high moisture content), resistance is relatively minor between pulling and cutting actions. An experienced hunter-butcher leaves few damage marks on the bones of a fresh carcass (e.g., Frison 1986:130, 1989:778; G. Haynes 1987). Once a carcass has stiffened, however, resistance increases significantly and a concentrated effort is needed to disarticulate and deflesh the carcass. More damage marks and multiple-stroke marks can be expected on bones from a stiffened carcass (e.g., Frison 1986:128–29; Saunders and Daeschler 1994:24; Shipman 1983, 1986; Shipman and Rose 1983).

In one set of localities (Table 3), carcass treatment is similar; these carcasses were processed while they were fresh (based on mark types, number of strokes per mark, and mark frequency). In the other set of localities (Table 3), the carcasses were treated much differently and received far more bone damage than those from the first set of localities. This difference is related to carcass condition. Carcasses in the second set of localities already were stiffened when processed, whereas those in the first set of localities were fresh.

This situation, however, does not necessarily indicate that the animals in the first set of localities were hunted and those in the second set were found carcasses that were scavenged. The mammoth in the first set of localities could have been found when they were dying or very shortly after death before becoming stiffened. The onset and duration of rigor mortis depends on a variety of factors, including ambient temperature, metabolic state at time of death, body composition, and amount of skeletal muscle (Gill-King 1997:98–99). Early decomposition encompasses the waning of rigor through bloating to the release of gases (Bass 1997:183–84; Clark et al. 1997:161; Galloway 1997:141; Gill-King 1997:99). For carcasses of modern elephants in Kenya, early decomposition takes place within 18 days, with rigor mortis setting in within a day and extensive bloating by the fourth day (Coe 1978:74–75, 84). The decay process slows in cold temperatures (Bass 1997:185; Clark et al. 1997:160; Gill-King 1997:94). The larger body mass of mammoth, coupled with cooler temperatures in the late Pleistocene, indicate a delay in the onset of rigor mortis that would have extended the period of early decomposition.

Although no method has been found to determine whether a carcass processed when fresh was that of a hunted animal or a found fresh carcass, inferences can be drawn from gender and behavior. Isolating a member of the herd would be a very difficult and dangerous undertaking; on the other hand, sexually mature males would occur as isolated individuals or in small unstructured groups (Douglas-Hamilton and Douglas-Hamilton 1975; G. Haynes 1991). The carcass of a male was likely that of a hunted animal, whereas that of an individual female was more likely a found fresh carcass. Among the first set of localities, then, Schaefer, Hebior, and Duewall-Newberry mammoth would likely be hunted males. The Domebo female would likely have been a found carcass, but whether a fresh or stiffened carcass cannot be determined owing to the absence of marks on the accessible bones. Mammoth #2 at Blackwater Draw Locality No. 1 was a found fresh carcass of undetermined gender.

The second set of localities indicates scavenging of stiffened carcasses. Either gender could be represented. At Blackwater Draw Locality No. 1, the Cotter mammoth (Saunders and Daeschler 1994:10, 24) are found stiffened male and female carcasses. Lupo (1994:835) has argued that processing a stiffened carcass is not diagnostic of scavenging (contra Binford 1981, 1984), but instead may indicate delayed butchering of a hunted animal. The focus of their discussion is on mark placement and not characteristics of different types of marks. While Lupo's (1994) point is well taken, it is difficult to assess the relationship between the marks described for the mammoth and those noted on the Hadza or Nunamiut material. However, the latter seem quite different from the multiple-stroke and leverage marks noted on the mammoth.

Carcasses from the two localities that indicate targeted hunting of small family units have different cut marks and carcass treatment profiles. At Lubbock Lake, the mammoth were fresh when butchered; at Miami, they were stiffened. Miami, an apparent anomaly, presents an unresolved dilemma. The available data are conflicting; the age profile indicates a family unit (a targeted hunting event), but the few marks on the bones indicate butchering of found stiffened carcasses (a scavenging event). Either the association of the various mammoth of appropriate ages is fortuitous (the age profile masking an accumulation rather than catastrophic death), the cultural marks are not the result of resistant tissue, the marks are not cultural, or something else not yet discerned is indicated. Depending on the size of the hunting party, butchering five mammoth could take some days. Whether Miami represents delayed butchering of hunted animals (as per Lupo [1994]) depends on various factors including climate and local environmental conditions. A cool, moist climate provides the maximum delay. The climate at the end of the Pleistocene was cool year-round. Whether the climate was cool and moist or cool and dry has been debated. A Clovis-age drought across the North American grasslands and Southwest has been postulated (C. V. Haynes 1991), with that drought as the stress mechanism that induced widespread mammoth die-off around bodies of water during the Clovis period (G. Haynes 1991). However, no geologic evidence of such a drought on the Southern High Plains has been documented (Holliday 1997, 2000). Whether the Miami playa held water at the time of the animals' death could not be determined (Holliday, pers. comm. 2000).

A mammoth carcass, even with targeted body parts, would yield a sizeable amount of meat and fatty foot pads, much more than a hunter could carry or consume before spoilage. The same situation would hold for hunters culling a small herd. Although the cool climate would delay spoilage, this surplus would need to be shared, stored, or both. To share the surplus, a sufficient number of people would have to be gathered and food prepared. The glut of meat from a small herd, or even the surplus from a single mammoth—a high-risk animal—implies planning for the hunt and for distributing the meat. Although a found carcass could not be planned and would not pose a risk from the mammoth itself, a distribution system already in place would facilitate handling a large amount of meat that became available. However, the risk would be in the happenstance of discovery and in potential competition from large carnivores. Happenstance may have been ameliorated through information sharing, knowing the distribution and habits of game animals hunted, and visual clues such as watching the sky for congregation of carrion birds (e.g., the Hadaza, O'Connell et al. 1988).

Dire wolf (*Canis dirus*) is one of the large carnivores of the late-Wisconsinan grasslands. Little is known about dire wolf behavior beyond their being "hunters and scavengers" (Anderson 1984:55). Modern wolves (*C. lupus*) prefer to kill prey but will scavenge fresh carcasses, leaving the scavenged carcass more intact than that from a kill. When food is scarce, scavenged bones may be gnawed more extensively, partially consumed, and fractured (G. Haynes 1982; Mech 1966). If modern wolf behavior is an appropriate analog for dire wolf, the behavior pattern indicates that the greatest potential conflict in terms of scavenging is over a found fresh carcass. Furthermore, results from another study (Johnson 2000) indicate that the frequency of carnivore-damaged bone at the Wisconsin localities appears high compared with later mammoth locations on the grasslands. Either the density of large carnivores is greater during the general time period represented by the Wisconsin localities than for Clovis-age sites or the frequency of carnivore scavenging is higher. This situation, if valid, suggests a reduction in the encounter rate with large carnivores over found carcasses at the end of the late Wisconsinan, and, therefore, a reduction (although by no means an elimination) in risk. As carnivores are more likely to scavenge fresh to nearly fresh carcasses, partitioning is a strategy that would minimize direct competition.

While immediate nutritional needs would be met, the group sharing in the mammoth meat would have to be sufficiently large to consume the surplus before spoilage. This requirement implies either a very large group feasting for a day, or a smaller group banqueting for several days. Mammoth procurement to acquire meat must have been an activity whose aim went beyond satisfying immediate nutritional needs. It may have represented more of a socioeconomic and political endeavor, based on the potential quantity of meat and fat obtained, planning activities, and handling (butchering, food preparation) and transport efforts. Moreover, it involved high risk in terms both of danger and of locating a solitary animal or found carcass. Even one mammoth, whether hunted or scavenged, represents an enormous effort beyond the capability of a single hunter.

To store the surplus may have required special food preparation and the effort of more than one person. Dried meat spoils in humid surroundings, and climatic conditions on the Southern and Mexican grasslands were not conducive to extended storage. On the Northern grasslands, climatic conditions made it possible to store frozen products. Colby (Frison and Todd 1986) was interpreted as a frozen meat cache. Mastodon remains found in pond deposits in the Great Lakes region were interpreted as subaqueous meat caches deposited in cold to frozen waters of late autumn to early winter (Fisher 1984, 1987, 1988, 1989, 1995; Fisher et al. 1991). Late-Wisconsinan-age peoples on the Northern grasslands then possibly stored food in frozen meat caches.

If food storage serves as a buffer against food shortage, that shortage can be potential (may or may not happen) or real (regularly occurs during that season or time of the year). An annual food shortage implies seasonal availability of food resources; in this case, food storage is a survival mechanism. The cached mastodons have the same season of death, indicating a pattern of procurement and caching as a buffer against food shortage in late winter to early spring. However, mastodon that died off naturally, with a season of death from late winter to early spring, also have been identified with equal frequency. Apparently meat was cached even when mastodon were available year-round in the region. Although meat was cached for future consumption, it was not necessarily for the purpose of surviving a food shortage. The caching of a large quantity of meat from an animal that was available year-round may indicate a planned endeavor to feed a large gathering of people at a future social event. On the other hand, year-round availability does not necessarily mean uniform quality year-round. Winter stress (e.g., cold, precipitation, forage quality and availability) may negatively affect the condition and nutritional value of large game animals (cf. Speth and Spielman 1983). Another purpose for storing food, then, may have been to maintain a level of nutritional quality through the year.

Neither the late-Pleistocene climate nor the grass communities were homogeneous over this vast grasslands. In general, summer temperatures were cooler than today by several degrees Celsius, with a lower average annual temperature regime. Annual precipitation was greater, with more available surface moisture. Perhaps the most important regional difference was in seasonality. Southern sections experienced rainy and non-rainy periods, lacking seasonal extremes and extended freezing conditions. In the Northern section, periods of winter alternated with non-winter, with seasonal depressed extremes. The Southern climatic regime was a maritime one, heavily influenced by the Gulf of Mexico; the Northern climatic regime was more continental, heavily influenced by the Laurentide ice sheet and periglacial conditions. Although the climate was not equitable across the entire grasslands, a high-diversity environment appears to have been fostered throughout the grasslands by a stable general climate that lacked or had depressed seasonal extremes with increased moisture. The same types of mammoth procurement sites occurred throughout the grasslands, for the purpose of meat-fat acquisition and bone raw material. An additional site type of meat caching may have occurred on the Northern grasslands as a food storage mechanism.

One of the most critical gaps in the late-Wisconsinan record for this construct is the lack of biological data on the age, gender, and seasonal time of death for the mammoth. A second gap is the need for detailed taphonomic data on mammoth remains. These data sets are crucial for examining the roles of hunting and scavenging in the subsistence of these early grasslands peoples, and for providing clearer insight into mammoth procurement strategies. Another gap is the absence of comparable sites, an issue of sample bias. It is not known whether the same site types existed throughout the grasslands for thousands of years and have not been discovered yet, or whether some types did not come into being until later in the period.

Concluding Remarks

This construct is built on inferences drawn from bone damage and carcass treatment. These assessments can only be made through taphonomic analysis. Just as critical biological data have not been compiled through systematic analysis, neither have comparable taphonomic analyses been undertaken in all grasslands localities where association of mammoth with people is known or suspected. Therefore this construct is a first approximation of the role of mammoth in the lifestyle of North American grasslands peoples in the late Wisconsinan. A generalist hunter-gatherer lifestyle constructed for Clovis-age Southern Plains peoples is applicable to a late-Wisconsinan lifestyle practiced across the grasslands; i.e., a broad-spectrum meat-related subsistence based on encounter hunting. The primary large game animals were herd herbivores. Targeted hunting focused on mammoth. Mammoth procurement apparently had two main purposes that could be accomplished in two ways. A mammoth carcass represented meat-related subsistence and raw material economy. Both purposes could be achieved either by hunting or by scavenging found carcasses. Either small family units or isolated individuals were hunted; either fresh or stiffened found carcasses were scavenged. The main competitors in this predator-scavenger strategy were the large carnivores that practiced a similar strategy, although not active cropping of a small herd. Competition and risks from danger may have been lessened by partitioning both carcass resources and timing of access. Food sharing occurs on a broader basis than the immediate residential group or extended family. Food storage in the Northern grasslands may have been a survival strategy but may also have served a socioeconomic and political purpose through advanced planning.

The timing of when people incorporated mammoth into the diet breadth and in which mode (hunted or scavenged) is critical to establishing the role of mammoth procurement in late-Wisconsinan lifeways. Prior to ca. 13,000 years ago, mammoth limb bones were quarried for raw material; bones lack direct evidence of butchery for meat acquisition. As mammoth became incorporated into the diet, the procurement mode became that of scavenging a found carcass. The earliest that mammoth hunting may have taken place is ca. 12,000 years ago.

In order to examine in greater depth and strengthen the construct, several postulates can be posed regarding the relationship of late-Wisconsinan grasslands peoples with mammoth. First, if an economic purpose was the first use of a mammoth carcass, then scavenging a found carcass would have yielded the highest net gain in a cost-benefit analysis. The high-diversity environment would have supplied adequate food resources so that hunting a high-risk animal ranked low or was absent from a subsistence base. Second, if a subsistence purpose was the second use of a mammoth carcass, then scavenging a found carcass again would have the highest net gain in a cost-benefit analysis. At some time between 18,000 and 13,000 years ago, mammoth meat moved up the ranked list to be included in the diet. The scavenging procurement model was a passive strategy that minimized the risk of danger and competition; its drawbacks were uncertainty and compromised nutritional quality. Third, if time period and carcass treatment differences are valid, then a change in procuring mammoth from a passive scavenging mode to an active hunting mode would have yielded the highest net gain in a cost-benefit analysis. The prey animal was the same, but the approach, investment, return, and danger were much different. A number of factors may have caused an alteration in procurement strategy. Among these might be a change in technology or technological efficiency, a change in the climate or environment, a change in the pattern or abundance of the prey animal (i.e., variability of the resource and unpredictability), a lowering of cost (an increase in encounter rate increases the benefit of hunting that animal), use of collective hunting, and an increased need for a food surplus. Lastly, if a lineal sequence is valid, then as each procurement strategy was developed and adopted, it further expanded the cumulative strategy for procuring mammoth in the repertoire of late-Wisconsinan grasslands peoples. Mammoth procurement was for economic purposes and subsistence. Hunting single mammoth or small herds and scavenging were strategies applied to both endeavors.

Regardless of how the animal was acquired and whether the carcass was fresh or stiffened, the focus on meat acquisition was the same. The easily accessible legs and up to four feet were disarticulated in order to remove sizeable muscle masses and fatty foot pads (and possibly tendons). The aim was to obtain protein-lipid resources. In this respect, late-Wisconsinan grasslands hunter-gatherers conform to a similar pattern used by modern hunter-gatherers (Kaplan and Hill 1992:201).

Nutrient quality and stress, however, are not addressed by this construct although certain aspects are implied. The quality of the protein-lipid resource deteriorates over time after death of the animal, even in a nearly or completely frozen state (Smith et al. 1978). On the other hand, a found fresh carcass and that of a hunted animal would yield comparable nutrient quality if the animals were in a similar state of health at death. The issue of nutritional stress would need to take into account not only if people were under stress but also if mammoth were, either because of short-term seasonal factors or because of long-term effects of a deteriorating climate. Such stress in mammoth would adversely affect the nutrient quality of the protein-lipid resource (Speth and Spielman 1983), which in turn would reduce the benefit to hunter-gatherers. The quality level of a carcass would be of concern to the hunter-gatherers if the purpose of procurement was for subsistence. If the purpose was for bone quarrying, then nutrient quality would not be a factor.

People were on the North American grasslands prior to the terminal Pleistocene and had developed strategies to obtain resources represented by a mammoth carcass. Scavenging mammoth carcasses appears to have been part of a standard repertoire of strategies of late-Wisconsinan peoples for at least 7,000 years throughout the grasslands. These strategies by early peoples prior to Clovis times provided a continuity with and economic link to the Clovis culture. Mammoth procurement came to an end when mammoth became extinct. That part of the late-Wisconsinan repertoire and lifestyle was abandoned. The cultural response was transformation or adoption of new strategies.

Acknowledgments

The author greatly appreciates the opportunity provided by Drs. David Overstreet (Center for Archaeological Research, Marquette University, Milwaukee) to conduct the taphonomic analysis of the southeastern mammoth localities, funded by National Science Foundation grant SBR-9708616, and Joaquín Arroyo-Cabrales, Instituto Nacional de Antropología e Historia, Mexico City) to undertake the bone technology analysis of the Tocuila mammoth locality with him; and access to the Schaefer and Mud Lake mammoths at the Kenosha Public Museum granted by Dan Joyce (Curator); sharing of information, continued interest, and encouragement by these gentlemen and Ruth Blazina-Joyce; photographic work by James

A. Clark, Jr. (Center for Archaeological Research, Marquette University, Milwaukee) and Mary Bartlett (formerly with the Museum of Texas Tech University, Lubbock); and technical assistance by Tara Johnson (Museum of Texas Tech University, Lubbock). Any errors in interpretation, however, are those of the author. General funding for this study was through the Museum of Texas Tech University. This manuscript represents part of the ongoing Lubbock Lake Landmark regional research program into grasslands adaptations of the Americas.

References Cited

Agenbroad, L. D.
1984 New World Mammoth Distribution. In *Quaternary Extinctions. A Prehistoric Revolution*, edited by P. S. Martin and R. G. Klein, pp. 90–108. University of Arizona Press, Tucson.

Arroyo-Cabrales, J., E. Johnson, and L. Morett
2001 Mammoth Bone Technology at Tocuila in the Basin of México. Paper presented at the Conference on World of the Elephant, Rome.

Arroyo-Cabrales, J., L. Morett, E. Johnson, and O. J. Polaco
2000 Tocuila (México): An Archaeological or Paleontological Site? Paper presented at the 50th Congreso Internacional de Americanistas, Varsovia, Poland.

Arroyo-Cabrales, J., O. J. Polaco, and E. Johnson
1998 Late Quaternary Environments and Mammal Faunas in México. Plenary session paper presented at the American Quaternary Association meeting, Puerto Vallarta, México.

Arroyo-Cabrales, J., O. J. Polaco, E. Johnson, and A. F. Guzman
1999 The Distribution of the Genus *Mammuthus* in México. DEINSEA, Neatherlands (in press).

Anderson, A. D.
1975 The Cooperton Mammoth: An Early Man Bone Quarry. *Great Plains Journal* 14(2):130–73.

Anderson, E.
1984 Who's Who in the Pleistocene: A Mammalian Bestiary. In *Quaternary Extinctions. A Prehistoric Revolution*, edited by P. S. Martin and R. G. Klein, pp. 40–89. University of Arizona Press, Tucson.

Arbogast, A. F., and W. C. Johnson
1994 Climatic Implications of the Late Quaternary Alluvial Record of a Small Drainage Basin in the Central Great Plains. *Quaternary Research* 41:298–305.

Baker, V. R.
1983 Late-Pleistocene Fluvial Systems. In *Late-Quaternary Environments of the United States. The Late Pleistocene*, edited by S. C. Porter, pp. 115–29. University of Minnesota Press, Minneapolis.

Barry, R. G.
1983 Late-Pleistocene Climatology. In *Late-Quaternary Environments of the United States. The Late Pleistocene*, edited by S. C. Porter, pp. 390–407. University of Minnesota Press, Minneapolis.

Bass, W. M. III
1997 Outdoor Decomposition Rates in Tennessee. In *Forensic Taphonomy. The Postmortem Fate of Human Remains*, edited by W. D. Haglund and M. H. Sorg, pp. 181–86. CRC Press, Boca Raton.

Bement, L. C.
1986 Excavation of the Late Pleistocene Deposits of Bonfire Shelter, Val Verde County, Texas. University of Texas at Austin, Texas Archeological Survey, *Archeology Series Paper* 1:1–69.

Bettinger, R. L.
1980 Explanatory/Predictive Models of Hunter-Gatherer Adaptation. In *Advances in Archaeological Method and Theory*, edited by M. B. Schiffer, 3:189–255. Academic Press, New York.

1991 *The Hunter-Gatherers. Archaeological Evolutionary Theory*. Plenum Press, New York.

Binford, L. R.
1981 *Bones. Ancient Men and Modern Myths*. Academic Press, New York.

1984 *Faunal Remains from Klasies River Mouth*. Academic Press, Orlando.

Blum, M. D., R. S. Toomey III, and S. Valastro, Jr.
1994 Fluvial Response to Late Quaternary Climatic and Environmental Change, Edwards Plateau, Texas. *Palaeogeography, Palaeoclimatology, Palaeoecology* 108:1–21.

Blumenschine, R. J., and M. Selvaggio
1991 On the Marks of Marrow Bone Processing by Hammerstone and Hyaenas: Their Anatomical Patterning and Archaeological Implications. In Cultural Beginnings. Approaches to Understanding Early Hominid Life-ways in the African Savanna, edited by J. D. Clark. *Union Internationale des Sciences Préhistoriques et Protohistoriques Monographien Band* 19:17–32.

Bonfield, W.
1975 Deformation and Fracture Characteristics of the Cooperton Mammoth Bones. *Great Plains Journal* 14:158–64.

Boyd, R., and P. J. Richerson
1985 *Culture and Evolutionary Process*. University of Chicago Press, Chicago.

Bradbury, J. P.
1989 Late Quaternary Lacustrine Paleoenvironments in the Cuenca de Mexico. *Quaternary Science Reviews* 8:75–100.

1997 Sources of Glacial Moisture in Mesoamerica. *Quaternary International* 43/44:97–110.

Brown, R. B.
1985 A Summary of Late-Quaternary Pollen Records from Mexico West of the Isthmus of Tehuantepec. In *Pollen Records of Late Quaternary North American Sediments*, edited by V. M. Bryant and R. G. Holloway, pp. 71–93. American Association of Stratigraphic Palynologists Foundation, Dallas.

Bryant, V. M. Jr.
1977 A 16,000 Year pollen Record of Vegetatational Change in Central Texas. *Palynology* 1:143–156.

Bryant, V. M., Jr., and R. G. Holloway
1985 A Late-Quaternary Paleoenvironmental Record of Texas: An Overview of the Pollen Evidence. In *Pollen Records of Late-Quaternary North American Sediments*, edited by V. M. Bryant, Jr. and R. G. Holloway, pp. 39–70. American Association of Stratigraphic Palynologists Foundation, Dallas.

Caballero-Miranda, M.
1997 The Last Glacial Maximum in the Basin of Mexico: The Diatom Record between 34,000 and 15,000 Years BP from Lake Chalco. *Quaternary International* 43/44.

Caballero, M., and B. Ortega Guerrero
1998 Lake Levels Since about 40,000 Years Ago at Lake Chalco, near Mexico City. *Quaternary Research* 50:69–79.

Calpaldo, S. D., and R. J. Blumenschine
1994 Quantitative Diagnosis of Notches Made by Hammerstone Percussion and Carnivore Gnawing on Bovid Long Bones. *American Antiquity* 59(4):724–48.

Camper, H. A.
1991 Pollen Analysis of Patschke Bog. Unpublished master's thesis, Texas A&M University, College Station.

Carlson, D. L., D. G. Steele, and A. G. Comuzzie
1984 Mammoth Excavations at the Duewall-Newberry Site on the Brazos River in Texas, 1983. *Current Research in the Pleistocene* 1:63–64.

Cashdan, E.
1990 *Risk and Uncertainty in Tribal and Peasant Economies*. Westview Press, Boulder.

Clark, M. A., M. B. Worrell, and J. E. Pless
1997 Postmortem Changes in Soft Tissues. In *Forensic Taphonomy. The Postmortem Fate of Human Remains*, edited by W. D. Haglund and M. H. Sorg, pp. 151–164. CRC Press, Boca Raton, Florida.

Coe, M.
1978 The Decomposition of Elephant Carcasses in the Tsavo (East) National Park, Kenya. *Journal of Arid Environments* 1(1):71–86.

Corona-M, E., and J. Arroyo-Cabrales
1997 New Record for the Flamingo (*Phoenicopterus* cf. *P. ruber* Linnaeus) from Pleistocene-Holocene Transition Sediments in Mexico. *Current Research in the Pleistocene* 14:137–39.

Coupland, R. T.
1992 *Natural Grasslands. Introduction and Western Hemisphere*. Elsevier, Amsterdam

Dibble, D. S., and D. Lorrain
1968 Bonfire Shelter: A Stratified Bison Kill Site, Val Verde County. Texas Memorial Museum, Austin, *Miscellaneous Papers* 1:1–138.

Dillehay, T. D.
1997 *Monte Verde. A Late Pleistocene Settlement in Chile. Volume 2: The Archaeological Context and Interpretation*. Smithsonian Institution Press, Washington, D.C.

Douglas-Hamilton, I., and O. Douglas-Hamilton
1975 *Among the Elephants*. The Viking Press, New York.

Elias, S. A., and T. R. Van Devender
1990 Fossil Insect Evidence for Late Quaternary Climatic Change in the Big Bend Region, Chihuahuan Desert, Texas. *Quaternary Research* 34:249–61.

Epps, L. W.
1973 A Geologic History of the Brazos River. *Baylor Geologic Studies Bulletin* 24:5–44.

Fisher, D. C.
1984 Taphonomic Analysis of Late Pleistocene Mastodon Occurrences: Evidence of Butchery by North American Paleo-Indians. *Paleobiology* 10:338–57.

1987 Mastodont Procurement by Paleoindians of the Great Lakes Region: Hunting or Scavenging? In *The Evolution of Human Hunting*, edited by M. H. Nitecki and D. V. Nitecki, pp. 309–421. Plenum Press, New York.

1988 Season of Death of the Hiscock Mastodonts. In Late Pleistocene and Early Holocene Paleoecology and Archeology of the Eastern Great Lakes Region, edited by R. S. Laub, N. G. Miller, and D. W. Steadman. *Bulletin of the Buffalo Society of Natural Sciences* 33:115–25.

1989 Meat Caches and Clastic Anchors: The Cryptic Record of Paleoindian Subsistence in the Great Lakes Region. Geological Society of America, *Abstracts with Program* 21:A234.

1995 Experiments on Subaqueous Meat Caching. *Current Research in the Pleistocene* 12:77–79.

Fisher, D. C., B. T. Lepper, and P. E. Hooge
1991 Taphonomic Analysis of the Burning Tree Mastodont. *Current Research in the Pleistocene* 8:88–92.

Fredlund, G. G.
1995 Late Quaternary Pollen Record from Cheyenne Bottoms, Kansas. *Quaternary Research* 43:67–79.

Fredlund, G. G., and P. J. Jaumann
1987 Late Quaternary Palynological and Paleobotanical Records from the Central Great Plains. In Quaternary Environments of Kansas, edited by W. C. Johnson, Kansas Geological Survey, *Guidebook Series* 5:167–78.

Fredlund, G. G., and L. L. Tieszen
1997 Phytolith and Carbon Isotope Evidence for Late Quaternary Vegetation and Climate Change in the Southern Black Hills, South Dakota. *Quaternary Research* 47:206–17.

Fredlund, G. G., R. B. Johnson, G. S. Porter, T. A. Revane, H. K. Schmidt, D. F. Overstreet, and M. Kolb
1996 Late Pleistocene Vegetation of Hebior Mammoth Site, Southeastern Wisconsin. *Current Research in the Pleistocene* 13:87–89.

Friedman, I.
1983 Paleoclimatic Evidence from Stable Isotopes. In *Late-Quaternary Environments of the United States. The Late Pleistocene*, edited by S. C. Porter, pp. 385–89. University of Minnesota Press, Minneapolis.

Frison, G. C.
1986 Mammoth Hunting and Butchering from a Perspective of African Elephant Culling. In *The Colby Mammoth Site. Taphonomy and Archaeology of a Clovis Kill in Northern Wyoming*, edited by G. C. Frison and L. C. Todd, pp. 115–34. University of New Mexico Press, Albuquerque.

1989 Experimental Use of Clovis Weaponry and Tools on African Elephants. *American Antiquity* 54:766–84.

Frison, G. C., and L. C. Todd (editors)
1986 *The Colby Mammoth Site. Taphonomy and Archaeology of a Clovis Kill in Northern Wyoming*. University of New Mexico Press, Albuquerque.

Galloway, A.
1997 The Process of Decomposition: A Model from the Arizona-Sonoran Desert. In *Forensic Taphonomy. The Postmortem Fate of Human Remains*, edited by W. D. Haglund and M. H. Sorg, pp. 139–50. CRC Press, Boca Raton, Florida.

Gill-King, H.
1997 Chemical and Ultrastructural Aspects of Decomposition. In *Forensic Taphonomy. The Postmortem Fate of Human Remains*, edited by W. D. Haglund and M. H. Sorg, pp. 93–108. CRC Press, Boca Raton, Florida.

González-Quintero, L.
1986 Análisis Polínicos de los Sedimentos. In Tlapacoya: 35,000 Años de Historia del Lago de Chalco, edited by J. L. Lorenzo and L. Mirambell. Instituto Nacional de Antropología e Historia, *Colección Cientifica* 155:157–66.

Graham, R. W.
1986 Taxonomy of North American Mammoths. In *The Colby Mammoth Site. Taphonomy and Archaeology of a Clovis Kill in Northern Wyoming*, edited by G. C. Frison and L. C. Todd, pp. 165–69. University of New Mexico Press, Albuquerque.

Graham, R. W., and E. L. Lundelius, Jr.
1987 Late Quaternary Mammalian Faunas and Paleoenvironments of the Southwestern Plains of the United States. In Late Quaternary Mammalian Biogeography and Environments of the Great Plains and Prairies. *Illinois State Museum Scientific Papers* 22:24–87.

1994 FAUNMAP: A Database Documenting Late Quaternary Distributions of Mammal Species in the United States. Illinois State Museum, *Scientific Papers* 25(2):1–690.

Haas, H., and J. Banewicz
1980 Radiocarbon Dating of Bone Apatite using Thermal Release of CO_2. *Radiocarbon* 22:537–44.

Hall, S. A.
1990 Channel Trenching and Climatic Change in the Southern U.S. Great Plains. *Geology* 18:342–45.

1991 Late Glacial Grassland at the Aubrey Clovis Site, North Texas. *Geological Society of America Abstracts with Program* 23(4):29.

2001 Palynology of the Aubrey Clovis Site. In The Archaeology and Paleoecology of the Aubrey Clovis Site (41DN479) Denton County, Texas, edited by C. R. Ferring, pp. 79–88. Center for Environmental Archaeology, University of North Texas, Denton. Report to the U.S. Army Corps of Engineers, Fort Worth District.

Hall, S. A., and S. Valastro, Jr.
1995 Grassland Vegetation in the Southern Great Plains during the Last Glacial Maximum. *Quaternary Research* 44:237–45.

Hannus, L. A.
1989 Flaked Mammoth Bone from the Lange/Ferguson Site, White River Badlands Area, South Dakota. In *Bone Modification*, edited by R. Bonnichsen and M. Sorg, pp. 395–412. Center for the Study of the First Americans, University of Maine, Orono.

1997 Mammoth Bone Flake Tools from the Lange/Furguson Site, South Dakota. In Proceedings of the 1993 Bone Modification Conference, Hot Springs, South Dakota, edited by L. A. Hannus, L. Rossum, and R. P. Winham. Augustana Collece, Sioux Falls, Archeology Laboratory, *Occasional Publication* 1:220–35.

Haragan, D. R.
1983 *Blue Northers to Sea Breezes: Texas Weather and Climate.* Hendrick-Long Publishing Co., Dallas.

Hassan, A. A., J. D. Termine, and C. V. Haynes, Jr.
1977 Mineralogical Studies on Bone Apatite and Their Implications for Radiocarbon Dating. *Radiocarbon* 19:364–74.

Hayden, B.
1981 Subsistence and Ecological Adaptations of Modern Hunter/ Gatherers. In *Omnivorous Primates: Gathering and Hunting in Human Evolution*, edited by R. S. O. Harding and G. Teleki, pp. 344–421. Columbia University Press, New York.

Haynes, C. V., Jr.
1975 Pleistocene and Recent Stratigraphy. In *Late Pleistocene Environments of the Southern High Plains*, edited by F. Wendorf and J. J. Hester, pp. 57–96. Fort Burgwin Research Center, Southern Methodist University, Dallas.

1991 Geoarchaeological and Paleohydrological Evidence for a Clovis-Age Drought in North America and Its Bearing on Extinction. *Quaternary Research* 35:438–50.

1995 Geochronology of Paleoenvironmental Change, Clovis Type Site, Blackwater Draw, New Mexico. *Geoarchaeology* 10:317–88.

Haynes, G.
1982 Utilization and Skeletal Disturbances of North American Prey Carcasses. *Arctic* 35(2):266–81.

1983 A Guide for Differentiating Mammalian Carnivore Taxa Responsible for Gnaw Damage to Herbivore Limb Bones. *Paleobiology* 9(2):164–72.

1987 Elephant-Butchering at Modern Mass-kill Sites in Africa. *Current Research in the Pleistocene* 4:75–77.

1991 *Mammoths, Mastodonts, & Elephants. Biology, Behavior, and the Fossil Record.* Cambridge University Press, Cambridge.

Hester, J. J.
1972 *Blackwater Locality No. 1. A Stratified, Early Man Site in Eastern New Mexico.* Fort Burgwin Research Center, Southern Methodist University, Dallas.

Hemmings, E. T.
1970 Early Man in the San Pedro Valley. Unpublished doctoral dissertation, University of Arizona, Tucson.

Holen, S. R.
1994 Sites without Lithics: Mammoth Bone Processing Sites in the Late Wisconsinan Loess of Nebraska. Paper presented at the 27th annual meeting of the Canadian Archaeological Association, Edmonton.

1996 The Lovewell Mammoth: A Late Wisconsinan Site in North-Central Kansas. *Current Research in the Pleistocene* 13:69–70.

1999 Late Pleistocene Bone Technology in the North American Mid-Continent. Paper presented at the Clovis and Beyond Conference, Santa Fe.

2004 Taphonomy of Two Last Glacial Maximum Mammoth Sites in the Central Great Plains of North America: A Preliminary Report. *Quaternary International* (in press).

Holliday, V. T.
1987 A Reexamination of Late-Pleistocene Boreal Forest Reconstructions for the Southern High Plains. *Quaternary Research* 28:238–44.

1995 Stratigraphy and Paleoenvironments of Late Quaternary Valley Fills on the Southern High Plains. *Geological Society of America Memoir* 186:1–136.

1997 *Paleoindian Geoarchaeology of the Southern High Plains.* University of Texas Press, Austin.

Holliday, V. T., C. V. Haynes, Jr., J. L. Hofman, and D. J. Meltzer
1994 Geoarchaeology and Geochronology of the Miami (Clovis) Site, Southern High Plains of Texas. *Quaternary Research* 41:234–44.

Huber, J. K., and G. Rapp, Jr.
1992 Results of a Pollen Study from Southeast Kenosha County, Wisconsin. In Archaeological Studies on the Southeast Wisconsin Uplands, edited by D. F. Overstreet. Great Lakes Archaeological Press, Milwaukee, *Case Studies in Great Lakes Archaeology* 1:74–131.

Hughes, E.
1984 Blackwater Draw Locality No. 1: A Case Study of Conservation, Collection Management, and Site Data Reconstruction Techniques. Unpublished master's thesis, Texas Tech University, Lubbock.

Imbrie, J., A. McIntyre, and T. C. Moore, Jr.
1983 The Ocean around North America at the Last Glacial Maximum. In *Late-Quaternary Environments of the United States. The Late Pleistocene*, edited by S. C. Porter, pp. 230–36. University of Minnesota Press, Minneapolis.

Johnson, E.
1985 Current Developments in Bone Technology. In *Advances in Archaeological Method and Theory*, edited by M. B. Schiffer, 8:157–235. Academic Press, New York.

1986 Late Pleistocene and Early Holocene Vertebrates and Paleoenvironments on the Southern High Plains. *Géographie Physique et Quaternaire* 40:249–61.

1987 *Lubbock Lake. Late Quaternary Studies on the Southern High Plains.* Texas A&M University Press, College Station.

1989 Human Modified Bones from Early Southern Plains Sites. In *Bone Modification*, edited by R. Bonnichsen and M. Sorg, pp. 431–72. Center for the Study of the First Americans, University of Maine, Orono.

1991 Late Pleistocene Cultural Occupation on the Southern Plains. In *Clovis: Origins and Adaptations*, edited by R. Bonnichsen and K. L. Turnmire, pp. 215–36. Center for the Study of the First Americans, Oregon State University, Corvallis.

1995 Site Formation and Disturbance Processes at Lubbock Lake (Southern High Plains, USA) during the Terminal Pleistocene. In *Ancient Peoples and Landscapes*, edited by E. Johnson, pp. 315–40. Museum of Texas Tech University, Lubbock.

2000 Taphonomy of Late Pleistocene Mammoth in Southeastern Wisconsin. Report submitted to the Great Lakes Archaeological Center, Milwaukee, Wisconsin.

Johnson, E., and V. T. Holliday
1989 Lubbock Lake: Late Quaternary Cultural and Environmental Change on the Southern High Plains. *Journal of Quaternary Science* 4(2):145–65.

1997 Analysis of Paleoindian Bonebeds at the Clovis Site: New Data from Old Excavations. *Plains Anthropologist* 42(161):329–52.

Johnson, E., P. Lewis, and R. Strauss
2000 A Measured Approach to Marks on Bones. Paper presented at the annual meeting of the American Association of Physical Anthropologists, San Antonio.

Johnson, E., L. Litwinionek, and V. T. Holliday
1994 The Sand Creek Mammoth Site, Llano Estacado of Texas. *Current Research in the Pleistocene* 11:70–72

Johnson, E., L. Morrett, and J. Arroyo-Cabrales
2001 Late Pleistocene Bone Technology at Tocuila, Basin of México. *Current Research in the Pleistocene* 18:83–85.

Johnson, E., and P. Shipman
1986 Scanning Electron Microscope Studies of Bone Modification. *Current Research in the Pleistocene* 3:47–48.

Johnson, W. C., and B. Logan
1990 Geoarchaeology of the Kansas River Basin, Central Great Plains. In Archaeological Geology of North America, edited by N. P. Lasca and J. Donahue. Geological Society of America, *Centennial Special* 4:267–99.

Joyce, D. J., and R. Blazina-Joyce
2004 A Chronology for Culturally Modified Proboscidea in Southeastern Wisconsin. *The Wisconsin Archaeologist* 85 (in press).

Kaplan, H., and K. Hill
1992 The Evolutionary Ecology of Food Acquisition. In *Evolutionary Ecology and Human Behavior*, edited by E. A. Smith and B. Winterhalder, pp. 167–201. Aldine De Gruyter, New York.

Keene, A. S.
1983 Biology, Behavior, and Borrowing: A Critical Examination of Optimal Foraging Theory in Archaeology. In *Archaeological Hammers and Theories*, pp. 137–55. Academic Press, New York.

Kelly, R. L.
1995 *The Foraging Spectrum*. Smithsonian Press, Washington, D.C.

Kreutzer, L. A.
1986 Cultural vs. Natural Bone Deposition in Feature Area 2-1, Lubbock Lake Site. Unpublished master's thesis, University of Washington, Seattle.

1988 Megafaunal Butchering at Lubbock Lake, Texas: A Taphonomic Reanalysis. *Quaternary Research* 30:221–231.

Leonhardy, F. C.
1966 Domebo: A Paleo-Indian Mammoth Kill in the Prairie-Plains. *Contributions of the Museum of the Great Plains* 1:1–53.

Lozano-García, M. Socorro, and B. Ortega-Guerrero
1998 Late Quaternary Environmental Changes of the Central Part of the Basin of Mexico; Correlation between Texcoco and Chalco Basins. *Review of Palaeobotany and Palynology* 99:77–93.

Lozano-García, M. Socorro, B. Ortega-Guerrero, M. Caballero-Miranda, and J. Urrutia-Fucugauchi
1993 Late Pleistocene and Holocene Paleoenvironments of Chalco Lake, Central Mexico. *Quaternary Research* 40:332–42.

Lundelius, E. L., Jr.
1972 Vertebrate Remains from the Gray Sand. In *Blackwater Locality No. 1. A Stratified Early Man Site in Eastern New Mexico*, edited by J. J. Hester, pp. 148–63. Fort Burgwin Research Center, Southern Methodist University, Dallas.

Lundelius, E. L., Jr., R. W. Graham, E. Anderson, J. Guilday, J. A. Holman, D. W. Steadman, and S. D. Webb
1983 Terrestrial Vertebrate Faunas. In *Late-Quaternary Environments of the United States. The Late Pleistocene*, edited by S. C. Porter, pp. 311–53. University of Minnesota Press, Minneapolis.

Lupo, K. D.
1994 Butchering Marks and Carcass Acquisition Strategies: Distinguishing Hunting from Scavenging in Archaeological Contexts. *Journal of Archaeological Science* 21(6):827–37.

Lyman, R. L.
1994 *Vertebrate Taphonomy*. Cambridge University Press, Cambridge.

Maglio, V. J.
1973 Origin and Evolution of the Elephantidae. *Transactions of the American Philosophical Society of Philadelphia, New Series* 63(3):1–149.

May, D. W., and S. R. Holen
1993 Radiocarbon Ages of Soils and Charcoal in Late Wisconsinan Loess, South-Central Nebraska. *Quaternary Research* 39:55–58.

Madole, R. F.
1991 Colorado Piedmont Section. In Quaternary Nonglacial Geology: Conterminous U.S., edited by R. B. Morrison. Geological Society of America, *The Geology of North America* K-2:456–462.

Maher, L. J.
1982 The Palynology of Devils Lake, Sauk County, Wisconsin. Quaternary History of the Driftless Area, pp. 119–135. University of Wisconsin Extension, Geological and Natural History Survey, Madison.

McAndrews, J. H.. and L. J. Jackson
1988 Age and Environment of Late Pleistocene Mastodont and Mammoth in Southern Ontario. In Late Pleistocene and Early Holocene Paleoecology and Archeology of the Eastern Great Lakes Region, edited by R. S. Laub, N. G. Miller, and D. W. Steadman. *Bulletin of the Buffalo Society of Natural Sciences* 33:161–72.

Mech, L. D.
1966 The Wolves of Isle Royale. U.S. National Park Service, Fauna of the National Parks of the United States, *Fauna Series* 7:1–210.

Metcalfe, S. E., F. A. Street-Perrott, R. A. Perrott, and D. D. Harkness
1991 Palaeolimnology of the Upper Lerma Basin, Central Mexico: A Record of Climatic Change and Anthropogenic Disturbance since 11,600 yr B.P. *Journal of Paleolimnology* 5:197–218.

Metcalfe, S. E., A. Bimpson, A. J. Courtice, S. L. O'Hara, and D. M. Taylor
1997 Climate Change at the Monsoon/Westerly Boundary in Northern Mexico. *Journal of Paleolimnology* 17:155–171.

Metcalfe, S. E., S. L. O'Hara, M. Caballero, and S. J. Davies
2000 Records of Late Pleistocene-Holocene Climatic Change in Mexico—A Review. *Quaternary Science Reviews* 19:699–721.

Morett, L. A., J. Arroyo Cabrales, and O. J. Polaco
1998a Tocuila, A Remarkable Mammoth Site in the Basin of Mexico. *Current Research in the Pleistocene*15:118–20.

1998a El Sitio Paleontológico de Tocuila. *Arqueología Mexicana,* 5(30):57.

Nordt, L. C., T. W. Boutton, C. T. Hallmark, and M. R. Waters
1994 Late Quaternary Vegetation and Climate Changes in Central Texas Based on the Isotopic Composition of Organic Carbon. *Quaternary Research* 41:109–20.

O'Connell, J., K. Hawkes, and B. Jones
1988 Hadza Scavenging: Implications for Plio-Pleistocene Hominid Subsistence. *Current Anthropology* 29:356–63.

Olsen, S. L., and P. Shipman
1988 Surface Modification on Bone: Trampling versus Butchery. *Journal of Archaeological Science,* 15:535–53.

Overstreet, D. F.
1996 Still More on Cultural Contexts of Mammoth and Mastodon in the Southwestern Lake Michigan Basin. *Current Research in the Pleistocene* 13:36–38.
1998 Late Pleistocene Geochronology and the Paleoindian Penetration of the Southwestern Lake Michigan Basin. *The Wisconsin Archeologist* 79(1):28–52.

Overstreet, D. F., and T. W. Stafford, Jr.
1997 Additions to a Revised Chronology for Cultural and Non-Cultural Mammoth and Mastodon Fossils in the Southwestern Lake Michigan Basin. *Current Research in the Pleistocene* 14:70–71

Overstreet, D. F., D. J. Joyce, K. F. Hallin, and D. Wasion
1993 Cultural Contexts of Mammoth and Mastodont in the Southwestern Lake Michigan Basin. *Current Research in the Pleistocene* 10:75–77.

Porter, S. C. (editor)
1983 *Late-Quaternary Environments of the United States. The Late Pleistocene.* University of Minnesota Press, Minneapolis.

Reeves, C. C., Jr.
1965 Pleistocene Climate of the Llano Estacado. *Journal of Geology* 73:181–188.
1966 Pleistocene Climate of the Llano Estacado II. *Journal of Geology* 74:642–47.

Rousseau, D. D., and G. E. Kukla
1994 Late Pleistocene Climate Record in the Eustis Loess Section, Nebraska, Based on Land Snail Assemblages and Magnetic Susceptibility. *Quaternary Research* 42:176–87.

Ruhe, R. V.
1983 Depositional Environment of Late Wisconsin Loess in the Midcontinental United States. In *Late-Quaternary Environments of the United States. The Late Pleistocene,* edited by S. C. Porter, pp. 130–37. University of Minnesota Press, Minneapolis.

Saunders, J. J.
1980 A Model for Man-Mammoth Relationships in Late Pleistocene North America. *Canadian Journal of Anthropology* 1:87–98.

Saunders, J. J., and E. B. Daeschler
1994 Descriptive Analyses and Taphonomical Observations of Culturally-Modified Mammoths Excavated at "The Gravel Pit," near Clovis, New Mexico in 1936. *Proceedings of the Academy of Natural Sciences of Philadelphia* 145:1–28.

Schmutz, E. M., E. L. Smith, P. R. Ogden, M. L. Cox, J. O. Klemmedson, J. J. Norris, and L. C. Fierro
1992 Desert Grassland. In *Natural Grasslands. Introduction and Western Hemisphere,* edited by R. T. Coupland, pp. 337–62. Elsevier, New York.

Sellards, E. H.
1938 Artifacts Associated with Fossil Elephant. *Bulletin of the Geological Society of America* 49:999–1010.
1952 *Early Man in America.* University of Texas Press, Austin.

Shipman, P.
1981a Applications of Scanning Electron Microscopy to Taphonomic Problems. In The Research Potential of Anthropological Museum Collections, edited by J. B. Griffin and N. A. Rothschild. *Annals of the New York Academy of Science* 376:357–85.
1981b *Life History of a Fossil: An Introduction to Taphonomy and Paleoecology.* Harvard University Press, Cambridge.
1983 Early Hominid Lifestyles: Hunting and Gathering or Foraging and Scavenging. In Animals and Archaeology: Hunters and Their Prey, edited by J. Clutton-Brock and C. Grigson. *British Archaeological Reports International Series* 163:31–49.
1986 Scavenging or Hunting in Early Hominids: Theoretical Framework and Test. *American Anthropologist* 88:27–43.
1988 Actualistic Studies of Animal Resources and Hominid Activities. In Scanning Electron Microscopy in Archaeology, edited by S. L. Olsen. *British Archaeological Reports International Series* 452:261–85.
1989 Altered Bones from Olduvai Gorge, Tanzania: Techniques, Problems, and Implications of Their Recognition. In *Bone Modification,* edited by R. Bonnichsen and M. H. Sorg, pp. 317–34. Center for the Study of the First Americans, University of Maine, Orono.

Shipman, P., and J. J. Rose
1983 Early Hominid Hunting, Butchering, and Carcass-Processing Behaviors: Approaches to the Fossil Record. *Journal of Anthropological Archaeology* 2:57–98.
1984 Cutmark Mimics on Modern and Fossil Bovid Bones. *Current Anthropology* 25:116–17.

Siebe, C., P. Schaaf, and J. Urrutia-Fucugauchi
1999 Mammoth Bones Embedded in a Late Pleistocene Lahar from Popocatépetl Volcano, near Tocuila, Central México. *Geological Society of America Bulletin* 111(10):1550–62.

Sims, P. L.
1988 Grasslands. In *North American Terrestrial Vegetation,* edited by M. G. Barbour and W. D. Billings, pp. 265–86. Cambridge University Press, Cambridge.

Smith, E. A.
1991 Risk and Uncertainty in the "Original Affluent Society": Evolutionary Ecology of Resource-Sharing and Land Tenure. In *Hunters and Gatherers 1. History, Evolution and Social Change,* edited by T. Ingold, D. Riches, and J. Woodburn, pp. 222–51. Berg, New York.

Smith, E. A., and B. Winterhalder
1992 Natural Selection and Decision-Making: Some Fundamental Principles. In *Evolutionary Ecology and Human Behavior,* edited by E. A. Smith and B. Winterhalder, pp. 25–60. Aldine De Gruyter, New York.

Smith, G. I., and F. A. Street-Perrott
1983 Pluvial Lakes of the Western United States. In *Late-Quaternary Environments of the United States. The Late Pleistocene,* edited by S. C. Porter, pp. 190–212. University of Minnesota Press, Minneapolis.

Speth, J. D., and K. Spielmann
1983 Energy Source, Protein Metabolism, and Hunter-Gatherer Subsistence Strategies. *Journal of Anthropological Archaeology* 2(1):1–31.

Steele, D. G., and D. L. Carlson
1989 Excavation and Taphonomy of Mammoth Remains from the Duewall-Newberry Site, Brazos County, Texas. In *Bone Modification,* edited by R. Bonnichsen and M. Sorg, pp. 413–30. Center for the Study of the First Americans, University of Maine, Orono.

Todd, N. E., and V. L. Roth
1996 Origin and Radiation of the Elephantidae. In *The Proboscidea. Evolution and Palaeoecology of Elephants and Their Relatives*, edited by J. Shoshani and P. Tassy, pp. 193–202. Oxford University Press, Oxford.

Toomey, R. S., M. D. Blum, and S. Valastro, Jr.
1993 Late Quaternary Climates and Environments of the Edwards Plateau, Texas. *Global Planetary Change* 7:299–320.

Torrence, R.
1983 Time-Budgeting and Hunter-Gatherer Technology. In *Prehistoric Hunters and Gatherers in Europe*, edited by G. Bailey, pp. 11–22. Cambridge University Press, New York.

1989 *Time, Energy, and Stone Tools*. Cambridge University Press, New York.

Van Devender, T. R.
1986 Pleistocene Climates and Endemism in the Chihuahuan Desert Flora. In *Second Symposium on Resources of the Chihuahuan Desert Region, United States and Mexico*, edited by J. C. Barlow, A. M. Powell, and B. N. Timmerman, pp. 1–19. Desert Research Institute, Alpine, Texas.

1990 Late Quaternary Vegetation and Climate of the Chihuahuan Desert, United States and Mexico. In *Packrat Middens. The Last 40,000 Years of Biotic Change*, edited by J. L. Betancourt, T. R. Van Devender, and P. S. Martin, pp. 104–33. University of Arizona Press, Tucson.

Van Devender, T. R., and T. L. Burgess
1985 Late Pleistocene Woodlands in the Bolson de Mapimi: A Refugium for the Chihuahuan Desert Biota? *Quaternary Research* 24:346–53.

Waters, M. R., and L. C. Nordt
1995 Late Quaternary Floodplain History of the Brazos River in East-Central Texas. *Quaternary Research* 43:311–19.

Wilson, L. R.
1966 Palynology of the Domebo Site. In Domebo. A Paleo-Indian Mammoth Kill in the Prairie-Plains, edited by F. C. Leonardy. *Contributions of the Museum of the Great Plains* 1:44–50.

Winterhalder, B., and C. Goland
1997 An Evolutionary Ecology Perspective on Diet Choice, Risk, and Plant Domestication. In *People, Plants, and Landscapes*, edited by K. J. Gremillion, pp. 123–60. University of Alabama Press, Tuscaloosa.

Winterhalder, B., and E. A. Smith
1981 *Hunter-Gatherer Foraging Strategies*. University of Chicago Press, Chicago.

1992 Evolutionary Ecology and the Social Sciences. In *Evolutionary Ecology and Human Behavior*, edited by E. A. Smith and B. Winterhalder, pp. 3–23. Aldine De Gruyter, New York.

Wright, H. A., and A. W. Bailey
1982 *Fire Ecology*. Wiley, New York.

Late-Glacial Ice-Marginal Adaptation in Southeastern Wisconsin

David F. Overstreet

Abstract

Human-mammoth interaction in southeastern Wisconsin has been radiocarbon-dated to between 13,500 and 12,500 RCYBP. Chronology indicates that local populations were occupying and ostensibly had adapted to an ice-marginal habitat. Previous Great Lakes Paleoindian studies have not focused on ice-marginal habitats, as they were formerly thought to be devoid of human presence. Existing models of Paleoindian adaptive strategies for the Great Lakes area are reviewed, the ongoing research at four mammoth processing sites and Chesrow complex habitation sites in geographic proximity are summarized, and alternative ice-marginal strategies are presented. It is also suggested that the Chesrow complex should not be considered unique, but rather is one component of a widespread late-glacial unfluted biface tradition in the eastern United States and Canada.

Introduction

Little attention has been directed to developing models of human adaptation to late-Pleistocene ice-marginal habitats in the Great Lakes region. In large part this is a function of historical accident. Also of critical importance has been the strength of paradigms developed in the North American grassland ecosystem. For example, characterizations of Great Lakes Paleoindians as big-game hunters were conveniently imported and regionally applied. In the American southwest, such characterization was based on firm data from numerous sites including Clovis, Domebo, Lehner, Naco, and others (Cotter 1937; Haury et al. 1953, 1959; Leonhardy 1966; Sellards 1938). This was not the case in the Great Lakes, but geographic proximity of fluted points and proboscidian remains proved sufficient. In spite of any contextual data, the "Mason-Quimby line" as an index of human-mammoth involvement was widely accepted (Mason 1958; Quimby 1958, 1960).

Following several decades of prominence for the "Mason-Quimby line," emphases shifted; and more contemporary models focused on the distribution of high-quality toolstone and the assumption that caribou rather than megafauna, the Kimmswick and Boaz mastodons notwithstanding (Graham et al. 1981; Palmer and Stoltman 1976), were key to understanding and explicating Paleoindian adaptive strategies. Exploitative territories were defined by distance from quarry sites, and caribou were said to be the focus of subsistence (cf. Deller and Ellis 1988, 1992; Gramly 1982; Storck 1984, 1988). Much was made of the single caribou element from the Holcombe Beach site (Cleland 1965). MacDonald (1968), Funk et al. (1970), Funk (1972, 1976), Ritchie and Funk (1973), Cleland (1976), Gramly (1982), and Curran and Grimes (1989), among others, in one way or another, all buttressed a Great Lakes–region model of caribou-reliant Paleoindian economies.

Until quite recently (e.g., Shott 1993; Storck and Speiss 1994), Great Lakes Paleoindian adaptive strategies, all in a post-11,500 RCYBP framework, paid homage to the toolstone-defined territory in which caribou provided primary sustenance. Shott succinctly states: "Concluding that caribou probably were an economic staple of the economy is neither original nor satisfactory as a characterization of the Paleo-Indian adaptation" (1993: 13). So, too, does he address the limitations of the toolstone = territory assumption:

> Accurate raw material identification is especially important however, in Paleo-Indian studies, both because eastern North American Paleo-Indians typically exhibited strong and specific preferences in stone (Ellis 1984:38–51, 1989; Goodyear 1989) and because distances between natural and archaeological occurrences provide a minimum measure of distance traveled by groups in their annual

Center for Archaeological Research, Marquette University, 1659 N. Jackson Street, Milwaukee, WI 53202; e-mail: david.overstreet@marquette.edu rounds (Binford 1979; Hester and Grady 1977; Storck 1983; Tankersley 1990). These measures should not be identified with territory size, however, since forager mobility practices are sufficiently complex and organized to seriously complicate the equation (Shott 1986a: 141) and chert procurement can be 'embedded' (Binford 1979:259) in other activities and movements (Shott 1993:14).

It is apparent from this brief summary that traditional models of Paleoindian adaptation in the region would be of limited utility as a point of departure to develop even rudimentary frameworks for interpreting humankind-land relationships in the late-Pleistocene ice-marginal habitat. Of paramount importance of course are the chronological parameters, and it is assumed here that the ice-front positions at ca. 13,500 and 12,500 RCYBP are the defining north limits of the habitat under investigation. Previous models of adaptation fall short of this interval by at least a millennium. They are flawed also in the contexts of caribou reliance and the toolstone = territory perspectives.

The taxon under consideration, the Chesrow complex (Overstreet 1993a), is viewed by the author as a late-Pleistocene Paleoindian manifestation situated in close proximity to the frontal position of the continental glacier. Chesrow, or more correctly, the mammoth carcasses with which the complex is inferred to be associated, have a bimodal chronometric distribution (Overstreet 1998). The distribution of these radiocarbon dates derived from purified bone collagen is not viewed as sampling error, but rather as a reflection of rapidly reorganizing ecosystems in the study area at the end of the last ice age. Over the short course of a few centuries, for example, ice-front positions of the Lake Michigan lobe of the continental glacier retreated and advanced as much as 515–595 km (320–370 miles). Inundation of inhabited landscapes by glacial meltwaters, burial by till and outwash, reorganization of drainage systems, and temperature averages and extremes likely had effects equal to or of greater profundity than the position of the ice front. How earliest Paleoindians utilized these rapidly changing landscapes is a question as difficult to understand as the ecological dynamics at play in the Great Lakes at the terminal Pleistocene. But utilize them they did, first at about 13,500 RCYBP and subsequently, following an approximate 1,000-year hiatus, at about 12,500 RCYBP.

By necessity, these questions call for abandoning traditional paradigms and developing new approaches. Fraught with many significant limitations, as are all current Great Lakes Paleoindian datasets, I do so with great caution, objectivity, and with considerable trepidation. At the same time, at least in view of my past experience, this is a less arduous and more rewarding task than attempting to integrate the new data into the frameworks of the old paradigms. To this end, the ensuing narrative provides background to the current research, a frank appraisal of recognized limitations (my own as well as those stated by others), and presentation of a model that will serve to guide future research.

Research History

Archaeology, Geomorphology, and Taphonomy

The Chesrow complex was conceptually fabricated as a Great Lakes Paleoindian manifestation of uncertain origin, distribution, and age (Overstreet 1993a, 1993b) (Figure 1). Excavations at the type station and nearby components such as the Lucas, Scott, Cabbage Patch, and WisPark XXXI sites revealed inter-

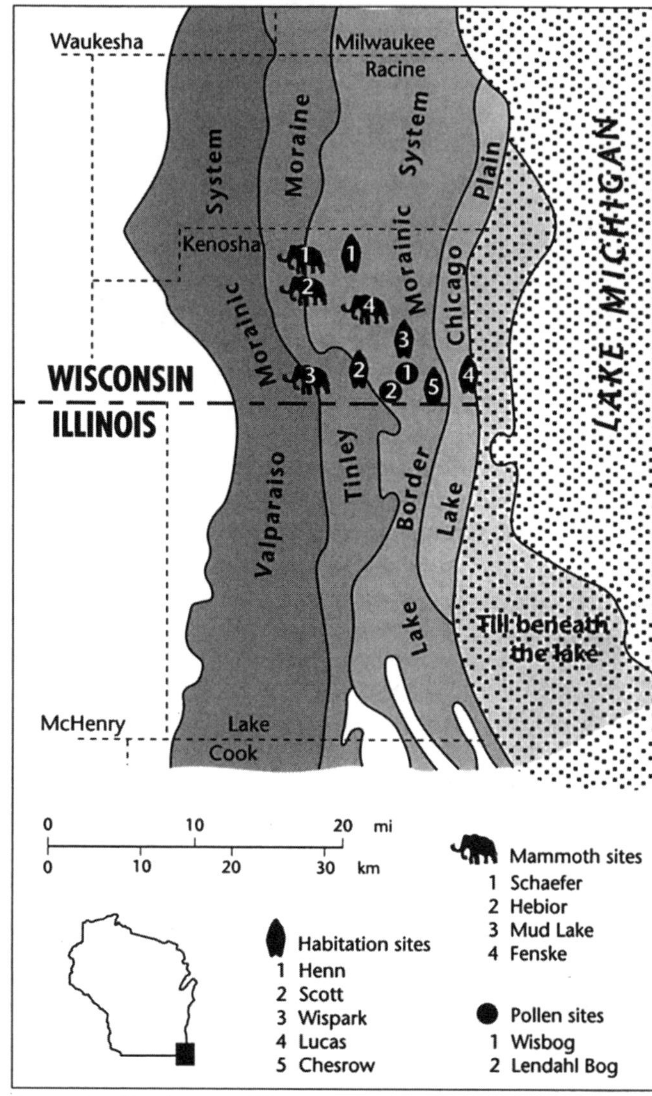

Figure 1. Southeastern Wisconsin locations of mammoth sites, Chesrow complex habitation sites, and pollen recovery sites.

nally consistent lithic assemblages, whose hallmark was a rather thick basally thinned or sometimes quasi-fluted, frequently heat-treated projectile point (Figure 2). Basal and/or lateral edge grinding occurs inconsistently and is nowhere as prominent or regular as in other Great Lakes Paleoindian lithic assemblages. Other tool types in the assemblages include delicate gravers, stone awls, end scrapers (rarely), bifacial drills, ad hoc flake tools best described as "cutters," "pseudo limaces," and elongate hammerstones. Notably, all are rendered from locally available, very poor quality toolstone derived from cobble-sized residual cherts in the local till. Exotic toolstone is virtually nonexistent. Raw materials were thus available wherever the till was exposed locally along the shores of local streams, rivers, and lakeshores. There is no evidence of on-anvil or bipolar reduction techniques, and the

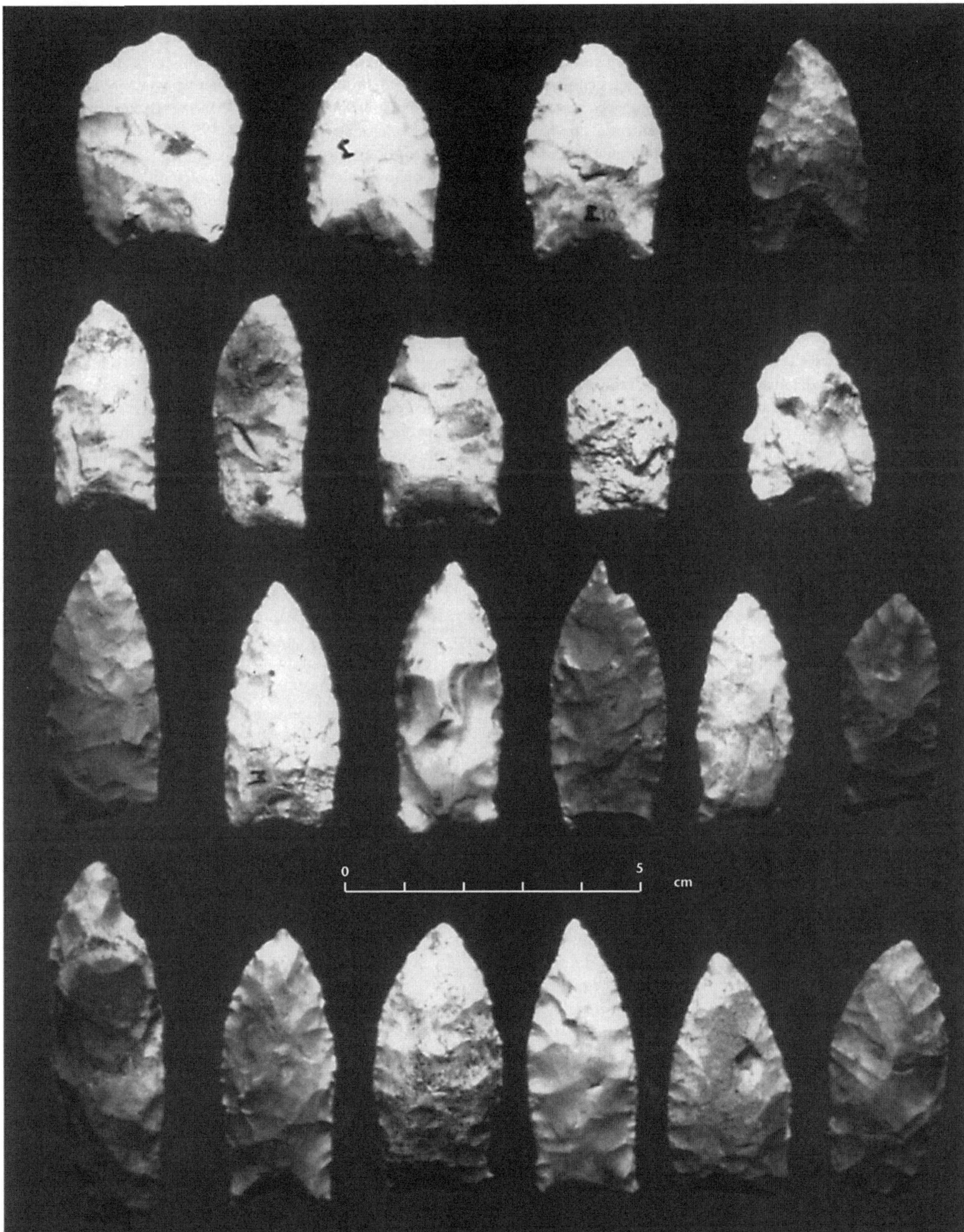

Figure 2. Chesrow-complex projectile points, Scott site (47 Kn 197).

debitage is devoid of bipolar cores (so-called wedges) and bipolar flakes.

There is considerable heterogeneity in the sample of Chesrow projectile points (Figure 2). Some appear to be "waisted" or weak-shouldered; others, particularly shorter reworked specimens, have a subtle bilateral declivity almost reminiscent of a notch. The basal configuration is often but not always concave. This form is likely a function of attempts at basal thinning; and the process often leaves projecting ears, upon which remnants of striking platforms are sometimes present.

Because the Chesrow site is situated on the lakeward flank of a Calumet Stage beach, the site was initially assumed to postdate 11,800–11,200 RCYBP (Hansel et al. 1985; Schneider and Hansel 1990), the age of the beach strand representing the maximum possible age of the complex. As analyses proceeded and other sites were investigated during the late 1980s through the early 1990s, contradictory data left this initial geochronological estimate for the Chesrow complex open to alternative interpretations. Bretz (1951, 1959) had suggested the water plane of glacial Lake Chicago was at an elevation of 189 m (620 ft) above mean sea level (amsl) on two occasions, with Calumet I sometime subsequent to the Glenwood high ca. 12,900–12,700 RCYBP and prior to Calumet II, ca. 11,800–11,200 RCYBP. A water-rolled drill from the Chesrow site could point to occupation during the Calumet I era (Overstreet 1998). This reasoning was also affected by analyses of the landscape position of the nearby Lucas site at an elevation of less than 184 m (605 ft) amsl. At this elevation Lucas would have been inundated during the Calumet II stage and obviously could not have been contemporaneous with Chesrow if the latter were inhabited during Calumet II times (11,800–11,200 RCYBP).

Reaction to the suggested chronology for the Chesrow complex (Overstreet 1993a) was prompt and generally unfavorable. Mason (1995), primarily on typological grounds, stated:

> In my view, quite apart from the venture in 'paleosociology,' the Chesrow complex (assuming it is a valid taxon and not a collapsed or mixed amalgam of disparate ages) ought to fit near the end of the Paleo-Indian tradition, perhaps falling temporally into the Early Archaic period. But not all the evidence points in the same direction, as Overstreet painstakingly makes clear (1995: 241).

Stoltman (1995), who also was not particularly enamored of an early chronology for the Chesrow complex, rejected the possibility of Clovis contemporaneity and characterized the complex as the "last gasp" of the fluted-point tradition. This chronological assessment was contradicted by the results of purified collagen accelerator mass spectrometer (AMS) radiocarbon assays from four local mammoth bonebeds (Dallman et al. 1996; Overstreet and Stafford 1997).

During the 1990s two mammoth sites, Schaefer and Hebior, were excavated. Both were intimately associated with chipped-stone tools (Overstreet 1996, 1998; Overstreet et al. 1993, 1995). Bifacial implements associated with the two mammoths, however, were diagnostic of no particular archaeological culture. Two additional sites, Fenske and Mud Lake, had yielded bones, identified as either mammoth or mastodon, that had been curated at local museums for many years. Preliminary excavations indicated depositional contexts very similar to those at the Schaefer and Hebior sites. Taphonomic analyses of the two complete and two incomplete mammoth were carried out by E. Johnson (2000, this volume; Johnson et al. 2000). Putative butchery marks (Overstreet 1996; Overstreet et al. 1993, 1995) were confirmed by her detailed taphonomic analyses.

Microwear analyses of chipped-stone tools associated with the Schaefer and Hebior mammoths, along with other selected bifacial implements from the Chesrow site, were also carried out (Yerkes and Weinberger 1997). As a result of this analysis, two bifaces associated with the Hebior mammoth were found to show meat and hide polish (Yerkes and Weinberger 1997).

Stratigraphic and geomorphic contexts at the Schaefer and Hebior mammoth sites are straightforward. Analyses of sediments and landscape position (Overstreet and Kolb 2003), supplemented by E. Johnson's taphonomic analyses (2000), support depositional integrity of both bonebeds; both were deposited in low-energy environments with no evidence of water transport or other processes of translocation. Elements in both bone piles, and those of the partial Fenske and Mud Lake specimens, show varying degrees of solution pitting, root etching, trampling, carnivore gnawing, and human modification in the form of cut and pry marks. Compiling additional data from excavation, specialized analyses, and radiocarbon dating fostered modifications in the initial responses taken by Stoltman (1995) and Mason (1995). Mason (1997), armed with new data from the ongoing research, adopted a decidedly different view of the Chesrow-mammoth construct from that espoused in his earlier pronouncement:

> The unresolved questions about the taxonomic integrity, age, cultural affiliation, and the geological and faunal associations of the Chesrow complex should not detract from the importance its discovery holds for Wisconsin prehistory and Paleo-Indian studies in a much wider context. On the contrary, not only has this research yielded the only solid instances of mammoth-mastodon-human interaction within the boundaries of the state, the distinctive cultural remains, whether or not they all relate to a single archaeological complex as has been proposed, increase the known variability of Late Pleistocene-Early Holocene artifacts. Of major interest is the metrically signaled increase in the time depth of the regional archaeological record, and, if corroborated by other relevant information, its consequential implications for a pre-Clovis occupation (1997:97–98).

Stoltman's (1995) position was also significantly ameliorated with the advent of additional data. In a more contemporary review of Paleoindian adaptive strategies, Stoltman noted:

> Recent research in Kenosha County has added a new dimension to the problem of dating the Chesrow/Price complex. From the time of its first formal recognition, Overstreet (1993b:82–90), noting the absence of Clovis and other fluted points in the immediate vicinity, has cautiously advocated the Chesrow complex, whatever its absolute age, as the most likely local complex to have produced cut marks and other signs of butchery observed on local mammoth and mastodon skeletons (e.g. Overstreet 1996; Overstreet et al. 1993; Overstreet et al. 1995). As this research in Kenosha County has progressed in the early 1990s, the recovery of definite (though not culturally diag-

> nostic) stone artifacts in association with mammoth remains at the Schaefer (Overstreet et al. 1995) and Hebior sites (Overstreet 1996) has heightened interest in the Chesrow complex. Complicating the issue is the recent release of what appear to be highly reliable (i.e., AMS) radiocarbon dates that place the two apparently butchered mammoths in a reliable temporal context, i.e., the interval 12,000–12,500 RCYBP (Overstreet and Stafford 1997). The surprising implications of these new radiocarbon assays, assuming all the available evidence can be accepted at face value, is that humans must have been on the landscape of southeastern Wisconsin in pre-Clovis times! Needless to say, these are, indeed, important and exciting implications, and, if Overstreet and his colleagues are correct in their assumptions, they raise the possibility for serious consideration that the presumed Late Paleoindian Chesrow complex may, in fact, be a pre-Clovis phenomenon (1998: 63–64).

And, in conclusion, he states:

> The most exciting new discoveries pertaining to the late glacial human settlement in Wisconsin derive from the interdisciplinary research coordinated by David Overstreet in Kenosha County. Starting with the recognition of suspected cut marks on mammoth and mastodon bones in the collections of the Kenosha Museum in the 1990's, subsequent field work has uncovered two mammoth carcasses with which probable stone tools (although none is so far culturally diagnostic) have been found in direct association (Overstreet et al. 1993, 1995). Most exciting about these finds, however, is their recent direct dating (AMS dating of bone collagen) that place them in the interval 12,000–12,500 RCYBP (Overstreet and Stafford 1997), raising the formerly unthinkable possibility that there was a pre-Clovis occupation of Wisconsin. Who could these pre-Clovis people be? Overstreet and his colleagues suspect that the Chesrow complex fills the bill. Isn't it interesting how the unthinkable can suddenly become plausible (1998: 64)?

I am indebted to both Ronald Mason and James Stoltman, first for their staunch conservatism in defending existing paradigms. As it should be, both have demanded extraordinary evidence with multiple lines converging in the same direction before they would seriously entertain modifications to Great Lakes Paleoindian time-space systematics. More importantly, however, both have served as sounding boards and willingly have aided in identifying data needs; but both, once those needs were met, accepted new directions with scholarly excitement. In no instance has exploring these new directions been met with the acrimony that sometimes confronts nontraditional directions in Paleoindian studies.

Radiocarbon Chronology

More than 15 radiocarbon assays have been run either on bone from the Hebior, Schaefer, Mud Lake, and Fenske mammoths, or on wood found within the Schaefer bone pile (Table 1); and nine others are pending (D. J. Joyce, pers. comm. 2000). During the 1994 fieldwork at the Schaefer site, two samples of spruce, and one sample consisting of bone fragments from a femur that had been shattered during drain tile installation, were submitted for standard radiocarbon dating. These preliminary assays had two distinct purposes. First, we sought to determine if sufficient collagen had been preserved in the bone for dating purposes. Second, we sought to determine ostensible contemporaneity of the wood associated with the bone pile and the mammoth remains. I had noted, while working at Lendhal bog (see Figure 1), that the bog's former proglacial lake shoreline was thickly littered with wood that had been rafted up at the excavation locality. My assumption was that the wood in association with the Schaefer mammoth might very well have been rafted up at that locality in much the same manner. Kaiser (1994) has described this process for the buried Two Creeks forest, indicating that rising waters drowned segments of the forest prior to ice advance and till deposition. The results of the assays on spruce at Schaefer were 12,220 ± 100 RCYBP (Beta 62823) and 12,480 ± 130 RCYBP (Beta 62824). The standard bone collagen assay yielded a date of 10,960 ± 100 RCYBP (Beta 62822).

The disparity was troubling. First, the age of the Schaefer mammoth, based on standard bone collagen radiocarbon assay, was considered too recent, as the species had likely been extirpated from the region sometime prior to 11,000 RCYBP. Second, it seemed untenable that more than 1,300 years could have elapsed between deposition of the bone pile and the wood in such intimate association. I contacted Thomas W. Stafford Jr., then at INSTAAR, University of Colorado–Boulder, to inquire about the magnitude of error between standard collagen dates and the AMS techniques utilizing purified collagen. He informed me that the date was probably in error by at least 1,000 years (Stafford 1990; Stafford, pers. comm. 1995). Samples submitted to Stafford for collagen extraction and subsequent dating at the Lawrence Livermore National Laboratory confirmed his predictions (Table 1). The Schaefer mammoth has a tight cluster of three assays at about 12,300 RCYBP. Hebior is slightly older, again tightly clustered at approximately 12,500 RCYBP. Fenske and Mud Lake are older still, with an assay average of about 13,500 RCYBP (Table 1). Taking the dates at face value, they are well beyond the range of Clovis and likely too old for the open spruce woodland said to persist in the locality until about 11,000 RCYBP.

Paleoecology

Local pollen data are largely uninformative for eras prior to 11,000–12,000 RCYBP. Analyses at several stations, namely Volo bog (King 1981) in northeastern Illinois, Wisbog (Huber and Rapp 1992), sites in south-central Wisconsin (Winkler 1985, 1988), and even sediments taken from intimate association with the Hebior mammoth (Fredlund et al. 1996) all speak to spruce dominance at the Pleistocene-Holocene transition. In spite of these studies, defining the limits of the ice-marginal habitat is hindered by significant data limitations. Furthermore, there is considerable disagreement regarding the composition, aerial extent, or even the existence of a tundra habitat. Finally, regional habitat reconstruction has relied almost exclusively on fossil pollen, with no systematic integration of data derived from macrofossils. Guthrie (1985; 1990) has cited limitations of this univariate approach to habitat reconstruction. For example, he notes:

> Thus, pollen-based portraits of the plant community in the Alaskan Pleistocene have been running at right angles to the clear pattern derived from macrofossils (Guthrie 1985). Although rarely the case, in this instance fossil mammals are a better indicator group of relative habitat distribution than pollen because the pollen rain from vegetation resorting into stripes may not reveal changes in habitat diversity (1990:51).

Table 1. ^{14}C dates from southeastern Wisconsin mammoth sites.[1]

Site/specimen	Standard ^{14}C date (RCYBP)	Lab no.	AMS date (RCYBP)▲	Lab no.	Tools present?
Schaefer mammoth, 47 Kn 252	10,960 ± 100, bone collagen	Beta-62822	12,310 ± 60	CAMS-30171	yes
	12,220 ± 80, spruce sample 4	Beta-62823	12.320 ± 50	CAMS-61135■	
	12,480 ± 130, spruce sample	Beta-62824	12.390 ± 40	CAMS-61143■	
	12,610 ± 80, unidentified wood	Beta-137412■			
	12,280 ± 80, unidentified wood	Beta-137413■			
Hebior mammoth, 47 Kn 265			12,590 ± 50	CAMS-61137	yes
			12,480 ± 60	CAMS-28303	
			12,520 ± 50	CAMS-24943	
Mud Lake mammoth, 47 Kn 246			13,440 ± 60	CAMS-36643	unknown
			13,530 ± 50	CAMS-61136	
Fenske mammoth, 47 Kn 240			13,470 ± 50	CAMS-36642	unknown
			13,510 ± 50	CAMS-61138	unknown

▲XAD-gelatin (KOH-collagen)
■samples submitted by Kenosha Public Museum

It is also problematic that all regional discussions of Paleoindian habitats begin at about 12,000 RCYBP with characterization as either open spruce parkland, closing spruce forest, or closed spruce-fir forest (cf. Cleland et al. 1998; Stoltman 1998). This time depth is insufficient in light of the radiocarbon chronology established for the Schaefer, Hebior, Fenske, and Mud Lake mammoth sites in southeastern Wisconsin noted in Table 1.

Perhaps a more accurate portrayal of the regional ice-marginal habitat during the period 17,000–20,000 RCYBP is provided by Graham (1990). While the onset of warming during the late glacial (14,000–10,000 RCYBP) is not likely a one-to-one analog for the full glacial period, it represents, I believe, a better characterization of local habitat than pollen-driven models. Guthrie (1990:50) and others have cited mammoth stomach contents as evidence indicating reliance on dry grasses, grasses representing climax groups rather than colonizers. This finding appears more consistent with a forb- and sedge-dominated temperate tundra-like habitat (Graham 1990:54–56) than spruce parkland or closing spruce forest, although this spruce cover may have been the case beginning about 12,000 RCYBP, marking the onset of degradation of mammoth habitat and ultimately resulting in regional extirpation of the species ca. 11,000 RCYBP. Cleland et al. (1998: 17) note the occurrence of a herb pollen zone throughout Michigan and the surrounding region at 14,300–12,000 RCYBP. This pollen zone has yet to be formally identified in southeastern Wisconsin. However, Fredlund et al. (1996) report higher frequencies of sedges and forbs from the Schaefer site than do Huber and Rapp (1992) at nearby Wisbog. Clearly more attention needs to be directed to both pollen and macrofossils from the 14,000–12,000 RCYBP interval, but continued characterization of earliest regional Paleoindian adaptation in either the open spruce parkland or closing spruce forest habitat as though this interpretation were empirically demonstrated is ill advised.

This observation is particularly true given the recent overview (Kapp 1999) of the paleoecology at 13,800 RCYBP of Michigan's lower peninsula, which lies at the same latitude as southeastern Wisconsin but east rather than west of the Lake Michigan lobe. Kapp (1999) indicates an expansive landscape dominated by tundra, flanked on the south, at approximately the base of current Lake Michigan, by wetlands and boreal parkland. Kapp's characterization is based on information derived from 18 paleoecological sites in Michigan's lower peninsula and northern Indiana. He states:

> I conclude that at 13,800 RCYBP in Michigan virtually all of the drained landscape was either cloaked with tundra-like vegetation or had scattered stands of pioneer trees (junipers, aspen, ash, spruce) and sun-tolerant shrubs (willow, silverberry, crowberry) (Figure 2.4). Intense water erosion and solifluction affected steep slopes and glacial end moraines. The interlobate and morainal hill country (such as the upland ridges extending from Jackson through Oakland Counties and the interior uplands from Newaygo and Wexford to Ogemaw Counties) certainly would have sustained open, treeless tundra-like communities because of the existence of exposed, drained sites (1999:45).

It is assumed, but not necessarily demonstrated, that similar conditions would have existed in the locality of the four mammoth sites in southeastern Wisconsin. Again, this interpretation is more consistent with Graham's (1990) temperate tundra-like environments prior to the onset of late-glacial times. At this juncture, though, it is not clear if the southeastern Wisconsin locality was either forb- or sedge-dominated.

Summary

The trajectories of evidence appear to converge in a single direction. Data acquired from archaeological, geomorphological, and taphonomic investigations indicate two periods of mammoth

[1] The ^{14}C dates reported here for the Fenske site were taken from a humerus fragment said to have been derived from that site along with a complete femur. Subsequent to compilation of the Fenske site dates in Table 1, the Kenosha Public Museum secured additional dates from the femur and these long-unpublished dates are considerably younger (ca. 11,000 RCYPB) than the dates reported here. As the specimens were curated in regional museums for many years, there are multiple possibilities: (1) both the complete femur with multiple unreported dates in possession of the Kenosha Public Museum and the fragmentary humerus are, in fact, from the Fenske Site, but are from separate animals of different ages; (2) the dated humerus fragment is from Mud Lake and not Fenske as we originally had been led to believe; or (3) the humerus fragment is from a locality other than the Fenske site.

utilization, one as early as 13,500 RCYBP and again at 12,400 RCYBP. The interval between these eras coincides with the Glenwood II phase of glacial Lake Chicago, formed during the early Port Huron advance (Hansel et al. 1985). During this time much of the landscape occupied by Chesrow complex people would have been inundated beginning as early as 13,000 RCYBP and persisting until sometime after 12,700 RCYBP, making it unsuitable for habitation. Johnson's (2000, this volume) results from taphonomic analyses indicate human involvement with the Fenske, Mud Lake, Schaefer, and Hebior mammoth carcasses. Her efforts, supplemented by those of Overstreet and Kolb (2000), mitigate against any notion that the excavated bone piles were translocated by water, ice, or other means. The intimate association of chipped-stone tools (Figure 3) with the Schaefer and Hebior bone piles (Overstreet 1996, 1998; Overstreet et al. 1995), augmented by the evidence of use of those tools for processing meat and hides (Yerkes and Weinberger 1997), leaves little doubt that mammoth played a role in local subsistence. Directly dating the mammoth remains employing the XAD purified collagen protocols (Stafford 1988, 1990; Stafford et al. 1987) supplements the multiple assays on wood associated with the bone piles. While these should be considered as archaeological facts, the association of the Chesrow complex with the mammoth bone piles is less well anchored and in need of confirmation by chronometric means. Nonetheless, at this point Chesrow remains the most viable candidate for human-mammoth interaction in southeastern Wisconsin, first following the Glenwood I phase after 13,600 RCYBP (Hansel et al 1985), and again following Glenwood II after 12,700–12,900 RCYBP.

Figure 3. Lithic implements. **A,** from the Hebior (47 Kn 265) site; **B,** from the Schaefer (47 Kn 252) site.

Limitations

There are three significant limitations that must be resolved before the Chesrow complex may be more fully documented and accurately interpreted, but none are insurmountable. First, we lack a refined understanding of the nature of the local habitat during the 13,500–12,500 following interval indicated for the four mammoth sites. Second, the association of the Chesrow complex with two excavated mammoth bonebeds is open to alternative interpretation. Third, long-held typological beliefs and conventional wisdom regarding the chronology of basally thinned projectile points that may be accommodated within such types as Hi-Lo, Quad, Chesrow, and Price will be most difficult to overcome. This will not occur unless greater focus is placed on confirming the absolute ages and undisturbed archaeological contexts of these types.

The late-glacial habitat is likely more akin to mammoth steppe (Guthrie 1990) or temperate tundra (Graham 1990) than to spruce parkland or closing spruce forest (cf. Fredlund et al. 1996; Huber and Rapp 1992; Stoltman 1998). Systematic and detailed pollen studies for the basin in which the Schaefer and Hebior mammoth remains were excavated have not yet commenced. Excellent pollen preservation (Fredlund et al. 1996) will yield important results to assist in answering this question; but pollen alone, as Guthrie (1985, 1990) points out, will be insufficient. Future habitat reconstruction will focus on developing detailed pollen zones for the region, but will be buttressed by intensive collecting and study of macrofossils. At the Hebior site environs, stratified forest floors with mats of plant debris and Pleistocene insects have been located in test excavations. A buried stump field with trees in growth position also is located within the basin that harbored the Schaefer and Hebior mammoths. Bivalves and snails occur in the Pleistocene lake clays at the Schaefer site and were also observed at Lendahl Bog. All these information sources need to be exploited and systematically integrated in order to accurately reconstruct the late-glacial habitat in southeastern Wisconsin. Since localities have already been isolated where appropriate data may be recovered, future research can readily determine accurate habitat characterization. Until that occurs, any model of ice-marginal adaptation will remain at the level of testable hypothesis. Finally, additional fieldwork at the Mud Lake and Fenske sites is critical in order to confirm the archaeological contexts, to expand Johnson's (2000, this volume) bone modification work on those specimens, and to secure additional site-specific paleoecological data.

Another significant limitation is our inability to determine chronometric dating for Chesrow-component habitation sites. Of course this problem remains true for almost all Paleoindian sites in the Great Lakes area. My surmise has been, and continues to be, that Chesrow complex populations are responsible for the butchered mammoth remains at the Schaefer, Hebior, Mud Lake, and Fenske mammoth sites (Overstreet 1993a, 1996, 1998). Both Mason (1995, 1997) and Stoltman (1998) have mentioned this problem. Mason (1997) has explicitly addressed the difficulty in affiliating the few chipped-stone tools found with mammoth remains to the Chesrow complex and has also questioned the integrity of the complex itself. He notes:

> But whatever the explanation for this unusual collection of artifact types, none of the artifacts actually associated with extinct animals can yet be shown to belong to Chesrow or for that matter to any other identified cultural complex. Attributing all of these things to a single proposed Chesrow category is by default (1997:94).

Stoltman, of like mind, states:

> However, until a Chesrow/Price point is found directly associated with a mammoth, or until a Chesrow site is independently dated by radiocarbon, the association between mammoths and the Chesrow complex remains on the inferential level only (1998:64).

Overstreet (1993a, 1993b) had noted the presence of both Clovis-like and Folsom-like projectile points some 56 km (35 miles) west of the Hebior and Schaefer mammoths on the Darien moraine, a landscape derived from recession of the Green Bay lobe. Mason (1997) indicated that perhaps these mammoth processing sites were within the hunting territories of Clovis and Folsom groups known to have occupied areas on the Darien moraine. This is possible, of course; but to verify this surmise would require considerable revisions in both Clovis and Folsom chronologies.

In my opinion, the ostensible association of the Chesrow complex with the Hebior and Schaefer mammoth remains rises above the level of default; rather, there is direct but limited evidence. Lithic assemblages from the Chesrow, Scott, and Lucas sites, for example, all share strong characteristics. The selection of raw materials is overwhelmingly from pebble-cobble sources in the local till. Bifacial reduction of this small-sized, poor-quality toolstone often results in cortical remnants remaining on the surfaces of bifacial and unifacial implements (Overstreet 1993a). Moreover, refined bifaces are often fashioned from heat-treated materials. Chipped-stone tools from Schaefer and Hebior: (1) are derived from local cobbles in glacial till; (2) are heat-treated; and (3) have cortical remnants on their surface. These characteristics are typical of the Chesrow complex but quite atypical of either Clovis- or Folsom-related manifestations in the region. This observation does not negate the need for radiocarbon dating of Chesrow complex habitation sites, nor does it confirm the Chesrow-mammoth association. The data do, however, raise the suggested association beyond the level of default, conjecture, or simple geographic proximity.

Suggesting that Hi-Lo, Price, Quad, Chesrow, Miller Lanceolate, and several other related projectile point styles represent a broad pattern of late-Pleistocene, pre-Clovis unfluted bifaces will doubtless raise some eyebrows. It is clear, however, that aside from antiquated typological frameworks, there are no substantive empirical data from which to argue they are not. The affinities of these styles with Chesrow points are inescapable. In fact, most reactions I have had to the proposed early chronology for Chesrow points have been based on typological grounds. Mason is firm in his assignment of Hi-Lo as late Paleoindian. He states, in his recent summary of Holcombe and Hi-Lo: "Although a few archaeologists might disagree about the former as the latest expression of the early Paleo-Indian stage, no one has qualms about the classification of Hi-Lo points" (1997:106). Schott, perhaps less adamant than Mason (1997) is equally clear that Hi-Lo is likely assignable to a Plano horizon; but he notes: "Although there are no radiocarbon dates for Hi-Lo sites, the Dalton placement suggested by Goodyear (1982) probably applies to Hi-Lo as well" (1999:75). Unfortunately, the most recent synthesis of Hi-Lo provides no specific clarification of Hi-Lo chronology, but the research was not designed to do so. Ellis and Deller (1982), working with existing Ontario collections, focused their analyses on technological and morphological attributes. Well aware of the chronological ambiguity, they conclude:

> It is not our purpose here to debate whether or not Hi-Lo can be conceived of as a Late Paleo-Indian or an Early Archaic industry. Instead, in the following, we are concerned only with demonstrating that Hi-Lo is a temporally early industry in the Great Lakes area (1982:15).

Just how early remains to be demonstrated, as is also the case with Price (Stoltman 1998) and Quad points (Bell 1960). With regard to the former, Stoltman was unable to accurately date Price points from the Governor Dodge rockshelter in southwestern Wisconsin's driftless area because they were mixed in the lowest level with Archaic and Paleoindian materials; but he estimated the age at "roughly 10,500 to 10,000 B.P." (1998: 63). His reasoning was that the Price/Chesrow materials may be as old as the late-Paleoindian (Agate Basin) artifacts with which they were mingled. This interpretation is not unreasonable, but it is just as possible that they could represent the oldest occupation at Governor Dodge rockshelter and thus predate the Agate Basin occupation by a considerable magnitude.

Whatever the absolute age of this constellation of lanceolate bifaces, it seems obvious, at least to me, that all (Chesrow, Price, Quad, Hi-Lo) are typologically comparable. It is just as obvious that the absolute age of all is uncertain. However, they share characteristics with the Miller Lanceolate at Meadowcroft, radiocarbon-dated to the interval 11,300–12,000 RCYBP (Adovasio et al. 1988), Tankersley et al. (1987) and Haynes (1987) notwithstanding; and also with the pre-Clovis-level bifaces at Cactus Hill (McAvoy and McAvoy 1997). Collectively, then, if we objectively assess the available empirical evidence rather than retreat to the position of wrestling with purely typological assumptions, there are more data in support of the early chronology than in contradiction. By accepting the evidence of course we raise some implications regarding the origins of Clovis technology, but I see no significant restraints to carrying forward the hypothesis that fluted points are a logical outgrowth of these widespread basally thinned forms throughout much of the eastern United States and Canada.

These limitations are imposing. The means for fine tuning

the late-glacial habitat and securing chronometric dating of the Chesrow complex, particularly at the Lucas site, where there is ample evidence of burning (Overstreet 1994), are at hand; and resolutions of these limitations are restrained only by time and adequate resources. Known site locations at Hebior, Schaefer, Lendahl Bog, Mud Lake, and Fenske have significantly reduced the size of the haystack in which "needles" may be sought. Overcoming habitual reliance on typologies unsupported by contextual or chronometric evidence and carrying forward the hypothesis that Eastern fluted points may be derived from an earlier, basally thinned lanceolate projectile point tradition will likely prove to be the greater chore.

A Model of Ice-Marginal Adaptation

Setting aside for the moment the question of the Chesrow complex–mammoth association and the existence of fluted-point precursors, whoever modified the mammoth remains at Fenske, Mud Lake, Hebior, and Schaefer were people of the tundra rather than the spruce parkland. Perhaps they also periodically, or even habitually, exploited the natural resources of the edge or interface between tundra and the boreal forest, located a short distance to the south. If they were habitual tundra residents, it seems most logical that their primary prey species would have been the woolly mammoth, musk oxen, and caribou within that zone circumscribed to the north by the ice-front and to the south by wetlands and boreal forest. Any answers are elusive, of course, because we lack adequate information to provide much of a portrayal of their lifeways. Were they effective big-game hunters, or were they simply opportunistic foragers who periodically scavenged moribund prey? No airtight case can be made for active predation of either mammoth or mastodon in the eastern United States. In addition to the four southeastern Wisconsin mammoth sites, there are examples of human involvement at Boaz (Palmer and Stoltman 1976), at Michigan and Indiana sites investigated by Fisher (1984, 1987, 1988, 1989, 1995), and Kimmswick (Graham et al. 1981). Many of these, however, likely occurred subsequent to the southeastern Wisconsin events; but none can unequivocally be characterized as actively hunted animals, as in the case of the Naco mammoth (Haury et al. 1953). Such limited information leaves us with two alternative models, neither of which can be tested adequately at this time.

The bifaces recovered from the Hebior and Schaefer mammoths indicate only that bifacial technology in the region was relatively well developed by 12,500 RCYBP. None of the three bifaces, however, can be classified as projectile points. Those at Hebior represent bifacial knives, one fairly thin and well made, the other rather crude with a sinuous edge and a cortical remnant found on its thick base. Neither could have functioned as a projectile point in a Clovis-like technology. The implement fragment recovered from Schaefer is less informative; it simply represents the edge of a broken biface fashioned from a local heat-treated chert. Many Chesrow complex bifaces clearly would have functioned as projectile points. Were they employed in actively hunting the tundra fauna? That cannot be ascertained from current evidence. If in fact the Chesrow complex–mammoth association is borne out by future investigations, and if the people were habitual residents of that fairly narrow zone of nearly treeless dry grassland dominated by forbs and sedges, it seems logical that economic emphasis would have been placed on exploiting larger fauna, supplemented by small game such as arctic hare and to lesser degree the restricted range of available plant foods. If this scenario approaches reality, perhaps Chesrow complex people could appropriately be categorized as big-game hunters.

It is also interesting to note that if a tundra-specific adaptation is correct, Chesrow complex populations may have simply drifted north with the receding ice front during the Intra-Glenwood low phase (Hansel et al. 1985). The early Port Huron advance would have then displaced them to the south and west after about 13,300–13,000 RCYBP. Any sites that may have been occupied in northeastern Wisconsin during the Intra-Glenwood low would likely either have been destroyed by the advancing ice front where it scoured down to the underlying bedrock, or now lie buried beneath tens of feet of outwash or till deposited during the early Port Huron advance and recession to the Two Creeks low phase. Confirming the presence of habitation beneath Great Lakean till sets forth logistical problems I do not even wish to contemplate. Certainly the Chesrow and Lucas sites, and possibly the Fenske site would have been inundated by the waters of glacial Lake Chicago during the Glenwood II phase high stand. This scenario will remain plausible until such time as other mammoth are located, excavated, and accurately dated to that 1,200-year hiatus between Fenske/Mud Lake and Schaefer/Hebior.

A short distance south, perhaps no more than 63–80 km (40–50 miles) using Kapp's (1999) paleoecological reconstruction at 13,800 RCYBP, the tundra habitat gave way to extensive wetlands and boreal parkland. The late-glacial populations responsible for the butchery marks at the four southeastern Wisconsin mammoth sites could have habitually occupied this edge area or the boreal parkland itself, from which they periodically made forays into the tundra zone. Moribund prey might have been predictable (Fisher 1988) and thus expected to provide nutritional resources; but mammoth carcasses could also have provided bone and ivory for raw materials less readily available in the boreal parkland. In this instance, if opportunistic foraging/scavenging was the primary adaptive strategy, perhaps the tundra zone was visited only seasonally for specific resources. The territory in question would certainly be no more expansive than those suggested for later times based on the location of toolstone sources and the distribution of implements fashioned from that material. In such a scenario Mud Lake, Fenske, Hebior, and Schaefer might best be characterized as sites for procuring and processing food situated some distance from stone procurement and tool manufacturing sites and base camps in the manner Tankersley (1998) sets forth. Unfortunately, the exclusive reliance on local till-transported cherts makes this approach difficult. Based on Tankersley's model, the Chesrow site (Overstreet 1993b), at least from the data at hand, may be identified as a tool processing locality; Lucas, with a varied assemblage and high densities of calcined bone, may have been a base camp (Overstreet 1994).

At this juncture both models may be said to have equal plausibility. The generalized forager model is equally plausible as a more traditional big-game hunting adaptation. The Chesrow complex, however, should not be considered unique. As Meltzer (1988), Guthrie (1990), Graham (1990), Tankersley (1998), Turner et al. (1999), and others have pointed out, it is the rapidly changing climatic conditions and reorganization of plant and animal species in patterns having no modern analogs that are so unique. In this context there is utility in considering the Chesrow complex as one spatial unit in a much more expansive "Initial Eastern Biface Tradition." Within such a taxon I would include, owing to the distinct similarities Stoltman (1998:62–64) has already noted, the Wisconsin driftless area Price point complex. Farther afield, based on strong stylistic, chronological, and typological attributes, I would include the Hi-Lo complex (Ellis and Deller 1982; Fitting 1963a, 1963b), Miller Lanceolate projectile points (Adovasio et al. 1988), Quad point complex (Lewis 1960; Soday 1954), unfluted lanceolate projectile points from the Eppley rockshelter (Brush 1993), and the complex represented by the refined bifaces found stratigraphically below Clovis at Cactus Hill (McAvoy and McAvoy 1997). Other "fluted bifaces" such as those at Hardaway also come to mind (Coe 1964), and undoubtedly other chronometrically unanchored lanceolate bifaces could be identified from other localities. Cognizance of an early basally thinned Eastern biface tradition has some bearing on the larger-scale question of extinctions, but poses no new lines of inquiry relevant to the peopling of the Americas. It merely shifts the time depth and presents a heretofore unrecognized adaptive strategy that is but one small element of the diverse suite of Eastern Paleoindian economies. With regard to extinction models, the data here extend human-mammoth involvement to 13,500 RCYBP and thereby leave the "Blitzkreig model" wounded if not fatally flawed. Mammoth predation may have had a significantly longer tenure than Martin (1984) anticipated in formulating his overkill model; if so, the megafauna predation history component of that equation would be seriously damaged. If the existence of the tradition suggested here were ultimately verified, it would have little influence in determining either the source or timing cited in the current multiple hypotheses relating to when and from where the continent's first human populations originated.

Acknowledgments

In part, this research was supported by The National Science Foundation under grant #SBR9708616. Support was also provided by the National Park Service through survey and planning grants administered by the State Historical Society of Wisconsin. I am indebted to E. Johnson, Museum, Texas Tech University; R. Yerkes and J. Weinberger, Department of Anthropology, Ohio State University; G. Fredlund, Department of Geography, University of Wisconsin-Milwaukee, K. Sverdrup, Department of Geosciences, University of Wisconsin–Milwaukee; T. W. Stafford, Jr., Stafford Laboratories; M. Kolb, Strata-Morph Geoexploration; and A. F. Schneider, Professor Emeritus, UW–Parkside, for their interest and participation in the ongoing research. I have benefited immeasurably from previous Paleoindian research in the Great Lakes region carried out by R. Mason, Professor Emeritus, Lawrence University; J. Stoltman, Department of Anthropology, University of Wisconsin–Madison; and R. Salzer, Department of Anthropology, Beloit College. Their consistent willingness to entertain new ideas, point out limitations (sometimes more bluntly that I would have liked), and provide encouragement is deeply appreciated. None of them, however, should be held responsible for any of the constructs presented here and for which I carry sole responsibility. I am also grateful to R. Bonnichsen for the opportunity afforded to have been a poster-presenter at the 1999 Clovis and Beyond Conference, Santa Fe, New Mexico. Dialogue with other presenters and colleagues at the conference lent considerable impetus to the finished version of this paper, and the opportunity for its submittal is also acknowledged to R. Bonnichsen, the Center for the Study of the First Americans, and the reviewers and editors of this volume whose efforts on my behalf are deeply appreciated. Support for fieldwork and conference travel was provided by the Department of Social and Cultural Sciences and the Graduate School at Marquette University. Finally, I extend my appreciation to Mr. Dan J. Joyce, Kenosha Public Museum, for the use of previously unreported radiocarbon dates from the Schaefer mammoth site.

References Cited

Adovasio, J. M., A. T. Boldurian, and R. C. Carlisle
1988 Who Are Those Guys?: Some Biased Thoughts on the Initial Peopling of the New World. In *America Before Columbus: Ice-Age Origins,* edited by R. C. Carlisle, pp. 45–62. Ecology Monographs No. 12. Department of Anthropology, University of Pittsburgh.

Bell, R. E.
1960 *Guide to the Identification of Certain American Indian Projectile Points.* Special Bulletin No. 2., Oklahoma Anthropological Society.

Bretz, J. H.
1951 The Stages of Lake Chicago—Their Causes and Correlations. *American Journal of Science* 249:401–29.
1959 The Double Calumet Stage of Lake Chicago. *Journal of Geology* 67:675–84.

Brush, N.
1993 Twelve Thousand Years of Human Occupation at the Eppley Rockshelter. *Current Research in The Pleistocene* 10:5–6.

Cleland, C. E.
1965 Barren Ground Caribou *(Rangifer arcticus)* from an Early Man Site in Southeastern Michigan. *American Antiquity* 30:350–51.
1976 The Focal-Diffuse Model: An Evolutionary Perspective on the Prehistoric Cultural Adaptations of the Eastern United States. *Midcontinental Journal of Archaeology* 1 (1):59–76.

Cleland, C. E., M. B. Holman, and J. A. Holman
1998 Mason-Quimby Line Revisited. *The Wisconsin Archeologist* 79 (1):8–27.

Coe, J. L.
1964 The Formative Cultures of the Carolina Piedmont. *Transactions of the American Philosophical Society,* N.S., Part 5. Philadelphia.

Cotter, J. L.
1937 The Occurrence of Flints and Extinct Animals in Pluvial Deposits near Clovis, New Mexico. Part IV-Report on Excavation at the Gravel Pit, 1936. *Proceedings of the Academy of Natural Sciences at Philadelphia* 89:1–16.

Curran, M. L., and J. R. Grimes
1989 Ecological Implications for Paleoindian Lithic Procurement Economy in New England. In *Eastern Paleoindian Lithic Resource Use,* edited by C. J. Ellis and J. C. Lothrop, pp. 41–74. Westview Press, Boulder, Colorado.

Dallman, J. E., D. F. Overstreet, and T. W. Stafford, Jr.
1996 A Revised Chronology for Mammoth and Mastodon Fossils in the Southwestern Lake Michigan Basin. *Current Research in the Pleistocene* 13:10–13.

Deller, D. B., and C. A. Ellis
1988 Early Paleo-Indian Complexes in Southwestern Ontario. In *Late Pleistocene and Early Holocene Paleoecology and Archaeology of the Eastern Great Lakes Region,* edited by R. Laub, N. Miller and D. Steadman, pp. 251–263. Bulletin 33, Buffalo Society of Natural Sciences

1992 *Thedford II- A Paleo-Indian Site in the Ausable River Watershed of Southern Ontario.* Memoirs, Museum of Anthropology, University of Michigan, No. 24. Ann Arbor.

Ellis, C. J., and D. B. Deller
1982 Hi-Lo Materials from Southwestern Ontario. *Ontario Archaeology* 38:3–22.

Fisher, D. C.
1984 Taphonomic Analysis of Late Pleistocene Mastodon Occurrences: Evidence of Butchery by North American Paleo-Indians. *Paleobiology* 10:338–357.

1987 Mastodont Procurement by Paleoindians of the Great Lakes Region: Hunting or Scavenging? In *The Evolution of Human Hunting,* edited by M. H. Nitecki and D. V. Nitecki, pp. 309–421. Plenum Press, New York.

1988 Season of Death of the Hiscock Mastodonts. In *Late Pleistocene and Early Holocene Paleoecology and Archeology of the Eastern Great Lakes Region,* edited by R. S. Laub, N. G. Miller, and D. W. Steadman, pp. 115–125. Bulletin of the Buffalo Society of Natural Sciences 33.

1989 Meat Caches and Clastic Anchors: The Cryptic Record of Paleoindian Subsistence in the Great Lakes Region. *Geological Society of America, Abstracts with Program* 21:A234.

1995 Experiments on Subaqueous Meat Caching. *Current Research in the Pleistocene* 12:77–79.

Fitting, J. E.
1963a The Hi-Lo Site: A Late Paleo-Indian Site in Western Michigan. *The Wisconsin Archeologist* 44 (2):87–96.

1963b An Early Post-Fluted Point Tradition in Michigan: A Distributional Analysis. *The Michigan Archaeologist* 9 (2):21–25.

Fredlund, G. G., R. B. Johnson, G. S. Porter, T. A. Revane, H. K. Schmidt, D. F. Overstreet, and M. F. Kolb
1996 Late Pleistocene Vegetation of the Hebior Mammoth site, Southeastern Wisconsin. *Current Research in the Pleistocene* 13:87–89.

Funk, R. E.
1972 Early Man in the Northeast and the Late-Glacial Environment. *Man in the Northeast* 4:7–39.

1976 *Recent Contributions to Hudson Valley Prehistory.* Memoir No. 22. New York State Museum, Albany.

Funk, R. E., D. W. Fisher, and E. M. Reilly, Jr.
1970 Caribou and Paleo-Indian in New York State: A Presumed Association. *American Journal of Science* 268:181–186.

Graham, R. W.
1990 Evolution of New Ecosystems at the End of the Pleistocene. In *Megafauna and Man: Discovery of America's Heartland* edited by L. D. Agenbroad, J. I. Mead, and L. W. Nelson, pp. 54–60. The Mammoth Site of Hot Springs, South Dakota, Inc., Scientific Papers, Volume 1. Hot Springs, South Dakota.

Graham, R. W., C. V. Haynes, D. L. Johnson, and M. Kay
1981 Kimmswick: A Clovis-Mastodon Association in Eastern Missouri. *Science* 213:1115–1117.

Gramly, R. M
1982 *The Vail Site: A Palaeo-Indian Encampment in Maine.* Bulletin No. 30, Buffalo Society of Natural Sciences, Buffalo, NY.

Guthrie, R. D.
1985 Woolly Arguments against the Mammoth Steppe—A New Look at Palynological Data. *Quarterly Review of Archaeology* 6:9–14.

1990 Late Pleistocene Faunal Revolution. In *Megafauna and Man: Discovery of America's Heartland,* edited by L. D. Agenbroad, J. I. Mead, and L. W. Nelson, pp. 42–53. The Mammoth Site of Hot Springs, South Dakota, Inc., Scientific Papers, Volume 1. Hot Springs, South Dakota.

Hansel, A. K., D. M. Mickelson, A. F. Schneider, and C. E. Larsen
1985 Late Wisconsinan and Holocene History of the Lake Michigan Basin. In *Quarternary Evolution of the Great Lakes,* edited by P. F. Karrow and P. E. Calkin, pp. 39–54. Geological Association of Canada, Special Paper No. 30.

Haury, E. W., E. Antevs, and J. F. Lance
1953 Artifacts with Mammoth Remains, Naco, Arizona. *American Antiquity* 19(1):1–14.

Haury, E. W., E. B. Sayles, and W. W. Wasley
1959 The Lehner Mammoth Site, Southeastern Arizona. *American Antiquity* 25:2–30.

Haynes, C. V.
1987 Clovis Origins Update. *The Kiva* 52(2):83–93.

Huber, J. K., and G. Rapp, Jr.
1992 Results of a Pollen Study from Southeast Kenosha County, Wisconsin. In *Archaeological Studies on the Southeastern Wisconsin Uplands,* edited by D. F. Overstreet, pp. 74–131. Case Studies in Great Lakes Archaeology No. 1. Great Lakes Archaeological Press, Milwaukee.

Johnson, E.
2000 Taphonomy of Late Pleistocene Mammoth in Southeastern Wisconsin. Report prepared and submitted to National Science Foundation in partial fulfillment of Grant #SBR9708616. Copy on file at Archaeological Research Center, Marquette University.

Johnson, E., P. J. Lewis, and R. Strauss
2000 A Measured Approach to Marks on Bone. Paper presented at the annual meeting of the American Association of Physical Anthropologists, San Antonio, April 2000.

Kaiser, K. F.
1994 Two Creeks Interstade Dated through Dendrochronology and AMS. *Quaternary Research* 42(2):288–298.

Kapp, R. O.
1999 Michigan Late Pleistocene, Holocene, and Presettlement Vegetation and Climate. In *Retrieving Michigan's Buried Past,* edited by J. R. Halsey and M. Stafford, pp. 31–58. Cranbrook Institute of Science Bulletin 64. Bloomfield Hills, MI.

King, J. E.
1981 Late Quaternary Vegetational History of Illinois. *Ecological Monographs* 51(1):43–62.

Leonhardy, F. C.
1966 *Domebo: A Paleo-Indian Mammoth Kill in the Prairie-Plains.* Contributions of the Museum of the Great Plains, 1:1–53.

Lewis, T. M. N.
1960 Editor's notes: The Guinn Collection. *Tennessee Archaeologist* 16(1):54–61.

MacDonald, G. F.
1968 Debert: A Paleo-Indian site in Central Nova Scotia. *National Museum of Canada, Anthropology Papers,* No. 16.

Martin, P. S.
1984 Prehistoric Overkill: The Global Model. In *Quaternary Extinc-*

tions: A Prehistoric Revolution, edited by P. S. Martin and R. G. Klein, pp. 354–403. University of Arizona Press. Tucson.

Mason, R. J.
1958 *Late Pleistocene Geochronology and the Paleo-Indian Penetration of the Lower Michigan Peninsula.* Anthropological Papers, Museum of Anthropology, University of Michigan, No. 11. Ann Arbor.

1995 Review: *Chesrow, A Paleoindian Complex in the Southern Lake Michigan Basin,* by D. F. Overstreet. *The Wisconsin Archeologist* 76(1-2):239–42.

1997 The Paleo-Indian Tradition. *The Wisconsin Archeologist* 78 (1–2): 78–111.

McAvoy, J. M., and L. D. McAvoy
1997 *Archaeological Investigations of Site 44SX202, Cactus Hill, Sussex County, Virginia.* Virginia Department of Historic Resources, Research Report Series No. 8.

Meltzer, D. J.
1988 Late Pleistocene Human Adaptations in Eastern North America. *Journal of World Prehistory* 2:1–52.

Overstreet, D. F.
1993a *Chesrow, A Paleoindian Complex in the Southern Lake Michigan Basin.* Case Studies in Great Lakes Archaeology, Number 2. Great Lakes Archaeological Press. Milwaukee.

1993b FY 1992 Historic Preservation Survey and Planning Grant-Early Holocene Megafauna Exploitation, Kenosha County, Wisconsin. *Great Lakes Archaeological Research Center, Inc., Reports of Investigation No. 325.* Milwaukee.

1994 FY 1995 Historic Preservation Survey and Planning Grant—Ice Age Landscapes of Southeastern Wisconsin. *Great Lakes Archaeological Research Center, Inc., Reports of Investigation No. 366.* Milwaukee.

1996 Still More on Cultural Contexts of Mammoth and Mastodont in the Southwestern Lake Michigan Basin. *Current Research in the Pleistocene* 13:36–38.

1998 Late Pleistocene Geochronology and the Paleoindian Penetration of the Southwestern Lake Michigan Basin. *The Wisconsin Archeologist* 79(1):28–52.

Overstreet, D. F., D. J. Joyce, K. Hallin, and D. Wasion
1993 Cultural Contexts of Mammoth and Mastodont in the Southwestern Lake Michigan Basin. *Current Research in The Pleistocene* 10:75–77.

Overstreet, D. F., D. J. Joyce, and D. Wasion.
1995 More on Cultural Contexts of Mammoth and Mastodont in the Southwestern Lake Michigan Basin. *Current Research in The Pleistocene* 12:40–42.

Overstreet, D. F., and M. F. Kolb
2003 Geoarchaeological Contexts for Late Pleistocene Archaeological Sites with Human-modified Woolly Mammoth Remains in Southeastern Wisconsin. *Geoarchaeology: An International Journal,* vol. 18, no. 1, 91–114.

Overstreet, D. F., and T. W. Stafford, Jr.
1997 Additions to a Revised Chronology for Mammoth and Mastodon Fossils in the Southwestern Lake Michigan Basin. *Current Research in the Pleistocene* 14:70–72.

Palmer, H. A., and J. B. Stoltman
1976 The Boaz Mastodon: A Possible Association of Man and Mastodon in Wisconsin. *Midcontinental Journal of Archaeology* 1(2):163–77.

Quimby, G. I.
1958 Fluted Points and Geochronology of the Lake Michigan Basin. *American Antiquity* 23(3):247–54.

1960 *Indian Life in the Upper Great Lakes.* University of Chicago Press.

Ritchie, W. A., and R. E. Funk
1973 *Aboriginal Settlement Patterns in the Northeast.* Memoir No. 20, New York State Museum and Science Service. Albany, New York.

Schneider, A. F., and A. K. Hansel
1990 Evidence for Post–Two Creeks Age of Type Calumet Shoreline of Glacial Lake Chicago. In *Late Quaternary History of the Lake Michigan Basin,* edited by A. F. Schneider and G. S. Fraser, pp. 1–8. Geological Society of America, Special Paper No. 251.

Sellards, E. H.
1938 Artifacts Associated with Fossil Elephant. *Bulletin of the Geological Society of America* 49:999–1010.

Shott, M. J.
1993 *The Leavitt Site—A Parkhill Phase Paleo-Indian Occupation in Central Michigan.* Memoirs, Museum of Anthropology, University of Michigan, No. 25. Ann Arbor.

1999 Early Archaic: Life after the Glaciers. In *Retrieving Michigan's Buried Past,* edited by J. R. Halsey and M. D. Stafford, pp. 71–82. Cranbrook Institute of Science Bulletin 64. Bloomfield Hills, MI.

Soday, F. J.
1954 The Quad Site, A Paleo-Indian Village in Northern Alabama. *Tennessee Archaeologist* 10(1):1–20.

Stafford, T. W., Jr.
1988 Accelerator ^{14}C Dating of Late Pleistocene Megafauna. *Current Research in the Pleistocene* 5:41–43.

1990 Late Pleistocene Megafauna Extinctions and the Clovis Culture: Absolute Ages Based on Accelerator ^{14}C Dating of Skeletal Remains. In *Megafauna and Man: Discovery of America's Heartland,* edited by L. D. Agenbroad, J. I. Mead, and L. W. Nelson, pp. 118–122. The Mammoth Site of Hot Springs, South Dakota, Inc., Scientific Papers, Volume 1. Hot Springs, South Dakota.

Stafford, T. W., Jr., A. T. Jull, K. Brendel, R. Duhamel, and D. Donahue
1987 Study of Bone Radiocarbon Dating Accuracy at the University of Arizona NSF Accelerator Facility for Radioisotopic Analysis. *Radiocarbon* 29:24–44.

Stoltman, J. B.
1995 Paleoindian Adaptive Strategies in Wisconsin. Paper presented at the annual meeting of the Society for American Archaeology, May 3–7, 1995.

1998 Paleoindian Adaptive Strategies in Wisconsin during Late Pleistocene Times. *The Wisconsin Archeologist* 79(1):53–67.

Storck, P. L.
1984 Glacial Lake Algonquin and Early Paleo-Indian Settlement Patterns in Southcentral Ontario. *Archaeology of Eastern North America* 12:286–298.

1988 The Early Paleo-Indian Occupation of Ontario: Colonization or Diffusion? In *Late Pleistocene and Early Holocene Paleoecology and Archaeology of the Eastern Great Lakes Region,* edited by R. Laub, N. Miller, and D. Steadman, pp. 243–250. Bulletin 33, Buffalo Society of Natural Sciences.

Storck, P. L., and A. E. Spiess
1994 The Significance of New Faunal Identifications Attributed to an Early Paleoindian (Gainey Complex) Occupation at the Udora Site, Ontario, Canada. *American Antiquity* 59(1):121–42.

Tankersley, K. B.
1998 Variation in the Early Paleoindian Economies of Late Pleistocene North America. *American Antiquity* 63(1):7–20.

Tankersley, K. B., C. A. Munson, and D. Smith
1987 Recognition of Bituminous Coal Contaminants in Radiocarbon Samples. *American Antiquity* 52(2):318–29.

Turner, M. D., E. J. Zeller, G. A. Dreschoff, and J. C. Turner
1999 Impact of Ice-related Plant Nutrients on Glacial Margin Environments. In *Ice Age Peoples of North America,* edited by R. Bonnichsen and K. L. Turnmire, pp. 42–77. Oregon State University Press. Corvallis, Oregon.

Winkler, M. G.
1985 *Late-Glacial and Holocene Environmental History of South-Central Wisconsin: A Study of Upland and Wetland Ecosystems.* Institute for Environmental Studies, University of Wisconsin, Madison.
1988 Effect of Climate on Development of Two Sphagnum Bogs in South-Central Wisconsin. *Ecology* 69(4):1032–43.

Yerkes, R. W., and J. W. Weinberger
1997 Microwear Analysis of Lithic Artifacts from the Hebior (47 Kn 265) and Schaefer (47 Kn 252) Mammoth Sites and Two "Stacked Bifaces" from the Chesrow Site (47 Kn 40), Kenosha County, Wisconsin. Report prepared in partial fulfillment of National Science Foundation Grant #SBR9708616. Copy on file at Archaeological Research Center, Marquette University.

III. Paleoamerican Prehistory

Archaeological Evidence, South America

The Ignored Continent: South America in Models of Earliest American Prehistory

Ruth Gruhn

Abstract

The venerable Clovis-first model of the peopling of the Americas was testable: It could be disproven by evidence of archaeological sites older than Clovis, or contemporary with Clovis but significantly different in character. South American archaeological sites presented such evidence, but they were long ignored or discounted. The history of development and entrenchment of the Clovis-first model in North America was a strong contributing factor. This paper discusses the treatment of three early South American sites—Taima-taima, Tibitó, and Monte Verde—by supporters of the Clovis-first model. A review of other early South American sites demonstrates that by the time of the Clovis horizon in North America, all major environmental zones of the southern continent were already occupied by well-adapted populations maintaining diverse subsistence systems and distinctive technologies unrelated to Clovis. Evidence from South America has not only disproven the Clovis-first model; it shows that the standard North American evolutionary model is inappropriate for the southern continent, for the earliest complexes are Paleoarchaic in character, not Paleoindian. North American models may be confounded by South American data once again, if evidence for South American sites earlier than 20,000 yr B.P. is verified.

Introduction

For most of the 20th century, the model of earliest American prehistory accepted by all but a few North American archaeologists envisioned the earliest people as specialized big-game hunters with an upper-Paleolithic technology, pursuing late-Pleistocene megafauna across the Bering land bridge into the interior of North America through an ice-free corridor east of the Rockies. According to this model, the early hunters, identified with the distinctive Clovis complex, reached the continental United States no earlier than ca. 11,500 radiocarbon years ago; and these "Paleoindians" very rapidly spread throughout both continents of the Americas, eventually giving rise to localized, more sedentary populations based upon a generalized hunting/collecting "Archaic" economy (cf. Jennings 1983; Fiedel 1987; Fagan 1987).

For all the shortcomings of the basic assumptions behind the notion of the Clovis complex as the first manifestation of human colonization of the Americas (cf. Alsoszatai-Petheo 1986; Whitley and Dorn 1993; Gruhn 1997; Bryan and Gruhn 2000), the "Clovis-first" model was testable. As formulated, it could be disproven by archaeological evidence of any New World cultural complex which was dated earlier, or which was contemporary but significantly different in character.

South American archaeological sites presented such evidence, but they were long ignored or discounted. In this paper I shall examine the treatment of early South American sites by supporters of the Clovis-first model. Following a very brief review of the historical development of the Clovis-first model in North America, I shall discuss the implications of this model for an understanding of the early prehistory of South America. To illustrate how these implications influenced North American researchers, I shall review the criticisms they raised about several early South American sites. Then, in a survey of other early sites in different regions of South America, I shall show that the Clovis-first model and the North American evolutionary model of Paleoindian to Archaic stages are not applicable in the southern continent. Others have made this point (cf. Ardila and Politis 1989; Dillehay et al. 1992; Kipnis 1998; Dillehay 1999, 2000: 6-8). In this paper I shall elaborate upon it with an historical perspective.

The Development of the Clovis-first Model

The development of the Clovis-first model began in 1927 with the discovery near Folsom, New Mexico, of a sophisticated lithic

Department of Anthropology, University of Alberta, Edmonton AB T6G 2H4, Canada; e-mail: rgruhn@gpu.srv.ualberta.ca technology featuring fluted projectile points in close association with now-extinct megafauna, large bison, in a late-Pleistocene geological context (Figgins 1927). By the early 1930s, the stratigraphic record at Clovis, New Mexico (cf. Hester 1972), and several kill sites on the Great Plains demonstrated a predecessor complex featuring larger, partially fluted projectile points in association with remains of mammoths. Clovis points were soon identified from sites all over the Great Plains and eastern North America, and eventually a scatter of such points were also recorded in the West (cf. Wormington 1957). The report of the excavations at the Lindenmeier Site, an extensive Folsom campsite in northern Colorado, also in the 1930s (Roberts 1935), confirmed the picture of the fluted-point people, soon termed the Paleoindians, as big-game hunters with a sophisticated lithic technology. By the late 1960s, the radiocarbon dating method had indicated the age of the Clovis complex as late Pleistocene, ca. 11,500–11,000 RCYBP (radiocarbon years before present), and the Folsom complex, which followed on the Great Plains, was dated ca. 11,000–10,500 RCYBP (Haynes 1969).

As the years went by, detailed archaeological, geological, and paleoenvironmental studies at a number of Clovis and Folsom sites on and adjacent to the Great Plains in the United States refined the picture of the earliest Americans, the Paleoindians, as nomadic, specialized, and highly successful big-game hunters with a sophisticated upper-Paleolithic type of technology including finely flaked points, blades, gravers, and bone points or foreshafts. On that basis, in the early 1970s a dramatic model of the peopling of the Americas was proposed by the paleoenvironmental specialist Paul S. Martin and his associate, J. F. Mosimann (Martin 1973; Mosimann and Martin 1975). They considered that a population of specialized upper-Paleolithic hunters coming into a vast uninhabited region replete with an abundance of unwary big game would multiply at an extremely high rate of growth and continuous population expansion; they would spread at great speed over both American continents, in their progress wiping out the many species of American megafauna known to have become extinct at the end of the Pleistocene. Using the assumed very high rate of population growth, Mosimann and Martin were able to show through computer simulation that it was theoretically possible for a population exiting the southern end of the ice-free corridor in North America by 11,500 RCYBP to expand to the Straits of Magellan within a thousand years. Martin (1973) pictured the expansion front crossing the U.S./Mexico border by 11,150 RCYBP and the Isthmus of Panama by 10,930 RCYBP, arriving at the Straits of Magellan at the southern end of South America by 10,500 RCYBP.

Thus by the 1970s the Clovis-first model was firmly established in North American archaeology. The empirical database in support of the model was solid; its premises were internally consistent; and the model presented a unique, dramatic, and colorful picture of the first colonization of the Americas. It became the norm among North American archaeologists and in textbooks of New World prehistory, where reported earlier sites might be mentioned, but usually did not fare well. The earliest prehistory of South America, assumed to be the same as North America, was often ignored in discussions of the early peopling of the New World. As late as 1987, a major book on the peopling of the Americas (Fagan 1987) devoted only 8 pages of a total of ca. 260 pages to South America, mainly to discount reported early sites of pre-Clovis age.

Implications of the Clovis-first Model for Early South American Prehistory

The empirical basis for the notion that early South American prehistory followed the North American model was the result of excavations in the 1930s by Junius Bird at Fell's Cave, just north of the Straits of Magellan, where stemmed "fishtail" projectile points were recovered in the basal occupation zone in association with bones of extinct horse and ground sloth (Bird 1938; Hyslop 1988: 134–201). It was noted that the stem portions of some of the fishtail points were fluted. The radiocarbon dates of 11,000 ± 170 and 10,720 ± 300 RCYBP subsequently obtained for the basal occupation zone at Fell's Cave (Bird 1969) forced Mosimann and Martin (1975) to their assumption of an extremely rapid rate of population growth and expansion in order to maintain that a Clovis population could have arrived at the southern tip of South America within a thousand years after their ancestors exited the ice-free corridor into the continental United States.

As time went by, stemmed fishtail points, some of them with fluted stems, were found to be distributed at intervals down the western cordilleran chain from southern Mexico to southern Patagonia. Most were isolated surface finds or undated surface clusters, as at Turrialba in Costa Rica (Snarskis 1979) and Madden Lake in Panama (Bird and Cooke 1978); excavated and dated sites were few. El Inga, Ecuador, which produced a sizable artifact assemblage with stemmed fishtail points but also a variety of other stemmed and non-stemmed point styles, was radiocarbon dated to ca. 9000 RCYBP (Mayer-Oakes 1986). The major dated sites, however, were those far to the south—at the Straits of Magellan, Fell's Cave, with dates of ca. 11,000 RCYBP (Bird 1969), and Cueva del Medio at ca. 10,500 RCYBP (Nami 1989); in Patagonia, Piedra Museo (AEP-1) at ca. 10,400 RCYBP (Miotti 1995); on the Pampas, Cerro La China and Cerro El Sombrero at ca. 10,700 RCYBP (Flegenheimer and Zárate 1997).

Given the Martin-Mosimann model of very rapid population expansion, the distribution of stemmed fishtail projectile points could be seen as marking the trail of Clovis hunters from the southern United States to southernmost South America (Lynch 1983; Fiedel1987: 80–81). With this model, it seems to have mattered little to North American archaeologists that despite the distance to the Straits of Magellan, the stemmed fishtail points there were nearly contemporary with Clovis in North America; and that very few of the South American stemmed fishtail points are actually fluted, and those not by the same technique as Clovis (Politis 1991). It was held that there was a morphological and therefore a historical connection between Clovis points in North America and stemmed fishtail points in South America (Bird 1969; Lynch 1983; Fiedel 1987: 80; Morrow and Morrow 1999).

The concept of a *Paleoindian stage,* which was closely associated with the development of the Clovis-first model, was also extended to South America. According to the standard evolutionary model of North American prehistory (cf. Jennings 1983; Fiedel 1987; Fagan 1987), Paleoindian people were not only the first colonists; they represented a basal stage of historical development in which small population groups were highly nomadic, far-ranging specialized hunters who focused upon the pursuit of large game, utilizing distinctive, well-made bifacial lanceolate projectile points such as Clovis and Folsom. This Paleoindian stage was followed in the North American evolutionary sequence by the development of an *Archaic stage,* in which population units became larger, more localized, and more sedentary, with a more generalized broad-spectrum hunting/collecting/fishing economy; and the development of more regionalized and diversified lithic assemblages, which might include ground and polished implements and utensils as well as stemmed and notched projectile points. Following the Clovis-first model, this North American developmental sequence of the primary Paleoindian stage succeeded by the Archaic stage had to be applicable to South America (e.g., Lynch 1983, 1991b, 1998).

Several corollary restrictions followed from imposing the North American model of early prehistory upon South American archaeological remains. If the Clovis-first model was to hold up, obviously no South American site could be earlier than Clovis age; or given the distance from the United States, even approximately contemporary with Clovis, although the Martin-Mosimann proposal that specialized big-game hunters could have spread from the southern United States all the way to the Straits of Magellan within 600 years could be a theoretical response to the latter concern. Furthermore, according to the North American evolutionary model, the earliest archaeological site in any area of the southern continent could not show an Archaic economy or technology; the earliest sites had to be Paleoindian in character, with a focus on hunting extinct megafauna using well-made bifacial projectile points. There had to be something wrong with data reported from any South American site that contradicted these expectations of the age and character of the earliest colonization. I shall now review how several such early South American sites were dealt with by North American archaeologists.

Examples of North American Critiques of Early South American Sites

Taima-taima

The site of Taima-taima, at a waterhole in low hills close to the present Caribbean coast of Venezuela, was first explored by José Cruxent, who found abundant remains of Pleistocene megafauna including mastodon, glyptodon, ground sloth, and horse. At intervals during the 1960s, excavations were carried out at the site by Cruxent, who reported finding a few long, thick lanceolate El Jobo points and other lithic artifacts in a basal gray sand deposit in association with the fossil bones (Cruxent 1967). A series of radiocarbon dates on bones and sediment samples (Tamers 1971) indicated an age of ca. 14,000–12,000 RCYBP for the fossiliferous zone with artifacts; a much higher overlying deposit of black organic clay gave dates of about 10,000 RCYBP.

In 1970 Bryan and Gruhn visited the site. Bryan's general discussion of early South American sites (Bryan 1973), which included Taima-taima, drew critical comment from Lynch (1974) and Haynes (1974), who questioned the integrity of the stratigraphy and the validity of the early radiocarbon dates at Taima-taima. It was proposed that since there were artesian springs at the site, the stratigraphy was disturbed, the artifacts and fossil bones redeposited and mixed up, and the early-dated radiocarbon samples possibly contaminated.

In 1976 Bryan and Gruhn returned to Taima-taima; and working with Cruxent and the paleoenvironmental specialist Claudio Ochsenius for six weeks, gained more definite evidence of the antiquity and integrity of the site (Bryan et al. 1978; Ochsenius and Gruhn 1979; Gruhn and Bryan 1984). They uncovered the remains of a juvenile mastodon with two flaked-stone artifacts, a median fragment of a quartzite El Jobo point, and a utilized jasper flake in close direct association with the skeleton. The paleontologist Rodolfo Casamiquela noted evidence of butchering in the form of series of sharp cut marks on several ribs and the left humerus, as well as a pattern of systematic disarticulation of the skeleton. Also in direct association with the skeleton, which rested in the basal deposit of waterlogged grey sand just above the Miocene bedrock, were masses of masticated fragments of twigs, likely the stomach contents of the mastodon, which produced four radiocarbon dates of ca. 13,000 RCYBP.

As well, the 1976 excavation clarified the stratigraphy at Taima-taima. The basal gray sand zone (Unit 1) that enclosed the mastodon remains and artifacts was sealed by a cemented erosional surface, the Unit I/II disconformity, which retained traces of an eroded lateritic paleosol formed on Unit I, and upon which were found fragmentary remains of glyptodon, horse, sloth, *Macrauchenia,* and extinct tortoise, but no mastodon remains and no artifacts. Within the thick sterile cemented reddish sand of Unit II overlying the disconformity was another paleosol horizon; and above, a thick sterile white sand zone directly underlying the sterile black organic clay (Unit III), which had previously been dated ca. 10,000 RCYBP. In sum, the 1976 excavations confirmed the stratigraphic integrity and antiquity of the early archaeological horizon at Taima-taima.

While Taima-taima fit the North American model of the Paleoindian as big-game hunter, obviously there were big problems with it; the site was dated far too early, and the projectile points were not Clovis or conceivable Clovis derivatives. Critiques of the site over the years addressed these problems.

One critique asserted on typological grounds that El Jobo points could not possibly be that old because they resembled other narrow leaf-shaped points like the Lerma type in Mexico, there dated only as early as the early Holocene, after 10,000 RCYBP (Fiedel 1987: 81). The corollary is that such "Late Paleoindian" points must be intrusive down into the lower early-dated fossiliferous zone at Taima-taima, being derived from an imagined higher occupation horizon at the site by mixing within spring deposits. The assertions of spring disturbance and artifact intrusion continued long after descriptive reports of the 1976 site stratigraphy had been published and the artifact associations had refuted them. It was as if the excavations of 1976 had never taken place; in major influential papers, neither Dincauze (1984) nor Lynch (1990, 1991a) described the 1976 stratigraphy or mentioned the physical evidence of butchering of the juvenile mastodon with which the El Jobo point fragment was found.

In 1976, then, the Venezuelan site of Taima-taima produced field evidence that disproved the Clovis-first model. Nevertheless, critics were successful in placing the site in the "equivocal and controversial" category; and until recently, few North American archaeologists accepted the evidence from Taima-taima.

Tibitó

Tibitó is an open site in a small basin on the Sabana de Bogotá, an extensive ancient lake basin in the central highlands of Colombia. Discovery of mastodon bones in a ditch by local workmen led archaeologist Gonzalo Correal to excavate around a nearby large protruding boulder in search of an associated occupation site. He did find it there, and meticulous excavation of the activity areas around the boulder was reported in detail in a monograph published in Colombia (Correal 1981) and also in an article (Correal 1986) that was more widely accessible to North American archaeologists.

The Tibitó site produced a radiocarbon date of 11,740 ± 110 RCYBP; and fragmented horse and mastodon remains were well represented in the faunal assemblage, along with white-tailed deer. The remains of extinct species together with a terminal-Pleistocene date would indicate a proper Paleoindian site, but there were problems with that identification. The radiocarbon date was almost a millennium too early for the postulated arrival of Paleoindians in South America from the north. Of most concern, however, was the fact that no bifacial projectile points were to be found in the lithic assemblage; in fact, there was no evidence of bifacial flaking, even in the debitage. The assemblage of about 120 lithic artifacts consisted entirely of small amorphous chunks or flakes of chert with minimal edge retouch, an extremely simple lithic industry known in Colombia as the Abriense. Probably unaware of Correal's monograph, most of the critics of early sites in South America (e.g., Dincauze 1984; Owen 1984; Waters 1985) failed to mention Tibitó; but Lynch has dealt with it in several articles (Lynch 1990, 1991a, 1991b, 1998). The early radiocarbon date was not challenged (it was supported by the site's pollen profile, which placed the occupation zone in the Guantiva Interstadial, dated elsewhere in the Colombian highlands between 12,400 and 11,000 RCYBP). However, Lynch implied that the lack of any bifacial projectile points at the site might be only a sampling error, "the Paleoindian occupation too brief or too functionally restricted to include projectile points" (Lynch 1990:17), although bifacial points are extremely rare within the long-lived Abriense industry on the Sabana de Bogotá. The problem is that the Abriense industry should be "Early Archaic" (Lynch 1990: 17). Most recently Lynch (1998: 92) seems to envision an instant conversion of Paleoindians to an Archaic lithic technology upon their first entrance into the Sabana de Bogotá, although he does allow that the Abriense industry may have a separate origin.

The Tibitó site did not fulfill the expectations of North American archaeologists. Although it featured remains of extinct species, the association with a very simple unifacial lithic industry was difficult to reconcile with the North American concept of the Paleoindian stage.

Monte Verde

The early South American site which completely failed the expectations of the North American Paleoindian model was Monte Verde, located in the temperate rain forest of south-central Chile (Dillehay 1989, 1997). Radiocarbon-dated between 13,000 and 12,500 RCYBP, Monte Verde II was a settlement with substantial constructed dwellings occupied continuously through most of the year and supported principally by plant foods; and it featured an extremely simple lithic industry.

By a stroke of luck, the site had been preserved under a later peat deposit, so it contained many perishables—wooden frameworks for the contiguous dwellings, secured by stakes and cordage; fragments of hide cover; and a diversity of wooden artifacts including lances and digging sticks, mortars containing remains of edible plants, a firedrill hearth, hafts for stone tools; and useful or edible parts of a wide variety of plants derived from forest, wetland, Andean, and coastal environments. Associated were some modified bone and tusk fragments derived from seven individual mastodons, possibly scavenged. There were several bifacial lithic implements, including two long, narrow lanceolate projectile points; but hundreds of the lithic artifacts found on the settlement were simply naturally faceted or sharp-edged stones that had been utilized.

The site, first discovered in 1977, was excavated and analyzed in meticulous detail over a 10-year period, with contributions by 40 specialists in geology, soils science, petrology, palynology, botany, paleobotany, paleontology, zoology, fossil insects, genetics, chemistry, bone modification, wood modification, textile analysis, lithic technology, use-wear analysis, trace residue analysis, and radiocarbon dating. In sum, the research at Monte Verde was a thorough, modern, indeed state-of-the-art interdisciplinary archaeological project. The final report on the Monte Verde site came to a total of 1,400 pages in two volumes (Dillehay 1989, 1997). It is probably the most exhaustive report on any archaeological site of its size. In January 1997, a few months before the second volume of the report was published, a delegation of experts, which included staunch supporters of the Clovis-first model, examined the artifacts and visited the site; and arrived at a consensus that the site was indeed valid (Meltzer et al. 1997). With that, at long last the Monte Verde site was accepted by most archaeologists as conclusive field evidence of the failure of the Clovis-first model.

But in fact the issue was still not settled, as generalized skepticism and specific critiques of the site continued to be advanced by determined supporters of the Clovis-first model. The matter came to a head in October 1999 with publication of a detailed and accusatory critique of the final report as an insert in a popular archaeology magazine and its Web site by Stuart Fiedel (Fiedel 1999), answered in equal detail by Dillehay and associates on another Web site (Dillehay et al. 1999).

The heart of the Fiedel critique was the assertion, also made by others (e.g., Lynch 1990, 1991a; Haynes 1999) that only a handful of undoubtable, unequivocal, unambiguous artifacts were recovered at the Monte Verde site: the two lanceolate bifacial point, a slender polished slate lance or perforator, a large quartzite biface, several grooved pebbles, a basalt core, and several choppers. Everything else—the structures and features, all other lithics, all the bone and wood specimens—was asserted to be non-artifactual, equivocal, or not clearly demonstrated to be artifacts. Following an earlier speculation published by Lynch (1990), it was thought by Fiedel and Haynes that the few unquestionable lithic artifacts were intrusive down into an earlier natural accumulation of wood, bone, and stones from an imagined later "Archaic" occupation in the area. To account for the exotic plant material found on the site, Fiedel suggested that the plant remains were conveyed there from distant locales by far-ranging mastodons that had used the site as a latrine or a deathbed.

Monte Verde II was so completely different from North American expectations of what a site its age should look like that it is not surprising that some supporters of the Clovis-first model could not recognize it. Haynes (1999:19) spoke of "bizarre archaeological associations." West (1999) titled his supporting comment on Fiedel's critique, "The Inscrutable Monte Verde." Questioned West rhetorically: "Why does the picture presented of this occupation at the end of the Pleistocene accord not at all with what was happening anywhere else in the world 13,000 years ago? Why is there nothing of a corresponding nature or time found elsewhere in the New World? "(West 1999: 15). For Lynch (1991a), on the basis of its age and character, the Monte Verde site would have to be *pre*-Paleoindian, apparently for him an unsupportable concept:

> I am not ready to accept, even now, that *pre*-Paleoindians settled year-round at this site. The site may have been used in several seasons, but I continue to doubt that Monte Verde was a year-round permanent settlement (Pino Quivira and Dillehay 1988: 188) in a Paleoindian or hypothetical earlier cultural context. Paleoindians elsewhere were nomadic. Any predecessors likely lacked permanent abodes as well. p. 353

The Monte Verde site, then, was completely incompatible with the Clovis-first model and the corollary notion that the first peoples in South America had to be nomadic, specialized big-game hunters. Not only did the Monte Verde site destroy the Clovis-first model of the earliest colonization of the New World; it demonstrated that the North American evolutionary model of a basal Paleoindian stage followed by an Archaic stage was not applicable in South America.

Late-Pleistocene Adaptations in South America

At about the same time Monte Verde II was becoming widely accepted, the prominent publication of two other early South American sites also contributed to a different picture of the early prehistory of the southern continent. Pedra Pintada, deep in the Amazon basin, showed that foragers had adapted to the tropical rain forest by the end of the Pleistocene (Roosevelt et al. 1996). On the west side of South America, QJ-280, on the bank of the Quebrada Jaguay of southern Peru, showed that at the same time, by ca. 11,000 RCYBP, a specialized maritime economy had been established on the Pacific Coast (Sandweiss et al. 1998).

Pedra Pintada (Roosevelt et al. 1996; Roosevelt 1998) is a large rockshelter in a sandstone bluff, now ca. 10 km north of the Amazon River near the town of Monte Alegre, Pará, Brazil. Excavations in 1990–91 revealed a stratified site, with the lower occupation zones overlying sterile sand and separated from an upper ceramic horizon by sterile deposits. Radiocarbon dates on carbonized plant remains indicated that occupation of the rockshelter began ca. 11,100 RCYBP. Food refuse reflected exploitation of a variety of tropical forest fruits and nuts, riverine fish and shellfish, and small mammals and reptiles. A great deal of flaking debitage was recovered, but only about two dozen formal tools, the lithic assemblage featuring a simple stemmed point and fragments of other small bifaces, as well as unifacial artifacts including *limaces.* Spatters and fragments of red pigment within the early occupation zone indicated that the earliest people at the site were responsible for the pictographs on the walls of the rockshelter.

The early people at Pedra Pintada were generalized foragers, supported by an efficient hunting/collecting/fishing economy utilizing a variety of forest and riverine resources. It is clear that in South America the tropical rain forest, long considered a formidable environment completely inhospitable to early hunter-gatherers (e.g., Bailey et al. 1989) was in fact already populated by such people by the end of the Pleistocene.

A specialized maritime economy has long been known to characterize the preceramic history of the desert coast of Peru; but until the mid-1990s, archaeologists did not realize how far back in time this adaptation to the rich marine resources of the area extended. In 1996 excavations at Site QJ-280 near the Quebrada Jaguay produced radiocarbon dates ranging from ca. 11,000 to ca. 10,000 RCYBP for the lowest occupation level (Sandweiss et al. 1998).

QJ-280 is located on the west bank of the Quebrada Jaguay, a long incised valley on the southern coast of Peru. The food refuse preserved in the lowest occupation zone was solely marine or littoral in origin; fish, shellfish, and seabirds were exploited, with a focus on certain genera of fish and clams. The size of the fish suggested the use of nets. Postholes indicated a structure in the lowest occupation level, although flakes of obsidian from a high volcano in the Andes 130 km away suggested transhumance. Lithic artifacts, however, were only a few simple unifacial tools. Most of the material culture was likely perishable. By means of a simple lithic technology and a specialized maritime economy, then, people had adapted to the desert Pacific coast of South America by the end of the Pleistocene.

Less publicized in North America was archaeological research in southern Argentina which showed that by 13,000 radiocarbon years ago, about the same time as the Monte Verde II settlement, people had moved east of the southern Andes into the steppes of Patagonia. From 1990, research has been carried out by Laura Miotti and her associates in a locality known as Piedra Museo, where the coincidence of a small paleolake, extensive grasslands rich with game, convenient sources of good toolstone, and habitable rockshelters offered an optimal environment for early people (Miotti 1992, 1995; Miotti and Cattaneo 1997; Miotti and Salemme 1999; Miotti et al. 1999).

The key site in the Piedra Museo locality is a large, well-stratified rockshelter termed AEP-1 (Alero El Puesto), situated at the base of a rock outcrop overlooking the paleolake and surrounding grasslands. The lowest stratigraphic unit, Unit 6, consisted of a gray sand deposit directly over bedrock. In 1996 charcoal from Unit 6 produced a date of 12,890 ± 90 RCYBP, which corresponds with a date of 12,600 ± 600 RCYBP for the earliest occupation (Level 11) at the site of Los Toldos, ca. 50 km away (Cardich 1978); other dates from the lower part of Unit 6 were 11,000 ± 65 RCYBP on charcoal and 10,925 ± 65 RCYBP on bone. Pleistocene species constituted 44 percent of the identifiable faunal remains from Unit 6 at AEP-1: *Lama gracilis* (a small extinct camelid), *Hippidion saldiasi* (horse), and mylodontids (sloth) (Miotti et al. 1999). Approximately two dozen lithic artifacts were recovered from Unit 6, mainly flaking debitage and retouched flakes. No stone projectile points were recovered, but there were several bifacial thinning flakes and a highly polished bone point (Miotti and Cattaneo 1997).

At AEP-1, Unit 6 is separated from overlying Units 5 and 4 by a thin zone of calcium carbonate. A radiocarbon date of 10,400 ± 80 RCYBP was obtained on bone from Unit 5, and there are a few bones of *Lama gracilis* and *Hippidion,* although most of the faunal remains are of modern species of guanaco and rhea. Units 5 and 4 yielded many more lithic artifacts, including two stemmed fishtail points as well as unifacial knives and scrapers (Miotti 1995).

Miotti and her associates interpret the early archaeological record at AEP-1 as reflecting the first colonization of the grasslands on the central plateau of Patagonia at the end of the Pleistocene. They note that the pioneer people who left material in Unit 6 at AEP-1 ca. 13,000 radiocarbon years ago hunted a diversity of fauna; later people in the area focused upon only a few species, guanaco and rhea, as the Pleistocene fauna died out. It is clear that the early population, which had colonized the grasslands of Patagonia by the end of the Pleistocene, adjusted readily to the ensuing environmental changes in the region, in time becoming more specialized hunters (Miotti and Salemme 1999).

By the end of the Pleistocene, foragers were well adapted to the uplands of interior eastern Brazil, a region of broad valleys and cuestas covered by savanna-parkland vegetation *(cerrado)* or xerophytic scrub forest *(caatinga)* rich in edible plants and game. By 1990, early occupation levels in about half a dozen stratified cave or rockshelter sites in the states of Pernambuco, Piauí, Bahia, Goías, Minas Gerais, and Mato Grosso had produced radiocarbon dates of ca. 11,000 RCYBP or older (Schmitz 1987; Kipnis 1998: Table 1). Best described at present is the early occupation at the Lapa do Boquete in north-central Minas Gerais, where Level VIII, rich in charcoal and ash, produced five radiocarbon dates ranging from 11,440 ± 240 RCYBP to 12,070 ± 170 RCYBP (Kipnis 1998: Table 1).

The Lapa do Boquete (Prous 1991; Prous et al. 1992; Fogaça and Lima 1991) is a large rockshelter in a limestone formation overlooking the Rio Peruaçu, a tributary of the Rio São Francisco. Faunal and floral remains from the basal dated occupation level, Level VIII, indicate a generalized foraging economy, with exploitation of a variety of medium- to small-sized game, fish, freshwater mussels, land snails, and palm nuts and other fruits. The associated artifact assemblage features unifacial knives, scrapers, and *limaces.* As well, it includes *quebra-cocos* (nut-cracking stones), hammerstones, paint palettes, and bone artifacts. Similar lithic assemblages are found at other late-Pleistocene and early-Holocene sites in eastern Brazil (Schmitz 1987). Stone projectile points are very rare in eastern Brazil; and it can be presumed that projectile tips were made of wood or bone, as among ethnographic native groups in the area.

By the late Pleistocene, then, foragers were well established throughout the uplands of eastern Brazil. A diversified hunting/collecting economy is indicated, associated with unifacial flake tool industries. This region of South America has also produced evidence for a very early settlement of the southern continent, to be described later in this paper.

Even the high Andes were occupied by the end of the Pleistocene. The cave site of Pachamachay in central Peru (Rick 1980) is at an elevation of about 4300 m on the *puna* (alpine grasslands). The basal occupation level, Level 32, overlying a sterile deposit, produced a radiocarbon date of 11,800 ± 930 RCYBP. It appears that human adaptation to high-altitude Andean grasslands was completed very soon after the area was deglaciated. The early population, likely highly mobile, survived on the *puna* by hunting camelids, using bifacial triangular or leaf-shaped projectile points.

Thus the archaeological record for South America at the close of the Pleistocene, by ca. 11,000 radiocarbon years ago, shows foragers in the forests of Chile and Brazil, hunters in the grasslands of Argentina and the high Andes, and fisherfolk on the Pacific coast. In fact, all the major environmental zones of South America were already occupied by well-adapted populations maintaining diverse subsistence systems and distinctive regional technological traditions.

Present Implications of South American Archaeology for North American Models of the Early Settlement of the New World

Not only have early South American archaeological sites disproven the Clovis-first model of the peopling of the New World; they have demonstrated that the North American evolutionary model of a primary Paleoindian stage of nomadic specialized big-game hunters succeeded by an Archaic stage of localized broad-spectrum foragers does not apply in South America. Put simply, it is now clear that models of development devised for the northern continent cannot be projected to the southern continent. The early prehistory of each continent must be considered unique.

North American theories concerning the initial settlement of the New World may continue to founder on data from South America. After Monte Verde II was confirmed for most archaeologists in 1997, some used the Monte Verde radiocarbon dates to push back the date of initial entry across the Bering Straits only a few thousand years earlier than the old "Clovis barrier" of ca. 12,000 radiocarbon years ago; that is, to accommodate Monte Verde in far-distant Chile at ca. 12,500 RCYBP, the date of initial entry into Alaska could now be placed between ca. 15,000 and 13,500 RCYBP (cf. Fagan 1997; Dixon 1999:254–255). That dating would pose a problem in that the ice-free corridor east of the Rocky Mountains was closed at its northern end by the Laurentide ice sheet between ca. 30,000 and 11,000 RCYBP (Lemmen et al. 1994); so a coastal entry route would be required. But again, South American data challenge North American models; for two archaeological sites, one in south-central Chile and one in northeast Brazil, suggest that people had arrived in South America over 35,000 radiocarbon years ago.

One of these sites is at the Monte Verde locality itself. Across Chinchihuapi Creek from the 12,500 RCYBP settlement of Monte Verde II described above, a test pit in 1985 discovered a deeper occupation level, designated Monte Verde I, in a sand deposit (Dillehay and Collins 1988; Dillehay 1989, 1997). Charcoal fragments found in clay lenses, possibly hearths, gave radiocarbon dates of ca. 34,000 RCYBP, corresponding with the geological interpretation of the sedimentary context. Associated were 26 lithic artifacts, including a well-flaked basalt core with edge retouch, use-wear, and an organic residue identified genetically as proboscidean. The researchers at Monte Verde and other archaeologists who have examined the artifacts have no doubt of the cultural nature of the material recovered; but obviously a definite archaeological site in south-central Chile dated at ca. 34,000 RCYBP has tremendous implications and is very difficult for most archaeologists to consider seriously at this time. As Michael Collins (1999:15) puts it:

> What Dillehay and I have been reluctant to do is take a dogmatic stance on what we consider absolutely to be a cultural manifestation, but one of such age that its implications go far beyond what we are willing to engage on so little evidence. We felt it better to simply report what was found.

More research on this occupation level at Monte Verde is planned for the future; but in the interval, archaeologists devising models of the initial settlement of the New World have to keep it in mind.

Another very early South American site that is proving very difficult for North American archaeologists to accept at present is the Toca do Boqueirão da Pedra Furada, a large stratified rockshelter near the town of São Raimundo Nonato in the state of Piauí in northeast Brazil, which has produced a series of radiocarbon dates going back beyond 40,000 years. Archaeological excavations were carried out at this site between 1978 and 1988 as part of a major research project in the area under the direction of Niède Guidon and associates (Guidon 1986; Guidon and Delibrias 1986; Delibrias et al. 1989; Parenti 2001; Pessis 1996).

The Toca do Boqueirão da Pedra Furada is a long, high overhang at the head of a large recess in a high escarpment of consolidated sandstone. The natural deposits in the sheltered area, which consist of fine particles and angular fragments of sandstone derived by slow disintegration of the shelter walls, extended to bedrock at a depth of approximately 5 m. Throughout the lower 4 m of deposit, quartz flake and pebble tools were recovered in association with a number of large hearths and charcoal concentrations, which produced a stratified series of more than a dozen finite radiocarbon dates ranging from ca. 32,000–17,000 RCYBP, with several indefinite dates in underlying deposits suggesting an age of more than 40,000 years for the initial occupation of the shelter.

In 1993 a field conference sponsored by Guidon and her associates brought archaeologists from North America, South America, and Europe to the site (cf. Pessis 1996). Many of the North Americans could not accept a Pleistocene occupation at Pedra Furada. In a paper published in a major journal (Meltzer et al. 1994), three of the archaeologists advanced hypotheses to provide possible natural explanations for what appeared to be hearths and artifacts in the Pleistocene deposits. The charcoal concentrations within the shelter, they thought, following an assertion made in earlier critiques of the site (e.g., Lynch 1990), might possibly have resulted from forest fires, despite Guidon's failure to find any charcoal horizons in test trenches outside the sheltered area. In their view, the flaked quartz pebbles and cobbles might possibly be attributed to natural breakage in falls of stones from a conglomerate stratum exposed in a salient in the cliff high above the site, the broken stones then introduced into the shelter and further flaked in an assumed high-energy fluvial environment. It was calculated that over a time span of 50,000 years, natural flaking on cobbles, mimicking lithic technology, could have occurred frequently enough to account for the ca. 600 specimens classified as artifacts by Guidon and her associates. And, forgetting the Pleistocene archaeological record for Australia (cf. White and O'Connell 1982, Mulvaney and Kamminga 1999), the archaeologists wondered why, if *Homo sapiens sapiens* was involved, there was so little technological change observable in the lithic assemblages over 40,000 years or more of Pleistocene deposition at Pedra Furada.

The crucial issue at Pedra Furada is whether or not there were mechanisms for natural flaking at the site that might possibly result in the patterns of flake removal observed on the specimens classified as artifacts. Because of the depth of the overhang, quartz cobbles from the conglomerate stratum high on the cliff face above the site could not fall directly into the sheltered area; and the salient below which a cone of fallen quartz cobbles and pebbles has accumulated is ca. 10 m to the east of the shelter. In 1993, when the North American archaeologists visited the site, most of the sandy deposit within the sheltered area had been removed. The exposed stratigraphic sections they observed, outside the sheltered area in which the artifacts were found, were much coarser in texture, giving rise to the notion that a high-energy fluvial environment existed within the site. However, the researchers and expert visitors (including the author of this paper) who saw the site in the 1980s, before completion of the excavations, concluded that the sediments within the sheltered area were derived from slow weathering of the sandstone walls, not a fluvial environment capable of flaking quartz.

Guidon and her associates (Guidon and Arnaud 1991; Guidon et al. 1996) have responded to the criticisms of Pedra Furada; but, as usual during the long controversy as to whether or not a reported pre-Clovis site is valid, once seemingly possible alternative explanatory hypotheses were advanced by critics, despite response by the researcher, the site was classified as equivocal and thereby discounted. Consequently, many people have now written off Pedra Furada as a possibly very early site. This situation may change with the publication of a dissertation on the excavations, the stratigraphy, and the artifact analysis by Fabio Parenti, successfully defended in Paris in 1993 (Parenti 2001).

Two archaeological sites, then, hint that people first arrived in South America more than 35,000 years ago. This dating would provide sufficient time depth for gradual population growth and regional adaptation, giving rise to the diversity of cultural traditions to be seen in each of the major environmental areas of the southern continent by the end of the Pleistocene.

In this paper I have shown that by 11,000 radiocarbon years ago, the time of emergence of the Clovis horizon in North America, the southern continent was already settled and South American populations were already highly diversified in economy and technology. Assuming a Bering Strait entry point for the first peoples into the New World, one would still expect the earliest archaeological sites to be in North America; and perhaps now a concerted search for archaeological remains in the paleolandscapes of the late Pleistocene may locate them. Perhaps, however, if the entry route was coastal, littoral-adapted populations spread down to South America before expanding beyond the western mountain ranges or around the Gulf of Mexico into the heartland of North America (Gruhn 1994, 1997:102; also cf. Dixon 1999:fig. 2.7). Hopefully more research on both continents will clarify the present picture and lead to new productive models of the first peopling of the Americas.

References Cited

Alsoszatai-Petheo, J.
1986 An Alternative Paradigm for the Study of Early Man in the New World. In *New Evidence for the Pleistocene Peopling of the Americas,* edited by A. L. Bryan, pp. 15–23. Center for the Study of the First Americans, University of Maine, Orono.

Ardila, G., and G. Politis
1987 Nuevos datos para un viejo problema: investigación y discusión en torno del poblamiento de la América del Sur. *Revista del Museo del Oro* 23:3–45. Bogotá.

Bailey, R. C., G. Head, M. Jenike, B. Owen, R. Rechtman, and E. Zechenter
1989 Hunting and Gathering in Tropical Rain Forest: Is It Possible? *American Anthropologist* 91:59–82.

Bird, J. B.
1938 Antiquity and Migrations of the Early Inhabitants of Patagonia. *The Geographical Review* 28:250–75.

1969 A Comparison of South Chilean and Ecuadorian "Fishtail" Projectile Points. *The Kroeber Anthropological Society Papers* 40:52–71.

Bird, J. B., and R. Cooke
1978 The Occurrence in Panama of Two Types of Paleo-Indian Projectile Points. In *Early Man in America from a Circum-Pacific Perspective,* edited by A. L. Bryan, pp. 263–73. Department of Anthropology, University of Alberta, Edmonton.

Bryan, A. L.
1973 Paleoenvironments and Cultural Diversity in Late Pleistocene South America. *Quaternary Research* 3:237–55.

Bryan, A. L., R. Casamiquela, J. M. Cruxent, R. Gruhn, and C. Ochsenius
1978 An El Jobo Mastodon Kill at Taima-taima, Venezuela. *Science* 200:1275–77.

Bryan, A. L., and R. Gruhn
2000 Observations on the Final Demise of the Clovis-first Model. In *Arqueología, Historia, y Antropología: In Memoriam, José Luis Lorenzo Bautista,* edited by Jaime Litvak and Lorena Mirambell, pp. 85–101. Colección Científica no. 415, Instituto Nacional de Antropología e Historia, Mexico City.

Cardich, A.
1978 Recent Excavations at Lauricocha (Central Andes) and Los Toldos (Patagonia). In *Early Man in America from a Circum-Pacific Perspective,* edited by A. L. Bryan, pp. 296–300. Department of Anthropology, University of Alberta, Edmonton.

Collins, M. B.
1999 Reply to Fiedel, Part II. In Monte Verde Revisited: *Discovering Archaeology* Special Report, pp. 14–15. In *Scientific American DiscoveringArchaeology* 1 (6), November–December.

Correal Urrego, G.
1981 Evidencias *Culturales y Megafauna Pleistocénicas en Colombia.* Banco de la República, Bogotá.

1986 Apuntes sobre el medioambiente pleistocénico y el Hombre Prehistórico en Colombia. In *New Evidence for the Pleistocene Peopling of the Americas,* edited by A. L. Bryan, pp. 115–32. Center for the Study of the First Americans, University of Maine, Orono.

Cruxent, J. M.
1967 El Paleo-Indio en Taima-taima, Estado Falcón, Venezuela. *Actas Científicas Venezolanos* Suplemente 3:3–17.

Delibrias, G., N. Guidon, and F. Parenti
1989 The Toca do Boqueirão da Pedra Furada: Stratigraphy and Chronology. In *Early Man in the Southern Hemisphere,* edited by J. R. Prescott, pp. S 3–S 11. Supplement to *Archaeometry: Australasian Studies* 1988. Department of Physics and Mathematical Physics, University of Adelaide.

Dillehay,T. D.
1989 *Monte Verde: A Late Pleistocene Settlement in Chile. Vol. I: Paleoenvironment and Site Context.* Smithsonian Institution Press, Washington D.C.

1997 *Monte Verde: A Late Pleistocene Settlement in Chile. Vol.* 2: *The Archaeological Context and Interpretation.* Smithsonian Institution Press, Washington D.C.

1999 The Late Pleistocene Cultures of South America. *Journal of Evolutionary Anthropology* (6):206–216.

2000 *The Settlement of the Americas: A New Prehistory.* Basic Books, New York.

Dillehay, T. D., G. Ardila C., G. Politis, and M. C. C. Beltrão
1992 Earliest Hunters and Gatherers of South America. *Journal of World Prehistory* 6 (2):145–204.

Dillehay, T. D., and M. B. Collins
1988 Early Cultural Evidence from Monte Verde in Chile. *Nature* 332:150–52.

Dillehay, T. D., M. B. Collins, M. Pino, J. Rossen, J. Adovasio, C. Ocampo,

X. Navarro, P. Rivas, D. Pollack, A. G. Henderson, J. Saavedra, P. Sanzana, P. Shipman, M. Kay, G. Muñoz, A. Karathanasis, M. Cibull, and R. Geissler
1999 On Monte Verde: Fiedel's confusions and misrepresentations. www.uky.edu/projects/monteverde/

Dincauze, D. F.
1984 An Archaeological Evaluation of the Case for Pre-Clovis Occupations. In *Advances in World Archaeology,* vol. 3, edited by F. Wendorf and A. Close, pp. 275–323. Academic Press, New York.

Dixon, E. J.
1999 *Bones, Boats, and Bison: Archaeology and the First Colonization of Western North America.* University of New Mexico Press, Albuquerque.

Fagan, B. M.
1987 *The Great Journey: The Peopling of Ancient America.* Thames and Hudson, New York.
1997 The First Americans. Review of *Monte Verde: A Late Pleistocene Settlement in Chile. Archaeology* 50 (2):60–63. March/April.

Fiedel, S. J.
1987 *Prehistory of the Americas.* Cambridge University Press, Cambridge.
1999 Artifact Provenience at Monte Verde: Confusion and Contradictions. In Monte Verde Revisited: *Discovering Archaeology* Special Report, pp. 1–12. In *Scientific American Discovering Archaeology* 1 (6). November/December.

Figgins, J. D.
1927 The Antiquity of Man in America. *Natural History* 27 (3):229–39.

Flegenheimer, N., and M. Zárate
1997 Considerations on Radiocarbon and Calibrated Dates from Cerro La China and Cerro El Sombrero, Argentina. *Current Research in the Pleistocene* 14:27–28.

Fogaça, E., and M. A. Lima
1991 L' Abri du Boquete (Brésil). Les premiéres industries lithiques de l'Holocene. *Journal de la Sociètè de Américanistes* 77:111–23.

Gruhn, R.
1994 The Pacific Coast Route of Initial Entry: An Overview. In *Method and Theory for Investigating the Peopling of the Americas,* edited by R. Bonnichsen and D. G. Steele, pp. 249–56. Center for the Study of First Americans, Oregon State University, Corvallis.
1997 Language Classification and Models of the Peopling of the Americas. In *Archaeology and Linguistics: Aboriginal Australia in Global Perspective,* edited by P. McConvell and N. Evans, pp. 99–110. Oxford University Press, Melbourne.

Gruhn, R. and A. L. Bryan
1984 The Record of Pleistocene Megafaunal Extinction at Taima-taima, Northern Venezuela. In *Pleistocene Extinctions,* edited by P. S. Martin and R. G. Klein, pp. 128–37. University of Arizona Press, Tucson.

Guidon, N.
1986 Las unidades culturales de São Raimundo Nonato—Sudeste del Estado de Piauí—Brasil. In *New Evidence for the Pleistocene Peopling of the Americas,* edited by A. L. Bryan, pp. 157–71. Center for the Study of the First Americans, University of Maine, Orono.

Guidon, N., and M. B. Arnaud
1991 The Chronology of the New World: Two Faces of One Reality. *World Archaeology* 23 (2):167–78.

Guidon, N., and G. Delibrias
1986 Carbon-14 Dates Point to Man in the Americas 32,000 years ago. *Nature* 321:769–71.

Guidon, N., A-M Pessis, F. Parenti, M. Fontugue, and C. Guérin
1996 Nature and Age of the Deposits in Pedra Furada, Brazil: Reply to Meltzer, Adovasio & Dillehay. *Antiquity* 70:408–21.

Haynes, C. V.
1969 The Earliest Americans. *Science* 166:709–715.
1974 Paleoenvironments and Cultural Diversity in Late Pleistocene South America: A Reply to Alan L. Bryan. *Quaternary Research* 4:378–80.
1999 Monte Verde and the Pre-Clovis Situation in America. In Monte Verde Revisited: *Discovering Archaeology* Special Report, pp. 17–19. In *Scientific American Discovering Archaeology* 1 (6), November/December.

Hester, J. J.
1972 *Blackwater Locality no. 1: A Stratified Early Man Site in Eastern New Mexico.* Fort Burgwin Research Center Publication no. 8. Southern Methodist University, Dallas.

Hyslop, J. (editor)
1988 *Travels and Archaeology in South Chile,* by J. B. Bird. University of Iowa Press, Iowa City.

Jennings, J. D. (editor)
1983 *Ancient Native Americans.* W. H. Freeman and Company, San Francisco.

Kipnis, R.
1998 Early Hunter-gatherers in the Americas: Perspectives from Central Brazil. *Antiquity* 72:581–92.

Lemmen, D. S., A. Duk-Rodkin, and M. Bednarski
1994 Late Glacial Drainage Systems along the Northwestern Margin of the Laurentide Ice Sheet. *Quaternary Science Reviews* 13:805–28.

Lynch, T. E
1974 The Antiquity of Man in South America. *Quaternary Research* 4:356–77.
1983 The Paleo-Indians. In *Ancient South Americans,* edited by J. D. Jennings, pp. 87–137. W.H. Freeman and Company, San Francisco.
1990 Glacial-Age Man in South America: A Critical Review. *American Antiquity* 55 (1):12–36.
1991a Lack of Evidence for Glacial-Age Settlement of South America: Reply to Dillehay and Collins and to Gruhn and Bryan. *American Antiquity* 56 (2):348–55.
1991b Paleoindians in South America: A Discrete and Identifiable Cultural Stage? In *Clovis Origins and Adaptations,* edited by R. Bonnichsen and K. Turnmire, pp. 255–59. Center for the Study of the First Americans, Oregon State University, Corvallis.
1998 The Paleoindian and Archaic Stages in South America: Zones of Continuity and Segregation. In *Explorations in American Archaeology: Essays in Honor of Wesley R. Hurt,* edited by M. G. Plew, pp. 89–100. University Press of America, Lanham, MD.

Martin, P. S.
1973 The Discovery of America. *Science* 179:969–74.

Mayer-Oakes, W. J.
1986 El Inga: A Paleo-Indian Site in the Sierra of Northern Ecuador. *Transactions of the American Philosophical Society* 76 (4).

Meltzer, D. J., J. M. Adovasio, and T. D. Dillehay
1994 On a Pleistocene Human Occupation at Pedra Furada, Brazil. *Antiquity* 68:695–714.

Meltzer D. J., D. K. Grayson, G. Ardila, A. W. Barker, D. F Dincauze, C. V. Haynes, F. Mena, L. Nuñez, and D. J. Stanford
1997 On the Pleistocene Antiquity of Monte Verde, Southern Chile. *American Antiquity* 62 (4):659–63.

Miotti, L.
1992 Paleoindian Occupation at Piedra Museo Locality, Santa Cruz Province, Argentina. *Current Research in the Pleistocene* 9:27–30.
1995 Piedra Museo Locality: A Special Place in the New World. *Current Research in the Pleistocene* 12:36–38.

Miotti, L., and R. Cattáneo
1997 Bifacial/Unifacial Technology ca. 13,000 Years Ago in Southern Patagonia. *Current Research in the Pleistocene* 14:62–65.

Miotti, L., and M. Salemme
1999 Biodiversity, Taxonomic Richness and Specialists-generalists during Late Pleistocene/Early Holocene Times in Pampa and Patagonia (Argentina, Southern South America). *Quaternary International* 53/54:53–68.

Miotti, L., M. Vazquez, and D. Hermo
1999 Piedra Museo: Un *yamnagoo* pleistocénico de los colonizadores de la Meseta de Santa Cruz: el estudio de la arqueofauna. *Actas de las Terceras Jornadas de Arqueologia de la Patagonia,* 1999:113–136.

Morrow, J. E., and T. A. Morrow
1999 Geographic Variation in Fluted Projectile Points: A Hemispheric Perspective. *American Antiquity* 64 (2):215–30.

Mosimann, J. E., and P. S. Martin
1975 Simulating Pleistocene Overkill by Palaeo-Indians. *American Scientist* 63 (3):304–13.

Mulvaney, J., and J. Kamminga
1999 *Prehistory of Australia.* Smithsonian Institution Press, Washington D.C.

Nami, H.
1989 New Dates for the Paleoindian Societies of South America. *Current Research in the Pleistocene* 6:18–19.

Ochsenius, C., and R. Gruhn (editors)
1979 *Taima-taima: A Late Pleistocene PaleoIndian Kill Site in Northernmost South America—Final Reports of 1976 Excavations.* South American Quaternary Documentation Program, Universidad Francisco de Miranda, Coro, Venezuela. Reprinted in 1992 by the Center for the Study of the First Americans, Oregon State University, Corvallis.

Owen, R. C.
1984 The Americas: The Case against an Ice-Age population. In *The Origins of Modern Humans,* edited by F. H. Smith and F. Spencer, pp. 517–63. Liss, New York.

Parenti, F.
2001 Le Gisement Quaternaire de la Toca do Boqueirão da Pedra Furada (Piauí, Brésil), Stratigraphique, Chronologie, Evolution C:ulturelle. Editions recherche sur les civilisations, Paris.

Pessis, A.-M. (editor)
1996 Proceedings of the International Meeting on the Peopling of the Americas, São Raimundo Nonato, Piauí, Brasil. *Revista* da *Fundação Museu do Homen Americano* vol. 1, no. 1. São Raimundo Nonato.

Politis, G.
1991 Fishtail Projectile Points in the Southern Cone of South America: An Overview. In *Clovis Origins and Adaptations,* edited by R. Bonnichsen and K. Turnmire, pp. 287–301. Center for the Study of the First Americans, Oregon State University, Corvallis.

Prous, A.
1991 Fouilles de l'Abri du Boquete, Minas Gerais, Brésil. *Journal de la Sociètè des Americanistes* 77:77–109.

Prous, A., M. A. Lima, E. Fogaça, and M. E. Brito
1992 A industria lítica da Camada VIII da Lapa do Boquete, Vale do Rio Peruaçu, MG (Brasil). *Anais III do Congresso ABEQUA, Belo Horizonte,* 1992, pp. 342–62.

Rick, J. W.
1980 *Prehistoric Hunters of the High Andes.* Academic Press, New York.

Roberts, F. H. H., Jr.
1935 A Folsom Complex: Preliminary Investigations at the Lindenmeier Site in Northern Colorado. *Smithsonian Miscellaneous Collections* 94 (4).

Roosevelt, A. C.
1998 Paleoindian and Archaic Occupations in the Lower Amazon, Brazil: A Summary and Comparison. In *Explorations in American Archaeology: Essays in Honor of Wesley R. Hurt,* edited by M. G. Plew, pp. 165–91. University Press of America, Lanham, MD.

Roosevelt, A. C., M. Lima da Costa, C. Lopes Machado, M. Michab, N. Mercier, H. Vallada, J. Feathers, W. Barnett, M. Imazio da Silveira, A. Henderson, J. Sliva, B. Chernoff, D. S. Reese, J. A. Holman, N. Toth, and K. Schick
1996 Paleoindian Cave Dwellers in the Amazon: The Peopling of the Americas. *Science* 272:373–84.

Sandweiss, D. H., H. McInnis, R. I. Burger, A. Cano, B. Ojeda, R. Paredes, M. C. Sandweiss, and M. D. Glasscock
1998 Quebrada Jaguay: Early South American Maritime Adaptations. *Science* 281:1830–32.

Schmitz, P. 1.
1987 Prehistoric Hunters and Gatherers of Brazil. *Journal of World Prehistory* 1:126–61.

Snarskis, M. l.
1979 Turrialba: A Paleo-Indian Quarry and Workshop Site in Eastern Costa Rica. *American Antiquity* 44:125–38.

Tamers, M.
1971 IVIC Natural Radiocarbon Measurements VI. *Radiocarbon* 13:32–44.

Waters, M.
1985 Early Man in the New World: An Evaluation of the Radiocarbon-dated Pre-Clovis Sites in the Americas. In *Environments and Extinctions,* edited by J. I. Mead and D. J. Meltzer, pp. 125–43. Center for the Study of Early Man, University of Maine, Orono.

West, F. H.
1999 The Inscrutable Monte Verde. In: Monte Verde Revisited: *Discovering Archaeology* Special Report, pp. 15–16. In *Scientific American Discovering Archaeology* 1 (6), November/December.

White, J. P., and J. F. O'Connell
1982 *A Prehistory of Australia, New Guinea, and Sahul.* Academic Press, Sydney.

Whitley, D. S., and R. I.Dorn
1993 New Perspectives on the Clovis vs. pre-Clovis controversy. *American Antiquity* 58 (4):626–47.

Wormington, H. M.
1957 *Ancient Man in North America.* 4th edition. Denver Museum of Natural History Popular Series no. 4.

Hunting and Butchering Events at the Pleistocene/ Holocene Transition in Piedra Museo: An Example of Adaptation Strategies of the First Colonizers of Patagonia

Laura Miotti[1] and Mónica Salemme[2]

Abstract

This paper is an analysis of the archaeofaunal materials recovered from the Pleistocene and Holocene stratigraphic layers of the rockshelter known as the Alero El Puesto 1 site (AEP-1), Piedra Museo locality, situated in the central plateau of Santa Cruz Province, Argentina, 47° 53′ 42″ S, 67° 52′ 04″ W. The faunal remains yielded information on the paleoenvironmental evolution of this enclosed basin, the taphonomy, the subsistence systems, and use of the Patagonian region by the earliest hunter-gatherer populations.

Two strata in the site were defined. The upper one is an eolian unit (U1); the lower is a paleosol containing five units (from top to bottom U2–U6) based on the sedimentological characteristics. Radiocarbon dating confirmed two occupational events. Unit 6, dated between 12,890 ± 90 RCYBP (AA-20125) and 10,925 ± 65 RCYBP (OXA-8528), represents the initial Pleistocene occupation in southernmost Patagonia. Unit 4/5, dated between 10,400 ± 80 RCYBP (AA-8428) and 9230 ± 105 RCYBP (LP-949), represents the stage of effective colonization of the region. The Holocene context of Unit 2, dated at 7670 ± 110 RCYBP (LP-450), is related to the stage of territorial consolidation of the local hunter-gatherer societies.

Taxonomic analysis reveals greater biodiversity and a higher proportion of extinct taxa in U6 compared with U4/5. Likewise, U5 contains more extinct species than U4. However, there are no extinct species in Unit 2, where *Lama guanicoe* dominates the spectrum. The disappearance of taxa from U6 to U4 suggests two possibilities, sample bias or differential extinction of Pleistocene fauna. Based on comparative studies of other archaeological contexts of similar antiquity in the Patagonian region, the second hypothesis seems more reliable.

Remains of U6 and U4/5 relate to different depositional events, chronologically separated, as suggested by radiocarbon dating, different depths below the surface of the site, and changes in the sedimentology of the units. Both depositional zones contain archaeological remains that are interpreted as two different hunting events in the human colonization of the region.

According to the faunal assemblage, stratigraphic position, and radiocarbon dating, U2 represents the latest occupation in the site, during the early/middle Holocene.

Introduction

During the last decade, archaeological investigations in the Patagonian region have focused on regional archaeology; and most studies have investigated the relationship between the hunter-gatherers and the fauna of the Pleistocene/Holocene transition. In other words, paleoenvironments, human colonization, mass extinction, and the role of technology in this peopling process are considered key elements in human dispersal and consolidation of territory. However, during recent years archaeologists have been trying to explain why bone accumulations in the archaeological sites show high variability in skeletal parts. What does the overrepresented presence of any taxa mean? Were human hunter-gatherers the only accumulating agents? Put simply, why are these bones here? As Lyman (1994:168) states, "Large ISD [index of skeletal disjunction] or IFD [index of fragment of disjunction] values denote significant dispersal, but do not identify the agent(s) or process(es) responsible for that dispersal." Nevertheless, some

[1] Departamento de Arqueología, Facultad de Ciencias Naturales y Museo, Universidad de La Plata, La Plata, Argentina. e-mail: lmiotti@museo.fcnym.unlp.edu.ar

[2] CADIC–CONICET, Ushuaia, Argentina. e-mail: msalemme@infovia.com.ar

indicators on bone surfaces like cutmarks, hackle, helical fractures, and frequency of dissolution pitting and punctures enable us to infer patterns in the process of accumulation, in which the principal agent could have been human.

To answer these and other questions, this paper presents the results of our zooarchaeological research at Piedra Museo locality. In Patagonia, there are just a few sites with archaeofaunas of high resolution (Figure 1), but these archaeofaunas are proof of the differential use of space and resources by hunter-gatherer societies of the Pleistocene/Holocene transition—groups that were among first colonizers of the Americas (Miotti and Salemme 1999).

The spatial analysis (intra- and intersite) conducted by us in the central plateau of Santa Cruz province is linked with ethnoarchaeological, taphonomic, and experimental studies (Cáttaneo 1999; Hermo and Vazquez 1999; Miotti 1996, 1998; Miotti and Salemme 1988; Salemme and Miotti 1998). Likewise, our interdisciplinary approach incorporates studies of palynology, geomorphology, paleopedology, and paleontology; all these perspectives are taken into account in interpreting paleoenvironments (Borromei 2003; Miotti and Salemme 1999; Pardiñas 1999; Rabassa 1997; Zárate et al. 2000).

Environments and Setting

Extra-andean Patagonia, apparently homogeneous, actually comprises a wide spectrum of environments and resources. It is assumed that there were also significant regional ecological differences in the past. The studied area (Figure 1) is the major and continuous plateau that stretches from the Colorado River through the provinces of Río Negro, Neuquén, Chubut, Santa Cruz, and Tierra del Fuego, and includes the XII and XIII Chilean regions. Much but not all of this steppe biome constitutes the Patagonian physiographic province. Although frequently used synonymously, the steppe biome and the Patagonian plateau have different boundaries. At least two major ecological regions are recognized, the steppe and the woodland; each has its own subset of topographic and climatic conditions, soil types, precipitation and wind regimes, grass species associations, and faunal communities. A mosaic of forest islands exists within the steppe.

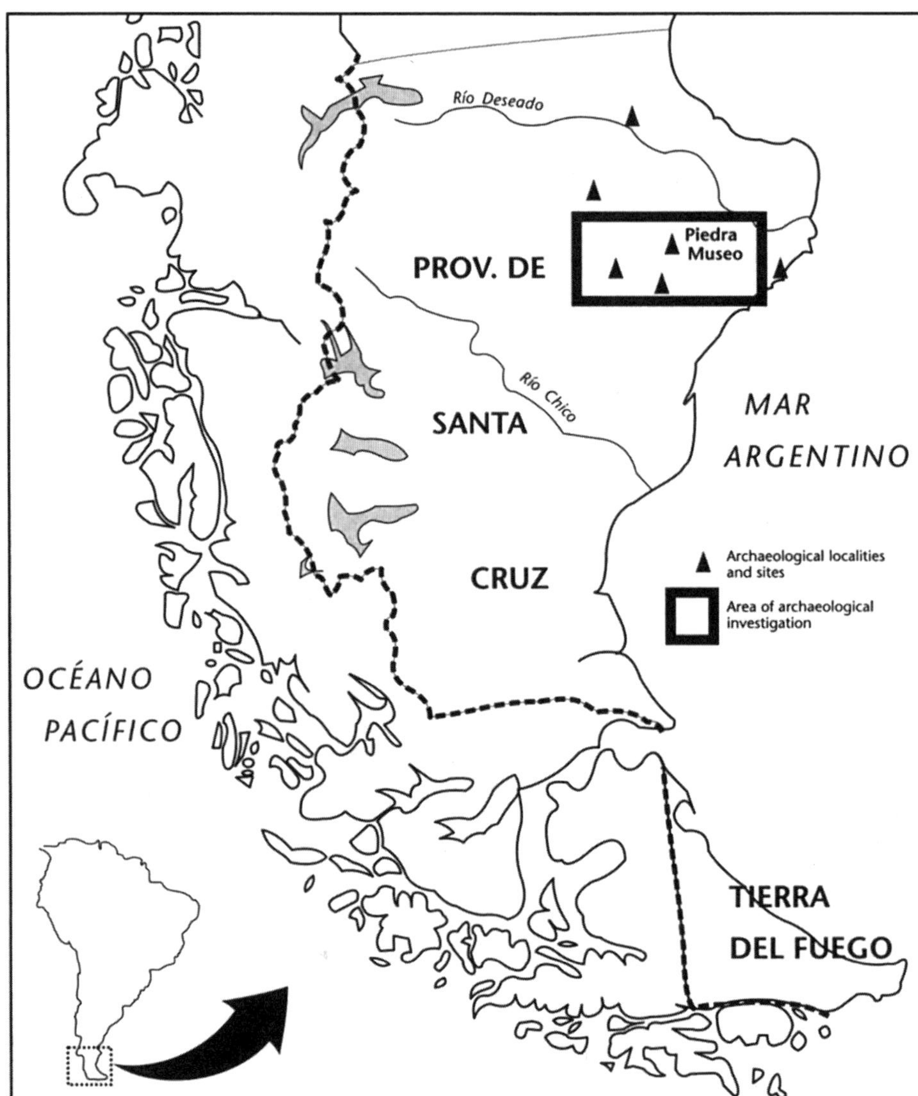

Figure 1. Patagonian sites of high resolution dating to Pleistocene/Holocene transition.

Within the steppe, the central plateau of Santa Cruz province is an independent geologic block with a specific environment; this physiographic structure is a volcanic plate known as Nesocratón del Deseado (De Giusto et al. 1980, 1982, 1994). Toward the west the basaltic plateau is crosscut by deep ravines, which toward the east dwindle to ephemeral shallow creeks; interior drainage basins (*bajos*) and lagoons are very frequent; volcanic cones and tuff formations produce a hilly landscape rather than a monotonous plateau of Patagonian steppe.

The Locality

Several open-air sites and rockshelters have been found in the Piedra Museo locality, which is situated in the central plateau, across the Zanjón Elornia, which flows to Laguna Grande (Figure 2). The AEP-1 site lies in an enclosed basin, a real topographic trap for animals.

AEP-1 (Figure 3) features two main archaeological contexts, the older dated between 12,800 and 9200 CALYBP, and the younger occupation 7600 CALYBP on average (Table 1). The excavated area is about 50 m^2 (Figure 4), which represents 80 percent of the surface under the rockshelter.

Six stratigraphic units were defined based on the sedimentological characteristics (texture, color, structures, granulometry, and limits). The preservation and spatial distribution of bones and bone modifications in each layer also contributed to separate this deposit (Zárate et al. 2000). It is an edaphic profile where there is a notable unconformity only between Unit 1 and 2; the other limits are gradual and imperceptible (Zárate et al. 2000).

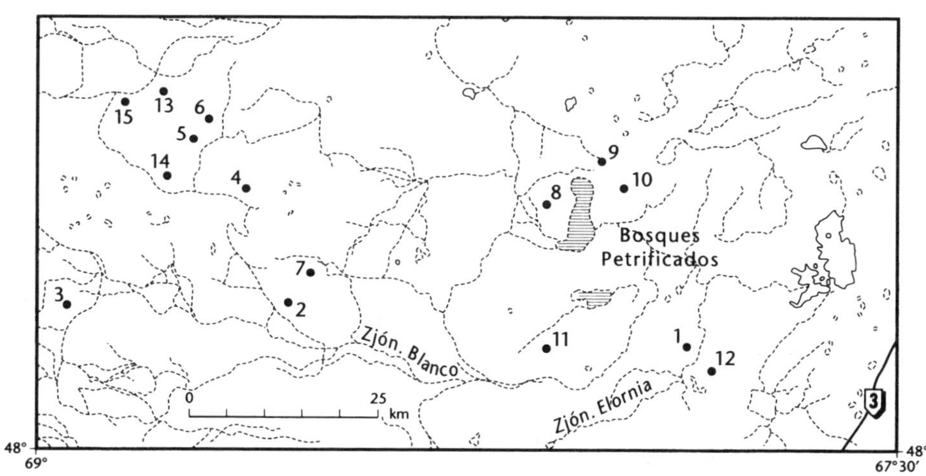

Figure 2. Topographic map of the study area in the Central Plateau of Santa Cruz province, Argentina, showing main investigated archaeological sites: **1,** Piedra Museo; **2,** Cueva Tito del Valle; **3,** Cueva Maripe; **4,** Cueva de la Hacienda; **5,** Cueva Moreno; **6,** La Leonera; **7,** Cueva Cerro Alto; **8,** Cerro Madre e Hija; **9,** El Cuadro; **10,** Talleres Bosque Petrificado; **11,** Pedrero Blanco; **12,** Cerros Colorados; **13,** Cerro Botellon; **14,** La Paloma; **15,** Cerro Bonete.

AEP-1 is the first site in the extra-andean Patagonian region of Argentina where fluted projectile points, called fishtail projectile points (FPPs), were found in the oldest major component associated with a high-resolution faunal assemblage and bone instruments. Radiocarbon analyses by the AMS method date the earliest human occupations of the site between 13,000 and 11,000 CALYBP (Table 1).

Taxonomic Diversity and Richness

The faunal assemblages of pedological units 4, 5, and 6 are very significant; they constitute the base of the sequence, corresponding to the colonization phase of the Pleistocene/Holocene transition (Miotti and Salemme 1999). They include remains of extinct megamammals and a species of flightless bird (*Rhea americana*) that does not inhabit this region at present, associated with guanaco (*Lama guanicoe*) and a smaller rheid (*Pterocnemia pennata*), both living today in the area. The taxonomic diversity and richness in the three units decreases from the bottom of the sequence to the top; moreover, the frequency of species decreases as indicated in percent NISP (Table 2 and Figure 5). Different taxa were identified through morphological and morphometrical analysis; for extinct horses, camelids, and rheids, multivariate analysis was also performed (Alberdi et al. 2001; Miotti ms; Salemme and Miotti 1998). The taxa recorded in AEP-1 were:

- ***Hippidion saldiasi***, an American extinct horse, is represented by a mandible, lumbar vertebrae, metapodials, and phalanges. Only one phalanx belongs to a young individual. A distal humerus from U6 has been radiocarbon dated at 10,925 ± 65 RCYBP (OXA-8528) (see Table 1); this specimen showed cutmarks, helical fractures with a point of impact, and hackles on the surface of the fracture. All these features, which are excellent indicators of human butchering, support the hypothesis that native horses were used as a profitable resource.
- ***Lama (Vicugna) gracilis*** is a small extinct camelid that occurs from stratigraphic units U6 to U4 in different frequencies. Pieces from fore and hind limbs are the best represented. This species was also recorded in other Patagonian archaeological sites, some of them nearby, and in Pampean paleontological sites (Cardich and Miotti 1983; Massone et al. 1998; Menegaz et al. 1989; Mengoni Goñalons 1976–80; Miotti 1998). A distal portion of a radius-ulna (AEP1Bs352) shows cutmarks and hackles, and a proximal radius-ulna (AEP1DwG10) shows cutmarks close to the joint. These specimens (Figure 6) suggest this species, as well as the horses, was used in human subsistence (Miotti 1995, 1998; Miotti et al. 1996)
- **Rheidae** (American ostrich) is represented by two species. The phalanges constitute 98 percent of the total NISP of these

Table 1. Radiocarbon datings of AEP1 rockshelter.

Stratigraphic unit	Date code	Sample material	Taxon	Date (RCYBP)	Calibrated date (CALYBP) (95%)◆
2 middle	OS-23745■	charcoal	indet.	7470 ± 140	8650 – 8100
2 bottom	LP450▼	bone	*Lama guanicoe*	7670 ± 110	8850 – 8200
4 top	LP 949▼	bone	*Lama guanicoe*	9230 ± 105	10,740 – 10,260
4 bottom	LP 859▼	bone	*Lama guanicoe*	9710 ± 105	11,350 – 10,700
5 bottom	AA-8428■	bone	Camelidae	10,400 ± 80	12,900 – 11,800
6 top	OXA8527■	bone	*Lama guanicoe*	10,390 ± 70	12,900 – 11,800
6 middle	GRA9837■	charcoal	indet.	10,470 ± 65	12,900 – 12,000
6 bottom	OXA8528■	bone	*Hippidion saldiasi*	10,925 ± 65	13,200 – 12,870
6 bottom	AA-27950■	charcoal	indet.	11,000 ± 65	13,250 – 12,800
6 bottom	AA-20125■	charcoal	indet.	12,890 ± 90	16,000 – 14,400

◆OxCal v3.3 Bronk Ramsey (1999) Oxford University program
▼Standard method
■AMS method

flightless birds. *Pterocnemia pennata* (*choique* or *ñandú petiso*) could be clearly distinguished from *Rhea americana* (*ñandú de las pampas*) by morphological analysis of anatomical units and multivariate analysis (Salemme and Miotti 1998).

- ***Mylodon* sp.** (ground sloth) is poorly represented; the only pieces recorded are a few ribs and a portion of skull (NISP in Table 2). The primitive nature of the bone tissue in these megamammals contributes to poor bone preservation, which makes it difficult to detect bone modifications by anthropic or non-anthropic agents. A sample from a rib was processed for radiocarbon dating, but unfortunately it did not contain collagen. At present, another sample of *Mylodon* bone, recovered close to the charcoal samples, that dated 12,890 ± 90 RCYBP (AA-20125) is still under analysis. In his latest report, Dr. Douglas Donahue informed us that "if the bone is really about 12,000 years old, it may not be well enough preserved to allow a radiocarbon measurement on its collagen. We will find out" (D. Donahue, pers. comm. 1998). Donahue also confirmed that sample AA-20125 was measured correctly, reinforcing his assessment of the great antiquity of the samples from U6.
- ***Lama guanicoe*** (guanaco) is a gregarious mammal; it is the species most frequently found throughout the sequence at AEP-1, as well as in the rest of sites all over Patagonia (Miotti and Salemme 1999). Every skeletal part of this extant camelid is found in different proportions in each of these stratigraphic units, integrating bone piles. The guanaco is also the best-represented species in the Holocene occupation of AEP-1. There are abundant cutmarks and intentional fractures on skeletal parts of guanaco, not only on long bones but also on vertebrae (Miotti 1999). This evidence has been verified in U6 as well as U4/5 and U2.

Figure 3. Entrance to AEP-1 Rockshelter.

- ***Canis* sp.** (canid) is represented by just a few bones of the distal leg. There are no marks or other evidence of anthropic action, such as high frequencies of selective skeletal parts or several individuals of the same or different species. Only one specimen of end tibia of a small carnivore (like wild cat or fox) shows a sign of puncture (F. Ramírez Rozzi, pers. comm.). However, the presence of this bone has been interpreted as introduced naturally within the matrix, that is, the result of a postdepositional event.
- **Medium-sized Aves.** There are 35 specimens of different long bones of birds in Unit 4/5 and other 6 in Unit 6 (see Table 2). Even though those remains could not be identified taxonomically to a specific level, the sizes range between partridge-like and wild geese–like. These remains of avifauna do not show marks or cutmarks that indicate the accumulator agents.
- **Rodentia.** Bones of different species of rodents, apparently introduced in the deposit by natural agents (i.e., owl pellets), became useful paleoenvironmental indicators. There is no evidence of serious disturbance by rodents in the site formation processes. However, some bones show marks of gnawing. The extremely low incidence of rodent remains (1.5 percent) suggests at least postdepositional restructuring of the bone materials at the site.

Spatial Distribution of Species and Anatomical Units of Prey

Note that in a previous paper (Table 2 in Miotti et al. 1999), the taxonomic abundance was presented separately for units 4, 5,

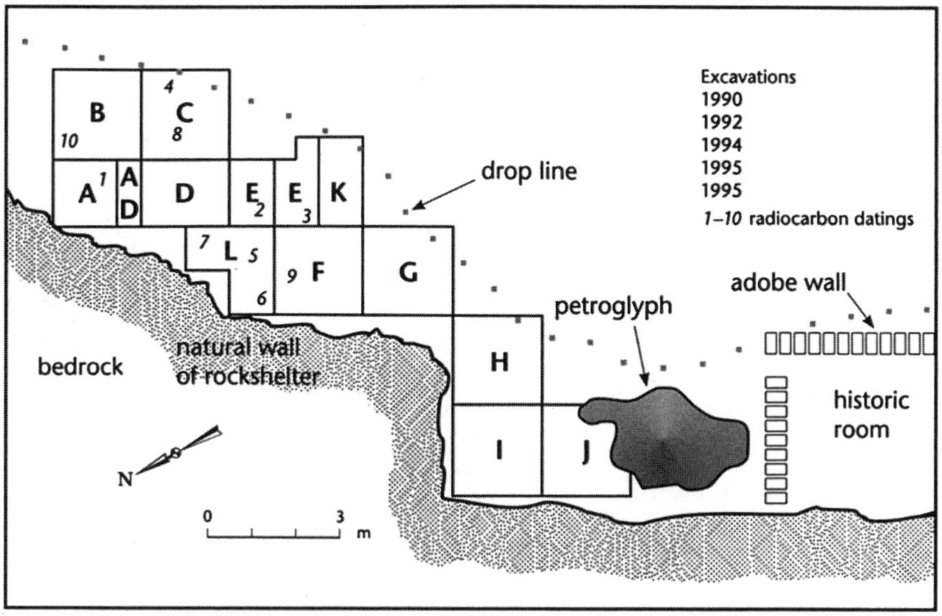

Figure 4. Plan of excavated area of AEP-1. Italic numbers indicate the distribution of radiocarbon datings (see Table 1).

and 6. However, after successfully refitting (mechanical and anatomical) long bones from both units and matching paired bones of the same individual, we decided to combine units 4 and 5 as evidence of a unique event or several events that occurred within a brief period. This decision is confirmed by this report, which documents trends of disappearance or decrease of taxa in these Paleoindian occupations (Table 2 and Figure 5). The scatterplot presented in previous papers (Miotti 1996; Miotti et al. 1999) shows that bones are especially concentrated in three different locations in each pedological unit. Note, for example, the bone plot of unit 4/5 and 6 (Figure 7).

A very detailed analysis has been made of remains of *Lama guanicoe*, such as different indices of economic utility (percent FUI and percent MGUI), survival and mineral density, and Spearman's correlation coefficients (the ratio of percent MAU:Bone Mineral Density, and the ratio of percent MAU:percent MGUI. From these results we infer the bone piles resulted from hunting activities and represent a locus of procuring and dismembering prey (Miotti et al. 1999).

The functionality of the site during this event is supported by the analysis of fractures, cutmarks, and other pre- and post-depositional and post-burial bone modifications (Miotti et al. 1995; see Figure 7). In addition, the low incidence of pitting and punctures (less than 1.5 percent) suggests patterns of human activities. In the Patagonian region during the 19th century, chroniclers recorded evidence in the accumulations of dismembered prey of similar behavior by aborigines. Claraz (1988) describes how Aonikenk or Tehuelches used topographic traps for communal hunting. A special place mentioned by Claraz and known as Yamnagoo, which means "paradise for the hunters," resembles Piedra Museo, with an enclosed basin with shallow lagoon, surrounded by low hills, abundant grasses, and rich faunal resources.

Continuing the intra-site analysis, around 0.7 m above the oldest archaeological component of AEP-1 in Unit 2 was recovered another stratigraphic archaeological context where only remains of

Table 2. Taxonomic abundance of Units 4/5 and 6.

	Unit 4/5		Unit 6		Units 4/5 and 6	
Taxon	NISP	Ratio (%)	NISP	Ratio (%)	NISP	Ratio (%)
medium-sized Aves	35	4.9	6	6.6	41	5
Rheidae	93	13	10	11	103	12.6
large mammals	35	4.9	22	25	57	7.1
Canis sp.	7	1	0	0	7	0.85
Mylodon sp.	0	0	5	5.5	5	0.62
Lama guanicoe	540	75	32	35.5	572	71
Lama gracilis	3	0.4	9	10	12	1.5
Hippidion saldiasi	4	0.7	6	6.6	10	1.2
Total	717		90		807	

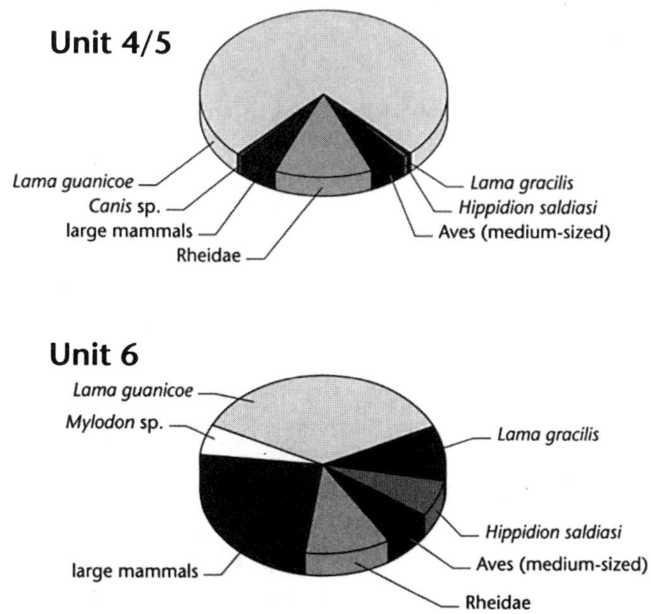

Figure 5. Taxonomic diversity and richness in the earliest Units 4/5 and 6, demonstrated by relative numbers of principal fauna.

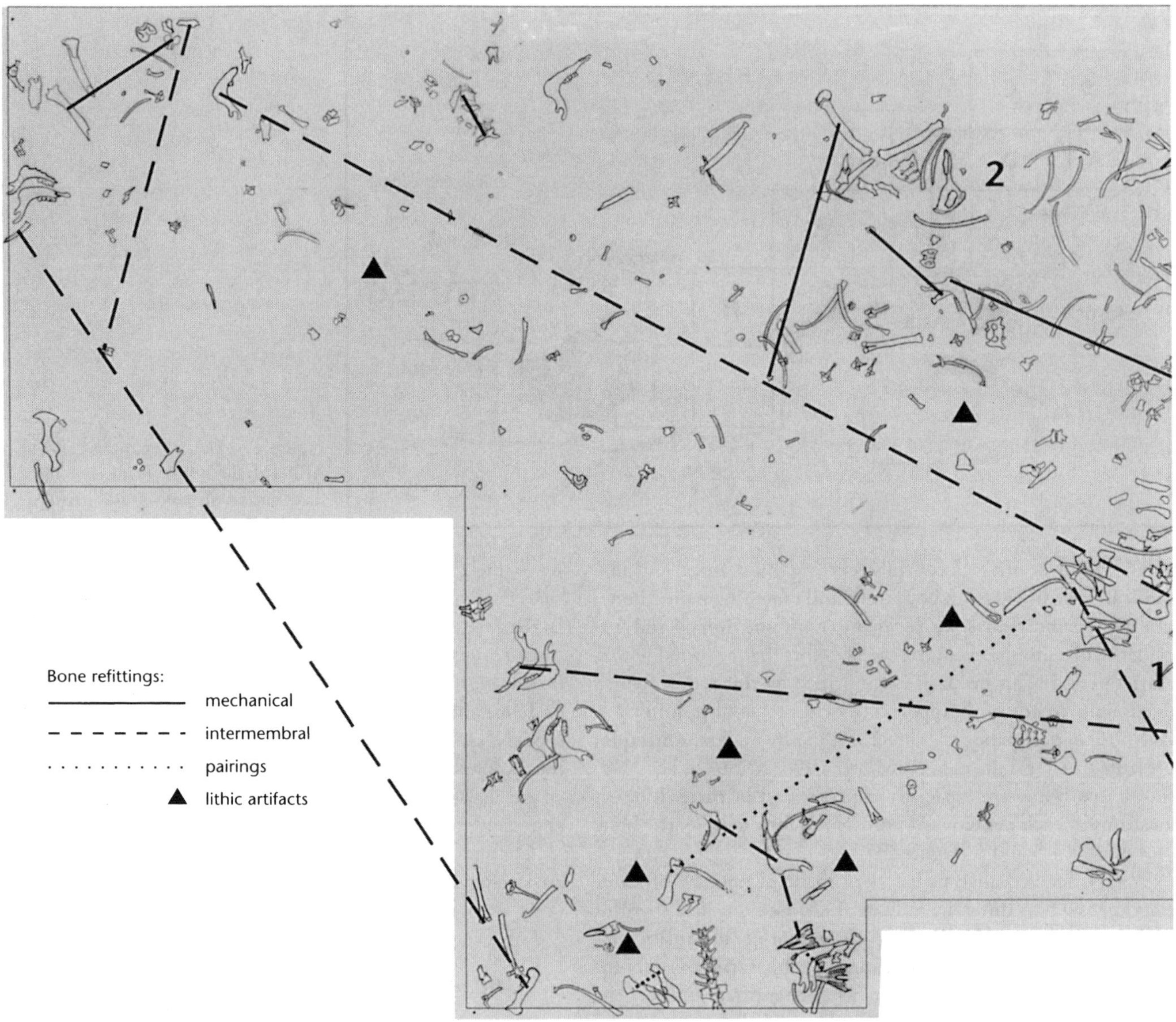

Figure 7. Scatterplot of Units 4/5 and 6 (after Giardina et al. 2000).

living fauna were recorded: guanaco, canids, *ñandú petiso* (*Pterocnemia pennata*), armadillos (*Zaedyus* sp.), and scarce fragments of marine mollusk shells (*Mytilus sp.*) associated with lithic artifacts and sophisticated bone tools reflecting intensive use of bone as raw material. This context has been dated between 7700 and 7400 CALYBP (Table 1). The lithic technology in this Unit 2 is similar to other regional assemblages like those of Fell III, late Toldense at Los Toldos and La Martita caves, and regional level Rio Pinturas II, all of them dating to early Holocene (Miotti 1996, 1998).

The Regional Level: The Intersite Analysis

A similar locus of kill and butchering occurs only in one other locality in the Southern Cone associated with FPPs and mastodon bones from a hunting event, the Tagua Tagua site in central Chile (Núñez et al. 1994). Moreover, the early component at AEP-1 could be related to the Nivel 11 and the initial-Toldense industries from Los Toldos and El Ceibo localities (*sensu* Cardich 1987). Although FPPs have not been recorded at these sites, all the other artifacts appear to display similar lithic technology. Likewise, extinct species of megamammals were associated in both site contexts. Furthermore, *Rhea americana* has been found in Los Toldos Cave 3 in association with the Toldense (Cardich and Miotti 1983; Salemme and Miotti 1987; Tambussi and Tonni 1985), confirming that the "big *ñandú*" was present up to ca. 9000 CALYBP on the central plateau of Santa Cruz.

Other sites identified as kill sites, kill and butchering sites, or multiple-activity sites are located in the Magellan Basin on the Chilean side (Fell's Cave, Palli Aike Cave, Cerro Sota, Cueva del Medio I and III, and Mylodon Cave), as described by Miotti and Salemme (1999). In these contexts extinct species except *Rhea americana* are associated. The lithic industry is characterized by unifacial and bifacial technology; however, fishtail projectile points are absent from Mylodon Cave. The occupations at these sites date between 12,000 and 8500 CALYBP.

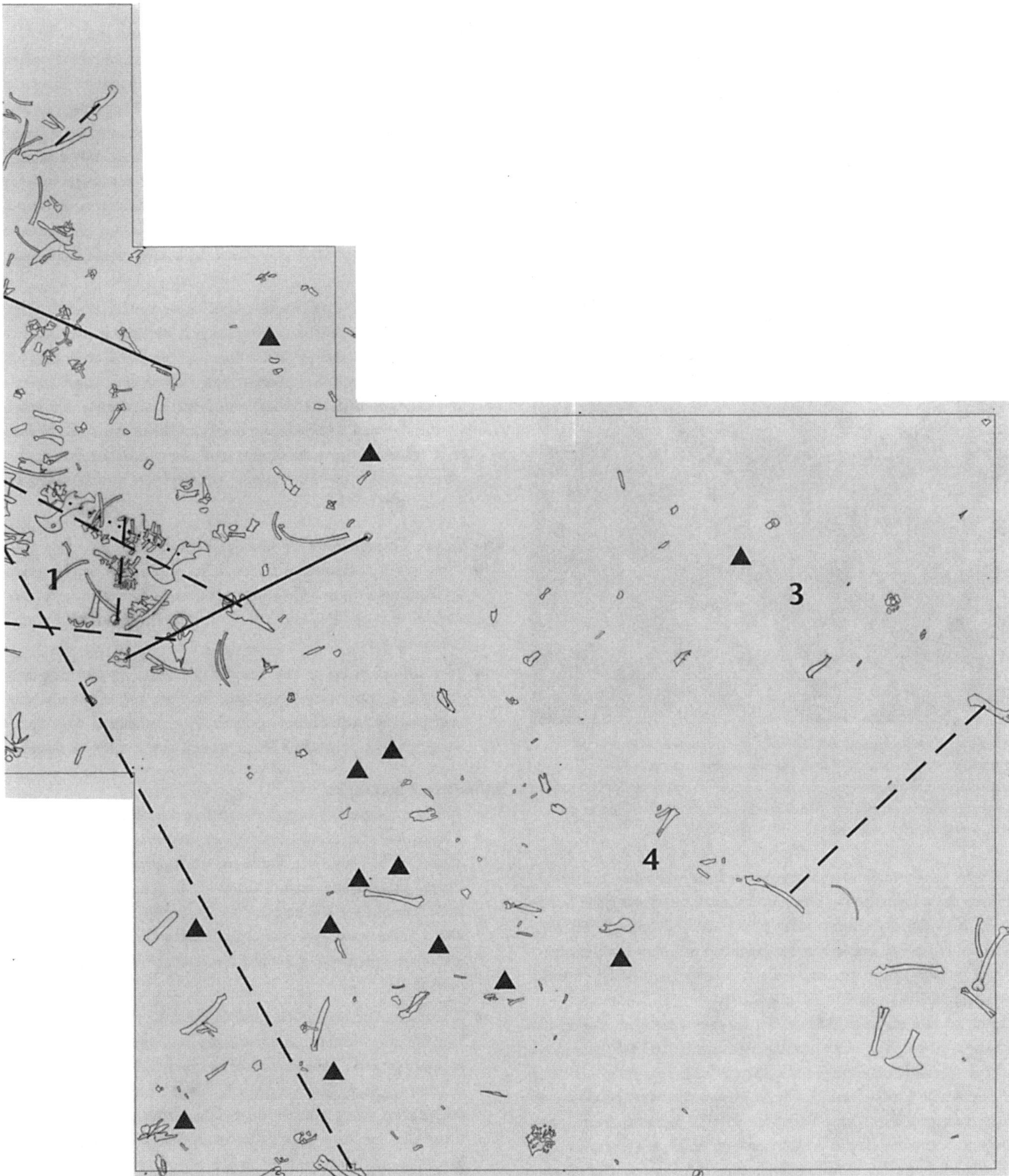

Thus, a similar environmental panorama can be found in different areas in Patagonia by the time of the Initial Colonization Phase. The archaeological associations suggest that hunter-gatherers were moving to resources (raw materials, water, prey, etc.) following "forager" strategies, that is, moving camps recurrently. Colonizers coexisted with the last megamammals under strong environmental stress at the end of the Pleistocene and the beginning of the Holocene. In any case, those societies learned how to

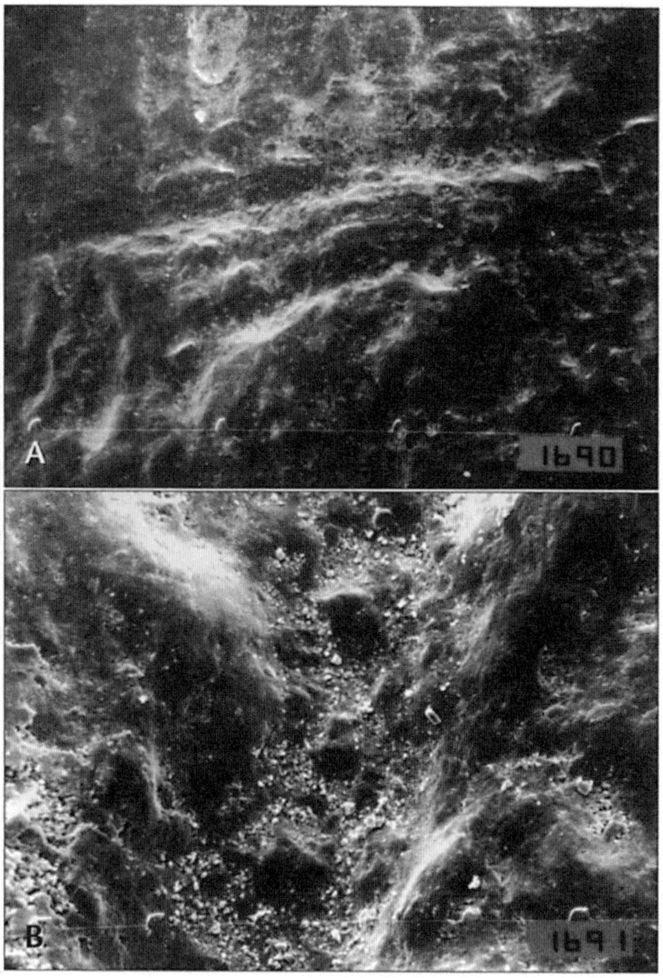

Figure 6. Microphotographs (MEB) of cutmarks on proximal epiphysis of radio-ulna of *Lama gracilis* (extinct camel). **A,** 35X magnification. **B,** transverse section of same cutmark at 200X magnification; evidence of microstriations on the right wall indicates the cutmark was made with the edge of a lithic tool.

live under such conditions, using generalized strategies and building their own social environments. In contrast, during the Territorial Consolidation Phase after 8500 CALYBP, people became "collectors," exercising a more specialized strategy that focused on one or two main resources; in this interval, resources were brought back to a residential base camp.

One of the authors (Miotti 1996) theorizes that the social landscape of the region was changing during the Holocene, and that the relationships among the hunter/gatherer groups during the Territorial Consolidation Phase were supported by alliances and exchange (Miotti and Salemme 1999). Items such as spear points and marine shells, which serve to mark territoriality, are likely candidates for items to exchange. Moreover, despite increasingly dense populations, groups continued to be mobile as late as the end of the Pleistocene. The plasticity of these societies, combined with the benefits of high technology, a communication network (through the technology and the ideas expressed in the rock art), and the existence of vacant ecological niches, contributed to the success of territorial occupation.

Conclusions

The Lower Component of AEP-1 Rockshelter

- The surrounding landscape in the Piedra Museo locality suggests this "situational locus" (in the sense of Binford 1983) was very different from today's. Hunter-gatherers would have found resources year round, thanks to a paleo lagoon in the center of a smooth hilly area that concentrated abundant game and subsistence resources on its boundaries, and habitation in nearby rockshelters. All these landscape features could have provided an excellent place for hunting activities: the ambush, killing, and butchering of game.
- Distributional, taphonomic, and bone modification analyses at Piedra Museo Locality suggest the lower component of AEP-1 rockshelter was a locus of confined activities related primarily to butchering megamammals, large American ostrich, and camelids, which were hunted at the edge of the lagoon approximately 50 m from the rockshelter. The relationship with nearby and contemporaneous open-air sites indicates major raw materials were procured over a local range of 6–20 km.

The Upper Component of AEP-1 Rockshelter

- The upper component contains only remains of living autochthonous mammals like guanaco and canids, the smaller *ñandú,* armadillos, and scarce fragments of marine shells (*Mytilus* sp.).
- Not only is bone in this context evidence of hunting and processing activities, it also indicates the use of bone as raw material for manufacturing tools. The associated lithic technology is similar to Fell III, assigned to the early Holocene.

Subsistence Strategies

The evidence reinforces our previous hypotheses:

- The faunal structure, stratigraphic position, and radiocarbon dating of U2 confirm the latest occupation in the site occurred during the early/middle Holocene, by the time of the Consolidation Phase in the conquest of Patagonian territory. The strategies had dramatically changed from generalist to specialist; guanaco became the most popular resource.
- The social and ecological landscape of the region was changing. Alliances and exchange maintained relationships among pioneer groups, demonstrated by the distribution of allochthonous goods like spear points and marine shells. The population was probably increasing, although mobility continued to be very high. The recorded sites suggest more places were inhabited by more people or repeatedly for longer periods.
- For the first interval of colonization in Patagonia and Pampa regions, a generalized subsistence strategy was practiced because of the availability of many species and the rich paleoenvironment (Miotti and Salemme 1999). Taxonomic analysis of AEP-1 reveals greater biodiversity and a higher

proportion of extinct taxa in U6 than in U4/5. Whether U4/5 is a single depositional event or the result of successive events, a few sequential hunting events is the most parsimonious hypothesis for explaining the decrease of extinct species from level U5 to U4.

Disappearance of Taxa

- The disappearance of taxa in the zooarchaeological record from U6 to U4/5 bears two possible explanations: 1) sample bias; or 2) differential extinction in the long-term evolutionary sequence, which led to the "massive extinctions" of Pleistocene fauna. The comparative studies of other archaeological contexts of similar chronology from the Patagonian region favor hypothesis (2).
- In contrast, there are no records of extinct species in U2, where *Lama guanicoe* dominates the faunal spectra. Unit 2, which dates to the early/middle Holocene (Table 1), represents the Territorial Consolidation Phase. By this time, the locale was occupied as an area of multiple activities that involved butchering, discarding the remains of prey and probably other resources, extracting marrow, and processing and eating foods (in the sense of Marshall 1993). The presence of only living species in the area and the high incidence of bone fracture suggest that the intrasite space was used differently than in the Paleoindian occupations, when the major activity would have been butchering large units of prey.

In sum, in southernmost Patagonia in the late Pleistocene, the rockshelter seems to have been the locus of a few tasks related to brief hunting and butchering events; in the Holocene, intra- and intersite space appears to have been used for other kinds of activities, mostly at residential campsites or multiple-activity sites. We caution, however, that since many of the studied contexts have a very complex taphonomic history and low or even very low archaeological resolution (coarse-grained in the sense of Lyman 1994), these interpretations may be biased.

The totality of the evidence—the discrete size of sites; the great variability of the technological contexts, with low tool standardization and high artifactual and raw material variability intra- and intersite (Scheinsohn 1999); great biodiversity; intrasite areas of few activities contrasting with intersite areas of multiple activities—indicates that the earliest hunter-gatherer strategies during the Pleistocene/Holocene transition in the Patagonian central plateau were of a generalized or forager nature.

Piedra Museo, because of the integrity and high resolution of the archaeological record (in the sense of Lyman 1994), is an excellent example of an occupation of the initial colonization phase. Evidence of a highly mobile hunter-gatherer group that exploited animal species of widely varying ethology and habits is the outstanding aspect of Piedra Museo. The integrity and resolution of its archaeofaunal evidence show that postdepositional processes are less important than the accumulation and preservation of archaeological remains.

Piedra Museo has been interpreted as a special place (Miotti 1995) where colonizers practiced an opportunistic strategy, probably hunting prey around the nearby lagoon or scavenging dead animals, partially butchering them in the nearby rockshelter and transporting pieces to another, unknown place. Lithic raw materials were obtained from local sources, not farther than 20 km. The hunter-gatherers were foragers until environmental conditions changed because of factors such as the increasing evolutionary success of gregarious species (camelids and rheids) and dwindling supplies of fresh water; their increasing population also encouraged a change in subsistence strategy.

Finally, we observe that the Piedra Museo complex and the El Ceibo, Los Toldos, and 3T localities (and possibly others) constitute a unified sociocultural system. The aggregate sites could also be related through rock art, which also describes a different distribution. We postulate a differential use of space whose variability became more pronounced during the Territorial Consolidation Phase, as is shown in the Upper Component at AEP-1 during the middle Holocene.

Acknowledgments

A previous version of this paper was presented in the Symposium "High Resolution Faunas at the Pleistocene/Holocene Boundary" (ICAZ '98, Victoria, Canada) and by one of the authors (L.M.) at the Clovis and Beyond Conference (Santa Fe, N. Mex. 1999). Funds from National Geographic Society partially supported this research (Grant #6371-98).

References Cited

Alberdi, M. T., L. Miotti, and J. L. Prado
2001 The South American Horses in the Patagonian Ecosystems during the Pleistocene-Holocene Transition. (Ed. Plenum). Cap.: *Hippidion saldiasi* Roth, 1899 (Mammalia, Perissodactyla) at the Piedra Museo Site (Patagonia): Their Implication for the Regional Economy and Environmental Reconstructions. *Journal of Archaeological Science,* 28:411–19.

Binford, L.
1983 *In Pursuit of the Past.* Thames and Hudson, London.

Borromei, A. M.
2003 Palynology at Piedra Museo Locality, Santa Cruz Province, Argentina. In *Where the South Winds Blow: Ancient Evidence of Paleo South Americans,* edited by L. Miotti, M. Salemme, and N. Flegenheimer, pp. 113–20. Center for the Study of the First Americans, Texas A&M University, College Station.

Cardich, A.
1987 Arqueología de Los Toldos y El Ceibo (provincia de Santa Cruz, Argentina). In *Investigaciones paleoindias al sur de la línea ecuatorial,* (Núñez, L. & B. Meggers, eds.). Estudios Atacameños, No. 8:98–117, Chile.

Cardich, A., and L. Miotti
1983 Recursos faunísticos en la economía de los cazadores-recolectores de Los Toldos (provincia de Santa Cruz, Argentina). *Rev. Relaciones* XV:147–57. Sociedad Argentina de Antropología, Buenos Aires.

Cattáneo, G. R.
1999 Organización de la tecnología en la Patagonia centro meridional: el caso de la localidad Piedra Museo, pcia. de Santa Cruz. *XII Congreso Nacional de Arqueología Argentina,* Actas III:16–22. La Plata.

Claraz, J.
1988 *Diario del viaje de Jorge Claraz al río Chubut por la ruta del centro.* (Rodolfo Casamiquela, ed.) Ediciones Marymar. Buenos Aires.

Di Giusto, J. M., C. A. Di Persia, and E. Pezzi
1980 Nesocratón del Deseado. *Geología Regional Argentina,* vol. II:1389–1430. Academia Nacional de Ciencias, Córdoba.

Giardina, M., D. Germo, L. Miotti, and M. Vazquez
2000 Resolución e integridad arqueológica del Componente Inferior de Piedra Museo. In *Guia de Campo de la visita a las localidades argueólogicas,* edited by L. Miotti, R. Paunero, M. Salemme, and G. R. Cattáneo, pp. 79–83. taller Internacional del INQUA "La Colonización del Sur de América del Sur durante la Transición Pleistoceno/Holoceno." La Plata–Santa Cruz.

Hermo, D., and M. Vázquez
1999 Cuánto que caminamos!!! Primeros resultados de las prospecciones en Cerro Vanguardia y Monumento Natural Bosques Petrificados. *XII Congreso Nacional de Arqueología Argentina,* Actas III:475–83. La Plata.

Lyman, R. L.
1994 *Vertebrate Taphonomy.* Plenum Press.

Marshall, L.
1993 Food Sharing and the Faunal Record, in *From Bones to Behavior: Ethnoarchaeological and Experimental Contributions to the Interpretation of Faunal Remains,* edited by J. Hudson, pp. 228–246. Center for Archaeological Investigations, Occasional Paper 21. Southern Illinois University at Carbondale.

Massone, M., A. Prieto, D. Jackson, and M. Arroyo
1998 Los cazadores tempranos y el fuego: nuevos antecedentes del sitio Tres Arroyos 1. *IV Jornadas de Arqueología de la Patagonia,* Resúmenes:22–23. Río Gallegos.

Menegaz, A., F. Goin, and E. Ortiz Jaureguizar
1989 Análisis morfológico y morfométrico multivariado de los representantes fósiles y vivientes del género *Lama* (Artyodactila, Camelidae). Sus implicancias sistemáticas, biogeográficas, ecológicas y biocronológicas. *Ameghiniana,* 26 (3-4):153–72. Buenos Aires.

Mengoni Goñalons, G.
1976-80 Los materiales óseos de la Cueva 2 de Los Toldos (expedición Menghin) y una aproximación a la metodología de análisis de restos faunísticos. *Rev. RUNA,* XIII (1-2):59–68. Buenos Aires.

Miotti, L.
1995 Piedra Museo Locality: A Special Place in the New World. *Current Research in the Pleistocene,* 12:37–40.

1996 Piedra Museo (Santa Cruz). Nuevos datos para la ocupación pleistocénica en Patagonia. *Arqueología. Sólo Patagonia* (J. Gómez Otero, ed.):27–38. Cenpat, Puerto Madryn.

1998 *Zooarqueología de la Meseta Central y la Costa de la provincia de Santa Cruz. Un enfoque de las estrategias adaptativas y los paleoambientes.* H. Lagiglia (ed.). Museo de Historia Natural. San Rafael, Mendoza.

1999 Los primeros americanos, la visión del problema desde la arqueología de Patagonia. Simposio *El Paleoindio en América del Sur.* (L. Núñez Atencio, ed.), Universidad Católica del Norte, Chile. In Press.

m.s. Análisis multivariado de camélidos. *Lama gracilis* en sitios arqueológicos de Patagonia meridional. *Unpublished report.*

Miotti, L., E. Johnson, E. and M. Gutiérrez
1995 One or Several Kill/Butchery Events at the End of Pleistocene in Piedra Museo Locality? Taphonomic Study of AEP1, Lower Component. Paper presented in *Simposio El Limite Pleistoceno/Holoceno. Intercongreso INQUA.* 1995. Unpublished.

Miotti, L., and M. Salemme
1988 De fracturas óseas modernas y arqueológicas: una hipótesis alternativa. *Revista de Arqueología e Historia Regional de Cuyo,* Vol. 2:41–48, 1988, Universidad Nacional de Cuyo. Mendoza.

1999 Biodiversity, Taxonomic Richness and Specialists-generalists during the Late Pleistocene/Early Holocene Times in Pampa and Patagonia (Argentina, South America). *Quaternary International,* 53/54:53–68. Pergamon Press.

Miotti, L., M. Vázquez, and D. Hermo
1999 Piedra Museo. Un yamnagoo pleistocénico de los colonizadores de la Meseta de Santa Cruz. El estudio de la arqueofauna. *Soplando en el Viento,* pp. 113–36. Bariloche.

Miotti, L., M. Salemme, D. Hermo, M. Vázquez, N. Carden, M. Giardina, and L. Magnin
1999 Aguada del Cuero, un escalón más en la arqueología de la meseta de Santa Cruz. *XIII Congreso Nacional de Arqueología Argentina.* Octubre, Universidad Nacional de Córdoba. Resumen:393.

Núñez Atencio, L., J. Varela, R. Casamiquela, V. Schiappacasse, H. Niemeyer, and C. Villagrán
1994 Cuenca de Taguatagua en Chile: el ambiente del Pleistoceno superior y ocupaciones humanas. *Revista Chilena de Historia Natural* 67:503–19. Chile.

Panza, J. L.
1982 Informe Geológico de las Hojas: 55a Gobernador Moyano y 54c Cerro Vanguardia, provincia de Santa Cruz. Servicio Geológico Nacional. (M.S.)

1994 Hoja Geológica 4969-II. Tres Cerros. Provincia de Santa Cruz. Dirección Nacional del Servicio Geológico. Secretaría de Minería de la Nación. Boletin No. 213.

Pardiñas, U.
1999 Los roedores muroideos del Pleistoceno tardío y Holoceno en la Región Pampeana (Sector Este) y Patagonia (República Argentina): Aspectos taxonómicos, importancia bioestratigráfica y significación paleoambiental. Ph.D. dissertation, Facultad de Ciencias Naturales y Museo, Universidad Nacional de La Plata, pp. 283.

Rabassa, J.
1997 Geological and Stratigraphical Observations at AEP-1, Piedra Museo Locality. *Unpublished report.*

Salemme, M., and L. Miotti
1987 Zooarchaeology and Palaeoenvironments: Some Examples from Patagonian and Pampean regions (Argentina). *Quaternary of South America & Antarctic Peninsula,* Vol. 5:33–57. Balkema Publishers, Rotterdam.

1998 The Status of Rheids in Patagonia: Environmental Approach and Economic Interpretation during the Late Pleistocene/Early Holocene Transition. *ICAZ '98,* Abstracts: 249. Victoria, Canada.

Scheinsohn, V.
1999 *Explotación de materias primas óseas en la Isla Grande de Tierra del Fuego.* Ph.D. Dissertation. Facultad de Filosofía y Letras, Universidad de Buenos Aires. Buenos Aires.

Tambussi, C. and E. Tonni
1985 Aves del sitio arqueológico Los Toldos, cañadón de las Cuevas, Provincia de Santa Cruz (República Argentina). *Ameghiniana* 22 (1-2):69–74. Buenos Aires.

Zárate, M., A. Blasi, and J. Rabassa
2000 Geoarqueología de la localidad Piedra Museo. In L. Miotti, R. Paunero, M. Salemme, and G. R. Cattáneo (Eds.), *Guía de Campo de la visita a las localidades arqueológicas,* pp. 56–64. Taller Internacional del INQUA "La Colonización del Sur de América durante la Transición Pleistoceno/Holoceno." La Plata–Santa Cruz.

IV. Paleoamerican Prehistory

Genetic Evidence

Tracking Genes through Time and Space: Changing Perspectives on New World Origins

Theodore G. Schurr

Abstract

Over the past several decades, researchers have obtained considerable genetic data from Native American and Siberian populations for the purpose of elucidating the nature and timing of the peopling of the New World. Most of these studies have shown a genetic linkage between populations from northern Asia and the Americas. However, because of using different molecular markers, not all of them have presented the same picture of biological variation in these populations. At the same time, comparing the colonization models generated from different genetic data sets has yielded new insights into the genetic relationships between Siberian and Native American groups. This paper provides a brief historical overview of these biogenetic studies, and also discusses the ways in which future research may further illuminate our understanding of the colonization of the Americas.

Introduction

As evident from the numerous papers presented in this volume, the colonization of the New World is an anthropological issue that is still being vigorously debated. All the earliest inquiries and much of the current research into this question have focused on the linguistic, archaeological, and osteological evidence from northern Asia and the Americas, with various models of the colonization process being proposed to explain the patterns of variation seen in these data sets. These models provided certain temporal benchmarks against which subsequent research results could be measured, and predicted population relationships that could be tested against new empirically derived associations.

In the mid-1960s, researchers began using biochemical methods to analyze molecular variation in Siberian and Native American populations, primarily protein polymorphisms such as red cell enzymes, serum proteins, and blood group markers. With these protein polymorphism data, they were able to define biological relationships among aboriginal populations of Siberia and the Americas in novel ways. New quantitative methods developed to deal with population genetic data further refined the assessment of population variation and structure, particularly issues such as effective population sizes, founder effects, and mutation rates. However, protein variants provide an indirect picture of the actual genetic variation in human populations because they are the products encoded by genes, not the gene sequences themselves. Hence, the underlying mutations responsible for these polymorphisms remained unknown.

With the advent of new technologies to assess genetic variation at the DNA sequence level, it became possible to determine the underlying genetic variants responsible for the known protein polymorphisms. These advances also expanded the number and types of genetic loci that could be analyzed in human evolutionary studies. A number of these loci encoded proteins involved in normal physiological or immunological functions of the body, such as the hemoglobin, HLA, and immunoglobulin genes. Because of their functional importance, the distribution of the alleles present at these loci appears to have been strongly influenced by selective forces, along with the usual population genetic forces such as drift, gene flow, and mutation.

An enormous boost for human genetic studies came in the mid-1980s with the invention of the polymerase chain reaction (PCR). With this method, researchers could enzymatically amplify specific regions of genes, or even whole gene sequences, a million-fold or more, thereby generating an abundance of DNA for use in various genetic studies, something that was previously

Department of Anthropology, University of Pennsylvania, 325 University Museum, 33rd and Spruce Streets, Philadelphia, PA. 19104-6398; e-mail: tgschurr@sas.upenn.edu possible only through the more involved process of molecular cloning. The PCR further permitted the analysis of genetic markers and genes that were formerly difficult to characterize, as well as the large-scale screening of population samples for these markers in far less time than required to conduct assays of classical genetic markers.

These trends were greatly enhanced in the last decade by the development of new technologies to assess genetic variation. Among these developments were automated sequencing methods for rapidly characterizing gene sequences, and computational methods for ascertaining population relationships and genetic substructure of populations. In addition, studies associated with the Human Genome Project (HGP) revealed whole new classes of genetic markers in different regions of the sex chromosomes and autosomes that could be used for population genetic studies. These markers included restriction site polymorphisms (RSPs), short tandem repeat (STR) polymorphisms, and single nucleotide polymorphisms (SNPs), most of which are effectively neutral in their overall genetic effects. The current and rapidly expanding genetic database of molecular markers pertinent to anthropological genetic studies is a direct reflection of these developments. These technological advances have further permitted the identification of the molecular mutations that create the protein polymorphisms in the classical genetic systems (e.g., blood group markers, serum proteins, erythrocyte enzymes, HLA and immunoglobulin alleles) that were previously used in population genetic studies.

In the following pages, I discuss a number of the major genetic systems that have been employed in studies of the peopling of the Americas over the past several decades. Although briefly considered because of space constraints, fuller explanations of the nature of these genetic systems and the mechanisms that generate polymorphisms in them can be found in the cited papers and references appearing therein. This group of genetic systems is by no means an exhaustive compilation, but does represent the primary set that has been utilized for most population genetic studies. It includes uniparentally and biparentally inherited markers, rapidly and slowly evolving genomes, and functional and non-functional loci. Together, they provide a comparative framework in which to evaluate the genetic diversity of Native American populations and its implications for the peopling of the New World. As will become apparent, these new molecular data have allowed researchers to re-examine colonization models based on archaeological, linguistic, osteological, and anatomical evidence, as well as test models of population relationships based on published protein polymorphism and nuclear genetic data sets.

Classical Protein Polymorphisms

The earliest studies of non-osteological biological affinities of Native Americans focused on the blood group antigen, serum protein, and erythrocyte enzyme variation in these populations. Blood group markers, or antigens, are proteins and glycoproteins located on the cell membranes of erythrocytes (red blood cells) (Avent 1997; King 1994). These antigens further represent five functional categories, including transporters and channels, receptors for exogenous ligands, adhesion molecules, enzymes, and structural proteins (Avent 1997; Cartron et al. 1998; King 1994). Many of them were initially discovered in transfusion experiments directed towards understanding immunogenetic complementation between mothers and their offspring or blood donors and recipients, and are also medically important in treating autoimmune anemia and in organ transplants (Avent and Reid 2000; Reid and Yazdanbakhsh 1998; Suto et al. 2000). A number of the macromolecules that confer blood group antigenicity are also receptors for pathogenic organisms, and as a result these pathogens may have played a role in the evolution of blood group polymorphisms (e.g., the Duffy glycoprotein [a chemokine receptor] and the malarial parasite *Plasmodium vivax*) (Daniels 1997; Hamblin and Di Rienzo 2000). At present, a total of 23 different blood groups markers have been identified on erythrocytes. Some of the more commonly known markers include the ABO, Rhesus (RH), and Duffy (Df) systems, with the majority of these having recently been molecularly characterized and cloned (e.g., ABO system, Yamamato et al. 1990; RH system, Ikemoto et al. 1996). Based on these studies, most blood group polymorphisms are caused by missense mutations that produce amino acid substitutions, but some are generated through gene deletion, single base deletion, or recombination (Daniels 1997).

Serum proteins and erythrocyte enzymes perform different but no less important physiological functions in the body. Among other functions, they are involved in the immune response (Langlois and Delanghe 1996), iron metabolism (Brock et al. 1986; Rouault and Klausner 1997), antioxidant protection (Krsek-Staples and Webster 1993; Langlois and Delanghe 1996), lymphocyte proliferation (Brock et al. 1986; Rouault and Klausner 1997), cellular differentiation (Moriwaki et al. 1999), glycogen metabolism (Whitehouse et al. 1998), adenine homeostasis (Nakazawa et al. 1990), and the elimination of immune complexes (Volanakis 1990). As a consequence, mutations in the genes coding these polypeptides can lead to serious biochemical diseases, particularly those affecting normal erythrocyte function. Erythrocytes themselves generate high-energy adenosine triphosphate (ATP) by anaerobic glycolysis, as well as cycle oxidized and reduced nicotinamide adenine dinucleotide phosphate (NADP) through the aerobic pentose phosphate shunt pathway (Valentine et al. 1985). Mutations that produce defects in the erythrocyte enzymes involved in the pentose phosphate shunt typically cause hemolysis resulting from the oxidative denaturation of hemoglobin, while those affecting enzymes participating in anaerobic glycolysis cause hemolytic anemia (Valentine et al. 1985). An example of one of these enzymopathies is glucose-6-phosphate (G-6-P) deficiency, which is known to be clinically, biochemically, and molecularly heterogeneous across the world. G-6-P catalyzes the first step in the pentose phosphate pathway, and generates NADPH to protect erythrocytes against oxidizing agents and in reductive biosynthesis reactions (Au et al. 2000; Notaro et al. 2000).

A large number of other erythrocyte enzyme loci have been analyzed for variation in Siberian and Native American populations. In addition to G-6-P, other loci include adenylate kinase (AK) phosphoglucomutase (PGM), carbonic anhydrase II (CAII), peptide hydrolase (PEP), 6-phosphogluconate dehydrogenase (6-PGD), adenosine deaminase (ADA), acid phosphatase (AcP), esterase D (EsD), glyoxalase (GLO), cholinesterase-2 (CHE2), and galactose-1-phosphate uridyltransferase (GALT). Quite a few serum protein loci have also been surveyed for genetic variation, including haptoglobin (Hp), transferrin (Tf), group-specific component (Gc), human complement component C3 (CŌ3), ceruloplasmin (Cp), and albumin (Al). Because so many of these protein loci have been studied in Native American and Siberian populations over the past three decades, it is impossible to adequately summarize all of the patterns of variation that have been observed at each of these loci in these populations. However, one can provide a general overview of the results and implications of these studies as they pertain to the peopling of the New World, which follows below.

In the 1960s and 1970s, researchers conducted a large number of studies of variation at these classical protein polymorphism loci in Native Americans (e.g., Barrantes et al. 1990; Mestriner et al. 1976, 1980; Mohrenweiser et al. 1979; Neel 1973, 1978, 1980; Neel, Gershowitz et al. 1977; Neel, Tanis et al. 1977; Neel et al. 1980; Salzano et al. 1972; Salzano, Gershowitz et al. 1977; Salzano, Neel et al. 1977; Salzano et al. 1980; Salzano and Callegari-Jacques 1980; Spielman et al. 1979; Tanis et al. 1973, 1977). These studies provided numerous insights into the population diversity and structure of unacculturated native populations. Many of them revealed a number of unique alleles, or "private polymorphisms," at different genetic loci in various populations in Central and South America. Private polymorphisms appear to have arisen after the colonization of the New World, and apparently reflect the influences of selection, genetic drift, or isolation on many of these populations since their initial settlement in various regions of the Americas. The continued discovery of these kinds of variants at the same genetic loci in newly characterized Native American populations, especially those from South America (e.g., Dipierri et al. 1998; Luiselli et al. 2000), confirms these trends. Furthermore, these studies indicated that the biological diversity in South American Indians resulted from a process of fission-fusion model of village propagation, one which led to the isolation of subpopulations and their subsequent genetic differentiation, with limited local gene flow occurring thereafter (Neel 1980; Salzano et al. 1980; Sokal et al. 1986; Ward et al. 1975).

Initial studies of classical protein markers in North American Indians revealed pronounced north-south clines in markers frequencies (Suarez et al. 1985). In particular, Eskimo populations appeared to be genetically distinct from Amerindians on the basis of lower ABO*O and higher BO*B frequencies (Suarez et al. 1985). There was also evidence of the association of allele frequencies and the linguistic affiliations of these populations (Spuhler 1979), and the possible geographic patterning of these traits (O'Rourke et al. 1985; Spuhler 1979; Suarez et al. 1985). These allelic distributions were further interpreted as showing the effects of natural selection operating in different ecological zones, as well as the influences of migration and admixture (O'Rourke et al. 1985; Suarez et al. 1985).

Looking more broadly across the New World, there was somewhat greater genetic heterogeneity among South Amerindians compared with North Amerindians, with Central Amerindians being much less diverse than either one of them (O'Rourke and Suarez 1986; O'Rourke et al. 1992). O'Rourke et al. (1992) also found no evidence of north-south clines for specific erythrocyte antigen alleles in or geographic structuring of allelic distributions in South American Indians. They attributed this pattern to the stochastic effects of genetic drift in geographically isolated populations. However, Salzano and Callegari-Jacques (1988) observed significant north-south clines for blood group markers, and found that these data separated South American populations into three major groups: Tropical Forest, Paleo-American, and Andean language groups. Salzano et al. (1991) and Callegari-Jacques et al. (1994) also noted differences between tribes living north and south of the Amazon River, suggesting this river might have constituted a barrier to north-south gene flow. These apparent discrepancies could also reflect differences in the number of genetic markers and the number of populations used in the two studies.

There were also substantial differences in genetic structure in North, Central, and South Amerindian populations. Both North and Central American populations showed evidence of isolation by distance, whereas South American groups did not (O'Rourke et al. 1992). This difference was attributed to low migration among South American groups. Low migration rates could have produced the extreme differentiation between adjacent groups because genetic drift shapes allele frequencies such that geographically separated, unrelated groups appear to be more similar than they actually are. On the other hand, O'Rourke et al. (1992) found little evidence that migration had been attenuated in these populations, as the genetic data revealed considerable proto-historical migrations within the continent.

Compared with North and South American groups, Central Amerindians were much less diverse (O'Rourke et al. 1992). In fact, the Central American pattern of gene-geographic relationships was intermediate between North and South America, but more similar to that seen in North America. Barrantes et al. (1990) also observed geographic structuring of erythrocyte allele frequency data in Central America, and attributed this pattern to the linguistic diversity of the region. It was also hypothesized that the geographic constriction in Central America, as well as increased population density and differentiation in Central America, may have resulted in blockage of population movement southward from North America (Barrantes et al. 1990; O'Rourke et al. 1992). As a result, genetic drift operating in small migrant groups present in South America before later population movement from Central America may have contributed to the spatial heterogeneity of genetic traits in the southern continent.

Other trends in population associations emerged through the analysis of data from circumarctic populations. Analyzing a variety of erythrocyte and serum protein markers in Alaskan and

Siberian Eskimos, Crawford et al. (1981) and Ferrell et al. (1981) showed a good fit between the genetic structure, geographic distribution, linguistic affiliation (Yupik versus Inupik), and ethnohistory of these populations. Similarly, differences between the Yupik, Inupik, and the Koniaq Eskimos and Aleuts were observed by Crawford and Enciso (1982) and Majumder et al. (1988). Crawford and Enciso (1982) further noted some separation between Coast and Reindeer Chukchi populations, with the former being closer to Eskimo and Aleut populations and the former showing a certain closeness to Nganasans and other Samoyedic-speaking populations.

Comparable differences between Na-Dené (Athapaskan) Indians and Eskimo-Aleuts were noted by Scott (1979) and Harper (1980) in their analysis of protein polymorphism data. Based on these data, Harper (1980) proposed that ancestral Native Americans (Amerindians) originated around 19,000 yr B.P. and entered the New World about this time. Na-Dené Indians, and the Bering Sea Mongoloids from which Eskimo-Aleut populations evolved, diverged somewhat later, around 15,000 yr B.P., with the Bering Sea Mongoloids emerging as a distinct entity at roughly 10,200 yr B.P. In addition, using genetic information and archaeological data from Alaska and the Aleutian Islands, Harper (1980) suggested that the Eskimos and Aleuts diverged around 9000 yr B.P., with the Inupiaq and Yupik-speaking Eskimo populations splitting around 5000 yr B.P..

Among northeast Siberian populations, the Chukchi and the Asiatic Eskimos showed significant allelic heterogeneity in blood group marker and serum protein distributions among their constituent subgroups, as well as between them (Sukernik et al. 1981). These findings supported the interpretation that genetic drift had played an essential role in shaping the genetic variation of Siberian and circumarctic peoples (Sukernik et al. 1981). They further suggested that northeastern Siberian populations have lived relatively isolated from one another, despite greater contact between different groups in historical times, with each experiencing a relatively unique genetic history since they began inhabiting this region.

This general pattern was also observed in later studies of native Siberian populations. One analysis of blood group markers and serum protein polymorphisms revealed clusters of eastern Siberian groups that reflected their geographic location or linguistic affiliation (Dubrova et al. 1993). These clusters included Turkic-, Mongolic-, Tungusic-, and Chukotko-Kamchatkan-speaking populations, with members of the latter cluster including Reindeer and Coast Chukchi and Siberian Eskimos. A similar analysis of classical protein polymorphisms in native populations from west, central, and eastern Siberia showed the same overall trends but also more complex relationships among these groups (Posukh et al. 1998). Eastern Siberian (Tungusic/Yukagir) and Samoyedic populations formed distinct clusters, with Nganasans showing similarities to eastern Siberian groups. In addition, northwestern native populations (Komi, Nentsi, Entsi) and ethnic Russians formed a grouping. In contrast, Altaic populations were equidistantly separated from all these clusters and showed affinities with Mongol groups, largely due to their having considerable frequencies of both European and Asian genetic markers (Posukh et al. 1998). These data suggested that the distribution of genetic markers among these partially isolated indigenous populations have been primarily shaped by genetic drift effects and gene flow, and to some extent by local and regional events in their recent history.

HLA Genes

Many aspects of Native American origins and affinities have been elucidated through the characterization of variants of the major histocompatibility locus antigen (HLA) system. The HLA system is the most polymorphic genetic system in humans, and has been extensively typed in numerous world populations (e.g., Bodmer et al. 1998). This system is highly polymorphic because balancing selection maintains a few allelic lines over very long periods of time (Harpending et al. 1998; Hedrick and Black 1997; Watkins et al. 1992). The great variation in its genetic markers also makes the HLA system very useful for determining genealogical relationships between populations.

HLA antigens belong to three classes based on their immunological properties, two of which have been frequently used in studies of human biological variation. Class I proteins, which are found on every mammalian cell, are called the transplantation antigens, since they are responsible for accepting or rejecting foreign tissue. The human class I proteins include the classical transplantation antigens HLA-A, -B, and -C (Bodmer et al. 1998; Browning and McMichael 1996). By contrast, class II proteins, which are found on the surfaces of B and T lymphocytes and macrophages, facilitate communications between cells that are involved in executing the immune response. The human class II region is arranged into four subregions: DR, DQ, DZ/DO, and DP (Bodmer et al. 1998; Browning and McMichael 1996). Recent molecular typing methodologies have permitted the further definition of subtypes within both class I and II genes, leading to the development of a new nomenclature for HLA variants that differs from the one based on immunocomplementation and protein electrophoresis (Bodmer et al. 1998).

Some general trends in the pattern of HLA variation in Native American populations have been seen in these studies. To begin with, Native American populations exhibit an overall reduction of the number of HLA alleles relative to other non-Amerindian populations. However, the same alleles and allelic lineages were missing in a number of Amerindian groups, suggesting that they shared a common set of HLA allotypes (Erlich et al. 1997). The occurrence of a number of unique alleles in various populations in the Americas also suggested that they have arisen since the colonization of the Americas, much like the "private polymorphisms" in blood group marker and serum protein systems in South American Indian populations. Together, these data suggested that the First Americans experienced a population bottleneck as they entered the New World (Erlich et al. 1997).

There was also evidence of larger-scale population dynamics within the American continents. For example, Rothhammer et al. (1997) studied HLA-A, -B, and -C types from 39 South American Indian populations and observed significant longitude and latitude clines in allele frequencies, which were viewed as reflecting ancient migration patterns through that continent. Similarly,

Black (1991) noted some geographic structuring of HLA alleles in South American Indian groups, and attributed this pattern to recent population histories, linguistic affiliations, or genetic drift effects. On the other hand, in a comparison of Venezuelan and Brazil tribes, Layrisse et al. (1995) noted a closer relationship between two linguistically but geographically distant Venezuelan tribes, the Bari and the Warao, compared with two culturally different Brazilian populations. Therefore, while language and geography have influenced the distributions of HLA-A, -B and -C alleles, they are not absolute predictors of genetic relationships between different South American Indian groups.

In addition, various studies have indicated that the HLA-B locus has evolved much more rapidly in Central and South American Indian populations than in North American tribes. For example, Iwanaga et al. (1997) detected the presence of new HLA-B alleles in the Kuna of Panama, Marcos et al. (1999) noted a variety of new HLA-B alleles and some new HLA-A alleles in Argentinean and Brazilian Amerindian populations, Martinez-Laso et al. (1996) observed new HLA-B alleles in the Venezuelan Bari, and Watkins et al. (1992) detected several new HLA-B alleles in the Brazilian Waorani. In contrast, the Navajo had limited HLA-B diversity comparable to other North American Amerindian tribes (Garber et al. 1996), and the Zuni were noted to have HLA-A and -B alleles that were also present in Asian populations (Watkins et al. 1992). These data have been interpreted as showing that South American populations have been under considerable selection pressure for increased allelic diversity (Erlich et al. 1997; Watkins et al. 1992).

Among HLA class II genes, the DRB1 locus has been most widely surveyed in Native American populations. Most studies of this locus indicate that Native American populations have a reduced number of alleles relative to non-Amerindian groups (Cerna et al. 1993; Erlich et al. 1997; Monsalve et al. 1998; Yunis et al. 1994). The same appears to be true for the DQA1 locus, as two alleles were common among the Navajo, Pueblo, and Sioux Indian groups (Scholl et al. 1996) and in two Alaskan native populations (Walkinshaw et al. 1996). As for HLA class I loci, there is also evidence of the emergence of Amerindian-specific alleles from the ancient DRB lineage (Erlich et al. 1997). For instance, Mack and Erlich (1998) found several new variants within the Brazilian Ticuna, including the new DRB1*0807 allele, Cerna et al. (1993) found the unique DRB1*0417 in the Brazilian Xavante, Layrisse et al. (1997) found the DRB1*0807 allele in the Venezuelan Yucpa, and Lozáro et al. (1999) found a number of novel alleles in the Brazilian Terena. Similarly, Erlich et al. (1997) identified new DRB1 alleles in several Amerindian groups from Brazil, Argentina, Ecuador and Canada. These results again point towards selection driving the creation of new HLA class II alleles, with most evidence indicating a functionally significant, more rapid evolution of class I compared with class II loci (e.g., Lozáro et al. 1999).

Based on combinations of these new molecular variants at class I loci, Monsalve et al. (1999) observed several major HLA haplotypes in Native American populations. One was common to Amerindian populations, including the Northwest Coast Salishans (Monsalve et al. 1999). The second was present at high frequencies in Athapaskans (Monsalve et al. 1998), occurred in a few Amerindian groups (Williams et al. 1994), and appeared in some eastern Siberian and East Asian populations (Grahovac et al. 1998). The third was common in both Salishan and Amerindian populations, as well as other Asian groups (Monsalve et al. 1999), whereas the fourth occurred at modest frequencies in Northwest Coast Indians (Monsalve et al. 1999) and northeast Asian populations (Grahovac et al. 1998). These data clearly showed the genetic links between North-East Asians and Native Americans and, in revealing specific patterns of haplotype sharing amongst these same populations, suggested genetic differences among Native American populations (Eskimo-Aleut, Na-Dené Indians, Amerindians).

Monsalve et al. (1999) also used a maximum-likelihood analysis to examine the phylogenetic relationships among HLA class II loci data from Native American and Asian populations (Cerna et al. 1993; Gao et al. 1991; Grahovac et al. 1998; Imanishi et al. 1992; Layrisse et al. 1995; Lozáro et al. 1999; Mack and Erlich 1998; Monsalve et al. 1998, 1999; Trachtenberg et al. 1996; Yunis et al. 1994). This analysis revealed that Northwest Coast Indians and Na-Dené Indians grouped together outside of the main Amerindian branch, and were intermediate between Asians and Amerindians. In addition, South American Indian groups generally clustered together away from North and Central American Indian groups. Eastern Siberian populations clustered together and formed a larger branch with central Siberians (Evenks, Kets), but were separated from East Asian populations. Monsalve et al. (1999) also noted greater interpopulational differences in Amerindian populations relative to Asian groups, probably because of the effects of genetic drift, which played a major role in shaping allelic distribution. Furthermore, geographic proximity was a major factor in the clustering of specific populations, as was linguistic affiliation to a lesser extent.

Variations in class II loci have also been examined in Siberian populations. One analysis of DRB1, DQA1, DQB1, DPA1 and DPB1 loci revealed a number of interesting features in seven central and eastern Siberian populations (Grahovac et al. 1998). While there was high polymorphism at most loci, no new alleles were detected. However, this high polymorphism generated numerous class II haplotypes, more than observed in other world populations. Some of these were shared with other non-Siberian groups but most were unique to these Siberian populations and, in some cases, represented novel combinations of DQA1 and DQB1 alleles. These data led Grahovac et al. (1998) to suggest that the colonizers of Siberia brought with them a set of haplotypes common to Eurasian populations that have since undergone extensive recombination to form these new haplotypes, ones which are still rapidly turning over. In addition, the allele frequencies at the DRB1 locus divided eastern and central Siberian populations, with the former showing clear genealogical relationships to Native Americans, a pattern also seen by Krylov et al. (1995).

Immunoglobulin Genes

The elucidation of genetic relationships among Native American tribal groups with immunoglobulin genes has also allowed the testing of hypotheses concerning their ancestry. Immunoglobulin proteins are antibodies secreted into the blood by the B-lymphocytes (B cells) in response to the presence of an antigen. Each immunoglobulin molecule comprises a tetramer of two identical light (L) chains and two identical heavy (H) chains, the combination of which generates an immunoglobulin with several discrete domains. Each individual protein chain consists of two principal regions, the variable (V) and constant (C) regions. The variable domain of the light chain (VL) is responsible for recognizing the antigen, with the production of various VL domains of differing specificities creating the ability to respond to diverse antigens. Accordingly, the total number of variable regions for both the light and heavy chains number in the hundreds (Alexander et al. 1985; Marchalonis and Schluter 1989; Sitnikiva and Su 1998). In addition, different classes of immunoglobulins have different effector functions, the class of which is determined by the heavy chain C region. Immunological responsiveness is maintained through the somatic recombination of the V, C, and joining-segment gene segment in B-lymphocytes (Alexander et al. 1985; Marchalonis and Schluter 1989; Sitnikiva and Su 1998). Because of their integral role in the human immune function, immunoglobulin genes have likely been under some form of selection during the expansion of modern humans into the Eurasia and the Americas.

The immunoglobulin data from Williams et al. (1985) were initially viewed as best supporting the tripartite migration hypothesis of Greenberg et al. (1986). This study showed three distinct sets of GM allotypes among Amerindians (GM*A G [GM1;21] and GM*X G [GM1,2;21]), Na-Dené Indians (GM*A G, GM*X G, and GM*A T [GM1,2;21]), and Aleut-Eskimos (GM*A G and GM*A T). Although there was a very large standard error estimated with the genetic distances on which this model was based, the Na-Dené Indians grouped closer to Eskimos and Chukchi than to Amerindians. However, Szathmary (1993) asserted that this data set actually shows Eskimos to have GM*X G allotype irrespective of admixture with other groups, and questioned whether the Eskimos were clearly demarcated from Na-Dené Indians and Amerindians with this genetic system.

Subsequent analyses of GM allotype data in Native American populations led to a further refinement of the tripartite migration model (Schanfield 1992; Schanfield et al. 1990). These analyses detected four main groupings of Native Americans, Eskimo-Aleuts, Na-Dené Indians, North and Central Amerindians, and South Amerindians, although these divisions were not entirely unambiguous. In addition, the Inupiaq Eskimos clustered closer to Yupik Eskimos and Canadian Athapaskans, and then to the Chukchi, whereas Alaskan Athapaskans clustered with Uralic-speaking Siberians (Schanfield et al. 1990). Both studies also suggested that the northern Asian haplotype GM*A T was polymorphic only in Eskaleut, Na-Dené Indians, and some northern Amerindian groups, being seen only sporadically in other Amerindian groups, although this interpretation was not universally accepted (Szathmary 1993).

These patterns led to the proposal that four separate migrations to the New World were responsible for the pattern of GM variation present in these aboriginal groups (Schanfield et al. (1990); Schanfield (1992). The first migration brought only GM*A G and GM*X G haplotypes, as reflected by the data from South American Indians (Black and Pandey 1997; Dugoujon et al. 1995; Schanfield 1992). The second migration to enter North America had high frequencies of GM*A G, and low frequencies of GM*X G and GM*A T, as seen in North and Central Amerindians. The third migration, represented by Na-Dené Indians, had high frequencies of GM*A G, and moderate frequencies of GM*X G and GM*A T, whereas the fourth, represented by Eskimo-Aleuts, had only GM*A G and GM*A T. These results were generally supported by a later analysis of GM variation in around 50 different Native American populations, which revealed a strong north-south gradient in GM allotypes frequencies from the Bering Strait to South America (Dugoujon et al. 1995), and the distinctiveness of Eskimos and Na-Dené Indians compared to other Amerindian groups.

Similar studies of nuclear genetic variation in aboriginal populations of Siberia and northern North America revealed close genetic relationships among certain northern Asian and Native American groups. To begin with, the GM*A G, GM*X G, and GM*A T allotypes were present in most eastern Siberian and East Asian populations (Crawford and Enciso 1982; Dubrova et al. 1993; Posukh et al. 1990, 1998; Schanfield et al. 1980, 1990; Sukernik et al. 1978, 1981). Hence, the major GM haplotypes present in Native American groups are widespread through Northern Asia. In particular, the GM*A T haplotype variant was observed among the Chukchi, Eskimos, Na-Dené Indians, and northern Amerindian populations of North America, suggesting a common origin for this haplotype among circumarctic populations, as well as the occurrence of gene flow between Na-Dené Indians and northern Amerindians. In addition, "private polymorphisms" and "lineal effects"observed in Native American groups (Neel 1980) were also detected in the Sel'kups and Forest Nentsy of western Siberian lowlands, including unique tribal-specific GM haplotypes such as IG1G-deleted, which varied significantly in frequency within their tribal boundaries (Sukernik et al. 1992). Furthermore, the population structure and genetic variation of the Forest Nentsy, Sel'kups, Nganasans, Yukagirs, Evens, Chukchi, and Siberian Eskimos also implied that aboriginal populations had undergone substantial genetic differentiation prior to their subsequent expansion into new environments (Sukernik et al. 1980, 1981, 1986a, b, 1992).

Synthetic Maps of Multiple Genetic Loci

The analysis of multiple nuclear genetic loci in Native American and Asian populations showed some of the same associations revealed in more focused studies of blood group marker or HLA loci variation, but also some intriguing discrepancies with them. These studies used allele frequency data from multiple nuclear genetic loci from many populations to estimate genetic distances between them, and then employed phylogenetic or principal components analysis to ascertain the evolutionary relationships amongst them. In one such analysis of 60-plus loci, Cavalli-Sforza et al. (1994) observed that Athapaskan Indians and North American Eskimos clustered together on the same branch but were not closely related to each other, and both were clearly separated from other Amerindian populations. In addition, North American Indian populations grouped together at some distance from those in the U.S. Southwest and Central America. There was also some clustering of Central and South American tribes along geographic lines, as well as some mixing of groups across linguistic boundaries.

Looking more broadly in the circumarctic region, Cavalli-Sforza et al. (1994) noted several distinct patterns of population relationships across the Bering Strait. The Siberian Eskimos grouped with the Coast Chukchi and Koryaks in a single branch that was positioned some distance from their North American counterparts. Even more interesting was the separation of Reindeer and Coastal Chukchi populations, and the association of the former population with the Mongols of southeastern Siberia. Thus, this analysis revealed splits among linguistically related populations such as the Siberian and Alaskan Yupik Eskimos, divisions within the same ethnic population, as in the case of the Reindeer and Maritime Chukchi, as well as associations between groups that had previously been seen in other genetic and non-genetic data sets (e.g., Na-Dené Indian closeness to Alaskan Eskimos).

The analysis of Asian and Siberian populations was equally as informative (Cavalli-Sforza et al. 1994). The results revealed a small cluster that encompassed northeastern Siberians, including the Koryaks, Chukchi, Tungus (Evens), and a North Turkic grouping (Yakuts, Tuvans, Altayans, Dolgans). There was also a northern mongoloid subcluster containing East Asian and northeastern Siberian populations, within which Bhutanese, Tibetans, northern Han Chinese, Koreans, and Japanese grouped together, then Mongols and Uralic Siberians, then Reindeer Chukchi. Principal components mapping of these allele frequencies showed roughly the same patterns of population associations: Eurasian (West and South Asia), Northeast and East Asia, and Southeast Asia. Within these general groupings were clusters containing Mongols, Ainu, Koreans, Japanese; (2) Tibetans, northern Tungus, northern Han Chinese; (3) Nentsy, Samoyed, Nganasan and, surprisingly, Reindeer Chukchi. When Arctic populations were considered alone, some of these population associations changed. The Eskimos clustered with Na-Dené Indians; Uralic-speaking groups clustered together; the Lapps (Saami) formed a small distinct cluster, as did northwestern Siberian (non-Russian) groups; and all central, northeastern, and southeastern Siberia groups clustered together.

In their analysis of multiple nuclear genetic loci, Nei and Roychoudhury (1993) observed somewhat different sets of population relationships among Native American and Asian populations. Native Americans clustered together away from a greater Asian cluster, with northeast Asians being closest to New World populations. In addition, North American Indians were separated from South American Indian populations, and both of these groups appeared distinct from Eskimos. The Eskimos were also positioned closer to North American Indians than South American groups. These differences in population clustering may reflect the use of different statistical and tree-building methods used by Nei and Roychoudhury (1993) and Cavalli-Sforza et al. (1994).

Alu Polymorphisms

The pattern of variation observed in Native Americans with polymorphic Alu insertions was somewhat different than for other genetic systems. Alu elements belong to the largest family of short interspersed repetitive elements (SINEs) in humans, with over 500,000 copies being present in each haploid genome (Batzer et al. 1991, 1996). While distributed throughout the genome of primate species, one Alu subfamily has been found to be largely human specific, having been inserted into the genome of the human lineage between 200,000 and 6,000,000 years ago (Batzer and Deininger 1991; Batzer et al. 1996). A minority of these are transcriptionally active (Matera et al. 1990) and undergo amplification, or retroposition, into other genomic locations (Wallace et al. 1991). Most Alu insertions are stable once situated in the genome because there is no mechanism for removing them, except for intrachromosomal deletion. In addition, because of the low rate of *de novo* insertions reaching polymorphic levels (Batzer and Deininger 1991; Batzer et al. 1991), the likelihood of the same insertion occurring in the same place more than once is extremely small. Therefore, the ancestral state of an Alu element is its absence at a particular position (-), with the derived state being its presence (+). It follows from this general pattern that Alu elements are inherited by offspring from parents through the simple Mendelian segregation of codominant + and - alleles (Novick et al. 1995).

Using a set of five Alu elements from different chromosomes, Novick et al. (1998) screened 24 native populations from North, Central, and South America for these markers. Their results supported an Asian origin of ancestral Native American populations, but not a multiple-migration hypothesis for the emergence of Eskimo-Aleuts, Na-Dené and Amerind linguistic groups. Instead, these data were interpreted as supporting a single migration to the New World followed by partial isolation and genetic drift. Based on heterozygosity values for these Alu loci, Novick et al. (1998) also noted the mixing of some North, Central, and South American populations apparently in both north-to-south and south-to-north directions. There was also a deficiency of heterozygotes in Native American populations. This finding likely reflected the variable degree of inbreeding in these populations or the lack of gene flow among them.

Interestingly, the maximum likelihood (ML) analysis of genetic distances generated from Alu haplotype data did not reveal clusters of Native American populations that conformed to the linguistic groupings of Greenberg et al. (1986). Instead, the clusters in the ML tree generally reflected the geographic location of the populations. However, a few tribal groups were positioned with geographically distant populations, such as the South American Wayuu with the North American Sioux, the South American Inca with the Greenland Eskimos, and the North American

Navajo and Zuni with the South American Karitiana. These anomalies were attributed to non-native admixture causing deviations from expected branching patterns based on known cultural and linguistic affiliations (Novick et al. 1998). While noting these apparent discrepancies, Novick et al. (1998) suggested that this pattern of genetic variation in the Americas was produced by a single migration to the New World followed by partial isolation and genetic drift. They further noted a close phylogenetic relationship between mainland Han Chinese and Native Americans, particularly the Mayans, and suggested possible recent gene flow from China to the New World.

Mitochondrial DNA

Much of the more recent molecular research into the colonization of the New World has employed the mitochondrial DNA (mtDNA). The human mtDNA has a number of distinct properties that make it an invaluable tool for molecular anthropological studies, including its maternal inheritance (Case and Wallace 1981; Giles et al. 1980), rapid mutation rate (Brown et al. 1979; Horai et al. 1995; Miyata et al. 1982; Stoneking et al. 1986; Wallace et al. 1987), and lack of recombination (Clayton 2000; Merriwether et al. 1991; Shadel and Clayton 1998). Because of these properties, one can reconstruct patterns of accumulated sequence changes in mtDNAs along branching female lineages with relatively minimal ambiguity. Furthermore, many of the mutations detected in human mtDNAs correlate with the geographic origin of the population in which they occur (e.g., Ballinger et al. 1992; Chen et al. 1995, 2000; Torroni et al. 1996, 1998; Torroni, Lott et al. 1994). This mutational pattern may provide a detailed record of ancient migration patterns based on their distribution in human populations, since the presence of specific mutations in groups distantly located likely reflects population movement or contacts in prehistoric times. Finally, being a haploid genome, the mtDNA is very sensitive to stochastic processes such as genetic drift and founder effects that result from geographic isolation, migration, or population splits (Livshits and Nei 1990; Nei 1987; Nei and Livshits 1989). For this reason, variation in mtDNA sequences often contains genetic signals of these past events.

Studies of mtDNA variation in modern Native American populations have shown that their mtDNAs belong to five different haplogroups, or mtDNA lineages, of related genotypes, which have been designated A–D and X (Bailliet et al. 1994; Batista et al. 1995; Brown et al. 1998; Easton et al. 1996; Forster et al. 1996; Ginther et al. 1993; Horai et al. 1993; Kolman et al. 1995; Kolman and Bermingham 1997; Lorenz and Smith 1996, 1997; Merriwether et al. 1994, 1995; Santos et al. 1994; Santos et al. 1996; Schurr et al. 1990; Smith et al. 1999; Torroni et al. 1992; Torroni, Chen, et al. 1994; Torroni, Neel, et al. 1994; Torroni, Schurr, et al. 1993; Ward et al. 1991, 1993, 1996). These findings have been substantiated by the results of similar analyses of ancient Native American populations (Fox 1996; Hayes 1999; Kaestle 1995, 1997; Merriwether et al. 1994; Monsalve et al. 1996; O'Rourke et al. 2000; Parr et al. 1996; Ribeiro-Dos-Santos et al. 1996; Stone and Stoneking 1998). Thus, the presence of five distinct haplogroups in the New World is no longer in question. In addition, some of these studies detected the presence of "Other" (non-haplogroup A–D) haplotypes in different Native American groups (Bailliet et al. 1994; Lorenz and Smith 1996; Merriwether et al. 1994, 1995; Ribeiro-Dos-Santos et al. 1996; Smith et al. 1999; Torroni, Schurr et al. 1993; Ward et al. 1991). The presence of "Other" haplotypes in Native Americans was significant because they could potentially represent previously unidentified founding mtDNA lineages that were brought to the New World during its initial phase of colonization. On the other hand, these haplotypes could reflect the occurrence of non-native admixture in Native American populations that has occurred in a post-Columbian context. Thus, the definition of these kinds of haplotypes was critical for resolving the number of founding lineages that were brought to the Americas.

One possible founding mtDNA lineage was observed among South American Indians in the form of "X6/X7" haplotypes (Easton et al. 1996; Merriwether et al. 1994, 1995). X6/X7 haplotypes had the +DdeI 10394 and +AluI 10397 sites (DdeI/AluI sites) seen in haplogroup M, a macrolineage that encompasses 55–70 percent of all Asian mtDNAs (Ballinger et al. 1992; Torroni, Miller, et al. 1994; Torroni, Schurr et al. 1993), but otherwise lacked the diagnostic RFLPs of haplogroups C and D, while differing between themselves by the presence or absence of the HaeIII 16517 site. Based on their mutational properties, and because putatively similar mtDNAs were identified in other Native American populations (Merriwether et al. 1994, 1995; Torroni, Schurr et al. 1993), X6/X7 haplotypes were proposed to represent an additional founding lineage of Asian origin (Easton et al. 1996; Merriwether et al. 1994, 1995). However, when the CR sequences of X6/X7 mtDNAs were analyzed with those from other Native American populations, they clustered among Amerindian CR sequences from haplogroups C and D (Schurr 1998; Schurr and Wallace 1999; Stone and Stoneking 1998). These results suggested that X6/X7 mtDNAs were autochthonous haplotypes that derived from haplogroup C and D mtDNAs after the peopling of the Americas rather than haplotypes belonging to an additional founding lineage.

The great majority of the remaining "Other" mtDNAs were likely acquired through historical non-native gene flow. Several studies have revealed slight but non-negligible European admixture in North American Indian groups by the presence of haplogroups H, J, and K (Scozzari et al. 1997; Smith et al. 1999; Torroni, Schurr et al. 1993). With further RFLP marker screenings, additional North American populations with "Other" mtDNAs may also exhibit the same kinds of haplotypes. These studies have also revealed the presence of African mtDNAs in Native American populations (Huoponen et al. 1997; Smith et al. 1999; Torroni, Chen, et al. 1994c). African haplotypes have also been found in admixed populations of North, Central, and South America, such as Black Caribs (Monsalve and Hagelberg 1997), Cubans (Torroni et al. 1995), Mexican-Americans (Green et al. 2000), and Afro-Brazilian and Afro-Uruguayan populations (Bortolini et al. 1997; Bravi et al. 1999). Because most of these admixed populations also possess Amerindian mtDNAs, they may be thought of as reservoirs of Native American genes belonging to groups that no longer exist as distinct tribal entities.

Even after eliminating haplotypes originating in African or European populations, some "Other" mtDNAs in Native American populations have not yet been assigned to a known haplogroup. Most of these "Other" mtDNAs were identified in ancient populations (e.g., Hauswirth et al. 1994; Parr et al. 1996; Ribiero-dos-Santos et al. 1996), with a minority being detected in modern populations (Bailliet et al. 1994). However, none of these "Other" mtDNAs were screened for the RFLP markers defining haplogroups M or X. Furthermore, in most cases, the portion of the CR region that was sequenced in ancient samples was not of comparable length to that typically obtained from modern samples. As a result, the limited data from these samples make it impossible to know exactly what their genealogical status actually is.

The accumulation of genetic data from a large number of Native American populations from all continental regions has revealed that mtDNA haplogroups are differentially distributed in Americas. While haplogroups A–D are found throughout the Americas (Bailliet et al. 1994; Batista et al. 1995; Brown et al. 1998; Easton et al. 1996; Forster et al. 1996; Ginther et al. 1993; Horai et al. 1993; Kolman et al. 1995, 1997; Lorenz and Smith 1996, 1997; Merriwether et al. 1994, 1995; Santos et al. 1994; Santos et al. 1996; Schurr et al. 1990; Smith et al. 1999; Torroni et al. 1992; Torroni, Chen, et al. 1994; Torroni, Neel, et al. 1994; Torroni, Schurr, et al. 1993; Ward et al. 1991, 1993, 1996), haplogroup X seems to be confined to North America (Brown et al. 1998). In addition, haplogroup A appears at high frequencies in North and Central America, but diminishes in frequency in South America, whereas haplogroups C and D appear at roughly inverse frequencies in the New World. By contrast, haplogroup B occurs at fairly similar frequencies in the Americas, although appearing at high frequencies in the American Southwest. Whether this pattern reflects the consequences of multiple colonization events, or instead the process of settlement and regional differentiation of a single ancestral population, is still being explored. However, it has apparently been stable over many thousands of years, at least in North America (O'Rourke et al. 2000).

These studies have further shown that all circumarctic populations have predominantly haplogroup A and D mtDNAs (Saillard et al. 2000; Schurr et al. 1999; Shields et al. 1993; Starikovskaya et al. 1998; Torroni, Schurr, et al. 1993; Ward et al. 1991, 1993). Amongst these populations, the Siberian and Alaskan Eskimos appeared genetically very similar to one another by having higher frequencies of haplogroup A, while the Aleuts differed from Eskimo groups by having mostly haplogroup D mtDNAs (Hayes 1999; Saillard et al. 2000; Schurr, et al. 1993; Schurr et al. 1999; Shields et al. 1993; Starikovskaya et al. 1998; Torroni, Merriwether et al. 1994; Ward et al. 1991, 1993). In addition, both sets of Eskimo groups had a very low frequency of haplogroup C mtDNAs. This finding suggested that haplogroup C could represent a founding mtDNA lineage in Eskimo-Aleut populations, although those in the Siberian Eskimos may have been obtained through gene flow with the Chukchi (Schurr et al. 1999; Starikovskaya et al. 1998). On the other hand, known gene flow from Europeans (Russians) into the Alaskan Eskimos and Aleuts, and from the Chukchi into the St. Lawrence Eskimos (Crawford et al. 1981; Ferrell et al. 1981), as well as tribal interactions with Northwest Coast Amerindian populations (Szathmary 1993), make it likely that the very low frequencies of haplogroup B and "Other" mtDNAs in Alaskan Eskimos and Aleuts (Merriwether et al. 1994) were acquired through gene flow with non-Eskimo-Aleut populations.

There has also been considerable regional differentiation in haplogroup distribution in Siberia and Asia. Mitochondrial DNA data have been used to suggest that Mongolia and the Lake Baikal region are the source area for ancestral Native Americans, because populations from these regions have polymorphic frequencies of haplogroups A–D (Derenko et al. 1999, 2000; Kolman et al. 1996; Merriwether et al. 1996; Sukernik et al. 1996). However, it is now known that haplogroup A–D mtDNAs are found together in populations originating as far west as the Altai Mountain region to Japan and Korea in the east, with Mongolia and the Lake Baikal region falling in between the two extremes (Ballinger et al. 1992; Derenko et al. 1999, 2000; Horai et al. 1996; Kolman et al. 1996; Merriwether et al. 1996; Schurr et al. 2000; Sukernik et al. 1996; Torroni, Miller et al. 1994). Thus, if the presence of all four founding haplogroups in an Asian population is the sole criterion for identifying the potential source area of ancestral Native Americans, then the range of possibilities has now been expanded beyond Mongolia per se.

Haplogroups A–D actually represent a minority of mtDNA lineages in Siberian and East Asian populations, with various other Eurasian and Asian haplogroups comprising the rest of their mitochondrial gene pool. In fact, most Siberian populations have only haplogroups A, C and D (Derenko et al. 1999, 2000; Schurr et al. 1999; Starikovskaya et al. 1998; Sukernik et al. 1996; Torroni, Sukernik et al. 1993). With few exceptions, Siberian groups lack haplogroup B mtDNAs, and those which do have these haplotypes inhabit the southern margin of Siberia adjacent to Mongolia and northern China (Derenko al. 1999, 2000; Kolman et al. 1996; Petrishchev et al. 1993; Schurr et al. 1999; Shields et al. 1992, 1993; Starikovskaya et al. 1998; Sukernik et al. 1996). In addition, Siberian and East Asian populations appear to lack haplogroup X mtDNAs (Horai et al. 1996; Schurr et al. 1999; Starikovskaya et al. 1998; Sukernik et al. 1996; Torroni, Sukernik et al. 1993), raising questions about the origin of this mtDNA lineage in Siberia and the Americas. Various other Eurasian and Asian haplogroups constitute the rest of the Siberian mitochondrial gene pool. This distribution probably reflects population dynamics in Siberia before and after the colonization of the New World, such as the expansion of Paleoasiatic speakers in northeast Asia (Schurr et al. 1999; Starikovskaya et al. 1998), the spread of Tungusic-speaking populations throughout east and central Siberia (Schurr et al. 1999; Starikovskaya et al. unpublished data; Torroni, Sukernik et al. 1993), the spread of Uralic speakers in northern Asia (Sajantila et al. 1995), and the northward expansion of Turkic speakers into Siberia (Yakuts; Torroni et al. 1998).

Based on these Siberian and Asian mtDNA data, several different models for the peopling of the New World have been developed. Some researchers have suggested that ancestral Amerindian populations brought at least haplogroup A, C, and D mtDNAs from Siberia during the initial colonization(s) of the New World, with haplogroup B possibly representing a second independent migration from East Asia to the Americas (Schurr et al. 1999; Starikovskaya et al. 1998; Torroni, Sukernik et al. 1993). In addition, it was suggested that haplogroup X might also represent a separate migration from somewhere in Eurasia, given its absence in Siberia and much of the Americas (Brown et al. 1998). However, other researchers argue that haplogroups A–D were brought to the New World in a single migratory event (Kolman et al. 1996; Merriwether et al. 1994, 1995, 1996). Indeed, statistical analyses and pairwise mismatch simulations using haplogroup A–D data tend to support a single migration to the New World (Bonatto and Salzano 1997; Stone and Stoneking 1998), although these studies did not account for the RFLP haplotype differences in Amerinds, Na-Dené Indians, and Eskimo/Aleuts.

The antiquity of these haplogroups and the timing of their initial entry into the New World have also been keenly debated. Recent studies have provided ages for haplogroups A–D and X of between 35,000 and 20,000 yr B.P., based on several different statistical estimates (Bonatto and Salzano 1997; Brown et al. 1998; Forster et al. 1996; Schurr et al. 1999; Stone and Stoneking 1998). Similar divergence estimates have been obtained for these lineages in both Asia/Siberia (Horai et al. 1996; Schurr et al. 1999; Starikovskaya et al. 1998; Torroni, Sukernik et al. 1993) and Europe (Torroni et al. 1996, 1998). Additional support for these data comes from the fact that Native American and Siberian populations lack any mtDNAs in common, even those that appear to be identical on the basis of their RFLP haplotypes due to having different CR sequences (Schurr et al. 1999; Starikovskaya et al. 1998; Torroni, Schurr et al. 1993). These findings imply that estimates of haplogroup divergence reflect the genetic diversity that has accumulated in the American branches of these mtDNA lineages, hence, the time at which modern humans first entered the Americas.

However, other investigators have argued that the ca. 25,000 yr B.P. divergence times for these haplogroups of the New World are overestimates of the time at which the colonization process took place, and instead represent the age of the haplogroup divergence in Asia, hence, that of the common ancestral population (Shields et al. 1993; Ward et al. 1991). Using a different coalescent method, Shields et al. (1993) proposed a "late" entry time (12,000–14,000 yr B.P.) of ancestral Amerindians to the New World rather than an "early" entry time, along with the late expansion of northern populations in the circumarctic region (5,000–7,000 yr B.P.). This late entry time for ancestral Amerindians falls within the lower bounds of other divergence time estimates for Native American haplogroups (Bonatto and Salzano 1997; Stone and Stoneking 1998). However, it was based on data taken from Northwest Coast Amerindian and circumarctic populations who are known to share a number of CR sequences from haplogroups A, C and D, and thus could be an underestimate of this date (Lorenz and Smith 1997; Schurr et al. 1999; Shields et al. 1993; Starikovskaya et al. 1998; Ward et al. 1993). On the other hand, the late entry time for circumarctic populations is consistent with other estimates of mtDNA diversity in these same groups (Saillard et al. 2000; Starikovskaya et al. 1998).

In addition, it has been proposed that the ancient haplogroup divergence dates are inflated because they did not consider that multiple variants (i.e., founding haplotypes) from haplogroups A–D were brought to the Americas by ancestral Amerindian populations, namely, haplotypes A1/A2, B1/B2, C1/C2, and D1/D2 (Bailliet et al. 1994; Easton et al. 1996; Merriwether et al. 1994, 1995). However, the hypermutability of the HaeIII 16517 site that is used to delineate these haplotypes (Ballinger et al. 1992; Chen et al. 1995; Torroni et al. 1992; Torroni et al. 1996, 1998; Torroni, Chen, et al. 1994; Torroni, Miller, et al. 1994; Torroni, Neel, et al. 1994; Torroni, Schurr, et al. 1993; Torroni, Sukernik, et al. 1993) raises serious questions about their validity as founding mtDNAs. Therefore, it is reasonable to exclude them from these calculations until stronger evidence for their existence is established.

Y Chromosome

Many new and exciting insights into the peopling of Siberia and the New World have been recently obtained through studies of Y-chromosome lineage distribution in these two regions. The Y chromosome exhibits strict paternal transmission and encompasses large regions of non-recombining sequence, thus offering the possibility for studies of male migration. Although the Y-chromosome sequence evolution rate is slower than that of the mtDNA, it contains several different types of polymorphic systems that have different mutational mechanisms and rates. These include single-nucleotide polymorphisms (SNPs), small insertions or deletions (indels), and short tandem repeat (STR) polymorphisms, also known as microsatellites, with the latter mutating more rapidly than the other two kinds of polymorphisms. All these polymorphisms, or genetic markers, can be used to construct compound haplotypes that are informative for tracing male migration and ascertaining phylogenetic relationships among various populations (Hammer 1995; Jobling and Tyler-Smith 1995).

Because of the variety of polymorphic systems present in the Y chromosome, researchers have adopted a genealogical approach based on a hierarchical analysis of different marker systems to identify paternally inherited haplotypes (Jobling and Tyler-Smith 1995; de Knijff et al. 1997; Lell et al. 1997; Santos et al. 1996). In this approach, sets of Y chromosomes are divided into distinct lineages, or haplogroups, defined by one to several infrequently occurring "biallelic" (present or absent) polymorphisms, such as SNPs or indels. These haplogroups are further assayed for diversity using more variable loci such as STRs, in which multiple alleles of different sizes are present within a population. The resulting "compound haplotypes" comprising biallelic and multiallelic markers are then compared with those occurring in various world populations to reconstruct the genetic prehistory of these groups. By combining the different marker systems, one can obtain additional information about the demographic history of these haplogroups while minimizing the effect of recurrent mutation in the multiallelic polymorphic systems.

Over the past five years, various research groups have analyzed different sets of biallelic and multiallelic markers to characterize Y-chromosome variation in Siberian and Native American populations. However, only a subset of these biallelic markers have been especially effective for defining the Y haplotypes and paternal lineages present in them, including the DYS287 (M1), DYS199 (M3), M9, DYS7C, RBF5/Tat (M46), M17, and RPS4Y (M130) loci (see descriptions below). In addition, a number of Y-chromosome STRs have been surveyed for variation in these populations, including the DYS19, DYS388, DYS389, DYS390, DYS391, DYS392, and DYS393 loci (de Knijff et al. 1997; Kayser et al. 1997; Lell et al. 1997, 1998, 1999; Santos et al. 1999; Santos, Gerelsaikhan, et al. 1996). These loci represent both tri- and tetranucleotide repeat motifs, as well as different kinds of repeat motifs within an STR class, e.g., $(ATA)_n$ versus $(GTA)_n$ within trinucleotide repeats. For this reason, Y-chromosome variation will be discussed in terms of the haplotypes defined by these informative biallelic and microsatellite markers. Because large numbers of both individuals and populations have been screened for these markers using PCR-based assays, it is possible to make direct comparisons of marker and haplotype frequencies within them.

All studies to date have indicated that at least six major paternal haplogroups are present among Siberian populations. These include the M9, M9/M13, M9/M17, M9/DYS7C/M46, M1 and RPS4Y lineages, although the characterization of additional SNPs will likely reveal further subdivisions within them. As shown below, not all these Y lineages were disseminated into the Americas, and those that were may not have been brought to the New World through a single population expansion. To draw out these distinctions more clearly, each lineage and its distribution in Siberia and the Americas will be separately described.

It should also be stated here that the data sets generated in studies of Y-chromosome variation are usually not entirely equivalent because different arrays of SNPs and STRs have been used in different studies. As a consequence, the nomenclatural system used to define individual Y haplotypes and the paternal lineages to which they belong varies according to a particular study or set of related studies (Hammer et al. 1997, 1998; Karafet et al. 1997, 1999; Lell et al. 1997, 1998, 1999; Santos et al. 1996; Santos et al. 1999). The system used in this paper focuses on the presence or absence of specific biallelic markers in Native American and Siberian Y-chromosomes, and the Y lineages that they define, rather than any particular published nomenclature.

The M9 lineage, which is defined by a C→G transversion at the M9 locus, defines one of the two major branches in human Y-chromosome lineage evolution. It essentially divides all non-African Y-chromosomes from African chromosomes, which, in turn, are defined by either the M1 polymorphism or the ancestral A1 haplotype (Hammer et al. 1997; Underhill et al. 1997). Based on existing data sets, most East Asian and Siberian Y haplotypes exhibit the M9 G allele, hence represent the forms derived from ancestral African haplotypes; and the same general pattern is also seen in Native American populations (Hammer et al. 1997; Karafet et al. 1999; Lell et al. 1998, 1999; Underhill et al. 1997). In the case of Asian, Siberian, and Native American Y haplotypes, the M9 polymorphism is associated with the M3, M46/DYS7C, and M17 SNPs, but not the RPS4Y or M1 markers. Overall, the M9 lineage comprises a minority of the Y haplotypes seen in northeastern Siberians (Hammer et al. 1997; Karafet et al. 1999; Lell et al. 1998, 1999; Underhill et al. 1997).

Defined by an additional C→T transition at the M3 locus, the M9/M3 lineage occurs in a significant proportion of Native American Y chromosomes (Bianchi et al. 1996, 1998; Karafet et al. 1997, 1999; Lell et al. 1997, 1998, 1999; Santos et al. 1999; Underhill et al. 1996), but has yet to be found in any African or European populations (Karafet et al. 1999; Santos et al. 1999; Underhill et al. 1996, 1997). The M9/M3 lineage is present at significant frequencies in most Native American populations, ranging in frequency from 0–56 percent in North America, 57–85 percent in Central America, and 48–100 percent in South America, thereby revealing a general north-to-south cline within the New World (Bianchi et al. 1996, 1998; Karafet et al. 1997, 1999; Lell et al. 1997, 1998, 1999; Santos et al. 1999; Underhill et al. 1996). New STR data also revealed significant differences in both Y lineage and haplotype distributions between North/Central and South American Amerindian populations, suggesting different population histories for the two major continental regions (Bianchi et al. 1998; Karafet et al. 1999; Lell et al. 1998, 1999; Ruiz-Linares et al. 1999; Santos et al. 1999).

In contrast, among Siberian and Asian populations, the M9/M3 lineage has only been observed in the Siberian Eskimos and the Chukchi (Karafet et al. 1997, 1999; Lell et al. 1997, 1998, 1999; Underhill et al. 1996), along with a single Even individual (Karafet et al. 1997, 1999). Given that southeastern Siberia or Mongolia has been proposed as a potential geographic source for ancestral Native American populations (Kolman et al. 1996; Merriwether et al. 1996; Neel et al. 1994; Sukernik et al. 1996), and that the M9 lineage has been suggested to have arisen in this same geographic region (Karafet et al. 1999; Lell et al. 1999; Santos et al. 1999), the absence of the M9/M3 lineage in nearly every Siberian ethnic group and central Asian populations implies that it evolved in the ancestral Native American population(s) after leaving this part of northern Asia, either in Beringia or in the Americas.

The origination of the M9/M3 lineage before the first New World colonization event is also consistent with its presence in all the major linguistic subdivisions of Native Americans. Based on this evidence, it appears that the M3 SNP was maintained in the ancestral Beringian population(s), which subsequently became separated from the populations living further south in the Americas due to glacial barriers (Rogers 1986; Rogers et al. 1991, 1992). This Beringian population(s), in turn, became ancestral to modern-day Eskimo-Aleuts and Chukchi, who are postulated to have had a Beringian origin based on their genetic affiliation with other circumpolar populations (Forster et al. 1996; Schurr et al. 1999; Shields et al. 1993; Starikovskaya et al. 1998; Szathmary 1984, 1985, 1993). Thus, the presence of the M3 SNP in the Chukchi may be attributable to its presence in the ancestral population of the Paleoasiatic-speaking groups of northeastern Siberia (Lell et al. 1997, 1998, 1999), although others have suggested it resulted from back-migration and gene flow from Alaskan Eskimos westward across the Bering Strait (Karafet et al. 1997, 1999).

Not surprisingly, considerable effort has been made to estimate the age of the M9/M3 lineage, given that it signals the entry of ancestral populations into the New World. An initial study provided an age estimate for this lineage of 30,000 yr B.P., based on the linkage of the M3 C→T mutation and the DYS19 alleles (Underhill et al. 1996) and an autosomal microsatellite mutation rate of 1.5×10^{-4} (Weber and Wong 1993). However, an alternative mutation rate of 2.1×10^{-3} for the DYS19 locus gave an age for the M9/M3 lineage of 2147 yr B.P. (Underhill et al. 1996). Although this lineal age severely underestimated the date for the colonization of the New World, it raised the possibility that the M3 marker arose in an ancestral Beringian or American population after the initial entry of human populations into the New World. Hammer et al. (1998) arrived at a similar conclusion, as their nested cladistic method of Y haplotype analysis provided a M3 mutation age of ca. 10,000 yr B.P. By contrast, Forster et al. (2000) recently estimated a 20,000 yr B.P. age for the M3 mutation based on a mutation rate of 2.6×10^{-4} mutations/20 years for slowly evolving Y STRs. Thus, there are some discrepancies in dating the entry of ancestral Y chromosome lineages into the Americas.

These differences in age estimates probably reflect the different mutational rates of different kinds (tri-, tetra-, penta-) of Y-chromosome STRs (de Knijff et al. 1997; Kayser et al. 1997). The mutation rate of a given STR can also vary considerably depending on the size (number of repeats of the variable block) of the founder allele at each STR locus (Carvalho et al. 1999). On the other hand, Heyer et al. (1997) reported that Y STRs have mutation frequencies comparable to those on the autosomes, roughly equal to that originally published by Weber and Wong (1993). In addition, recent work suggests that STRs have a more rapid mutation rate than previously estimated, meaning that modern human populations spread into different parts of the Old and New Worlds later than estimated in earlier genetic studies (Thomson et al. 2000). Judging from these data sets, more focused studies of STR evolution in Amerindian populations are clearly needed to address the issue of the date(s) of the initial colonization of the New World from a Y-chromosome perspective.

In this context, it should also be mentioned that several other studies of Y-chromosome variation in Siberian and Native American populations used a different set of biallelic and STR loci to characterize Y haplotype variation, including the alphoid heteroduplex (ah) type II gene (Bianchi et al. 1997, 1998; Pena et al. 1995; Santos et al. 1996; Santos et al. 1999). These studies also revealed the presence of a major founding Y haplotype or lineage in Native American populations defined by the ahII/DYS19/M3 markers, one which almost certainly is synonymous with the M9/M3 lineage, and showed that its ancestral form originated somewhere in central-east Siberia. In addition, Bianchi et al. (1998) estimated the age of the ancestral Native American-specific Y lineage of 22,270 yr B.P., a date that was consistent with the age estimates of Underhill et al. (1996) and Forster et al. (2000). Thus, most analyses of Y chromosome variation in Native American populations suggest an "early" rather than a "late" entry of ancestral populations into the New World.

The M46/DYS7C lineage is defined by two different mutations, a T→C transition at the RBF5 locus, and a 50-bp deletion at the DYS7C locus. In fact, the M46 lineage is now known to subsume all the DYS7C haplotypes seen in the same groups (Karafet et al. 1999; Lell et al. 1998, 1999, 2000). Based on the distribution of this lineage, it appears that there has been a substantial paternal genetic contribution of Asians to northern European populations such as the Finns and Saami, as well as an expansion of Uralic-speakers eastward into northern Asia (Jobling et al. 1996; Karafet et al. 1999; Lahermo et al. 1998; Lell et al. 1998, 1999, 2000; Santos et al. 1999; Zerjal et al. 1997). Recent studies also suggest that the derived M46-C allele arose in southeastern Siberia, since this is where the ancestral form of the M46/DYS7C lineage appears (Karafet et al. 1999; Lell et al. 1998, 1999, 2000; Zerjal et al. 1997), with the ancestral populations bearing these haplotypes then influencing the Kets, Altayans and Buryats, as well as Mongolians, Yakuts and Chinese Han populations (Karafet et al. 1999; Lell et al. 1998, 1999, 2000; Santos et al. 1999; Zerjal et al. 1997).

Interestingly, the derived M46/DYS7C haplotypes are common in northeastern Siberian groups, including the Chukchi and Siberian Eskimos, but occur far less frequently in Amur River groups (Karafet et al. 1999; Lell et al. 1998, 1999, 2000). In addition, they are completely absent in Native American populations with the exception of the Navajo, in which a single haplotype bearing the DYS7C mutation has been detected (Karafet et al. 1999; Lell et al. 1998, 1999, 2000; Santos et al. 1999; Zerjal et al. 1997). This distribution suggests that the M46/DYS7C lineage arose in Siberia after the initial settlement of the New World, possibly only 4,000–2,000 years ago (Hammer et al. 1998; Zerjal et al. 1997), perhaps with the expansion of reindeer-herding culture in northern Asia and Eurasia (Zerjal et al. 1997).

The RPS4Y (M130) lineage is defined by a C→T transition occurring at np 711 within the 7th exon of the RPS4Y (ribosomal protein S4 gene on the Y-chromosome) locus (Bergen et al. 1996, 1999). Unlike some of the other Y haplogroups, the M130 lineage appears to be quite ancient and widespread in East and Southeast Asia, appearing in populations as far apart as Australia and Chukotka (Bergen et al. 1996, 1999; Karafet et al. 1999; Lell et al. 1998, 1999, 2000). Based on the STR patterns for its Y haplotypes in Siberian populations, the M130 lineage appears to have arisen in East Asian populations (Bergen et al. 1999; Lell et al. 1998, 1999, 2000), rather than the Altai-Sayan/Lake Baikal region, where nearly all other Y lineages are thought to have evolved. This interpretation is supported by the high frequencies and diversity of haplotypes in the M130 lineage in the Amur River region and northeastern Siberia (Lell et al. 1998, 1999, 2000). In fact, while sharing founding haplotypes, both Kamchatkan and Amur River populations have distinct sets of RPS4Y haplotypes, implying the considerable divergence of this haplogroup in East Asia.

Based on STR allele distributions, this lineage appears to have spread westward from the Russian Far East into the Baikal region, where it appears at significant frequencies in Central Asian and Central Siberian populations (Bergen et al. 1999; Karafet et al. 1999; Lell et al. 1998, 1999, 2000). However, in the Americas, the M130 lineage has only been seen in the Na-Dené-speaking Tanana and Navajo and Amerindian Cheyenne (Bergen et al. 1996, 1999; Karafet et al. 1999; Lell et al. 1998, 1999). Because the STR allele sizes for the Navajo haplotype are consistent with those present in M130 haplotypes in Siberian groups (Lell et al. 1998, 1999, 2000), it appears that this Y lineage was passed into Na-Dené Indian and Amerindian populations through the secondary expansion of Beringian populations into the New World.

The M1 lineage is defined by an Alu element insertion at Yq11, which has also been called the Y Alu Polymorphic element (YAP) (Hammer 1994). This lineage appears at intermediate to high frequencies (46–78 percent) in African populations and at low frequencies (0–11 percent) in European populations (Hammer 1994; Lahermo et al. 1998; Spurdle et al. 1994). It has also been detected at low to moderate frequencies in Central and East Asian populations, such as the Japanese (Hammer and Horai 1995), Koreans (Kim et al. 1998), Tibetans, Mongolians, Yakuts, Altayans and Tuvinians (Hammer et al. 1997; Karafet et al. 1997, 1999; Lell et al. 1997, 1998, 1999, 2000). However, the Asian M1 haplotypes differ from those in African populations by having longer 3' oligo (dA) tails on the Alu element, and also lack the DYS271 (M2) SNP, an A→G transition that results in an *Nla*III site loss, and which has been found only on African M1 chromosomes (Seielstad et al. 1994).

In the New World, the M1 lineage has been observed in several Native American populations, the Mixe from southern Mexico (Karafet et al. 1999; Lell et al. 1997), the Seminoles of Florida (Huoponen et al. 1997), and several Central and South American populations (Karafet et al. 1999). The two Mixe with the M1 haplotypes also exhibited DYS1 alleles previously observed in other non-Native American populations (Lell et al. 1997), and the Central and South American Indian groups exhibited additional SNPs that typically occur in African M1 haplotypes. Therefore, it appears that these Native American populations acquired their M1 haplotypes through intermarriage with persons having non-native ancestry.

In addition to the above-mentioned SNPs, a number of other SNPs have recently been shown to be phylogenetically informative (Underhill et al. 1997). These mutations mostly define the interior portion of the Y-chromosome phylogeny, i.e., the subbranches of the M9 major lineage. Among those informative for Siberians and Native Americans is the M17 marker, a 1-bp deletion, which defines the M9/M17 lineage. This lineage is present at low frequencies in a small, but not insignificant, number of Siberian populations and occurs at the highest frequency among the Itel'men, whereas it is absent from the neighboring Koryaks (Lell et al. 1998, 1999, 2001). However, the M9/M17 lineage is completely absent from Native American populations, with the exception of the Guaymi (Ngöbé), a Chibcha-speaking tribe from Costa Rica (Lell et al. 1998, 1999, 2000). Because the STR pattern for this Guaymi M9/M17 haplotype is consistent with those present in Siberian groups, it appears to have been brought to the Americas through a secondary expansion of ancient Asian peoples, rather than with the initial immigrants to the New World.

A few Y haplotypes found in Native American populations did not fit into any of the known paternal lineages, hence were placed into an "Other" category. However, it is likely that additional screenings for newly discovered SNPs will further define their origins and affinities, hence the regional source of these haplotypes. In this regard, screening Asian/Siberian and Native American populations for the STR markers observed in Southeast Asian and Chinese populations (Parra et al. 1998; Su et al. 1999) will likely illuminate the origins of Siberian of Native American populations.

From a Y-chromosome perspective, there are clear linkages between Siberia and the Americas, as well as population dynamics within Siberia itself, that have led to the distribution of various paternal lineages within it. These linkages can be summarized as follows. First, at least six different paternal lineages have been identified in Siberia and the Americas using biallelic and STR markers (M9, M9/M3, M9/DYS7C/M46, M9/M17, M130, and M1). Only two of these (M9 and M9/M3) fundamentally contributed to the initial peopling of the New World, either through single (Bianchi et al. 1998; Santos et al. 1999; Underhill et al. 1996) or multiple (Karafet et al. 1999; Lell et al. 1998, 1999, 2000) migration events. Three of the other major Y lineages present in Siberia either arose after the colonization of the New World, being distributed almost exclusively in Siberia (M9/DYS7C/M46, M9/M17), or else originated outside of the geographic region from which ancestral Native American populations bearing the M9 and M9/M3 lineages evolved (M130). One or more later, secondary expansions of human groups from Beringia into the Americas brought with them a different set of M9 haplotypes and the M9/M17 lineage, with the M130 lineage possibly being contributed from the Amur River region. On the other hand, M9/DYS7C/M46 haplotypes appear not to have been introduced into the New World at any appreciable frequency, being dispersed in Siberia and Eurasia during the Neolithic. Finally, the M1 haplotypes present in Native American populations were acquired through non-native admixture, rather than being contributed by founding Asian populations.

Overview of Data

Given all these genetic data, what general patterns of biological affinities and origins for Native Americans can be ascertained? To begin with, all these studies support the Asian origin of Native American populations. Most of them have suggested a region extended from the Altai Mountains to southeastern Siberia and northern China as the potential source area(s) for ancestral

Native American populations. However, the mtDNA data also suggest a possible Eurasian genetic influence with presence of haplogroup X in North America, and Y chromosome data suggest both south-central and eastern Siberia as the source area for ancestral Y lineages brought to the New World. Other data sets suggest possible population contacts with Asian populations from China (Novick et al. 1998) or Polynesia (Cann and Lum 1996; Lum et al. 1994) after the peopling of the New World, but these putative contacts have not been supported by other genetic data (Bonatto et al. 1996). On the other hand, various data sets further reveal post-Columbian (historical) admixture with peoples of European and African descent (e.g., Bortolini et al. 1997; Bravi et al. 1997; Huoponen et al. 1997; Rodriguez-Delfin et al. 1997; Torroni, Chen, et al. 1994), but these data are less important for reconstructing the prehistory of the New World.

While there is a general consensus about the source area for ancestral Paleoamerican populations, there is much less agreement among these studies about the number of migratory events bringing these populations to the New World. The number of possible migrations that have been proposed range from one (Bianchi et al. 1998, 1999; Merriwether et al. 1994, 1995; Monsalve et al. 1998; Novick et al. 1998; Santos et al. 1999), two (Karafet et al. 1999), three (Williams et al. 1985), to four (Horai et al. 1993; Lell et al. 1998, 1999, 2000; Schanfield et al. 1990; 1992; Schurr and Wallace 1999; Torroni, Schurr et al. 1993), depending on the genetic data and the definition of "migration" being used in a particular study. The primary debate taking place in these studies is whether the "Amerind" linguistic/genetic stock represents a single expansion event, or instead the consequence of multiple population movements into the New World. In this respect, should certain molecular data sets be correct in indicating that multiple migrations contributed to the genetic diversity of Amerindian populations, then Amerind cannot represent a single linguistic stock as previously asserted by Greenberg (1987). Regardless of the number of migrations that reached the New World, however, all these data sets are consistent in showing the distinctiveness and the great haplotypic diversity of members of the so-called Amerind linguistic group relative to those from the Eskimo-Aleut and Na-Dené groups, hence the earlier immigration of ancestral Amerindians to the New World.

Perhaps the most controversial aspect of recent molecular studies is the early colonization dates for the Americas that they have suggested. Most archaeological studies have dated this event to between 18,000 and 12,500 yr B.P., whereas genetic studies have provided somewhat older dates ranging from 35,000 to 15,000 yr B.P. (Bianchi et al. 1998; Cavalli-Sforza et al. 1994; Forster et al. 2000; Schurr et al. 1999; Torroni, Schurr et al. 1993; Underhill et al. 1996), again depending on the genetic system being used. The later times have been associated with the entry of the Clovis lithic culture in North America (Greenberg et al. 1986; Merriwether et al. 1994, 1995), while the earlier ones have been used to support the presence of pre-Clovis lithic traditions in North and South America (Schurr and Wallace 1999; Torroni, Neel et al. 1994; Torroni, Schurr et al. 1993; Underhill et al. 1996). Ongoing analyses of the mutation rates of the various genetic systems may ultimately narrow this time range, or perhaps push the initial entry time closer to the dates associated with various archaeological dates in the Americas. Irrespective of the initial entry time, most data sets are concordant in dating the emergence or expansion of the Eskimo-Aleuts and Na-Dené Indians at between 10,000 and 5000 yr B.P., well after ancestral Paleoindians arrived in the New World.

Moreover, all these genetic systems have shown that settlement of the Americas after their initial colonization was not a static process. Although most studies show general genetic divisions between North and Central versus South America, there is also evidence for both north-south and south-north population contact and gene flow between these regions. Underlying these broader patterns is considerable evidence of regional and tribal differentiation of populations in the major continental regions of the New World, particularly South America, with these patterns undoubtedly being shaped by geographic isolation, linguistic differentiation, and genetic drift. In the case of immunogenetic variation, selection appears to have played an especially significant role in producing the variety of protein polymorphism marker, HLA and immunoglobulin haplotypes detected in Siberian and Native American populations. There is also suggestive evidence of selection having acted on specific mtDNA (Torroni et al. 1997) and Y chromosome (Jobling et al. 1998) haplogroups in human populations. Hence, the distribution of maternal and paternal lineages may have been influenced by selection while being disseminated across the globe.

Finally, it should be clear from the discussion of these molecular anthropological studies that analyzing a large number of different genetic markers in human population provides a more complete picture of genetic variation in these groups. We have seen somewhat analogous but not identical views of the peopling of the New World originating from the use of multiple genetic systems. These differences are, in part, due to the varying rates of mutation and mechanisms of sequence evolution that underlie each genetic system, as well as their unique inheritance patterns. Ultimately, some molecular markers may be more informative than others. However, all of them will contribute to the overall understanding of the genetic variation present in Siberian and Native American populations. In this regard, the use of new STR, SNP and RSP data from autosomal loci will further help to delineate these population relationships and genetic diversity.

References Cited

Alexander, A., S. Rosen, and J. Buxbaum
1985 Human Immunoglobulin Genes in Health and Disease. *Clinical Immunology Review* 4:31–94.

Au, W., S. Gover, V. M. Lam, et al.
2000 Human Glucose-6-Phosphate Dehydrogenase: The Crystal Structure Reveals a Structural NADP(+) Molecule and Provides Insights into Enzyme Deficiency. *Structure Fold Des* 8:293–303.

Avent, N. D.
1997 Human Erythrocyte Antigen Expression: Its Molecular Basis. *British Journal of Biomedical Science* 54:16–37.

Avent, N.D., and M. E. Reid
2000 The Rh Blood Group System: A Review. *Blood* 95:375–87.

Bailliet, G., F. Rothhammer, F. R. Carnese, et al.
1994 Founder Mitochondrial Haplotypes in Amerindian Populations. *American Journal of Human Genetics* 54:27–33.

Ballinger, S. W., T. G. Schurr, A. Torroni, et al.
1992 Southeast Asian Mitochondrial DNA Analysis Reveals Genetic Continuity of Ancient Mongoloid Migrations. *Genetics* 130: 139–52.

Barrantes, R., P. E. Smouse, H. W. Mohrenweiser, et al.
1990 Microevolution in Lower Central America: Genetic Characterization of the Chibcha-speaking Groups of Costa Rica and Panama, and a Consensus Taxonomy Based on Genetic and Linguistic Affinity. *American Journal of Human Genetics* 46:63–84.

Batista, O., C. J. Kolman, and E. Bermingham
1995 Mitochondrial DNA Diversity in the Kuna Amerinds of Panama. *Human and Molecular Genetics* 4:921–29.

Batzer, M. A., and P. Deininger
1991 A Human-Specific Subfamily of Alu Sequences. *Genomics* 9:481–487.

Batzer, M. A., S. S. Arcot, J. W. Phinney, et al.
1996 Genetic Variation of Recent Alu Insertions in Human Populations. *Journal of Molecular Evolution* 42:22–29.

Batzer, M. A., V. A. Gudi, J. C. Mena, et al.
1991 Amplification Dynamics of Human-Specific (HS) Alu Family Members. *Nucleic Acids Research* 19:3619–23.

Bergen, A., C.-Y. Wang, K. Jefferson, et al.
1996 RSP4Y and STR Evolution on the Y-Chromosome: Identification of a Distinct Paternal Lineage with Asian and Amerindian Populations. *American Journal of Human Genetics* Supplement 59:A174.

Bergen, A.W., C.-Y. Wang, J. Tsai, et al.
1999 An Asian-Native American Paternal Lineage Identified by RPS4Y Resequencing and Microsatellite. *Haplotyping Annals of Human Genetics* 63:63–80.

Bianchi, N. O., C. I. Catanesi, G. Bailliet, et al.
1998 Characterization of Ancestral and Derived Y-chromosome Haplotypes of New World Native Populations. *American Journal of Human Genetics* 63:1862–71.

Bianchi, N. O., G. Bailliet, C. M. Bravi, et al.
1997 Origin of Amerindian Y-chromosomes as Inferred by the Analysis of Six Polymorphic Markers. *American Journal of Physical Anthropology* 102:79–89.

Black, F. L.
1991 Reasons for Failure of Genetic Classifications of South Amerind Populations. *Human Biology* 63:763–74.

Black, F. L., and J. P. Pandey
1997 Evidence for Balancing of KM, but GM, Alleles by Heterotic Advantage in South Amerinds. *Human Genetics* 100:240–44.

Bodmer, J. G., S. G. Marsh, E. D. Albert, et al.
1999 Nomenclature for Factors of the HLA System, 1998. *Tissue Antigens* 53:407–46.

Bonatto S. L., and F. M. Salzano
1997 Diversity and Age of the Four Major mtDNA Haplogroups, and Their Implications for the Peopling of the New World. *American Journal of Human Genetics* 61:1413–23.

Bonatto, S. L., A. J. Redd, F. M. Salzano, et al.
1996 Lack of Ancient Polynesian-Amerindian Contact. *American Journal of Human Genetics* 59:253–56.

Bortolini, M. C., M. A. Zago, F. M. Salzano, et al.
1997 Evolutionary and Anthropological Implications of Mitochondrial DNA Variation in African Brazilian Populations. *Human Biology* 69:141–59.

Bravi, C. M., M. Sans, G. Bailliet, et al.
1997 Characterization of Mitochondrial DNA and Y-chromosome Haplotypes in a Uruguayan Population of African Ancestry. *Human Biology* 69:641–52.

Brock, J. H., T. Mainou-Fowler, and L. M. Webster
1986 Evidence that Transferrin May Function Exclusively as an Iron Donor in Promoting Lymphocyte Proliferation. *Immunology* 57:105–10.

Brown, M. D., S. H. Hosseini, A. Torroni, et al.
1998 Haplogroup X: An Ancient Link between Europe/Western Asia and North America? *American Journal of Human Genetics* 63:1852–61.

Brown, W. M., M. George, and A. C. Wilson
1979 Rapid Evolution of Animal Mitochondrial DNA. Proceedings of the National Academy of Sciences USA 76:1967–71.

Browning, M., and A. McMichael (editors).
1996 HLA and MHC: Genes, Molecules, and Function. BIOS Scientific Publishers, Herndon, VA.

Callegari-Jacques, S. M., F. M. Salzano, J. Constans, et al.
1993 GM Haplotype Distribution in Amerindians: Relationship with Geography and Language. *American Journal of Physical Anthropology* 90:427–44.

Callegari-Jacques, S. M., F. M. Salzano, T. A. Weimer, et al.
1994 Further Blood Genetic Studies on Amazonian Diversity—Data from Four Indian groups. *Annals of Human Biology* 21:465–81.

Cann, R. L., and J. K. Kum
1996 Mitochondrial Myopia: Reply to Bonatto et al. *American Journal of Human Genetics* 59:256–57

Cartron, J.-P.
1994 Defining the Rh Blood Group Antigens: Biochemistry and Molecular Genetics. *Blood Reviews* 8:199–212.

Cartron, J.-P., P. Bailly, C. Le Van Kim, et al.
1998 Insights into the Structure and Function of Membrane Polypeptides Carrying Blood Group Antigens. *Vox Sanguinis* 74 Suppl 2:29–64.

Carvalho-Silva, D. R., F. R. Santos, M. H. Hutz, et al.
1999 Divergent Human Y Chromosome Microsatellite Evolution Rates. *Journal of Molecular Evolution* 49:204–14.

Case, J. T., and D.C. Wallace
1981 Maternal Inheritance of Mitochondrial DNA Polymorphisms in Cultured Human Fibroblasts. *Somatic Cell Genetics* 7:103–108

Cavalli-Sforza, L. L., P. Menozzi, and A. Piazza
1994 The History and Geography of Human Genes. Princeton University Press.

Cerna, M., M. Falco, H. Friedman, et al.
1993 Differences in HLA Class II Alleles of Isolated South American Indian Populations from Brazil and Argentina. *Human Immunology* 37:231–230.

Chen, Y. S., A. Olckers, T. G. Schurr, et al.
2000 Mitochondrial DNA Variation in the Southern African Kung and Khwe and Their Genetic Relationships to Other African Populations. *American Journal of Human Genetics* 66:1362–83.

Chen, Y.-S., A. Torroni, L. Excoffier, et al.
1995 Analysis of mtDNA Variation in African Populations Reveals the Most Ancient of All Human Continent-specific Haplogroups. *American Journal of Human Genetics* 57:133–49.

Clayton, D. A.
2000 Vertebrate Mitochondrial DNA—A Circle of Surprises. *Experimental Cellular Research* 255:4–9.

Crawford, M. H., and V. B. Enciso
1982 Population Structure of Circumpolar Groups of Siberia, Alaska, Canada and Greenland. In *Current Developments in Anthropological Genetics,* Vol. 2. Ecology and Population Structure, edited by M. H. Crawford and J. H. Mielke. pp. 51–91. Plenum Press, New York.

Crawford, M. H., J. H. Mielke, E. J. Devor, et al.
1981 Population Structure of Alaskan and Siberian Indigenous Communities. *American Journal of Physical Anthropology* 55:167–85.

Daniels, C.
1997 Blood Group Polymorphisms: Molecular Approach and Biological Significance. *Transfusion and Clinical Biology* 4:383–90.

de Knijff, P., M. Kayser, C. Caglià, et al.
1997 Chromosome Y Microsatellites: Population Genetic and Evolutionary Aspects. *International Journal of Legal Medicine* 110:134–40.

Derenko, M. V., I. K. Dambueva, B. A. Maliarchuk, et al.
1999 Structure and Diversity of the Mitochondrial Gene Pool of the Aboriginal Population of Tuva and Buriatia from Restriction Polymorphism Data. *Genetika* 35(12):1706–12.

Derenko, M. V., B. A. Malyarchuk, I. K. Dambueva, et al.
2000 Mitochondrial DNA Variation in Two South Siberian Aboriginal Populations: Implications for the Genetic History of North Asia. *Human Biology* 72(6):945–73.

Dipierri, J. E., E. Alfaro, V. L. Martinez-Marignac, et al.
1998 Paternal Directional Mating in Two Amerindian Subpopulations Located at Different Altitudes in the Northwest of Argentina. *Human Biology* 70:1001–10.

Dubrova, Y. E., L. V. Bogatyrjova, and E. I. Pushkina
1993 Genetic Affinities of Buryat Populations and Other Mongoloids of Siberia. *Human Heredity* 43:82–85.

Dugoujon, J.-M., B. Mourrieras, M.-T. Senegas, et al.
1995 Human Genetic Diversity Immunoglobulin GM Allotypes, Linguistic Data, and Migrations of Amerindian Tribes. *Human Biology* 67:231–49.

Easton, R. D., D. A. Merriwether, D. E. Crews, et al.
1996 mtDNA Variation in the Yanomami: Evidence for Additional New World Founding Lineages. *American Journal of Human Genetics* 59:213–25.

Erlich, H. A., S. J. Mack, T. Bergstrom, et al.
1997 HLA Class II Alleles in Amerindian Populations: Implications for the Evolution of HLA Polymorphisms and the Colonization of the New World. *Hereditas* 127:19–24.

Ferrell, R. E., R. Chakraborty, H. Gershowitz, et al.
1981 The St. Lawrence Island Eskimos: Genetic Variation and Genetic Distance. *American Journal of Physical Anthropology* 55:351–58.

Forster P., R. Harding, A. Torroni, et al.
1996 Origin and Evolution of Native American mtDNA Variation: A Reappraisal. *American Journal of Human Genetics* 59:935–45.

Forster, P., A. Ršhl, P. Lünnermann, et al.
2000 A Short Tandem Repeat-based Phylogeny for the Human Y Chromosome. *American Journal of Human Genetics* 67:182–96.

Fox, C. L.
1996 Mitochondrial DNA Haplogroups in Four Tribes from Tierra del Fuego–Patagonia: Inferences about the Peopling of the Americas. *Human Biology* 68:855–71.

Gao, X., Y. Sun, J. An, et al.
1991 DNA Typing for HLA-DR and -DP Alleles in a Chinese Population Using the Polymerase Chain Reaction (PCR) and Oligonucleotide Probes. *Tissue Antigens* 38:24–30.

Garber, T. L., S. N. McAdam, L. M. Butler, et al.
1996 HLA-B Alleles of the Navajo: No Evidence for Rapid Evolution in the Nadene. *Tissue Antigens* 47:143–46.

Giles R. E., H. Blanc, H.M. Cann, and D.C. Wallace
1980 Maternal Inheritance of Human Mitochondrial DNA. Proceedings of the National Academy of Sciences of the United States of America 77:6715–19.

Ginther, C., D. Corach, G. A. Penacino, et al.
1993 Genetic Variation among the Mapuche Indians from the Patagonian Region of Argentina: Mitochondrial DNA Sequence Variation and Allele Frequencies of Several Nuclear Genes. In *DNA Fingerprinting: State of the Science*, edited by S. D. J. Pena, R. Chakraborty, J. T. Epplan, and A. J. Jefferies, pp. 211–19. Berkhauser Verlag. Basel

Grahovac, B., R. I. Sukernik, C. O'hUigin, et al.
1998 Polymorphism of the HLA Class II Loci in Siberian Populations. *Human Genetics* 102:27–43.

Green, L. D., J. N. Derr, and A. Knight
2000 mtDNA Affinities of the Peoples of North-Central Mexico. *American Journal of Human Genetics* 66:989–98.

Greenberg, J. H.
1987 Language in the Americas. Stanford University Press, Stanford.

Greenberg, J., C. G. Turner, II, and S. L. Zegura
1986 The Settlement of the Americas: A Comparison of the Linguistic, Dental and Genetic Evidence. *Current Anthropology* 4:477–97.

Hamblin, M. F., and A. Di Rienzo
2000 Detection of the Signature of Natural Selection in Humans: Evidence from the Duffy Blood Group Locus. *American Journal of Human Genetics* 66:1669–79.

Hammer, M. F.
1994 A Recent Insertion of an Alu Element on the Y Chromosome Is a Useful Marker for Human Population Studies. *Molecular Biology and Evolution* 11:749–61.
1995 A Recent Common Ancestry for Human Y Chromosomes. *Nature* 378:376–378.

Hammer, M., and S. Horai
1995 Y Chromosomal DNA Variation and the Peopling of Japan. *American Journal of Human Genetics* 56:951–62.

Hammer, M. F., A. B. Spurdle, T. Karafet, et al.
1997 The Geographic Distribution of Human Y Chromosome Variation. *Genetics* 145:787–805.

Hammer, M. F., and S. L. Zegura
1996 Role of the Y Chromosome in Human Evolutionary Studies. *Evolutionary Anthropology* 5(4):116–34.

Hammer, M. F., T. Karafet, A. Rasanayagam, et al.
1998 Out of Africa and Back Again: Nested Cladistic Analysis of Human Y Chromosome Variation. *Molecular Biology and Evolution* 15:427–41.

Harpending, H. C., M. A. Batzer, M. A. Gurven, et al.
1998 Genetic Traces of Ancient Demography. Proceedings of the National Academy of Sciences USA 95:1961–67.

Harper, A. B.
1980 Origins and Divergence of Aleuts, Eskimos and American Indians. *Annals of Human Biology* 7:547–54.

Hauswirth, W. W., C. D. Dickel, D. J. Rowold, et al.
1994 Inter- and Intrapopulation Studies of Ancient Humans. *Experientia* 50:585–91.

Hayes, M. G.
1999 Mitochondrial DNA Variation of Prehistoric Eastern Canadian Arctic Inhabitants. *American Journal of Physical Anthropology* Supplement 28:148.

Hedrick, P. W., and F. L. Black
1997 Random Mating and Selection in Families against Homozygotes for HLA in South Amerindians. *Hereditas* 127:51–58.

Heyer, C., J. Puymirat, P. Dieltjes, et al.
1997 Estimating Y Chromosome Specific Mutation Frequencies Using Deep Rooted Pedigrees. *Human Molecular Genetics* 6:799–803.

Horai, S., and K. Hayasaka
1990 Intraspecific Nucleotide Sequence Differences in the Major Non-coding Region of the Human Mitochondrial DNA. *American Journal of Human Genetics* 46:828–42.

Horai, S., K. Hayasaka, R. Kondo, et al.
1995 Recent African Origin of Modern Humans Revealed by Complete Sequences of Hominoid Mitochondrial DNAs. Proceedings of the National Academy of Sciences of the USA 92:532–36.

Horai, S., R. Kondo, Y. Nakasawa-Hattori, et al.
1993 Peopling of the Americas, Founded by Four Major Lineages of Mitochondrial DNA. Molecular Biology and Evolution 10:23-47.

Horai, S., K. Murayama, K. Hayasaka, et al.
1996 mtDNA Polymorphism in East Asian Populations, with Special Reference to the Peopling of Japan. *American Journal of Human Genetics* 59:579–90.

Huoponen, K., A. Torroni, P. R. Wickman, et al.
1997 Mitochondrial and Y Chromosome-Specific Polymorphisms in the Seminole Tribe of Florida. *European Journal of Human Genetics* 5:25–34.

Ikemoto, S., S. Iwamoto, S. Tsuchida, et al.
1996 Molecular Genetic Basis of Red Cell Markers and Its Forensic Applications. *Forensic Science International* 80:147–61.

Imanishi, T. T. Akaza, A. Kimura, et al.
1992 Allele and Haplotype Frequencies for HLA Complement Loci in Various Ethnic Groups. In HLA 1991b. Proceedings of the 11th Histocompatibility Workshop and Conference, vol. 1, edited by K. Tsuji, K. Tokunaga, T. Sasazuki, pp. 1065–1220. Oxford University Press, Oxford.

Iwanaga, K. K., M. Eberle, C. J. Kolman, et al.
1997 Further Diversification of the HLA-B Locus in Central American Amerindians: New B*39 and B*51 Alleles in the Kuna of Panama. *Tissue Antigens* 50:251–57.

Jobling, M. A., and C. Tyler-Smith
1995 Fathers and Sons: The Y Chromosome and Human Evolution. *Trends in Genetics* 11:449–56

Jobling, M. A., V. Samara, A. Pandya, et al.
1996 Recurrent Duplication and Deletion Polymorphisms on the Long Arm of the Y Chromosome in Normal Males. *Human Molecular Genetics* 5:1767–75.

Jobling, M. A., G. Williams, K. Schiebel, et al.
1998 A Selective Difference between Human Y Chromosomal DNA Haplotypes. *Current Biology* 8:1391–94.

Kaestle, F.
1995 Mitochondrial DNA Evidence for the Identity of the Descendants of the Prehistoric Stillwater Marsh Population. In *Bioarcheology of the Stillwater Marsh: Prehistoric Human Adaptation in the Western Great Basin,* edited by C. S. Larsen and R. L. Leyy, pp. 737–49. New York, Anthropological Papers of the American Museum of Natural History.

Kaestle, F. A.
1997 Molecular Analysis of Ancient Native American DNA from Western Nevada. *Nevada Historical Society Quarterly* 40:85–96.

Karafet, T., S. L. Zegura, J. Vuturo-Brady, et al.
1997 Y-Chromosome Markers and Trans-Bering Strait Dispersals. *American Journal of Physical Anthropology* 102:301–14.

Karafet, T. M., S. L. Zegura, O. Posukh, et al.
1999 Ancestral Asian Source(s) of New World Y-chromosome Founder Haplotypes. *American Journal of Human Genetics* 64:817–31.

Kayser, M., C. Caglià, D. Corach, et al.
1997 Evaluation of Y-chromosomal STRs: A Multicenter Study. *International Journal of Legal Medicine* 110:125–33.

Kim, W., D. J. Shin, S. A. You, et al.
1998 Y-specific DNA Polymorphism of the YAP Element and the Locus DYS19 in the Korean Population. *Journal of Human Genetics* 43:195–98.

King, M. J.
1994 Blood Group Antigens on Human Erythrocytes—Distribution, Structure and Possible Functions. *Biochimica et Biophysica Acta* 1197:15–44.

Kolman, C. J., and E. Bermingham.
1997 Mitochondrial and Nuclear DNA Diversity in the Choco and Chibcha Amerinds of Panama. *Genetics* 147:1289–1302.

Kolman, C. J., E. Bermingham, R. Cooke, et al.
1995 Reduced mtDNA Diversity of the Ngöbé Amerinds of Panama. *Genetics* 140:275–83.

Kolman, C. J., N. Sambuughin, and E. Bermingham
1996 Mitochondrial DNA Analysis of Mongolian Populations and Implications for the Origin of New World Founders. *Genetics* 142:1321–34.

Krsek-Staples, J. A., and R. O. Webster
1993 Ceruloplasmin Inhibits Carbonyl Formation in Endogenous Cell Proteins. *Free Radical Biology and Medicine* 14:115–25.

Krylov, M., S. Erdesz, L. Alexeeva, et al.
1995 HLA Class II and HLA-B27 Oligotyping in Two Siberian Native Populations. *Tissue Antigens* 46:382–86.

Lahermo, P., M.-L. Savontaus, P. Sistonen, et al.
1999 Y Chromosomal Polymorphisms Reveal Founding Lineages in the Finns and Saami. *European Journal of Human Genetics* 7:447–58.

Langlois, M. R., and J. R. Delanghe
1996 Biological and Clinical Significance of Haptoglobin Polymorphism in Humans. *Clinical Chemistry* 42:1589–1600.

Layrisse, Z., Y. Guedez, E. Dominguez, et al.
1995 Extended HLA Haplotypes among the Bari Amerindians of the Perija Range: Relationship to Other Tribes Based on Four-Loci Haplotype Frequencies. *Human Immunology* 44:228–35.

Layrisse, Z., M. Matos, S. Montagnani, et al.
1997 The Novel HLA-DRB1*0807 Allele Is Present in a Carib Amerindian Population: The Yucpa. *Tissue Antigens* 49:427–28.

Lell, J. T., M. D. Brown, T. G. Schurr, et al.
1997 Y Chromosome Polymorphisms in Native American and Siberian Populations: Identification of Founding Native American Y Chromosome Haplotypes. *Human Genetics* 100:536–43.

Lell, J. T., T. G. Schurr, R. I. Sukernik, et al.
1998 Identification and Characterization of Siberian and Native American Y-Chromosome Lineages by Microsatellite Analysis. *American Journal of Human Genetics* Supplement 63:A213.

Lell, J. T., T. G. Schurr, R. I. Sukernik, et al.
1999 Y Chromosome Haplotypes Reveal Distinct Migration Patterns in Siberia and the Americas. *American Journal of Human Genetics* Supplement 65:A41.

Lell, J. T., R. I. Sukernik, Y. B. Starikovskaya, et al.
2002 The Dual Origins and Siberian Affinities of Native American Y Chromosomes. *American Journal of Human Genetics* 70(1): 192–206.

Livishits, G., and M. Nei
1990 Relationships between Intrapopulational and Interpopulational Genetic Diversity. *Annals of Human Biology* 17:510–17.

Lorenz, J. G., and D. G. Smith
1996 Distribution of Four Founding mtDNA Haplogroups among Native North Americans. *American Journal of Physical Anthropology* 101:307–23.
1997 Distribution of Sequence Variations in the mtDNA Control Region of Native North Americans. *Human Biology* 69:749–76.

Lozáro, A. M., M. E. Moraes, C. Y. Marcos, et al.
1999 Evolution of HLA-Class I Compared to HLA-Class II Polymorphism in Terena, a South American Indian Tribe. *Human Immunology* 60:1138–49.

Luiselli, D., L. Simoni, E. Tarazona-Santos, et al.
2000 Genetic Structure of Quechua-speakers of the Central Andes and Geographic Patterns of the Gene Frequencies on South Amerindian Populations. *American Journal of Physical Anthropology* 113:5–17.

Lum, J. K., O. Rickards, C. Ching, et al.
1994 Polynesian Mitochondrial DNAs Reveal Three Deep Maternal Lineage Clusters. *Human Biology* 66:567–90.

Mack, S. J., and H. A. Erlich
1998 HLA Class II Polymorphism in the Ticuna of Brazil: Evolutionary Implications of the DRB1*0807 Allele. *Tissue Antigens* 51:41–50.

Majumder, P. P., W. S. Laughlin, and R. E. Ferrell
1998 Genetic variation in the Aleuts of the Pribilof Islands and the Eskimos of Kodiak Island. *American Journal of Physical Anthropology* 76:481–88.

Marchalonis, J. J., and S. F. Schluter
1989 Evolution of Variable and Constant Domains and Joining Segments of Rearranging Immunoglobulins. *FASEB Journal* 3:2469–79.

Marcos, C. Y., M. A. Fernández-Viña, A. M. Lozáro, et al.
1999 Novel HLA-A and HLA-B Alleles in South American Indians. *Tissue Antigens* 53:476–85.

Martinez-Laso, J., Z. Layrisse, E. Gomez-Casado, et al.
1996 A New HLA-B15 Allele (B*1522) Found in Bari-Motilones Amerindians in Venezuela: Comparison of Its Intron 2 Sequence with Those of B*1501 and B*3504. *Immunogenetics* 43:108–9.

Matera, A. G., U. Hellmann, and C. W. Schmid
1990 A Transpositionally and Transcriptionally Competent Alu Subfamily. *Molecular and Cellular Biology* 10:5424–32.

Merriwether, D. A., A. G. Clark, S. W. Ballinger, et al.
1991 The Structure of Human mtDNA Variation. *Journal of Molecular Evolution* 33:543–55.

Merriwether, D. A., F. Rothhammer, and R. E. Ferrell
1994 Genetic Variation in the New World: Ancient Teeth, Bone, and Tissue as Sources of DNA. *Experientia* 50:592–601.
1995 Distribution of the Four-Founding Lineage Haplotypes in Native Americans Suggests a Single Wave of Migration for the New World. *American Journal of Physical Anthropology* 98:411–30.

Merriwether, D. A., W. W. Hall, A. Vahlne, et al.
1996 mtDNA Variation Indicates Mongolia May Have Been the Source for the Founding Population for the New World. *American Journal of Human Genetics* 59:204–12.

Mestriner, M. A., A. L. Simoes, and F. M. Salzano
1980 New Studies on the Esterase D Polymorphism in South American Indians. *American Journal of Physical Anthropology* 52:95–101.

Mestriner, M. A., F. M. Salzano, J. V. Neel, et al.
1976 Esterase D in South American Indians. *American Journal of Human Genetics* 28:251–61.

Miyata, T., H. Hayashida, R. Kikuno, et al.
1982 Molecular Clock of Silent Substitution: At Least Six-fold Preponderance of Silent Changes in Mitochondrial Genes over Those in Nuclear Genes. *Journal of Molecular Evolution* 19:28–35.

Mohrenweiser, H., J. V. Neel, M. A. Mestriner, et al.
1979 Electrophoretic Variants in Three Amerindian Tribes: The Baniwa, Kanamari, and Central Pano of Western Brazil. *American Journal of Physical Anthropology* 50:237–46.

Monsalve, J. V., and E. Hagelberg
1997 Mitochondrial DNA Polymorphisms in Carib People of Belize. Proceedings of the Royal Society of London, Series B 264: 1217–24.

Monsalve, M. V., F. Cardenas, F. Guhl, et al.
1996 Phylogenetic Analysis of mtDNA Lineages in South American Mummies. *Annals of Human Genetics* 60:293–303.

Monsalve, J. V., G. Edin, and D. V. Devine
1998 Analysis of HLA Class I and Class II in Na-Dene and Amerindian Populations from British Columbia, Canada. *Human Immunology* 59:48–55.

Monsalve, J. V., A. Helgason, and D. V. Devine
1999 Languages, Geography and HLA Haplotypes in Native American and Asian Populations. Proceedings of the Royal Society of London, Biology, Series B 266:2209–16.

Moriwaki, Y., T. Yamamoto, and K. Higashino
1999 Enzymes Involved in Purine Metabolism—A Review of Histochemical Location and Functional Implications. *Histology and Histopathology* 14:1321–40.

Nakazawa, A., M. Yamada, H. Tanaka, et al.
1990 Gene Structures of Three Vertebrate Adenylate Kinase Isozymes. Progress in Clinical and Biological Research 344:495–514.

Neel, J. V.
1973 "Private" Genetic Variants and the Frequency of Mutation among South American Indians. Proceedings of the National Academy of Sciences, USA. 70:3311–15.
1978 Rare Variants, Private Polymorphisms, and Locus Heterozygosity in Amerindian Populations. *American Journal of Human Genetics* 30:465–90.

1980 Isolates and Private Polymorphisms. In Population Structure and Genetic Disorders, ed. A. Eriksson, pp. 173–93. Academic Press, London.

Neel, J. V., R. J. Biggar, and R. I. Sukernik
1994 Virologic and Genetic Studies Relate Amerind Origins to the Indigenous People of the Mongolia/Manchuria/Southeastern Siberia Region. Proceedings of the National Academy of Sciences, USA 91:10737–41.

Neel, J. V., H. Gershowitz, H. W. Mohrenweiser, et al.
1980 Genetic Studies of the Ticuna, an Enigmatic Tribe of Central Amazonas. *Annals of Human Genetics* 44:37–54.

Neel, J. V., H. Gershowitz, R. S. Spielman, et al.
1977 Genetic Studies of the Macushi and Wapishana Indians. II. Data on 12 Genetic Polymorphisms of the Red Cell and Serum Proteins: Gene Flow Between Tribes. *Human Genetics* 37:207–19.

Neel, J. V., R. J. Tanis, E. C. Migliazzi, et al.
1977 Genetic Studies on the Macushi and Wapishana Indians. I. Rare Genetic Variants and a "Private Polymorphism" of Esterase A. *Human Genetics* 36:81–107.

Nei, M.
1987 Molecular Evolutionary Genetics. Columbia University Press, New York.

Nei, M., and G. Livhsits
1989 Genetic Relationships of Europeans, Asians and Africans and the Origins of Modern Homo sapiens. *Human Heredity* 39:276–81.

Nei, M., and A. K. Roychoudhury
1993 Evolutionary Relationships of Human Populations on a Global Scale. *Molecular Biology and Evolution* 10:927–43.

Notaro, R., A. Afoloyan, and L. Luzzatto
2000 Human Mutations in Glucose 6-Phosphate Dehydrogenase Reflect Evolutionary History. *FASEB Journal* 14:485–94.

Novick, G. E., C. C. Novick, J. Yunis, et al.
1995 Polymorphic Human Specific Alu Insertions as Markers for Human Identification. *Electrophoresis* 16:1596–601.
1998 Polymorphic Alu Insertions and the Asian Origin of Native American Populations. *Human Biology* 70:23–39.

O'Rourke, D. H., and B. K. Suarez
1986 Patterns and Correlates of Genetic Variation in South Amerindians. *Annals of Human Biology* 13:13–31.

O'Rourke, D. H., M. G. Hayes, and S. W. Carlyle
2000 Spatial and Temporal Stability of mtDNA Haplogroup Frequencies in Native North America. *Human Biology* 72:15–34.

O'Rourke, D. H., A. Mobarry, and B. K. Suarez
1992 Patterns of Genetic Variation in Native America. *Human Biology* 64:417–34.

O'Rourke, D. H., B. K. Suarez, and J. D. Crouse
1985 Genetic Variation in North Amerindian Populations: Covariance with Climate. *American Journal of Physical Anthropology* 67:241–250.

Osipova, L. P., and R. I. Sukernik
1983 Immunoglobulin Allotypes in Aboriginal Populations of the Taimir Peninsula. *Journal of Immunogenetics* 10:11–16.

Parr, R. L., S. W. Carlyle, and D. H. O'Rourke
1996 Analysis of Fremont Amerindians of the Great Salt Lake Wetlands. *American Journal of Physical Anthropology* 99:507–18.

Parra, E., M. D. Shriver, A. Soemantri, et al.
1999 Analysis of Five Y-Specific Microsatellite Loci in Asian and Pacific Populations. *American Journal of Physical Anthropology* 110:1–16.

Pena, S. D. J., F. R. Santos, N. O. Bianchi, et al.
1995 A Major Founder Y-Chromosome Haplotype in Amerindians. *Nature Genetics* 11:15–16.

Petrishchev, V. N., A. B. Kutueva, and Y. G. Rychkov
1993 Deletion-Insertion Polymorphism in mtDNA Region V in Ten Mongoloid Populations of Siberia. *Genetika* 29:1196–1203.

Posukh, O. L., L. P. Osipova, Y. O. Kashinskaya, et al.
1998 A Genetic Study of the South Altaian Population of the Mendur-Sokkon Village, Altai Republic. *Russian Journal of Genetics* 34:92–98.

Posukh, O. L., V. P. Wiebe, R. I. Sukernik, et al.
1990 Genetic Study of the Evens, an Ancient Human Population of Eastern Siberia. *Human Biology* 62:457–65.

Reid, M. E., and K. Yazdanbakhsh
1998 Molecular Insights into Blood Groups and Implications for Blood Transfusions. *Current Opinion in Hematology* 5:93–102.

Ribeiro-Dos-Santos, A. K. C., S. E. B. Santos, A. L. Machado, et al.
1996 Heterogeneity of Mitochondrial DNA Haplotypes in Pre-Columbian Natives of the Amazon Region. *American Journal of Physical Anthropology* 101:29–37.

Rodriguez-Delfin, L., S. E. B. Santos, and M. S. Zagos
1997 Diversity of the Human Y Chromosome of South American Amerindians: A Comparison with Blacks, Whites and Japanese from Brazil. *Annals of Human Genetics* 61:439–48.

Rogers, R. A.
1986 Language, Human Subspeciation, and Ice Age Barriers in Northern Siberia. *Canadian Journal of Anthropology* 5:11–22.

Rogers, R. A., L. A. Rogers, and L. D. Martin
1992 How the Door Opened: The Peopling of the New World. *Human Biology* 64:281–302.

Rogers, R. A., L. A. Rogers, R. S. Hoffmann, et al.
1991 Native American Biological Diversity and the Biogeographic Influence of Ice Age Refugia. *Journal of Biogeography* 18:623–30.

Rosser Z. H., T. Zerjal, M. E. Hurles, et al.
2000 Y-chromosomal Diversity in Europe Is Clinal and Influenced Primarily by Geography, Rather Than by Language. *American Journal of Human Genetics* 67(6):1526–43.

Rothhammer, F., C. Silva, S. M. Callegari-Jacques, et al.
1997 Gradients of HLA Diversity in South American Indians. *Annals of Human Biology* 24:197–208.

Rouault, T., and R. Klausner
1997 Regulation of Iron Metabolism in Eularyotes. *Current Topics in Cellular Regulation* 35:1–19.

Ruiz-Linares, A., D. Ortiz-Barrientos, M. Figueroa, et al.
1999 Microsatellites Provide Evidence for Y Chromosome Diversity among the Founders of the New World. Proceedings of the National Academy of Sciences, USA 96:6312–17.

Saillard, J., P. Forster, N. Lynnerup, et al.
2000 mtDNA Variation among Greenland Eskimos: The Edge of the Beringian Expansion. *American Journal of Human Genetics* 67:718–26.

Sajantila, A., P. Lahermo, T. Anttinen, et al.
1995 Genes and Languages in Europe: An Analysis of Mitochondrial Lineages. *Genome Research* 5:42–52.

Salzano, F. M., and S. M. Callegari-Jacques
1980 Demographic and Genetic Relationships among Brazilian Wapishana Indians. *Annals of Human Biology* 7:129–38.

1988 South American Indians: A Case Study in Evolution. Oxford: Clarendon Press.

Salzano, F. M., S. M. Callegari-Jacques, and J. V. Neel.
1980 Genetic Demography of the Amazonian Ticuna Indians. Journal of Human Evolution 9:179–91.

Salzano, F. M., F. L. Black, S. M. Callegari-Jacques, et al.
1991 Blood Genetic Systems in Four Amazonian Tribes. *American Journal of Physical Anthropology* 85:51–60.

Salzano, F. M., H. Gershowitz, P. C. Junquiera, et al.
1977 Blood Group and H-Lea Salivary Secretion of Brazilian Cayapo Indians. *American Journal of Physical Anthropology* 36:417–26.

Salzano, F. M., J. V. Neel, H. Gershowitz, et al.
1977 Intra- and Intertribal Genetic Variation within a Linguistic Group: The Ge-speaking Indians of Brazil. *American Journal of Physical Anthropology* 47:337–47.

Salzano, F. M., J. V. Neel, L. R. Weitkamp, et al.
1972 Serum Proteins, Hemoglobins and Erythrocyte Enzynes of the Brazilian Cayapo Indians. *Human Biology* 44:443–58.

Salzano, F. M., F. Pagés, J. V. Neel, et al.
1978 Unusual Blood Genetic Characteristics among the Ayoreo Indians of Bolivia and Paraguay. *Human Biology* 50:121–36.

Santos, M. R., R. H. Ward, and R. Barrantes
1994 mtDNA Variation in the Chibcha Amerindian Huetar from Costa Rica. *Human Biology* 66:963–77.

Santos, F. R., T. Gerelsaikhan, B. Munkhtuja, et al.
1996 Geographic Differences in the Allele Frequencies of the Human Y-linked Tetranucleotide Polymorphism DYS19. *Human Genetics* 97:309–13.

Santos, F. R., A. Pandya, C. Tyler-Smith, et al.
1999 The Central Siberian Origin for Native American Y-Chromosomes. *American Journal of Human Genetics* 64:619–28.

Santos, S. E. B., A. K. C. Ribeiro-Dos-Santos, D. Meyer, et al.
1996 Multiple Founder Haplotypes of Mitochondrial DNA in Amerindians Revealed by RFLP and Sequencing. *Annals of Human Genetics* 60:305–19.

Santos, F. R., L. Rodriguez-Delfin, S. D. Pena, et al.
1996 North and South Amerindians May Have the Same Major Founder Y Chromosome Haplotype. *American Journal of Human Genetics* 58:1369–70.

Schanfield, M. S.
1992 Immunoglobulin Allotypes GM and KM Indicate Multiple Founding Populations of Native Americans: Evidence of at Least Four Migrations to the New World. *Human Biology* 64:381–402.

Schanfield, M. S., T. E. Alekseyeva, and M. H. Crawford
1980 Studies on the Immunoglobulin Allotypes of Asiatic Populations. VIII. Immunoglobulin Allotypes among the Tuvinians of Siberia. *Human Heredity* 30:343–49.

Schanfield, M. S., M. H. Crawford, J. B. Dosseter, et al.
1990 Immunoglobulin Allotypes in Several North American Eskimo Populations. *Human Biology* 62:773–89.

Scholl, S., B. Budowle, K. Radecki, et al.
1996 Navajo, Pueblo, and Sioux Population Data on the Loci HLA-DQA1, LDLR, GYPA, HGBB, D7S8, Gc, and D1S80. *Journal of Forensic Sciences* 41:47–51.

Schurr, T. G.
1998 Population Genetic Variation in Indigenous Peoples of Northeastern Siberia: Prehistoric and Historic Influences on Genetic Diversity. Ph.D. Dissertation, Department of Anthropology, Emory University.

Schurr, T. G., and D. C. Wallace
1999 mtDNA Variation in Native Americans and Siberians and Its Implications for the Peopling of the New World. In *Who Were the First Americans?: Proceedings of the 58th Annual Biology Colloquium,* edited by R. Bonnichsen. pp. 41–77. Center for the Study of the First Americans, Oregon State University, Corvallis.

Schurr, T. G., S. W. Ballinger, Y. Y. Gan, et al.
1990 Amerindian Mitochondrial DNAs Have Rare Asian Variants at High Frequencies, Suggesting They Derived from Four Primary Maternal Lineages. *American Journal of Human Genetics* 46:613–23.

Schurr, T. G., Y. B. Starikovskaya, R. I. Sukernik, et al.
2000 Mitochondrial DNA Diversity in Lower Amur River Populations, and Its Implications for the Genetic History of the North Pacific and the New World. American Journal of Physical Anthropology Supplement 30:274–75.

Schurr, T. G., R. I. Sukernik, E. B. Starikovskaya, et al.
1999 Mitochondrial DNA Diversity in Koryaks and Itel'men: Ancient and Recent Population Expansions and Dispersals in Okhotsk-Bering Sea Region. *American Journal of Physical Anthropology* 108:1–40.

Scott, E. M.
1979 Genetic Diversity of Athapascan Indians. *Annals of Human Biology* 6:241–247.

Scozzari, R., F. Cruciani, P. Santolamazza, et al.
1997 mtDNA and Y-chromosome-Specific Polymorphisms in Modern Ojibwa: Implications About the Origin of Their Gene Pool. *American Journal of Human Genetics* 60:241–44.

Seielstad, M., E. Bekele, M. Ibrahim et al.
1999 A View of Modern Human Origins from Y Chromosome Microsatellite Variation. *Genome Research* 9:558–67.

Seielstad, M., L. Hebert, A. Lin, et al.
1994 Construction of Human Y-chromosomal Haplotypes Using a New Polymorphic A to G Transition. *Human Molecular Genetics* 3:2159–61.

Shadel, G. S., and D. A. Clayton
1998 DNA Maintenance in Vertebrates. *Annual Review of Biochemistry* 66:409–35.

Shields, G. F., K. Hecker, M. I. Voevoda, et al.
1992 Absence of the Asian-specific Region V Mitochondrial Marker in Native Beringians. *American Journal of Human Genetics* 50:758–765.

Shields, G. F., A. M. Schmiechen, B. L. Frazier, et al.
1993 mtDNA Sequences Suggest a Recent Evolutionary Divergence for Beringian and Northern North American Populations. *American Journal of Human Genetics* 53:549–62.

Sitnikova, T., and C. Su
1998 Coevolution of Immunoglobulin Heavy- and Light-Chain Variable Region Gene Families. *Molecular Biology and Evolution* 15:617–625.

Smith, D. G., R. S. Malhi, J. Eshleman, et al.
1999 Distribution of Haplogroup X among Native North Americans. *American Journal of Physical Anthropology* 110:271–84.

Sokal, R. R., P. E. Smouse, and J. V. Neel
1986 The Genetic Structure of a Tribal Population, the Yanomama Indians. XV. Patterns Inferred by Autocorrelation Analysis. *Genetics* 114:259–87.

Spielman, R. S., E. C. Migliazza, J. V. Neel, et al.
1979 The Evolutionary Relationships of Two Populations: A Study of the Guaymi and the Yanomama. *Current Anthropology* 20:377–88.

Spuhler, J. N.
1979 Genetic Distances, Trees and Maps of North American Indians. In *The First Americans: Origins, Affinities and Adaptations,* W. S. Laughlin and A. B. Harper, eds. pp. 135–83. Gustav Fischer, New York.

Spurdle, A. B., M. F. Hammer, and T. Jenkins
1994 The Y Alu Polymorphism in Southern African Populations and Its Relationship to Other Y-specific Polymorphisms. *American Journal of Human Genetics* 54:319–30.

Starikovskaya, Y. B., R. I. Sukernik, T. G. Schurr, et al.
1998 Mitochondrial DNA Diversity in Chukchi and Siberian Eskimos: Implications for the Genetic Prehistory of Ancient Beringia. *American Journal of Human Genetics* 63:1473–91.

Stone, A. C., and M. Stoneking
1998 mtDNA Analysis of a Prehistoric Oneota Population: Implications for the Peopling of the New World. *American Journal of Human Genetics* 62:1153–70.

Stoneking, M., K. Bhatia, and A. C. Wilson
1986 Rate of Sequence Divergence Estimated from Restriction Maps of Mitochondrial DNAs from Papua New Guinea. Cold Spring Harbor Symposium of Quantitative Biology 51:433–39

Su, B., J. Xiao, P. Underhill, et al.
1999 Y-Chromosome Evidence for a Northward Migration of Modern Humans into Eastern Asia during the Last Ice Age. *American Journal of Human Genetics* 65:1718–24.

Suarez, B. K., J. D. Crouse, and D. H. ORourke
1985 Genetic Variation in North Amerindian Populations: The Geography of Human Genes. *American Journal of Physical Anthropology* 67:212–32.

Sukernik, R. I., and L. P. Osipova
1982 GM and KM Immunoglobulin Allotypes in Reindeer Chukchi and Siberian Eskimos. *Human Genetics* 61:148–53.

Sukernik, R. I., T. M. Karaphet, and L. P. Osipova
1978 Distribution of Blood Groups, Serum Markers and Red Cell Enzymes in Two Human Populations from Northern Siberia. *Human Heredity* 28:321–27.

Sukernik, R. I., L. P. Osipova, and M. S. Schanfield
1992 Distribution of GM Allotypes and Deleted IGHG1 Haplotypes in the Sel'kups in Comparison to the Forest Nentsi of Northwestern Siberia. *Experimental and Clinical Immunogenetics* 9:15–23.

Sukernik, R. I., S. V. Lemza, T. M. Karaphet, et al.
1981 Reindeer Chukchi and Siberian Eskimos: Studies on Blood Groups, Serum Proteins, and Red Cell Enzymes with Regard to Genetic Heterogeneity. *American Journal of Physical Anthropology* 55:121–28.

Sukernik, R. I., L. P. Osipova, T. M. Karaphet, et al.
1980 Studies on Blood Groups and Other Genetic Markers in Forest Nentsi: Variation among the Subpopulations. *Human Genetics* 55:397–404.

Sukernik, R. I., L. P. Osipova, T. M. Karaphet, et al.
1986a Genetic and Ecological Studies of Aboriginal Inhabitants of Northeastern Siberia. I. GM Haplotypes and Their Frequencies in Ten Chukchi Populations: Genetic Structure of Reindeer Chukchi. *Genetika* 22:2361–68.

Sukernik, R. I., L. P. Osipova, T. M. Karaphet, et al.
1986b Polymorphic Blood Systems, Immunoglobulin Allotypes, and Other Genetic Markers in Asian Eskimos: Genetic Structure of Bering Sea Eskimos. *Genetika* 22:2369–80.

Sukernik, R. I., T. G. Schurr, Y. B. Starikovskaya, et al.
1996 Mitochondrial DNA Variation in Native Siberians, with Special Reference to the Evolutionary History of American Indians: Studies on Restriction Endonuclease Polymorphism. *Genetika* 32:432–39.

Suto, Y., Y. Ishikawa, H. Hyodo, et al.
2000 Gene Organization and Rearrangements at the Human Rhesus Blood Group Locus as Revealed by Fiber-FISH analysis. *Human Genetics* 106:164–71.

Szathmary, E. J. E.
1984 Peopling of Northern North America: Clues from Genetic Studies. *Acta Anthropologia* 8:79–109.
1985 Peopling of North America: Clues from Genetic Studies. In *Out of Asia: Peopling the Americas and the Pacific,* edited by R. L. Kirk and E. J. E. Szathmary, pp. 79–104. Journal of Pacific History, Inc., Canberra.
1993 Genetics of Aboriginal North Americans. *Evolutionary Anthropology* 1:202–20.

Tanis, R. J., J. V. Neel, and R. Torrez de Aranz
1977 Two More "Private" Polymorphisms of Amerindian Tribes: LDHB GUA-1 and ACP1 B GUA-1 in the Guaymi in Panama. *American Journal of Human Genetics* 29:419–30.

Tanis, R. J., J. V. Neel, H. Dovey, et al.
1973 The Genetic Stucture of a Tribal Population, the Yanomama Indians. IX. Gene Frequencies for Seventeen Serum Protein and Erythrocyte Enzyme Systems in the Yanomama and Five Neighboring Tribes: Nine New Variants. *American Journal of Human Genetics* 25:655–76.

Thomson, R., J. K. Pritchard, P. Shen, et al.
2000 Recent Common Ancestry of Human Y Chromosomes: Evidence from DNA Sequence Data. Proceedings of the National Academy of Sciences USA 97:7360–65.

Torroni, A., H.-J. Bandelt, L. D'Urbano, et al.
1998 mtDNA Analysis Reveals a Major Late Paleolithic Population Expansion from Southwestern to Northwestern Europe. *American Journal of Human Genetics* 62:1137–52.

Torroni, A., M. D. Brown, M. T. Lott, et al.
1995 African, Native American and European Mitochondrial DNAs in Cubans from the Pinar del Rio Province and Implications for the Recent Epidemic Neuropathy in Cuba. *Human Mutation* 5:310–17.

Torroni, A., Y.-S. Chen, O. Semino, et al.
1994 mtDNA and Y-chromosome Polymorphisms in Four Native American Populations from Southern Mexico. *American Journal of Human Genetics* 54:303–18.

Torroni, A., K. Huoponen, P. Francalacci, et al.
1996 Classification of European mtDNAs from an Analysis of Three European Populations. *Genetics* 144:1835–50.

Torroni, A., M. T. Lott, M. F. Cabell, et al.
1994 mtDNA and the Origin of Caucasians: Identification of Ancient Caucasian-Specific Haplogroups, One of Which Is Prone to a Recurrent Somatic Duplication in the D-loop Region. *American Journal of Human Genetics* 55:760–76.

Torroni, A., J. A. Miller, L. G. Moore, et al.
1994 Mitochondrial DNA Analysis in Tibet: Implications for the Origin of the Tibetan Population and Its Adaptation to High Altitude. *American Journal of Physical Anthropology* 93:189–99.

Torroni, A., J. V. Neel, R. Barrantes, et al.
1994 A Mitochondrial DNA "Clock" for the Amerinds and Its Implications for Timing Their Entry into North America. Proceedings of the National Academy of Sciences USA 91:1158–62.

Torroni, A., M. Petrozzi, L. D'Urbano, et al.
1997 Haplotype and Phylogenetic Analysis Suggests That One European-specific mtDNA Background Plays a Role in the Expression of Leber Hereditary Optic Neuropathy by Increasing the Penetrance of the Primary Mutations 11778 and 14484. *American Journal of Human Genetics* 60:1107–21.

Torroni, A., T. G. Schurr, M. F. Cabell, et al.
1993 Asian Affinities and the Continental Radiation of the Four Founding Native American mtDNAs. *American Journal of Human Genetics* 53:563–90.

Torroni, A., T. G. Schurr, C.-C. Yang, et al.
1992 Native American Mitochondrial DNA Analysis Indicates That the Amerind and the Na-Dené Populations Were Founded by Two Independent Migrations. *Genetics* 130:153–62.

Torroni, A., R. I. Sukernik, T. G. Schurr, et al.
1993 mtDNA Variation of Aboriginal Siberians Reveals Distinct Genetic Affinities with Native Americans. *American Journal of Human Genetics* 53:591–608.

Trachtenberg, E. A., G. Keyeux, J. Bernal, et al.
1996 Results of Expedicion Humana. I. Analysis of HLA Class II (DRB1-DQA1-DQB1-DPB1) Alleles and DR-DQ Haplotypes in Nine Amerindian Populations from Colombia. *Tissue Antigens* 48:174–81.

Underhill, P. A., L. Jin, A. A. Lin, et al.
1997 Detection of Numerous Y Chromosome Biallelic Polymorphisms by Denaturing High Performance Liquid Chromatography. *Genome Research* 7:996–1005.

Underhill, P. A., L. Jin, R. Zemans, et al.
1996 A pre-Columbian Y Chromosome-Specific Transition and Its Implications for Human Evolutionary History. Proceedings of the National Academy of Sciences USA 93:196–200.

Valentine, W. N., K. R. Tanaka, and D. E. Paglia
1985 Hemolytic Anemias and Erthrocyte Enzymopathies. *Annals of Internal Medicine* 103:245–57.

Volanakis, J. E.
1990 Participation of C3 and Its Ligands in Complement Activation. *Current Topics in Microbiology and Immunology* 153:1–21.

Walkinshaw, M., L. Strickland, H. Hamilton, et al.
1996 DNA Profiling in Two Alaskan Native Populations Using HLA-DQA1, PM, and D1S80 Loci. *Journal of Forensic Sciences* 41:478–84.

Wallace, D. C.
1986 Mitotic Segregation of Mitochondrial DNAs in Human Cell Hybrids and the Expression of Chloramphenicol Resistance. *Somatic Cell and Molecular Genetics* 12:41–49.
1992 Diseases of the Mitochondrial DNA. *Annual Review of Biochemistry* 61:1175–1212.

Wallace, M. R., L. B. Andersen, A. Saulino, et al.
1991 A De Novo Alu Insertion Results in Neurofibromatosis Type 1. *Nature* 353:864–66.

Wallace, D. C., J. H. Ye, S. N. Neckelmann, et al.
1987 Sequence Analysis of cDNAs for the Human and Bovine ATP Synthase ?-subunit: Mitochondrial DNA Genes Sustain Seventeen Times More Mutations. *Current Genetics* 12:81–90.

Ward, R. H., B. L. Frazier, K. Dew-Jager, et al.
1991 Extensive Mitochondrial Diversity within a Single Amerindian Tribe. Proceedings of the National Academy of Sciences USA 88:8720–24.

Ward, R. H., H. Gershowitz, M. Layrisse, et al.
1975 The Genetic Structure of a Tribal Population, the Yanomama Indians. XI. Gene Frequencies for 10 Blood Groups and the ABH-Le Secretor Traits in the Yanomama and their Neighbors: The Uniqueness of the Tribe. *American Journal of Human Genetics* 27:1–30.

Ward, R. H., A. Redd, D. Valencia, et al.
1993 Genetic and Linguistic Differentiation in the Americas. Proceedings of the National Academy of Sciences USA 90:10063–67.

Ward, R. H., F. M. Salzano, S. L. Bonatto, et al.
1996 Mitochondrial DNA Polymorphism in Three Brazilian Indian Tribes. *American Journal of Human Biology* 8:317–23.

Watkins, D. I., S. N. McAdam, X. Lui, et al.
1992 New Recombinant HLA-B Alleles in a Tribe of South American Amerindians Indicate Rapid Evolution of MHC Class I Loci. *Nature* 357:329–33.

Weber, J. L., and C. Wong
1993 Mutation of Human Short Tandem Repeats. *Human Molecular Genetics* 2:1123–28.

Whitehouse, D. B., J. Tompkins, J. U. Lovegrove, et al.
1998 A Phylogenetic Approach to the Identification of Phosphoglucomutase Genes. *Molecular Biology and Evolution* 15:4456–62.

Williams, R. C., A. G. Steinberg, H. Gershowitz, et al.
1985 GM Allotypes in Native Americans: Evidence for Three Distinct Migrations across the Bering Land Bridge. *American Journal of Physical Anthropology* 66:1–19.

Williams, T. M., J. Wu, T. Foutz, et al.
1994 A New DRB1 Allele (DRB1*0811) Identified in Native Americans. *Immunogenetics* 40:314.

Yamamoto, F., H. Clausen, T. White, et al.
1990 Molecular Genetic Basis of the Histo-Blood Group ABO System. *Nature* 345:229–33.

Yunis, J. J., H. Ossa, M. Salazar, et al.
1994 Major Histocompatibility Complex Class II Alleles and Haplotypes and Blood Groups of Four Amerindian Tribes of Northern Colombia. *Human Immunology* 41:248–58.

Zerjal, T., B. Dashnyam, A. Pandya, et al.
1997 Genetic Relationships of Asians and Northern Europeans Revealed by Y-Chromosomal DNA Analysis. *American Journal of Human Genetics* 60:1174–83.

Zhang, S., M. Fernandez-Vina, M. Falco, et al.
1998 A Novel HLA-DRB1 Allele (DRB1*0417) in South American Indians. *Immunogenetics* 38:163.

Mitochondrial DNA Haplogroups of Paleoamericans in North America

David Glenn Smith[1], Ripan S. Malhi[2], Jason A. Eshleman[3], Frederika A. Kaestle[4], and Brian M. Kemp[5]

Abstract

Some studies of craniofacial traits of late-Pleistocene and early-Holocene human remains from the Americas have cited the presence of traits uncharacteristic of modern Native Americans as evidence for a separate migration to the New World whose source was outside Siberia. To test this hypothesis, we attempted to extract mitochondrial DNA (mtDNA) for genetic analysis from some of the earliest known human remains from the New World. The remains from only one of twenty sites represented by these remains yielded DNA for analysis. Together with previously reported studies, late-Pleistocene/early-Holocene remains provide evidence for three of the five haplogroups (haplogroups B, C and D) present in modern Native Americans approximately 10,000 years ago, notably excluding both the most common haplogroup in modern Native Americans (haplogroup A) and all other haplogroups absent in modern Native American populations. These results provide no support for hypotheses of more than a single migration to the New World, the one that brought the ancestors of all extant Native Americans to the New World, followed by a late population expansion in Beringia.

Evidence from archaeological, linguistic, and ethnohistorical studies have all been used to support hypotheses about the population prehistory of North America, including the circumstances pertaining to its colonization. The data that define boundaries among cultures, traditions, or languages have no isomorphic relationship with the Mendelian populations that practice them, but isolation by distance should ensure that morphological skeletal traits with high heritabilities closely reflect the phylogenetic relationships among living human populations and between living and prehistoric populations. Studies using a variety of methods (e.g., compare the typologies of Nelson 1998 and Neumann 1952) have shown (though not without criticism—e.g., see Keita and Kittles 1997) that craniofacial metric traits lacking adaptive significance provide phylogenies some of which are both relatively stable over time and highly geographically patterned, suggesting marked regional continuity in time or space; moreover, these studies lead to inferences similar to those based on discrete genetic markers (e.g., Sanghvi 1953; Spielman and Smouse 1976; Konigsberg and Ousley 1995). Thus, comparisons among prehistoric skeletal remains from different geographic regions and different time periods can provide valid clues about the settlement of the New World.

Early assessments that characteristic features of the skeletal remains of the first humans to reach the New World closely resemble those of modern Native Americans (Hrdlička 1937) have been challenged by recent craniometric studies. These studies of skeletal material from North America dating to late-Pleistocene or early-Holocene times have employed more sophisticated multivariate analyses than were used in earlier studies. They have concluded that at least some of the earliest Native Americans had longer, narrower cranial vaults and shorter and narrower faces with narrower and higher orbits than modern Native Americans (Steele and Powell 1992; 1994; Jantz and Owsley 1997; 2000).

[1]Department of Anthropology and California National Primate Research Center, University of California, Davis, CA 95616; e-mail: dgsmith@ucdavis.edu

[2]Department of Anthropology, University of California, Davis, CA 95616; e-mail: rsmalhi@ucdavis.edu

[3]Department of Anthropology, University of California, Davis, CA 95616; e-mail: jae@ucdavis.edu

[4]Department of Anthropology, Indiana University, Bloomington, IN 47405; e-mail: kaestle@indiana.edu

[5]Department of Anthropology, University of California, Davis, CA 95616; e-mail: bmkemp@ucdavis.edu

These and studies of certain South American materials (Neves and Puciarelli 1991) have led to conclusions that more than one single founder population (hence migration) colonized the Americas and that the earliest of these populations to arrive did not closely resemble modern Native Americans. Craniometric traits of the earliest Americans are actually said to resemble those of modern Europeans, Polynesians, or Ainu/Jomon people more closely than those of modern peoples of either the Americas or Northeast Asia (Steele and Powell 1992; Brace and Nelson 1999; Jantz and Owsley 2000; Powell and Neves 1999; Brace et al, 2001).

Due to geographically specific environmental and genotype by environment interaction effects on the phenotype, morphological differentiation among populations typically exceeds that based on discrete genetic traits (Spielman and Smouse 1976), which are more reliable indicators of shared, derived phylogenetic relationships. Regional differentiation of *Homo sapiens* worldwide, followed by subsequent homogenization through gene flow, might have been driven by glacial pulses and climatic events associated with them (Rogers et al. 1992; Sarich 1995). These events, whose nature and timing are unknown, might have exerted profound (and unknown) selective influences on morphology, which are responsible for the apparent temporal discontinuity in the morphology of Native Americans (Powell and Neves 1999; but see Lovvorn et al., 1999). For example, Sciulli and Mahaney (1991) found evidence that selection has driven reductions in tooth size in prehistoric populations in Ohio. Nasal height appears to be adapted to extreme climates (Carey and Steegmann 1981), and its clinal distribution is more marked than that of most other traits (Jantz et al. 1992). There is as yet no general agreement regarding whether the more widespread pattern of brachycephalization was introduced to the New World by a later migration or represents an in situ adaptation to changing environmental conditions.

Unlike many morphometric traits and products of protein coding loci, point mutations in human mitochondrial DNA (mtDNA), whose complete sequence is known (Anderson et al. 1981), have minimal influences on fitness (Avise 1994; but see Tanaka et al., 2000 for some evidence of selection on the mitochondrial genome). Moreover, mtDNA is solely maternally inherited and thus does not experience recombination (Giles et al. 1980; Merriwether and Kaestle, 1999). Because it evolves more rapidly than nuclear DNA (Brown et al. 1979), mtDNA provides phylogenetic information about closely related populations. This information comprises the unique series of accumulated mutations that link modern mtDNA matrilineages to their ancestral forms. With the automation of polymerase chain reaction (PCR) amplification it is now possible to study mtDNA extracted from prehistoric skeletal material (e.g., Hagelberg and Clegg 1993; Herrmann and Hummel 1993) which can be used to test for genetic discontinuities between ancient and modern populations in the Americas, implying multiple origins (hence migrations) (Kolman and Tuross 2000).

Modern Native Americans belong to one of five different matrilines, or mtDNA haplogroups (Schurr et al 1990; Forster et al. 1996), each comprising a unique set of mtDNA haplotypes that share a recent common ancestry. Each haplogroup can be characterized by the presence or absence of a nine-base pair (bp) deletion (in the case of haplogroup B), or by the gain or loss of specific restriction sites (in the case of haplogroups A, C, D, and X), and by corresponding mutations in the mtDNA control region (CR) (Horai et al. 1993; Brown et al. 1998;Torroni et al. 1993a). These five haplogroups and the restriction sites and CR mutations that are diagnostic of each of them are given in Table 1. Since these same haplogroups and a putative ancestral (Old World) haplotype representing each of them are also found in lower frequencies, together with additional haplogroups, in Asian populations (Ballinger et al. 1992; Torroni et al. 1993a; Bailliet et al. 1994; Kolman et al. 1996; Lum et al. 1998; Merriwether et al. 1996; Derenko et al. 2001), their presence in the Americas in high frequencies has been interpreted as the result of a founder effect (Wallace et al. 1985).

Table 1. Mutations in PCR-amplified mtDNA fragments defining haplogroups A, B, C, D, and X.

Haplogroup	Primer	Reference	CR Mutations	Enzyme	Restriction Site
A	L611	Parr et al. 1996	16290T,16319A	+Hae III	663
A	H743	Parr et al. 1996			
B	L8215	Handt et al. 1996	16217C (9-bp deletion)		
B	H8297	Handt et al. 1996			
C	L13257	Handt et al. 1996	16298C, 16327C	-Hinc II	13259
C	H13290	Handt et al. 1996		+Alu I	13262
D	L5120	Parr et al. 1996	16325C,16362C	-Alu I	5176
D	H5189	Handt et al. 1996			
X	L1651	Kaestle 1998	16278T	-Dde I	1715
X	H1776	Smith et al. 1999			
X	L14440	Brown et al. 1999		+Acc I	14465
X	H14591	Brown et al. 1999			

The geographical distribution of mtDNA haplogroups in North America is illustrated in Figure 1. Haplogroup A, the most common haplogroup in modern Native Americans, is especially common in the Arctic and Subarctic regions of the northwestern portion of North America and along the entire western coastal strip (i.e., Pacific shoreline) of North America to southern California, where much lower frequencies of haplogroups C and D are also found. All three of these haplogroups are also common in eastern Siberia. Haplogroup A is also common, albeit less, among Algonquian populations, including those of eastern North America, who might have originated on the Columbia Plateau (Denny 1991; Malhi 2000; Malhi et al. 2001; Goddard 1994). Haplogroup B, the second most common Native American haplogroup, is absent among Beringian populations (Shields et al. 1992) but predominates in the American Southwest (Lorenz and Smith 1994; Malhi et al. 2003), where haplogroup C is also fairly common. While present in Central America and the northern part of South America (Merriwether et al. 1995), haplogroup B generally declines in frequency to the east, north, and south of this region. Haplogroup D, like haplogroup C, is common in the southern portion of South America, but is one of the two least common haplogroups in North America with frequencies of 3–5 percent. It occurs east of the Coast Range in western

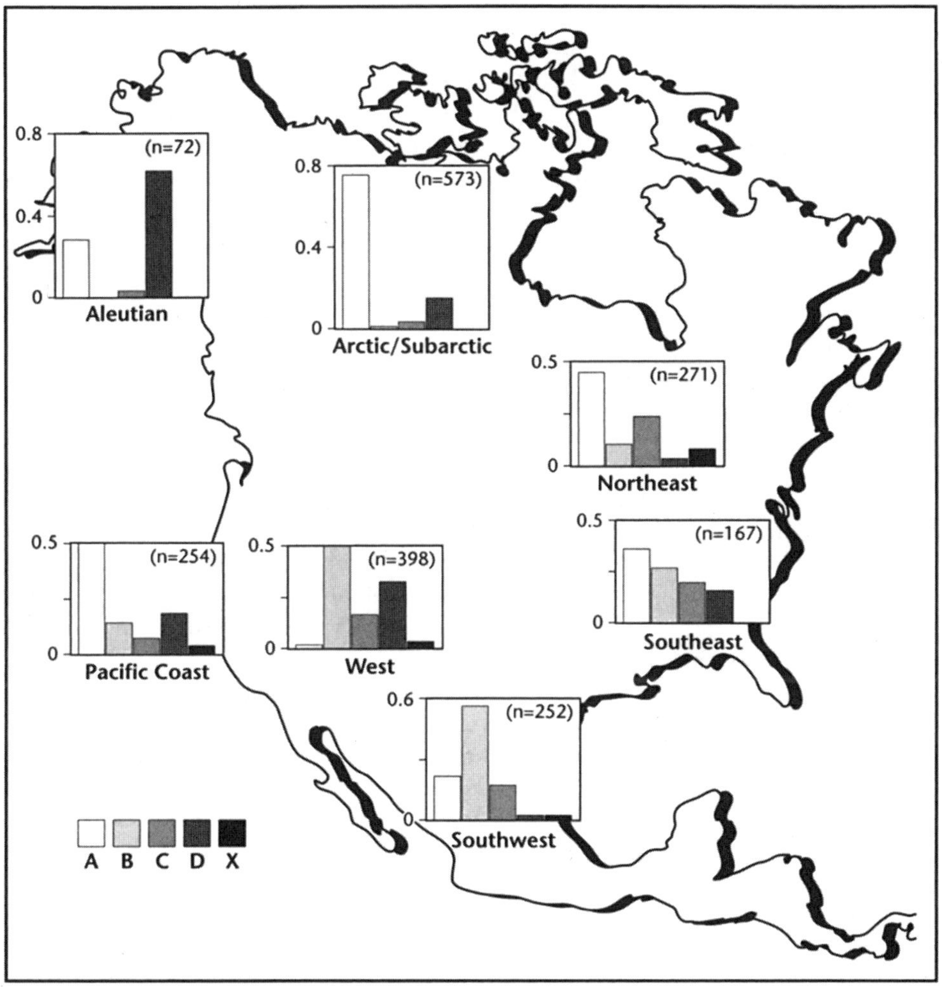

Figure 1. Mitochondrial lineage distribution of modern Native North Americans (Malhi et al. 2002).

North America, including Eskimo populations, and is especially common in the Aleut (Hayes and O'Rourke 2000). Haplogroup X, the least common of the five haplogroups, exhibits a broad geographic and language distribution in North America (Smith et al. 1999), but appears to be more common among Algonquian-speaking groups than elsewhere (Brown et al. 1998; Smith et al., 1999) and might be absent from South America.

The marked geographic variation of haplogroup frequency distributions in the North America, illustrated in Figure 1, and the similarity of these distributions among closely related tribes (Lorenz and Smith 1994; 1996; 1997; Malhi et al. 2001) suggest that random genetic drift has not sufficiently influenced the haplogroup frequency distributions of Native Americans to obscure ancestor/descendant relationships (but see Weiss and Smith [2003] for an exception to this). The mtDNA signature of some Asian population(s) from whose ancestors Native Americans also descend might once have resembled that of living Native Americans, but subsequently experienced the loss or gain of one or more haplogroups. Correspondingly, Native Americans might once have carried additional, presumably Asian-derived, haplogroups that later became extinct in the Americas. The occurrence of additional haplogroups (other than the five cited above, to one of which most, if not all, modern Native Americans belong) among the earliest Native Americans could indicate 1) a severe genetic bottleneck (after settlement of the New World) without a prior founder effect, or 2) evidence of a separate migration for which genetic evidence no longer survives. If such novel haplogroups are discovered in prehistoric Native Americans and are non-randomly distributed in space and time prior to the hypothetical bottleneck, the second of these two hypotheses is the more likely. Otherwise, it is unlikely that representative samples of the descendants of the founders of two separate and independent migrations to the New World, especially ones that arrived at different times or from different origins, would exhibit identical, or even very similar, mtDNA haplogroup frequency distributions (hereafter referred to as mtDNA signatures). Although it has also been argued that there were two founding haplotypes representing each haplogroup, based on the presence of haplotypes with and without the HaeIII restriction site at np 16517 in all five haplogroups (Bailliet et al 1994; Merriwether et al., 1995), this pattern is more likely due to the hypervariable nature of this nucleotide position. It seems to us unlikely that exactly the same mutation would be present in each random sample of each haplogroup represented in the founding population. Therefore, most, if not all, haplotype variation in the Americas probably arose after the colonization event. Thus, if the earliest Native Americans are not ancestral to modern Native Americans or, alternatively, represent only the first of a larger number of migrations to the Americas, the mtDNA signatures of the two are unlikely to be similar. Correspondingly, if the Americas were colonized by a single migration, the original colonists must have carried all five of the mtDNA haplogroups found in living Native Americans (and perhaps even more, if some were eventually lost in the Americas), and their haplogroup signature should not differ substantially from that of modern populations. Specific differences between the early prehistoric and modern geographic distributions of mtDNA signatures might also provide clues about major migration events or population expansions in the Americas (Kolman and Tuross 2000). In the present study we report the haplogroups of mtDNA extracted from some of the earliest skeletal remains known from North America and provide a preliminary assessment of whether or not they comprise a representative sample of the haplogroup distribution of modern Native North American populations.

Materials and Methods

Samples of skeletal remains acquired for this study are listed in Table 2 and include those documented to date to late-Pleistocene or early-Holocene times (see also Steele and Powell 1992; 1994). They represent a broad geographic sample from North America and a time span from approximately 10,000 yr B.P. to about 8000 yr B.P. Samples of teeth, cancellous bone protected by a dense cortical surface (such as carpals or metatarsals), or rib fragments (3–5 grams) were requested from museum or private individuals housing them because they have provided the best results in our lab using our protocols. Methods of extraction and amplification of ancient DNA studied here are modifications of those previously described (e.g., see O'Rourke et al. 1996; Kaestle 1997, 1998; Kaestle and Smith, 2001).

Table 2. Haplogroups of Paleoamerican and early-Archaic skeletal samples analyzed in this and other studies.

Site	Age (RCYBP)	Haplogroup	Reference
1 Arlington Springs, CA	10,000–11,000	■	Orr 1968; Johnson et al. 2000
2 Browns Valley, MN	9049	■	Jenks 1937
3 Central, WA	8000	■	Chatters and Taylor, pers. comm.
4 Clark Fork, ID (n=2)	8000	◆	Cook, pers. comm.
5 Cutler Sink (Miami), FL (n=3)	9620	■	Carr 1986
6 Horn Shelter (Waco), TX (n=2)	10,000	■	Young 1985
7 Hourglass Cave, CO	8000	B	Stone and Stoneking 1996
8 Kennewick, WA	8410	■	Taylor et al. 1998
9 Los Vaqueros, CA	7400	■	Meyer and Rosenthal 1997
10 Little Salt Springs, FL (n=2)	6000–10,000	▲	Clausen 1979
11 O-Y-K Cave, AK	9740	D	Dixon et al. 1997
12 Pelican Rapids, MN	7840	■	Jenks 1936
13 Sulfur Springs, AZ	8200–10,400	◆	Waters 1986
14 Sumatra, MT	Pleistocene?	▲	Lehren, pers. comm.
15 Vero Beach, FL	late Pleistocene	▲	Steward 1946
16 Warm Mineral Spgs, FL (n=3)	10,260	■	Clausen et al. 1975
17 Watsonville, CA (n=6)	7000–8000	◆	
18 Wilson-Leonard, TX	9500	■	Weir 1985
19 Windover, FL (n=12)	8000	X	Hauswirth 1994
19 Windover, FL (n=8)	8000	■	Smith et al. 2001
19 Windover, FL (n=4)	8000	■	Doran et al. 1986
20 Wizards Beach, NV	9200	C	Kaestle 1998; Kaestle and Smith 2001
21 Pintoscayoc, Argentina	9208	◆	Taylor, pers. comm.

■ sufficient DNA for amplification could not be extracted from this sample
▲ substance in DNA extraction inhibited amplification; sample still under study
◆ samples still under study

Surface contamination was removed from each sample by soaking in 0.6 percent bleach (albeit recent research indicates that much higher concentrations are far more effective) and cross-linking with UV light (254 nm; within 5 inches of UV source), each for a period of about 10 minutes. To remove calcium from these samples, approximately 0.25 gm of each fragment was incubated at 4°C for 3–4 days (and sometimes for up to two weeks) in sterile, filtered 0.5M EDTA, pH 8.0, the EDTA solution replaced daily. Teeth were lightly cracked (using a steel mortar and pestle) before being decalcified. The sample was then deproteinized overnight at 55°C (on a rocker or rotating platform incubator) in a solution containing 50 mM Tris, pH 8.0, 1 mM $CaCl_2$, 1 mM DTT, 0.5% Tween 20, and 1 mg/ml proteinase K.

DNA was extracted following the Yang protocol (Yang et al. 1998) using a Qiagen PCR purification kit after two successive extractions using an equal volume of (pH 8.0) phenol: chloroform:isoamyl alcohol (25:24:1), followed by an extraction with chloroform, then precipitated with two volumes of absolute ethanol and one-half volume of ammonium acetate. Five volumes of Qiagen PB Buffer were added to and mixed with the sample, loaded on the Qiagen column, and washed with an equal volume of Qiagen PE buffer, centrifuging the columns at 12,800 x g for 1 minute after each step; DNA was eluted from each column with 50–100 µl Qiagen EB buffer.

PCR reactions were performed in 25-ml volumes in a 1605 Air Thermo Cycler (Idaho Technology, Idaho Falls, Idaho). Reactions contained 3 ml ancient DNA extract, 1.5 units of Amplitaq Gold (Perkin Elmer) or Platinum Taq (Gibco) DNA polymerase to provide "hot start" conditions, and 20 ml of a master mix (UV cross-linked for 10 min to destroy contaminant DNA) containing 10X rapid Air Thermo Cycler buffer (50 mM Tris, pH 8.4, 1.5 mM $MgCl_2$, 20 mM NaCl, 500 mg/ml BSA), 0.6 mM each primer, 200 mM each dNTP and sterile ddH_20. Primers for amplifying fragments containing the restriction sites that identify haplogroups A (Parr et al. 1996), C (Handt et al. 1996), D (Parr et al. 1996; Handt et al. 1996), and X (Brown et al. 1998; Smith et al. 1999) and the 9-bp deletion that identifies haplogroup B (Handt et al. 1996) are those described elsewhere. After an initial 4-min denaturation step at 95°C to activate the "hot start" Taq, 35–40 cycles of amplification were conducted under the following conditions: denature at 95°C for 30 sec, anneal at 51°–55°C for 30 sec, and extend at 72°C for 30 sec, ending with a final 3-min extension at 72°C.

The following precautions, most of which are cited elsewhere (e.g., Kelman and Kelman 1999; Kaestle 1997; Kolman and Tuross 2000), were taken to minimize the influence of sample contamination on inferences drawn from the results of this study:

1. Gloves and laboratory coats were always worn when working with Native American DNA.
2. Skeletal samples were UV irradiated and bleached (10 percent v/v Clorox) on each surface for 10 minutes prior to the DNA extraction process.
3. Buffers, liquids, and reagents used were micro-filtered, UV irradiated, and tested for contamination before use with ancient DNA.
4. DNA extractions and PCR preparations were conducted in a bleach-sterilized, UV cross-linked, ½-inch-thick Plexiglas glove box. The box contains dedicated equipment and supplies and is housed in a controlled access, air filtered, and positively pressurized room dedicated solely to pre-PCR study of ancient DNA. No modern DNA samples are allowed in this pre-PCR ancient DNA laboratory.
5. Negative controls including all reagents except DNA template were prepared for each extraction step and amplification, and run beside each restriction digest to increase the likelihood that cases of contamination could be detected.
6. Every month the entire pre-PCR ancient DNA room and all its dedicated equipment are decontaminated using bleach and enzymatic decontaminants.

Successful amplification of a PCR fragment of the appropriate size was confirmed by separating 5 ml of each PCR product on a 6-percent polyacrylamide gel, staining with ethidium bromide, and visualizing a fragment of the appropriate size under UV light. Amplifications that were not successful on the first attempt were repeated, up to a maximum of two times. Samples that still consistently failed to amplify were mixed 50:50 with a modern (positive) control sample of DNA; subsequent failure to amplify the modern DNA was considered evidence that the prehistoric sample had been inhibited by some co-extracted substance. Serial dilutions of additional extracts from these samples were made and in some cases led to successful amplification of fragments for restriction analysis.

The remainder of the PCR product was digested overnight at 37°C with 10 units of the restriction enzyme required to characterize haplogroups A, C, D, and X, and the digests were electrophoresed on a 6-percent polyacrylamide gel. The gel was stained with ethidium bromide and photographed over a UV transilluminator, using an IS 2000 imaging system (Alpha Inotech, San Leandro, Calif.), to determine whether or not the fragment had been digested by the restriction enzyme at the appropriate site.

Haplogroup B was identified by the amplification of a 114-bp (rather than the typical 123-bp) fragment and visualized in an identical manner. Haplogroup C was identified by both the failure of the Hind II restriction enzyme to digest mtDNA at np 13,259 and the presence of an Alu I restriction site at np 13,262. Haplogroup D was identified by the absence of an Alu I restriction site at np 5,176. Haplogroup A was identified by the digestion (at np 663) of the appropriate PCR fragment by the Hae III restriction enzyme. Haplogroup X was identified by both the absence of the Dde I restriction site at np 1715 and the presence of an Acc I restriction site at np 14,465. Provisional haplogroups were assigned to each prehistoric sample based on this restriction analysis and compared with the frequency distribution of haplogroups of modern Native Americans (Lorenz and Smith 1996). The partial digestion of fragments of some samples with the appropriate restriction enzyme was regarded as potential sample contamination and the above analysis was repeated on an additional extraction from skeletal material representing that sample.

Results

At least one sample from each of the 21 archaeological sites listed in Table 2 was tested for its mtDNA haplogroup (12 samples from Windover were tested). Haplogroups that could be assigned to any of the samples are listed in Table 2 together with the haplogroups of samples from North America that were previously reported in the literature. Three of the samples cited in Table 2 had been previously reported to be members of one of the five modern Native American haplogroups. One of these, Hourglass Cave, belongs to haplogroup B (Stone and Stoneking 1996) and another, Wizards Beach, belongs to haplogroup C (Kaestle 1998; Kaestle and Smith 2001). The haplogroup assignments of both of these two samples were confirmed by the presence of specific mutations in the first hypervariable region (HVSI) of the mtDNA CR that invariably accompany the restriction sites used to identify their haplogroups.

Sufficient DNA for analysis could not be extracted from the samples from 12 of the remaining 19 sites despite repeated efforts. Samples from three other of these sites (Little Salt Springs, Sumatra, and Vero) exhibited DNA that was not a member of any of the five modern Native American haplogroups. Samples from these three sites were also used to amplify an mtDNA fragment containing the characteristic restriction site (Alu I and Dde I restriction sites at both 10397 and 10394) diagnostic of (macro) haplogroup M, which itself comprises the Asian-specific haplogroups C, D, G, and Z; none of these four samples could be assigned to (macro) haplogroup M. Because tests showed that most amplifications of these three samples were significantly inhibited, contamination is probably responsible for these results. Future studies are planned to overcome inhibition in these samples and screen them for membership in other Asian-specific haplogroups that might once have been present, but have since become extinct, in the Americas. It is quite possible that further analysis will show that the haplogroups and haplotypes belong to individuals known to have worked with these samples.

The results of the analysis of four of the samples (Clark Fork, Watsonville, Pintoscayoc, and Sulfur Springs) were inconclusive, and the samples are still under study. The sample from the sole remaining site yielded sufficient DNA to amplify fragments for restriction analysis. The characterization of one Windover sample as a member of haplogroup X is an interpretation by Smith et al. (1999) of one of eight HVS1 sequences reported by Hauswirth (1994). However, since none of the remaining seven sequences reported by Hauswirth exhibited CR sequences characteristic of any other Asian-derived haplogroup and might therefore reflect either contamination or sequencing errors, the assignment of one of those sequences to Haplogroup X was probably in error. Because haplogroup X is found in Europe at a frequency of about 3 percent, it is possible that contaminant DNA in the Windover sample was the source of this member of haplogroup X, which has otherwise not been reported from populations of southeastern North America (Malhi et al. 2001; Weiss and Smith 2003). Likewise, the characteristic CR mutations at np 16223 and np 16278 found in haplogroup X are also found without the associated restriction markers for haplogroup X. Thus, assessment of haplogroup X without restriction analysis is problematic. None of the 12 Windover samples analyzed as part of the present study contained sufficient DNA for analysis.

The only sample in Table 2 not previously assigned a Native American haplogroup that yielded sufficient DNA for restriction analysis was the sample from On-Your-Knees Cave. That sample was assigned to haplogroup D and, therefore, now represents the oldest known member of that haplogroup.

Discussion

Horai et al. (1993) speculated that each haplogroup of the New World was introduced by a separate founding population, but others have argued that a single migration event introduced all five of the New World founding haplogroups (Merriwether et al. 1995; Kolman et al. 1996; Lorenz and Smith 1997; Stone and Stoneking 1998). Curiously, only three of the five haplogroups found among modern Native Americans—haplogroups A, C, and D—are also found in modern populations of eastern Siberia (Torroni et al. 1993b; Schurr et al. 1999), the region through which North America is often presumed to have first been colonized (Goebel 1999; Dixon 2001). Mitochondrial haplogroups A, C, and D are found in populations on both sides of the Bering Strait that are widely believed to descend from the latest colonizers of the New World from Asia. While mitochondrial haplogroups B and X are absent from these populations (Shields et al 1993; Smith et al 1999), it is possible that separate migrations are responsible for introducing these two haplogroup sets into the New World and that some or all of the five haplogroups reached Beringia by a coastal route that by-passed the interior of eastern Siberia.

It has been argued that haplogroup B was introduced to the Americas by practitioners of the Clovis culture, perhaps through the Alberta Corridor, long after haplogroups A, C, and D arrived there (Starikovskaya et al. 1998) and began to diversify. This interpretation is based in part on estimates, derived from whole-genome restriction analyses, of the level of genetic diversity among the mtDNA haplotypes of New World members of haplogroup B, which was first reported to be lower than that of members of haplogroups A, C, and D (e.g., see Torroni et al. 1994; Starikovskaya et al. 1998; Schurr et al. 1999). However, this conclusion is not consistent with our own estimates of diversity based on CR sequences (e.g., see Lorenz and Smith 1997; Malhi et al. 2001; Malhi et al. 2002), which were independently confirmed, using a different sample, by Bonatto and Salzano (1997). The conflict between estimates of diversity based on RFLP and sequencing analysis is probably due either to the inherent level of diversity revealed by the two different methods or to the different level of heterogeneity in ethnic origin of the samples employed in the two different types of analyses.

Although present in low frequency in east Asian (Harihara et al. 1992) and central Asian (Comas et al. 1998) populations, haplogroup B is more common in Southeast Asia (Ballinger et al. 1992; Harihara et al. 1988). It is interesting that its distribution in Asia is similar to that of Sundodont dental traits (Turner 1985). Haplogroup B is particularly prevalent in Polynesian populations (Serjeantson et al. 1989; Merriwether et al. 1999a), whom the earliest prehistoric Native American skeletal material is sometimes said to resemble more closely than any other (e.g., Jantz and Owsley 2000). Some studies of mtDNA have provided data that have been interpreted as supporting a genetic relationship between Polynesia and the Americas (e.g., Cann 1994; Sykes et al. 1995). However, although haplogroup B is known to have been in North America for at least 10,000 years (Stone and Stoneking 1996), Polynesians with haplogroup B arrived in their current homeland about 3,500 years ago (Merriwether et al. 1999b), much too late to have first introduced haplogroup B to the Americas. Moreover, the specific CR mutations that characterize the modern Polynesian motif of haplogroup B (e.g., the 16247G and 16261C transitions) are absent in the Americas. Thus, the common ancestry of New World and Old World members of haplogroup B must be very ancient indeed, long predating the settlement of the Americas.

Haplogroup X is more common in European populations (Torroni et al. 1996), whose population some of the earliest prehistoric Native American remains are also said to resemble more closely than any other outside North America (Jantz and Owsley 2000). This observation has complemented recent popular publication (e.g., see Preston 1997) of speculations about cultural and genetic connections between prehistoric European and Native American populations, as critiqued by Straus (2000). However, since members of haplogroup X in both Europe and Asia lack the transition at CR site np 16213, which is characteristic of most members of haplogroup X in the Americas (Smith et al. 1999), the common ancestry between members of haplogroup X in the New and Old Worlds, just as for haplogroup B, must also be very ancient, probably long predating the colonization of the Americas. Recent whole-genome sequencing of the mitochondrial genomes of European and Asian members of haplogroup X has failed to identify an Old World ancestor of the founding haplotype of haplogroup X in the New World (Reidla et al. 2003). The Native American forms of both haplogroups B and X exhibit uniquely non-Asian mutations and present unimodal mismatch distributions, suggesting that neither was introduced to the New World more than once. Thus, the distributions of these five Native American haplogroups in the Old World suggest that three independent migrations might have introduced them to the Americas. However, since the distributions of these haplogroups do not coincide with the boundaries among Greenberg's (1986) three-language macrophylum, these haplogroup distributions do not support his three-migration hypothesis. Moreover, the contrasting levels of diversity of the Native

American haplogoups, if not due to sampling error, might be the result of the more limited early geographic distribution and later expansion of haplogroups B and X, due, respectively, to the Clovis and Algonquian expansions. This interpretation is consistent with the nonrandom geographic distributions of these haplogroups in the Americas.

Recently published sequences of the CR of mtDNA of some Central Asian populations (e.g., the Kazakh and the Uighurs) exhibit the CR mutations characteristic of four of the five Native American haplogroups, including haplogroup X (Comas et al. 1998), but the restriction analysis needed to confirm that all these Native American haplogroups exist in these populations has not been done. More recently, haplogroups A, B, C, D, and X, assigned both by restriction analysis and CR sequencing, have all been reported in an Altai population from southern Siberia (Derenko et al. 2001), where archaeological sites dating to at least 20,000 yr B.P. are found (Goebel 1999). The ancestors of this population represent the only known surviving possible source for peopling of the New World via a single migration. The region of southern Siberia just west of Lake Baikal, the hypothesized origin location of two of the most common Native American Y-chromosome haplotypes (Bortolini et al. 2003), might have provided the staging point for both the migration that introduced all five haplogroups into the Americas and haplogroups A, C, and D into eastern Siberia. Of course the (presumably Asian) ancestors of Native Americans might have lived in a location completely different from that of their modern Asian descendants when the Americas were first settled. Moreover, the existence of a particular modern Asian population with all five Native American haplogroups does not necessarily prove that the ancestors of this population were the sole source—in fact, any source—for New World colonization. The presence of all five Native American haplogroups in southern Siberia might have resulted from recent gene flow (as Reidla et al. 2003 hypothesized for haplogroup X); therefore, these haplogroups might not have been present in southern Siberia when the Americas were first colonized.

Most members of haplogroup A in the Americas exhibit a 16111T mutation that is present but very rare in eastern Siberian members of haplogroup A with the exception of several eastern Siberian populations. The homogeneous distribution of both this haplogroup A mutation and a minority of Native Americans of haplogroup A lacking this mutation in the Americas suggests that it, like haplogroups B and X, originated in one panmictic Beringian population soon after that region was first settled, then spread southward. If the few haplogroup A haplotypes of modern Native Americans that lack the 16111T mutation originated from reverse-mutations at np 16111 that occurred after the settlement of the New World, as suggested by their apparent derivation from haplotypes with the 16111T mutation in haplotype networks (e.g., Malhi et al. 2001), then all haplogroup A haplotypes in the Americas probably descend from a single founding haplotype. Haplogroup A is completely or nearly absent in western North America (although it is common in Central America and in the Northern portion of South America) except along the Pacific coast west of the front range (Eshleman et al. 2004) and in the American Southwest among southern Athapaskans (Malhi et al. 2003), who are recent arrivals to that region. In some of the rare instances in which haplogroup A occurs in other Southwestern groups, it exhibits characteristically Athapaskan- (and Eskimo–Aleut-) specific CR mutations (16192T, 16233G and 16331G), suggesting that its presence results from admixture with recent (i.e., within approximately the last 500 years) female Athapaskan immigrants. The restricted distribution of these CR mutations suggests that they are younger than the 16111T mutation, roughly coeval with the more recent expansion of Eskimo and Athapaskan peoples in North America (Torroni et al. 1993b; Shields et al. 1993) after the end of the Younger Dryas event. The modern distribution of haplogroup A in North America, illustrated in Figure 1, is consistent with the hypothesis that it assumed its present form relatively recently and expanded from Beringia along coastal North America after groups descending from earlier southward movement out of Beringia were well established east of the Coastal Range (Eshleman et al. 2004).

The presence of haplogroups B, C, and D in the prehistoric samples studied here is consistent with the foregoing interpretation of the modern distribution of the five Native American haplogroups. The paucity of haplogroup A during the early Holocene is consistent with a later expansion in the Northwest among Eskimo/Aleut and Athapaskan people bearing the distinctive CR mutation 16192T (Shields et al. 1993; Torroni et al. 1993b). However, this version of haplogroup A apparently did not spread southward. Haplotypes of haplogroup A shared between the Chumash of southern California and four different northwestern populations (Haida, Bella Coola, Nuu-Chah-Nulth, and Eskimo) are also consistent with a late expansion of haplogroup A down the Pacific coastline (Eshleman et al. 2004). That the modern unimodal mismatch distribution of haplogroup B approximates that of the unimodal mismatch distribution for haplogroup A (albeit studies based on restriction analysis have led to other conclusions [Schurr et al. 1999]) suggests both arrived in Beringia as part of the same migration to the Americas, but expanded at different times as a result of independent demographic events (Malhi et al. 2002). That the later expansion of haplogroup A included coastal areas of western North America, and that the original inhabitants were displaced to the interior by this expansion is suggested by the preponderance of haplogroup B in the American Southwest, the central part of the Columbia Plateau, and the western Great Basin, all of which have provided more marginal ecological habitats during the last few thousand years. The absence of the 16192T mutation southward down the coast of north America indicates that this mutation occurred late, not early, during this expansion.

The absence of haplogroup X among the early prehistoric samples studied could easily represent sampling effect, but is consistent with its higher frequencies only in groups that currently occupy, or (in the case of Algonquian-speaking populations) might once have occupied, the Northwest portion of North America. Otherwise, this haplogroup is widely distributed, albeit in low frequency, throughout North America (Smith et al 1999). The prehistoric occurrence of haplogroup X in North America (Malhi and Smith 2002) was recently demonstrated using restriction analysis and the identification of CR mutations that accompany the diagnostic restriction sites associated with haplogroup X in skeletal remains at a prehistoric site in northwestern North America. Thus, the present and prehistoric distributions of New World haplogroups and the presence of all five haplogroups in at least one modern southern Siberian region suggest that haplogroups A, B, C, D, and X were introduced to the New World by a single migration followed much later by a re-expansion of populations with high frequencies of haplogroup A (Forster et al. 1996). Recent studies of Y-chromosome haplotypes also provide evidence of a single migration occurring after the Last Glacial Maximum (Zegura et al. 2004). However, as with mtDNA, methods for calibrating diversity estimates to time the settlement of the Americas are controversial and not well substantiated.

In summary, we hypothesize that all five haplogroups reached the New World simultaneously after the Last Glacial Maximum. Very soon thereafter, the characteristic New World forms of haplogroups A (with the 16111T mutation) and X (with the 16213A mutation) emerged in Beringia and spread southward quickly as the population expanded. Recent interpretations of archaeological, paleoclimatic and geophysical data (Goebel 1999; Dixon 2001) suggest that this colonization of Beringia occurred over a western coastal route approximately 14,000 years ago, overlapping the recalibrated ages of the oldest of the samples analyzed in the present study. This date is coeval with the earliest evidence of Beringian brown bears in the contiguous United States (Leonard et al. 2000), a species well attested in Beringia at least 40,000 yr B.P. The distribution of mtDNA CR sequences of prehistoric and modern brown bears in North America suggests they followed a coastal route southward (Leonard et al 2000), rather than the Alberta corridor, and it is not clear why humans and brown bears should have penetrated south of the ice sheet at different times or by different routes. Much later the 16192T mutation in haplogroup A occurred and spread throughout Eskimo/Aleut and Athapaskan-speaking populations after they had begun to expand southward and eastward.

The present study provides no evidence that haplogroups other than A, B, C, D, and X were introduced to the New World or that some prehistoric remains in the Americas said not to resemble modern Native Americans belong to mtDNA haplogroups that are not common among modern Native Americans. Although genetic evidence of some population replacements has been reported in some regions (Kaestle 1997; 1998; Kaestle and Smith 2001; Schultz et al. 2002), evidence of temporal continuity of mtDNA haplogroup distributions in many regions of North America has emerged from studies that include both prehistoric and modern populations (e.g. Parr et al. 1996; Carlyle et al. 2000; Stone and Stoneking 1998; Malhi 2000), recently reviewed by O'Rourke et al. (2000). In some regions, such as the Aleutian Islands (Hayes and O'Rourke 2000), regional continuity of mtDNA haplogroup frequency distributions occurs in the presence of marked discontinuity in craniometric features, suggesting an ancestral/descendant relationship between the younger brachycephalic and the older dolichocephalic populations. The present study provides neither evidence that the earliest settlers of the Americas were not the direct ancestors of modern Native Americans nor that they first arrived in Beringia in more than a single migration. In the future the continued development of new, more effective techniques, such as the use of DNase to eliminate contamination of supplies, reagents and labware (Eshleman and Smith 2001) and the use of PTB to cleave crosslinks between ancient DNA and sugar molecules that prevent its extraction (Poinar et al. 1998; 2001), should increase the level of success reported in the present study.

Conclusions

The present study demonstrates that sufficient mtDNA to conduct phylogenetic analyses survives in some prehistoric remains from North America, but seldom in those of early Holocene age. Known skeletal material of early Holocene age has yielded evidence of the presence of three of the five modern Native American haplogroups, B, C, and D, but not haplogroup A, the most common among the modern Native Americans. Assuming the occurrence of a major expansion of haplogroup A-bearing groups in northwestern North America long after the arrival of this haplogroup in the Americas, the early Holocene distribution of Native American haplogroups cannot be said to differ from that of modern Native Americans.

The present study provides no support for the hypotheses that the first Americans were not the direct ancestors of modern Native Americans and that the New World was settled by more than a single migration from a relatively panmictic population.

References Cited

Anderson, S. et al.
1981 Sequence and Organization of the Human Mitochondrial Genome. *Nature* 290:457–65.

Avise, J. C.
1994 *Molecular Markers, Natural History and Evolution.* Chapman and Hall, New York.

Bailliet, G., F. Rothhammer, F. R. Carnese, C. M. Bravi, and N. O. Bianchi
1994 Founder Mitochondrial Haplotypes in Amerindian Populations. *American Journal of Human Genetics* 55:27–33.

Ballinger, S. W., T. G. Schurr, A.Torroni, Y. Y. Gan, J. H. Hodge, K. Hassan, K-H. Chen, and D. C.Wallace
1992 Southeast Asian Mitochondrial DNA Analysis Reveals Genetic Continuity of Ancient Mongoloid Migrations. *Genetics* 130: 139–52.

Bonatto, S. L., and F. M. Salzano
1997 Diversity and Age of the Four Major mtDNA Haplogroups, and Their Implications for the Peopling of the New World. *American Journal of Human Genetics* 61:1413–23.

Bortolini, M. C. et al.
2003 Y-chromosome Evidence for Differing Ancient Demographic Histories in the Americas. *American Journal of Human Genetics* 73:524–39.

Brace, C. L., and A. R. Nelson
1999 The Peopling of the Americas: Anglo Stereotypes and Native American Realities. *General Anthropology* 5:1–7.

Brace, C. L., A. R. Nelson, N.Seguchi, H. Oe, L. Sering, P. Oifeng, L. Yongyi, and D. Tumen
2001 Old World Sources of the First New World Inhabitants: A Comparative Craniofacial View. Proceedings of the National Academy of Sciences, USA 98:10017–22.

Brown, M. D., S. H. Hosseini, A. Torroni, H-J. Bandelt, J. C. Allen, T. G. Schurr, R. Scozzari, F. Cruciani, and D. C.Wallace
1998 mtDNA Haplogroup X: An Ancient Link between Europe/ Western Asia and North America? *American Journal of Human Genetics* 63:1852–61.

Brown, W. M., M. George, Jr., and A. C. Wilson
1979 Rapid Evolution of Animal Mitochondrial DNA. Proceedings of the National Academy of Sciences, USA 76:1967–71.

Cann, R. L.
1994 mtDNA and Native Americans: A Southern Perspective. *American Journal of Human Genetics* 55:7–11.

Carey, J. W., and A. T. Steegman, Jr.
1981 Human Nasal Protrusion, Latitude, and Climate. *American Journal of Physical Anthropology* 56:313–19.

Carlyle, S. W., R. L. Parr, M. G. Hayes, and D. H. O'Rourke
2000 Context of Maternal Lineages in the Greater Southwest. *American Journal of Physical Anthropology* 113:85–101.

Carr, R. S.
1986 Preliminary Report of Archaeological Excavations at the Cutler Fossil Site in Southern Florida. Paper presented at the 51st annual meeting, Society for American Archaeology, April 27, New Orleans.

Clausen, C. J., A. D. Cohen, C. Emiliani, J. A. Holman, and J. J. Stipp
1979 Little Salt Springs,Florida: A Unique Underwater Site. *Science* 203:609–14.

Clausen, C. J., H. K. Brooks, and A. B. Wesoloesky
1975 The Early Man Site at Warm Mineral Springs, Florida. *Journal of Field Archaeology* 2:191–213.

Comas, D., F. Calafell, E. Mateu, A. Perez–Lzuaun, E. Bosch, R. Martinez-Arias, J. Clarimon, F. Facchini, G. Fiori, D. Luiselli, D. Pettener, and J. Bertranpetit
1998 Trading Genes along the Silk Road: mtDNA Sequences and the Origin of Central Asian Populations. *American Journal of Human Genetics* 63:1824–38.

Denny, J. P.
1991 The Algonquian Migration from Plateau to Midwest: Linguistics and Archaeology. Papers of the 2nd Algonquian Conference, Carleton University Press, Carleton, Ottawa.

Derenko, M. V., T. Grzybowski, B. A. Malyarchuk, J. Czarny, D. Miscicka-Sliwka, and I. A. Zakharov
2001 The Presence of Mitochondrial Haplogroup X in Altaians from South Siberia. *American Journal of Human Genetics* 69:237–41.

Dixon, E. J.
2001 Human Colonization of the Americas: Timing, Technology and Process. *Quaternary Science Review* 20:277–99.

Dixon, E. J., T. Heaton, T. E. Fitfield, T. D. Hamilton, D. E. Putnam, and F. Grady
1997 Late Quaternary Regional Geoarchaeology of Southeast Alaska Karst: A Progress Report. *Geoarchaeology* 12:689–712.

Doran,G. H., D. N. Dickle, W. E. Ballinger, Jr., F. Agee, P. J. Laipis, and W. W. Hauswirth
1986 Anatomical, Cellular and Molecular Analysis of 8,000-yr.-old Human Brain Tissue from the Windover Archaeological Site. *Nature* 323:803–806.

Eshleman, J. A., and D. G. Smith
2001 Use of Dnase to Eliminate Contamination in Ancient DNA Analysis. *Electrophoresis* 22:4316–19.

Eshleman, J. A., R. S. Malhi, J. Johnson, F. A. Kaestle, J. Lorenz, and D. G. Smith
2004 Mitochondrial DNA and Prehistoric Settlements: Native Migrations on the Western Edge of North America. *Human Biology* 76:55–75.

Forster, P., R. Harding, A. Torroni, and H-J. Bandelt
1996 Origin and Evolution of Native American mtDNA Variation: A Reappraisal. *American Journal of Human Genetics* 59:935–45.

Giles, R. E., H. Blanc, H. M. Cann, and D. C. Wallace
1980 Maternal Inheritance of Human Mitochondrial DNA. Proceedings of the National Academy of Sciences, USA 77:6715–19.

Goebel, T.
1999 Pleistocene Human Colonization of Siberia and Peopling of the Americas: An Ecological Approach. *Evolutionary Anthropology* 8:208–29.

Goddard, I.
1994 The East-west Cline in Algonquian Dialectology. In *Actes. Cong. Algonq. 25th*, edited by W. Cowen, pp. 187–211. Carleton University Press, Carleton, Ottawa.

Greenberg, J. H., C. G. Turner, and S. L. Zegura
1986 The Settlement of the Americas: A Comparison of the Linguistic, Dental and Genetic Evidence. *Current Anthropology* 27:477–97.

Hagelberg, E., and J. B. Clegg
1993 Genetic Polymorphisms in Prehistoric Pacific Islanders Determined by Analysis of Ancient Bone DNA. Proceedings of the Royal Society of London B 252:163–70.

Handt, O., M. Krings, R. H. Ward, and S. Pääbo
1996 The Retrieval of Ancient Human DNA Sequences. *American Journal of Human Genetics* 59:368–76.

Harihara, S., N. Saitou, M. Hirai, T. Gojobori, K. S. Park, S. Misawa, S. B. Ellepola, T. Ishida, and K. Omoto
1988 Mitochondrial DNA Polymorphism among Five Asian Populations. *American Journal of Human Genetics* 43:134–43.

Harihara, S., M. Hirai, Y. Suutou, K. Shimizu, and K. Omoto
1992 Frequency of a 9-bp Deletion in the Mitochondrial DNA among Asian Populations. *Human Biology* 64:161–66.

Hauswirth, W. W., C. D. Dickel, D. J. Rowold, and M. A. Hauswirth
1994 Inter- and Intrapopulation Studies of Ancient Humans. *Experientia* 50:585–91.

Hayes, G., and D. H. O'Rourke
2000 Transitions in the North American Arctic: A Comparison of Archaeological, Skeletal and Ancient DNA Data. Paper presented at the 99th Annual Meeting, American Anthropological Association, San Francisco, November 15–19.

Herrmann, B., and S. Hummel
1993 *Ancient DNA: Recovery and Analysis of Genetic Material from Paleontological, Archaeological, Museum, Medical, and Forensic Specimens.* Springer-Verlag, New York.

Horai, S., R. Kondo, Y. Nakagawa-Hattori, S. Hayashi, S. Sonoda, and K. Tajima
1993 Peopling of the Americas Founded by Four Major Lineages of Mitochondrial DNA. *Molecular Biology and Evolution* 10:23–47.

Hrdlička, A.
1937 Early Man in America: What Have the Bones to Say? In *Early*

Man as Depicted by the Leading Authorities at the International Symposium at the Academy of Natural Sciences, Philadelphia, March 1937, edited by G. G. MacCurdy, pp. 93–104. J.B. Lippincott, London.

Jantz, R. L., and D. W. Owsley
1997 Pathology, Taphonomy, and Cranial Morphometrics of the Spirit Cave Mummy. *Nevada Historical Society Quarterly* 40: 62–82.
2001 Variation among Early North American Crania. *American Journal of Physical Anthropology* 114:146–55.

Jantz, R. L., D. R. Hunt, A. B. Falsetti, and P. J. Key
1992 Variation among North Amerindians: Analysis of Boas's Anthropometric Data. *Human Biology* 64:435–61.

Jenks, A. E.
1936 *Pleistocene Man in Minnesota: A Fossil Homo Sapiens.* University of Minnesota Press, Minneapolis.
1937 *Minnesota's Browns Valley Man and Associated Artifacts.* Memoirs of the American Anthropological Association No. 49. Menasha, Wisconsin.

Johnson, J. R., T. W. Stafford, Jr., H. O. Ajie, and D. P. Morris
2000 Arlington Springs Revisited. In *Proceedings of the Fifth California Islands Symposium,* edited by D. Brown, K. Mitchell, and H. Chaney. U.S. Minerals Management Service and Santa Barbara Museum of Natural History, pp. 541–45.

Kaestle, F. A.
1997 Molecular Archaeology: An Analysis of Ancient Native American DNA from Western Nevada. *Nevada Historical Society Quarterly* 40:85–96.
1998 Molecular Evidence for Prehistoric Native American Population Movement: The Numic Expansion. Ph.D. dissertation. Department of Anthropology, University of California, Davis.

Kaestle, F. A., and D. G. Smith
2001 Ancient Mitochondrial DNA Evidence for Prehistoric Population Movement: The Numic Expansion. *American Journal of Physical Anthropology* 115:1–12.

Keita, S., and R. A. Kittles
1997 The Persistence of Racial Thinking and the Myth of Racial Divergence. *American Anthropologist* 99:534–44.

Kelman, L. M., and Z. Kelman
1999 The Use of Ancient DNA in Paleontological Studies. *Journal of Vertebrate Paleontology* 19:8–20.

Kolman, C. J., N. Sambuughin, and E. Bermingham
1996 Mitochondrial DNA Analysis of Mongolian Populations and Implications for the Origin of New World Founders. *Genetics* 142:1321–34.

Kolman, C. J., and N. Tuross
2000 Ancient DNA Analysis of Human Populations. *American Journal of Physical Anthropology* 111:5–23.

Konigsberg, L. W., and S. D. Ousley
1995 Multivariate Quantitative Genetics of Anthropometric Traits from the Boas Data. *Human Biology* 67:481–98.

Leonard, J. A., R. K. Wayne, and A. Cooper
2000 Population Genetics of Ice Age Brown Bears. Proceedings of the National Academy of Sciences, USA. 97:1651–54.

Lorenz, J. G., and D. G. Smith
1994 Distribution of the 9 bp Mitochondrial DNA Region V Deletion among North American Indians. *Human Biology* 66:777–88.
1996 Distribution of Four Founding mtDNA Haplogroups among Native North Americans. *American Journal of Physical Anthropology* 101:307–23.
1997 Distribution of Sequence Variation in the mtDNA Control Region of Native North Americans. *Human Biology* 69:749–75.

Lovvorn, M. B., G. G. Gill, G. F. Carlson, J. R. Bozell, and T. L. Steinach
1999 Microevolution and the Skeletal Traits of a Middle Archaic Burial: Metric and Multivariate Comparison to Paleoindians and Modern Amerindians. *American Antiquity* 64:527–45.

Lum, J. K., R. L. Cann, J. J. Martinson, and L. B. Jorde
1998 Mitochondrial and Nuclear Genetic Relationships among Pacific Island and Asian Populations. *American Journal of Human Genetics* 63:613–24.

Malhi, R. S.
2000 Context of Maternal Lineages on the Columbia Plateau. Paper presented at the semi-annual Ancient DNA Conference. Manchester, England, July 12–14.

Malhi, R. S., and D. G. Smith
2002 Haplogroup X Confirmed in Prehistoric North America. *American Journal of Physical Anthropology* 119:84–86.

Malhi, R. S., B. A. Schultz, and D. G. Smith
2001 Distribution of Maternal DNA Lineages among Native American Tribes of Northeastern North America. *Human Biology* 73:17–55.

Malhi, R. S., J. A. Eshleman, J. A. Greenberg, D. A. Weiss, B. A. Schultz, B. M. Kemp, F. A. Kaestle, J. G. Lorenz, J. R. Johnson, and D. G. Smith
2002 The Structure of Diversity within Native North American mtDNA Haplogroups: Implications for the Prehistory of North America. *American Journal of Human Genetics* 70:905–19.

Malhi, R. S., H. M. Mortensen, J. A. Eshleman, B. M. Kemp, J. G. Lorenz, F. A. Kaestle, J. R. Johnson, C. Gorodezky, and D. G. Smith
2003 Native American mtDNA Prehistory in the American Southwest. *American Journal of Physical Anthropology* 120:108–24.

Merriwether, D. A., and F. A. Kaestle
1999 Mitochondrial Recombination? (Continued). *Science* 285:837.

Merriwether, D. A., F. Rothhammer, and R. E. Ferrell
1995 Distribution of the Four Founding Lineage Haplotypes in Native Americans Suggests a Single Wave of Migration for the New World. *American Journal of Physical Anthropology* 98:411–30.

Merriwether, D. A., W. W. Hall, A. Vahlne, and R. E.Ferrell
1996 mtDNA Variation Indicates Mongolia May Have Been the Source for the Founding Population for the New World. *American Journal of Human Genetics* 59:204–12.

Merriwether, D. A., F. A. Kaestle, B. Zemel, G. Koki, C. Mgone, M. Alpers, and J. Friedlander
1999a Mitochondrial DNA Variation in the Southwest Pacific. In *Genomic Diversity: Applications in Human Population Genetics,* edited by S. S.Papiha, R. Deka and R. Chakraborty, pp. 153–80. Plenum Publishers, New York.

Merriwether, D. A., J. S. Friedlaender, J. Mediavilla, C. Mgone, F. Gentz, and R. E. Ferrell
1999b Mitochondrial DNA Variation is an Indicator of Austronesian Influence in Island Melanesia. *American Journal of Physical Anthropology* 110:243–70.

Meyer, S. J., and J. S. Rosenthal
1997 Archaeological and Geoarchaeological Investigations at Eight Prehistoric Sites in the Los Vaqueros Reservoir Area, Contra Costa County, California. Report to the Contra Costa Water District. Anthropological Studies Center. Sonoma State University, Rohnert Park, California.

Nelson, A. R
1998 A Craniofacial Perspective on North American Indian Population Affinities and Relations. Ph.D. dissertation, University of Michigan, Ann Arbor.

Neumann, G.
1952 Archaeology and Race in the American Indian. In *Archaeology of Eastern United States,* edited by J. B. Griffin, pp. 13–34. University of Chicago Press, Chicago.

Neves, W. A., and H. M. Puciarelli
1991 Morphological Affinities of the First Americans: An Exploratory Analysis Based on Early South American Human Remains. *Journal of Human Evolution* 21:261–73.

O'Rourke, D. H., S. W. Carlyle, and R. L. Parr
1996 Ancient DNA:A Review of Methods, Progress and Perspectives. *American Journal of Human Biology* 8:557–71.

O'Rourke, D. H., M. G. Hayes, and S. W. Carlyle
2000 Spatial and Temporal Stability of mtDNA Haplogroup Frequencies in Native North Americans. *Human Biology* 72:15–34.

Orr, P. C.
1968 Prehistory of Santa Rosa Island. Santa Barbara Museum of Natural History, Santa Barbara.

Parr, R. L., S. W. Carlyle, and D. O'Rourke
1996 Ancient DNA Analysis of Fremont Amerindians of the Great Salt Lake Wetlands. *American Journal of Physical Anthropology* 99:507–18.

Poinar, H. N., M. Hofeiter, W. G. Spaulding, P. S. Martin, B. A. Stankiewicz, H. Bland, R. P. Evershed, G. Possuert, and S. Pääbo
1998 Molecular Coproscopy: Dung and Diet of the Extinct Ground Sloth *Nothrotheriops shasteusis. Science* 281:402–406.

Poinar, H. N., M. Kuch, K. D. Sobolik, I. Barnes, A. B. Stankiewicz, T. Kuder, W. G. Spaulding, V. M. Bryant, A. Cooper and S. Pääbo
2001 A Molecular Analysis of Dietary Diversity for Three Archaic Native Americans. Proceedings of the National Academy of Sciences, USA. 98:4317–25.

Powell, J. F., and W. A. Neves
1999 Craniofacial Morphology of the First Americans: Pattern and Process in the Peopling of the New World. *Yearbook of Physical Anthropology* 42:153–88.

Preston, D.
1997 The Lost Man. *The New Yorker* 16 June:70–81.

Reidla, M. et al.
2003 Origin and Diffusion of mtDNA Haplogroup X. *American Journal of Human Genetics* 73:1178–90.

Ribeiro-dos-Santos, A. K. C., S. E. B. Santos, A. O. Machado, V. Guapindaia, and M. A. Zago
1996 Heterogeneity of Mitochondrial DNA Haplogroups in Pre-Columbian Natives of the Amazon Region. *American Journal of Physical Anthropology* 101:29–37.

Rogers, R. A., L. A. Rogers, and L. D. Martin
1992 How the Door Opened: The Peopling of the New World. *Human Biology* 64:281–302.

Sanghvi, L. D.
1953 Comparison of Genetical and Morphological Methods for a Study of Biological Differences. *American Journal of Physical Anthropology* 2:385–404.

Sarich, V. M.
1995 Human Races Are Very Real and Very Young. Paper presented at the annual meeting of the American Association of Physical Anthropologists, March 28–April 1, Oakland, CA.

Schultz, B. A., R. S. Malhi, and D. G. Smith
2002 Examining the Proto-Algonquian Migration: Analysis of mtDNA. In Proceedings of the 32nd Algonquian Conference, Montreal, 2000, pp. 470–92. Carlton University Press, Ottawa, Canada.

Schurr, T. G.,S. W. Ballinger, Y-Y. Gan, J. A. Hodge, D. A. Merriwether, D. N. Lawrence, W. C. Knowler, K. M. Weiss, and D. C. Wallace.
1990 Amerindian Mitochondrial DNAs Have Rare Asian Mutations at High Frequencies, Suggesting They Derived from Four Primary Maternal Lineages. *American Journal of Human Genetics* 46:613–23.

Schurr, T. G., R. I. Sukernik, Y. B. Starikovskaya, and D. C. Wallace
1999 Mitochondrial DNA Variation in Koryaks and Itel'men: Population Replacement in the Okhotsk Sea-Bering Sea Region during the Neolithic. *American Journal of Physical Anthropology* 108:1–39.

Sciulli, P. W., and M. C. Mahaney
1991 Phenotypic Evolution in Prehistoric Ohio Amerindians: Natural Selection versus Random Genetic Drift in Tooth Size Reduction. *Human Biology* 63:499–511.

Serjeantson, S. W., J. F. Prior, and J. Trent
1989 An Asian-specific 9-bp Deletion of Mitochondrial DNA is Frequently Found in Polynesians. *American Journal of Human Genetics* 44:504–10.

Shields, G. F., K. Hecker, M. I. Voevoda, and J. K. Reed
1992 Absence of the Asian-specific Region V Mitochondrial Marker in Native Beringians. *American Journal of Human Genetics* 50:758–65.

Shields, G. F., A. M. Schmiechen, B. L. Frazier, A. Redd, M. I. Voevoda, J. K. Reed, and R. H. Ward
1993 mtDNA Sequences Suggest a Recent Evolutionary Divergence for Beringian and Northern North American Populations. *American Journal of Human Genetics* 53:549–62.

Smith, D. G., R. S. Malhi, J. G. Lorenz, J. A. Eshleman, and F. A. Kaestle
1999 Distribution of mtDNA Haplogroup X among Native North Americans. *American Journal of Physical Anthropology* 110:273–84.

Smith,D. G., B. K. Rolfs, F. A. Kaestle, R. S. Malhi, and G. H.Doran
2002 Serum Albumin Phenotypes and a Preliminary Study of the Windover mtDNA Haplogroups and Their Anthropological Significance. In *Windover Multidisciplinary Investigations of an Early Archaic Florida Cemetery,* edited by G. H. Doran, pp. 244–49. University of Florida Press, Gainesville.

Spielman, R. S., and P. E. Smouse
1976 Multivariate Classification of Human Populations. I. Allocation of Yanomama Indians to Villages. *American Journal of Human Genetics* 28:317–31.

Starikovskaya, Y. B., R. I. Sukernik, T. G. Schurr, A. M. Kogelnik, and D. C. Wallace
1998 mtDNA Diversity in Chukchi and Siberian Eskimos: Implications for the Genetic History of Ancient Beringia and the Peopling of the New World. *American Journal of Human Genetics* 63:1473–91.

Steele, D. G., and J. F. Powell
1992 Peopling of the Americas: Paleobiological Evidence. *Human Biology* 64:303–36.

1994 Paleobiological Evidence of the Peopling of the Americas: A Morphometric view. In *Method and Theory for Investigating the Peopling of the Americas,* edited by R. Bonnichsen and D. G. Steele, pp.141–63.

Steward, T. D.
1946 A Reexamination of the Fossil Human Skeletal Remains from

Melbourne, Florida. *Smithsonian Miscellaneous Collections* 106:1–28.

Stone, A. C., and M. Stoneking
1996 Genetic Analysis of an 8,000 year-old Native American Skeleton. *Ancient Biomolecules* 1:83–87.
1998 mtDNA Analysis of a Prehistoric Oneota Population: Implications for the Peopling of the New World. *American Journal of Human Genetics* 62:1153–70.

Straus, L. G.
2000 Solutrean Settlement of North America? A Review of Reality. *American Antiquity* 65:219– 26.

Sykes, B., A. Leiboff, J. Low-Beer, S. Tetzner, and M. Richards
1995 The Origin of the Polynesians: An Interpretation from Mitochondrial Lineage Analysis. *American Journal of Human Genetics* 57:1463–75.

Tanaka, M., J. S. Gong, J. Zhang, Y. Yamada, H. J. Borgeld, and K. Yagi
2000 Mitochondrial Genotype Associated with Longevity and Its Inhibitory Effect on Mutagenesis. *Mechanisms of Aging and Development* 116:65–76.

Taylor, R. E., D. C. Kirner, J. R. Southon, and J. C. Chatters
1998 Radiocarbon Dates of Kennewick Man. *Science* 280:1171–72.

Torroni, A., T. G. Schurr, M. F. Cabell, M. D. Brown, J. V. Neel, M. Larson, D. G. Smith, C. M. Vullo, and D. C. Wallace
1993a Asian Affinities and Continental Radiation of the Four Founding Native American mtDNAs. *American Journal of Human Genetics* 53:563–90.

Torroni, A., R. I . Sukernik, T. G. Schurr, Y. B. Starikovskaya, M. F. Cabell, M. H. Crawford, A. G. Comuzzie, and D. C. Wallace
1993b mtDNA Variation of Aboriginal Siberians Reveals Distinct Genetic Affinities with Native Americans. *American Journal of Human Genetics* 53:591–608.

Torroni, A., K. Huoponen, P. Francalacci, M. Petrozz, L. Morelli, R. Scozzari, D. Obinu, M-L. Savontaus, and D. C. Wallace
1996 Classification of European mtDNAs from an Analysis of Three European Populations. *Genetics* 144:1835–50.

Torroni, A., J. V. Neel, R. Barrantes, T. G. Schurr, and D. C. Wallace
1994 Mitochondrial DNA "Clock" for the Amerinds and Its Implications for Timing Their Entry into North America. Proceedings of the National Academy of Science, USA. 91:1158–62.

Turner, C. G., II
1985 The Dental Search for Native American Origins. "Out of Asia." In *Peopling of the Americas and the Pacific,* edited by R. Kirk and E. Szathmary, pp. 31–78. Journal of Pacific History, Canberra.

Wallace, D. C., K. Garrison, and W. C. Knowler
1985 Dramatic Founder Effects in Amerindian Mitochondrial DNAs. *American Journal of Physical Anthropology* 68:149–55.

Waters, M. R.
1986 Sulfur Springs Woman: An Early Human Skeleton from Southeastern Arizona. *American Antiquity* 51:361–65.

Weir, F. A.
1985 An Early Holocene Burial at the Wilson-Leonard Site in Central Texas. *Mammoth Trumpet* 2:2–3.

Weiss, D. A., and D. G. Smith
2003 Unexpected Patterns of Mitochondrial DNA Variation among Native Americans from the Southeastern United States. *American Journal of Physical Anthropology* 122:336–54.

Yang, D. Y., B. Eng, J. S. Waye, J. C. Dudar, and S. R. Saunders
1998 Improved DNA Extraction from Ancient Bones Using Silica-based Spin Columns. *American Journal of Physical Anthropology* 105:539–43.

Young, D. E.
1985 The Paleoindian Skeletal Material from Horn Rock Shelter in Central Texas. *Current Research in the Pleistocene* 2:39–40.

Zegura, S. L., T. M. Karafet, L. A. Zhivotovsky, and M. F. Hammer
2004 High Resolution SNPs and Microsatellite Haplotypes Point to a Single, Recent Entry of Native American Y Chromosomes into the Americas. *Molecular Biology and Evolution* 21:164–75.

V. Paleoamerican Prehistory

Skeletal Evidence

Appearance of the "Mongoloid Skeletal Trait Complex" in the Northwestern Great Plains: Migration, Selection, or Both?

George W. Gill

Abstract

Within the Northwestern Great Plains region of North America, recent studies in skeletal biology have shown a surprising degree of biological change from the Plains Archaic to the late-prehistoric period. Cultural and demographic changes such as a dramatic increase in interpersonal violence, changes in diet and subsistence, alterations in burial customs, and a precipitous drop in longevity are all clearly evidenced in the skeletal record. Accompanying these life-style changes are alterations in craniofacial morphology and in some postcranial traits as well. Many of the Archaic skeletons, especially from western Nebraska and eastern Wyoming, resemble certain recently examined Paleoindians (such as Spirit Cave and Kennewick). These early Americans lack the well-developed Mongoloid skeletal trait complex so common to late-prehistoric and modern American Indian populations and instead project a "vaguely Caucasoid" general appearance. Well documented are changes in skull form, cranial vault height, midfacial morphology, femur form, chin shape, and several other traits.

In nearly all cases these changes are in the direction of an intensification of the Mongoloid trait complex. Parallel changes occur in East Asia at the same time or at a somewhat earlier time. Could these changes in the Northwestern Plains, and elsewhere in North America, be merely a mirroring of Old World population dynamics, transplanted by migration? Or are the changes in the Northwestern Plains an adaptive response to the emerging new cultural conditions? Or are they a result of both?

Biological Anthropology, Department of Anthropology, University of Wyoming, Laramie, WY 82071; e-mail: ggill@uwyo.edu

Introduction

The Northwestern Plains region is a vast area of North America that covers over 200,000 square miles and includes a wide range of ecological zones and habitats (Frison 1991). Open plains, high mountain ranges, lesser uplifts, intermontane basins, foothills, and stream valleys are the primary microenvironments of the region. It encompasses all of the modern state of Wyoming, eastern Montana, western North Dakota, western South Dakota, western Nebraska, and the northeastern border of Colorado (Frison 1991). Southern Alberta and southern Saskatchewan are also included if one extends beyond the borders of the United States. Low rainfall and a short growing season as well as other factors of climate and terrain in the region precluded the development of large-scale agriculture in prehistoric times. Human groups were mobile and practiced a wide range of hunting and gathering subsistence activities.

As a result of this high degree of mobility of the prehistoric people, the number of skeletal remains recovered from the region is significantly less than those recovered from adjacent geographical areas (Gill 1991a); nevertheless a small body of bioarchaeological literature for the Northwestern Plains region has begun to accumulate. It started with Howells (1938) and Mulloy (1954), increased somewhat in the 1960s (Agogino 1961; Agogino and Galloway 1963; Bass and Lacy 1963; Birkby and Bass 1963; Steege 1960), and has expanded considerably in the last 30 years (see Scheiber and Gill 1996). At least one significant human skeletal collection has been developed within the region during this same 30-year period; it is presently housed at the University of Wyoming Department of Anthropology.

The University of Wyoming collection consists of over 200 skeletons of pioneer Whites, historic Chinese, frontier Blacks, and prehistoric, protohistoric, and early-historic Plains Indians. Records exist on another 200 skeletons from the Northwestern Plains region that have been studied at the University of Wyoming but are no longer curated there (owing to reburial, specimens returned to other agencies and institutions, etc.).

The University of Wyoming human remains (HR) series, as well as the files on returned or repatriated skeletons (that is, the database (DB) series) span temporally a period from 7,500 years ago, through the Archaic, the late-prehistoric period, and on through the protohistoric period and into the historic era. Modern remains exist as well (in the forensic case [FC] series, and the cadaver specimen [CS] collection) but these will not be treated within this study, nor will the historic Chinese, pioneer Whites, or frontier Blacks. The Amerindian skeletal remains are the central focus of this study.

Not all the Northwestern Plains region is represented by the skeletal sample. Skeletons come from all parts of Wyoming and much of Montana and western Nebraska, but none exist from western South Dakota or any part of Canada. One very ancient maxillary fragment comes from North Dakota, and a few specimens come from the northern border of Colorado. All time periods except the Paleoindian period are represented. The Anzick site skeletal material from Montana is from the Paleoindian period (Owsley 1999); but until the descriptions are complete on these fragmentary human remains, they will not become part of the University of Wyoming database. The Gordon Creek skeleton (Breternitz, et al. 1971) from a few miles south of the Wyoming border is a well-preserved Paleoindian female, but has been housed at the University of Colorado and is only presently being entered into the University of Wyoming database.

During the long span of time from the early Plains Archaic to the early historic period, dramatic biological changes occurred within the Northwestern Plains region. These include shifts in patterns of disease, injury, mortality, and dental pathology, as well as alterations in craniofacial morphology and changes in some postcranial skeletal traits.

By the early 1990s the University of Wyoming skeletal samples and skeletal database files had expanded to the point that definitive studies could begin to be made. A few attempts to define microevolutionary trends had been attempted earlier (Agogino and Galloway 1963; Gill 1974, 1981; Howells 1938), but samples were too small to allow definitive conclusions. Bioarchaeological research involving regional differences in stature began first (Miller 1990); then some preliminary comments on other regional physical differences, as well as temporal trends in craniofacial morphology (Gill 1991a). Dental pathology studies followed (Zitt 1992), and then an examination of demographic patterns (Scheiber and Gill 1996) and discrete cranial traits (Rogers 1996). More recent accomplishments include a thorough examination and classification of skeletal pathology (Miller 1999) and a comprehensive osteometric (largely craniometric) evaluation of the entire adult sample from the Northwestern Plains (Stuart 2000).

Few multiple burial complexes occur in the Northwestern Plains region, owing to the high mobility and low density of populations in the prehistoric past. University of Wyoming anthropologists, however, excavated and studied two sites with multiple interments and completed master's theses on them (Armstrong 1999; Davis 1992). These studies, which focused on skeletal demographics, taphonomy, and cultural practices, also serve as sources for this study.

While assessing craniofacial traits of some of the earliest skeletons from the Northwestern Plains (or adjacent areas), such as the Harlan County Lake skeleton from Nebraska (Baker 1989; Gill 1991a), the Sydney skeleton from western Nebraska (Lovvorn et al. 1999), the Smilden-Rostberg maxilla from North Dakota (Gill 1991b), and the Huntley and Willson crania from eastern Wyoming (Gill and Jimenez 1999, n.d.), it has been observed that these specimens lack almost entirely the "Mongoloid craniofacial trait complex" that is so characteristic of all later Northwestern Plains Indian populations. Some characteristics of the femur differ between the Archaic and late-prehistoric peoples as well.

Baker and I reported a preponderance of "Caucasoid-like" features on the Harlan County Lake cranium in 1989 and 1991 (Gill 1991a; Tibezar et al. 1989), but we were not the first skeletal biologists to notice such features on ancient American crania. For example, almost 50 years before, J. B. Birdsell (see Gill 1991a) referred to the few known Paleoindian crania of that time as "vaguely Caucasoid" in craniofacial morphology, and others commented on this as well. The perceptive physical anthropologists of Birdsell's day, however, were basically ignored, likely because sample sizes were so small and because most investigators could not imagine how significant numbers of Caucasoids could have reached the ancient New World.

Within a year after the Gill (1991a) discussion of the Baker-Gill findings on the Harlan County Lake skeleton, Steele and Powell (1992) reported on their own thorough study of all available Paleoindian skeletons in North America. They were impressed by the number of Caucasoid-like traits and other "non-Mongoloid" skeletal features found in their sample of earliest known Americans.

Since the important work of Steele and Powell in 1992, some additional well-preserved skeletons from the Paleoindian period have surfaced. Most notable among these are Buhl (Green et al. 1998), Spirit Cave (Jantz and Owsley 1997), and Kennewick (Powell and Rose 1999). Interestingly, two of these (Spirit Cave and Kennewick) reveal a "non-Mongoloid" craniofacial pattern of traits similar to that seen on some of the Archaic specimens from the Northwestern Plains (e.g., Harlan County Lake, Sydney, and Huntley and Willson). This observation prompted a personal examination of the Spirit Cave specimen in 1998, and also of a cast of the Kennewick Man skull and mandible in 1999. Using a basic forensic-style protocol, which combines a number of morphological assessments of the palate, zygomatic arches, nasal skeleton, and mandible along with the usual osteometric approaches, the non-Mongoloid pattern of craniofacial traits on these ancient skeletons was further substantiated (Gill 2001; Gill and Jimenez n.d.).

The question now is, What does this mean? Why did Mongoloid traits increase within populations in the Americas after Paleoindian times? Did biological changes accelerate during the transition from the Archaic to the late prehistoric? If so, were parallel cultural and demographic changes occurring at the same time that might suggest migration, followed by replacement or admixture? Or were there possibly radical alterations in the direction of natural selection, or even genetic drift?

Samples and Methods

The University of Wyoming skeletal sample of Amerindians consists of 170 specimens (with Whites, Asians, Blacks, and subadults removed). Of these, a sample of 100 complete or nearly complete adult crania exists for analysis. These have formed part of a recent craniometric study by Stuart (2000), which will be referred to occasionally in this study. The cranial data are drawn from human remains (HR) files, database (DB) files, and even some forensic case (FC) files (those FCs that happen to contain information on specimens of a prehistoric age).

Table 1 shows a temporal breakdown of the Northwestern Plains adult Amerindian skeletal sample (at least those 71 that have been adequately dated). Since the Northwestern Plains Amerindian sample seems to represent at least two separate populations, a Wyoming Basin (desert) population, and a High Plains population of grasslands hunters (Gill 1991a; Scheiber and Gill 1996), Stuart (2000) in her recent craniometric evaluation treated the two regional subsamples separately (High Plains n = 83; Wyoming Basin n = 17). Skeletons from both the Great Divide and the Green River Basins constitute the Wyoming Basin sample. The High Plains sample is made up largely of specimens from the plains of Montana and northern and eastern Wyoming; it also includes those specimens from the Laramie Basin and the Big Horn Basin, since they seem to belong morphologically more to the Plains groups (and also are outside the Red Desert environment). The age, sex, and race of each specimen have been determined utilizing standard visual and metric techniques (Bass 1995; Gill and Rhine 1990; Gill et al. 1988; Krogman and Iscan 1986; Reichs 1998; Stewart 1979). Other Plains skeletons (from the Northwestern Plains or immediately adjacent areas) that form a basis for discussion and conclusions, beyond those mentioned in the study by Stuart (2000), are the Harlan County Lake skeleton (Tibesar et al. 1989), the Sydney skeleton (Lovvorn et al. 1999) and the Smilden-Rostberg maxillary fragment (Gill 1991b).

Table 1. Time periods of the Northwestern Plains as defined by Frison (1991) and available adult crania from the Northwest Plains in good condition and well dated.

Temporal period	Dates (yr B.P.)	Adult crania (n)
Paleoindian	11,000–8000	–
Early Plains Archaic	8000–5000	2
Middle Plains Archaic	5000–3000	1
Late Plains Archaic	3000–1500	23
Late Prehistoric	1500–300	20
Protohistoric	300–150	25
Historic	150–present	–
Total		71

The three Paleoindian skeletons studied here (all three from the American West but none from the Northwestern Plains), are Spirit Cave (9300 yr B.P.), Wizards Beach (9200 yr B.P.) and Kennewick Man (9300 yr B.P.). Our study of the Kennewick specimen has been only through casts, photographs, and radiographs, since it has not yet been released by the U.S. Army Corps of Engineers for complete study.

The University of Wyoming standard three-page osteological checklist has been utilized in this study and other University of Wyoming osteological studies. It includes 20 standard craniofacial measurements defined by Howells (1973), 6 interorbital measurements defined by Gill (Gill et al. 1988), 12 vault measurements from Bass (1995), 7 mandibular measurements (Moore-Jansen et al. 1994), and 3 dimensions of the bony palate (Olivier 1969). In the most recent study of the Northwestern Plains sample, Stuart (2000) calculated eight commonly used craniofacial indices from these metrics to augment her metric comparisons with populations both to the west and east of the Northwestern Plains.

The morphological characteristics used in this study are a combination of traits, some of which are recorded on the University of Wyoming standard form and some not. These are cranial shape, malar projection, zygomaticomaxillary suture form, palate form, palatine suture shape, nose form, nasal sill formation, nasal spine projection, incisor form, jawbone shape, chin form, and projection and shape of the proximal femur. All these traits are commonly used by forensic osteologists in skeletal race attribution among modern *Homo sapiens* populations. They are defined, illustrated, and discussed in Bass (1995), Gill and Rhine (1990), and Gill (1995, 1998).

Results

Research results from the Northwestern Plains skeletal sample are emerging currently from a number of different areas (craniometrics, morphological analysis, demography, paleopathology and trauma, etc.). Rather than treat all these findings in depth, I will present an overview of the most important findings, particularly as they relate to original settlement, migration, drift, and selection.

Cranial and Skeletal Comparisons

Figure 1 shows two skulls (adult male and female) from the late Plains Archaic of eastern Wyoming. Figure 2 shows two skulls for comparison from two other populations common to the Northwestern Plains. The Willson and Huntley crania of the late Plains Archaic (Figure 1) reveal a number of traits common to individuals earlier in the Archaic and also to Paleoindians such as Spirit Cave Man from Nevada and Kennewick Man from the banks of the Columbia River in Washington. The frontier White in Figure 2 shows well-developed Caucasoid skeletal features typical of Europeans (and certain Asians) for comparison with and in contrast to the Paleoindian/Archaic specimens. The other skull in Figure 2, from the late prehistoric period of the Northwestern Plains, reveals the "Mongoloid craniofacial trait complex" so common to modern East Asians and late prehistoric-, protohistoric-, and historic-period Amerindians from the Northwestern Plains (and elsewhere across the Americas).

The left column of Table 2 lists the morphological traits of the skull and femur that characterize Paleoindians such as Spirit Cave and Kennewick Man and also certain Northwestern Plains Archaic specimens such as Willson and Huntley. Europeans and other Caucasoids (from West Asia, South Asia, and the Ainu of Japan) show many traits in common with these ancient Americans, as revealed by the list in the middle column. By contrast,

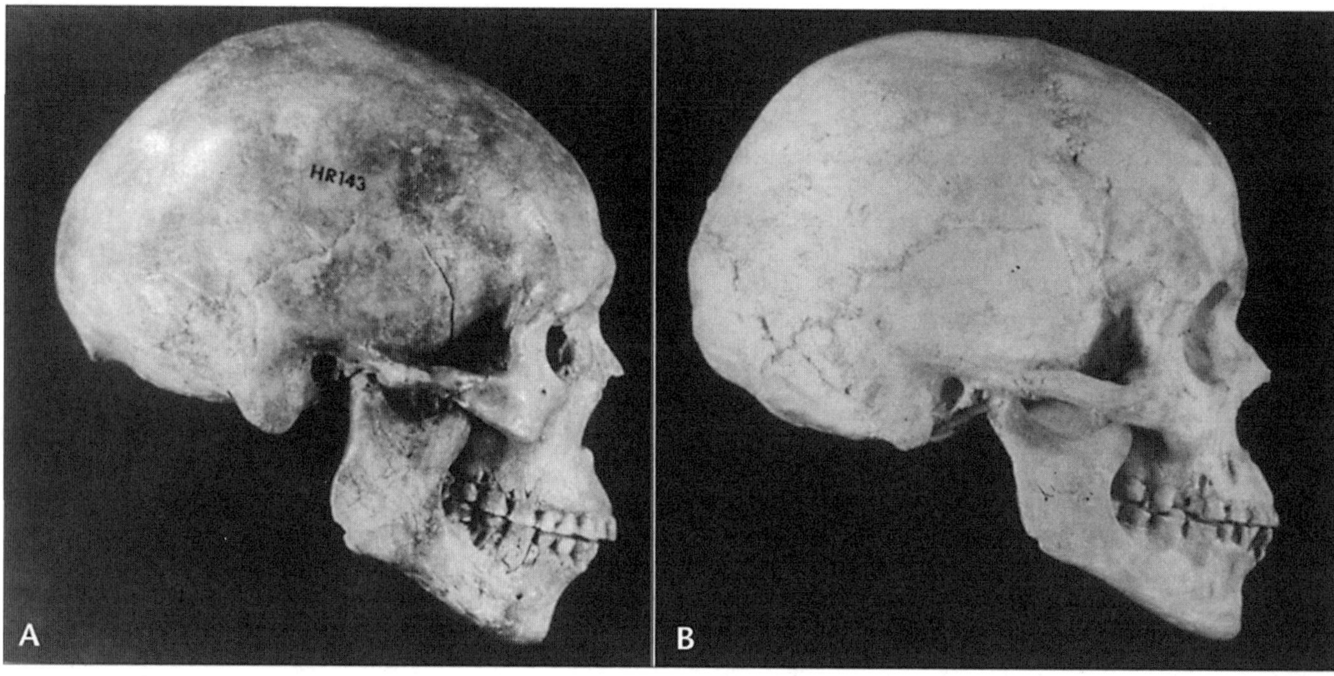

Figure 1. Male (**A**) and female (**B**) skulls from the late Plains Archaic in eastern Wyoming. This population preserved a number of Paleoindian traits such as long, high crania, alveolar (mouth) projection, square chins, and retreating cheekbones.

the late-period Amerindians from the Northwestern Plains (and elsewhere) listed in the right column show very few traits in common with the Paleoindians and the Europeans. Their proximal femora, for instance, are flat in anterior-posterior diameter (i.e., platymeric), while those of most Whites and Paleoindians are quite thick (see chart in Gill and Jimenez n.d.)

Even more impressive than the changes in femur morphology from the earlier to later temporal periods on the Plains are the changes in craniofacial morphology. The late-prehistoric and recent Northwestern Plains Indians have smaller, somewhat shorter crania with lower vault heights than earlier populations (or Europeans). In this feature modern East Asians differ somewhat from their cousins, the late Amerindians. Instead of a somewhat sloping, low-vaulted cranial profile, East Asians exhibit a high-vaulted large cranium that is even shorter in length than those of the late-period Amerindians. This results in a distinctive "globular" appearance to the East Asian cranium, clearly lacking among most Native American populations. East Asians also have, on average, a more reduced nasal skeleton than Amerindians. Otherwise the Mongoloid craniofacial complex manifests itself almost equally within these two closely related world populations.

Most Paleoindians and Archaic period skeletons show parabolic palates like those of Caucasoids. The majority of late-prehistoric skulls from the Northwestern Plains, on the other hand, show short, elliptic palates, as exhibited by other Mongoloid peoples. Exceptions from the Paleoindian period are Buhl and Wizards Beach. Not only do they show elliptic palates like the

Table 2. Morphological traits of Paleoindians and their Archaic-period descendants, compared with Europeans and other Caucasoids, and with late Northwestern Plains Indians and other Mongoloids.

Paleoindian/Archaic Northwest Plains	European/Caucasoid	Late Northwest Plains/Mongoloid
long skulls	long skulls	—
high skulls	high skulls	—
parabolic palates	parabolic palates	—
retreating malars	retreating malars	—
prominent nasals	very prominent nasals	—
concavo-convex nasals	—	concavo-convex nasals
curved zygomaticomaxillary suture	curved zygomaticomaxillay suture	—
nasion depression	nasion depression	—
dull nasal sill	—	—
reduced nasal spine	—	—
shovel-shaped incisors	—	shovel-shaped incisors
square, heavy jaw	—	square, heavy jaw
square, bilateral chin	square, bilateral chin	—
prominent chin	prominent chin	—
thick AP femur diameter	thick AP femur diameter	—

later Amerindians, they also show many other attributes of the Mongoloid craniofacial trait complex. This will be discussed more later on. Spirit Cave and Kennewick have parabolic forms that actually resemble the European morphology in the degree of anterior narrowing. Rhine (1990) has termed this narrow parabolic palate found among many American Whites "triangular." The posterior width, at the location of the third molars, is so great compared with the anterior breadth (at the canines) that it does create a basic triangle. Spirit Cave Man exhibits this narrow parabolic or triangular form of the palate, which is almost never found among late Amerindians. The Nebraska and eastern Wyoming Archaic specimens (as well as the 7500-year-old Smilden-Rostberg maxilla) all show clear parabolic forms (like Whites and East Polynesians), but not quite the narrow, triangular form of Spirit Cave and some of the more extreme Europeans.

Cheekbones are difficult to assess objectively; but with combinations of metric and visual approaches, comparisons can be made. It is obvious from the profiles of the Willson and Huntley skulls in Figure 1 that their malars are less robust and less forward projecting than those of the late-prehistoric Robbers Gulch male in Figure 2. Clearly Spirit Cave and Kennewick show the same retreating malars as these eastern Wyoming (and the western Nebraska) Archaic specimens. Here again Buhl and Wizards Beach, and some of the western Wyoming Archaic skeletons (from the Red Desert), have more prominent malars, resembling more those of the later Plains groups. In order to better test the temporal changes within the Northwestern Plains with regard to the placement of the cheekbones (i.e., the reduction of the facial midline in relation to the rest of the face and skull), Stuart (2000) tracked changes in five midfacial measurements (basion-nasion, basion-prosthion, porion-nasion, porion-subnasale, porion-prosthion). The High Plains male sample is the only one large enough to examine, and her results show a reduction in all of these from the late Plains Archaic to the late prehistoric, and further reduction in all but one measurement from the late prehistoric on to the protohistoric. Basion-nasion length, for instance, averages 102.8 mm in the late Plains Archaic, 101.7 mm in the late prehistoric, and 100.9 mm in the protohistoric. Our very small sample of Wyoming Basin males seems to show about the same degree of reduction from the early Plains Archaic to the late Plains Archaic and on to the late prehistoric (no protohistoric specimens available).

In order to see if lateral flare of the zygomatic arches correspondingly increased through time on the Northwestern Plains, I averaged bizygomatic breadth measurements of 12 males from the Archaic (pooled Wyoming Basin and High Plains) and compared it with the late-prehistoric/protohistoric mean (n = 18 males). The results are nearly identical, with an average breadth for both samples of 138 mm, thus showing that the changes through time involve an increase in facial flatness but not increases in facial breadth. To confirm this, both Stuart and I compared the upper facial index from the Archaic with that from late periods, utilizing slightly differently constituted Northwestern Plains samples. Neither one of us found significant change from early to late samples.

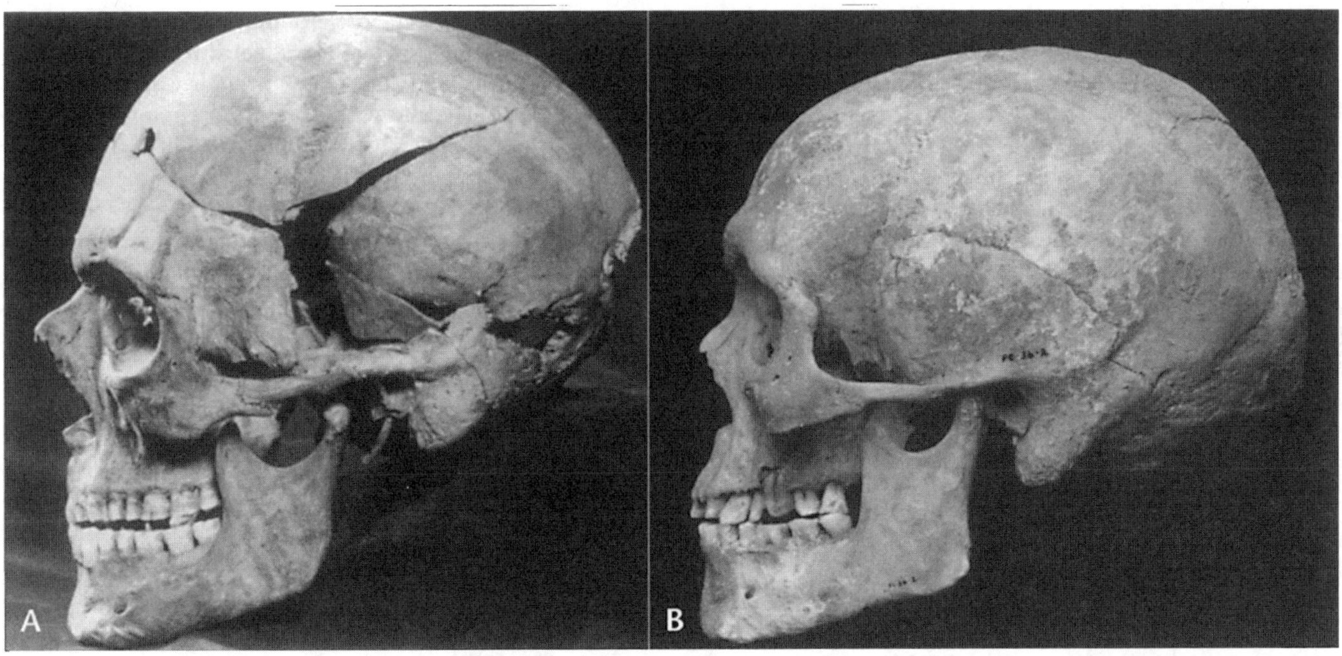

Figure 2. Pioneer White male skull (**A**) shows traits common to Europeans and other Caucasoids such as a long, high cranium, projecting nasal skeleton, and a square, projecting chin. A late-prehistoric male skull (**B**) exhibits traits common to recent Amerindians such as a low-vaulted skull, low nasal root, blunt rounded chin, and projecting cheekbones.

Much more dramatic changes are documented metrically for the cranial vault in the Northwestern Plains population. For years a drop in cranial vault height has been observed from Archaic to late prehistoric times (Gill 1974, 1981, 1991a). In my own recent calculations (including the western Nebraska Archaic skeletons), the male basion-bregma height decreases nearly 6.5 mm from the Plains Archaic (n = 10) to the late prehistoric/protohistoric (n = 21). In these tabulations once again the small Wyoming Basin sample was combined with the High Plains sample. The former (i.e., Archaic males) average 134.2 mm in height, and the late males 127.8 mm. This provides a noticeably different cranial profile for the average skull within these two samples (the ancient ones much more closely approximate the Caucasoid cranial profile). To substantiate this decrease in vault height, mean basion height indices have been calculated for all Northwestern Plains adults. Statistically significant differences have been found for both regions (High Plains and Wyoming Basin) and for both sexes within the Northwestern Plains from the Archaic to late temporal periods (Stuart 2000). In one of my own compilations, in which the two regional samples of males were pooled (and two individuals added to Stuart's original sample), I obtained an average of 82.87 for the mean basion height index of the Archaic males (n= 10) (which is at the middle of the medium range of vault height), and 79.01 for the late males (n = 20) (which is at the very low end of the medium range).

The length and breadth of the skull also change from the Archaic to the late prehistoric/protohistoric, but these are not quite as dramatic as the changes in vault height. The Archaic males from the Northwestern Plains have crania about 2.5 mm longer and 4 mm narrower than those of the late males. These differences result in an almost dolichocranic average for the Archaic period (75.03) and a high mesocranic average (78.07) for the late prehistoric/protohistoric. Females seem to show the same trend, but again sample sizes are too small to make meaningful comparisons.

Rather large and sometimes prominent nasal bones are common in the Northwestern Plains region throughout the various time periods, especially among males. The concavo-convex nasal profile common to most Amerindian populations is found among the Archaic specimens of the Northwestern Plains sample as well as within the late populations. Interestingly, Kennewick Man falls Caucasoid on nasal projection when the Gill simometer method (Gill et al. 1988) is applied (see Powell and Rose 1999). So does the Harlan County Lake specimen (see Tibesar et al. 1989). The nasal bones (and adjacent maxillary processes) of Spirit Cave Man look somewhat Caucasoid as well, but application of the simometer method shows it to be somewhat closer to Amerindians (i.e., nose projection in relation to the total midfacial area). In general it appears that the older the skull within the ancient American samples, the more likely it is to have a European-like nasal form and projection. Certainly more well-preserved specimens are needed, however, before this hypothesis can be fully tested.

Noticeable nasion depressions are common with Paleoindians and the Northwestern Plains Archaic individuals. These angled depressions between the top of the nasal bridge and the brow ridge, so common with Caucasoids and East Polynesians (especially males), are only rarely seen on late Amerindian or recent East Asian crania.

The nasal sills (i.e., the lower margins of the nasal aperture) of late Amerindians are generally quite medium in development (and sometimes even sharp); but those of the Northwestern Plains Archaic individuals (and the Paleoindians) are normally dull (i.e., smooth and rounded off) like those of Africans, Melanesians, and Polynesians. A recent study by Valentine (1999) shows that our Northwestern Plains Archaic sample has 89 percent dull or "guttered" (i.e., absent) nasal sills, and only 11 percent medium sills and no sharp sills. By protohistoric times that had changed to 8 percent sharp sills, 48 percent medium, and only 44 percent dull or guttered sills. Reduced nasal spines are also typical of the Paleoindian and Archaic samples. By later time periods within the Northwestern Plains, the majority show medium development of the nasal spine.

One trait of the Mongoloid craniofacial trait complex that does occur early in the Northwestern Plains area is shovel-shaped incisor teeth. These occur from the earliest Archaic skeletons on to the very late ones. This continuity is also true for some aspects of mandible form. Most Archaic individuals, as well as the later groups, have square, heavy jaws.

A feature of the mandible that is quite different, however, from ancient times to more recent in the Northwestern Plains is chin form. Nearly all Paleoindians and many Archaic specimens from the Northwestern Plains have prominent, square chins (sometimes bilateral) like those of Europeans (see Figure 3). By contrast, no late-prehistoric or recent Northwestern Plains Indians have this form. They show heavy, rounded chins (normally median and not bilateral) with reduced projection (i.e., neutral chins, non-prominent). With the prominent, bilateral chin form of the Paleoindians and Archaic individuals comes a marked curvature (cupping) as well. As illustrated in Figure 3, the curve starts from the bottom margin of the tooth row and extends inferiorly to the mental eminence. This condition among the earliest Americans, nearly identical to that among Europeans, is very different from the heavy, rounded chin form (without curvature) of later Amerindians.

Biological Indicators of Culture Change

During the same period of time in the Northwestern Plains that the dramatic morphological changes were occurring, other biological changes were happening too. Skeletal injuries increased dramatically from the Archaic to the late prehistoric, longevity decreased precipitously, and the incidence of dental caries changed as well.

As pointed out by Scheiber and Gill (1996), there is strong evidence for increased conflict among groups in the Northwestern Plains as the Archaic period drew to an end and the late-prehistoric period began. In the case of late-prehistoric interments, small projectile points (probably arrow points) are present (sometimes embedded in bone) in one third of the skeletons. Actual deaths by projectiles or other severe injuries are extensive through the late-prehistoric and protohistoric periods. This situation is in marked contrast to the very few injuries (or wounds from projectile points) observed in association with the Plains Archaic burials.

Further evidence of an escalation of conflict is the shift in mortality curves. The average age at death in the late Plains Archaic (n = 14) was 47.5 years; during the late prehistoric (n = 48)

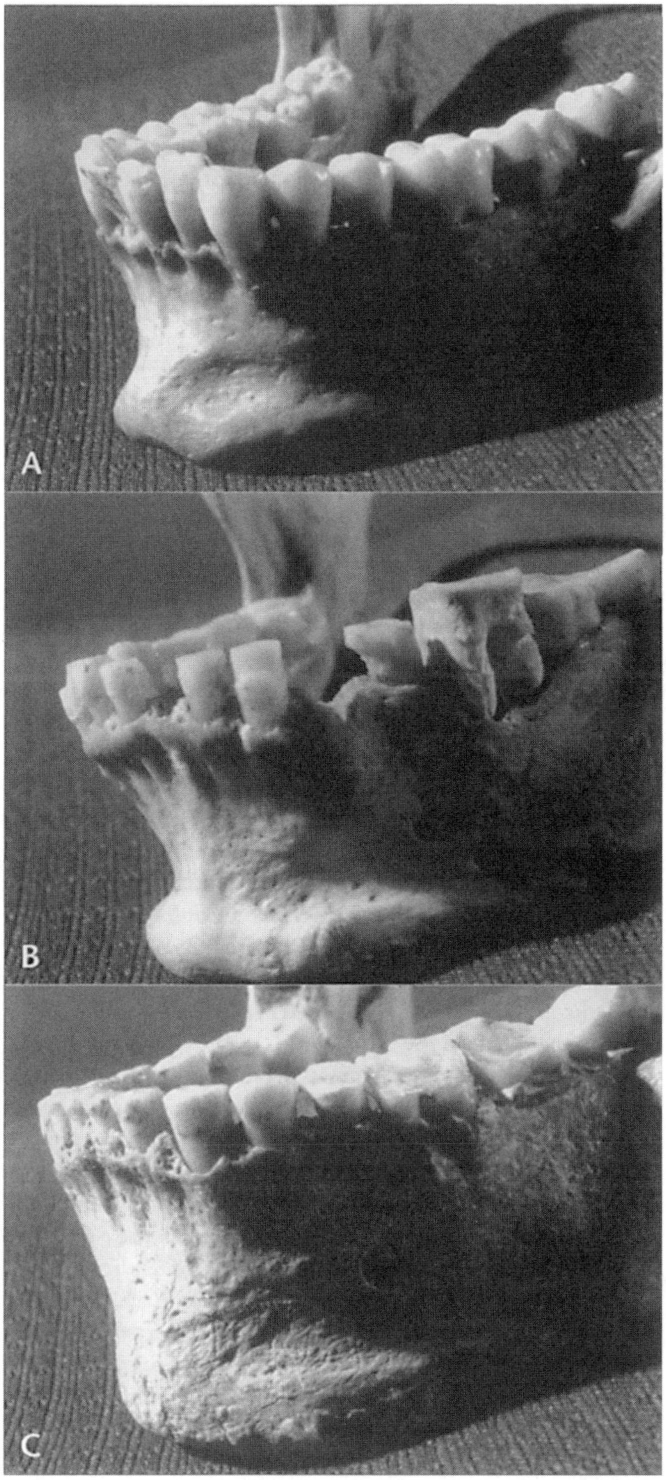

Figure 3. Chin form on three female mandibles. The pioneer White (**A**) and the late–Plains Archaic female (**B**) show strong curvature from the tooth row to the corner of a square, prominent chin. The protohistoric Amerindian (**C**) shows a blunt, heavy, rounded chin with little curvature below the tooth row.

it was 32 years. As we move on into the protohistoric (n = 55), the age at death drops even further to 27 years. The five-year decline in life expectancy from the late prehistoric to the protohistoric/historic does not seem so amazing. This period witnessed the advent of Euroamerican weapons and goods, the coming of the horse, and increased movement of nomadic tribes (due to dislocation, heightened competition for resources, etc.). What is amazing is that more than 1000 years earlier than this, some dramatic disruption occurred that caused three times that decrease in longevity. The dramatic increase in cranial injuries, scalpings and other bodily injuries, and embedded projectile points in conjunction with the 15½-year drop in longevity from the Archaic to the late prehistoric suggests a time of dramatic disruption and upheaval.

During the same period of prehistory, we are able to document a sharp decline in dental caries. Zitt (1992) records a caries frequency of 4.0 percent for the Plains Archaic (n = 225 teeth). The late prehistoric period shows just half as many dental cavities, with a caries frequency of 2.0 percent (n = 200 teeth); and the protohistoric yields little evidence of caries at all (0.2 percent frequency, n = 394 teeth) (Zitt 1992). According to Turner (1979) and other dental anthropologists, such changes indicate a decrease in carbohydrates and an increase in meat in the diet. Simply put, carbohydrate foods cause cavities and meat does not.

Discussion and Conclusions

Morphological and metric analyses demonstrate that during the Plains Archaic, at least some population elements within the Northwestern Plains still existed that were very similar to certain earlier Paleoindians of the western regions of North America and were likely descended from them or similar early populations. These Archaic peoples (e.g., Sydney, Harlan County Lake, Willson, Huntley) had long, rather high skulls and proximal femora that were thick in A-P diameter like those of Europeans, Asiatic Caucasoids, and many of the Paleoindians (e.g., Kennewick, Spirit Cave).

Other Caucasoid-like skeletal traits of ancient Americans (both the Paleoindians and their Archaic descendants) are parabolic (sometimes triangular) palates; retreating malars (cheekbones); prominent nasal skeletons (occasionally in the "White" sector metrically); a nasion depression; and square, prominent bilateral chins. These people lacked the so-called "Mongoloid skeletal trait complex" entirely except for square, heavy mandibles and shovel-shaped incisor teeth. Moreover, their crania sometimes exhibit the concavo-convex nasal profile that is so common to late prehistoric and modern American Indians (but not so common among modern East Asians).

Other distinctive traits found among Paleoindians and their Archaic descendants of the Northwestern Plains are *not* common to either the modern Caucasoids or Mongoloids. Among these are alveolar (mouth) prognathism (with a correspondingly long palate), dull (or absent, guttered) nasal sills, and reduced nasal spines. Quite interestingly, these are traits that some South Asian Caucasoids and Polynesians possess (also in combination with European-like skulls, chins, nasal skeletons, and cheekbones).

It is important to keep in mind that not all Paleoindian skeletons reveal this "vaguely Caucasoid" morphology. Wizards Beach, for instance, shows almost perfectly the Mongoloid craniofacial complex (except for chin form), as well as a proximal femur that is platymeric like those of late-period and modern Amerindians. According to Fenton (see Green et al. 1998), the Buhl woman is primarily "Mongoloid" in appearance as well. As we enter the Northwestern Plains Archaic, at least a third of the specimens (even more from the Wyoming Basin) show the Mongoloid trait complex (craniofacial as well as femur), and not the "vaguely Caucasoid" look of their neighbors.

By late-prehistoric times (1500–300 yr B.P.) in the Northwestern Plains, there are no individuals remaining with the complex of skeletal traits exhibited by Huntley, Willson, or Harlan County Lake (or Spirit Cave or Kennewick). By that time these "ancient ones," with their Caucasoid looks, have been completely replaced by people with low-vaulted, shorter skulls; heavy jaws with blunt, rounded chins; forward-projecting cheekbones; and femora with a distinctively flat (platymeric) shape to the proximal end of the shaft. These "new people" also possess (and also share with their modern East Asian cousins) short, often elliptic palates (with straight palatine sutures), widely divergent "angled" zygomaticomaxillary suture lines, a complete lack of nasion depression (smooth nasal root), medium nasal sills (and spines), and very robust shovel-shaped anterior teeth oriented with a more vertical edge-edge occlusion.

The corresponding changes in longevity and patterns of injury (from the Archaic to the late prehistoric) suggest the entry onto the Northwestern Plains of new people, perhaps with a new weapon technology and new subsistence strategies. Changes in the incidence of caries suggest a diet with more meat and fewer carbohydrates. This trend is quite noticeable from the Archaic to the late prehistoric, and again from the late prehistoric to the protohistoric. Scheiber and Gill (1996) suggest that these changes in diet, as well as patterns of injury from the Archaic to the late prehistoric, might correspond to the entrance into the Northwestern Plains of the bow and arrow. This development could have increased the efficiency in hunting bison and other game, heightened the competition for resources, and provided new weapons for warfare and raids. Changes in burial practices at this same time (see Scheiber and Gill 1996) also strongly support the idea of a new, more nomadic life-style.

During the protohistoric period, the introduction first of metal, then the horse, and eventually firearms would have intensified all these forces of change, resulting in even better weapons, more bison exploitation, and more rapid movement of peoples. Human populations probably continued to increase as well.

According to Brace (1995), the Mongoloid craniofacial trait complex evolved in response to working hides with the incisor teeth and other activities involving biting, holding, chewing, and stripping leather, dried meat, and other animal products. The entire biomechanics of the Mongoloid face (heavy shovel-shaped incisors; short, broad palate; heavy lower jaw; strong forward-projecting cheekbones) would seem to be a design for heavy use of the anterior dentition. Biomechanics evidence, then, seems to support Brace's hypothesis. If Brace is right, then the evolving cultural conditions on the Northwestern Plains would have favored the evolutionary development of (or intensification of) the basic Mongoloid trait complex.

The ultimate question for the biological anthropologist in all of this is, Precisely which force or forces caused the morphological changes on the Northwestern Plains? Was it entirely regional microevolution in place, without significant population migration into the region? Or was it a major migration of physically different people (Mongoloid peoples) accompanying the new cultural introductions?

Perhaps it would be useful to look at the Old World, since throughout much of East Asia a very similar process occurred, that is, an ancient Caucasoid population was replaced by expanding Mongoloid peoples. This process is probably documented best in Japan, where ancient Ainu hunters (a race of "White" people with Caucasoid skeletal features) were invaded from the mainland by the Yayoi people (a Mongoloid racial element) during the Bronze Age (Brace et al. 2001). Despite their technological advantage and greater numbers (since the Yayoi people were agriculturalists), replacement throughout Japan took over a thousand years, accomplished by warfare and conflict at times, and in other instances by intermarriage and assimilation. One great difference between Bronze Age Japan and the prehistoric Northwestern Plains of North America is that in the Plains both the indigenous people and those entering were hunter-gatherers, in presumably approximately equal numbers. However, if Brace is correct, the entering people in the Northwestern Plains would have had an adaptive physical advantage (heavy teeth, jaws, etc.) to the new life-style that they themselves introduced. This would not be the case with the Yayoi people of Japan, since their agricultural practices would have eliminated the need for a heavy masticatory complex. Thus the Old World example can only serve as a limited model for what happened in the Northwestern Plains and elsewhere within the New World.

I believe that *both* these factors, the migration of new people into the Northwestern Plains and the changing natural-selection forces brought about by the new life-style they introduced, led to the genetic/morphological changes we have documented through our osteology analyses. Had it been migration alone with total replacement, then the transition from the Archaic to the late prehistoric would have been even more abrupt, with no sharing of traits with late-Archaic individuals. Yet these shared traits do exist. Had the process been drift or selection without the entry of new people with additional Mongoloid traits, the process could not have moved so swiftly as it did. The fact that horticulturalists were moving up the major river systems about this time into South Dakota and Nebraska seems to have had limited effect upon this overall pattern of change occurring in the drier, colder, and more hostile Northwestern Plains.

As the highly successful Mongoloid peoples spread across northern Asia thousands of years ago, they eventually came to replace the Ainu and the other Ainu-like peoples of mainland Asia. In South China, the cousins of the Polynesians (i.e., those who stayed in China and did not sail eastward into Oceania) were close relatives of the Ainu (see Gill 2001) and were among the next to be replaced in Asia by Mongoloid peoples expanding out of northern China. Judging from some skeletal samples, this gradual process of assimilation and replacement lasted in East Asia until within the last few thousand years (Brace et al. 2001). It would seem that a somewhat similar process was occurring almost simultaneously in the Northwestern Plains and elsewhere in North America. The groups entering the Americas in subsequent waves of migration were increasingly more Mongoloid in appearance. This process of migration, with subsequent replacement and gene flow, mirrored in the New World what was happening almost at the same time in East Asia. Yet in the Northwestern Plains and much of western and northern North America, a hunting way of life was intensifying, not yielding to agriculture as in the Old World. This meant the migrants, although fewer in number, possessed traits that gave them a great advantage. Therefore in the Northwestern Plains and much of North America, selective forces favoring the Mongoloid trait complex augmented the effects of migration and gene flow, thus further accelerating the process of gene replacement. The new people who brought the new way of life also carried with them the biological traits necessary to adapt to it.

References Cited

Agogino, G. A.
1961 A Middle Prehistoric Burial from Laramie, Wyoming. *Plains Anthropologist* 6:202–204.

Agogino, G. A., and E. Galloway
1963 The Skulls from Torrington, Wyoming: A Re-Evaluation. *American Antiquity* 29:106–109.

Armstrong, A. K.
1999 *A Bioarchaeological Examination of the Korell-Bordeaux Site (48G054).* Master's Thesis, Department of Anthropology, University of Wyoming, Laramie.

Baker, S. J.
1989 Osteological Analysis. In *Archaeological and Osteological Analysis of Two Burial Sites along Harlan County Lake, Nebraska: Chronological and Evolutionary Implications,* by W. L. Tibesar, T. K. Larson, S. J. Baker, T.A. Christianson, K. H. Dueholm and P. H. Sanders. Report to U.S. Army Corps of Engineers. Kansas City District. Larson-Tibesar Associates, Laramie, Wyoming.

Bass, W. M.
1995 *Human Osteology: A Laboratory and Field Manual.* Fourth edition, Missouri Archaeological Society, Columbia, Missouri.

Bass, W. M., and D. C. Lacy
1963 Three Human Skeletons from the PK Burial, Sheridan County, Wyoming. *Plains Anthropologist* 8(21):142–57.

Birkby, W., and W. M. Bass
1963 Possible Shoshonean Skeletal Material from the Turk Burial Site, 48WA301. *Plains Anthropologist* 8:103–108.

Brace, C. L.
1995 *The Stages of Human Evolution.* Fifth edition, Prentice Hall: Englewood Cliffs, New Jersey.

Brace, C. L., A. R. Nelson, N. Seguchi, H. Oe, L. Sering, P. Qifeng, L. Yongyi, and D. Tumen
2001 Old World Sources of the First New World Human Inhabitants: A Comparative Craniofacial View. *Proceedings of the National Academy of Sciences* 98(17): 10017–22.

Breternitz, D. A., A. Swedlund, and D. Anderson
1971 An Early Burial from Gordon Creek, Colorado. *American Antiquity* 36(2):170–82.

Davis, D. P.
1992 *Archaeology and Human Osteology of the Benick Ranch Site: (48AB571).* Master's thesis, Department of Anthropology, University of Wyoming, Laramie.

Frison, G. C.
1991 *Prehistoric Hunters of the High Plains.* Second edition. Academic Press, New York.

Gill, G. W.
1974 Human Skeletons from Wyoming and Their Bearing on the Concept of Morphological Dating. In *Applied Geology and Archaeology: The Holocene History of Wyoming,* edited by M. Wilson, pp. 100–107. Geological Survey of Wyoming,Report of Investigations No. 10.

1981 Human Skeletal Populations of the Northwestern Plains: A Preliminary Analysis. In *Progress in Skeletal Biology of Plains Populations,* edited by R. J. Jantz and D. H. Ubelaker, pp. 57–70. *Plains Anthropologist.* Memoir 17.

1991a Human Skeletal Remains on the Northwestern Plains. In *Prehistoric Hunters of the High Plains,* Second edition, by G. C. Frison, pp. 431–447. Academic Press, New York.

1991b Human Osteological Material. In *The Smilden-Rostberg Site: An Early Archaic Component in the Northeastern Plains,* edited by T. K. Larson and D. M. Penny. Larson-Tibesar Associates, Laramie. Report submitted to KBM, Inc., Grand Forks.

1995 Challenge on the Frontier: Discerning American Indians from Whites Osteologically. *Journal of Forensic Sciences* 40(5):783–88.

1998 Craniofacial Criteria in the Skeletal Attribution of Race. In *Forensic Osteology.* Second edition, edited by K. Reichs, pp. 293–317. Charles C. Thomas, Springfield.

2001 Basic Skeletal Morphology of Easter Island and East Polynesia: With Paleoindian Parallels and Contrasts. In *Pacific 2000: Proceedings of the Fifth International Conference on Easter Island and the Pacific,* edited by C. M. Stevenson, G. Lee, and F. J. Morin, pp. 447–56. The Easter Island Foundation, Bearsville Press, Los Osos, California.

Gill, G. W., and C. J. Jimenez
1999 Paleoamerican Skeletal Features of the Huntley and Willson Crania from the Late Archaic of Eastern Wyoming. Poster presented at the Clovis and Beyond Conference, October 29–30, Santa Fe, New Mexico.

n.d. Paleoamerican Skeletal Features Surviving into the Late Plains Archaic. Manuscript accepted for publication in *New Directions in First American Studies,* edited by B. T. Lepper.

Gill, G. W., S. Hughes, S. Bennett, and B. M. Gilbert
1988 Racial Identification from the Midfacial Skeleton with Special Reference to American Indians and Whites. *Journal of Forensic Sciences* 13:92–99.

Gill, G. W., and S. Rhine, editors.
1990 *Skeletal Attribution of Race: Methods for Forensic Anthropology.* Maxwell Museum of Anthropology, Albuquerque, New Mexico.

Green, T. B., B. Cochran, T. W. Fenton, J. C. Woods, G. L. Titmus, L. Tieszen, M. A. Davis, and J. J. Miller
1998 The Buhl Burial: A Paleoindian Woman from Southern Idaho. *American Antiquity* 63:437–65.

Howells, W. W.
1938 Crania from Wyoming Resembling "Minnesota Man." *American Antiquity* 4:318–26.

1973 *Cranial Variation in Man: A Study of Multivariate Analysis of*

Patterns of Differences among Recent Human Populations. Harvard University Press, Cambridge, Massachusetts.

Jantz, R. L., and D. W. Owsley
1997 Pathology, Taphonomy and Cranial Morphometrics of the Spirit Cave Mummy. *Nevada Historical Society Quarterly* 40(1): 62–84.

Krogman, W. M., and Y. Iscan
1986 *The Human Skeleton in Forensic Medicine.* Second edition. Charles C. Thomas. Springfield, Illinois.

Lovvorn, M. B., G. W. Gill, F. G. Carlson, J. R. Bozell, and T. L. Steinacher
1999 Microevolution and the Skeletal Traits of a Middle Archaic Burial: Metric and Multivariate Comparison to Paleoindians and Modern Amerindians. *American Antiquity* 64)3):527–45.

Miller, J. C.
1990 *A Summary Evaluation of Stature in the Aboriginal American West.* Manuscript on file, Department of Anthropology, University of Wyoming, Laramie.

Miller, M.
1999 *The Paleopathology of Northwestern Plains Indians.* Master's thesis, Department of Anthropology, University of Wyoming, Laramie.

Moore-Jansen, P., S. D. Ousley, and R. L. Jantz
1994 *Data Collection Procedures for Forensic Skeletal Material.* Report of Investigations No. 48. Department of Anthropology, University of Tennessee, Knoxville.

Mulloy, W.
1954 The McKean Site in Northeastern Wyoming. *The Wyoming Archaeologist* 7:16–40.

Olivier, G.
1969 *Practical Anthropology.* Charles C. Thomas, Springfield, Illinois.

Owsley, D. W.
1999 *Human Skeletal Remains from the Anzick Site (24PA506), Park County, Montana.* Paper presented at the Fourth Biennial Rocky Mountain Anthropological Conference, October 2, 1999, Glenwood Springs, Colorado.

Powell, J. F., and J. C. Rose
1999 *Report on the Osteological Assessment of the "Kennewick Man" Skeleton.* http://cast.uark.edu/other/nps/nagpra/CENWW.97.Kennewick

Reichs, K. J., editor
1998 *Forensic Osteology: Advances in the Identification of Human Remains.* Charles C. Thomas. Springfield, Illinois.

Rhine, S.
1990 Non-metric Skull Racing. In *Skeletal Attribution of Race: Methods for Forensic Anthropology,* edited by G. W. Gill and S. Rhine, pp. 9–20. Maxwell Museum of Anthropology, Albuquerque, New Mexico.

Rogers, M. A.
1996 *An Analysis of Discrete Traits of the Northwestern Plains and Wyoming Basin Indians with a Comparison to Other Amerindian Populations.* Master's thesis, Department of Anthropology, University of Wyoming, Laramie.

Scheiber, L. L., and G. W. Gill
1996 Bioarchaeology of the Northwestern High Plains. In *Archaeological and Bioarchaeological Resources of the Northern Plains,* edited by G. C. Frison and R. C. Manifort, pp. 91–119. Arkansas Archaeology Survey Research Series No. 47, Fayetteville, Arkansas.

Steege, L. C.
1960 A Probable Middle Period Burial in Wyoming. *Plains Anthropologist* 10:82–84.

Steele, D. G., and J. F. Powell
1992 The Peopling of the Americas: The Paleobiological Evidence. *Human Biology* 63:301–36.

Stewart, T. D.
1979 *Essentials of Forensic Anthropology: Especially as Developed in the United States.* Charles C. Thomas, Springfield, Illinois.

Stuart, L. J.
2000 *Metric Analysis of Northwestern Plains Amerindian Skeletal Remains.* Master's Thesis. Department of Anthropology. University of Wyoming, Laramie.

Tibesar, W., L., T. K. Larson, S. J. Baker, T. A. Christiansen, K. H. Dueholm, and P. H.Sanders
1989 *Archaeological and Osteological Analyses of Two Burial Sites along Harlan County Lake, Nebraska: Chronological and Evolutionary Implications.* Report to U. S. Army Corps of Engineers, Kansas City District. Larson-Tibesar Associates, Laramie, Wyoming.

Turner, C. G.
1979 Dental Anthropological Implications of Agriculture among the Jomon People of Central Japan. *American Journal of Physical Anthropology* 51:619–36.

Valentine, C. C.
1999 *Microevolutionary Trends in Nasal Sill Development on Northwestern Plains Indian Crania.* Manuscript on file, Department of Anthropology, University of Wyoming, Laramie.

Zitt, J. O.
1992 *Prehistoric and Early Historic Subsistence Patterns and Dental Pathology on the Northern Great Plains.* Master's thesis, Department of Anthropology, University of Wyoming, Laramie.

Circumpacific Populations and the Peopling of the New World: Evidence from Cranial Morphometrics

Richard L. Jantz[1] and Douglas W. Owsley[2]

Abstract

The peopling of the New World is now considered to have been a more complicated process than previously imagined. Instead of or in addition to the Bering land bridge route, early populations may have arrived via a coastal route. Most of the evidence supporting a coastal route is archaeological and geological. In this paper we test the hypothesis of a coastal route using cranial morphometrics. Our hypothesis is that early populations from western North America should resemble early circumpacific Asian and modern Pacific populations rather than recent Native Americans. We test this hypothesis using four samples from early California sites dating from about 9000 to 1500 yr B.P. We also use six late-Pleistocene/early-Holocene fossil crania, three from western North America (La Brea, Buhl, and Spirit Cave) and three from Asia (Upper Cave 101 and 103, and Liujang). Early Californians are highly variable, but in general are more similar to modern Pacific populations, such as Polynesians, than they are to modern Native Americans. The individual crania are also highly variable, but those from Asia are not strongly differentiated from early Americans. In particular, the Spirit Cave Mummy and Upper Cave 101 could easily belong to the same population. Early American crania also share large cranial vault size with recent Polynesians and with early Asians, in contrast to recent Native Americans and Asians, which have smaller cranial vaults. We conclude that the morphometric evidence supports a coastal route.

[1]Department of Anthropology, University of Tennessee, 250 South Stadium Hall, Knoxville, TN 37996-0720; e-mail: rjantz@utk.edu

[2]Department of Anthropology, Mail stop 112, National Museum of Natural History, Smithsonian Institution, Washington, D.C. 20560: e-mail: owsley.douglas@nmnh.si.edu

The recent discovery or redating of skeletal remains in the Americas from 8000–10,000 RCYBP has rekindled interest in their contribution to discussions of the peopling of the New World. Morphometric dissimilarity between early crania and recent Native Americans has contributed to the perception that populating the New World was a more complex process than previously imagined. One component of that complexity is the idea that some of the early migrants may not have used a route over the Bering land bridge, but rather a sea route along the coast. The evidence is mainly archaeological or geological (Dixon 1999; Gruhn 1994), but morphology suggesting Pacific affinities of early crania could also support a coastal migration route. Recent craniometric studies showing that early-Holocene crania are more similar to Ainu, Polynesian, or southern Pacific Rim populations (Steele and Powell 1992) can be taken as evidence supporting a coastal route. In this paper we examine some of this evidence in greater detail, focusing on early skeletons from California that have received little attention in morphometric research, and on the few specimens from East or Southeast Asia for which we have data. We also review and reanalyze the cranial morphometrics of the Buhl burial because the few analyses to date have not resulted in a clear picture of this important skeleton.

Our hypothesis is that early crania from California, the western Great Basin, and Plateau should be more similar to modern Pacific populations, such as the Ainu, Polynesians, or Southwest Pacific islanders, than to recent Native Americans. We may also expect these early American crania to resemble early crania from the Asian Pacific rim, as the latter could represent source populations from which the earliest migrants were drawn.

Methods

As in our previous studies, we use two approaches. In the first, early or fossil crania are compared to recent (post-Neolithic) crania from Howells's (1989) world sample, supplemented with crania from historic Native American tribes. This approach is described by Owsley and Jantz (1999). In the second, individual fossil crania are compared with one another. Morphometric methods used in this paper have been outlined in Jantz and Owsley (2001) and are similar to those proposed by Van Vark (1995). Briefly, the method is based on Mahalanobis distances calculated between samples and between fossil crania. The pooled within-sample covariance matrix is used to obtain all distances, including those among individual fossil specimens. Principal coordinates are used to scale the distances along orthogonal axes for graphical presentation (Gower 1972). The null hypothesis concerning fossil crania is that they are drawn from the same population. This hypothesis can be tested using the single-population expectation from Defrise-Gussenhoven (1967).

Samples

Table 1 presents the circumpacific samples identified as useful for investigating the hypothesis of Pacific affinities of early Americans. We have had an opportunity to examine and measure early California crania from Tulare Lake, La Jolla, Karlo, and La Brea. The Tranquillity crania were extensively reported by Angel (1966), and his measurements are used here. Following is the salient information for each site:

Tulare Lake

The five burials from this site (TUL-90) have been reported by Warren and McKusick (1959), although no morphological interpretation was presented. The site is located in the southern San Joaquin Valley between Bakersfield and Fresno. Warren and McKusick place the site in the middle Horizon between 1500 and 3500 yr B.P., but no radiocarbon date is available. Tulare Lake is likely the latest site among the test samples used.

La Jolla

La Jolla refers to a complex of shell midden sites along the coast of southern California and Baja California. La Jolla sites indicate intermittent occupation from 7000 yr B.P. until a few hundred years ago (Shumway et al. 1961). The five burials used in the present analysis are from several sites, only one of which, the Scripps Estate site, has been extensively described (Shumway et al. 1961). Our sample has been assigned to La Jolla II. The crania are mineralized, sometimes markedly so, suggesting considerable antiquity. We believe it is likely that our sample dates from 5000–7000 yr B.P., but additional dating is required. Shumway et al. (1961) present no analysis of the burials, but do comment on the dolichocephaly and prognathism seen in some crania.

Karlo

The Karlo site (LAS-7) site was described by Riddell (1960). It is located in Secret Valley in northeastern California. The Karlo site apparently dates to about 3000 yr B.P., although the Karlo period extends from 2400–4000 yr B.P. Krantz (1960) described the five Karlo crania, noting their large size, dolichocephaly, large and laterally prominent brow ridges, large faces, alveolar prognathism, low noses, and little chin projection.

Tranquillity

The Tranquillity site (FRE-48) is in the central San Joaquin Valley, about 24 miles west of Fresno. Angel (1966) has described the six Tranquillity skeletons, but there is still a critical issue about their date. Angel was convinced that the crania belonged to a late-Pleistocene or early-Holocene temporal context. Not only were the skeletons associated with extinct fauna (*Camelops, Equus,* and *Bison*), the chemical analysis by Heizer and Cook (1952) supported that association. However, a conventional radiocarbon date on two skeletons (burials 4 and 5) in Angel (1966) yielded an age of 2550 ± 60 RCYBP (Berger et al. 1971). Berger et al. argue that the radiocarbon date calls the fluorine date into question. This conflict should be resolved by AMS dating. In the meantime, we are suspicious of the radiocarbon date because of the chemical analysis and because Angel describes the skeletons as heavily mineralized, suggesting antiquity. Angel describes the crania as robust, long, and extremely high, with somewhat pinched foreheads, broad faces, and low-rooted and broad noses. He characterized them as "proto mongoloid."

Table 1. Sites, sample sizes and dates of crania used in this study.

Specimen	Sample (n)	Date (yr B.P.)	Reference
Tranquillity	6	9000?	Angel 1964, Berger et al. 1971
La Jolla	5	2550? 9000?	Shumway et al. 1961
Karlo	5	2500–4000	Riddell 1960
Tulare Lake	5	1500–3500	Warren and McCusick 1959
La Brea	1	9000	Berger 1975
Buhl	1	10,275	Green et al. 1998
Spirit Cave	1	9000	Tuohy and Dansie 1997
Upper Cave 101	1	26,000–28,000	Weidenreich 1938–39; Chen et al. 1989
Upper Cave 103	1	26,000–28,000	Weidenreich 1938–39; Chen et al. 1989
Liujang	1	late Pleistocene	see Pope (1992)

La Brea

The La Brea skeleton was recovered along with extinct fauna from the La Brea tar pits in Los Angeles and has been given a conventional radiocarbon date of 9000 ± 80 RCYBP (Berger 1975). Kroeber (1962) assessed the La Brea skull as resembling modern Chumash Indians, and this characterization has been carried through to recent times. The skull was placed in the "narrow-nosed subtype" of the Californian physical type. Our initial evaluation of the La Brea skull shows that it differs markedly from all our recent samples (Jantz and Owsley 1999).

Buhl

The Buhl skeleton from Idaho is extremely important because of its antiquity and excellent preservation. The loss to science brought about by its reburial is compounded because no comprehensive morphological assessment of the skeleton was performed while it was available. No images other than black-and-white photographs were made. Based on their limited morphological assessment of the skeleton and with no analytical evaluation, Green et al. (1998) concluded that it falls within the range of variation of recent Native Americans. Fenton and Nelson (1998) are more explicit:

> It was determined that Buhl registers most closely with the Amerind, or New World samples, practically to the exclusion of the rest of the world. Buhl does not show any affinities with the European samples and indicates only a very tenuous connection with the Ainu. The results of our analysis show the Buhl woman to have been an American Indian. We believe that, in general, she is not morpho-

logically distinct from later North American Indians. It is our contention that the Buhl woman is a representative of an early Paleoindian migration that is more closely related to later groups of American Indians than to European or Asian populations.

An even more recent assessment of Buhl's morphometric affinities (Neves and Blum 2000) supporting the Fenton-Nelson interpretation also has some problems. Their two-dimensional principal component plot comparing Buhl with recent world populations shows Buhl to be rather strongly differentiated from recent Native Americans and Siberians in the direction of Polynesians. For whatever reason, they have chosen not to emphasize this aspect of their results.

Lovvorn et al. (1999) classify the Buhl cranium as Ainu using FORDISC 2.0 (Ousley and Jantz 1996). Classification as Ainu suggests the cranium bears little similarity to recent Native Americans. Lovvorn et al. (1999) apparently excluded Polynesians from FORDISC's comparative framework. Lovvorn et al. use 16 measurements, but do not specify which ones. We applied FORDISC using the 16 variables we consider likely candidates for their measurement set and including all of Howells's (1989) female samples. Our results show that the four populations to which Buhl is most similar are the Moriori, Mokapu, Hawaiians, and Easter Islanders, all Polynesians. The Ainu are the fifth most similar group. Using different measurement subsets did not materially alter the results.

Obviously Buhl is a priority for systematic comparative analysis because of earlier suggestions that it differs little from recent Native Americans. If that is, in fact, its morphological pattern, Buhl would be almost unique among early skeletons, and this would add further complexity to the fossil record.

Spirit Cave

The now well-known Spirit Cave (Scave) mummy, discovered in 1940, was recently dated to 9415 ± 25 RCYBP (Tuohy and Dansie 1997). Its taphonomy, pathology, and cranial morphometrics have been reported by Jantz and Owsley (1997). The Spirit Cave cranial vault is large, long, and narrow; the face is vertical and not markedly robust.

Upper Cave 101 and 103

The crania from Upper Cave (UC), or Shandingdong, of Zhoukoudian of northern China have been controversial since Weidenreich's (1938/1939) assessment. Although Weidenreich's characterization of Upper Cave 101 and 103 as "Primitive Mongoloid" and "Eskimoid" respectively was typological, the two crania have recently been shown to be so different from one another that it is unlikely they come from the same population (Cunningham and Wescott 2000). Dates for the Upper Cave crania have variously been considered late Pleistocene-early Holocene (Kamminga and Wright 1988), but more recent AMS dates from animal bones yield a range of 29,000–24,000 RCYBP (Chen et al. 1989).

Morphometric relationships have been extensively considered, but little agreement has emerged. Howells has emphasized their Amerindian character (Howells 1983;1995). Kamminga and Wright (1988) emphasized their Australian affinities. Van Vark and Dijkema (1988) and Cornell (1998) emphasize the uniqueness of Upper Cave compared with recent *Homo sapiens*, although both identify Pacific affinities in the skulls. There is substantial agreement that the Upper Cave crania have little in common with recent Chinese crania.

Liujang

Howells (1983) describes the Liujang skull of southeastern China as "definitely of a generalized Mongoloid character" and "somewhat un-Chinese" in appearance, particularly in the short face and low orbital openings (p. 298). Howells later (1995) identifies European affinities in the skull. Unlike the Upper Cave crania, the Liujang skull falls easily within the range of variation of recent *Homo sapiens* skulls. The date of this cranium remains indefinite. Woo (cited in Coon 1963, p. 467) considered it upper Pleistocene geologically, although a uranium date puts it at 67,000 years (Yuan et al. 1986). Its similarity to recent populations makes the early date suspect (Pope 1992).

Morphometric Relationships of the California Samples

To better illustrate relationships, the world reference populations were grouped into major geographical regions: Europe, subsaharan Africa, East Asia, Southwest Pacific, Polynesia, and America. Populations especially relevant to New World origins that do not easily group into major geographic categories are retained as separate samples. These include the Ainu (Japan), Buriats (Siberia), and Eskimos (New World Arctic). Measurement sets available for the individual crania differ in the number of variables because of the degree of preservation or the source of the data, which limit the number of variables that can be used in comparing crania. The most limiting set is for the Tranquillity crania, followed closely by the set available for the Buhl skull. We have therefore carried out two analyses, one that focuses on the California samples and another more extensive analysis that omits Tranquillity, but includes the other samples. Figure 1 shows a two-dimensional plot portraying the distances among the modern samples and the early California samples.

The first axis can be seen as a major dimension on which modern populations are arrayed. On one end of the axis with low scores are African and Southwest Pacific populations. On the opposite end with high scores are the Siberian Buriats and American Indians. East Asia, Europe, and Polynesia are more centrally located on the graph, although Polynesia tends toward the Africa–Southwest Pacific end and Europe and East Asia tend toward the American Indian–Siberian end of the axis. The early California samples exhibit almost as much variation along the first axis as do modern populations.

Measurement contributions to these two axes permit description of the major metric features responsible for the observed differentiation. Variables contributing to the first axis are vault breadth and face height. Populations with high scores, such as Native Americans and the extreme Buriats, have wide vaults and high faces. Those on the opposite end, Southwest Pacific and Africa, have long, narrow vaults and low faces. Among early Californians, the Tranquillity crania are characterized by narrow

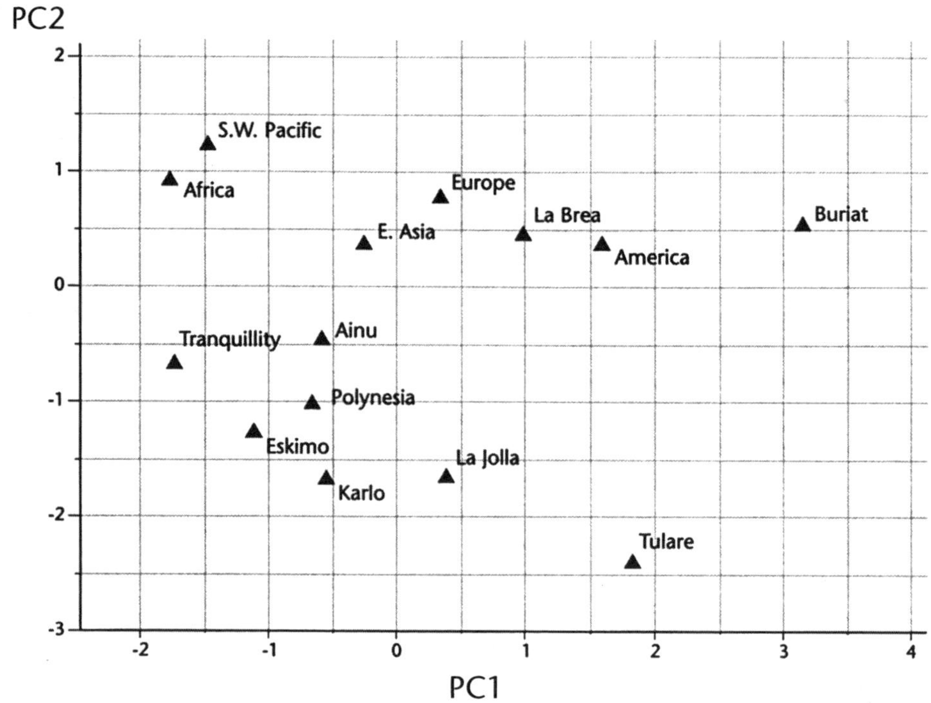

Figure 1. Two-dimensional principal coordinates plot showing relationships of recent human populations and early Californians (including Tranquillity).

vaults and low faces. Tulare Lake, on the other hand, is characterized by wider vaults and higher faces.

The second axis primarily expresses the distinctiveness of the California samples, except for La Brea, from most recent samples. The recent Polynesians, Eskimos, and to a slightly lesser extent Ainu are also differentiated on the second axis. Tulare Lake has a first-axis score approximately the same as recent Native Americans, but is differentiated from them on the second axis. Tranquillity has a first-axis score corresponding to Africa and Southwest Pacific, but is likewise differentiated from them on the second axis. Karlo and La Jolla assume positions very close to Polynesians and Eskimos. Nearly all variables contribute in the same direction to the second axis, making it a reflection of general size. Variables defining facial forwardness and vault height contribute most prominently to this axis. Groups with low scores on this axis can therefore be characterized as generally large, but with more pronounced facial forwardness and high vaults. These features are seen in modern Polynesians, Ainu, and Eskimos and are a consistent description of the early Californians, with the exception of La Brea.

The La Brea skull is not distinctive on these first two axes, where it assumes a position in the vicinity of Native Americans. This skull is distinctive on the third axis (not shown), which to a considerable extent reflects the small size of the face. We deal with the La Brea skull in greater detail in the next section.

Morphometric Relationships of Early Californians, Americans, and Asians

In this analysis all samples presented in Table 1 except Tranquillity are used. Omitting Tranquillity allows use of a larger, but still limited variable set.

General Analysis

Figure 2 is the two-dimensional plot of these samples in relation to recent human populations. The axes are dominated by the fossil crania, which occupy the low and high positions on both axes. Their inclusion has the effect of compressing the modern populations to the center, which reflects their similarity to one another and the distinctiveness of some of the fossils. On the first axis La Brea occupies an extreme positive position and Upper Cave 103 an extreme negative. The extreme position of Upper Cave 103 distorts some of the interesting population arrangements on the first axis. Somewhat more moderate negative scores are associated with the recent populations of the Pacific rim: Polynesians, Eskimos, Ainu, and Southwest Pacific populations. Upper Cave 101, Buhl, and Spirit Cave share this general location to varying degrees. On the second axis La Brea is the positive extreme and Buhl and Upper Cave 103 are the extreme negative values.

The morphological pattern expressed by axis 1 relates to vault length, especially the parietal contribution to length, vault breadth, and face height. The crania of Upper Cave 103, and to a lesser extent Upper Cave 101 and Buhl, can be characterized as long and narrow with short faces. Buriats, Native Americans, and La Brea represent the opposite condition. Axis 2 reflects to some extent general size, but emphasizes upper facial forwardness, vault breadth, especially at the base, and nose height. Low-scoring populations on axis 2 can first be characterized as generally large, and more subtly as upper facially forward with vaults wide at the base compared with the maximum cranial breadth, and high noses relative to facial height.

La Brea is clearly a metric anomaly, bearing little resemblance to any modern human population or any other fossil. The other fossil crania can generally be described as large, but the La Brea cranium is small with a high face height relative to size. This

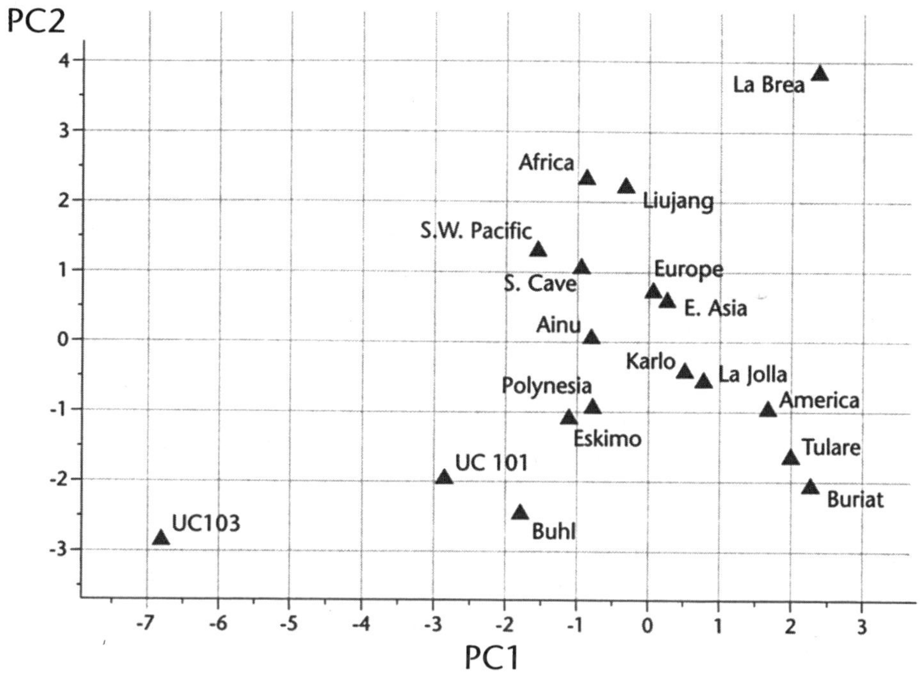

Figure 2. Two-dimensional principal coordinate plot showing relationships of recent human populations, early Californians (excluding Tranquillity), and individual fossil crania.

cranium would be an atypical member of any population examined here.

Relationships of Early Groups to Recent Populations

Figure 2 gives an overall summary of intergroup relationships, but fails to account for important details. Table 2 presents the full Mahalanobis squared distance of each early group from the modern regional populations, with distances arranged in descending order. Decreasing similarity to Native Americans is seen in three California samples, Tulare Lake, La Jolla, and Karlo, where Native Americans are in first, second, and third positions respectively. The California samples are also similar to East Asia, Ainu, and Polynesia, in marked contrast to the early fossil crania, where the Native American rank ranges from fifth (La Brea) to eighth (UC103 and Spirit Cave). The three groups most similar to the early fossils are overwhelmingly Pacific Rim populations, although Europe is most similar to Spirit Cave and Liujang.

Table 2. Mahalanobis distances (D^2) between early samples and recent regional populations.

Tulare	Amerindian 5.67	Polynesian 20.90	Ainu 22.02	E. Asia 23.32	Buriat 24.21	European 24.62	S.W. Pacific 34.05	African 34.19	Eskimo 34.23
Karlo	Ainu 26.41	E. Asia 27.43	Amerindian 27.59	European 28.75	Polynesian 29.45	Eskimo 30.22	S.W. Pacific 36.07	Buriat 39.43	African 41.84
La Jolla	E. Asia 15.94	Amerindian 18.42	European 19.38	Polynesian 21.33	Ainu 22.39	S.W. Pacific 25.84	Eskimo 28.06	Buriat 30.04	African 32.18
La Brea	E. Asia 46.10	European 47.03	African 52.68	Ainu 53.63	Amerindian 55.59	S.W. Pacific 57.81	Polynesian 66.85	Eskimo 69.84	Buriat 76.23
UC103	Eskimo 63.71	S.W. Pacific 65.77	Polynesian 65.82	Ainu 70.80	European 79.62	E. Asia 80.71	African 86.81	Amerindian 92.68	Buriat 108.78
UC101	Ainu 37.99	Polynesian 42.37	European 45.71	Buriat 50.00	S.W. Pacific 51.02	Amerindian 51.77	E. Asia 53.64	African 54.89	Eskimo 57.40
Buhl	Polynesian 33.44	S.W. Pacific 40.35	European 40.64	E. Asia 42.28	Ainu 43.68	Amerindian 45.19	Eskimo 51.15	African 51.85	Buriat 56.45
Sp. Cave	European 22.52	Ainu 25.89	E. Asia 26.79	Polynesian 27.82	Eskimo 30.66	S.W. Pacific 30.72	African 31.66	Amerindian 33.57	Buriat 54.17
Liujang	European 23.44	African 28.36	S.W. Pacific 33.27	E. Asia 33.59	Ainu 34.18	Amerindian 38.60	Polynesian 41.45	Buriat 54.21	Eskimo 55.73

Relationships among Early Fossils

The Mahalanobis distance (D) matrix giving the distance between crania in each pair is presented in Table 3. Significance levels designate whether each pairwise distance is greater than would be expected if the two crania were drawn from a single population. The single-population criterion used here is the pooled-within-sample covariance matrix from Howells's (1989) world samples with Native American additions as described in Jantz and Owsley (2001). These samples are local populations, so this test will be conservative in permitting the conclusion that two crania are from the same population.

Table 3. Distances (D) between individual fossil specimens.

	La Brea	Buhl	Sp. Cave	UC101	Liujang	UC103
La Brea	0.000					
Buhl	10.617▲	0.000				
Sp. Cave	6.768	8.318■	0.000			
UC 101	10.896▲	9.054▲	6.802	0.000		
Liujang	7.877■	9.074▲	5.498	7.368	0.000	
UC 103	12.150▲	8.594▲	9.357▲	8.965▲	10.254▲	0.000

■P < 0.05 ▲P < 0.01

The extreme position of Upper Cave 103 is evident in its high distances from all other crania and low probability that it could have been drawn from the same population as any of the others. Buhl's relationship to other crania is almost as distinctive. Spirit Cave, by contrast, is the most centrally located skull, differing only from Upper Cave 103 and Buhl. Spirit Cave, Upper Cave 101, and Liujang are all similar enough to each other to have been drawn from the same population.

Size

The foregoing analyses have shown that early Californians exhibit a great deal of variation over the 5000–6000 years represented. A common feature is the large size of the cranial vault. It is useful to examine size variation explicitly in relation to world variation. We limit size to cranial vault size, quantified as the geometric mean of cranial length, width, and height. Cranial length and breadth have the usual definitions, but vault height is defined as vertex radius. This procedure allows use of crania with the bregma measurement missing, as is often the case in the California series. Size data were centered on world sex-specific averages and expressed as standard scores.

Table 4 shows the distribution of cranial vault size in worldwide samples, arranged from largest to smallest. These are generally presented in broad regional groupings, although Buriats, Eskimos, and Ainu were retained as separate samples either because they represent the only sample from a general region (Siberia for the Buriats and the Arctic for the Eskimos) or because the group is generally considered different from the majority population of its region (Ainu in Asia). North American Indian populations were grouped into smaller, culture-area regions in order to determine whether significant variation exists among regions.

Table 4. Cranial vault size (in standard units) of recent human populations.

group	n	X	S.D.
Buriat	109	1.037	0.860
Ainu	86	0.844	0.741
Polynesia	371	0.816	0.810
Europe	317	0.261	0.749
Eskimo	108	0.225	0.741
E. Asia	400	-0.111	0.807
Historic Plains	213	-0.196	0.754
California (Santa Cruz)	102	-0.447	0.737
S.W. Pacific	298	-0.520	0.856
Numic	22	-0.604	0.521
Africa	373	-0.752	0.985
Southwest U.S. (Pueblo)	43	-0.800	0.706

The largest vaults belong to Buriats, followed closely by the Ainu and Polynesians. These three groups have vault sizes considerably larger than the other groups: more than 0.8 standard deviations above the world mean, with a marked difference relative to the next largest groups, Europeans and Eskimos. All other groups have vault sizes less than the world mean. Amerindians are uniformly below the world mean, ranging from only slightly below (historic Plains), to markedly small (Santa Cruz, Great Basin, and Southwest). Human populations are strongly differentiated on vault size; the minimum Fst among five regions (Europe, Africa, East Asia, Polynesia, and S. W. Pacific) is 0.179, which is greater than the Fst for the average cranial measurement (Relethford 1994).

Table 5 presents vault-size data for individual fossil crania, their mean, and the mean of the early Californians. The largest cranial vaults belong to Upper Cave, particularly Upper Cave 101. Except for La Brea, the early American crania are considerably larger than the world average. Taken as a group, vault sizes of the American and Asian fossil crania are well above the world average and are similar to the Ainu and Polynesians. The early Californians are also very large.

Discussion

An unanswered question at this point is the taxonomic relevance of cranial size. Size is often regarded as a nuisance variable to be removed. Various removal techniques are available (Jungers et al. 1995), but in cranial morphometrics Darroch and Mosimann shape variables (Darroch and Mosimann 1985) or C-scores (Howells 1989) are commonly used. Many of the recent morphometric analyses of early American crania have attempted to eliminate size (e.g., Neves and Blum 2000; Powell and Neves 1999; Powell and Rose 1999) because size is considered less relevant than shape. However, size is patterned in time and space and may reflect as much of a genetic component of variation as shape does.

Table 5. Cranial vault size (in standard units) of early American and Asian crania.

Group/site	n	X	S.D.
Upper Cave 101	1	2.003	—
Upper Cave 103	1	1.013	—
Spirit Cave	1	0.980	—
Buhl	1	0.941	—
Liujang	1	0.863	—
La Brea	1	-0.183	—
Fossil mean	6	0.936	0.693
Early Californians	26	1.000	0.980

Howells's (1989) analysis of his world cranial database using PENSIZE clearly demonstrates the patterning in size. Although PENSIZE, as used by Howells, reflects the contributions all variables make to overall size, as opposed to vault size as used here, his size patterning is much the same as we observe. The distribution of large cranial vaults in recent populations includes Polynesians (including the Micronesian Guam sample in Howells's database),

Ainu, and Buriats. Except for Buriats, this is a Pacific distribution. We now see that this characteristic includes early-Holocene Americans from the western United States and early Californians as late as the Middle Horizon, perhaps 1500 B.C. No recent Native American sample can be characterized as large-vaulted; Table 4 shows all are below the world mean, some well below.

The meaning of vault-size variation is poorly understood and has not been subjected to systematic investigation. In most parts of the world, cranial capacity, which will be closely related to size as defined here, has become smaller since the end of the Pleistocene (Henneberg 1988; Henneberg and Steyn 1993). Henneberg (1988) ascribes the change in cranial capacity to generalized structural reduction or gracilization, perhaps driven by hormonal changes (Henneberg and Steyn 1995). Lahr and Wright (1996) have argued that size variation may cloud phylogenetic relationships because of its relationship to robusticity. Vault size as defined here, however, seems to have only a weak relationship to robusticity. The robust Southwest Pacific populations and the more gracile subsaharan Africans are small-vaulted, while the large-vaulted Polynesians are robust, but less so than Southwest Pacific populations. Beals et al. (1984) found a strong correlation between climate and cranial capacity, putting it into the framework of adaptation. In the present data, the climate model fits only loosely. Buriats and Eskimos both qualify as cold-climate populations, but differ considerably in vault size. Although Houghton (1990) has argued that Polynesian physique in general is cold-adapted, it is hard to imagine cold stress sufficient to explain their markedly large vaults.

The large vault size seen in early-Holocene Americans and early Californians must unfortunately be left unexplained, but may in part be traced to the original founding populations. What is required at this point is a model that would explain the distribution of vault size in recent human populations that also accounts for the change in vault size in western North Americans. Models from which we might choose would presumably include the possibility that vault-size variation reflects an important component of genetic variation.

The primary question addressed in this paper is whether morphometric relationships support hypotheses that early populations came to America via a coastal route. If early populations came via this route, they presumably came out of coastally adapted populations along the western Pacific rim. There is not much evidence to illustrate what these early populations might have looked like, although those who have examined early Asian crania describe them as unspecialized with sundadont dental features (Lahr 1995). They are typically large and low-faced and lack the facial flatness associated with recent East Asians. Using modern populations as a comparative framework, the early Asians in our analysis (UC101, UC103, Liujang) generally align with Pacific populations, although Liujang is most similar to Europe and Africa. A similar result for this skull was obtained by Howells (1995). These early Asian crania would generally be atypical members of modern populations.

Another early-Holocene skeleton from Malaysia has been judged similar to Australian Aboriginals, and to Mesolithic populations from Malaysia and Flores (Matsumura and Zuraina 1999). Although we have not used the Minatogawa crania from Japan, they too have been characterized as possessing low faces and orbits and are metrically similar to Liujiang (Suzuki 1982). Evidence from cranial morphology supports the idea that this generalized S. E. Asian population represents the morphological background of Polynesians and the Jomon-Ainu lineage in Japan (Hanihara 1992). A Southeast Asian or South Chinese origin for Polynesians is seen in both mtDNA (Melton et al. 1995) and Y chromosome data (Bing et al. 2000). A convincing argument can therefore be made that the early western Pacific Rim populations were generalized in cranial morphology, features still seen in such modern groups as Polynesians and Ainu, and also characterizing early-Holocene American crania from western North America.

The California samples, Tranquillity, La Jolla, Karlo, and Tulare Lake, are highly variable, but to some varying degree also show evidence of having been derived from these generalized Asian circumpacific populations. Tranquillity, which we are considering the earliest pending a reliable AMS date, is the most distinctive and bears the least resemblance to recent Native American populations. Our results provide quantitative support to earlier investigators who observed the absence of Amerindian morphology in both Tranquillity (Angel 1966) and Karlo (Krantz 1960) and who mention "Protomongoloid" or "Australoid" morphology. The metric expression of these observations in the present analysis does not align any of them with modern aboriginal Australians, although Tranquillity comes closest. Rather, early California samples are removed from modern samples, including Amerindians, by virtue of longer, narrower cranial vaults, greater upper facial forwardness, and lower facial height. The California sample with the most recent date, Tulare Lake, exhibits the greatest morphometric similarity to recent Amerindians, but, like the other California samples, has greater vault size than recent Native Americans.

The similarity of early crania from California and the western U.S. to early crania from circumpacific East Asia and to contemporary Pacific populations is most parsimoniously explained by a circumpacific movement of people responsible for at least a part of the early population of America. A finding of more intense expression of the generalized circumpacific Asian morphology in California than further inland would provide further support for a coastal route. The present evidence is probably too limited to allow that conclusion, but future research should be directed at that question.

Acknowledgments

The California collections reported on in this study were examined at the University of California at Los Angeles (Tulare Lake), the San Diego Museum of Man (La Jolla), and the University of California at Berkeley (Karlo); the La Brea skeleton was made available for study by the Page Museum in Los Angeles. We especially thank Wendy Teeter at UCLA; Rose Tyson, Curator at the San Diego Museum of Man; Leslie M. Freund at UC–Berkeley; and Christopher Shaw, the Collections Manager at the Page Museum. Dr. David Hunt, Collections Manager at the National Museum of Natural History, measured several of the specimens used in this study. Sandy Schlactmeyer provided editorial guidance, and Rebecca Kardash assisted with the preparation of this manuscript.

References Cited

Angel, J. L.
1966 Early Skeletons from Tranquility, California. Smithsonian Contributions to Anthropology, Vol. 2, No. 1. Washington, D.C.

Beals, K. L., C. L. Smith, and S. M. Dodd
1984 Brain Size, Cranial Morphology, Climate and Time Machines. *Current Anthropology* 25:301–30.

Berger, R.
1971 New Radiocarbon Dates Based on Bone Collagen of California Paleoindians. University of California, Berkeley, Archaeological Research Facility, Department of Archaeology, Contributions No. 12, pp. 43–49.
1975 Advances and Results in Radiocarbon Dating: Early Man in America. *World Archeology* 7:174–84.

Bing, S., P. Underhill, J. Martinson, S. Nilmani, S. T. McGarvey, M. D. Sriver, J. Chu, P. Oefner, R. Chakraborty, and R. Deka
2000 Polynesian Origins: Insights from the Y Chromosome. *Proceedings of the National Academy of Science* 97:8225–28.

Chen, T., R. E. M. Hedges, and S. Yuan
1989 Accelerator Radiocarbon Dating for Upper Cave at Zhoukoudian. *Acta Anthropologica Sinica* 8:216–21.

Coon, C. S.
1963 *The Origin of Races*. Alfred A. Knopf, New York.

Cornell, D. L.
1998 The Morphometric Relationship of Upper Cave 101 and 103 to Modern Homo Sapiens. Unpublished M.A. thesis, University of Tennessee, Knoxville.

Cunningham, D.L., and D. J. Wescott
2000 Within-group Human Variation in the Asian Pleistocene: An Assessment of the Three Upper Cave Crania. *American Journal of Physical Anthropology* Supplement 30:132.

Darroch, J. N., and J. E. Mosimann
1985 Canonical and Principal Components of Shape. *Biometrika* 72:241–52.

Defrise-Gussenhoven, E.
1967 Generalized Distance in Genetic Studies. *Acta Genetica et Medicae Gemellologiae* 17:275–88.

Dixon, E. J.
1999 *Bones, Boats and Bison: Archeology and the Colonization of Western North America*. The University of New Mexico Press, Albuquerque.

Fenton, T. W., and A. R. Nelson
1998 Biological Affinities of the Buhl Woman: One of the Oldest Paleoindian Skeletons. *American Journal of Physical Anthropology* Supplement 26:81–82.

Gower, J. C.
1972 Measures of Taxonomic Distance and Their Analysis. In *The Assessment of Population Affinities in Man*, edited by J. S. Weiner and J. Huizinga, pp. 1–24. Clarenden Press, Oxford.

Green, T. J., B. Cochran, T. W. Fenton, J. C. Woods, G. L. Titmus, L. Tieszen, M. A. Davis, and S. J. Miller
1998 The Buhl Burial: A Paleoindian Woman from Southern Idaho. *American Antiquity* 63:437–56.

Gruhn, R.
1994 The Pacific Coast Route of Initial Entry: An Overview. In *Method and Theory for Investigating the Peopling of the Americas*, edited by R. Bonnichsen and D. G. Steele, pp. 249–56. Center for the Study of the First Americans, Oregon State University. Corvallis.

Hanihara, T.
1992 Dental and Cranial Affinities among Populations of East Asia and the Pacific: The Basic Populations in East Asia, IV. *American Journal of Physical Anthropology* 88:163–82.

Heizer, R. F., and S. F. Cook
1952 Fluorine and Other Chemical Tests of Some North American Human and Fossil Bones. *American Journal of Physical Anthropology* 10:289.

Henneberg, M.
1988 Decrease of Human Skull Size in the Holocene. *Human Biology* 60:395–405.

Henneberg, M., and M. Steyn
1995 Diachronic Variation of Cranial Size and Shape in the Holocene: A Manifestation of Hormonal Evolution? *Rivista di Antropologia* 73:159–64.

Houghton, P.
1990 The Adaptive Significance of Polynesian Body Form. *Annals of Human Biology* 17:19–32.

Howells, W. W.
1983 Origins of the Chinese People: Interpretations of the Recent Evidence. In *The Origins of Chinese Civilization*, edited by D. N. Keightley, pp. 297–317. University of California Press, Berkeley.
1989 *Skull Shapes and the Map*. Papers of the Peabody Museum of Archaeology and Ethnology, Harvard University, vol. 79. Cambridge, MA.
1995 *Who's Who in Skulls*. Papers of the Peabody Museum of Archaeology and Ethnology, Harvard University, vol. 82. Cambridge, MA.

Jantz, R. L., and D. W. Owsley
1997 Pathology, Taphonomy and Cranial Morphometrics of the Spirit Cave Mummy. *Nevada Historical Society Quarterly* 40:62–84.
1999 Circumpacific Populations and the Peopling of the New World. Paper presented at Clovis and Beyond Conference, Santa Fe, NM, Oct. 30.
2001 Variation among Early North American Crania. *American Journal of Physical Anthropology* 114:146–55.

Jungers, W. L., A. B. Falsetti, and C. E. Wall
1995 Shape, Relative Size, and Size Adjustments in Morphometrics. *Yearbook of Physical Anthropology* 38:137–61.

Kamminga, J., and R. V. S. Wright
1988 The Upper Cave at Zhoukoudian and the Origin of the Mongoloids. *Journal of Human Evolution* 17:739–67.

Krantz, G.
1960 Description of Human Remains from the Karlo Site, Appendix A of *The Archaeology of the Karlo Site (LAS-7), California*, edited by F. Riddell, pp. 93–99. The University of California Archaeological Survey, Department of Anthropology, University of California, Berkeley, CA.

Kroeber, A. L.
1962 The Rancho La Brea Skull. *American Antiquity* 27:416–17.

Lahr, M. M.
1995 Patterns of Modern Human Diversification: Implications for Amerindian Origins. *Yearbook of Physical Anthropology* 38:163–198.

Lahr, M. M., and R. V. S. Wright
1996 The Question of Robusticity and the Relationship between Cranial Size and Shape in *Homo Sapiens*. *Journal of Human Evolution* 31:157–91.

Lovvorn, M. B., G. W. Gill, G. F. Carlson, J. R. Bozell, and T. L. Steinacher
1999 Microevolution and the Skeletal Traits of a Middle Archaic

Burial: Metric and Multivariate Comparisons to Paleoindians and Modern Amerindians. *American Antiquity* 64:527–45.

Matsumura, H., and M. Zuraina
1999 Metric Analysis of an Early Holocene Human Skeleton from Gua Gunung Runtuh, Malaysia. *American Journal of Physical Anthropology* 109:327–40.

Melton, T., R. Peterson, A. J. Redd, N. Saha, A. S. Sofro, J. Martinson, and M. Stoneking
1995 Polynesian Genetic Affinities with Southeast Asian Populations as Identified by mtDNA Analysis. *American Journal of Human Genetics* 57:403–14.

Neves, W. A., and M. Blum
2000 The Buhl Burial: A Comment on Green et al. *American Antiquity* 65:191–93.

Ousley, S. D., and R. L. Jantz
1996 FORDISC 2.0: Personal Computer Forensic Discriminant Functions. University of Tennessee, Knoxville.

Owsley, D. W., and R. L. Jantz
1999 Databases for Paleoamerican Skeletal Biology Research. In *Who Were the First Americans?*, edited by R. Bonnichsen, pp. 79–96. Center for the Study of the First Americans, Oregon State University, Corvallis.

Pope, G. G.
1992 Craniofacial Evidence for the Origin of Modern Humans in China. *Yearbook of Physical Anthropology* 35:243–98.

Powell, J. F., and W. A. Neves
1999 Craniofacial Morphology of the First Americans: Pattern and Process in the Peopling of the New World. *Yearbook of Physical Anthropology* 42:153–88.

Powell, J. F., and J. C. Rose
1999 Chapter 2. Report on the Osteological Assessment of the "Kennewick Man" Skeleton (CENWW97.Kennewick), In *Report on the Non-destructive Examination, Description, and Analysis of the Human Remains from Columbia Park, Kennewick, Washington* [Oct. 1999]. National Park Service. htpp://www.cr.nps.gov/aad/kennewick/

Relethford, J. H.
1994 Craniometric Variation among Modern Human Populations. *American Journal of Physical Anthropology* 95:53–62.

Riddell, F. A. (editor)
1960 *The Archaeology of the Karlo Site (LAS-7), California.* Reports of California Archaeological Survey, No. 53, Department of Anthropology, University of California, Berkeley, CA.

Shumway, G., C. L. Hubbs, and J. R. Moriarty
1961 Scripps Estates Site, San Diego, California: A La Jolla Site Dated 5460 to 7370 years before the Present. *Annals of the New York Academy of Sciences* 93:37–132.

Steele, D. G., and J. F. Powell
1992 Peopling of the Americas: Paleobiological Evidence. *Human Biology* 64:303–36.

Suzuki, H.
1982 Chapter 2. Skulls of the Minatagowa Man. In *The Minatagowa Man: The Upper Pleistocene Man from the Island of Okinawa*, edited by H. Suzuki and K. Hanihara, pp. 7–49. The University Museum, The University of Tokyo, Bulletin No. 19. Tokyo, Japan.

Tuohy, D. R., and A. J. Dansie
1997 New Information regarding Early Holocene Manifestations in the Western Great Basin. *Nevada Historical Society Quarterly* 40:24–53.

Van Vark, G. N.
1995 The Study of Hominid Skeletal Remains by Means of Statistical Methods. In *Biological Anthropology: The State of the Science*, edited by N. T. Boaz and Linda D. Wolfe, pp. 71–90. International Institute for Human Evolutionary Research, Bend, OR.

Van Vark, G. N., and J. Dijkema
1988 Some Notes on the Origin of the Chinese People. *Homo* 39:143–48.

Warren, C. N., and M. B. McKusick
1959 A Burial Complex from the Southern San Joaquin Valley. In *Annual Report, Archaeological Survey, Department of Anthropology-Sociology.* University of California, Los Angeles.

Weidenreich, F.
1938/1939 On the Earliest Representatives of Modern Mankind Recovered on the Soil of East Asia. *Peking Natural History Bulletin* 13:161–74.

Yuan, S., T. Chen, and S. Gao
1986 Uranium Series Chronological Sequence of Some Paleolithic Sites in South China. *Acta Anthropologica Sinica* 5:179–90.

Patterns of Craniometric Variation and Geographical Distribution in North America: A Historical Comparison

A. Russell Nelson

Abstract

Georg Neumann's *Archaeology and Race in the American Indian* (Neumann, 1952) presented a brief overview of prehistoric population distributions in North America in the context of the broad archaeological overview in *Archaeology of Eastern United States* (Griffin 1952). In the nearly 50 years since those publications, the study of human variation has shifted its emphasis from typology to population distribution. A systematic craniometric examination of broad-scale New World population relations was attempted by Nelson (1998). This paper, derived from a portion of that work, discusses the perspectives, samples, and methodologies employed, and compares the results obtained with the conclusions reached by Neumann.

Comparing the results derived by two very different methods—subjective visual assessment versus computerized statistical analysis of multivariate data—more than simply describes two ways of looking at population distribution in North American prehistory, for this work affirms that although typological essentialism as an academic discipline has rightly been relegated to the dustbin of history, the morphological variation on which it was based is nevertheless real. Human anatomical variation is quantifiable for the purposes of study. Although we no longer indulge in microtaxonomic dalliances such as the pursuit of racial essences, nonetheless we can expect to find answers to many archaeological questions by increasing our understanding of broad prehistoric population relations.

Introduction

Comparative morphometrics has seen a resurgence as an analytic tool in exploring biological relations among groups of people through time and over geographical space (Brace 1989, 1990, 1992, 1993; Droessler 1981, Ishida 1993, 1997; Jantz 1973, 1998, 1998b; Kamminga 1988; Key 1983; Li Y. 1991; Pietrusewski 1984, 1990, 1992) since Howells's (Howells 1973, 1989, 1995) explorations of broad-scale population relations beginning in the 1970s. Many studies have focused on the relationships themselves; others have worked to refine methodologies for making morphometric comparisons (Albrecht 1992; Jungers 1995; Relethford 1994a, 1994b). Increasingly powerful computer technology has made this quantitative approach to the data a practical reality. However, many of our anthropological forebears collected vast quantities of metric data (for example, Hrdlička, 1927b), some of which are only recently being rediscovered and analyzed (Ousley 1995, Szathmary 1995). One of the last physical anthropologists prior to the widespread use of computer technology who attempted to deal with broad patterns of variation among Amerind groups from a comparative morphological perspective was Georg Neumann (1908–1971) (Hall 1997). His chapter, "Archaeology and Race in the American Indian," in Griffin's *Archaeology of Eastern United States* (1952) is the outline for a large-scale project he unfortunately never completed.

A generation and a half later, many aspects of these issues are still relevant and are getting the attention of scientists who are using comparative morphology to systematically explore interpopulational biological relationships on a continental or multicontinental scale. Other lines of evidence also contribute to this process. Ancient-DNA studies are giving us clues into prehistoric population distributions, and mitochondrial—and Y chromosome—DNA studies have presented an outline and many details of contemporary populations; extrapolating to prehistory is an ongoing process. The study of linguistic family distributions has been used as an analytical tool to draw conclusions for comparison with the picture presented by archaeological site distributions (Greenberg 1986), although other, very different perspectives have also been developed (Nichols 1990, 1992, 1995). In any event,

Museum of Anthropology, University of Michigan, Ann Arbor, MI 48109-1079. Department of Anthropology, University of Wyoming, Laramie, WY 82071; e-mail: arnelson@umich.edu single lines of inquiry need to be considered in the broader perspective of current understanding.

This paper presents the results of cluster and discriminant function analysis of craniometric data collected from archaeological assemblages in museums and anthropology departments across the U.S. and parts of Canada. This material is a sampling of Amerind inhabitants, particularly of North America, of the prehistoric to early-historic period; it summarizes analyses described by Nelson (Nelson 1998). The data for this study come from the University of Michigan Museum of Anthropology (UMMA) Craniofacial Database (Brace 1989, 1990, 1993; Nelson 1998). This analysis includes a comparative assessment of Neumann's Amerind varieties on the basis of the seven variables he uses (Neumann 1952). Few if any contemporary physical anthropologists ascribe to the typological perspectives Neumann espoused for most of his professional life. However, his acuity as an observer of human skeletal variation produced a legacy of information, which stands as a remarkable contribution to the contemporary study of population distribution and prehistoric population dispersals.

Before making any comparisons it is important to consider expectations and assumptions. As a morphologist, Neumann needed to organize his observations into some sort of meaningful order. The computer cluster program using the variable set performs the same function mathematically. However, it is important to recognize that the samples examined are not the same except in a general sense. For example, the UMMA Plains samples are sets of measurements taken from the remains of historic and prehistoric Northern Plains people, both males and females; among these samples are measurements from historic Crow and Blackfoot, as well as prehistoric and protohistoric Arikara. Neumann's Plains ("Lakotid") samples, on the other hand, are measurements taken from 63 adult male individuals of "various bands of Dakota of the northern Plains" described by Hrdlička (Hrdlička 1927b). We assume that both these samples are representative of the larger population of people living on the Northern Plains in the late-prehistoric, protohistoric, and early-historic periods. Furthermore we assume both these samples of Plains people are representative, morphologically, of their parent or sister populations in the Old World areas from which they presumably came at some time in the distant past. Consequently, we expect that Neumann's groupings approximate some of the groupings produced by the UMMA data if variations observed by the eye correspond to quantified craniofacial dimensions.

In *Archaeology and Race in the American Indian,* Neumann describes eight Amerind "varieties." He emphasizes that he is not presenting *types* in the sense that particular individuals having "certain metrical or indicial combinations" are selected out of larger populations; rather, his samples are "relatively homogeneous units, from a number of archaeological horizons." However, by designating type specimens and using only males in his comparisons, he seems to define a selectively drawn sample. By defining "varieties" in this manner, Neumann seems to effectively establish types among his samples with specific spatial and temporal characteristics.

Of Neumann's eight varieties, two (Otamid and Iswanid) represent Archaic groups; two (Walcolid and Lenapid) represent Central Valley and northeastern Mississippian and Woodland groups; one (Ashiwid) represents the Southwestern Pueblos; one (Lakotid) the Plains; one (Deneid) Athabascans; and one (Inuid) the Eskimo groups spread across the far northern Arctic portion of the New World. This picture gives a reasonable representation of the range of geography encompassed by North America, although some of the regions, notably the Pacific coast, Great Basin, and Southern Plains, present significant questions.

In order to compare the UMMA material with Neumann's material, it is necessary first to examine how this set is organized compared with Neumann's method, then to compare the results. Our tests were run using cluster analysis of multivariate craniometric data. Neumann organized his results by visually assessing comparative morphology; he made some use of craniometric data analysis, and fortunately published variable descriptions and means for his groupings. Finally, the cluster groups were compared with Neumann's varieties using discriminant function and general distance analysis.

Methods: Unit of Analysis

In order to compare details of craniofacial morphology across regions to infer population relationships, samples from given contexts are used as the basic units of analysis. Each individual within a sample is represented by a set of 24 or fewer numbers corresponding to the battery of craniofacial measurements assembled by C. L. Brace (Brace 1989, 1990, 1993) in the course of developing this database.

Samples represent a region (e.g., Massachusetts, approximately in the late Woodland), an archaeological provenience (e.g., Indian Knoll or Moundville), or a component of archaeological sites (e.g., the middle-Woodland component of the Klunk Mounds in Illinois, distinct from the late-Woodland component of that same site). I did not feel it was necessary to further restrict the representation to temporal spans within a given horizon; for example, I did not distinguish early, middle, or late periods at a site such as Indian Knoll or Eva where there were long periods of occupation. Samples such as the material representing "Massachusetts" (# 556) were tested with samples of known provenience before being used. This sample, from the collections of the Peabody Museum at Harvard University, comprises East Coast Woodland material donated by local farmers, constables, and physicians; the Peabody has long been a locally known repository for such materials. Discriminant function comparison with other Eastern Woodland material, such as Seneca samples of secure provenience from upstate New York (see below), suggests these individuals represent a relatively homogeneous population that inhabited this region in late-prehistoric times, a few hundred years prior to contact. In general, the sample represents the local prehistoric population whose remains are frequently encountered in the course of construction or other ground disturbance. When individuals are compared with other regional samples in discriminant function testing, Massachusetts and Maine tie most closely with samples from the Northeast Woodlands. Among these are samples of secure provenience from the Kleinburg, Uxbridge, and Ossossane Ossuaries in Ontario, and the Tram, Adams, Harscher, and Dutch Hollow sites in upstate New York (see "Materials" below). In other cases a single individual may represent a sample listing. This is the case for much of the Wyoming Plains and Nevada Great Basin samples.

Methods: Variables

Table 1, modified from Brace and Hunt (Brace 1990), lists the measurements used in this study. This project utilizes published methodologies (Brace 1990, 1992, 1993) and an existing variable list[1]. The sheer size of the database at the time the present author began working with it made it impractical to change the data-collection regimen to any major degree. Hence the measurements used here are the 24 assembled by Brace. Neumann used seven variables, which correspond to seven of the dimensions in the UMMA set. The comparisons presented here will be based on those variables held in common.

Table 1. Craniofacial measures and computed variables in the UMMA dataset (from Brace and Hunt 1990).

Variable no.	Description	Source■
1	nasal height	Martin # 55
2	nasal bone height	Martin # 56(2)
3	piriform aperture height	Martin # 55(1)
4	nasion prosthion length	Martin # 48
5	nasion basion	Martin # 5
6	basion prosthion	Martin # 40
7	superior nasal bone width	Martin # 57(2)
8	simotic width	Brace and Hunt, 1990
9	inferior nasal bone width	Martin # 57(3)
10	nasal breadth	Martin # 54
11	simotic subtense	Howells, 1973
12	inferior simotic subtense	Brace and Hunt, 1990
13	FOW subtense at nasion	Brace and Hunt, 1990
14	MOW subtense at rhinion	Woo and Morant, 1934
15	bizygomatic breadth	Martin # 54
16	glabella opisthocranion	Martin # 1
17	maximum cranial breadth	Martin # 8
18	basion bregma	Martin # 17
19	basion rhinion	Brace and Hunt, 1990
20	width at 13 (fmt fmt)	Brace and Hunt, 1990
21	width at 14	Woo and Morant, 1934
22	IOW subtense at nasion	Woo and Morant, 1934
23	width at 22 (fmo fmo)	Woo and Morant, 1934
24	minimum nasal tip elevation	Brace and Hunt, 1990

■Martin numbers from Martin, 1928

"Regions" and "Horizons"

In the course of examining broad-scale population relationships it is often necessary to summarize geographic regions or periods of time by condensing closely related clusters of samples into composite samples for comparison. These summaries over space and time are referred to as "regions" and "horizons."

Clusters are generated using samples of secure provenience as the baseline unit of comparison. These clusters and the matrices from which they are generated are then examined for patterns of morphological similarity indicated by distance figures from the matrix. In the next step, samples clustering together are grouped and examined for patterns of geographic distribution (regions). They may also be grouped according to temporal provenience (horizons).

Methods: Cluster Analysis

"The basic data for cluster analysis is a set of *N* entities (for example, people) on which we have recorded *p* measurements. This initial choice of the particular set of measurements constitutes a frame of reference within which to establish the clusters" (Everitt 1974:48). Cluster analysis is applied as a general way of exploring relations between samples. It is most useful for blocking out sections, or relationships, within a matrix of distance measures. Using cluster analysis is an interactive and exploratory process. Analysis of any body of data naturally requires running many similar data files and experimenting in order to become familiar with the data.

Samples whose mean variable profiles match most closely will produce smaller distance measures and link together into primary clusters, and secondary or tertiary "macro-clusters." A macro-cluster is here defined as a second- or third-order grouping of primary clusters. The operating expectation is that overall patterning among these macro-clusters should roughly describe collections of morphometrically similar (ergo related) groups of the analyzed samples.

Howells (1984:6) states that cluster dendrograms do nothing more than illustrate arrangements of data and do not represent phylogenies "however well the two seemingly or actually correspond." This is very true—a given dendrogram can only graphically illustrate the results of analysis of the specific data being analyzed. If a different set of crania is analyzed, the resulting clusters describe a different set of morphological relationships. Likewise, different combinations of variables may produce different analytic results.

Although the dendrogram is in fact nothing more than a graphical depiction of a particular arrangement of data, those data are details of human skeletal anatomy. The anatomical detail in the phenotype of each individual in the sample was produced through parental inheritance with some general environmental effect presumably induced by the conditions of the developmental trajectory of growth to maturity. The environmental conditions are to some extent regional, but more important are idiosyncratic polygenic configurations that cluster regionally and thus effectively indicate breeding distance (Wright 1943, 1946).

In this project, standardized data are analyzed using the SPSS cluster algorithm (Norušis 1994:267) and Wards method (Everitt 1974:15; Ward 1963). Individual data elements are coded into strata and standardized to C-scores, as discussed in Brace et al.

[1] The variable list was originally set up to test for differences between Japanese and Ainu samples (Brace et al., 1989, 1990). Brace and co-workers discovered it worked for a variety of other population samples and conditions, but after 24 variables they did not add more because of the back-tracking and extra labor involved with going back to the collections and re-measuring each specimen (Brace and Hunt, 1990).

(1989:98–99). This technique, called "double centering" (Jungers 1995), eliminates the factor of sheer size from the analysis.[2] Following data standardization, the routine sums the male and female mean figures by strata and divides by 2 to calculate the midsex (arithmetic) means for each sample. The cluster algorithm operates on these midsex mean C-score data to generate the distance matrix.

The Wards method is a hierarchical cluster technique. Hierarchical techniques either link or divide in the "most efficient step, in some defined sense, at each stage in the progressive subdivision or synthesis of the population" (Everitt 1974:8). The Wards method is agglomerative rather than divisive, and is a variance-minimizing method well suited for analyzing biological data (*ibid.*, p. 62).

Methods: Discriminant Function

Discriminant function analysis is a technique widely used to classify samples in operations such as determining skeletal sex and population affinity. In this study, discriminant function analysis compares a pseudo-individual (Neumann's mean figures) with groups produced by cluster analysis. Albrecht (1992) discusses two approaches to calculating population affinity, the "restricted" approach that involves comparisons within a closed universe, and the "unrestricted" approach that involves comparisons in an open-ended universe.

The Restricted Approach to Group Allocation

The restricted approach to population affinity uses discriminant function routines widely available in statistical packages. In this project, the SPSS version (Norušis, 1994:278) is used. Comparative samples are first assembled based on the results of cluster analysis. At the outset, the subject individual has an equal *a priori* likelihood of belonging to any of the comparative groups, whose aggregate represents the complete universe of available options for allocation. The relative probability of the individual's belonging to each of these groups is calculated mathematically, then proportioned such that the sum of all calculated relative probabilities equals 1. The resultant probabilities are the *a posteriori* probabilities of inclusion, otherwise known as posterior probabilities.

This method of comparative analysis has been criticized recently (Howells 1995) on the grounds that results based on posterior probabilities mean that anything put into the system will be put in a cluster somewhere, *even if it is not human*. Brace et al. (1993) anticipated this critique and note that it is up to the researcher to be certain that samples run for comparison are appropriate to questions being asked. This type of discriminant analysis, used by Brace et al. (1993), can be useful in limited applications. If the comparative samples, as well as the test case, are known to be related and the question is one of relative degrees of relatedness, the restricted approach can be legitimately employed because the assumptions inherent in the approach are met. Furthermore, if the region being tested is sufficiently well sampled that the aggregate of the comparison groups does indeed reasonably approximate the universe of possible choices, then the inherent assumptions are again met.

In light of these qualifications, the case can be made that the restricted approach is appropriate in the present study, where comparisons are made of Amerind material, because this situation is quite unlike that discussed in 1973 by Brace and coworkers (Brace 1973), and criticized by Albrecht (1992). In that paper the affinity of *Homo habilis*, a post-*Australopithecine* fossilized hominid, was in question. In that instance, as in many paleoanthropology studies, many thousands of years and many thousands of miles may separate specimens in a relatively sparse fossil record. On the other hand, in our study of the New World there is not a separation of a thousand years and 1500 miles between any specimen and its nearest neighbor, and data from over 2000 individuals span only 10,000 years. All these people are modern *Homo sapiens*, and all derive from groups that came from somewhere in Asia. While this database certainly doesn't include every single one of these people, it nevertheless samples a wide geographic and temporal spread of New World skeletal material and is broadly inclusive of the range of Amerind craniofacial variation. Justification for the use of posterior probabilities in discriminant function analyses increases with the thoroughness of sampling. In our comparisons with the work of Neumann, we use this technique because the comparisons are with recent (Archaic and later) population samples and reasonably complete coverage can be assumed.

Typicality: The Unrestricted Approach to Group Allocation

Albrecht (1992) contrasts his discussion of the restricted approach, used to infer taxonomic group membership for unknown fossils, with what he terms the unrestricted approach. He discusses typicality probability, the measure of how typical of the comparative sample a specimen may be. This, he points out, is the "more realistic situation in physical anthropology given the possibilities for temporal, spatial, and taxonomic variation among living primates, previously described fossil species, and yet to be discovered fossil taxa" (p. 58). The fundamental utility of the unrestricted approach is that it may suggest group affinities with one, two, several, *or none* of the comparative samples used. Since these probabilities do not have to sum to 1, their aggregate does not constitute an essentially closed (hence restricted) system. They are produced in the same manner as posterior probabilities except that the final step, proportioning and standardizing to 1, is eliminated. This places the unknown specimen within the context of other members of the comparison samples by assessing its

[2] Jungers et al. (1995) criticize this method because they contend that size plays an important role in shape. In the case of interspecific comparisons (they give the extreme example of comparing a gorilla and a mouse lemur), which are their primary concern, this is very true. Converting a mouse lemur into something the size of a gorilla involves logarithmic scaling. In the case of comparing human samples, where sexual dimorphism and intersample body size differences do not even approach the orders of magnitude addressed by their study of primates, logarithmic scaling does not improve the interpretation of results beyond that attainable by arithmetic scaling. Thus C-score standardization is quite acceptable for a study like this.

position in proportion to other members residing within the same distance to the center of the comparative group.

Because this approach admits the possibility of no clear-cut group membership, it is considered the more appropriate method for assessing fossil material from very distant temporal horizons, as in Albrecht's (1992) review of Brace et al. (1973). In the New World, where the analogous situation would be examining Paleo-indian and single Archaic specimens, assigning typicality probabilities is an appropriate analytic tool. The geographic and temporal spans encompassed by the entire period of New World habitation are an order of magnitude short of those under consideration in the study of human evolution in the Old World. Nonetheless, there exists the possibility of a broad range of human variation in the New World at the Paleoindian horizon (Jantz 1998b) for which analogous comparative samples may not be extant at this time. Because of this, and until such time as areas in Siberia and Northeast Asia can be more fully explored, an appropriate technique admits the possibility of non-inclusion.

This is accomplished by producing a Mahalonobis D^2 matrix of distances (in this case using the SAS PROC-DISCRIM procedure), which includes the subject individual in the series of samples (as in Albrecht, 1992:65). The distances are read from the question individual, then looked up on a chi-square table; the degrees of freedom correspond to the number of variables in the measurement battery that were analyzed when deriving the matrix.

Materials

Neumann (1952) published the sample means for seven variables. Fortunately, all these variables are present in the UMMA series, and they form the basis of the comparison with his material reported here. The variable list used for this comparison is restricted to the following seven dimensions:

1. nasal height (Martin #55)
2. nasion-prosthion length (Martin #48)
3. nasal breadth (Martin #54)
4. bizygomatic width (Martin #45)
5. glabella-opisthocranion length (Martin #1)
6. maximum cranial breadth (Martin #8)
7. basion-bregma (Martin #17)

This set of dimensions records the length, breadth, and height of the skull, the height of the face from prosthion to nasion, the height of the nose from the floor of the nasal sill to nasion at the frontonasal suture, and the maximum width of the nasal aperture. It is a small set of generalized variables.

The collections from which the comparative samples are drawn (the UMMA dataset) represent archaeological assemblages from sites across the New World, with a time depth of approximately 7,000–8,000 years. Table 2 presents a list of samples representing New World populations used in the cluster analysis.

Results of the Cluster Analysis

The dendrogram (Figure 1) illustrates the results of the comparison of New World material from the UMMA dataset with

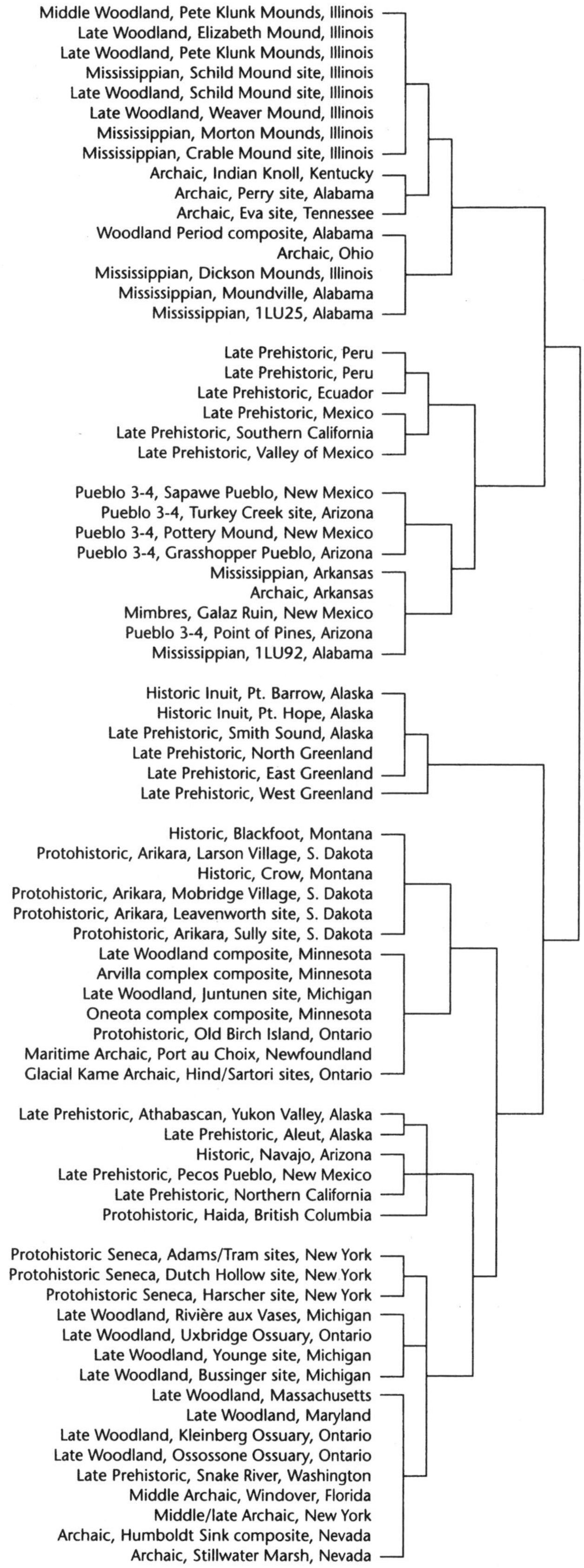

Figure 1. Dendrogram presenting results of the cluster analysis of New World samples.

Table 2. Samples used to generate Figure 1.

UMMA #	Site and region	Collection location	Horizon	n
204	Bussinger site, Michigan	UMMA	Late Woodland	9
756	Port au Choix	St John's, Newfoundland	Maritime Archaic	27
215	Old Birch Island	UMMA	Historic	4
221	Juntunen	UMMA	Late Woodland	10
230	Younge	UMMA	Late Woodland	9
281	Rivière aux Vases	UMMA/SUNY Alb	Late Woodland	25
116	Kleinberg	Erindale, Ontario	Late Woodland	25
742	Uxbridge	Guelph, Ontario	Late Woodland	22
118	Ossossone	Erindale, Ontario	Late Woodland	12
119	Ball site, Ontario	Erindale, Ontario	Late Woodland	1
712	Morrison Mound, Minnesota	Hamline U, St Paul, Minnesota	Late Woodland	4
720	Round Mound, Minnesota	Hamline U, St Paul, Minnesota	Late Woodland	9
558	Adams site, New York	Rochester Museum, New York	Protohistoric Seneca	21
559	Tram site, New York	Rochester Museum, New York	Protohistoric Seneca	12
560	Harscher site, New York	Rochester Museum, New York	Protohistoric Seneca	19
556	Massachusetts composite	Peabody Museum, Harvard U, Mass.	Massachusetts Late Woodland	48
725	Red Lake, Minnesota	U Minnesota, Minneapolis	Arvilla Complex	4
726	Warner, Minnesota	U Minnesota, Minneapolis	Arvilla Complex	6
727	Peter Lee site, Minnesota	U Minnesota, Minneapolis	Arvilla Complex	2
728	Snake River Mounds, Minnesota	U Minnesota, Minneapolis	Arvilla Complex	2
729	Harstadt Mound, Minnesota	U Minnesota, Minneapolis	Arvilla Complex	1
717	Rushford site	Hamline U, St. Paul, Minnesota	Oneota	2
718	Vosberg site	Hamline U, St. Paul, Minnesota	Oneota	7
719	Hogback site	Hamline U, St. Paul, Minnesota	Oneota	6
129	Sully site, South Dakota	U Tennessee, Knoxville	Protohistoric Arikara	16
561	Dutch Hollow, New York	Rochester Museum, New York	Protohistoric Seneca	9
655	Pete Klunk Mound, Illinois	Indiana U, Bloomington	Middle Woodland	34
138	Elizabeth site, Illinois	Illinois State Museum	Late Woodland	19
656	Pete Klunk Mound, Illinois	Indiana U, Bloomington	Late Woodland	15
658	Schild Mound Group, Illinois	Indiana U, Bloomington	Late Woodland	12
650	Weaver Mound, Illinois	Indiana U, Bloomington	Late Woodland	9
761	Eva site, Tennessee	U Tennessee, Knoxville	Middle Archaic	42
382	Valley of Mexico	Smithsonian Institution, Wash., D.C.	Late Prehistoric/ Historic	14
122,123	Indian Knoll, Kentucky	U Kentucky, Lexington	Riverine Archaic	45
250	Perry site, Alabama	U Alabama, Tuscaloosa	Riverine Archaic	10
739	Hind, Ontario	U of Guelph, Ontario	Glacial Kame Archaic	7
740	Hind, Ontario	U of Guelph, Ontario	Glacial Kame Archaic	1
741	Sartori, Ontario	U of Guelph, Ontario	Glacial Kame Archaic	2
245	Pickwick Basin, Alabama	U Alabama, Tuscaloosa	Woodland	7
246	Wheeler Basin, Alabama	U Alabama, Tuscaloosa	Woodland	8
134	Dickson Mounds, Illinois	Dickson Museum	Mississippian	23
659	Schild Md Group, Illinois	Indiana U, Bloomington	Mississippian	20
645	Morton Mounds, Illinois	Indiana U, Bloomington	Mississippian	2
648	Morton Mounds, Illinois	Indiana U, Bloomington	Mississippian	17
137	Crable site, Illinois	Dickson/Indiana U, Bloomington	Mississippian	17

Neumann's data. Sample clusters in the upper half of the arrangement are rooted in Central and South America and the Southern (and Southwestern) portion of the present-day U.S. Those in the lower half of this first division of the samples generally represent northern North America. At the initial level there are as many as 19 clusters consisting of one to eight samples. Many of these can be collapsed into macro-samples representative of the region and general horizon indicated in this initial assessment. At this point some care must be taken to assemble comparative samples into relatively uniform sample sizes. Macro-samples are assembled on the basis of sample size as well as geographic representation. Groupings with lower inter-sample distances are preferred in order to keep composite group variance as low as possible. Although in some cases it may be important to use samples relatively tightly circumscribed in time and space, in other cases it may be more important to concentrate on the geographical distribution of the samples. Even in the New World, with abundant sources of skeletal material, sometimes perfectly circumscribed comparative samples simply cannot be drawn, and to do the work we need to consider what is available.

Since this project is principally concerned with North America, we eliminated the Mesoamerican and South American groupings. The cluster algorithm suggests six broad group-

Table 2 (continued).

UMMA #	Site and region	Collection location	Horizon	n
657	Arkansas composite	Indiana U, Bloomington	Mississippian	22
247	Moundville, Alabama	U Alabama, Tuscaloosa	Mississippian	9
249	Pickwick Basin LU 25, Alabama	U Alabama, Tuscaloosa	Mississippian	6
251	Pickwick Basin LU 92, Alabama	U Alabama, Tuscaloosa	Mississippian	5
410	Blackfoot, Montana, and Alberta, Canada	Field Museum	Historic	32
381	Crow, Montana	American Museum of Natural History	Historic	15
413	Crow, Montana	American Museum of Natural History	Historic	2
685	Humboldt Sink, Nevada	Nevada State Museum, Carson City	Great Basin Archaic	15
125	Larson site, South Dakota	U Tennessee, Knoxville	Arikara: Plains Protohistoric	30
126	Mobridge site, South Dakota	U Tennessee, Knoxville	Arikara: Plains Protohistoric	30
128	Leavenworth site, South Dakota	U Tennessee, Knoxville	Arikara: Plains Protohistoric	15
253	Peru, South America	Peabody Museum, Harvard U, Mass.	Late Prehistoric	18
508	Peru, South America	Smithsonian Institution, Wash., D.C.	Late Prehistoric	45
689	Stillwater Marsh, Nevada	Nevada State Museum, Carson City	Great Basin Archaic	7
379	Athabascan, Yukon Valley, Alaska	Field Museum	Historic	6
391	Athabascan, Yukon Valley, Alaska	Smithsonian Institution, Wash., D.C.	Historic	42
378	Aleut, Alaska	Field Museum	Historic	15
386	Aleut, Alaska	Field Museum	Historic	1
507	Mexico	Field Museum	Historic	17
505	Ecuador	Smithsonian Institution, Wash., D.C.	Historic	11
425	Ecuador	Smithsonian Institution, Wash., D.C.	Historic	1
372	Inuit: Point Barrow, Alaska	Field Museum	Historic	41
373	Inuit: Point Hope, Alaska	Field Museum	Historic	41
375	Inuit: Smith Sound, Alaska	Field Museum	Historic	9
504	Navajo, from Arizona	Smithsonian Institution, Wash., D.C.	Historic	16
380	Canyon du Chelly, Arizona	Field Museum	Historic	8
730	Galaz Ruin, New Mexico	U Minnesota, Minneapolis	Mimbres	24
666	Grasshopper Pueblo, Arizona	Arizona State Museum	Puebloan	13
665	Point of Pines Pueblo, Arizona	Arizona State Museum	Puebloan	9
667	Turkey Creek Pueblo, Arizona	Arizona State Museum	Puebloan	5
260	Southern California	Fowler Museum, UCLA	Prehistoric	26
487	Malibu, Southern California	Lowie, U California, Berkeley	Prehistoric	10
252	Northern California	Lowie, U California, Berkeley	Prehistoric	24
261	Central California: Kern Co.	Lowie, U California, Berkeley	Prehistoric	10
392	Maryland composite	Smithsonian Institution, Wash., D.C.	Woodland	27
817	Snake River, Washington	Central Washington State U	Late Prehistoric	14
148	Windover site, Florida	Florida State U, Tallahassee	Early Archaic	38
570	Pecos Pueblo, New Mexico	Peabody Museum, Harvard U, Mass.	Late Prehistoric	13
376	Inuit: North Greenland	Field Museum	Prehistoric	12
285	Inuit: Greenland	U of Zurich, Switzerland	Prehistoric	10
377	Inuit: West Greenland	Field Museum	Prehistoric	4
774	Stratton Wallace site, Ohio	Ohio State U, Columbus	Glacial Kame Archaic	7
764	Kay's Landing site, Tennessee	U Tennessee, Knoxville	Middle Archaic	8
409	Haida: Washington	Field Museum	Historic	51
771	Kirian Treglia site, Ohio	Ohio State U, Columbus	Glacial Kame Archaic	11

ings. Two are in the U.S. South, where sites along the broad drainages of the Mississippi and Missouri rivers south to Alabama tie closely together. The close links across this region suggest a similarity presumably arising out of relatively stable population continuity going back at least to the middle Archaic. Neumann's Iswanid type was based on material from this area. It is interesting to note the position of the Mississippian culture horizon (corresponding presumably to Neumann's Walcolid variety) and links to other sites within this array of data. Those connections suggest ties across the Texas Plateau to the other major division of the Southern macro-cluster, the Southwest. These samples in the Arizona and New Mexican Pueblan areas should correspond to Neumann's Ashiwid type; these in turn indicate ties back to the Valley of Mexico and south to samples from Peru and Ecuador.

The Northern macro-cluster is slightly more complex. At least four broad divisions suggest population relationships potentially reflecting migratory pulses across Beringia (Nelson, 1998); considering these groupings in terms of broad linguistic/geographic regions lends some support to this conjecture. Inuit samples are distributed across the northern periphery of the New World, from Alaska to Greenland. They cluster closely, indicating a narrow range of morphological variation. This high degree of relatedness suggests a recent and rapid dispersion across the Inuit range.

These samples should contain some of the information abstracted by Neumann for his Inuid type.

South of the Inuit distribution, across the northern part of what is today the United States and the southern part of Canada, is the next cluster grouping depicted in Figure 1. Extending from the high Northern plains across the upper Great Lakes and northeast to Newfoundland is an array of samples representing northern Great Lakes people and subsequent relations west to the Plains. Depending upon whether one wishes to lump or split, this sub-cluster could be broken in two: an Eastern, or northern Great Lakes cluster; and a Western, or Plains cluster. The Eastern cluster includes the Maritime Archaic people of Port au Choix, Newfoundland, as well as the Glacial Kame Archaic people of southern Ontario from the Hind and Sartori sites. Neumann's Lakotid type could be expected to have been generalized from data in a Plains grouping like this, though Lakotid ties to the Northern Great Lakes and Northern Archaic samples would not be at all unexpected.

The third Northern cluster represents chiefly Athabascan people arriving within the past few thousand years and the route taken by their probable descendants down the western edge of North America into the U.S. Southwest. This cluster of samples should contain much of the information abstracted by Neumann to create his Deneid variety. The loose grouping in this cluster as well as the range of geography represented by these samples appears to reflect the Athabascan migration to the Southwest as well as their long-term association with nearby Aleutian Island groups.

Below the Athabascan group cluster is the final sub-cluster, the Northeast Woodlands, with samples representing southern Michigan, New York, Massachusetts, Maryland, and, interestingly enough, one of the oldest large burial sites in the New World, the Archaic sample from Windover, Florida. This cluster also contains both of the Nevada Great Basin samples, from the Humboldt Sink and the Stillwater Marsh. Although the connection between the Northeast Woodlands and the Nevada Great Basin would seem tenuous at best, this one has been consistent and stable, suggesting relatively close biological ties between these two groups. It is interesting to note among the northeast Woodland groups a sample of relatively recent material from along the Snake River in Washington, from collections at Central Washington University in Ellensburg. Connections among so many geographically distant samples as these are curious; however, the distance matrix suggests close morphological resemblances for these groups and, by association, some underlying polygenic relationship. Perhaps the unifying causation in this lies in the relative proximity of the Athabascan cluster to the Northeast Woodland cluster. This link, like the Puebloan-Mississippian link across the Texas Plateau, may derive from a common origin for these people. Neumann (1952) discussed what he termed the "shatter" effect, the marginalizing of one group of people to the peripheries of a range postulated to have been moved into by a subsequent group. However, without research directed specifically at this distribution of samples and the archaeology that goes with it, it is useless to speculate on the origin of this apparent link.

Analyzing clusters by a combination of proximity and temporal horizon yields 14 samples, which are compared with Neumann's mean figures for the eight groupings in his analysis. Table 3 lists samples and sample sizes; Table 4 lists generalized distance figures (D^2); and Table 5 lists chi-squared values. Unrestricted-type probabilities (Albrecht 1992) can be determined using these two tables. Table 6 lists posterior probabilities for comparing typicality figures with posterior probability figures.

Table 3. Samples compared with Neumann (1952) mean figures.

Site#	Sample	Sites	n
1	Central Woodland	Klunk mounds (M-L Wood), Elizabeth mound (L Wood), Schild mound (L Wood)	76
2	Mississippian	Dickson Mounds, Moundville, 1LU25, 1LU59, 1MA48	53
3	Central Archaic	Eva, Perry, Indian Knoll, Kirian Treglia	108
4	Pueblan	Pottery Mound, Sapawe, Grasshopper, Turkey Creek	54
5	Inuit	Point Barrow, Point Hope	82
6	Plains	Blackfoot, Crow, Larson site	79
7	N. Great Lakes	Minn: Woodland, Oneota, Arvilla; Mich: Juntunen, Old Birch Island	58
8	Northern Archaic	Port au Choix, Hind, Sartori	37
9	Na-Dene	Athabascan, Navajo, Northern California	88
10	N.E. Archaic	Windover, Lamoka Lake, Frontenac Island, Horseshoe Pond, Cole Gravel pit	61
11	Great Basin	Humboldt Sink, Stillwater Marsh	22
12	N.E. Woodland	Massachusetts, Kleinburg, Ossossane	85
13	Seneca	Adams, Tram, Harscher, Dutch Hollow	61
14	S.E. Great Lakes	Rivière aux Vases, Bussinger, Younge, Uxbridge	65
	Total		929

Discussion

At this point it is important to restate the purpose of the analysis pursued here and the goal pursued by the late Georg Neumann, and to note the difference between the two methodologies. Neumann was a pre-Darwinian typologist who believed that human variation could be catalogued by defining an essence to describe a given human group, in effect an identifying label. The goal was to create a subspecific typology analogous to "racial" identification. The flaws of such an approach have been pointed out often enough (see as well Brace 1996), but it is also true that those who sought to define such characteristics were often very acute observers. Georg Neumann was such a one.

The current study deals with patterns of morphology discernible in different human groups, patterns more obviously evident in some samples than in others. Unlike Neumann's typological approach, however, there is no assumption of a defining group essence. Similarities and differences simply suggest degrees of relatedness. Correlating the clusters in terms of their geographic location merely indicates where such human morphological configurations are found. Although the purpose of our study is fundamentally different from the attempt to define human "types," and although the two approaches start from very different assumptions about variations in human morphology—their significance and manner of distribution—it is nevertheless true that both approaches depend on discernible human physical features.

Table 4. Generalized distances between Neumann means and comparative samples, for comparison using seven variables.

	Ashiwid	Deneid	Inuid	Iswanid	Lakotid	Lenapid	Otamid	Walcolid
Central Woodland	4.95	9.77	11.48	2.59	13.06	2.36	10.76	2.94
Mississippian	11.46	11.99	18.79	5.16	17.12	5.08	19.18	2.64
Central Archaic	6.29	8.95	8.58	0.96	13.40	3.38	11.89	3.87
Pueblan	8.39	10.32	20.23	6.71	16.46	7.26	19.22	7.25
Inuit	4.25	6.78	2.41	3.75	8.40	5.53	6.88	8.32
Plains	1.71	3.67	9.81	7.18	2.71	5.44	9.45	8.19
North Great Lakes	5.61	3.94	12.25	9.00	2.97	8.09	13.65	10.47
Northern Archaic	4.15	3.16	9.02	7.44	3.22	8.17	12.96	11.14
Na-Dene	2.88	5.26	7.59	2.52	9.00	3.26	9.76	4.72
Northeast Archaic	5.11	6.83	8.03	2.77	9.06	3.14	10.18	4.80
Great Basin	0.95	4.79	6.79	5.01	4.89	4.44	9.29	6.55
Northeast Woodland	2.63	5.74	7.01	3.22	6.19	2.19	7.01	4.33
Seneca	5.61	11.07	10.16	6.02	11.12	1.66	5.47	4.39
Southeast Great Lakes	3.83	5.57	12.55	8.04	4.60	5.03	11.33	7.55

This paper is neither a typological exercise nor a defense of the typology espoused by Georg Neumann. It is, however, a recognition of his skill as a morphologist and of the fact that some of the bases for his assessments of human form are consistent with the relationships indicated by the clusters documented here.

Assembling comparative samples from the samples provided by the algorithm presents options at every turn. A purely mechanical process would interpret the samples determined by the algorithm as composites and perform discriminant function analysis directly on them. On one hand, it may be significant that core Archaic sites such as Eva, Indian Knoll, and Perry located in Tennessee, Kentucky, and Alabama tie closely with subsequent Woodland and even Mississippian sites in the same region; however, it is inappropriate to lump all these horizons together in a big summary sample and compare it with groups specifically sorted on the basis of temporal and cultural horizons. On the other hand, if we circumscribe comparative samples too narrowly, a great deal of regional representation is lost; Neumann cautions against this in *Archaeology and Race* (1952:29). There are remarkable similarities between the groups visually organized by Neumann (1952) and those grouped statistically by Nelson (1998). There are also some marginal agreements between the two systems, as well as some surprising differences.

Table 5. Chi-square values.

X^2	.005	.01	.025	.05	.10	.25	.50	.75	.90	.95
D^2	1.7	2.09	2.7	3.3	4.2	5.9	8.3	11.4	14.7	16.9

The mean figure for Neumann's Otamid group doesn't seem to be very close to any of the comparison samples. It comes closest to the Seneca sample: its Mahalanobis D^2 distance finds it typical of approximately 75 percent of that grouping; posterior probabilities suggest the Otamid configuration would be found in less than 40 percent of the Seneca, based on the sample used. Neumann believed there was a basic difference between the Otamid and Lenapid groups; he suggested the Lenape had their closest affinity to hunters northwest of the Great Lakes and arrived at the Northeastern Woodland scene relatively recently, replacing Otamid peoples who were already there. The Lenapid mean actually also finds its closest link with the Seneca; typicalities suggest they are virtually identical (its typicality score of 1.66 places the Lepanid within 99.5 percent of the sample). Such a robust typicality figure makes us expect a strong posterior probability,

Table 6. Posterior probabilities for comparing Neumann means with cluster samples.

	Ashiwid	Deneid	Inuid	Iswanid	Lakotid	Lenapid	Otamid	Walcolid
Central Woodland	.03650	.00839	.00706	.12990	.00150	.14335	.02637	.19343
Mississippian	.00141	.00281	.00018	.03591	.00020	.03666	.00039	.22342
Central Archaic	.01859	.01261	.03015	.29441	.00125	.08653	.01493	.12241
Pueblan	.00649	.00635	.00009	.01642	.00027	.01234	.00038	.02231
Inuit	.05178	.03748	.66821	.07263	.01550	.02927	.18587	.01305
Plains	.18580	.17761	.01637	.01310	.26799	.03119	.05138	.01428
North Great Lakes	.02847	.16791	.00528	.00582	.23909	.00924	.00676	.00528
Northern Archaic	.05537	.23190	.02474	.01178	.20734	.00816	.00905	.00339
Na-Dene	.10290	.08019	.04958	.13427	.01148	.09151	.04367	.07944
Northeast Archaic	.03362	.03655	.03994	.11982	.01104	.09832	.03533	.07783
Great Basin	.27205	.10156	.07410	.03865	.08983	.05099	.05542	.03204
Northeast Woodland	.11667	.06308	.06644	.09536	.04669	.15888	.17445	.09871
Seneca	.02623	.00441	.01374	.02339	.00397	.20523	.37608	.09461
Southeast Great Lakes	.06413	.06915	.00413	.00855	.10382	.03833	.01994	.01978

and in fact the posterior probability profile approximates the typicality profile.

The highest posterior probability scores are for the Central Woodland (.143), Northeast Woodland (.159), and Seneca (.205) samples, in accordance with the typicality figures (2.4, 2.2, and 1.7, respectively). The difference between the smallest and largest Otamid D^2 is about 13.8 (Seneca versus Pueblo), while the same difference for the Lenapid grouping is about 6.5 (Seneca versus Northern Archaic); thus the D^2 distances are much less for the Lenapid than for the Otamid, and their overall spread is smaller by approximately half. Compare these results with the Inuid grouping, where the D^2 to the closest sample, the Inuit, is about 2.4, and to the most distant sample, the Pueblo, is 20.2; the difference of 17.8 is probably the widest spread in the chart. The corresponding posterior probability, .668, is also the highest (Table 6). All of this suggests a high degree of morphological homogeneity among the Inuit and also illustrates the differing expressions of typicality versus posterior probability figures.

In comparisons among other groups, the Walcolid sample, taken from an Illinois, or Central Valley Mississippian series, corresponds closely to the Central Woodland and Mississippian samples from the UMMA set and falls incrementally distant from the Central Valley Archaic samples, as would be expected. The Iswanid mean is nearly identical with the Central Valley Archaic and Woodland samples, with the Mississippian sample being the slightly divergent one. On the Plains, the Lakotid mean falls closest to those comparative samples representing the Plains, Northern Great Lakes, and Northern Archaic cluster, and farthest from the Eastern Woodland, Central Valley and Puebloan groups.

Comparisons achieve some depth—and pique curiosity—in the examples of the Ashiwid (Desert Archaic) and Deneid (Athabascan). One of the greatest distances between the Lakotid mean and any other group is that to the Pueblo (16.5); yet the corresponding comparison between the Ashiwid and the UMMA Plains sample yields a very short distance (1.7) that suggests virtual identity. Furthermore, the closest D^2 ties for the Ashiwid are not to the Pueblan Southwest, from which they are reputed to have come, but rather with the Nevada Great Basin sample (0.95) from the Humboldt Sink and Stillwater Marsh. Neumann selected the sample he uses from an Arizona Basket Maker series defined by Hooton. Basket Maker is an Archaic expression in the Southwest, while the UMMA Pueblan samples used here (Sapawe and Pottery Mound, N.M., and Grasshopper and Turkey Creek, Ariz.) are more recent in the Puebloan sequence. It is not at all improbable, given the transitions and population movements cited by Neumann, that the comparison here is between different groups of people. This being the case, then it is also reasonable to deduce from the D^2 figures that there is not a representative of the later Pueblan sequences in Neumann's scheme.

The D^2 distance of less than 1 between the Ashiwid group and the Great Basin sample may suggest an origin or illustrate a distribution of the people who were Basket Makers. More than 9000 years of continuity in the Great Basin has been documented by recent work from the Nevada State Museum (Dansie 1997). Consequently the close biological connection between the Great Basin Archaic and the Desert Archaic indicated here supports the picture of regional population distributions suggested by Nelson (Nelson 1998) and Brace and Nelson (Brace in press). If Neumann is correct in his assertion that the Plains people may represent a secondary, or derived group, then the very short D^2 distance from the Ashiwid mean to the Plains sample may suggest a source for some incipient Plains peoples in the Great Basin and Southwest. Nelson (Nelson 1998) also found striking resemblances between individuals from Plains Archaic sites and this Great Basin sample. Finally, the almost identical distances between the Ashiwid and both the Na-Dene and Northeastern Woodland samples is noteworthy, since the Michigan data have repeatedly suggested this relationship. It is also confirmed by the placement of the Na-Dene and Northern Woodland macro-clusters (Figure 1); the Northern Woodland macro-cluster contains the Seneca, Southeast Great Lakes, and Northeast Woodland samples used in this comparison.

In the second problematic example, that of the Deneid, Neumann chose a specifically limited series to represent his Athabascan grouping. After eliminating Yukon Valley, Salishan, and Haida series, he chose 32 individuals (also Haida and Salish) that he felt most closely approximated the Athabascan craniofacial configuration. In the UMMA sample, which doesn't show any particularly close affinity to Neumann's, the Athabascan cluster consists of Yukon Valley Athabascans, Aleuts, Northern Californians, Navajo, and Haida, as well as the Pecos sample. Of those, the Athabascans, Northern Californians, and Navajo were selected on the basis of their Euclidean distances as a sample representing that cluster area. Neumann's Deneid sample actually finds its closest match with the Northern Great Lakes, Northern Archaic, and Plains cluster materials in a comparative profile that very closely parallels that of his Lakotid sample. The D^2 figures here, however, suggest that Neumann was probably not looking at just another basic Plains configuration. The closest Lakotid ties to the three samples contributing to that cluster are to the Plains groups; the farthest are to the Archaic samples from Ontario and Newfoundland. On the other hand, in the Deneid comparison, the closest ties are to the Archaic samples, and the most distant are to the Plains samples of the three samples representing that cluster (Figure 1). At the time of Neumann's publication (1952), Port au Choix would not be excavated for nearly a decade and a half and would not be published for over two decades (Tuck 1976). One can't help but wonder if Neumann would have considered this to be an Athabascan site.

Conclusions

The search for patterns of correspondence in data derived from our own broad range of phenotypic expression has driven the development of analytic methodologies in physical anthropology for a century or more. Over the course of that time the end purpose of defining those patterns has shifted greatly, from the divination of racial types and essences to the exploration of settlement patterns in prehistory and the correlations between phenotypic and genotypic information (see, for example, Brace et al., 2001).

Regardless of the typological framework in which he was trained and worked most of his life, Georg Neumann was a practically unparalleled observer with a respectable grasp of the diversity and distribution of prehistoric peoples in the New World. He was concerned with a broad picture of the distribution of the data at a time when analyzing and processing multivariate data was such a tedious undertaking that it was more practical to base assessments on visual observations. Considering the highly subjective nature of visual observations, the present comparison using computer-generated groupings produces remarkable concurrences. Cross-referencing Neumann's groups with computer-derived cluster groupings is a useful and instructive method of examining data that may no longer be accessible. The D^2 figures for Neumann's Walcolid, Iswanid, Lenapid, Lakotid, and Inuid means fall just where we would have been expected, and the UMMA samples used for comparison agree with the facial configurations Neumann evidently observed. The intriguing comparisons are with the Ashiwid and Deneid means, which we might have expected to tie closely with one cluster, but which instead appear to connect more closely with unexpected groups.

The potential for solid, productive research based on assessing craniofacial metric data has been demonstrated in study after study. Physical anthropologists are recognizing this potential, and students are picking up the mantle and making some real contributions (Ozolins 1999). This shows promise of becoming an exiting and dynamic part of the overall discipline. Jantz and Owsley (Jantz 1998), speaking of Paleoindian studies, note that the

> major problem facing synthesis of Paleoamerican craniometric variation concerns lack of comparable data. There has not yet been any systematic metric analysis of early American cranial or skeletal remains using a comprehensive set of variables.

The key phrase in this statement is "lack of comparable data." In fact, however, there are abundant variable lists, many of them derived or borrowed whole cloth from Howells's pioneering work in broad-scale analysis (Howells 1973). What has been lacking to date is a database with the breadth to adequately address New World population relation issues. For this project, the University of Michigan Museum of Anthropology craniofacial database has been utilized and expanded.

It is unfortunate that biological anthropology is hitting its stride when so many collections have either been destroyed or are in imminent jeopardy. Jantz and Owsley (1998) are right: there is a lack of comparable data, at least in many databases. Data are available. They are sitting on the shelves and in the cabinets of our universities and museums. They are available for scholarly research, at least for the time being.

References Cited

Albrecht, G.
1992 Assessing the Affinities of Fossils Using Canonical Variates and Generalized Distances. *Human Evolution* 7(1):49–69.

Brace, C. L.
1996 A Four Letter Word Called Race. In *Race and Other Misadventures: Essays in Honor of Ashley Montagu in his Ninetieth Year.*, edited by L. T. Reynolds and L. Lieberman, pp. 106–41. General Hall, Inc., Dix Hills, NY.

Brace, C. L., and K. D. Hunt
1990 A Non-racial Craniofacial Perspective on Human Variation: (A)ustralia to (Z)uni. *American Journal of Physical Anthropology* 82:341–60.

Brace, C. L., and A. R. Nelson
in press Francthi: Between Europe and Asia; or, *Qui Erant Danai, et per Crania Cogni Sunt?* In *Volume Commemorating J. Lawrence Angel*, edited by D. C. Cook. Indiana University Press, Bloomington, IN.

Brace, C. L., and D. P. Tracer
1992 Craniofacial Continuity and Change: A Comparison of Late Pleistocene and Recent Europe and Asia. In *The Evolution and Dispersion of Modern Humans in Asia.*, edited by T. Akazawa, K. Aoki, and T. Timura, pp. 439–71. Hokusen-Sha, Tokyo.

Brace, C. L., M. L. Brace, and W. R. Leonard
1989 Reflections of the Face of Japan: A Multivariate Craniofacial and Odontometric Perspective. *American Journal of Physical Anthropology* 78:93–113.

Brace, C. L., P. E. Mahler, and R. B. Rosen
1973 Tooth Measurements and the Rejection of the Taxon "Homo habilis." *Yearbook of Physical Anthropology* 16:50–68.

Brace, C. L., A. R. Nelson, N. Seguchi, H. Oe, L. Sering, Pan Q., Li Y., and D. Tumen
2001 Old World Sources of the First New World Inhabitants: A Comparative Craniofacial View. PNAS (USA) 98(17):10017–22.

Brace, C. L., D. P. Tracer, L. A. Yaroch, J. Robb, K. Brandt, and A. R. Nelson
1993 Clines and Clusters versus "Race": A Test in Ancient Egypt and the Case of a Death on the Nile. *Yearbook of Physical Anthropology* 36:1–31.

Dansie, A., and D. R. Tuohy
1997 Nevada State Museum Prehistoric Human Burial Study—Overview, Issues and Proposals. Nevada State Museum, Carson City, NV.

Droessler, J.
1981 *Craniometry and Biological Distance: Biocultural Continuity and Change at the Late Woodland Interface.* Research Series 1. Center for American Archaeology, Evanston, IL.

Everitt, B.
1974 *Cluster Analysis.* Reviews of Current Research 11. Heinemann Educational Books.

Greenberg, J. H., C. G. Turner, and S. L. Zegura
1986 The Settlement of the Americas: A Comparison of the Linguistic, Dental, and Genetic Evidence. *Current Anthropology* 27:477–97.

Griffin, J. B. (editor)
1952 *Archaeology of Eastern United States.* University of Chicago Press, Chicago, IL.

Hall, R. L.
1997 Neumann, Georg Karl (1908–1971). In *History of Physical Anthropology: An Encyclopedia*, edited by F. Spencer, pp. 731–32. vol. 2. 2 vols. Garland Publishing Co., New York, NY.

Howells, W. W.
1973 *Cranial Variation in Man: A Study by Multivariate Analysis of Patterns of Difference among Recent Human Populations.* Papers of the Peabody Museum of Archaeology and Ethnography 67. Harvard University, Cambridge, MA.

1984 Introduction. In *Multivariate Statistical Methods in Physical Anthropology*, edited by G. N. Van Vark, and W. W. Howells, pp. 1–11. D. Reidel Publishing Co., Dordrecht, Netherlands.

1989 *Skull Shapes and the Map: Craniometric Analysis in the Dispersion of Modern Homo.* Papers of the Peabody Museum of Archaeology and Ethnography 79. Harvard University, Cambridge, MA.

1995 *Who's Who in Skulls: Ethnic Identification of Crania from Measurements.* Papers of the Peabody Museum of Archaeology and Ethnography 82. Harvard University, Cambridge, MA.

Hrdlička, A.
1927b Cataloge of Human Crania in the United States National Museum Collections: Thye Algonquian and Related Iroquois; Siouan; Caddoan; Salish and Sahaptin; Shoshonean and California Indians. *Proceedings of the US National Museum* 69(5):1–169.

Ishida, H., and Y. Dodo
1993 Nonmetric Cranial Variation and the Populational Affinities of the Pacific Peoples. *American Journal of Physical Anthropology* 90:49–57.

1997 Cranial Variation in Prehistoric Human Skeletal Remains from the Marianas. *American Journal of Physical Anthropology* 104(3):399–410.

Jantz, R. L.
1973 Microevolutionary Change in Arikara Crania: A Multivariate Analysis. *American Journal of Physical Anthropology* 38(1):15–26.

Jantz, R. L., and D. W. Owsley
1998 Pathology, Taphonomy, and Cranial Morphometrics of the Spirit Cave Mummy. *Nevada Historical Society Quarterly* 40(1): 62–84.

1998b How Many Populations of Early North Americans Were There? *American Journal of Physical Anthropology, Supplement* 26:128.

Jungers, W. L., A. B. Falsetti, and C. E. Wall
1995 Shape, Relative Size, and Size-Adjustments in Morphometrics. *Yearbook of Physical Anthropology* 38:137–61.

Kamminga, J., and R. V. S. Wright
1988 The Upper Cave at Zhoukoudian and the Origins of the Mongoloids. *Journal of Human Evolution* 17(8):739–67.

Key, P. J.
1983 *Craniometric Relationships among Plains Indians: Culture-Historical and Evolutionary Implications.* Reports of Investigations 34. University of Tennessee Dept. of Anthropology, Knoxville, TN.

Li Y., C. L. Brace., J. Gao, and D. P. Tracer
1991 Dimensions of Face in Asia in the Perspective of Geography and Prehistory. *American Journal of Physical Anthropology* 85:269–79.

Nelson, A. R.
1998 *A Craniofacial Perspective on North American Indian Population Affinities and Relations.* Doctoral dissertation, University of Michigan.

Neumann, G. K.
1952 Archaeology and Race in the American Indian. In *Archaeology of Eastern United States*, edited by J. B. Griffin, pp. 13–34. University of Chicago Press, Chicago, IL.

Nichols, J.
1990 Linguistic Diversity and the First Settlement of the New World. *Language* 66(3):475–521.

1992 *Linguistic Diversity in Space and Time.* University of Chicago Press, Chicago, IL.

1995 The Spread of Language around the Pacific Rim. *Evolutionary Anthropology* 3(6):206–15.

Norušis, M. J.
1994 *SPSS Professional Statistics 6.1.* SPSS Professional Statistics. SPSS Inc., Chicago, IL.

Ousley, S. D.
1995 Relationships between Eskimos, Amerindians, and Aleuts: Old Data, New Perspectives. *Human Biology* 67(3):427–58.

Ozolins, E. G.
1999 Are Paleoindians Too Variable to Be from One Population: A Test of a Single Migration Origin. *American Journal of Physical Anthropology* Supplement 28:216.

Pietrusewski, M.
1984 *Metric and Non-metric Cranial Variation in Australian Aboriginal Populations Compared with Populations from the Pacific and Asia.* Occasional Papers in Human Biology 3. Australian Institute of Aboriginal Studies, Canberra.

1990 Craniofacial Variation in Australasian and Pacific Populations. *American Journal of Physical Anthropology* 82:319–40.

Pietrusewski, M., Y. Li, X Shao, and N. Q. Quyen
1992 Modern and Near Modern Populations in Asia and the Pacific: A Multivariate Craniofacial Interpretation. In *The Evolution and Dispersion of Modern Humans in Asia*, edited by T. Akazawa, K. Aoki and T. Kimura, pp. 439–71. Hokusen-Sha, Tokyo, Japan.

Relethford, J. H.
1994a Craniometric Variation among Modern Human Populations. *American Journal of Physical Anthropology* 95(1):53–62.

Relethford, J. H. **and** H. C. Harpending
1994b Craniometric Variation, Genetic Theory, and Modern Human Origins. *American Journal of Physical Anthropology* 95:249–70.

Szathmary, E. J. E.
1995 Overview of the Boas Anthropometric Collection and Its Utility in Understanding the Biology of Native North Americans. *Human Biology* 67(3):337–344.

Tuck, J. A.
1976 *Ancient People of Port au Choix: The Excavation of an Archaic Indian Cemetery in Newfoundland.* Memorial University of Newfoundland Social and Economic Studies 17. Memorial University, St. John's, NF.

Ward, J. H.
1963 Hierarchical Grouping to Optimize an Objective Function. *Journal of the American Statistical Association* 58:236–244.

Wright, S.
1943 Isolation by Distance. *Genetics* 28(2):114–38.

1946 Isolation by Distance under Diverse Systems of Mating. *Genetics* 1(1):38–59

Nearsightedness in Paleoamerican Research: Historical Perspective and Contemporary Analysis

Douglas W. Owsley[1] and Richard L. Jantz[2]

Abstract

Both increased analytical capability and compilation of an extensive reference database for morphometric research enable the testing of old ideas versus new theories regarding the origin of the First Americans. Measurements from four ancient skulls—Browns Valley, Minnesota Woman, Spirit Cave, and Wizards Beach—are compared with corresponding data for 34 population samples including nine Native American groups from the western half of North America. The craniofacial morphology of Browns Valley and Minnesota Woman are quite similar; together with Spirit Cave, they fall outside the range of variation of all modern groups, especially that of American Indians. This interpretation is very different from the traditional view of relatedness and close morphological similarity championed by Hrdlička more than half a century ago. Historical perspective focusing on differing interpretations of Minnesota Woman reaffirms the need for farsightedness in determining the disposition of ancient remains in accordance with the Native American Graves Protection and Repatriation Act (NAGPRA). Preservation of culturally unaffiliated human remains for future study is essential for advancing knowledge of ancient American prehistory.

Human bones and teeth represent such a remarkable archive of prehistoric ways of life that now we commonly expect the professional examination of a skeleton to provide information about age, sex, physical size and stature, and sometimes even the cause of death. Some bones reveal personal information such as habitual activities as well as the aches, pains, and illness of life (Owsley and Jantz 2000). Often even ancestry and population affiliation can be established (Gill and Rhine 1990; Owsley 1999).

Until the 1980s, we had been looking at this information with a very narrow vision: one, or only a few skeletons at a time. Now the computer allows us to view the information from groups of skeletons, giving us a much wider vision. The potential for breakthrough discovery is limited only by the availability of data.

Expert analyses of population data can discern not only trends in demography, health, diet, general lifestyle, and mortuary practices, but also patterns of sociocultural interaction, gene flow, microevolutionary change, migrations, and origins. These changes can be viewed through time and in different regions because more sophisticated investigative procedures and research designs utilize advances in technology, analytical methods, and systematic approaches that emphasize the development of computerized databases (for example, Brace and Tracer 1992; Howells 1973, 1995; Larson and Milner 1994; Ortner and Aufderheide 1991; Owsley and Rose 1997; Saunders and Katzenberg 1992; Turner 1987, 1990).

Our work has added to these databases by intensively collecting osteological data for numerous groups from different time periods and geographical areas. We have developed and used a standardized recording format for dental and bone inventories, pathological conditions, and measurements of crania and postcrania (Owsley and Jantz 1994, 1999a, 1999b). This approach expands our vision by maximizing the comparability of data and facilitating direct comparisons across samples. Data have been entered for well over 7,000 African-American, Euro-American, and Native American skeletons from North America. Our computerized database includes especially extensive records for prehistoric and historic populations from the Plains and Great Basin. Ancient remains are extremely rare, but this database does include detailed information for more than 20 early-Archaic and Paleoamerican skeletons including important discoveries from Minnesota and Nevada. (Jantz and Owsley 2001).

Although this database was initiated for studying Holocene

[1]Department of Anthropology, Mail stop 112, National Museum of Natural History, Smithsonian Institution, Washington, D.C. 20560: e-mail: owsley.douglas@nmnh.si.edu

[2]Department of Anthropology, University of Tennessee, 250 South Stadium Hall, Knoxville, TN 37996-0720; e-mail: rjantz@utk.edu

populations from the western half of North America, its volume and breadth provide a structure that enables systematic analysis of ancient remains either individually or statistically within a comparative framework. Our objective in this presentation is to illustrate part of the database's potential to expand our vision of ancient peoples by focusing on one component, the craniometric data. In doing so, it is important not only to give historical perspective as to how an earlier generation of anthropologists looked at ancient burials, but also to address a central concern of contemporary researchers involved in these studies; namely, the restriction of our vision that is caused by reburial of crucial remains before all the issues are sorted out. We intend to illustrate the effects of these untimely reburials by example; but first, let us provide some background on morphometric research as it relates to the peopling of the Americas.

Recent Analytical Insight

We are involved in testing old ideas and new theories against traditional assumptions that the First Americans crossed the Bering land bridge connecting Siberia to North America about 11,500 years ago. First Americans have been referred to as relating to the Clovis culture because their technology was first identified in Clovis, N.M. Their designation as Paleoindian was linked to the belief that biologically they were related to northeast Asians and were directly ancestral to modern Native Americans (Bonnichsen 1999; Bonnichsen and Schneider 2001).

Bioanthropological studies initiated during the 1990s (Jantz and Owsley 1997; Steele and Powell 1992, 1993, 1994) and accurate dating of early sites have challenged this model and suggest that "Paleoamerican" may be a better designation than "Paleoindian" for this population. Tantalizing evidence indicates that multiple migrations of pre- and post-Clovis arrivals involving diverse populations may be a more accurate assessment of the events involved in peopling the New World (Bonnichsen 1999; Brace et al. 2001; Jantz and Owsley 2001).

One way of examining questions of origin and relationships is through morphometric analysis. Because craniofacial morphology is under strong genetic control, size and shape provide a basis for tracing biological affinity and population-historical relationships through time. This approach uses statistical methods to compare the metric data of a fossil with samples representing recent American Indian groups and other world populations. It addresses the question of whether the fossil falls within the range of variation of recent populations; and if so, to which group it is most similar.

To illustrate the power of this technique, we incorporated four well-dated fossil crania into an analysis. Two are from Nevada: Spirit Cave, dated at 9415 years RCYBP; and Wizards Beach, dated at 9200 RCYBP. Two are from Minnesota: Browns Valley, dated at 8700 RCYBP; and Minnesota Woman, dated at 7900 RCYBP (Dixon 1999; Kirner et al. 1997; O'Connell and Myster 1996; Tuohy and Dansie 1997). The extensive reference database we have used consists of cranial measurements compiled by Howells (1995) and Jantz (1997) for 34 world populations including nine American Indian groups from western North America.

The Spirit Cave Mummy was found in a small rockshelter near Fallon, Nevada, and excavated by Sydney and Georgia Wheeler in 1940 while conducting salvage archaeology for the Nevada State Parks Commission (Dansie 1997; Wheeler 1997). The Wheelers systematically excavated areas of the cave, carefully recorded their findings, and documented the site with photographs. They removed the mummy from the cave in its mat wrappings, and it has been stored for half a century at the Nevada State Museum.

The skeleton from Wizards Beach at Pyramid Lake was found during low lake levels in 1968 by avocational archaeologist Peter Ting, and was dug by Robert Stevenson of the Nevada Archaeological Survey (Dansie 1997). It is also stored at the Nevada State Museum.

The bones from Browns Valley, Minnesota, were noticed by a layman interested in archaeology as gravel was unloaded in his driveway in 1933. Mr. Jensen picked the bones out of that load, rescued more bones from the next load, located the probable source in the quarry, and notified Professor Albert E. Jenks of the find. Jenks was able to excavate the site in 1934. He found another tooth and 17 additional bone fragments, some of which fitted the original discovery (Jenks 1937).

Minnesota Woman was found in 1931 by highway workers who were digging a roadbed with machinery (Figures 1a and 1b). They reported the find immediately, dug out the skeleton, and kept it safe until it could be collected by Albert Jenks's staff. A year later, Dr. Jenks reexamined the site but did not find any additional bones or artifacts (Jenks 1936).

The cranial data on these four discoveries were subjected to canonical analysis, with the fossils treated in the same way as the other samples and given equal weight. Canonical analysis is a multivariate statistical procedure that displays Mahalanobis distances in low-dimensional space. The 25 measurements used in this comparison were selected to express the size and shape of the cranial vault in the sagittal plane (e.g., heights), vault and face breadths, and facial forwardness and prognathism.

Figure 2 shows the results displayed in two independent (orthogonal) axes. The first axis, which represents 23 percent of the total group variation, separates the Minnesota fossils, and even Spirit Cave to some extent, from Africa and South West Pacific populations. It highlights the uniqueness of Browns Valley and Minnesota Woman relative to all modern populations. This axis reflects variation in vault base width, nose width, frontal flatness, and upper facial forwardness at nasion. Browns Valley, Minnesota Woman, and, to a lesser degree, Spirit Cave differ from other world populations, especially those of Africa and the South West Pacific, by having wider vault bases, narrower noses, flatter frontal bones, and upper facial forwardness.

The Minnesota fossils are also extreme on the second axis. This axis represents 15 percent of the total sample variation, and it separates most modern American Indians and Siberians from the Minnesota fossils. This axis reflects variation in face height including the nasal and orbital heights, vault breadth, and parietal length. Browns Valley and Minnesota Woman differ from modern populations, and especially American Indians, in having low faces, narrow vaults, and long parietal bones. Indian cra-

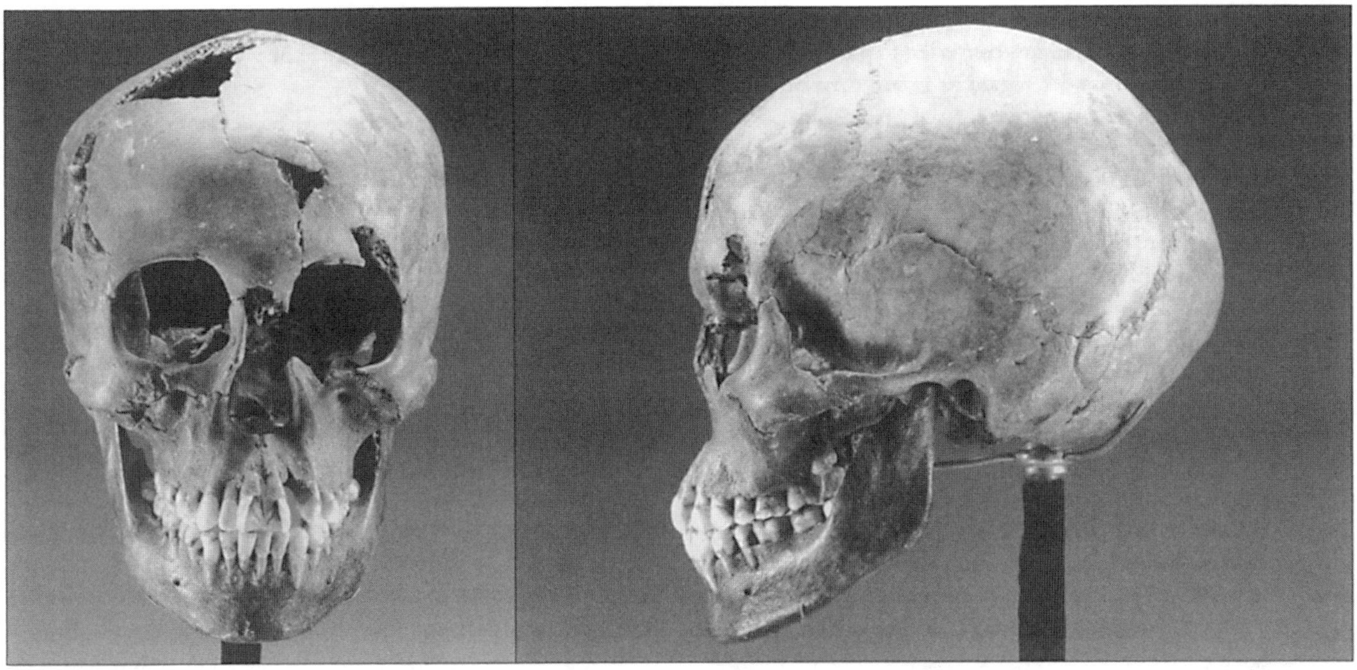

Figure 1. Frontal and left lateral views of the skull of Minnesota Woman.

nia are characterized by longer and wider faces and broader and shorter (more rounded) cranial vaults.

The Minnesota fossils are extreme on the first two axes of variation, and are so different from modern populations that their inclusion tends to compress heterogeneity among groups within the modern series. In fact, it makes little sense to ask which groups they are most similar to. Spirit Cave falls a bit closer, and is nearest the Polynesian-Eskimo part of the two-dimensional space. Spirit Cave's uniqueness is not fully represented on these two axes, but it appears on subsequent axes. As demonstrated by Jantz and Owsley (1997), the Spirit Cave skull falls outside the range of variation of modern samples; and, in particular, it shows no affinity to American Indian samples. Spirit Cave's closest biological distance is to the Ainu of Japan. In contrast, the Wizards Beach cranium classifies as most similar to modern Indians.

Overall, the Paleoamerican sample exhibits a great deal of heterogeneity as reflected by the wide spread of the fossil crania within the multivariate distribution. Because the fossils were treated in the same way as the modern samples and given equal weight, it could be argued that the procedure has emphasized unique features of these individual crania. However, when considering typicality probabilities of group membership, only Wizards Beach falls

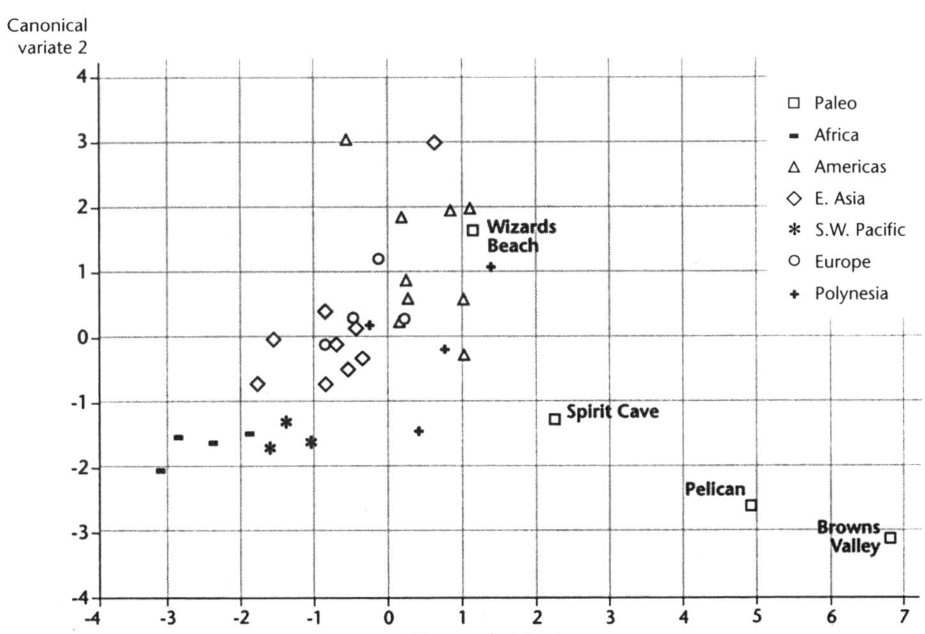

Figure 2. Canonical plot of 34 world population samples and four well-dated ancient American fossils. The world populations are coded by general region.

within the range of variation of modern groups. Typicality probabilities are Mahalanobis distances converted to probabilities, and there are no assumptions with regard to group membership. In this regard, the craniofacial morphology of Spirit Cave, Browns Valley, and Minnesota Woman falls outside the range of variation of all modern groups. These specimens have a very low probability of belonging to any of the groups represented here.

To reiterate what we have seen from this analysis: some of the oldest humans identified on the continent have skeletal features that are distinctly different from modern-day American Indians. In fact, early American crania are usually different from any modern crania. To the extent that they do resemble modern populations, they show greater similarity to certain recent Old World populations including Polynesians, Europeans, and the Ainu than to recent American Indians.

Whether individual skeletons or specific early groups were directly related to later peoples is yet to be learned. Early migrants may have been replaced through competition, or changed through gene flow by later arrivals. It is important to point out here that what we have learned so far has only been possible because we have had the skeletal material available for study.

Hrdlička's Assessment

To understand how important that availability has been, consider the historical record and the debate over one of those early skulls: Minnesota Woman. A detailed monograph entitled *Pleistocene Man in Minnesota, a Fossil Homo Sapiens* was produced by Albert Jenks in 1936. This detailed study described the discovery, geoarchaeological context, and skeletal morphology of Minnesota Woman. In his opinion the skull exhibited a number of features that are different from modern Native Americans, and some of these characteristics seemed primitive. Specifically, Jenks (1936:80) wrote:

> The face of Minnesota Man shows closest relationship to the Mongoloid and White groups. . . . It is closest to the Mongoloids in face length and bigonial width, and to the Whites and Mongoloids in total facial and upper facial height, and it lies between Whites and Mongoloids in zygofrontal index.

In addition, Jenks (1936:60) noted various features that seemed primitive including "the backward extension of the skull" and the "house type" shape of the posterior vault.

Jenks's report was reviewed by the eminent Smithsonian scientist Aleš Hrdlička, who attempted "to prove that the characteristics of the Minnesota skeleton are of modern Sioux type and that the skeletal remains therefore represent a modern Sioux burial" (Jenks 1938:332). Hrdlička compared Minnesota Woman's features and measurements one at a time with a sample of 41 historic Sioux crania. His conclusion was that

> the Minnesota skull, both in the face and the vault, are so close to those prevalent in the female Sioux skulls that there is conveyed to the observer a forcible impression of close similarity, and in not one feature of significance does the Minnesota skull show anything substantially different (Hrdlička 1937:192).

Jenks correctly asserted that the individual measurements could not be compared separately, but that simultaneous evaluation of the entire constellation of features was necessary. In effect, he recognized the importance of evaluating the total morphological pattern.

> In his [Hrdlička's] comparisons with Sioux crania he does not show that the primitive features of the Minnesota skull are those of the Sioux crania generally. Instead, he finds that one or a few Sioux individuals share one primitive feature with Minnesota Man, others share a second primitive trait, etc. (Jenks 1938:332).

Jenks had noted distinctive differences between the Minnesota discoveries and those of modern American Indians. Hrdlička, believing in the relatively recent arrival of humans across the Bering land bridge, concluded that only anatomically modern humans could have inhabited the New World and was compelled to minimize the significance of any of the observed differences.

Given his preeminence, it is not surprising that Hrdlička's view of how America was peopled and his intransigence in accepting conflicting data effectively discouraged studies of Paleoamerican skeletons for decades. The intensity of Hrdlička's (1937:199) conviction is evident in his writings:

> Should [an] experienced professional physical anthropologist be given the task of proving by its somatic characteristics the Minnesota skull as ancient or even really primitive, he would have to give up in despair. The task would be impossible.

This interpretation held sway for more than half a century: "A. Hrdlička argued for decades that no Paleoamerind remains fall outside the range of variation found in living Amerinds. This has yet to be refuted" (Wolpoff 1980:325).

If Minnesota Woman had not been available for study and measurement using the Howells system, we would now be repeating, with conviction, Hrdlička's statement that she was Sioux. Hrdlička may have been nearsighted in not realizing that future scientists could develop techniques and methodology that would make his hard-won skills and abilities obsolete. Fortunately, painstaking measurements and the multivariate analysis that we have described show that some of Professor Jenks's interpretations had merit, i.e., that the cranial morphology of Minnesota Woman and Browns Valley are different from those of modern populations, including those of contemporary Native Americans.

A Disappearing Legacy

If Hrdlička was nearsighted about Minnesota Woman, we have been blinded. At this time we can learn no more about her and her place in the peopling of the Americas because, regrettably, Minnesota Woman was reburied in October 1999. The haste to rebury ancient remains before these issues are adequately sorted out is of great concern. What these skeletons can tell us about ancient environments, health, nutrition, lifestyles, and the ancestry of the earliest peoples in the Americas is their legacy to current and future generations. They are the ultimate opportunity to learn from our elders.

This is a national concern. In June 1999 the State of Minnesota completed an inventory of human remains and associated funerary objects as required under NAGPRA. They listed "the physical remains of a minimum of 1,059 individuals of Native

American ancestry," and approximately 306 objects "reasonably believed to have been placed with or near individual human remains at the time of death or later as part of the death rite or ceremony" (National Park Service 1999a:43221). All these remains were termed culturally unaffiliated; or, in their words, "there is no relationship of shared group identity which can be reasonably traced between these Native American human remains and associated funerary objects and any present-day Indian tribe or group" (National Park Service 1999a:43221).

These unaffiliated remains represented nine thousand years of diverse cultures in Minnesota and had the potential of increasing knowledge of this rich heritage. On 28 September 1999, they were handed over to a coalition of Sioux tribes, a group whose identifiable antiquity is at best one thousand years in this region. All the material, including Minnesota Woman and Browns Valley, was reburied in a mass grave in Sisseton, South Dakota, on October 2.

But even that act may not be the end of the damage done to bioarchaeology by the Minnesota decision. The National Park Service (1999b) has drafted "principles of agreement" that cite the Minnesota example as a model to follow when other states and institutions need to determine how to dispose of culturally unidentifiable remains. The Department of the Interior seems to consider reburial as the most desirable outcome even for skeletons and artifacts that are not related to current Indian groups.

There are similar concerns with regard to the fate of the Spirit Cave remains. The Northern Paiute have claimed this mummy, which we have shown here as being most like the Ainu of Japan. But their claim compels the Bureau of Land Management to defer requests to conduct DNA research on this individual and on other ancient remains under their control. Our concern is clearly in order, as aptly expressed by Steele (1997) in a letter to the Nevada State Museum:

> Like other Paleoamerican remains, the two Nevada individuals are distinguished by their relatively long and narrow crania, and their relatively narrower and shorter face than I have seen in more recent prehistoric and historic North American Indians. Given the distinctiveness of the features of the early Holocene remains from North America I am convinced that assuming a biological relationship of the Paleoamericans to more recent Native Americans simply on the proximity of their sites of recovery to location of living Native American populations is not warranted. Further, I believe that determining the biological relationship of Spirit Cave and Pyramid Lake individuals to living Native Americans from Nevada and surrounding areas can only be determined by detailed comparative metrical analyses of the cranial remains, and DNA analyses, and such studies have yet to be completed.

These concerns are being most dramatically played out in the Kennewick lawsuit, in which eight scientists challenged the Corps of Engineers decision to allow reburial of an ancient skeleton without allowing opportunities for scientific study by multiple disciplines and points of view. Such concerns involve not only remains, but also archaeological sites. In the Kennewick case, rather than allow a careful geoarchaeological assessment of the discovery site, the Corps chose to bury the site under tons of rubble.

It occurs to us that, in a way, we are repeating history. In the 1930s Hrdlička, through his position and force of argument, not only retained his nearsighted belief that America was peopled by modern Asians, but he also persuaded others to see it his way for more than half a century. We have had 20 years of clearer vision made possible by remarkable advances in technological and scientific capability. We fear that nearsightedness is returning through the application of NAGPRA to all remains, even ones that cannot be attributed to current Indian groups.

We have shown that through craniometrics we can point out those remains that are so unusual as to merit further study. In the intervening 60 years since Hrdlička, we have seen the development of powerful dating techniques, such as accelerator mass spectrometry, which allow accurate dating of archaeological discoveries. We have the statistical capability and technology to manipulate more information simultaneously. What we need now is the vision to preserve ancient remains so future generations can learn from them.

Acknowledgments

Sandy Schlachtmeyer provided editorial assistance and Rebecca Kardash assisted with the preparation of this manuscript. Figures 1a and 1b were photographed by Chip Clark and printed by Brenda McLain. Figure 2 was prepared by Robert Ruch.

References Cited

Bonnichsen, R.
1999 An Introduction to *Who Were the First Americans*? In *Who Were the First Americans*? edited by R. Bonnichsen, pp. 79–96. Center for the Study of the First Americans, Oregon State University, Corvallis.

Bonnichsen, R., and A. L. Schneider
2001 The Case for a Pre-Clovis People. *American Archaeology* 5(4): 35–39.

Brace, C. L., and D. P. Tracer
1992 Craniofacial Continuity and Change: A Comparison of Late Pleistocene and Recent Europe and Asia. In *The Evolution and Dispersal of Modern Humans in Asia*, edited by T. Akazawa, K. Aoki, and T. Kimura. Hokusen-Sha, Tokyo.

Brace, C. L., A. R. Nelson, N. Seguchi, H. Oe, L. Sering, P. Qifeng, L. Yongyi, and D. Tumen
2001 Old World Sources of the First New World Human Inhabitants: A Comparative Craniofacial View. *Proceedings of the National Academy of Sciences* 98(17):10017–22.

Dansie, A.
1997 Early Holocene Burials in Nevada. Overview of Localities, Research and Legal Issues. *Nevada Historical Society Quarterly* 40(1):4–14.

Dixon, E. J.
1999 *Bones, Boats, & Bison. Archeology and the First Colonization of Western North America*. The University of New Mexico Press, Albuquerque.

Gill, G. W., and S. Rhine (editors)
1990 *Skeletal Attribution of Race*. Maxwell Museum of Anthropology, Anthropological Paper No. 4, University of New Mexico, Albuquerque.

Howells, W. W.
1973 *Cranial Variation in Man*. Papers of the Peabody Museum of Archaeology and Ethnology, Vol. 67, Harvard University, Cambridge, MA.

1995 *Who's Who in Skulls.* Papers of the Peabody Museum of Archaeology and Ethnology, Vol. 82, Harvard University, Cambridge, MA.

Hrdlička, A.
1937 The Minnesota "Man." *American Journal of Physical Anthropology* 22(2):175–99.

Jantz, R. L.
1997 Cranial, Postcranial, and Discrete Trait Variation. In *Bioarchaeology of the North Central United States*, edited by D. W. Owsley and J. C. Rose, pp. 240–47. Arkansas Archaeological Survey Research Series No. 49. Arkansas Archeological Society, Fayetteville.

Jantz, R. L., and D. W. Owsley
1997 Pathology, Taphonomy, and Cranial Morphometrics of the Spirit Cave Mummy. *Nevada Historical Society Quarterly* 40(1):62–84.

2001 Variation among Early North American Crania. *American Journal of Physical Anthropology* 114(2):146–55.

Jenks, A. E.
1936 *Pleistocene Man in Minnesota: A Fossil Homo Sapiens.* University of Minnesota Press, Minneapolis.

1937 *Minnesota Browns Valley Man and Associated Burial Artifacts.* American Anthropological Association, Memoir 49. Menasha, WI.

1938 Minnesota Man: A Reply to a Review by Dr. Aleš Hrdlička. *American Anthropologist* 40:328–36.

Kirner, D. L., R. Burky, K. Selsor, D. George, R. E. Taylor, and J. R. Southon
1997 Dating the Spirit Cave Mummy: The Value of Reexamination. *Nevada Historical Society Quarterly* 40(1):54–56.

Larson, C. S., and G. R. Milner
1994 *In the Wake of Contact.* Wiley-Liss, New York.

National Park Service, Department of the Interior
1999a Notice of Inventory Completion for Native American Human Remains and Associated Funerary Objects from the State of Minnesota in the Possession of the Minnesota Indian Affairs Council, Bemidji, MN. *Federal Register* 64(152):43211–22.

1999b Notice of Draft Principles of Agreement Regarding the Disposition of Culturally Unidentifiable Human Remains—Extended Date for Comments. *Federal Register* 64(145):41135–36.

O'Connell, B. H., and S. M. Myster
1996 *Paleo-Indian and Early Archaic Human Remains in Minnesota.* Presented at the 5th Annual Plains Conference, Iowa City, Iowa.

Ortner, D. J., and A. C. Aufderheide (editors)
1991 *Human Paleopathology.* Smithsonian Institution Press, Washington.

Owsley, D. W.
1999 From Jamestown to Kennewick: An Analogy Based on Early Americans. In *Who Were the First Americans*, edited by R. Bonnichsen, pp. 127–40. Center for the Study of the First Americans, Oregon State University, Corvallis.

Owsley, D. W., and R. L. Jantz (editors)
1994 *Skeletal Biology in the Great Plains.* Smithsonian Institution Press, Washington.

1999a Databases for Paleo-American Skeletal Biology Research. In *Who Were the First Americans?* edited by R. Bonnichsen, pp. 79–96. Center for the Study of the First Americans, Oregon State University, Corvallis.

1999b A Systematic Approach to the Skeletal Biology of the Southern Plains. In *Bioarcheology of the South Central United States*, edited by J. C. Rose, pp. 171–83. Arkansas Archeological Survey Research Series No. 55, Fayetteville.

2000 Biography in the Bones. *Discovering Archaeology* 2(1):56–58.

Owsley, D. W., and J. C. Rose (editors)
1997 *Bioarcheology of the North Central United States.* Arkansas Archeological Survey Research Series No. 49, Fayetteville.

Saunders, S. R., and M. A. Katzenberg (editors)
1992 *Skeletal Biology of Past Peoples: Research Methods.* Wiley-Liss, New York.

Steele, D. G.
1997 Letter to A. Dansie, Nevada State Museum, Carson City.

Steele, D. G., and J. F. Powell
1992 Peopling of the Americas: Paleobiological Evidence. *Human Biology* 64(3):303–36.

1993 Paleobiology of the First Americans. *Evolutionary Anthropology* 2(4):138–44.

1994 Paleobiological Evidence of the Peopling of the Americas: A Morphometric View. In *Method and Theory for Investigating the Peopling of the Americas,* edited by R. Bonnichsen and D. G. Steele, pp. 141–63. Center for the Study of the First Americans, Oregon State University, Corvallis.

Tuohy, D. R., and A. J. Dansie.
1997 New Information regarding Early Holocene Manifestations in the Western Great Basin. *Nevada Historical Society Quarterly* 40(1):24–53.

Turner, C. G.
1987 Late Pleistocene and Holocene Population History of East Asia Based on Dental Variation. *American Journal of Physical Anthropology* 73(3):305–21.

1990 Major Features of Sundadonty and Sinodonty, including Suggestions about East Asian Microevolution, Population History, and Late Pleistocene Relationships with Australian Aboriginals. *American Journal of Physical Anthropology* 82(3):295–317.

Wheeler, S. M.
1997 Cave Burials near Fallon, Nevada. *Nevada Historical Society Quarterly* 40(1):15–23.

Wolpoff, M. H.
1980 *Paleoanthropology*, first edition. Knopf, New York.

VI. Conclusions

Where Are We Going? Public Policy and Science

Alan L. Schneider[1] and Robson Bonnichsen[2]

Introduction

The papers presented in this book illustrate the remarkable progress that has been made in First Americans studies over the past three decades. New research tools such as DNA analyses, AMS radiocarbon dating, and advanced statistical programs for analyzing data have made it possible to pursue lines of inquiry that were scarcely imagined when the Clovis-First theory was initially proposed. New, more refined models for explaining human settlement of the Americas are being explored. New questions are being asked, and old explanations are being reexamined (Bonnichsen, this volume). In short, First Americans studies are maturing.

Alongside these advances, however, have come other developments that are not as positive for the future of First Americans research. Science and its explanations for understanding the world are being challenged by critics who advocate nonempirical ways of explaining empirical phenomena. These critics claim that such nonempirical explanations should be given parity with (if not priority over) scientific explanations, and that all pursuit of knowledge should be subordinate to the political or other ideological agendas currently favored by the critics (Harding 1997; Johnson 1996; Tsosie 1999). Study of the human past (both historic and prehistoric) is not exempt from such demands for a reordering of priorities. Research involving indigenous human skeletal remains, sites, and artifacts has become particularly controversial. If the critics of science have their way, future advances in understanding the peopling of the Americas will come to depend more on political considerations than on hard work and scientific techniques.

[1]1437 S.W. Columbia, Suite 200, Portland, OR 97201; e-mail: alslawoffices@uswest.net

[2]Director, Center for the Study of the First Americans, deceased December 2004. Address all inquiries to Michael R. Waters, Director, Center for the Study of the First Americans, Department of Anthropology, Texas A&M University, College Station, TX 77843-4352; e-mail: mwaters@tamu.edu

These external threats to First Americans research are the subject of this paper. The first two sections examine how government policies and museum practices are being affected by the controversy over who should control the past. Some of the effects include the enactment of disposition (i.e., repatriation) statutes that have resulted in the reburial of many human skeletal remains, including some of the oldest remains ever discovered in the United States. Many of the remaining government and museum skeletal collections are in peril of similar treatment. In addition, researchers are beginning to experience difficulties in gaining access to federally controlled collections, sites, and data. Depending upon the institution involved, similar difficulties can be encountered in gaining access to some state and museum collections.

The next two sections of this paper examine how the principal scientific associations and various segments of the American public have responded to these challenges to scholarly investigation of the past. For all practical purposes, the scientific associations have been passive spectators in the battle over control of the American past. To date, they have displayed little effective leadership in this area. Reporters, editors, publishers, and the public in general seem interested in First Americans issues and are generally sympathetic to the interests and goals of researchers. So far, however, such interest has not translated into positive action at the political level.

The final section offers some general observations on the political and cultural challenges now confronting First Americans researchers, and some recommendations on what researchers must do if they wish to prevent their position from deteriorating any further.

Part I. Government Policy

Unlike many European countries, the United States does not have a unified national policy for managing the nation's prehistoric heritage. Instead, legal authority to regulate such matters is divided between the federal government and the 50 states. On the state level, regulatory jurisdiction may be further subdivided between the state government and various local governments (counties, cities, etc.). In addition, decision-making authority within each level of government is shared by legislatures, administrative agencies (e.g., Army Corps of Engineers, Department of the Interior), and courts. ers from different levels and branches of government interact with one another in many direct and indirect ways, thereby creating a complicated feedback system for formulating and applying government policy. As a result, decisions made at one level or by one branch of government can affect ing in other parts of the system.

Because of these factors, First Americans researchers in the United States are confronted with a legal environment consisting of an array of policies that are often inconsistent, vague and difficult to interpret, and subject to unpredictable change. The following discussion focuses primarily on policies and programs stemming from federal laws, with occasional references to some pertinent state laws. County and other local government laws are not separately discussed since they are usually patterned after federal or state models. Discussed are those policies and laws relating to:

- disposition (or ownership) of human skeletal remains and other archaeological items
- study of collection materials
- access to information
- study of archaeological sites
- funding of research.

Disposition of Remains

As demonstrated by some of the papers in this book, ancient human skeletal remains can provide important information about the peopling of the Americas. Such remains are the most direct source of information about the biology of early New World populations—their genetics, skeletal characteristics, age at time of death, sex, diseases, etc. Despite their scientific importance, however, some of the oldest and best-preserved remains have been transferred to the possession of native tribal claimants who have no known affiliation to the remains in question. In each case, the remains were buried in a location known only to the participating tribes. Depending on future political and legal developments, many of the remains that are still left in federal and state collections could suffer the same fate.

Background

The repatriation movement in this country was initially an informal process that concentrated largely on items that were clearly affiliated to a modern tribe. For example, a war god was voluntarily returned by the Denver Art Museum in 1978 to the Zuni, and in 1989 Stanford University transferred more than 500 Ohlone human remains to modern tribal members (DuBoff 1992). In 1990, the process became more formal with the passage of federal legislation on the subject. Disposition or repatriation laws were subsequently enacted in many states.

Terminology

The word "repatriation" is often used today to refer to all situations in which ownership of human remains or other cultural items is transferred to a tribal claimant regardless of the circumstances of the transfer. However, the word "repatriation" implies that something is being returned to a group to which the object once belonged. It is debatable whether such a relationship exists in many of the so-called "repatriations" that have occurred to date. This article will use the terms "disposition," "transfer of possession," or something equivalent when referring to custody changes for remains or items that have not been shown to be culturally affiliated to the receiving tribe.

Federal Statutory Structure

At the federal level, the principal statute regulating these matters is the Native American Graves Protection and Repatriation Act of 1990, 25 U.S.C. §3001 et. seq. ("NAGPRA"). Under NAGPRA, culturally affiliated tribes can claim possession of Native American human remains and other "cultural items" that either: (a) are found on federal or tribal land after 16 November 1990; or (b) were held in a federal or museum collection on 16 November 1990 (excluding the Smithsonian Institution, which is subject to a separate repatriation law). Cultural affiliation, however, is not the only basis for a claim under NAGPRA. In the case of items found on federal or tribal land after 16 November 1990, proof of a cultural relationship is not needed if the discovery occurred on a tribe's tribal land (e.g., on its reservation) or on land that the tribe has been judicially determined to have aboriginally occupied at the time of European contact. In addition, the Department of the Interior is considering proposed regulations to allow for the disposition and burial of culturally unaffiliated remains. See "Proposed Regulations" below. For more details of NAGPRA see Schneider 1995.

State Statutory Structure

Practices at the state level vary from jurisdiction to jurisdiction. Some states allow claims only for items found on land owned by state or local government agencies, while others extend the claims process to include "Native American" or "Indian" remains and cultural items found on private land. Some states require proof of a cultural relationship to the items being claimed, while others dispense with any need to prove such a relationship. An example of the latter is Oregon's reburial laws, which allow a tribe to claim any "Native Indian" human remains or funerary object found in the tribe's area of interest as established by the State Office of Historic Preservation (Oregon 1993a, 1993b). Such areas of interest can be quite arbitrary and in some cases may be based on nothing more than suspected settlement or foraging patterns at the time of European contact.

Government Transfers

Government agencies have already transferred to tribal claimants some of the oldest skeletons found in the United States. As far as the authors can determine, in none of these cases was there any credible evidence to establish any legitimate cultural affiliation between the remains and the claiming tribe. The transferred skeletal remains were:

Federal Hourglass Cave, a 7700- to 7900-year-old partial skeleton was transferred under NAGPRA by the U.S. Forest Service to the Southern Ute Tribe in 1993 (Hall 1997). Affiliation appears to have been inferred solely from the location of discovery. However, geography alone is not sufficient to establish a valid NAGPRA claim unless there has been a prior (and final) determination by the Court of Claims or the Indian Claims Commission that the claiming tribe aboriginally occupied the discovery area (25 U.S.C. 3002(a)(2)(c)). Only limited study of this skeleton was conducted before it was transferred and buried.

Idaho Buhl Woman, a 10,675-year-old nearly complete skeleton, was transferred by Idaho state authorities to the Shoshone-Bannock Tribe in December 1991 (Wisner 1992). The only known connection between the skeleton and this tribe was the chance fact that it was found in the vicinity of the tribe's present reservation. Thomas Green, Idaho State Archaeologist, later declared in a private communication that Idaho law does not require proof of "*any* degree of cultural affiliation" (Green 1992; underlining in original). Subsequent reanalysis of cranial measurements recorded by the only investigator of the skeleton indicates that it does not resemble modern American Indians (Jantz and Owsley, this book).

Minnesota In late 1999, the Minnesota Indians Affairs Council transferred the skeletal remains of 1,059 individuals to a coalition of Upper Midwest tribes. Shortly thereafter, all the remains were buried on a tribal reservation in South Dakota. Included in the transferred remains was a partial skeleton known as Browns Valley Man, which has been dated to approximately 8900 years RCYBP (Minnesota Indian Affairs Council 1999, p. 43215; Steele and Powell 1994). Important (and possibly irreplaceable) information about the biological affinities of early New World populations may have been lost with the burial of this skeleton. Its cranium is morphologically quite different from modern American Indians and many other early New World individuals such as Spirit Cave, Wizard Beach, and Kennewick Man (Owsley and Jantz 1999). The only reported connection between this skeleton and the tribal coalition was the fortuitous fact that coalition members happen to have occupied or used the area of the present state of Minnesota many thousands of years after Browns Valley Man died. Even the Minnesota ers conceded that "there is no relationship of shared group identity" that could be traced between any of these remains and any present-day Indian tribe or group (Minnesota Indian Affairs Council 1999, p. 43221). Minnesota law was cited as justification for the transfer.

Also included in the 1999 Minnesota transfers was a partial skeleton known as Minnesota Woman (or Pelican Rapids). Like Browns Valley Man, the morphology of this skeleton's cranium (approx. 7,800 years old) does not link closely to any known North American population, whether modern or prehistoric (Owsley and Jantz 1999). The Council's NAGPRA notice stated that the skeleton (and its associated funerary objects) "are associated with the Archaic Tradition, a broad archeological classification which cannot be associated with any present-day Indian tribe or group" (Minnesota Indian Affairs Council 1999, p. 43216). Also included in the transfer were numerous other Archaic-age skeletal remains, one example being Sauk Valley Man (Minnesota Indian Affairs Council 1999, p. 43218). The latter remains have been dated to approximately 4360–4190 yr B.P. (Myster and O'Connell 1997).

Other Tribal Claims

Tribal claims also have been made for other ancient skeletal remains still held in federal or state collections. Two prominent examples of such claims are:

Kennewick Man In September 1996, the U.S. Army Corps of Engineers announced plans to transfer custody of this nearly complete 8,400-year-old skeleton to a coalition of five local Indian groups (Chatters 2001, p. 53, 89). The Army Corps decision was subsequently set aside by a federal court for lack of evidence and procedural errors (*Bonnichsen et al. v. U.S.* 1997b). On 21 September 2000, the Department of the Interior and the Army Corps reaffirmed the earlier vacated decision (*Bonnichsen et al. v. U.S.* 2000a, 2000b). The local tribal coalition was said to be culturally affiliated to the skeleton because of the location of the discovery and inferences derived from tribal creation accounts. These new decisions were challenged in court and set aside. For more information about the case see Postscript at the end of this article. Copies of court documents can be found at the Friends of America's Past Web site (**www.friendsofpast.org**).

Spirit Cave The Fallon Paiute-Shoshone Tribe has claimed this 9,400-year-old skeleton on the basis of geography and tribal creation accounts. On 26 July 2000, the Bureau of Land Management rejected the Tribe's claim. In a 114-page report reviewing available biological, archaeological, anthropological, linguistic, oral tradition and other data, BLM concluded that the remains "are unaffiliated with any modern individual, tribe, or other group" (BLM 2000a, p. 66). In a companion document, one of the authors of the BLM report separately wrote: "The antiquity and rarity of Spirit Cave and other Paleoamerican remains and artifacts are critically important to the biological history of contemporary Homo Sapiens. These facts make Spirit Cave Man and other Paleoamerican remains National Patrimony" (Damadio 2000). The NAGPRA Review Committee later issued an opinion supporting the tribal claim (NAGPRA Review Committee 2002). This decision, however, has no operative legal effect. Under current NAGPRA regulations, decisions of the Review Committee are not binding on federal agencies or any other party (35 C.F.R. 10.16(b)).

Recent State Legislation

Several states have recently passed laws that are particularly noteworthy in their potential effect on the preservation of archaeological collections. Legislation adopted in California in 2001 provides that tribal oral histories, documentation, and testimony "shall not be afforded less evidentiary weight" than other lines of evidence when resolving repatriation disputes (California 2001, §8016(h)). This new law also allows tribal groups to nominate seven of ten voting members of the Oversight Commission charged with enforcing California's repatriation laws (§§8025 and 8026). In early 2002, Maine enacted a law requiring that all "Indian human remains" in the possession of any private person must be transferred to a "repatriation organization" appointed exclusively by native groups (Maine 2002). Transfer is required regardless of the age of the remains in question and does not depend upon any proof of cultural affiliation. Montana adopted a law in 2001 that permits tribes to claim human remains and funerary objects that are held in private as well as state collections (Montana 2001). Although the lithic artifacts found with the 10,800-year-old Anzick burial are expressly excluded from the application of this new law, no exemption was given for the Anzick remains themselves or for any other collection (Montana 2001 at 22-3-921).

Proposed Regulations

The National Park Service has proposed that the Secretary of the Interior adopt new NAGPRA regulations to authorize the disposition of culturally unaffiliated human remains and funerary objects (Department of Interior 2002). One versoin proposed in 2002 would allow such remains and objects to be transferred to any tribe that once occupied, or now has a "cultural relationship" to, the area where the remains or objects were found. If the geographic origin of the remains or object is unknown, it could be given to any tribe that has a cultural relationship to the region where the possessing museum or federal agency is located. The premise behind these proposed regulations seems to be the concept that all prehistoric human remains and funerary objects found in the United States should be buried regardless of their ethnic affinities or scientific values (Department of Interior 2002a, section entitled "Supplementary Information"). The author of the proposed regulations, C. Timothy McKeown of the National Park Service, has openly advocated that NAGPRA should be interpreted in accordance with tribal wishes. For example, in one article, he wrote: "The substantive information upon which claims to [cultural] property are made will come from the tribes, rather than being dictated to them" (Hutt and McKeown 1999, pp. 388–89). He was one of the advisors consulted by the Army Corps of Engineers when it reached its controversial (and later overturned) decision to give the Kennewick Man skeleton to tribal claimants based solely upon geography and the age of the skeleton. He reportedly advised the Army Corps that it "couldn't have made another choice" (*Bonnichsen et al. v. U.S.* 1996b).

Access to Collections

A fundamental reason for gathering the collections of skeletal remains and other archaeological items that were compiled by government agencies over the past 150 years was to provide study materials for scholarly research. Prior to NAGPRA, it was generally agreed that such collections should be made available to all qualified scientists and students. The only limitations typically placed on access (or handling by) researchers were those reasonably needed to protect a collection. These assumptions are now being challenged. Some American Indian and academic critics of science now demand that tribal consent be obtained for study of any human remains or artifact that a tribe considers culturally sensitive. Some also contend that all reports and data relating to such items should be treated as the exclusive property of interested tribes and should not be released without their permission. Such claims are beginning to influence government decisions concerning access to collections and data.

Federal Practices

Some federal officials have begun to impose restrictions on access to all remains and artifacts that are (or might become) the subject of a tribal claim. NAGPRA is the most common justification given for such actions. The restrictions imposed can take many forms, ranging from a complete ban on study to complicated requirements that are difficult to meet (such as the need to obtain consent from all potentially interested tribes). Study requests may also encounter interminable delays while paperwork is processed and policies are debated by agency ers. The Kennewick Man case provides an example of how such tactics can be used to frustrate research. Government representatives initially refused to give a definite "yes" or "no" response to the scientists' request to study the skeleton (*Bonnichsen et al. v. U.S.* 1997a). Later when new administrative proceedings were ordered by the court, the scientists' study request was put on hold by the Department of the Interior while it completed its affiliation investigation (*Bonnichsen et al. v. U.S.* 1999b, note 10, p. 12). Government representatives even refused to state whether they would allow study of the skeleton if it was determined not to be Native American within the meaning of NAGPRA, or if it could not be affiliated to any present-day tribe (*Bonnichsen et al. v. U.S.* 1999c, p. 9; 1999d, p. 35). Finally, after four years, the scientists were told that they would not be permitted to study the skeleton unless tribal consent was obtained (*Bonnichsen et al. v. U.S.* 2000b, p. 6). Government documents on file in the case indicate that the refusal to permit study of the Kennewick skeleton reflects decisions made at the highest agency levels in Washington, D.C. In other cases, study restrictions originate at the local level. An example of the latter is the decision by the Nevada State Director of the Bureau of Land Management to prohibit further "invasive testing" of Spirit Cave Man because of "the cultural sensitivity of the materials" (BLM 2000b).

State Policies

Conditions are equally unsettled at the nonfederal level. It is not uncommon for state agencies to limit or prohibit scholarly study of items claimed to be culturally sensitive. The items most commonly restricted are human remains. For example, University of Nebraska officials refused to allow study of their culturally unaffiliated human remains prior to transfer and burial of the remains (Grew 1998). It is sometimes claimed (incorrectly) that NAGPRA requires tribal consent for study of all potentially repatriable items, even those that have not been culturally affiliated. Requirements imposed (or thought to be imposed) by state repatriation and reburial statutes also may be offered as a justification for restrictions on study. Researchers' rights are particularly vulnerable to adverse legislation at the state level. State legislatures tend to be more susceptible to pressure from local interest groups, and tribal influence at the state legislative level is growing. For example, Montana tribes reportedly employed eight lobbyists to work for passage of the new Montana law noted in the preceding subsection. Some state laws seem specifically designed to prevent scientific study of human remains. The new burial law enacted by Maine in early 2002 repealed an existing state statute that allowed "persons skilled in the anthropological and archaeological fields" to study remains for up to one year before they could be transferred to native claimants (compare Maine 2002 and 1993).

Access to Information

Restrictions on Access

On occasion NAGPRA has been used as a pretext to deny researchers access to information about items held in government collections. One such attempt was made in the Kennewick Man case. There the government argued that CT scan data of the skeleton could be withheld from scientists because, among other things, such data are said to be considered sensitive by the tribes who opposed its release (*Bonnichsen et al. v. U.S.* 2001a). It required a court order to obtain release of the data (*Bonnichsen et al. v. U.S.* 2001b). Tribal demands for restrictions on release of data are often couched in ownership terms. For example, in the Kennewick Man case, Yakama representatives asserted that distribution of data obtained from study of the skeleton raised "issues of intellectual property" (*Bonnichsen et al. v. U.S.* 1998b). Colville representatives argued that all information concerning the skeleton was the "exclusive property" of the tribal claimants and should be "kept closed and reburied with the skeleton" (*Bonnichsen et al. v. U.S.* 1998a, 1999a).

Government Monopolization of Studies

The Department of the Interior has taken the position that NAGPRA limits only the study of skeletal remains by members of the general scientific community, but does not bar government agencies from conducting studies of their own. According to Dr. Francis McManamon, Interior's Chief Consulting Archaeologist, such government-directed studies are sufficient to "allow good science to go forward" (Department of Interior 1999). The authors disagree. Such a policy is not consistent with good science. The issues that interest First Americans researchers may not be a priority for government officials who often have other objectives. In addition, government officials may not have the expertise needed to plan and coordinate complex interdisciplinary research projects. Furthermore, studies that are conducted in secret and are not submitted to peer review and independent verification are the antithesis of the scientific method. Finally, monopolistic control over research opportunities would enable government officials to promote the theories they favor and discourage investigation of theories they oppose. Such a desire to promote personally favored theories of New World human origins appears to have influenced selection of the studies conducted by the government in the Kennewick Man case (*Bonnichsen et al. v. U.S.* 2001c, note 48, p. 31).

Access to Sites

The ability of researchers to seek out and investigate early sites is critical to future progress in First Americans studies. New data from carefully investigated sites are essential to answering important questions about the peopling of the Americas, such as when humans entered different parts of the New World and the adaptive strategies they used after arrival. Obtaining access to sites, however, may become increasingly difficult as archaeological matters become more politicized and regulated. One concern is that tribal challenges to the investigation of sites said to be sacred or otherwise culturally sensitive could impact the issuance of needed government permits. In addition, research interests are often jeopardized by cultural resource management policies that disfavor excavation of sites.

Permit Requirements

The permit system for regulating activities affecting archaeological sites has become an almost universal practice in this country. At the federal level, the Archaeological Resources Protection Act of 1979 (ARPA) requires that a permit be obtained before any activity (including scientific research) is conducted on federal or tribal land that would alter or damage an archaeological site older than 100 years. Permits are issued by the federal agency or tribe that manages the land in question. Most states have enacted comparable laws requiring a permit for activities affecting archaeological sites located on land owned by the state or any local government. Many of these state laws also apply to sites located on private land. The original justification for such permit laws was to protect archaeological sites so they would be available for future scientific investigation. That purpose is now only one of many objectives promoted by the permit system. It is not unusual now for scientific interests to be given a low priority when determining the permissible uses that can be made of a site.

A NAGPRA Example

The Kennewick Man case is an example of how the permit system can be used to block legitimate scientific study of an important early site. In August 1997, five scientists applied to the U.S. Army Corps of Engineers for a permit to investigate the terrace overlooking the river beach where the Kennewick Man skeleton was found (*Bonnichsen et al. v. U.S.* 2001c, p. 33). Their objectives were to establish the terrace's age and geologic characteristics, and to gather data concerning the skeleton's taphonomic history. Local tribes objected, claiming that the terrace was a sacred site (*Bonnichsen et al. v. U.S.* 2001c, Extract of Record, pp. 292–93). Despite a lack of evidence verifying any prior religious or other cultural use of the terrace, the Army Corps accepted the tribes' objections and refused to issue the requested permit. In April 1998, after conducting only a limited survey of the area (Chatters 2001, pp. 155–56), the Army Corps buried the beach and adjacent bank under 600 tons of rock, fill, and organic materials (*Bonnichsen et al. v. U.S.* 2001c, p. 35; Science 1998).

A State Law Example

In May 2000, one of the authors (Bonnichsen) applied for an Oregon permit to investigate a site on private land near Corvallis, Oregon. Excavation of a cattle watering pond at the site had unearthed an assortment of Pleistocene faunal remains, including fragments of ground sloth, mammoth, cervid, and canid. Also found was a diaphysis and distal end of a human humerus from a small individual. The age, ethnicity, and original depositional context of this fragment are unknown. Tribal objections have prevented further investigation of the intact sediments around the pond and any radiocarbon dating or other testing of the humerus fragment. In the authors' opinion, such objections are a misuse of the Oregon law that allows tribes the right to veto scientific investigation of an Indian "burial" located on public or private land (Oregon 1993a). Without more data, it is mere speculation at this point to conclude that this humerus fragment is "Indian" within the generally accepted meaning of that term or that the site is actually a burial location.

CRM Policies

Cultural resource management (CRM) considerations play a major role in the permit process and in the formulation of government land-use plans. Policies favored by CRM personnel can determine which sites will be available (or unavailable) for research and the conditions under which they can be investigated. Of particular concern for First Americans studies is the growing trend in CRM quarters to view in situ preservation as a preferred alternative to data recovery through controlled excavations (regardless of how limited). Although in situ preservation is a legitimate and often necessary technique, it must be balanced with other considerations. Often too little recognition is given to the fact that researchers must be allowed to excavate sites if progress is to be made in understanding the American past. Land managers also often forget that all sites are subject to ongoing deterioration from natural or human causes, and that in many cases excavation may be necessary to ensure that essential data are not lost.

Research Funding

Funding for First Americans research has never been generous. Most researchers must rely upon financial support from a variety of sources, and few have the level of funding they consider adequate for the work that needs to be done. Even when government funds are only a modest fraction of a project's total financing, they can provide the critical difference that enables the project to go forward. Loss of even a small part of these resources would severely impact the ability of researchers to develop more refined explanations for the peopling of the Americas.

Thus far the debate over access to the past does not appear to have caused any appreciable reduction in the amount of government funding available for First Americans research. However, the situation could change at any time. Government officials who are willing to accept nonempirical interpretations of the past over scientific interpretations could gain greater influence over future funding decisions. In addition, as control of the past becomes more politicized, officials who are otherwise sympathetic to research may decide to choose expediency over principle when faced with potentially controversial funding decisions.

Other Comments

Many government policy-makers seem to view the pursuit of knowledge as a secondary (or even irrelevant) consideration when resolving questions over who should be given access to the past. One proponent of such a view is the NAGPRA Review Committee, which has repeatedly advocated that all culturally unaffiliated remains be transferred to native claimants for burial regardless of the scientific importance of the remains (NAGPRA Review Committee 2002; Department of Interior 2002a, section entitled "Supplementary Information"). According to the Review Committee, scientists and members of the public should not be allowed even to participate in the decision-making process without tribal consent (NAGPRA Review Committee 2000, p. 36463). On the other hand, there have been some positive developments such as BLM's decision on the Spirit Cave skeleton and the so far successful struggle to save the Kennewick Man skeleton. It is too early to tell which of these views of the past—one in favor of scientific inquiry, one opposed—will be more influential. Given the nature of the American political system, a tidy all-encompassing resolution of the issues raised by these conflicting viewpoints may be difficult to achieve.

Part II. Museum Activities

In addition to government agencies, NAGPRA also applies to all institutions of "higher learning" and other organizations that receive federal funds and have possession or control of Native American cultural items as defined in the Act. NAGPRA refers to all these institutions as "museums." This specialized usage of the word "museum" can be misleading since it includes organizations not commonly thought of as museums. However, to avoid even more confusion, this paper will follow that usage of the term.

Museums as so defined are subject to NAGPRA with respect to all Native American cultural items (i.e., human remains, funerary objects, sacred objects, and objects of cultural patrimony) that were in their possession or control as of 16 November 1990. The source of such cultural items is irrelevant. NAGPRA controls their ownership and disposition regardless of whether they were found on federal, state, or private land, and in many cases regardless of whether they were acquired by gift, purchase, exchange, or other means. The National Park Service has argued that NAGPRA should be extended to cover all collections acquired by museums after 16 November 1990 (Department of Interior 2002b). Museums that are not owned by the federal government are also subject to state-enacted disposition laws. Such state laws may expand the types of items that can be claimed and the types of persons or organizations that can pursue claims. However, state law cannot prohibit or limit claims that would otherwise be valid under NAGPRA. The California and Montana statutes discussed earlier are examples of the kinds of expansive laws permitted by NAGPRA.

To date, most museums seem to have treated their repatriation obligations in a responsible manner. Most have been careful in processing tribal ownership claims, and most have generally cooperated with researchers requesting access to collection materials. However, there have been some notable exceptions.

Disposition of Remains

Nebraska

In late 2000, the remains of 330 individuals of "unknown" age were given by the University of Nebraska to a coalition of 14 tribes. The University's NAGPRA notice in the Federal Register stated that the cultural affiliation of these 330 remains could not be established (Nebraska 2000). An earlier inventory of the remains noted that many were too fragmentary to assign any racial or temporal designation (Nebraska 1998). The sole reason given for transferring the remains to the coalition was the claim that these tribes had "historic or aboriginal cultural ties to the area of the present State of Nebraska" (Nebraska 2000). However, aboriginal occupation of an area (by itself) is a justification for awarding custody of human remains under NAGPRA only in the case of new discoveries on federal or tribal land after 16 November 1990 and only where such occupation has been confirmed in a judicial judgment (25 U.S.C. 2002(a)(2)(c)). Moreover, the University's NAGPRA notice conceded that most of the remains were found in "an unknown location by person(s) unknown under unknown circumstances." The University's disposition of these remains was reportedly approved by the Secretary of the Interior (Department of Interior 2002a), thereby releasing the University from the need to comply with the requirement that museums "must retain possession" of all culturally unidentifiable human remains until regulations have been adopted for their disposition (NAGPRA Regulations 1995, 10.9(e)(6)). Priscilla Grew, NAGPRA Coordinator for the University, has since declared that all the University's culturally unidentified human remains have been given to tribal claimants (Grew 2001).

Oregon

Another dubious transfer was a 1999 decision by the Oregon State Museum of Anthropology involving a nearly complete skeleton found by L. S. Cressman in 1940. It, along with the remains of 142 other individuals of unknown age, was given to a coalition of Oregon tribes (Oregon, 1999). Although never radiocarbon-dated, this skeleton was believed to be more than 6800 years old because of its reported association with one of the Mount Mazama ash deposits (Hill 2000). The Museum's NAGPRA notice indicated that "affiliation" of the skeleton was based solely upon the inferred geographic distribution of the Molalla Indians "since precontact times." As previously noted, such a geographic coincidence is a far cry from the proof required for a valid custody claim under NAGPRA. Moreover, the Oregon NAGPRA notice admitted that cultural affiliation with the Molalla Indians was only "possible."

Improper Decision Making

It is difficult to find any reasonable justification for the transfers discussed above (or others like them). They were not compelled by any requirement imposed by NAGPRA, and such ill-conceived decisions have potential harmful consequences far beyond the particular circumstances involved in these incidents. Among other things, they create precedents that put pressure on other museums to disregard the requirements of the law and the museums' trust responsibilities when faced with similar transfer demands, and they encourage decision making based on emotion rather than reason and evidence. Furthermore, the balance between the needs of science and the interests of native claimants is made meaningless when important determinations are left to the personal whim of museum decision makers. It is particularly disturbing that the Department of the Interior has sought to promote such questionable practices. The Secretary has reportedly recommended the disposition of culturally unaffiliated remains to tribal claimants in 21 cases (Department of Interior 2002a). Some of these cases (such as Nebraska) involved multiple remains.

Access to Collections

Depending on the institution involved, researchers can experience difficulties in obtaining access to museum collections of skeletal remains and other objects said to be culturally sensitive. Requests to study such items can be met with unreasonable conditions and delays. For example, one researcher has reported that he was denied permission to study a museum collection of skeletal remains unless he first obtained the written consent of over 20 Indian tribes and groups (D. Owsley pers. comm.). This condition was impossible to meet since at least some of the tribes and groups were known to be opposed to research on human skeletal remains. Study restrictions such as these are shortsighted and inconsistent with the objectives of NAGPRA. Even the NAGPRA Review Committee has acknowledged that additional data can sometimes help to affiliate remains that would otherwise be culturally unidentifiable (NAGPRA Review Committee 2000).

Part III. Scientific Associations

Professional scientific associations can play a significant role in determining whether researchers will have the freedom and support needed to pursue their investigations. The positions these associations take (or do not take) on issues of interest are generally seen by other segments of society as reflecting the views of their members. In addition, they usually have greater access than their individual members to editors and publishers, government officials, and other shapers of public policy, and they are usually better able to coordinate and implement the measures needed to influence government decisions. Moreover, through their publications and other forums, these associations can encourage particular areas of research and can influence members' opinions on political, ethical, and other issues.

Two associations that claim to have a special right to speak for archaeologists are the Society for American Archaeology (SAA) and the American Anthropological Association (AAA). Discussed here are their respective responses to the battle over control of the past and their interest (or lack thereof) in First Americans studies. Also included are comments on the American Association for the Advancement of Science (AAAS), the Society for Archaeological Sciences (SAS) and the American Association of Physical Anthropologists (AAPA).

The Society for American Archaeology

Organization and Objectives

SAA currently has more than 6600 members (SAA 2001a). Membership is open to all persons (whether or not professional archaeologists) who support the Society's objectives (SAA 1997, Art. IV, Section 1). SAA's governing body is a board consisting of four (sometimes five) executive officers and six elected at-large members. SAA was organized to promote and stimulate "research in the archaeology of the American continents" and to encourage public "appreciation" for archaeological research (SAA 1997, Art. II). Despite this professed objective, SAA has been neither forceful nor consistent in its support of research interests against those who seek to politicize the past.

Views on Repatriation

SAA's position on repatriation and related issues has been disjointed. SAA has declared that it "opposes universal or indiscriminate reburial of human remains, either from ongoing excavations or from extant collections" (SAA 1986). Nonetheless, despite the obvious flaws in NAGPRA, SAA was one of the statute's original supporters and claims to have "lobbied hard for the bill's passage" (SAA 1998). SAA made no concerted or public effort to challenge the transfer and burial of the Hourglass Cave skeleton, Browns Valley Man, Minnesota Woman, or any of the other reburied skeletal remains discussed in the preceding section on Government Policy. On the state level, SAA did lobby against the new California repatriation law mentioned earlier. However, it did not oppose passage of the equally flawed Maine and Montana statutes. SAA did support the proposed Hastings amendments to NAGPRA (SAA 1998), which, among other things, would have eliminated claims based solely upon aboriginal occupation of a discovery location and would have clarified the study rights of researchers. These amendments, however, died in committee in 1998.

Kennewick Man Issues

SAA's position in the Kennewick Man case seems equally confused. At the eleventh hour, it filed an amicus brief challenging the Secretary of the Interior's interpretation of what is meant by NAGPRA's definition of cultural affiliation and his decision to give oral creation accounts preference over scientific data in determining the skeleton's affiliation (SAA 2001b). On the other hand, SAA's brief supported Interior's claim that age alone is enough (i.e., if older than A.D. 1492) to classify a skeleton as Native American for NAGPRA purposes. Such a position would exclude science from a critical step in the NAGPRA process and would allow for the "repatriation" of remains that have no demonstrated relationship to modern native people. SAA has said that scientific interests in human remains must be recognized and that conflicting claims over the treatment of remains must be resolved reasonably on a case-by-case basis (SAA 1986). Nonetheless, its amicus brief did not contest the government's arguments that independent researchers can be excluded from study of federal skeletal collections. Moreover, SAA commended the government for its investigations of the skeleton even though they were incomplete and were never subjected to peer review or independent verification. Also perplexing was SAA's failure to challenge the government's decision to bury the skeleton's discovery site despite its incomplete investigation of the site and its failure to make any assessment of the potential damage to the site's contents from the burial and subsequent landscaping.

Information Forums

Papers and posters on new excavations and other relevant First Americans topics are frequently presented at SAA's Annual Meetings. Its journal *American Antiquity* also publishes occasional articles on First Americans topics. SAA's monthly newsletter to members has published articles and letters on issues or events relating to the ongoing struggle over who will control access to the past. Most of these represent the personal views of the writers and do not reflect any official policy on the part of SAA's leadership.

Observations

SAA's support for such a flawed statute as NAGPRA is difficult to reconcile with its stated objective of promoting and stimulating archaeological research. Equally difficult to explain is its inconsistent support of research interests against those who seek to control access to the past for political or other ideological reasons. SAA's record on these matters may reflect an underlying ambivalence on the part of its leaders with respect to the role of archaeology in modern society. If so, First Americans researchers cannot expect to receive forceful support from SAA in future controversies over scientific investigation of American prehistory.

The American Anthropological Association

Organization and Objectives

AAA has more than 10,000 members (AAA 2001). Any person who has "expressed an interest" in anthropology may join the association (AAA 2001). Business affairs are managed by a 15-member Executive Board with input from a 30-member Section Assembly. AAA's objectives include "to promote the science of anthropology, to stimulate and coordinate the efforts of American anthropologists" and "to publish and encourage the publication of matter pertaining to anthropology" (AAA 2001). Part of AAA's mission is to speak for anthropology to other fields, to federal and state governments, and to the public (AAA 2001).

Response to Topical Issues

AAA has done little to support archaeological research interests in the controversies that have arisen during the past two decades. For example, it did not endorse the proposed Hastings amendments to NAGPRA. Likewise, it did little, if anything, to stop passage of the new California, Maine, and Montana repatriation statutes. Like SAA, the leadership of AAA did not challenge transfer and burial of the Hourglass Cave skeleton, Browns Valley Man, Minnesota Woman, or any of the other premature dispositions discussed in the section on Government Policy. AAA did not intervene to defend research interests in the Kennewick Man lawsuit.

Information Forums

AAA's Annual Meetings are not a significant venue for disseminating information on First Americans issues. However, its journal *American Anthropologist* does publish occasional articles on subjects relating to First Americans topics. In addition, AAA's monthly newsletter has featured numerous articles and letters on the Kennewick Man controversy. Its coverage of the subject has been relatively well balanced between supporters and opponents of the government's decisions in the case.

Observations

AAA's lack of support for research interests in the struggle over access to the past is not surprising. AAA is dominated by cultural anthropologists whose interests and agendas are often quite divergent from, and sometimes diametrically opposed to, those of science-oriented archaeological researchers. Some of these differences can be seen in AAA's Code of Ethics (1998) and Statements on Ethics (1986). For example, AAA's Code of Ethics states: "Anthropological researchers should obtain in advance the informed consent of persons . . . identified as having interests which might be impacted by the research." Under current conditions, such a requirement would be impossible to meet for most First Americans research given the number of people who are adamantly opposed to any investigation of the past that conflicts with their religious or political beliefs.

Other Associations

AAAS

The American Association for the Advancement of Science claims to be the world's largest general science organization with more than 138,000 members (AAAS 2001a). Objectives of AAAS include "improving understanding of the relationship between the law and science" and "monitoring ethical, legal, and social issues related to science and technology and bringing them to the attention of scientists, engineers, policymakers, and the public" (AAAS 2001b). Notwithstanding these goals, AAAS has been largely passive in the ongoing public debate over the interplay between government policy and archaeological research in the United States. It did not intervene in the Kennewick Man case or challenge any of the dubious dispositions of skeletal remains discussed in the section on Government Policy. To its credit, the Association's Section H (Anthropology) did express support for the proposed Hastings amendments to NAGPRA (Brace 1998). In addition, the AAAS weekly journal *Science* has carried more than a dozen news stories and updates on developments in the Kennewick Man case.

SAS

The Society for Archaeological Sciences was founded in 1977 to provide a forum for communication among scholars who apply physical science methods to archaeological questions (SAS 2001). It currently has approximately 400 members. Among the Society's purposes are "to advance relations with government agencies and the public in general" (SAS 1984). Nonetheless, SAS has not intervened significantly in any of the developments or other issues discussed in the section on Government Policy.

AAPA

The American Association of Physical Anthropologists has more than 1,800 members and claims to be the world's leading professional organization for physical anthropologists (AAPA 2002a). Membership is open to professional physical anthropologists, advanced graduate students, and other qualified biologists (AAPA 2002b). Like the other professional scientific associations, AAPA has generally been a passive spectator in the various anti-research developments and controversies discussed in the section on Government Policy. However, it did support the Hastings amendments to NAGPRA (SAA 1997). It also issued a forceful rebuke of Interior Secretary Babbitt's decision on cultural affiliation of the Kennewick Man skeleton. Among other things, AAPA observed that

> using the Secretary's logic, anything is possible and there is no basis for evaluating competing ideas with current evidence. Using the Secretary's logic, as the antiquity of a set of remains increases, so does the ease with which the preponderance of evidence is shifted to cultural affiliation with a federally recognized tribe using the flimsiest of arguments based on geographic proximity, folklore, and oral traditions (AAPA 2000 at p.2).

AAPA also declared "where cultural affiliation is absent, repatriation claims have no moral foundation." (AAPA 2000, p. 1.)

Leadership Void

First Americans studies are something of an orphan in the world of scientific associations. SAA claims to represent the interests of archaeological research, but has been neither consistent nor effective in supporting the interests of First Americans researchers against critics who oppose scientific investigation of the past. AAA seems unconcerned about the increasing restrictions that are being imposed on researcher access to prehistoric sites, collections, and materials. AAAS and SAS also seem largely indifferent to the situation, and AAPA's occasional responses to developments affecting researcher rights can best be described as "too little, too late." The reasons for this lethargy probably vary from association to association. Whatever the reasons, it is a problem that archaeologists and other scientists interested in American prehistory must confront. More effective leadership on the part of the professional scientific associations is essential if First Americans researchers hope to withstand the ongoing challenges that are being raised to the legitimacy and value of their work.

Part IV. Public Attitudes

Public opinion is a powerful force in American society. It can determine which activities or causes will receive government and institutional support, and which ones will be marginalized or proscribed. Public attitudes toward First Americans studies are difficult to gauge. To the authors' knowledge, no survey has ever been conducted on the topic. Nonetheless, any assessment of the external forces affecting First Americans studies would be incomplete without some consideration of public opinion. Unlike many other scientific disciplines, the study of prehistory can seldom be justified by reference to revenues generated, lives saved, new foods produced, or other tangible benefits. As a result, public goodwill is critical to the future of First Americans research.

First Americans Coverage

One rough measure of interest in a subject is the amount of attention it receives from the print and broadcast media and in other venues of public communication. By this measure, public interest in First Americans studies has increased significantly over the past decade. The attention that has been given to First American topics is unprecedented for a discipline devoted to the study of American prehistory.

Magazines

Popular archaeology and science magazines have featured a number of recent articles on First Americans topics. Some examples are *American Archaeology* (Fall 2001 and Winter 2001–02); *Discover* (Feb. 1999); *Discovering Archaeology* (Feb. 2000); *National Geographic* (Dec. 2000); *New Scientist* (17 Oct. 1999); *Scientific American* (Feb. 1999 and Sept. 2000). National mass-circulation magazines have also shown an interest in the peopling of the Americas. *U.S. News & World Report* made the subject its main cover-page story on 12 October 1998. *Newsweek* did the same on 16 April 1999 and followed with a shorter article on 15 November 1999. A 14-page article entitled "The Diffusionists Have Landed" appeared in *Atlantic Monthly* (Jan. 2000).

Newspapers

Over the past five years, many newspapers (both large and small) have carried articles on First Americans topics. Some of them are: *Los Angeles Times, New York Times, Oregonian, Sacramento Bee, Seattle Post-Intelligence, Tucson Citizen,* and *Washington Post.* A recent study by C. Loring Brace and other researchers comparing New World and Asian cranial characteristics was reported by the *Associated Press,* and articles on the study appeared in numerous newspapers (front page on some). A few years ago such a specialized study would have received attention only from scientists.

Other Indications of Interest

The past three years have seen the publication of several First Americans books written for nonspecialists: *Bones, Boats & Bison* (Dixon 1999); *The Settlement of the Americas* (Dillehay 2000); *Bones: Discovering the First Americans* (Dewar 2002). A 60-minute show on the peopling of the Americas aired on NOVA on 15 February 2000.

Significance of Coverage

The amount of coverage given to First Americans topics by the print and broadcast media suggests that there is a growing interest in the subject on the part of a broad spectrum of the American public. It can be presumed that editors, publishers, and producers have some sense of what interests the public, and that they will usually choose their stories to match those interests. The quality of the coverage also suggests that editors, writers, and producers appreciate the importance of the subject and are fascinated with the methods used to investigate it. Such persons could be important allies for researchers since they are uniquely positioned to influence public opinion and command the attention of legislators and other policymakers.

Kennewick Man Coverage

Another indication of the growing interest in First Americans studies is the enormous amount of attention that has been given to the Kennewick Man case. Few archaeological stories in the past 50 years have generated as much media coverage as this case. In addition to its more sensational elements (e.g., adversarial litigation, religious fundamentalism, agency defiance of Congress), the Kennewick Man case touches on a legitimate unsolved scientific mystery—who were the First Americans and where did they come from.

Scope of Interest

Discovery of the skeleton and the ensuing lawsuit generated hundreds of articles in newspapers and magazines, reports on television and radio, five books, and a number of Internet Web sites. Five members of Congress (three representatives and two senators) wrote letters to the U.S. Army Corps of Engineers protesting its refusal to permit scientific study of the skeleton and its discovery site (*Bonnichsen et al. v. U.S.* 1996a, 1996c). Publicity generated by the case was a major stimulus for introducing the proposed Hastings amendments to NAGPRA. For more examples of the widespread attention that has been given to the case, see the Appendix at the end of this paper.

Effect on Public Attitudes

The Kennewick Man case has undoubtedly helped to raise public awareness of First Americans studies and the restrictions being imposed on researchers. According to a poll conducted by Harris Interactive in early 2000 (SAA 2000), the chief sources of information about archaeology for members of the public are television (56 percent), magazines (33 percent), and newspapers (24 percent). Individuals who use these sources for information are not likely to have escaped exposure to the Kennewick Man case over the past five years. What they will have learned is likely to be positive for First Americans researchers. Many of the articles and other reports on the case have been favorable to the goals of First Americans researchers. For example, the *Washington Times* stated in an editorial:

> This is a landmark case, with profound implications. Should Kennewick Man be forever lost to research for, in essence, fear of what he might teach us about our past, freedom of inquiry will have lost another battle to the forces of political correction—a reason to fear for the future of knowledge itself (27 April 2001).

Another writer commented that

> No single person or group owns the past; we all do, collectively. And it is only through a spirit of scientific inquiry that we may learn the answer to that fascinating question: How did the New World come to have such people in it? (*National Review* 25 June 2001).

Observations

Public interest in the work of First Americans researchers has increased substantially over the last decade. What was once a relatively unknown field of research has now become a topic for discussion in newspapers, magazines, books, and television programs. This increased level of public interest can be attributed, at least in part, to two factors. First, the new ferment in scientific thinking about the peopling of the Americas has added an element of excitement that was missing under the conventional Clovis-First model. The American public enjoys a mystery, and the story of human colonization of the Americas is now seen as abounding with unsolved mysteries. Second, the Kennewick Man case has increased public awareness of First Americans issues and provided a vivid example of how study of the past can be held hostage to political and religious forces.

So far, however, the public has not been engaged in the debate over who should control American prehistory. According to the 2000 Harris Interactive poll, a majority of Americans (60 percent) believe in the value of archaeological research (SAA 2000). Most also believe that archaeology should be protected and promoted through appropriate laws. There is no reason to think that this desire to support archaeology does not extend to First Americans studies and the need to maintain researchers' access to the past. The problem is not a lack of public goodwill, but rather a failure by the scientific community to mobilize it effectively (see e.g., Lepper, in press).

Part V. Concluding Comments

Despite all of the advances that have been made in scientific techniques over the past two decades, the climate in the United States for scholarly investigation of American prehistory has deteriorated appreciably. NAGPRA and equivalent state laws have resulted in the burial of many important skeletal remains, and a similar fate could happen to many of the other ancient remains still in federal, state and museum collections. Access to government and museum collections and to archaeological sites on public land is becoming increasingly problematic, and demands are being made for tribal control over data in federal records. Because of these obstacles, researchers are becoming more and more dependent upon collections and sites owned by private parties. As the laws adopted in Maine and Montana demonstrate, however, not even these resources are entirely safe from anti-research legislation.

Whether they like it or not, First Americans researchers are participants in a contest between two competing views of the roles of science and public policy. One view, which stems from the libertarian traditions of the Western Enlightenment, asserts that knowledge should be the common heritage of all people and that individuals should not be regulated in what they think, say, or question. The other view, as characterized by some of the more extreme versions of postmodernism and identity politics, asserts that knowledge is merely another form of power and accordingly should be controlled for political and ideological purposes. There is little room for compromise or coexistence between these two views of knowledge because they proceed from diametrically opposed premises and objectives. The outcome of the contest between these views will have as much bearing on the future of First Americans studies as will technological advances.

The struggle taking place over control of and access to materials and sites relating to American prehistory is part of a worldwide phenomenon. Archaeology and related disciplines are favorite targets of those who reject the scientific method and all that it represents. The critics of science argue that knowledge of the past should be controlled and used to promote indigenous rights or other political/religious objectives. Such arguments have led to restrictions on the study of human prehistory in a number of countries (e.g., Australia, Canada, Israel). In Tasmania, for example, archaeological fieldwork cannot be conducted unless aboriginal representatives conclude that it is relevant to their particular interests (Mulvaney and Kamminga 1999, p. 396). Similar restrictions are sought in the United States and justified by claims of "cultural integrity" and "human rights" (Tsosie 1999, p. 654). Knowledge of the human past is seen not as the common heritage of all people but as the exclusive property of those who claim special "caretaker" status.

For the most part, First Americans researchers (and archaeologists in general) have remained silent while this battle over the past unfolds. Such silence is perplexing since the fate of their profession is at stake. The assault on science cannot be dismissed as a mere passing fad that will be inevitably die of its own accord if ignored long enough. Ignorance, intolerance, and superstition are powerful forces that do not die easily (Sagan 1996). If not resisted, they can become the prevailing norm (at least for a time), and the havoc they can create can take generations to correct. Some researchers believe that the assault on science can best be defused through conciliation and compromise. They are mistaken. The hard-core critics of the scientific method do not believe in compromise. Their goal is to suppress all scientific interpretations of American prehistory that do not further their preferred religious and political interests. In the view of one commentator, such subordination of science to special minority interests is appropriate since "the discipline of science, like that of history, is not neutral" (Tsosie 1999, p. 630).

Thus far, most scientists have chosen to leave the defense of archaeological research to the elected leadership of the professional scientific associations. After all, isn't that what they were elected to do? Such a response by researchers is understandable, but misguided. To date, there has been little effective leadership on these issues by the professional associations. They seem unable, or perhaps unwilling, to articulate and defend the scientific method against those who seek to substitute their version of the truth for open, empirically based study of the past. In many cases, it has been the leaders of the professional associations who have advocated compromises that are gradually destroying independent scholarly study of the American past one costly piece at a time.

First Americans researchers cannot afford to remain on the sidelines any longer while the fate of their discipline is being determined. They must begin to speak out and let the rest of society know what researchers do and why it is important. They must make themselves available to reporters and magazine writers, and they must take advantage of every opportunity to speak to amateur archaeological societies, business groups and other interested audiences. They must write more articles for popular magazines, and they must begin to take notice of and (where appropriate) challenge what is written in school textbooks about American prehistory. First Americans researchers must demand more effective leadership from their professional associations, and if that leadership is not forthcoming, they must develop it within their own community.

Of critical importance is the need to forge alliances with potential supporters in native communities. The critics of science seek to create the impression that all Americans of Indian ancestry are opposed to scientific investigation of the past. We do not believe that is true. From what we have seen, many Native Americans support the goals and methods of scientific research provided it is done in a respectful and sharing way (see e.g., Lippert 1997). To date, these potential supporters of science have been largely silent in the debate over control of the past. They must be encouraged to speak out, since their views would be extremely influential with legislators and other government ers.

The situation confronting First Americans researchers is far from hopeless. As BLM's decision in the Spirit Cave Man case demonstrates, there are still some bastions of rational thought in the government agencies that have jurisdiction over public lands and collections. The Kennewick Man case is another reason for guarded optimism. If this particular skeleton can be saved, it will provide a strong precedent in favor of a more balanced interpretation of NAGPRA and possibly other federal and state laws affecting research interests. In addition, large segments of the American public appear to be interested in what First Americans researchers do and are sympathetic to their research goals. If properly encouraged and mobilized, such public support could help to stem, and perhaps eventually reverse, the tide of anti-science legislation.

There are those who believe that science is outmoded, oppressive, and imperialistic. They are wrong. Science is one of the most liberating forces ever developed in the long history of humankind. The scientific method is an open-minded way of viewing the world that rejects authoritarian thought control in favor of proof based on evidence and logic. Science frees minds rather than enslaving them. It is a legacy worth preserving.

Acknowledgment

The authors would like to thank Cleone Hawkinson for her assistance in the preparation of this paper.

Appendix

Examples of the range of attention that has been given to the Kennewick Man case include the following:

Print Media

The authors have counted more than 35 U.S. daily newspapers that carried articles on the case over the past seven and a half years. Some examples are *Baltimore Sun, Chicago Tribune, Christian Science Monitor, Denver Post, Los Angeles Times, New York Times, Salt Lake Tribune, Wall Street Journal, Washington Post,* and *USA Today.* Many had multiple articles, and many tried to inform their readers about issues relating to human settlement of the Americas and how scientists investigate this subject. In addition, more than 30 magazines printed stories or reports on the case, including *Academic Questions, American Archaeology, Chronicle of Higher Education, Discovering Archaeology, Economist, National Review, New Yorker, Reason. Science News,* and *U.S. News and World Report. Archaeology* magazine carried eight reports or updates on the case, and *Science* carried 14.

Broadcast Media

The case has been featured on a number of national television news programs, including "60 Minutes" and "The News Hour with Jim Lehrer." A story on the case entitled "The Earliest Immigrants" was shown by the Discovery Channel on at least four occasions in 2000 and 2001. Local television coverage in the Pacific Northwest has been extensive. National Public Radio and Canada's CBC network have carried frequent reports on the case. Local radio coverage in the region has been frequent.

Books

The skeleton and the controversy over its study and disposition have figured prominently in five books, and several others are reported to be in preparation. Of the existing books, the most detailed and well researched are *Ancient Encounters* (Chatters 2001) and *No Bone Unturned* (Benedict 2003).

Internet

There are numerous Web sites dedicated to the case (or some aspect of it). Examples include Friends of America's Past (**www.friendsofpast.org**) and Tri-City Herald (**www.kennewickman.com**). The Department of the Interior also has a Web site on the case (**www.nps.gov/aad/kennewick**), which is now back online after a court-ordered shut down of all DOI Web sites.

International Interest

Foreign coverage of the Kennewick Man case has been extensive, with numerous reports appearing in newspapers and magazines and on radio and television. A few examples are *The Bulletin* (Australia, 17 October 2000), *Financial Times* (London, 13 September 1997), *The Guardian* (London, 3 September 1997), *Jakarta Post* (23 April 1998), *London Times* (6 June 1997), and *Royal Review* (Canada, 16 October 1996).

–Originally submitted 10 June 2002
Postscript added 15 March 2004

Postscript

On 30 September 2002, the U.S. District Court set aside the Secretary of the Interior's decision to award ownership of the Kennewick Man skeleton to the tribal coalition (Bonnichsen et. al. v. U.S., 2002). The court's opinion enjoined any transfer of the skeleton to the tribal claimants and ordered that the suing scientists be allowed to study it (id. at 1166).

The district court's decision represents the first definitive ruling on many important issues relating to the interpretation and application of NAGPRA. One issue of particular importance for First American researchers is the court's rejection of the Department of the Interior's (and SAA's) contention that NAGPRA automatically applies to all pre-Columbian artifacts and human remains found in the United States. The court found that this interpretation is inconsistent with the objectives and plain language of the statute which require "at a minimum, a cultural relationship between the remains and other cultural items and a present-day tribe, people, or culture indigenous to the United States" (id. at 1138). In the case of Kennewick Man, proof of such a relationship was not established. His culture is unknown, and it cannot be presumed without proof that he was necessarily ancestral to modern American Indians (id. at 1138). "The record would not support a finding that the ancestors of the American Indians were the only people here in prehistoric times, or that only one culture existed throughout prehistoric times. Congress did not create a presumption that items of a particular age are 'Native American' [for purposes of NAGPRA]" (id. at 1138).

The court declined to decide the question of whether scientists have a constitutional right to study government-owned skeletal remains and other primary research materials. Instead, the court decided the issue of study on statutory grounds, ruling that the Archeological Resources Protection Act (ARPA) gives scientists a right to obtain access to ancient skeletal remains such as the Kennewick Man skeleton. "Allowing study is fully consistent with applicable statutes and regulations, which are clearly intended to make archaeological information available to the public through scientific research" (id. at 1167). The court cautioned, however, that the right to access is not absolute (id. n. 74 at 1166) and is subject to "the normal terms and conditions routinely imposed when studies of objects subject to ARPA are carried out" (id. at 1167).

The Army Corps's decision to bury the skeleton's discovery site was sharply criticized by the court. It found that the Army Corps violated the National Historic Preservation Act by failing to consider the views of interested parties (i.e., the scientists), by failing to mitigate the potential loss of archaeological data, and by failing to consider the potentially negative effects of the project (id. at 1163–64). The Army Corps had attempted to justify its actions by arguing that it covered the site in order to prevent further erosion. The court rejected this argument and concluded that the primary reason for the project was to prevent further investigation of the site (id. at 1125, 1133, 1163).

The court also found that the decision-making process followed by the government was fatally flawed. Government ers improperly sided with the tribal claimants, secretly provided them with information while denying such information to the scientists, and deprived the scientists of an opportunity to submit materials and comments on important issues (id. at 1133–34). While government ers have some flexibility in structuring the process to be followed for resolving disputed issues, "regardless of the particular method used to reach a decision, the decision-making process must be fair to all affected parties" (id. at 1133).

The court's opinion also contains a lengthy discussion of what is required to prove cultural affiliation under NAGPRA. Among other things, the court ruled that tribal origin accounts are not sufficient to establish cultural affiliation with skeletal remains as old as Kennewick Man. Even if one assumes that "the oral traditions help to establish a 'cultural continuity . . . extending into the prehistoric past,' the narratives do not help to establish how far into the 'prehistoric past' such continuity extends. The 9,000 years between the life of the Kennewick Man and the present is an extraordinary length of time to bridge with evidence of oral traditions" (id. at 1155). For a more detailed description of the court's decision, see Lepper 2002. The complete text of the opinion can be viewed at the Friends of America's Past Web site (www.friendsofpast.org).

On 4 February 2004, a three-judge panel of the Ninth Circuit Court of Appeals affirmed the district court's decision (Bonnichsen et. al. v. U.S. 2004). Like the district court, the Ninth Circuit found that there was insufficient evidence to demonstrate that the Kennewick Man skeleton is Native American within the meaning of that term as it is used in NAGPRA. The case was remanded to the district court for enforcement of its order permitting the scientists to proceed with their studies of the skeleton. The Ninth Circuit's opinion reaches a number of conclusions of potential importance to First Americans researchers.

The Ninth Circuit agreed with the district court's ruling that the scientists in the case had the requisite standing to challenge the actions of the Army Corps of Engineers and the Department of the Interior. According to both courts, the jurisdictional language of NAGPRA is broad enough to permit lawsuits such as this to prevent over-enforcement of the statute (id. at 1594).

Like the district court, the Ninth Circuit rejected the argument that NAGPRA automatically applies to all pre-Columbian cultural items found in the United States. The Ninth Circuit found that such arguments were not consistent with either the text or the purpose of the statute (id. at 1596–98, 1601). Instead, the term "Native American" as used in NAGPRA must be interpreted as applying only to remains and other objects that have a "special and significant genetic or cultural relationship to some presently existing indigenous tribe, people, or culture" (id.

at 1603). If followed by other courts, the Ninth Circuit's (and the district court's) interpretation of the term Native American should protect other paleo-remains (such as Arlington Springs or Spirit Cave) from ill-founded NAGPRA claims or from disposition under any future regulations that are adopted for the disposition of culturally unaffiliated remains. Like Kennewick Man, the relationship (if any) of such paleo-remains to living American Indians is unclear, and it may be many years before any reasonable conclusions can be reached in that regard. Similar problems of proof will exist for many younger (e.g., Archaic) remains. State-controlled skeletal remains, however, will still be subject to disposition under state laws, which in many cases require a lower level of proof.

Like the district court, the Ninth Circuit was critical of Secretary Babbitt's acceptance of oral tradition as sufficient to establish a relationship between the skeleton and the tribal claimants. The court's opinion states:

> Because oral accounts have been inevitably changed in context of transmission, because the traditions include myths that cannot be considered as if factual histories, because the value of such accounts is limited by concerns of authenticity, reliability, and accuracy, and because the record as a whole does not show where historical fact ends and mythic tale begins, we do not think that the oral traditions of interest [here] were adequate to show the required significant relationship of the Kennewick Man's remains to the Tribal Claimants. As the district court observed, 8340 to 9200 years between the life of Kennewick Man and the present is too long a time to bridge merely with the evidence of oral traditions (id. at 1607).

The Ninth Circuit received amici (i.e., friends of the court) briefs supporting the scientists from the Ethnic Minority Council, Ohio Archaeological Council, Pacific Legal Foundation, Society for American Archaeology and Texas Historical Commission. Amici brief were also filed by Drs. Ives Goddard and William Shipley (on linguistics) and by Drs. Harry G. Custred and Andrei Simic (on oral tradition). SAA's brief effectively dismantled any claims that the skeleton can be culturally affiliated to the tribal claimants or any other existing group. However, it did not address the pivotal issue of what is meant by the term Native American. For a more detailed description of the opinion, see Lepper 2004. The complete text of the opinion can be found at Web site www.friendsofpast.org.

References Cited

AAA (American Anthropological Association)

1986 Statements on Ethics, Principles of Professional Responsibility, May 1971 (as amended through November 1986).

1995 Commission to Review the AAA Statements on Ethics, Final Report, 16 September 1995.

1998 Code of Ethics of the American Anthropological Association, June 1998.

2001 A Brief History of the American Anthropological Association, AAA website for 4 December 2001

AAAS (American Association for the Advancement of Science)

2001a About AAAS, AAAS Web site for 4 December 2001.

2001b Scientific Freedom, Responsibility & Law, Mission Statement, AAAS Web site for 4 December 2001.

AAPA (American Association of Physical Anthropologists)

1997 Statement of Phillip L. Walker at committee hearings on H.R. 2893.

2000 Statement by the American Association of Physical Anthropologists on the Secretary of Interior's Letter of 21 September 2000 Regarding Cultural Affiliation of Kennewick Man, 20 October 2000.

2002a First page, AAPA Web site for 1 April 2002.

2002b Membership in the AAPA, AAPA Web site for 1 April 2002.

Benedict, J.

2003 *No Bone Unturned*, Harper Collins. New York, NY.

BLM (Bureau of Land Management)

2000a Determination of Cultural Affiliation of Ancient Human Remains from Spirit Cave, Nevada, Bureau of Land Management, Nevada State Office, 26 July 2000.

2000b BLM Makes Spirit Cave Man Determination, BLM Web site for 15 August 2000.

Bonnichsen et al. v. United States, Civil No. 96-1481-JE, U.S. District Court, District of Oregon.

1996a Letter dated 6 September 1996 from Senator Patty Murray to Mr. Richard Apple, Office of the Chief of Engineers, U.S. Army Corps of Engineers. At pp. 9363–64 in U.S. Army Corps of Engineers Administrative Record filed with the Court.

1996b e-mail dated 17 September 1996 from Don Curtis CENPW-DE at NPW (Army Corps of Engineers) to Col. Bart B. Bohn. Obtained in response to Plaintiffs' First Request for Production. Copy on file with the authors.

1996c Letter dated 11 October 1996 from Senator Slade Gorton and Representatives Doc Hastings, Jack Metcalf, and George K. Nethercutt, Jr. to Lt. Gen. Joe Ballard, Commander, U.S. Army Corps of Engineers. At pp. 9129–30 in U.S. Army Corps of Engineers Administrative Record filed with the Court.

1997a Affidavit of Paula A. Barran, Plaintiffs' Motion for Order Granting Access to Study, 11 March 1997.

1997b Court opinion reported at 969 F. Supp. 628 (D. Or. 1997).

1998a Letter dated 3 April 1998 to Lt. Gen. Joe N. Ballard, U.S. Army Corps of Engineers, from the Confederated Tribes of the Colville Reservation. At p. DOI 02720 in the Department of the Interior Administrative Record filed with the court.

1998b Letter dated 31 July 1998 to Francis P. McManamon from the Confederate Tribes and Bands of the Yakama Indian Nation. At DOI p. 03368 in the Department of the Interior Administrative Record filed with the court.

1999a Letter dated 12 April 1999 to Secretary Bruce Babbitt from Office of the Reservation Attorney, Confederated Tribes of the Colville Reservation, at p. DOI 04214 in the Department of the Interior Administrative Record filed with the court.

1999b Federal Defendants' Opposition to Plaintiffs' Motion for Immediate Response Re: Study Request, 16 August 1999.

1999c Plaintiffs' Reply (Motion for Immediate Response), 30 August 1999.

1999d Transcript of Motion Hearing on 14 September 1999.

2000a 21 September 2000 letter from Secretary of the Interior Bruce Babbitt to Secretary of the Army Louis Caldera, attached to Notice of Filing of Federal Defendants' Response to Plaintiff's Request to Study, 25 September 2000.

2000b Carl A. Strock, Brigadier General, USA, The U.S. Army Corps of Engineers Response to Requests for Scientific Study-Kennewick Man Human Remains, attached to Notice of Filing of Federal Defendants' Response to Plaintiffs' Request to Study, 25 September 2000.

2001a Federal Defendants' Response to Plaintiffs' Motion to Compel Production of Raw Computer Data, 22 January 2001.

2001b Order of court dated 5 February 2001.

2001c Plaintiffs' Memorandum in Support of Motion to Vacate Second Administrative Action, 16 April 2001.

2002 District Court Opinion. 217 F. supp. 2d 1116 (D. Or. 2002).

2004 Ninth Circuit Court of Appeals Opinion. Appeals nos. 02-35994 and 02-35996 (Feb. 4, 2004).

Brace, C. L.
1998 Letter dated 3 June 1998, sent as Chair of Section H (Anthropology) AAAS to Jonathan L. DeVaney and Congressman Doc Hastings.

California
2001 California Native American Graves Protection and Repatriation Act. In California Health and Safety Code §8010 et. seq.

Chatters, J. C.
2001 *Ancient Encounters,* Simon & Schuster. New York, NY.

Damadio, S. M.
2000 Spirit Cave Man: Biological Aspects, Bureau of Land Management, Washington D.C.

Department of the Interior
1998 Press Release, Office of the Secretary, 3 March 1999.

2002a Native American Graves Protection and Repatriation Act Regulations, Disposition of culturally unaffiliated human remains (draft presented at NAGPRA Review Committee meeting, 31 May–2 June 2002).

2002b Native American Graves Protection and Repatriation Act Regulations, Future Applicability (draft presented at NAGPRA Review Committee meeting, 31 May–2 June 2002).

Dewar, E.
2002 *Bones: Discovering the First Americans,* Carroll & Graf Publishers. New York, NY.

Dillehay, T. D.
2000 *The Settlement of the Americas,* Basic Books.

Dixon, E. J.
2000 Bones, Boats & Bison, University of New Mexico Press.

DuBoff, L. D.
1992 Protecting Native American Cultures, *Oregon State Bar Bulletin,* November 1992.

Green, T.
1992 Letter dated 4 March 1992, to C. Meighan.

Grew, P. C.
1998 Letter dated 11 November 1998 to A. R. Nelson.

2001 Statement (17 November 2001) presented to meeting of the NAGPRA Review Committee, Harvard Law School, Cambridge, Mass.

Hall, D. A.
1997 8000-Year-Old Cave Explorer Challenges Research Team, *Mammoth Trumpet* 12(1):10.

Hill, R. L.
2000 A Question for the Ages: Returning Ancient Remains, *Oregonian,* 25 October 2000.

Hutt, S., and C. T. McKeown
1999 Control of Cultural Property as Human Rights Laws. In 30 *Arizona State Law Journal.* Tempe, Arizona.

Johnson, G.
1996 Indian Tribes' Creationists Thwart Archaeologists, *New York Times,* 22 October 1996.

Lepper, B. T.
in press Public Policy, Academic Archaeology, and the First Americans. In *New Perspectives on First Americans,* edited by B. T. Lepper and R. Bonnichsen. Center for the Study of the First Americans, Texas A&M University, College Station.

2002 Judge Rules Scientists Can Study Kennewick Man. *Mammoth Trumpet* 18(1):1.

2004 Kennewick Man Decision Upheld! *Mammoth Trumpet* 19(2):1.

Lippert, D.
1997 In Front of the Mirror: Native Americans and Academic Archaeology. In *Native Americans and Archaeologists: Stepping Stones to Common Ground,* edited by N. Swider, K. E. Dongoske, R. Anyon and A. S. Downer, pp. 120–27. Altamira Press, Walnut Creek, California.

Maine
1993 Public Law 1993, C. 738, Pt. C, Section 5. Codified as 22 MRSA §2842-B.

2002 LD 1940. Codified as 22 MRSA §2842-B.

Minnesota Indian Affairs Council
1999 Notice of Inventory Completion for Native American Human Remains and Associated Funerary Objects from the State of Minnesota in the Possession of the Minnesota Indian Affairs Council, Bernidji, MN. In Federal Register 64(152): 43211, 9 August 1999.

Montana
2001 Montana Repatriation Act. In Montana Code Annotated at 22-3-901 et. seq.

Mulvaney, J., and J. Kamminga
1999 *Prehistory of Australia,* Smithsonian Institution Press. Washington and London.

Myster S. M. T., and B. O'Connell
1997 Bioarchaeology of Iowa, Wisconsin, and Minnesota. In *Bioarchaeology of the North Central United States,* edited by D. W. Owsley and J. C. Rose. Arkansas Archaeological Survey Research Series. Fayetteville, AR.

NAGPRA Regulations
1995 Native American Graves Protection and Repatriation Act Regulations, 43 Code of Federal Regulations Part 10. In Federal Register 60 (232): 62134, 4 December 1995.

NAGPRA Review Committee
2000 Recommendations regarding the Disposition of Culturally Unidentifiable Native American Human Remains. In Federal Register 65 (111): 36462, 8 June 2000.

2002 Native American Graves Protection and Repatriation Review Committee Findings and Recommendations regarding Human Remains and Associated Funerary Objects from Spirit Cave in Nevada. In Federal Register 67 (69): 17463, 10 April 2002.

Nebraska
1998 University of Nebraska, Inventory of Culturally Unaffiliated Remains, 30 September 1998.

2000 Inventory Completion for Native American Human Remains and Associated Funerary Objects in the Possession of the University of Nebraska–Lincoln, Lincoln, NE. In Federal Register 65(191): 58803, 2 October 2000.

Oregon
1993a Oregon Revised Statutes §97.750.

1993b Oregon Revised Statutes §358.940.

1998 Notice of Inventory Completion for Native American Human Remains and Associated Funerary Objects in the Possession of the Oregon State Museum of Anthropology, University of Oregon, Eugene, Oregon. In Federal Register 64(12): 36035, 2 July 1999.

Owsley, D. W., and R. L. Jantz
1999 Databases for Paleo-American Skeletal Biology Research. In *Who Were the First Americans?* edited by R. Bonnichsen, Center for the Study of the First Americans, Oregon State University. Corvallis.

SAA (Society for American Archaeology)
1986 Statement concerning the Treatment of Human Remains, May 1986.
1997 Society Bylaws, revised October 1997.
1998 Statement of Vin Steponaitis at committee hearings on H.R. 2893.
1998 Poll Finds Public Support of Archaeology, Release dated 17 March 2000, SAA website.
2001a About SAA, SAA website for 4 December 2001.
2001b Memorandum of Law in Support of the Society for American Archaeology's Amicus Curiae Submission, *Bonnichsen et al. v. United States,* Civil No. 96-1481-JE, U.S. District Court, District of Oregon.

SAS (Society for Archaeological Sciences)
1984 By-Laws of Society for Archaeological Sciences, revised 15 May 1984.
2001 SAS website for 27 November 2001.

Sagan, C.
1996 The Demon-Haunted World: Science as a Candle in the Dark.

Schneider, A. L.
1995 NAGPRA and First Americans Studies, *Current Research in the Pleistocene* 12:117.

Science
1998 Kennewick Man's Trials Continue, Sciences 280:190, 10 April 1998.

Steele, D. G., and J. F. Powell
1994 Paleobiological Evidence of the Peopling of the Americas: Morphometric View. In *Method and Theory for Investigating the Peopling of the Americas,* edited by R. Bonnichsen and D. G. Steele, pp. 141–163, Center for the Study of the First Americans. Oregon State University, Corvallis.

Tsosie, R.
1998 Privileging Claims to the Past: Ancient Human Remains and Contemporary Cultural values, *Arizona State Law Journal* 31: 583.

Wisner, G.
1992 Idaho Burial Suggests Life of Hardships, *Mammoth Trumpet* 7(2):1.

Paleoamerican Origins: Models, Evidence, and Future Directions

Dennis Stanford[1], Robson Bonnichsen[2], Betty Meggers[3], and D. Gentry Steele[4]

Abstract

Following the discovery of "Paleolithic" projectile points associated with Ice Age animals near Clovis, New Mexico, scholars made the assumption that these artifacts were used by descendents of ancient hunters who migrated "out of Asia" across the Bering Land Bridge and traversed an ice-free corridor to populate the Americas. This unilineal theory of New World origins became so rooted in scientific and public thinking that it evolved into "fact" without the benefit of proof. This paper assesses the progress of more than three quarters of a century of research focused on resolving the questions of the peopling of the Americas and concludes that there is sufficient evidence of multiple migrations to rule out the unilineal "Clovis-first" descent theory.

The growing body of new interdisciplinary data has currently resulted in three colonizing scenarios to replace the Clovis-first model. These models include: multiple terrestrial migrations across Beringia into the Americas; expansion around the North Pacific Rim by Southeast Asian paleomariners; and a westward extension of Paleoeuropean sea mammal hunters exploiting the North Atlantic ice margin. We conclude that while these or even combinations of these models are logical possibilities to be tested, at present they only have minimal empirical supporting evidence.

We also examine variations in technology against a chronological framework to assess the existence of multiple cultural patterns that represent adaptations to the varied environments and dramatic climate changes encountered as the First Americans dispersed across the Western Hemisphere. These patterns include a long-lived megamammal bone-flaking technology that extended from Eurasia across Beringia and southward into North America as far as the Valley of Mexico. An early unifacially flaked stone tool pattern is common in South America and lasts until the beginning of the Holocene. This pattern is poorly represented in North America. The "signature" pattern of the Americas is the bifacially flaked stone projectile point.

While the bifacially flaked pattern is most common and diverse in North America, it also occurred southward to the very tip of the South American Cone. From its earliest occurrences, this pattern is so variable in technology and form that we considered the variations co-traditions, many of which overlapped time and space. These include pre-Clovis (Cactus Hill, Meadowcroft, and Page-Ladson), Clovis, Nenana, Goshen/Plainview, Western Stemmed points of North America, and El Jobo, Fishtail and Paiján points of South America. At present it is ambiguous if these co-traditions represent regional adaptations or have independent origins.

We conclude that the scientific database, though growing, is currently insufficient to explain the intricate web of social, biological, and environmental interactions that over untold millennia evolved into modern Americans. Therefore, it is imperative to study and preserve every important archaeological discovery if we ever hope to achieve an understanding of the rich story of the peopling of the Americas.

Unfortunately, the Native American Graves Protection and Repatriation Act (NAGPRA), which was intended to balance the interest of Native Americans, scientists, and the public, may in fact make it difficult to acquire critical data in the future. Satis-

[1] Department of Anthropology, Museum of Natural History 304, Smithsonian Institution, Washington, DC 20560.

[2] Director, Center for the Study of the First Americans, deceased December 2004. Address all inquiries to Michael R. Waters, Director, Center for the Study of the First Americans, Department of Anthropology, Texas A&M University, College Station, TX 77843-4352; e-mail: mwaters@tamu.edu

[3] Department of Anthropology, Museum of Natural History, Smithsonian Institution, Washington, DC 20560.

[4] Department of Anthropology, Texas A&M University, College Station, TX 77843-4352.

factory regulations have not been developed to insure fair and objective decisions regarding study of Paleoamericans, and many federal agencies and institutions have returned remains that have little if any affiliation with modern tribal groups. This situation became apparent with the discovery of the Kennewick Man, a Paleoamerican who died as a result of a violent confrontation over 9,000 years ago. The Corps of Engineers decided to summarily give Kennewick Man's remains to the local tribes for burial even though preliminary studies strongly suggested they were not related. Fortunately, this did not happen and Kennewick Man's contribution to the history of the First Americans will be preserved for the time being.

State and federal agencies as well as tribes and the general public should understand that many different tribal groups as well as non-Indians have a common heritage with prehistoric people such as Kennewick Man. Or, in the case of lineal extinction, there may be no living descendents. To accurately identify prehistoric social groups and understand their relationships to one another as well as to modern tribes, we need to develop and use new technologies and methods to test alternative hypotheses using scientific methods. By constructing better descent models and following rules of evidence, we can accurately resolve conflicts of ethnic and social affiliation. Certainly, because theoretical, technological, and analytical innovations will occur, it is imperative to hear the "Ancient Ones" who have patiently waited to tell the stories of their times and lives through the archaeological record.

Introduction

This is an exciting time in First Americans studies. New and significant archaeological, paleobiological, and environmental evidence is expanding our knowledge of early cultures and human populations. Advances in computer, accelerator mass spectrometry (AMS) radiocarbon dating, and molecular technologies have made new lines of evidence available and are increasing our ability to date accurately and interpret early cultural and biological remains over the last 50,000 years. These advances, plus new discoveries, have not only enhanced our abilities to learn more about the past but have contributed to the demise of the Clovis-first model (a model that dominated the field during the past 40-plus years) and are laying the foundation for new advances in our understanding of the initial peopling of the Americas.

The goal of First Americans research is to understand the origin and descent of the earliest cultures and biological populations to enter the New World. Our objectives are to determine where they come from, who they were, when and where they arrived, and how they changed once here. Establishing who, when, where, why, and how the Americas were peopled is essential for understanding the conditions and processes that led to the occupation of the hemisphere and to the development of subsequent cultures.

The field is now at a very interesting juncture. This is a time when First Americans studies lack a prevailing model or theory. After more than 100 years of research, we have neither definitive answers as to when, where, or how the Americas were first colonized, nor indisputable evidence of how many colonization events were involved. We also do not have a clear understanding of the processes that led to the myriad of diverse cultural and biological groups found throughout the New World at the time of European contact. Although we have more scientific evidence now than at any time in the past, we lack carefully articulated, well-defined models for explaining the available evidence. The Clovis and Beyond Conference held in Santa Fe, New Mexico, in October 1999 and the creation of this volume provide an opportunity to reflect on past and present directions of the field, as well as to consider the kinds of ways needed to move this active and dynamic field forward.

Before considering fresh approaches, a few comments are in order. The initial peopling of the Americas is not just another historical problem; it is a very basic problem, and its solution is important to many other fields. For example, information developed through the study of evidence left by the first Americans provides the factual, temporal, and to some extent the theoretical framework of Paleoamerican prehistory. This information can then be used by cultural anthropologists, archaeologists who specialize in later time periods, linguists, and geneticists to support and bolster cultural, linguistic, and biological theories about the origin, development, spread, and diversification of New World cultures, languages, and biological populations.

Finding solutions to the questions of American origins is truly a significant global problem. If we think people first came to the Americas only 11,500 RCYBP or alternatively more than 50,000 years ago, it makes an enormous difference in how we approach the study of the past, what we think we know, how we frame future Paleoamerican research projects, and what we ultimately infer about human development in the Americas. The scientific evidence for the antiquity of humans in the Americas is not only of importance to the scientific and humanist communities, but also to decision makers that grant research funds and who are the stewards of archaeological resources on federal, state, and municipality lands in the United States. If it is assumed that humans have only been here since the end of the last Ice Age, agency personnel will focus survey and management efforts on sites and landscapes dating to that time. If in fact the Americas were peopled as much as 50,000 RCYBP years ago, then scientists, researchers, and agency decision makers who accept that human occupation began only about 11,500 RCYBP will be overlooking the majority of the early archaeological record.

A substantial body of evidence presented in this volume and elsewhere (Bonnichsen and Turnmire 1999; Dillehay 2000a) strongly supports the proposition that humans were in North and South America well before 11,500 RCYBP. Exactly how much earlier is a focus of ongoing research. An important new development has been the demise of the Clovis-first model; its failure created a void that has resulted in three different responses by the professional community. First, and not unexpectedly, there are a few residual Clovis-first advocates (Fiedel this volume; Haynes 2002; Haynes this volume) who still cling tenaciously to the hope that some of the propositions of the Clovis-first model can be reformulated to correct the model, or that definitive proof will be found to provide credence to this "failed" model. Some "hard core" Clovis-first proponents believe "all pre-Clovis age sites" and biological evidence that supports a pre-11,500 RCYBP occupation must reflect flawed science. Therefore, all pre-Clovis sites and biological evidence are the focus of negative critiques (Fiedel 2004). The second type of response to the collapse of the Clovis-first model is by former Clovis-first advocates who have accepted the 12,500 RCYBP dates from Monte Verde. They speculate that there must have been a peopling event that occurred slightly earlier than Clovis, 13,000–14,000 RCYBP (Dixon 2001; Meltzer 2004). Scholars of this persuasion tend either to ignore or discount sites and evidence that support a pre-glacial maximum occupation of the Americas. The third response has been to create a series of new alternative models to explain the peopling of the Americas.

As Merriwether (2002) has observed, these models incorporate conflicting scenarios. The most prominent include the three-wave migration, based on linguistic, genetic, and archaeological data that correlate with the Amerind, Na-Dene, and Eskaleut languages described by Greenberg et al. (1987) and Greenberg (1986). The single-wave migration model proposes that the entry of humans was followed by the differentiation of languages and peoples (Forester et al. 1996; Merriwether et al. 1995). The North Atlantic Maritime model proposes that humans came to the New World by boats while hunting sea mammals along the southern edge of the north Atlantic ice sheet during the Last Glacial Maximum (Bradley and Stanford 2004; Stanford and Bradley 2002). Another popular migration model is the coastal north Pacific migration in which colonists made use of small boats to move along the Pacific Rim to more southern climates (Erlandson 2002; Fladmark 1979). Others have proposed that pre-Clovis pedestrian populations came to the Americas prior to the late-Wisconsinan closure of the Ice-free Corridor from the Mongolia/southern Siberia/Manchuria/Lake Baikal region of Northeastern Asia (Kolman et al. 1996; Merriwether et al. 1996; Schurr and Wallace 1999; Starikovskaya et al. 1998). Based on craniofacial evidence, others suggest early populations could have come from Japan (Brace et al. 2001), Africa, Australia, or Southeast Asia (Neves et al. 1993), presumably by watercraft.

The appearance of this new generation of models signals that the field of Early American studies is rapidly changing and several interesting new trends are emerging. First, no single model has been universally accepted as a replacement for the Clovis-first model. Second, many scholars now entertain the idea that more than one human population colonized the Americas and that more than one colonization model could be correct. Third, some models are not all-inclusive and are based on a single line of evidence. Fourth, it is evident that all the colonizing scenarios under consideration cannot be correct. Thus, the next logical step to move the field forward is to develop a blueprint for determining which lines of evidence are relevant and reliable and can be used to construct and test propositions that can be verified through the study of surviving empirical evidence. In the following pages, we review model-building procedures in First Americans studies based on the nature of the evidence.

Theory Building in First Americans Studies

Over the past few decades, First Americans studies have emphasized excavating and reporting individual sites with clear stratigraphy and undeniable artifacts found in an unambiguous dated context. While we applaud this rigorous approach, this focus alone will not resolve the fundamental question of how the Americas were peopled. We also need to develop testable models that draw on all lines of evidence available for reconstructing the cultural and biological adaptive processes that allowed human groups to move from other continents to the Americas and to cope with changing circumstances during their journey.

In seeking to understand Paleoamerican prehistory, we need to develop testable models that will allow us to move beyond the individual artifact, site, or skeleton to a more encompassing level of analysis. We must reach a level that will allow us to trace specific cultural groups and biological populations from site to site, from region to region, and between continents. Carefully articulated scientific models are needed to bring order to and to integrate the diverse lines of evidence from the fossil record. Consideration needs to be given to what constitutes relevant and reliable evidence and how evidence from the fossil record can be used to test model propositions.

The approach of using scientific models to explain the past has been in vogue since the late 1960s (Clarke 1968). Archaeologists have experimented with many different kinds of models covering a wide range of subjects, such as settlement and subsistence, lithic procurement, and trade networks. Indeed, these functional models have advanced our understanding of how human groups have adapted to local and regional environments. They do not, however, provide answers to basic origin and descent issues raised by specialists seeking to understand the peopling of the Americas.

We suggest our efforts should focus on creating descent models. Rather than using terms such as colonization or migration, which imply a directed goal, descent models emphasize the temporal, spatial, cultural, and biological relationships among different discrete archaeological and paleobiological data sets that are supported by evidence recovered from the fossil record.

From a model-building perspective, models can accommodate many kinds of problems using different temporal and spatial scales. They can accommodate research problems that range in geographic scope from local to intercontinental, or they can accommodate very short time periods of only a few years to millennium-scale events. They can incorporate either a single line of evidence, or complex interdisciplinary problems such as the peopling of the Americas that involve integrating many lines of cultural, biological, and environmental evidence in seeking to explain a peopling event. This latter category of models is the focus of the following discussion.

The next generation of descent models should seek to explain and to advance knowledge about the peopling of the Americas by formulating explicit propositions that can be tested by collecting and analyzing archaeological, paleobiological, and environmental evidence from the fossil record. Determining how the Americas were colonized will require the use of carefully defined scientific standards for classifying and reporting data. To demonstrate that a human group moved across space and time will entail comparing data from different local areas and regions, and even between regions in the Americas and other continents. Accuracy in model-building endeavors will be improved by drawing on all possible independent lines of evidence to assure greater reliability and accuracy. These models should be useful not only to researchers, but also to agency decision makers charged with managing cultural and biological resources that occur on public sector lands.

With the goal of moving beyond the debate between the Clovis and pre-Clovis camps, set up almost 50 years ago with the introduction of the Clovis-first model, the following discussion first focuses on outlining general premises used in scientific model building. Next, these concepts are used to assess: 1) where we have been; 2) where we are; and 3) where we should go in respect to modeling the peopling of the Americas.

Scientific Model-building Cycle

When confronted with the question of why we still do not have a definitive answer as to when and how the Americas were peopled, it appears as if the reasons may be more complex than simply a debate about timing. One group advocates a short chronology, explaining the peopling of the Americas by the Clovis-first model (Haynes this volume; Martin 1973; West 1996). Another view proposes a longer chronology, advocating that the Americas were peopled well before the end of the last Ice Age (Adovasio this volume, Bonnichsen and Lepper this volume, Bryan 1986, Dillehay 2000a, Gruhn this volume, Morlan 2003). Much of the debate against the long chronology has focused on invalidating chronological propositions by creating doubt about dates and by arguing the context, associated artifacts, and in some cases the credibility of site investigators (see Gruhn this volume).

There is also a lack of consensus among specialists as to what constitutes an appropriate scientific model and what is acceptable evidence. In considering the reasons for this disagreement, it will be useful to consider prior model-building efforts in First Americans studies.

Investigators who have contributed to our understanding of the peopling of the Americas have done so in essentially two different ways. These are the "descriptive" and the "scientific" model-building approaches. The distinction between these two facets of the field is not unique to First Americans studies and is common in other disciplines that seek to understand the past. It is the distinction between the "data people" and the "theory/model people." We suggest that both views will allow the field to advance and flourish. Some believe careful descriptive reporting of the evidentiary record will lead to the accumulation of enough facts that an understanding of the peopling of the Americas will emerge from all of the details. In all fairness, reports of this nature are often carefully done. They are notable, however, for their lack of explicit theoretical statements and propositions that can be tested against the fossil record. Descriptive reports add to the general corpus of knowledge about First Americans sites, artifacts, and environments. The down side of this approach is that there are infinitely more studies possible than can be done. Choices have to be made as to what is studied. If our goal is to learn about the peopling of the Americas, evidence important to understanding descent relationships may go unreported. Without theory/models, many significant observations may be neglected while others with little meaning may be overemphasized. In general, theories do not simply emerge from compiling data. First Americans studies is a particularly complex field, with basic data reported using various standards and languages, including English, Spanish, Portuguese, French, Japanese, Chinese, and Russian, and is based on a wide array of approaches.

The function of scientific model building in prehistory is to identify significant research problems. Scientific model building differs from the descriptive approach in that the latter is not theory/model–driven. Theories tend to be abstract and usually cannot be directly tested. Models, however, provide the bridge between theory and data. If properly designed, they can be used to explain past events and processes by providing hypotheses/propositions that can be tested against evidence from the fossil record. The principal components of the model include theory, models, data, and analysis.

Systems Theory

Most First Americans specialists use some form of systems theory to guide their interdisciplinary research projects. They construct models based on the general systems theory propositions that the Earth's climatic, atmospheric, oceanographic, geologic, paleoecologic, cultural, and human systems are linked as a series of subsystems. Subsystems can vary in scale from a societal level and focus down to the individual. Each subsystem has its own internal dynamics, and changes in linked systems can affect their operation. In recent years, many Quaternary specialists have proposed that not all subsystems are equal and that changes in the climatic system are of paramount importance in forcing change in linked systems. "New Archaeologists" first began to experiment with the use of systems theory during the 1960s. Soon thereafter, a theoretical debate arose between "culture history" and "process" schools of thought. Rather than being antithetical, the goals of the culture history and the systems theory approaches are complementary. Spokespersons for the new archaeology emphasized understanding the internal dynamics of systems responsible for the archaeological record. But little discussion was offered as to how the spatial and temporal boundaries of systems were to be defined. Without defining spatial and temporal boundaries, it is impossible to determine which remains belong to which culture. The new archaeology advocated the need to understand breaks in the archaeological record and emphasized changes in systems operations, including sudden shifts in internal dynamics, such as a new innovation or war with a neighboring group. In seeking to understand breaks in the archaeological record, it is also important to consider historical factors, such as the migration of a new population into the area or diffusion of new and significant ideas across existing populations. In evaluating breaks in stratified cultural records, e.g., continuity vs. discontinuity, system dynamics issues as well as historical factors must be considered.

Evolutionary Materialism

Scientific evolutionary materialism is an important approach to understanding the past that complements the systems approach. It is based on the premise that we live in a material world and that human cultural and biological adaptation can be understood in reference to specific material conditions that have shaped the human experience. Even if spiritual matters shaped human decision making, there is generally little or no surviving evidence to reconstruct this component of Paleoamerican prehistory.

First Americans specialists interested in reconstructing the descent of cultural and paleobiological systems through time must focus on the study of past cultural and human biological systems and their environmental context. Such systems may no longer exist, and there are no modern analogs. Specialists must therefore concentrate on investigating the cultural and biological products left behind that have survived in the fossil record. This is accomplished by studying ecofacts from past environments, cultural materials such as artifacts, and features and biological materials such as human skeletal remains and ancient DNA. By studying ancient environmental, cultural, and biological materials found in context and by referring to modern analogs, e.g., modern cultural and biological systems, prehistorians make inferences with respect to how specific systems and the groups of humans who lived in them and adapted to past environments have changed through time. The material evidence that survives through time limits our knowledge about past social and biological groups. We may in some cases never know about the "nonmaterial" aspects of the past; for example, we will probably never know about many aspects of Clovis life such as spiritualism, language, or music.

Models

The construction of descent models during the Paleoamerican period, e.g., models that seek to explain specific descent patterns, is based on the theory of evolutionary materialism and must be linked to the surviving evidentiary record.

Creating testable scientific descent models is a two-step process. The first entails constructing a logical scenario, e.g., system, with explicit propositions that define a cultural, biological, or cultural and biological descent system and is informed by facts available from the evidentiary record. The second step involves collecting and analyzing evidence that can be used to test and/or refute the model propositions.

A well-constructed and tested model should answer the basic questions of who, when, where, how, and why. At least some of the model propositions/hypotheses should be framed so that they can be tested against available empirical evidence from the fossil record. Ideally, testing model propositions should facilitate organizing existing data in new meaningful ways and provide predictive statements to guide future research. If these ideals are met, the construction and testing of descent models will enhance our understanding of the process that shaped the unique cultural and biological heritage of the Americas.

All descent models, regardless of their complexity, should incorporate propositions that specify the operational dynamics and linkages between subsystems that include the following dimensions: 1) unambiguous cultural or biological signature characteristics such as unique artifacts, features, skeletal traits or statistical patterns, and genetic markers; 2) time or temporal period covered; 3) the spatial area in which signature evidence occurs; 4) concordance of independent lines of dated evidence that provide proof for the model's temporal and spatial parameters; and 5) a clear chain of evidence, e.g., from artifacts to genetic traits, demonstrating a linkage between the temporal starting and end points proposed by the model.

1. Signature Evidence Signature evidence must include unique cultural or biological features or a unique association of cultural or biological traits that can be used to isolate cultural groups and biological populations. Signature evidence can be used to test model propositions about past cultural or biological events and to define both temporal and spatial dimensions of ancient cultural and biological patterns. If signature evidence is recovered that fails the test of temporal relevance, the modeler has several choices as how to proceed. Dating accuracy may be checked. If the date appears to be reliable or if additional dates support the first date, another option would be to determine whether signature evidence was accurately identified. A third option would be to modify the model to accommodate new dates. The final option would be to reject the model proposition and start over.

2. Time Time is the variable used to tie other lines of evidence together in a chronological framework. Other variables, such as signature evidence, can only be integrated into the model if they date to the time period specified by the model. For example, to identify prehistoric people, we must have signature cultural or biological evidence. But if that evidence cannot be precisely dated or associated with other lines of dated evidence, then such evidence will fail the "test of temporal relevance" and must be excluded from the model.

3. Spatial Parameters Space is an important dimension of descent models that can be used to test model propositions. If migration is proposed for a group, support for the migration should be found along the spatial corridor proposed by the model propositions. If in situ development is proposed for a specific locality, signature evidence for the group being discussed should not be found outside the area where the in-place development occurred.

4. Concordance First Americans scientists should use as many lines of evidence as possible, drawing on environmental, cultural, and biological evidence in seeking to understand descent issues from local to global-scale problems. The key element that facilitates integrating multiple lines of evidence is time.

5. Chain of Evidence The chain of evidence standard is very important for constructing future descent models. If a modeler proposes a culture or population movement from Asia to the Americas, signatory evidence must be found in Asia and America that was produced by the same group. The evidence in the Americas should be slightly younger than in Asia to accommodate time required to move from one continent to another. Likewise, establishing a link between modern tribes and ancient groups requires an unbroken chain of evidence. Both the Clovis-first and Kennewick Man claims fail to meet this fundamental standard.

Complex Records and Rules of Evidence

In constructing models, it is prudent to integrate as many concordant lines of evidence as possible to test model propositions. Additionally, consideration must be given to possible biases in the cultural, biological, and environmental evidentiary record that might affect interpretations.

Constructing Paleoamerican descent models involves integrating many different kinds of evidence. First Americans specialists draw on environmental, human, biological, and archaeological records produced by past oceanographic, climatic, geologic, soils, paleoecologic, paleoentomologic, paleontologic, human skeletal and genetic systems, and cultural/behavioral systems in their model-building activities. It also involves the use of rules of evidence for collecting and interpreting data so that other investigators can replicate or refute model propositions, predictions, and conclusions.

There is a distinction between fossil and evidentiary records. The fossil record includes remains or products from past cultural, biological, and environmental systems. By contrast, the evidentiary record is limited to the sample of evidence collected from the fossil record to test or refute specific model propositions.

Interpreting interdisciplinary evidence to form a picture of past events is a complex task requiring highly trained specialists. Each discipline that seeks to create new knowledge about the past has its own rules of evidence. Within any given discipline, there may be many different levels of rules (Fiebleman 1954). There are rules for transforming observations about empirical data into meaningful statements about past cultural, biological, and environmental systems. For example, identifying and interpreting a single artifact may involve rules for identifying and interpreting material, shape, technology, and use-wear information and transforming these observations into statements about human adaptation to past cultural systems. Because of the specialized nature of scientific data, specialists must make their assumptions explicit and articulate the limitations of their data.

Rules of evidence accommodate several important functions in the model building and testing process. Rules of evidence specify: 1) what data are collected; 2) methods used to collect the data; 3) definitions of data procedures, both classification and description; 4) rules for interpreting the evidentiary record; and 5) documenting the collecting, processing, and analytic rules in sufficient detail that others can replicate them.

Finally, we must recognize the limitations of our data. The fossil record of environmental, cultural, and human paleobiological remains may be so complex as to defy modeling efforts to develop an empirical foundation for understanding past events. In formulating descent models, our studies must focus on collecting and studying tangible material cultural remains that survive the rigors of time, e.g., houses, hearths, artifacts, and ecofacts such as food debris, skeletal remains, and ancient DNA. We have no way of reconstructing the nonmaterial aspects of culture. We must also anticipate that many mechanisms or processes can explain biological and cultural change through time, such as random genetic drift, migration, diffusion, and demographic shifts as well as other mechanisms that cannot be directly observed. Since fossil evidence is static, processes are inferred from the patterning of empirical evidence. Such inferences are often informed by modern analogs. Ideally, there should only be one process that can account for observed patterning, but in dealing with ancient skeletal and archaeological remains, the resolution of our classification systems is often not sufficiently refined to discriminate between competing hypotheses. In such cases, it may be necessary to develop comparative studies to refine the resolution of our interpretations. We should also anticipate that despite our best efforts, the evidentiary record might be too poorly preserved to yield the evidence we seek.

Where We Were

Clovis-first Model

Archaeologists have long been interested in the peopling of the Americas. Following the famous discoveries near Folsom, New Mexico (Figgins 1927), which established an association between humans and extinct mammals, many scholars deduced that the ancestors of these people probably crossed the Bering Land Bridge from Asia at the end of the last Ice Age. Scientific model building was introduced to First Americans studies in the early 1960s with the Clovis-first model, a detailed scenario conceived to explain the peopling of the western hemisphere and the extinction of many mammal species. This model dominated scientific and popular thinking until the mid 1990s.

The Clovis-first model met the scientific standards by including propositions that could be tested by collecting information from the evidentiary record. The key propositions included (cf. Bonnichsen and Schneider 1999): 1) a distinctive, highly efficient technology that allowed; 2) a single human population to migrate from Siberia to North America, a continent previously devoid of humans, about 11,500 RCYBP; 3) these people expanded throughout North and South America, arriving at Terra del Fuego approximately 10,600 RCYBP and 4) caused the extinction of many mammalian species.

The rules for testing model propositions against the archaeological and paleontological records were not formulated at the time the model was published. Nonetheless, specialists quickly recognized that evidence was required to support or refute the Clovis-first model propositions. The major points of contention can be summarized in five arguments.

1. Extinction An extensive literature has developed in response to the original proposal that Clovis hunters were responsible for the extinction of big-game animals in North and South America (Martin and Klein 1984; Martin and Wright 1967). A significant counter-model proposed that rapid global environmental change at the end of the last Ice Age caused dramatic restructuring at the biome level and the reorganization of plant and animal communities to which large mammals could not adapt (Graham and Lundelius 1984). Radiocarbon dating of bones from specific taxa (Mead and Meltzer 1985) has shown that the extinctions were more gradual than originally anticipated and took place over several thousand years, and that the environment may have been the causal agent (Grayson and Meltzer 2002). Haynes (2002), however, has proposed that both humans and environmental changes were causal factors.

2. Spatial Issues The Clovis-first model predicts that the Clovis homeland is in Siberia, but after more than 40 years of research, no evidence has been found in Siberia to support the Siberian homeland proposition (Bradley and Stanford 2004; Goebel 2004). Recent research at Ushki Lake, Kamchatka (Goebel et al. 2003), has demonstrated that the Ushki site is younger than anticipated and cannot be considered ancestral to Clovis. Fluted points have been encountered in Alaska along the flanks of the Brooks Range (Clark 1991), but obsidian hydration dates on these specimens, stratigraphic position (Reanier 1995), as well as the timing of deglaciation of the northern Brooks Range (Schweger 1969) suggest they are younger than Clovis and therefore cannot be regarded as ancestral. The model also predicts that fluted points should be younger in South America, but fishtail points in Argentina and Chile are approximately the same age as those in North America (Miotti et al. 2003).

3. Variability It has long been argued that Clovis represents a single culture produced by a single biological population (see discussion under the Clovis pattern). The signature calling card of this complex is the Clovis toolkit. However, this proposed kit is not known in its entirety from a single site. Rather, it is a composite construct based on tools associated with Clovis points from multiple sites over the entire continent.

The second and related argument is that there are different regional variants of Clovis. These include types such as East Wenatchee, Washington (Gramly 1993); Debert, Nova Scotia (MacDonald 1985); Ross County (Prufer and Baby 1963); and Gainey (Simons et al. 1984).

The variability in point styles and toolkits has led to the suggestion that Clovis was not a culture, but represents the diffusion of a technological complex across populations already in the Americas during a time of severe environmental stress (Frison and Bonnichsen 1996; Stanford 1978).

4. Co-traditions The Clovis-first model proposed that the Clovis big-game hunters were the first colonists to enter the Americas. Bryan (1979) in his Smith Creek Cave study was one of the first to propose that other North American cultures were contemporaneous with Clovis (Bryan and Touhy 1999). Some of the most important, which are discussed in greater detail in the next section, include the Nenana complex of Alaska, the Stemmed Lanceolate complex of the western United States, the Goshen/Plainview complexes of the Great Plains, and the El Jobo, Fishtail Point, and Itaparica complexes in South America.

5. Pre-Clovis The Clovis-first model predicted that sites older than Clovis would not be found. Now at least 50 sites found in North and South America that have pre-Clovis radiocarbon assays seriously challenge the Clovis-first model.

Summary

The Clovis-first model has collapsed under the weight of a substantial body of contrary evidence that refutes its basic premises. What is known about the spatial, temporal, and assemblage diversity does not support either the Clovis-first model or the revised models (for a contrasting view see Fiedel this volume; Haynes this volume). They do not explain the co-traditions in North and South America and do not account for pre-Clovis-age sites.

Although the Clovis-first controversy diverted a tremendous amount of time and energy from the overall goal of developing new knowledge about Paleoamerican prehistory, it had an important benefit. The exchange between specialists has led some down the model-constructing and testing path and toward formulating evidentiary standards that have helped move scientific model building forward.

Where We Are

During the Clovis-first era, many came to regard the Paleoamerican record as beginning with Clovis and all subsequent cultures, biological populations, and languages as descending from a single founding group (Bonnichsen et al. 1987). Gradually, a new multi-origin perspective is emerging; it envisions that the Americas were colonized multiple times by distinct cultural and biological groups and that each group adapted to the local and regional biological and environmental circumstances of its habitat.

This development has challenged the scientific community to create a new generation of descent models to explain the cultural and biological patterns in the archaeological and paleobiological records.

A general overview of some of the major cultural and biological patterns that evolved in the Americas during late-Pleistocene times begins with potential routes of migration now actively under consideration. This is followed by descriptions of lithic and bone technologies both earlier than and contemporary with Clovis. These data permit developing a new model for reconstructing the peopling of the hemisphere.

Archaeological Evidence and Models

Routes The principal routes that have been proposed to account for the peopling of the Americas are the Bering Land Bridge, the Pacific Rim, and the North Atlantic (Figure 1). Although similarities between cultural features of present-day populations in Australia, Melanesia, and Polynesia were cited in the 1920s as evidence for immigration across the southern Pacific to South America, Rivet (1943) and recent skeletal studies suggest possible affinities with Southeast Asian populations (Neves 1996),

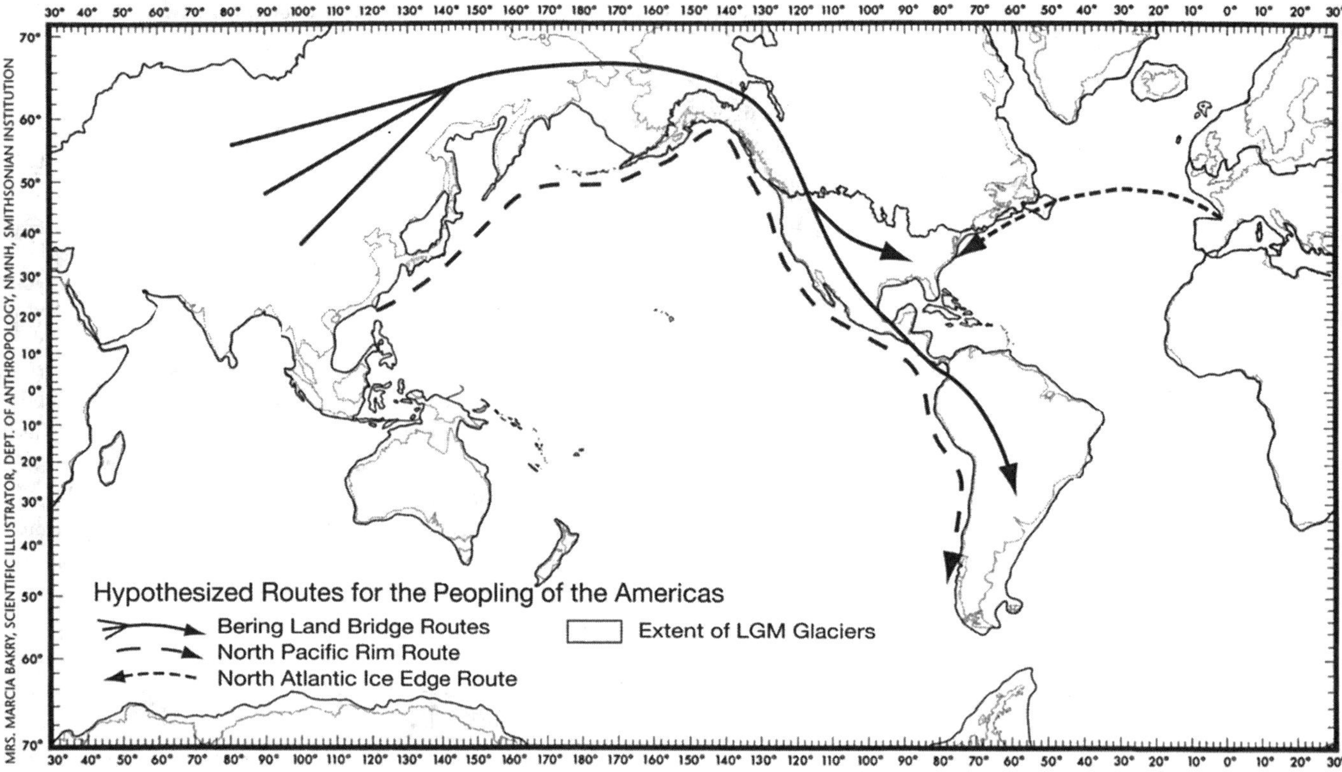

Figure 1. Principal routes currently proposed for the peopling of the Americas.

there is no archeological evidence to justify seriously considering this possibility at this time.

1. Bering Land Bridge Traditionalists who proposed a continental route from Siberia via the Bering Land Bridge envisioned that hunters walked from Asia to the Americas when sea levels were lower and the Bering Platform connected the continents. Presumably, Asian populations crossed Beringia and moved south and east through an Ice-free Corridor. Recent glaciological research suggests, however, that this route was blocked by ice from approximately 28,000 to 11,600 RCYBP (Bonnichsen et al., this volume; Jackson and Duk-Rodkin 1996; Mandryk and Rutter 1996 and papers therein).

The Bering Land Bridge model was challenged as researchers realized that radiocarbon-dated sites, such as Monte Verde, Chile (Dillehay 2000b); Meadowcroft, Pennsylvania (Adovasio and Page 2002); Cactus Hill, Virginia (McAvoy and McAvoy 1997); and mammoth sites in Wisconsin (Overstreet 2004) placed humans south of the ice sheet prior to the opening of the Ice-free Corridor.

Despite these changing perceptions, the Bering Land Bridge could have been used, however, prior to 28,000 and after 11,600 RCYBP. For this option to achieve model status, scientists need to address the questions: 1) who crossed the Land Bridge? 2) where did they come from? 3) where did they go after crossing the Land Bridge? and 4) what signature evidence did they leave in the archaeological and paleobiological records?

The oldest dated evidence from the Siberian side comes from the 11,000-year-old level at the Ushki site on the Kamchatka Peninsula (Goebel et al. 2003), which is younger than many Nenana sites in Alaska. However, the widespread occurrence of microblade industries in Japan (Akazawa 1999) and in eastern Siberia in the Lena and Aldan river regions suggests strong ties in the far north between Asia and the eastern Beringia during late-Glacial times (Mochanov and Fedoseeva 1996).

2. The North Pacific Rim What has traditionally been called the Pacific Rim or Coastal-entry hypothesis is a complex model that involves several different propositions. Hrdlička (1923) and Huesser (1960) were among the first to recognize the potential importance of this route. Fladmark (1978) was the first to articulate the Coastal-entry route as an alternative to the Ice-free Corridor route.

The Coastal-entry hypothesis was an idea before its time and did not draw much attention until the demise of the Clovis-first model. With the discovery of archaeological evidence from dated contexts in North and South America earlier than the opening of the Ice-free Corridor, archaeologists have focused on the North Pacific Rim as a logical alternative route for bringing people to the Americas. Proponents envision that early maritime-adapted populations using small boats could have coast-hopped along the Pacific Rim from Asia to the Americas just prior to the opening of the Ice-free Corridor between 14,500 and 14,000 radiocarbon years ago. This adjustment from 11,500 to 14,000 RCYBP presumably would have allowed humans to get to Monte Verde in late-Glacial times (Dixon 2001; Earlandson 2002; Fladmark 1978; Meltzer 2004).

Although there is considerable interest in the Pacific Rim route, this has not been articulated as a formal model with a series of testable propositions with strong supportive evidence. Such a model would need to specify: 1) a source population and homeland; 2) signature evidence with characteristics that can be used to define the colonizing group; 3) when humans came to the Americas; and 4) details of the spatial or geographical areas where evidence will be found to support the colonization event.

The Pacific Rim route could have been used during both the late Pleistocene and Holocene, and opinion varies as to when colonization might have occurred. Scholars such as Fladmark (1978) and Dixon (2001) suggest that south of the Queen Charlotte Islands, glacial ice extended into the Pacific Ocean with refugium conditions in the islands during the Last Glacial Maximum. Humans could have moved along Pacific Rim as deglaciation began to occur about 14,000 RCYBP. Others (e.g., Gruhn 1994) propose that colonization along the Pacific Rim route could have begun as early as 50,000 RCYBP. Differences of opinion also exist as to where the colonists may have originated. For example, Fladmark (1978) suggests they introduced the Dyuktai technology from Northeast Asia. Others (Brace et al. 2001; Steele and Powell 2002) believe the Americas were colonized by populations that came from south of Siberia, possibly from Southeast Asia, through Japan.

Although the route is logically appealing, archaeological evidence from sites along the Pacific Rim has been amazingly meager. The dearth of supporting evidence is related to sea level fluctuation. Since the Last Glacial Maximum, sea level has risen over 120 m and would have drowned landscapes along the shore and in river estuaries where sites would have been located. With the exception of the Canadian effort (Josenhans et al. 1997) to locate early sites off the coast of the Queen Charlotte Islands using remote sensing and dredging techniques, little effort has been dedicated to searching along submerged coastlines.

The earliest archaeological evidence along the Pacific Rim of North America comes from On Your Knees Cave on Prince of Wales Island, where a bifacial stemmed point associated with microblades, a bone point that has an AMS age of 10,300, and a human skeleton dated to 9330 and 9880 RCYBP were recovered (Dixon 1999). Other early sites include the Blade site on Anagula Island, which has a beginning occupation date of 8500 RCYBP (Ackerman et al. 1979), and the Ground Hog Bay and Hidden Falls sites in southeastern Alaska, where Ackerman (1996) found a microblade and heavy-chopper industry thought to date between 9000 and 9500 RCYBP.

Occupation of the Channel and Santa Rosa islands implies an early population using watercraft along the California coast. This is evidenced by an 10,600 RCYBP date from Daisy Cave, where humans were using marine resources, on the northern shore of San Miguel Island, 40 km off the Santa Barbara mainland coast (Erlandson 2002). Research by Johnson on remains found at the Arlington Springs site on Santa Rosa Island yielded an AMS date of 10,090 RCYBP from charcoal recovered in the same stratum that yielded human bones (Wenner et al. 1991). A radiocarbon assay of 10,090 RCYBP processed from pygmy mammoth remains excavated on Santa Rosa Island indicates that these animals were contemporaries of humans occupying the Channel Islands (Agenbroad et al. 1999).

Early shoreline sites have also been encountered in several places along the west coast of South America. Stothert (1985) has reported initial dates around 10,800 RCYBP for the Las Vegas complex in central Ecuador characterized by pebble-based pounders and hammerstone associated with utilized unifacial flakes. On the adjacent coast of Peru, Richardson (1978) excavated campsites of the Amotape complex dating 11,200 ± 115 to 8125 ± 80 RCYBP consisting of mangrove shells and bones of birds and small mammals, along with expedient tools.

On the south coast of Peru, excavations by Sandweiss et al. (1998) at Quebrada Jaguay encountered obsidian flakes, unifacial expedient tools, and a few unfinished bifaces in association with fish bones, marine mollusks, and crustaceans. Thirteen charcoal dates extend from 11,105 to 9850 RCYBP. Keefer et al. (1998) found a rich maritime fauna at nearby Quebrada Tacahuay associated with unifacial tools, a few bifaces, and flakes dating 10,770–10,530 RCYBP. The Ring site excavated by Richardson (1978) is another shell midden associated with an old shoreline and lagoon. The earliest cultural levels contain a few marine shells, abundant marine bird bones, and unifacial lithics. The shell dates suggest an initial occupation around 10,500 RCYBP. A shell midden in the same region investigated by Lavallée (2003) at Quebrada de los Burros contained burials dated 11,030 and 9790 RCYBP associated with unifacial and bifacial stone tools.

In the Los Vilos region of central Chile lithics are associated with extinct fauna in the upper deposits of raised marine terraces. Quebrada de Quereo, a settlement on the edge of a ravine that dates earlier than 11,000 RCYBP, contains horse and camelid bones possibly modified by humans (Núñez et al. 1994). A site in the same region, Quebrada El Membrillo, has yielded surface evidence of flaked stone associated with equid, camelid, and mylodon bones, some of which show cut marks. One vertebra yielded an age of 13,500 RCYBP.

Excavations at Monte Verde, along the banks of Chinchihauapi Creek in southern Chile, yielded an unusual archaeological record preserved under a late-Pleistocene bog (Dillehay 2000b). Radiocarbon-dated remains from the Monte Verde II occupation have an average date of 12,500 RCYBP. Cultural materials include flake tools typical of South American unifacial sites, extinct animal bones including mastodont and paleolama remains, a bi-pointed projectile point, plant remains, log frames and planks anchored by stakes, and walls covered by animal hide. The presence of salt, beach-rolled pebbles, and bitumen implies that Monte Verdeans either traveled to the coast or were part of a coastal exchange system.

In summary, archaeological evidence from the eastern Pacific indicates there were a series of regional adaptations for exploiting maritime fauna by 11,000 to 11,500 RCYBP in both North and South America. Thus the possibility cannot be excluded that terrestrial large-mammal hunters refocused their efforts on lucrative coastal resources as Pleistocene extinction took place at the end of the last Ice Age. It is equally possible that rapidly rising sea levels displaced coastal peoples, who relocated at higher elevations. Interestingly, there were two major spikes in sea level rise, one that brackets the period from 14,800 to 14,100 RCYBP and one that brackets the period from 12,600 to 12,100 RCYBP, and sea level rose at a rate of 5.1 m per century.

3. North Atlantic Ice Edge Another hypothesis is that hunters living along the pleniglacial coast of northern Spain and western France developed an economy utilizing maritime resources along the sea ice margin of the Polar front. Eventually, following the sea ice margin, they crossed the North Atlantic by watercraft (Bradley and Stanford 2004; Stanford and Bradley 2002).

These mariners explored the east coast of North America and established a foothold along the mid-Atlantic coastline. Through time their technology evolved into the Clovis complex and the population spread westward across North America. Evidence in support of this colonization event consists of unique similarities in complex biface and blade manufacture, as well as bone tools, and assemblage content found in Solutrean and Clovis assemblages. Radiocarbon dates from Solutrean sites in Spain overlap at the one-sigma level dates from the pre-Clovis sites of Cactus Hill, Meadowcroft level IV, and Page-Ladson in the eastern United States. Dates from the pre-Clovis sites also overlap at the one-sigma level Clovis dates, forming a time regression cline along an east-west axis that extends from Spain to Montana. Moreover, a fluted point found on the upper strand line of the Champlain Sea in Vermont, made of Ramah chert from northern Labrador, is unequivocal evidence of the sophisticated maritime abilities of early Americans in the Northeast (see Jodry this volume).

The North Atlantic Paleo-maritime model meets the criteria of a scientific model as outlined above and should be rigorously tested.

In summary, three routes have been considered logical possibilities for bringing humans to the Americas. These could have been used multiple times during the late Pleistocene or Holocene. At present, however, they have only minimal empirical supporting evidence.

Archaeological Patterns

Pre-Clovis Technologies Creating adequate models for the peopling of the Americas is ultimately dependent on evidence in the archaeological and biological records. Although we currently lack models to explain specific archaeological patterns/adaptive systems, we now know a great deal more about the North and South American archaeological records than we did when the Clovis-first model was constructed. Although there has been and still is a focus on Clovis (Fiedel 2002; Haynes 2002; Meltzer 2004), there are other significant issues that need to be taken into account in considering the peopling of the Americas. The following discussion summarizes the most important late-Glacial archaeological patterns in the Americas.

Paleoamerican sites earlier than Clovis are known as pre-Clovis even though they may have no historical relationship to Clovis. Figure 2 illustrates the spatial distributions of 28 North American sites that have yielded radiocarbon ages older than 11,600 RCYBP plotted as well as the glacial ice distribution and sea level positions during the Last Glacial Maximum (LGM). These locations were plotted rather than providing details for each individual site; the following discussion focuses on three pre-Clovis technical patterns: 1) early bone modification; 2) unifacially flaked stone; and 3) flaked-stone blades and bifaces.

1. Bone Modification Pattern Since the 1970s researchers reported early bone modification patterns in Old Crow Flats and the Blue Fish Caves in eastern Beringia (Bonnichsen 1979; Cinq-Mars 1979, 1990, 2001; Cinq-Mars and Morlan 1999; Irving et al. 1977; Jopling et al. 1981; Morlan 1980, 1986; Morlan and Cinq-Mars 1982; Morlan et al. 1990).

In a recent overview of bone modification evidence for human occupation in eastern Beringia during the last 40,000 radiocarbon years, Morlan (2003) has challenged the "Nay Sayers" by shifting the burden of proof. The Nay Sayers' various hypothetical "what if" scenarios argue that natural processes, not humans, altered Pleistocene-age bones (see Haynes 1991). However, Morlan asserts that the critics have failed to present any evidence to falsify the evidentiary record and that rather than subvert more time, energy, and effort to senseless debate, research efforts should focus on testing scientific hypotheses by empirical methods. He proposes that

> human occupation of eastern Beringia began during the mid-Wisconsin interstade shown by the recovery of cut marked bison bones and fractured proboscidean limb bones dated to between 25,000 and 42,000 RCYBP in the Old Crow Basin, northern Yukon Territory. This hypothesis can be falsified if the cut marks are restudied and found not to be the results of stone tool use; if localized impact scars can be shown to result from animal trampling; and if some series of natural processes can be shown to fracture bone in one plane and subsequently detach flakes from its cortex by force delivered in a different plane.

He also states,

> Human occupation of eastern Beringia began or continued during the early part of the late Wisconsin and persisted, at least intermittently, throughout the glacial period as shown by a variety of artifact types from the Bluefish Caves in northern Yukon. This hypothesis will be more difficult to falsify, because it rests on a wider variety of material evidence from an excavated context. One would need to identify natural processes that can fracture and flake mammoth bones, cut and polish caribou and sheep bones, and cause cut-mark mimics on many anatomical parts of many vertebrate taxa. The stone tools are undeniable artifacts, but it might be shown that they are younger than their stratigraphic associates (Morlan 2003).

Morlan's assessment is that no information has been presented to refute or falsify the evidence from Old Crow Flats that humans occupied eastern Beringia prior to the LGM. Twenty-three culturally modified bones that lack secure stratigraphic context have yielded ages that range from 25,200 to 42,000 RCYBP. Unlike the specimens from Old Crow Flats, a "Rosetta" bone core found in stratigraphic context at Blue Fish Cave II is an unquestionable artifact (Cinq-Mars and Morlan 1999). The core dates to 23,200 RCYBP, and the flakes removed from it average 23,500 RCYBP in age. The Blue Fish Caves mammoth bone core, which is identical to cores lacking context from Old Crow Flats, vindicates earlier

Figure 2 (opposite). Last Glacial Maximum, North America, showing the locations of selected pre-Clovis sites.

MRS. MARCIA BAKRY, SCIENTIFIC ILLUSTRATOR, DEPT. OF ANTHROPOLOGY, NMNH, SMITHSONIAN INSTITUTION

claims for an early bone-flaking pattern in eastern Beringia. In short, propositions supported by empirical evidence are not trumped by competing propositions that are not supported by empirical evidence. The burden of proof lies with the critics to develop evidence to refute or support competing hypotheses.

Bone-flaking technology is widely documented in Eurasia and occurs in the Acheulian of Italy, the Micoquien and Gravettian of Czechoslovakia, and the Dyuktai culture of Siberia (Morlan 2003). In addition to eastern Beringia, evidence for an early bone-flaking pattern exists at several sites south of the Laurentide sheet. The earliest is the McMinnville mammoth site, Oregon (Bonnichsen et al. 2004), where green bone fractures appear on the tusk sheath, mandible, and skull of a mammoth. Additionally, a bison tibia broken while green, with flaking of compact bone at its distal end, suggests the presence of human butchers. A single AMS date from collagen extracted from the skull bone yielded a radiocarbon age of 44,000 RCYBP.

The LaSena, Shaffert (Holen and May 2002), and Jensen sites (May and Holen 2005) in Nebraska and the Lovewell mammoth site (Holen 1996, 1997) in Kansas are buried in loess and have produced evidence of mammoth limb bones broken at mid-shaft by percussion blows. Mammoth limb bone cores and flakes are also present at these sites. Radiocarbon ages of these sites range from 14,000 to 19,000 RCYBP (May and Holen 1993). Mammoth long bones with mid-shaft breaks dating between 11,000 and 13,000 RCYBP were recovered at the Dutton and Selby sites in Colorado (Stanford 1979). At Lamb Spring near Denver, fractured mammoth bones were recovered along with a 33-lb boulder (Fisher 1992; Stanford et al. 1981b). The Sand Creek site, Texas, has also yielded flaked mammoth bone that dates to 13,450 RCYBP (Johnson et al. 1994; Johnson this volume). The Duewall Newberry locality in south-central Texas has produced flaked mammoth bone, but owing to poor bone preservation it has not been dated (Carlson and Steele 1992; Steele and Carlson 1990). The southernmost known occurrence of the early flaked-bone pattern is at the Tocuila site in the Valley of Mexico (Morett et al. 1998).

Several mammoth butchering/scavenging localities on the southwestern side of Lake Michigan in the Kenosha, Wisconsin, area, along what would have been the southern margin of the Laurentide Ice Sheet, have yielded pre-Clovis radiocarbon ages (Overstreet 2004; Overstreet this volume). Unlike mammoth localities associated with the early bone-flaking pattern, the Wisconsin mammoth limb bones were not broken in mid-shaft. Instead, they exhibit cut marks in the joint regions, suggesting dismemberment by flaked-stone tools. Cut marks have also been documented at the Hebior site, which dates about 12,500 RCYBP. Similarly, the Schaefer site has a date of 12,310 ± 60 RCYBP, and the Mud Lake site dates around 13,500 RCYBP (Overstreet and Stafford 1997). In addition, flakes and biface stone tools were recovered from the Hebior and Schaefer sites. The tantalizing evidence of cut marks suggests a butchering and meat processing strategy used by ice-marginal peoples, who relied on sharp-edged stone tools for dismembering and processing mammoth meat. This strategy differs from that at sites on the Central Plains.

Fluted points have been found associated with flaked mammoth bone at Owl Cave, Idaho (Miller 1989), the Lange-Ferguson site, South Dakota (Hannus 1989), and the Clovis-type site of Blackwater Draw, New Mexico (Hester 1972). This indicates that mammoths provided bone and ivory as raw materials for tool manufacture until their extinction.

Evidence of megamammal bone flaking extends from Eurasia through Beringia and south through the United States to the Valley of Mexico. Bone breaking is an indicator of a widespread technique used by early peoples to butcher animals; severing long bones in the mid-shaft section rather than cutting through joints exposes marrow in medullary cavities. Additionally, experimental tools produced by flaking elephant bone were used to extract bone from carcasses to create cutting, chopping, and scraping tools used in butchering and other activities (Stanford et al. 1981a). Available radiocarbon dates suggest that the bone-flaking pattern first appeared slightly earlier than 40,000 RCYBP and persisted through Clovis times, spanning a period of about 30,000 years.

2. Unifacially Flaked Stone Tools Another early technological pattern, although poorly represented, emphasizes unifacially flaked stone tools. Signature characteristics of this pattern include unifacially flaked cobbles or nodules, edge-trimmed bipolar spalls, and flakes. Sites with this tool-working strategy may include the upper pre-Clovis level at the Topper site in South Carolina (Goodyear and Steffy 2003), the early level at the Hueyatlaco site, Mexico, thought to have an age of more than 20,000 RCYBP (Irwin-Williams 1969), and 40,000–60,000 RCYBP levels at Pedra Furada in northeastern Brazil (Guidon et al. 1994). The occurrence of sites in South America that date to the eleventh millennium, including Tibitó in Colombia (Van der Hammen and Correal 2001), Las Vegas in Ecuador (Stothert 1985), and Itaparica in Brazil (Schmitz et al. 2004), suggests this technological system may have persisted until the beginning of the Holocene. These assemblages have yet to be fully investigated and described in the literature.

3. Flaked-stone Bifaces The third pre-Clovis technological pattern includes basally thinned projectile points (Figure 3), bifaces, and blades and flakes made from polyhedral and disk-shaped cores. These tools occur at the Cactus Hill site and may date as early as 16,900 RCYBP (Wagner and McAvoy 2004). Research still in progress at this site will lead to a detailed description of this important assemblage. Similar tool types come from Level IIA at Meadowcroft Rockshelter (Adovasio and Page 2002). Provided the radiocarbon assays of ca. 12,500 RCYBP adequately date the unfluted Suwannee points and butchered mammoth bone obtained at the Page-Ladson site in Florida (Dunbar and Hemmings 2004), they also fall into this pattern. The South American El Jobo lanceolate projectile points dating between 12,000 and 13,000 RCYBP at Taima-Taima (Gruhn this volume) and Monte Verde (Dillehay 2000a, b) also fall into the pre-Clovis bifacial technological pattern.

At this juncture, we know very little about the adaptive systems and lifeways of early peoples who flaked bone, created unifacially flaked and core tools, or practiced the biface and flake/

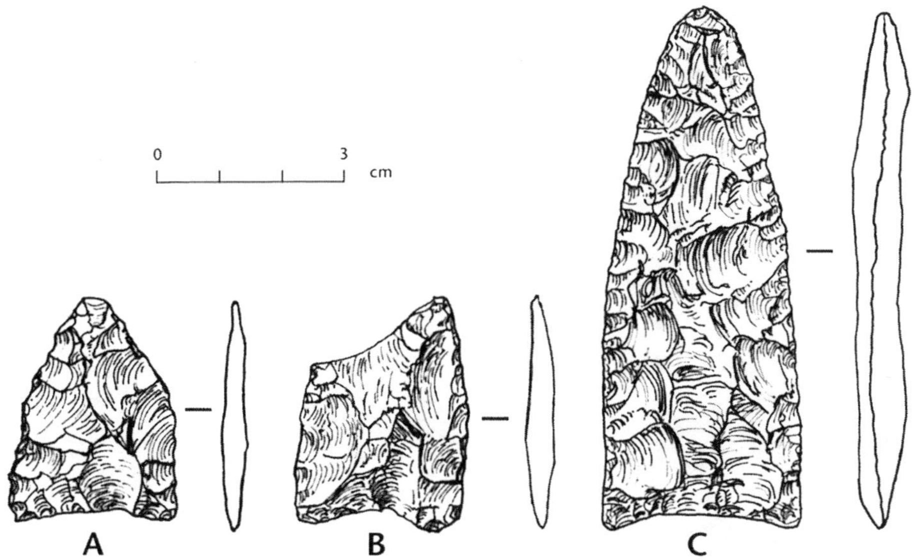

Figure 3. Pre-Clovis indented-base bifacial projectile points. **A, B,** Cactus Hill; **C,** Page-Ladson Suwannee.

blade industry that preceded Clovis. Nor do we know how these patterns relate to Clovis. Work in progress on pre-Clovis materials recovered from Virginia, South Carolina, Mexico, and Brazil should help fill these voids in our knowledge. While there are remaining doubts about some pre-Clovis sites mentioned above, there is little doubt about the presence of humans in the Americas before Clovis. The challenge that lies before the profession is to address those lingering doubts with new research using state-of-the art research methods.

North American Co-traditions

The Clovis-first model proposed that Clovis big-game hunters and their descendants, by using a new hunting technology, were able to spread throughout the hemisphere in about 1,000 years. An initial founding group of about 150 people and their descendants therefore created all the archaeological complexes in the Americas.

From a descent perspective, there are two serious problems with the Clovis-first model. The first, dealt with in detail in the previous section, is that there should be no archaeological evidence earlier than Clovis in the Americas. There is now ample evidence from North and South America, however, to document earlier colonization events. The second descent issue is that if Clovis is indeed the single first culture in the Americas, we should not find archaeological evidence of other complexes contemporaneous with Clovis. A significant body of archaeological evidence, however, indicates the presence of projectile point–using cultures that are coeval with Clovis in North and South America. In the following discussion we present archaeological evidence for multiple co-traditions in North and South America that are as old as if not older than Clovis; we then propose a diffusion model to explain the widespread appearance of divergent projectile-point patterns throughout the hemisphere during approximately the same time period.

Clovis

Clovis is the best known of the late Ice Age projectile point co-traditions in the Americas. Recent publications on the topic include studies by Collins (1999) and Haynes (2002) and reviews by Tankersley (2002) and Dixon (1999). For comparative purposes a base map of North America has been constructed (Figure 4) showing the locations of selected Clovis sites relative to the extent of glaciers and exposed continental shelves at approximately 11,000 RCYBP.

Clovis is generally considered to date between approximately 11,500 and 10,900 RCYBP. (Stanford 1999). It is typically represented as a normative culture in which all individuals share similar concepts, values, and beliefs replicated by each subsequent generation (Young and Bonnichsen 1984). Signature artifacts in the Clovis toolkit are typically made from very high quality stone selected to produce bifacial fluted projectile points, large bifaces, blades used for cutting, endscraping tools and flakes, gravers, and burins. Other implements include bone and ivory points and spear foreshafts. A bone shaft wrench was found at Murray Springs, Arizona (Haynes and Hemmings 1968), and at Blackwater Draw, New Mexico; an ivory billet (Saunders et al. 1991) and a sectioned tusk fragment (Saunders et al. 1990) point to the use of large bone and ivory items made from proboscidean remains. Mammoth bone cores and bone flakes are known from the Lange-Ferguson site (Hannus 1989), and expedient tools are reported from Lubbock Lake (Johnson 1987). At the Gault site in central Texas, mobile art (incised lines on limestone pebbles and cobbles) appears to be associated with the Clovis horizon (Collins et al. 1991), suggesting a symbolic component of Clovis life. Quarry, kill, settlement, and cache sites indicate a complex and variable organizational system.

The majority of Clovis artifacts in the Plains and the Southwest have been found at kill localities, near springs and at small playas (Stanford 1999). In most cases, such as at Colby and Lange-Ferguson, the primary activity appears to have been mammoth butchering. At Murray Springs, bison butchering is associated with mammoth butchering. By contrast, most Clovis artifacts in the East appear to be associated with larger settlement sites, such

MRS. MARCIA BAKRY, SCIENTIFIC ILLUSTRATOR, DEPT. OF ANTHROPOLOGY, NMNH, SMITHSONIAN INSTITUTION

as Shawnee-Minisink (McNett 1985), Williamson (Peck 1985), and Paw Paw Cove, Virginia, on the Delmarva Peninsula bordering Chesapeake Bay (Lowery 2003). Many of the Eastern sites, including Carson-Conn-Short (Broster et al. 1994) and Thunderbird (Gardner 1977), are located near quarry outcrops.

Examples of Clovis caching occur at the Drake Cache, Colorado (Stanford and Jodry 1988), Anzick (Jones and Bonnichsen 1994; Lahren and Bonnichsen 1974), East Wenatchee (Gramly 1993), Simon (Titmus and Woods 1985), and Fenn (Frison and Bradley 1999). The artifact content of caches varies, and little comparative research has been conducted on what are often exquisite Clovis artifacts to establish whether these unique associations are burial offerings, materials stored or cached for the future use, or items having ritual significance.

There are nagging problems when applying the normative approach to Clovis and other early New World projectile point–using cultures. The normative approach cannot accommodate variability and change through time. It presumes that all members of the society share the same norms and values and that these norms and values are transmitted to their descendants in each succeeding generation. The values governing the production and use of material culture, some of which are preserved in the archaeological record, are passed from generation to generation in much the same manner as nonmaterial aspects of culture.

Clovis-first specialists often implicitly adopt the normative view of Clovis culture and accommodate the problem of variability by ignoring, transcending, or averaging the differences away. For example, traditional Clovis-first advocates have argued that Clovis is a single culture and artifacts produced by Clovis peoples are essentially the same whether these groups lived in Nova Scotia or in California. Their projectile points are not identical, but similarities outweigh minor differences (Haynes 2002). Since variability is deemed to be unimportant, comparative studies of fluted-point artifact assemblages that would reveal differences and similarities are seldom undertaken.

The alternative to the normative approach is to assume that all members of society do not hold the same norms and values. Each generation does not replicate the preceding one. The study of variability within and among groups or artifact assemblages can provide important insights into how human decision making occurs through time and how humans adapt to changing sociocultural and environmental circumstances (Young and Bonnichsen 1984).

The first and most basic question is, What material signature characteristics identify Clovis? The single most diagnostic artifact is the fluted point. Although advocates of the Clovis-first culture would argue that similarities outweigh differences, they do not specify what they mean by similarity. Artifacts are complex, and their production involves the use of different domains of knowledge, including material, shape, technology, and functional use. Lithic analysts can identify and compare the material, shape, technology, and use-wear decisions made by ancient artisans to learn about the nature of inter- and intra-site variation. Unfortunately, most descriptions of fluted points and associated lithic industries focus primarily on form and pay limited attention to technology, such as the specialized trait of fluting, and almost no attention to use-wear patterns. These typological descriptive data are used in conjunction with chronology to assign sites to the Clovis complex. In short, the current archaeological definition of Clovis culture is a top-down intuitive normative construct based on fluted points and a list of traits of associated artifact types.

Figure 4 (opposite). North America at 11,000 RCYBP, showing the locations of selected Clovis and related sites.

Following the initial discovery of Clovis at Blackwater Draw during the early 1930s, additional discoveries at Southwestern sites such as Dent (Haynes et al. 1984; Stafford 1988), Naco (Haury 1953), Lehner (Haury 1956), Murray Springs (Haynes 1981), and others led to the recognition that Clovis represented an important widespread cultural pattern. Points from these sites set the standard for judging what is and is not Clovis, following the Krieger (1944) typological method. Since those early days of definition, many additional fluted-point sites have been found and significant advances have been made in how artifacts are analyzed.

It is now evident that there is not a single Clovis fluted-point type. Rather, there are several forms that do not conform well to "classic" fluted points from the Southern Plains and Southwest. Some of these variants, illustrated in Figure 5, include Debert (MacDonald 1968), Cumberland (Goodyear 1999), Barnes (Deller and Ellis 1988, 2001; Ellis and Deller 1998), and Gainey (Simons et al. 1984). Other than Debert, these regional variants have not been dated.

There is also significant variation in Clovis assemblages. All the artifacts considered to represent the Clovis toolkit have never been found together at a single site, and there are significant differences in the artifacts found at Clovis sites. None, however, is more profound than the presence or absence of blade cores and tools made from blades. For example, in the Southeast and Southwest macroblade cores and tools are common. By contrast, they are rare in the Northeast, along the southern limit of the Laurentide Ice Sheet in the Great Lakes region, in the Northern Plains, and in the West.

The question posed by fluted-point variants and assemblage variability is, What does this mean? Do all sites yielding Clovis points represent a single monolithic Clovis culture? Do the variants represent daughter or descendant populations of the original founding population? Or do the formal differences in tools and point form imply functional differences in regional adaptive systems?

There is evidence that mammoth hunting was not the only economic focus of groups who used fluted points. At the Kimmswick (Graham et al. 1981), and Hiscock sites (Laub and Haynes 1998), Clovis points were associated with mastodon remains. Elsewhere in the eastern United States and adjacent areas of Canada, the preferred prey species was likely caribou (Speiss et al. 1984); white-tailed deer were probably a favored prey in the Southeast. Investigations at the Lewisville site, Texas, recovered numerous burned remains of small mammals, reptiles, and amphibians in hearth features (Stanford 1999). The number of points

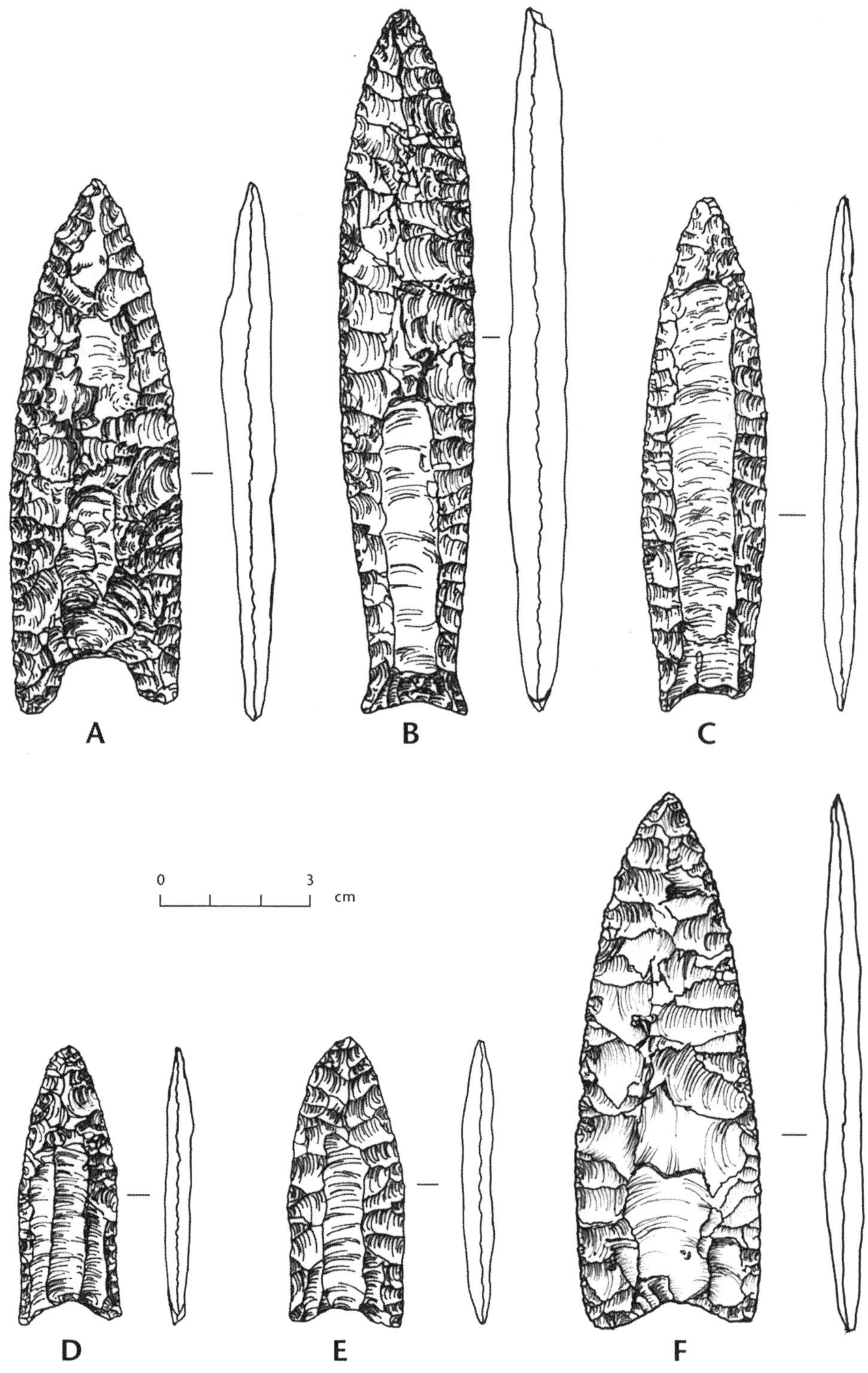

Figure 5. Examples of the range of variations of early fluted projectile points from North America. **A,** Debert; **B,** Cumberland; **C,** Barnes; **D,** Arctic fluted; **E,** Gainey; **F,** Clovis.

found at Eastern fluted-point localities tends to be substantially higher than at Southern Plains and Southwestern kill sites, suggesting more points were produced for hunting smaller prey. These variations suggest that Clovis does not represent a single adaptive pattern. Rather, it encompasses a set of poorly known regional adaptive systems.

The distribution of fluted points is instructive for understanding Clovis patterns. The map in Figure 5 shows Clovis-era glacial and sea level positions as well as significant pluvial lake systems. Examining the plotted distribution of the 62 fluted-point sites, including variants with stratigraphic contexts, makes several important dimensions of the Clovis pattern apparent.

If a founding population from Northeast Asia traveled down the Ice-free Corridor, we would expect fluted-point sites to be earlier in Alaska, the Yukon Territory, and the Ice-free Corridor. Fluted points from the Brooks Range, Alaska (Figure 5D), dated using the obsidian hydration method, are younger than radiocarbon dates from the "classic" Clovis sites (Clark 1984; Clark and Clark 1993). The only dates for fluted points found in the Ice-free Corridor come from Charlie Lake Cave (Fladmark et al. 1988) in British Columbia. Here, too, radiocarbon ages of 10,100 ± 100 and 10,770 ± 120 RCYBP suggest that these points are younger than "classic" Clovis. In short, the distribution of fluted-point sites in the Northwest does not conform to that expected for a colonizing population from Beringia.

The age typically cited for Clovis is 11,500–10,900 RCYBP. These dates, however, are based on sites from the Southern Plains and Southwest. If the dates for sites in the Northeast, such as 10,680 ± 400 RCYBP for Whipple (Curran 1984), 10,190 ± 300 RCYBP for Templeton (Moeller 1980), and 13 dates for Debert that average 10,590 ± 50 RCYBP (MacDonald 1968) are taken at face value, the inescapable conclusion is that Clovis did not end abruptly at 10,900 RCYBP, but may have continued until about 10,000 RCYBP or into the Holocene. However, radiocarbon ages from Northeastern fluted-point sites have been challenged because the charcoal may be from forest fires rather than human hearths (Bonnichsen and Will 1999). Nevertheless, it is possible that Clovis in the Northeast may bracket a time interval from 11,000 to as late as 10,000 RCYBP.

Likewise, it appears that fluted points may be more recent in the western United States than in the Plains and Southwest. Although there are no directly dated Clovis occupations in the Great Basin, five sites suggest a younger age in that area (Bryan and Touhy 1999). These include the Sunshine locality in eastern Nevada that yielded a fluted point about 20 cm below charcoal dated at 10,320 ± 50 RCYBP (Jones et al. 1996). At Danger Cave, Utah, two fluted points along with stemmed points were dated to 10,080 ± 130 RCYBP (Holmer 1986; Willig and Aikens 1988; Madsen and Rhode 1990). The Wasden site, also known as Owl Cave, Idaho, produced points with full-length flutes associated with mammoth bone. A refined collagen date of 10,920 ± 150 RCYBP on the bone (Miller and Dort 1978), along with an AMS charcoal date of 10,640 ± 85 RCYBP, led Miller to establish the occupation at Owl Cave between 10,900 and 10,600 RCYBP (Bryan and Touhy 1999). Connley Cave #5, near Fort Rock, Oregon, yielded a fluted point above a level dated at 9540 ± 260 RCYBP (Beck et al. 2004). The Henwood site located near Barstow, California, produced a fluted point in association with two Lake Mohave points and a hearth dated at 8470 ± 370 RCYBP (Douglas et al. 1988).

It has been argued that these Western fluted-point dates are in error (Beck and Jones 1997; Grayson 1993). This is especially true for the dates from the Henwood site, Danger Cave, and Connley Cave, where it appears that the points were recovered from mixed contexts. However, four recent bone dates that range between 10,525 ± 35 and 10,705 ± 35 RCYBP (T. Stafford pers. comm. 2005), from the Anzick Cache, southwestern Montana, support the idea that the Western fluted points may be younger than Clovis on the Southern Plains. Thus, the existing dates indicate that the Clovis pattern persisted longer in the Northeast, possibly in the Great Lakes region, the West, and even in Alaska than in the Great Plains and Southwest, where Clovis is best known.

In summary, Clovis has been regarded as the initial colonizing population of the Americas (Kelly and Todd 1988; Meltzer 2002). However, the existing archaeological data including the spatial distribution of fluted-point sites, and radiocarbon dates do not support the Clovis-first migration model. With only 25 dated Clovis sites in North America, it is impossible to determine exactly where the pattern originated.

An alternative explanation is that the Clovis pattern is simply the diffusion of a new technological complex across widely dispersed preexisting populations in North and South America at the end of the Ice Age. The low visibility of these earlier cultures in the archaeological record results from their technologies based on simple flaked-stone, bone, wood, and fiber tools (Bonnichsen and Schneider 1999; Frison and Bonnichsen 1996; Stanford 1978).

To move beyond the intuitively formulated normative Clovis culture concept and intuitively defined types, more rigorous and holistic studies of Clovis assemblages are necessary. It would be appropriate to focus on detailed descriptive studies of assemblages found in context using empirical analysis of material, shape, technology, and use. Such studies would evaluate similarities and differences from specific sites and would provide a rational foundation for higher-order constructs seeking to explain variation.

Nenana

The Nenana complex, originally defined by Powers and Hoffecker (1989), represents a hunting culture dated between 11,600 and 10,500 RCYBP (Dixon 2001), or approximately the same age as or slightly earlier than Clovis. What is known about Nenana is based on excavations during the past decade at a series of sites in the Tanana River drainage system of central Alaska. Sites include Owl Ridge (Hoffecker et al. 1996a) in the Teklanika River Valley; Moose Creek (Hoffecker 1996); Walker Road (Goebel et al. 1996); Dry Creek (Hoffecker et al. 1996b) located above the Nenana River; and Broken Mammoth (Holmes 1996; Yesner 1996); Mead (Hamilton and Goebel 1999; Dixon 2001); Swan Point (Holmes et al. 1996); and Healy Lake Village sites (Cook 1996), all located within the Tanana River Valley itself (Figure 6). Recent research has located sites in the southern Yukon Territory, suggesting that Nenana has a significantly wider distribution than suggested by the cluster of sites in central Alaska (Easton 2004).

All Nenana camps were small and generally located on bluffs with panoramic overviews. Sites consist of well-stratified loess, suggesting they were occupied by small groups of people for short periods of time. Occupants of these open-air sites may have lived in temporary tent-like structures, judging from the circular distribution of artifacts around a hearth at Walker Road, but the range of the Nenana settlement and subsistence patterns is poorly known.

Artifacts diagnostic of this complex are: (1) triangular and tear-shaped projectile points and knives (Figure 7A), (2) straight-

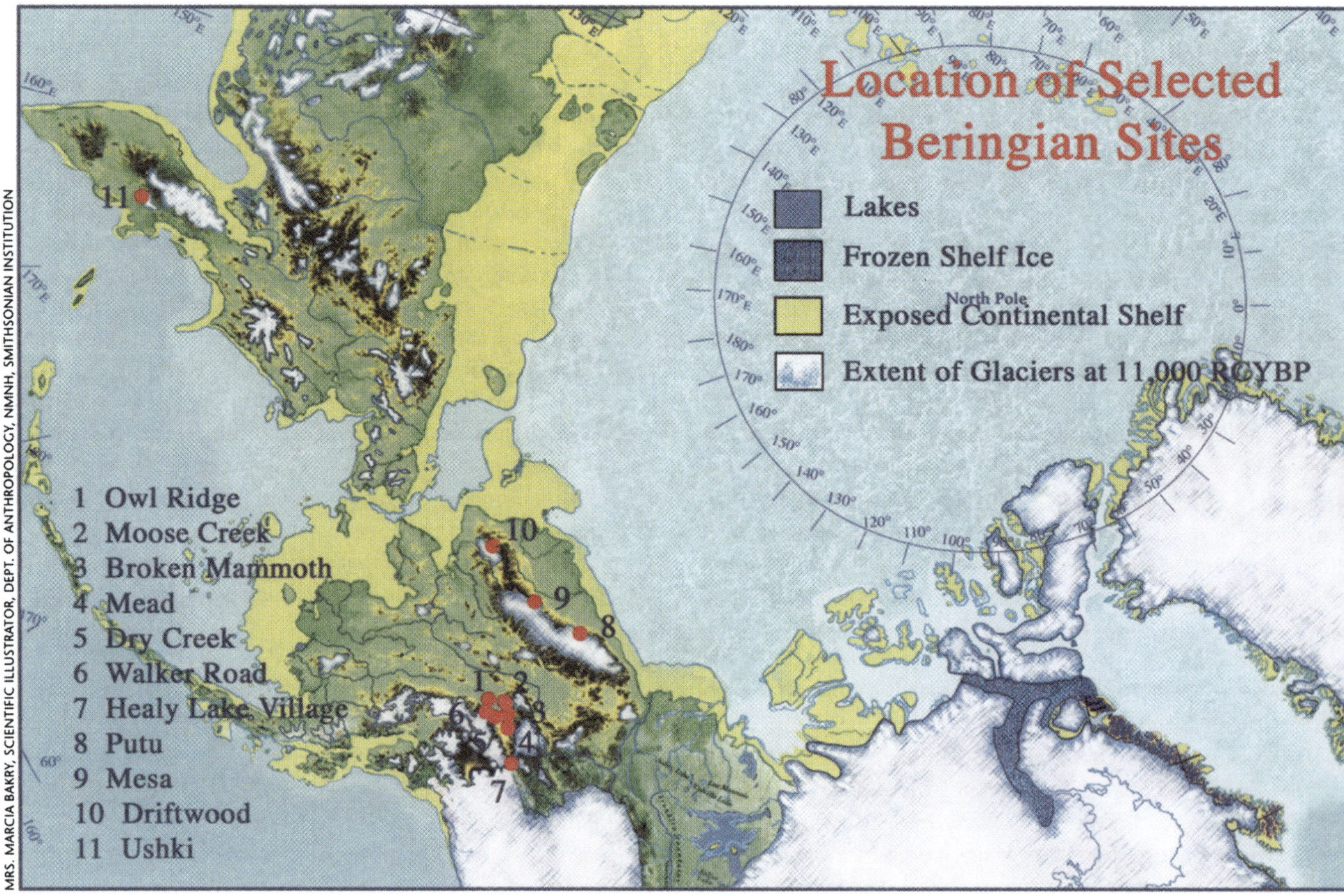

MRS. MARCIA BAKRY, SCIENTIFIC ILLUSTRATOR, DEPT. OF ANTHROPOLOGY, NMNH, SMITHSONIAN INSTITUTION

Figure 6. Beringia at 11,000 RCYBP, showing the locations of selected Paleoarctic sites.

or concave-based projectile points (Figure 7B), (3) perforators, (4) endscrapers, (5) burins, (6) hammerstones and anvil stones, and (7) unifacial knives and scrapers (Dixon 2001:283). While these types have become a normative "toolkit" for Nenana, it should be noted that bifacially flaked lanceolate projectile points (Figure 7C) occasionally occur at Nenana sites. In contrast, lanceolate bifaces are usually found at sites where microblade technology is absent. Moreover, the radiocarbon dates range between 10,000 and 11,000 RCYBP from sites such as Mesa (Kunz 2004), Putu (Alexander 1987), and Irwin's Sluiceway (Stanford and Gal n.d.), suggesting that the lanceolate points overlap in time with Nenana. It is not clear if all these point forms were used with a variety of Nenana weaponry systems during different seasonal/subsistence activities or if they represent independent co-traditions.

As originally defined, the Nenana complex lacked microblades, in contrast to the Denali complex that occurs stratigraphically above Nenana at the Dry Creek site. However, microblades produced using the Siberian (Dyuktai) techniques have been recovered from the Nenana levels at the deeply stratified site of Swan Point (Crass and Holmes 2004). The oldest cultural level at Swan Point is dated 11,600 RCYBP and contains "fossil" ivory dating 12,060 RCYBP that may have been scavenged from animals predating the occupation. Along with the microblade technology, the lithic assemblage includes split quartz pebbles, quartz pebble chopper-planes, burins, and red ocher.

The earliest firm date for the Nenana complex comes from the Tanana River Valley, Swan Point, and Broken Mammoth sites. Located near Swan Point, Broken Mammoth, also a deeply stratified site, contained four periods of cultural occupation. Nine radiocarbon dates from the oldest level, Zone IV, range from 11,800 to 11,000 RCYBP. Cultural materials from this zone include waste flakes, a quartz "chopper-scraper-plane," retouched flakes, biface thinning flakes, a scrap of fossil ivory, and ivory points. The lithic artifacts were made from wind-polished local quartz and chert cobbles. Interestingly, about 60 percent of the assemblage is composed of quartz that was used to manufacture expedient tools with little retouch, and some were used as cores that may have served as scraper planes. About 30 percent of the remaining assemblage is made from chert. A large number of the "finished" artifacts are cobble tools utilized in butchering activities (Yesner 2001).

Faunal evidence from the original excavations at Dry Creek suggests that Nenana peoples were big-game hunters whose prey species included elk and mountain sheep. Subsequent research at Broken Mammoth (Yesner 2001) indicates a more generalized opportunistic gathering economy whose activities included hunting water fowl, gathering eggs, and hunting or trapping small mammals. Elk, bison, and sheep were also important. The presence of mammoth ivory resulted from scavenging rather than hunting proboscideans.

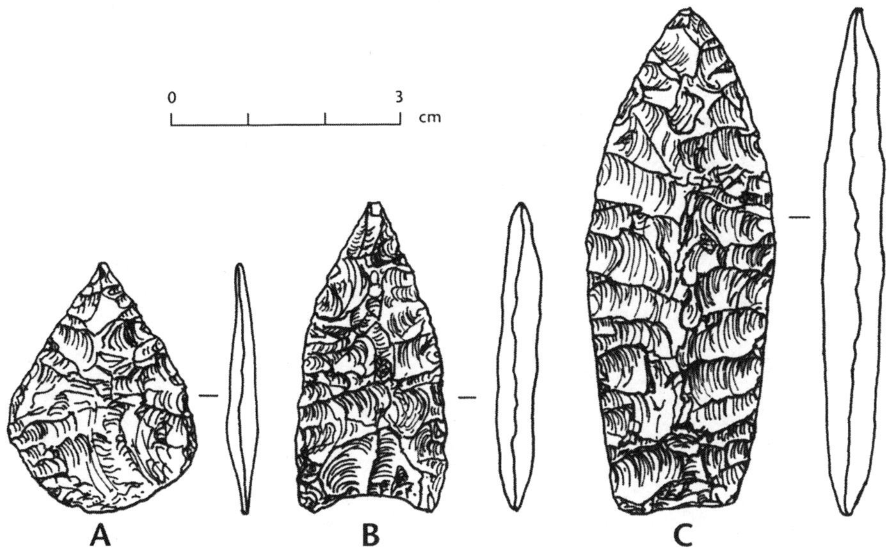

Figure 7. Paleoarctic bifaces. **A,** Nenana tear-shaped projectile point or knife; **B,** Nenana indented-base point; **C,** lanceolate projectile point.

In respect to the origin of the Nenana complex, Goebel (1999) observes that it was associated with the mammoth-steppe environment and suggests that the central Alaskan Nenana sites are technologically similar to Siberian middle upper-Paleolithic industries. Both are characterized by blade and bifacial technologies. However, it is difficult to link the two, given the temporal gap between the latest middle upper-Paleolithic sites in Siberia and the earliest sites in Beringia. Moreover, the recent discovery of Dyuktai microblade technology at Swan Point indicates that Nenana is an eastern extension of the late Paleolithic of Siberia.

Western Stemmed

Western North America has tremendous geographical, topographical, and ecological diversity that can be traced in part to environmental disruption brought about by the massive breakup of glacial conditions. Extreme environmental events, including volcanism and rapid climate change, led to the recession of the continental ice sheets and alpine glaciers, lowering of pluvial lakes, and catastrophic flooding and extensive deposition of glacial outwash and flood sediments in regional river valleys. Within this setting, human adaptation to environmentally diverse but geographically restricted settings led to the emergence of local/regional groups that used a variety of stemmed lanceolate projectile points usually made from highly siliceous chert and volcanic rocks. This type, designated the Western Stemmed series, is coeval with "classic" Clovis and continues into the early Holocene. Points of this type are far more common than fluted points.

The Western Stemmed points occur in the Great Basin, Plateau, and Snake River Plains regions of southern Idaho and the large interior valleys in California (Figure 8). They have been found at many surface locations, including near pluvial lake shorelines and along rivers. They are also known from stratified caves and rockshelters, including Marmes Rockshelter and Windust Caves, Washington (Rice 1972); Cougar Mountain Cave (Cowles 1959), Connley Caves, and Dirty Shame Rockshelter, Oregon (Hanes 1988a, b); Fort Rock Cave, Oregon (Bedwell 1973); Bison Rockshelter, Veratic Rockshelter (Swanson 1972), and Wilson Butte Cave (Gruhn 1961; Bryan and Touhy 1999) in Idaho; Redfish Lake, Overhand, Idaho; Smith Creek Cave, Nevada (Bryan 1979); Danger Cave, Utah (Jennings 1957); and Ventana Cave, Arizona (Haury 1975). Important open-air sites include Lind Coulee, Washington (Daugherty 1956); Dietz, Oregon (Willig 1988, 1990); and Hatwai and Cooper's Ferry, Idaho (Wisner 1998). Western Stemmed points have been found along fossil shorelines at many lakes including Tulare Lake (Wallace and Riddle 1988) and Borax Lake, California (Harrington 1948). Human burials associated with the Western Stemmed tradition include Buhl, Idaho (Green et al. 1998); Marmes, Washington (Fryxell et al. 1968); and possibly Fishbone Cave, Nevada (Orr 1956).

The signature artifact is not a single type, but rather a variety of lanceolate to stemmed projectile/knife types (Figure 9). Base forms vary from concave to convex, and some specimens have stems and shoulders. An underlying design feature of these projectile points is that they have thick cross sections relative to their width. This reduction scheme produced robust and versatile projectile points that could be resharpened multiple times. As noted by Bryan (1988), lanceolate and stemmed points represent a socketed hafting tradition, which is entirely different from Clovis. As with Clovis, several production grammars were used to make Western Stemmed artifact assemblages; hence there was not a single discrete culture, but variations through time and space. Our understanding of the formal and technological variability of Western Stemmed is largely unknown owing to the lack of comparative studies.

Other technological markers are stone crescents found throughout much of the western United States. Significantly, blade core technology is absent in Western Stemmed assemblages. However, ground-stone tools, including *manos* and metates, suggest gathering of plant produce supplemented their hunting economy. The importance of plants to these societies is under-

MRS. MARCIA BAKRY, SCIENTIFIC ILLUSTRATOR, DEPT. OF ANTHROPOLOGY, NMNH, SMITHSONIAN INSTITUTION

scored by the wealth of cordage and basketry, along with woven sandals and mats preserved in dry caves and rockshelters (Adovasio 1974).

In general, the Western Stemmed pattern represents a series of adaptations to local/regional resources. The Western Pluvial Lake tradition, first defined by Bedwell and Cressman (1971) and expanded by Bedwell (1973), envisioned an adaptive focus on transition zones or ecotones between lake environments and land where plants, fish, local game, and waterfowl could be exploited. Sedentary settlement was suggested by consistent occupation patterns at various caves and rockshelters. Beck and Jones (1988) note, however, that Western Stemmed points occur in other environmental contexts, suggesting a wider array of adaptive activities, including big-game hunting. Based on his work at Lind Coulee, Daugherty (1962) proposed a seasonal subsistence pattern that combined hunting locally available large game animals, such as camel, bison, mountain sheep, and deer, with exploiting riverside and lake resources such as anadromous fish and plants.

The Western Stemmed tradition is not a single discrete culture any more than is Clovis. As noted by Bryan (1988), lanceolate and stemmed points represent a hafting tradition that involves the use of socket hafts, entirely different from Clovis. As with Clovis, several production grammars were likely used to make Western Stemmed artifact assemblages.

The radiocarbon ages of 22 Western Stemmed point sites range from 11,600 to 7500 RCYBP, suggesting that it was a stable and long-lasting cultural pattern. Eleven sites have yielded radiocarbon ages greater than 10,800 RCYBP, including Clark's Flat, California; Connley Caves, Oregon; Cougar Mountain Cave, Oregon; Fishbone Cave, Nevada; Hatwai, Idaho; Lind Coulee and Marmes, Washington; Smith Creek, Nevada; and Wasden, Idaho. Several of these sites strongly support the hypothesis that the early part of the Western Stemmed pattern was coeval with Clovis (see Bryan and Touhy 1999).

Bryan (1979) was the first to recognize the importance of the Western Stemmed complex as a Clovis co-tradition during his work on Smith Creek Cave, Nevada. The Smith Creek excavations recovered a lithic assemblage of stemmed-point bases, Moriah points, scrapers, and gravers associated with faunal remains left behind by mountain sheep hunters. The earliest layer yielded dates of 11,690 ± 160 RCYBP on charcoal and 12,150 ± 120 RCYBP on wood. Additional AMS dates of 12,060 ± 450 RCYBP on camelid hair, 10,845 ± 250 RCYBP on bovid hair, and 10,420 ± 100 RCYBP on a piece of cordage (Bryan and Touhy 1999) suggest that the Western Stemmed pattern is coeval with Clovis.

The Cooper's Ferry site, located on a terrace about 15 m above the Salmon River near Grangeville, Idaho, produced a cache of points (Davis 2001; Wisner 1998). Four stemmed points that correspond to the Lind Coulee and Windust Phase artifacts were found in a circular pit about 80 cm in diameter that extended 3.2 m below the surface. There are eight radiocarbon dates from the pit, but Davis (2001) suggests that all but three dates are suspect because of contamination or provenience errors. He accepts dates of 11,410 ± 130 RCYBP and 11,370 ± 70 RCYBP for the Lind Coulee points, indicating that they are ancestral to later Windust Phase points. Davis also suggests that a reevaluation of the stratigraphy at Lind Coulee places Lind Coulee points stratigraphically below the Mount St. Helen's J tephra, which dates between 11,000 and 12,000 RCYBP. More work is needed at the site to fully evaluate this little-understood Lind Coulee feature.

Figure 8 (opposite). Locations of selected Western Stemmed sites.

Peak and Crew (1990) report a series of Western Stemmed points from the Stanislaw River, California, at the Clarks Flat site. The earliest radiocarbon date is 11,720 ± 145 RCYBP. Even though the date was not directly associated with the Stemmed Point series, it is indicative of the age of the occupation.

Fort Rock Cave, Oregon (Bedwell 1973; Bedwell and Cressman 1971) yielded a single charcoal date of 13,200 ± 70 RCYBP from Pleistocene gravels (level 10) containing a small assemblage of flakes, a *mano* fragment, two unnotched points, and a resharpened Western Stemmed point. Willig and Aikens (1988) indicate there is no evidence to disprove the possibility of vertical mixing (cf. Haynes et al. 1971); but neither is there evidence to support vertical mixing, and the date could identify an early Western Stemmed occupation.

Additional early dates for Western Stemmed points come from Oregon's Cougar Mountain Cave Number 2, which yielded an age of 11,950 ± 350 RCYBP (Bedwell and Cressman 1971), and from the bottom layer of a rockshelter on the western shore of Tule Lake, California, which yielded an age 11,450 ± 340 RCYBP on wood charcoal from a firepit (Beaton 1991). Dates from Fishbone Cave include 11,535 ± 500 RCYBP; on textiles, 10,900 ± 300 RCYBP; on plant remains, 11,200 ± 250 RCYBP; and on horse bone gelatin, 12,280 ± 520 RCYBP (Orr 1956).

The proposition that "classic" Clovis precedes the Stemmed Point series in the West has never been proven (Bryan and Touhy 1999). Provided that the pre-Clovis dates on Stemmed Point occurrences are accurate and that data suggesting that Clovis entered the Far West from the Plains at a time late in the Clovis period are correct, Stemmed Points may represent a long-lasting tradition that is coeval with Clovis and may even predate Clovis.

Goshen/Plainview

Unfluted, indented-base projectile points found in an occupation level below Folsom at the Hell Gap site in Wyoming were labeled Goshen by Irwin (1967). Although Goshen points are technologically similar to the Plainview points reported by Sellards et al. (1947), their stratigraphic position at Hell Gap suggested they were much older (see Campbell 1961; Stanford 1999). Later, because of typological considerations, Irwin (1971) abandoned the term Goshen in favor of Plainview (Figure 10).

Goshen once again achieved prominence in the literature when the Mill Iron site in Montana was investigated (Frison 1996). At this site two localities were excavated, a bison kill area and a campsite, both producing Goshen points. Radiocarbon assays suggested the occupation occurred between 10,840 ± 60 RCYBP

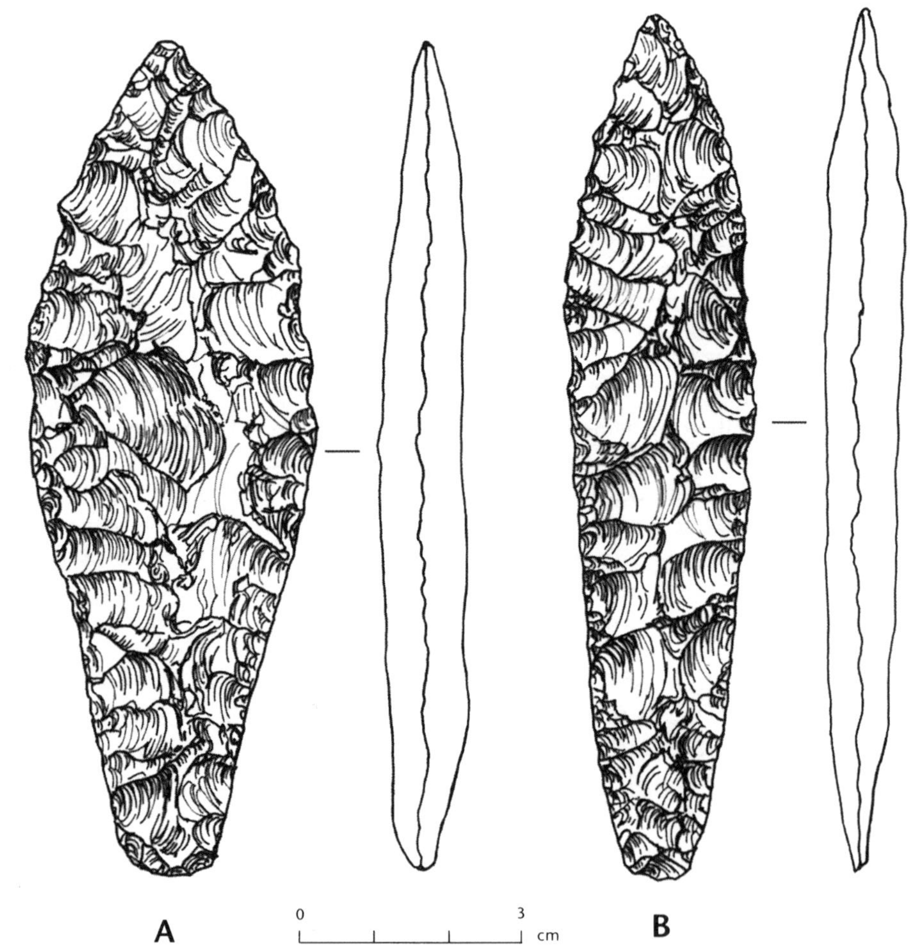

Figure 9. Examples of Western Stemmed projectile points. **A,** stemmed and shouldered; **B,** stemmed without shoulders.

and 11,360 ± 70 RCYBP (Frison 1996), supporting the pre-Folsom antiquity of the Goshen occupation unearthed at Hell Gap.

Indeed, a reanalysis of Hell Gap charcoal from the Goshen level resulted in a date of 10,955 ± 135 RCYBP (Frison 1996). Furthermore, Frison et al. (1996) suggest that the projectile points from the "Midland" level are in fact Goshen and that the cultural sequence at Hell Gap was Goshen-Folsom-Goshen. Hence, based on both its stratigraphic position and radiocarbon dates, Goshen appears to be a co-tradition with both Clovis and Folsom.

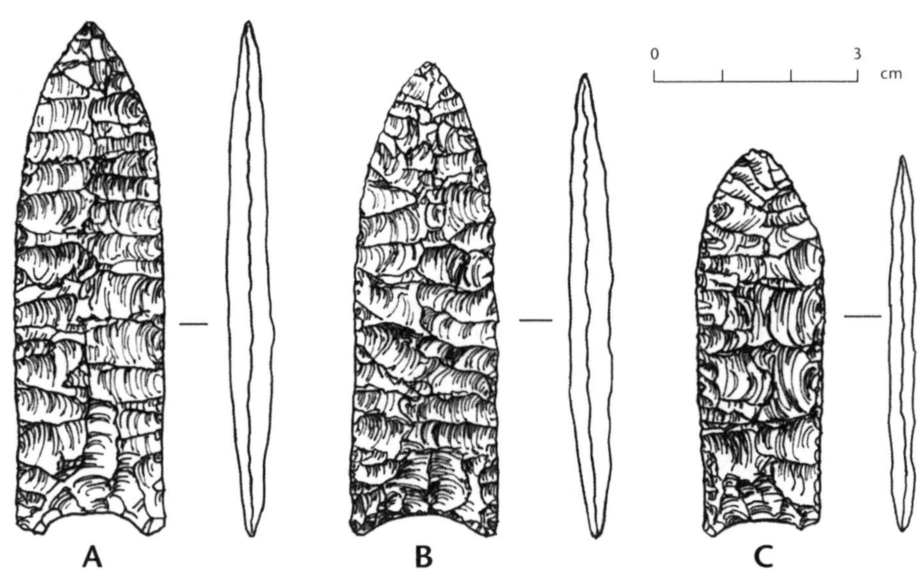

Figure 10. Early indented-base projectile points from the Western Plains and Rocky Mountains. **A,** Goshen; **B,** Plainview; **C,** Midland.

Midland points are technologically distinctive and are normally associated with Folsom sites.

Sellet (2001) refined the microstratigraphy at Hell Gap and corroborated the previous conclusion that the lowest level was pure Goshen. A Folsom occupation was wedged between Goshen and an occupation level with an undetermined point style. The next level contained a mix of Folsom and Agate Basin points; and in a final level both Folsom- and Goshen-style points were mixed. If Sellet's analysis is correct, it further supports the notion that multiple point styles were being used on the High Plains during the same time period.

A Goshen occupation occurred below Folsom at Carter/Kerr-McGee in Wyoming (Frison et al. 1996). Another Goshen bison bone bed was excavated at the Jim Pitts site in South Dakota (Donohue and Sellet 2001); it too was below a level where Folsom, Goshen, and even some Agate Basin artifacts were intermixed. Seven charcoal samples from the Goshen bone bed were assayed; if an anomalous young date with a large standard of error is ignored, the other six dates average slightly over 11,100 RCYBP. Kornfeld and Frison (Kornfeld and Frison 2000; Kornfeld et al. 1999) excavated another Goshen assemblage at Upper Twin Mountain in Middle Park, Colorado. Two bone dates of 8090 ± 60 and 10,240 ± 70 RCYBP from this site are considerably younger than the Northern Goshen sites, but they overlap Plainview dates from the Southern Plains. These dates may suggest that Goshen originated in the North and slowly moved southward onto the Southern Plains, where this style of point was in use for another two thousand years (Figure 11).

Because of sites and occupation levels where only a single point style occurred, one is led to believe that these specific artifacts are the products of identifiable social groups. However, when a site or living surface has a mixture of point styles, there are several ways to interpret past events. A mixture of types may result from sequential occupation by different social groups over a short time period, or an identified stratigraphic unit may have been redeposited with mixed artifacts from multiple occupations. On the other hand, it is also possible that coeval groups utilizing different artifact types came together in the same encampment for various activities. It would appear, however, that the Goshen/Plainview type was indeed a distinct technological form lasting from the Clovis time period until later Paleoindian times.

South America

North American archaeologists have generally discounted the antiquity of South American sites with initial dates exceeding 12,000 RCYBP because South American technologies do not conform to the North American stereotypes (Bryan and Gruhn 2003). Nevertheless, increasing numbers of early sites are well documented and their abundance, continental distribution, diversity of subsistence adaptations, and distinctive lithic complexes provide conclusive evidence for their pre-Clovis age. Figure 12 shows the locations of selected sites with initial dates exceeding 10,000 RCYBP. Ardila and Gerardo (1991), Dillehay (2000b), Dillehay et al. (1992), Lanata and Borrero (1999), Miotti (2003), Miotti et al. (2003), Scheinsohn (2003), and Zárate (1999) provide additional sites and dates. Descriptions of artifacts and regional interpretations can be found in Borrero (1999), Chauchat and Pelegrin (2004), Meggers and Miller (2003), Prous and Fogaça (1999), Kipnis (1998), Schmitz et al. (2004), and Suárez Sáinz (2000). Gruhn also provides an excellent overview (this volume).

The abundance and diversity of complexes with pre-Clovis dates in South America contrasts with their rarity in North America, where their antecedents should exist. They include rockshelters in Colombia, Peru, Chile, Argentina, and eastern Brazil, butchering sites on the pampas of Argentina and Patagonia, riverside camps in Amazonia, shell middens along the southern Pacific coast, and quarries and camps along powerline transects in several parts of Brazilian Amazonia. Several general factors are likely to be responsible for their relative abundance. First, much of South America escaped major remodeling by glaciations, inundation, erosion, and other physical processes during the Pleistocene and early Holocene, which repeatedly destroyed early North American archaeological sites (Wilson and Burns 1999). Second, caves and rockshelters suitable for habitation, which are widespread in South America, were occupied seasonally or continuously prior to the adoption of agriculture and have intact stratigraphic records. Third, major regions attractive to Paleoamericans, especially the Argentine pampas, Patagonia, the *cerrado* of northeastern Brazil, and the rain forest of Amazonia have remained little disturbed by modern settlement, roads, and agriculture; consequently sites are preserved and accessible.

The South American Database

The 55 sites shown on Figure 12 give a misleading impression of the number of sites recorded and the number of dates available. For example, RS-1-50 is one of 21 sites of the Uruguay and Ibicui phases, which have 20 dates extending from 12,770 ± 220 to 8585 ± 115 RCYBP (Miller 1987). Abrigo do Sol, a rockshelter in south-central Brazil, has 34 dates, 6 from outside the present drip line extending from 14,700 ± 195 to 7950 ± 115 RCYBP, 18 from the drip line extending from 12,300 ± 95 to 5730 ± 60 RCYBP, and 10 from inside the drip line extending from 7695 ± 65 to 115 ± 55 RCYBP (Miller 1987). Boquête, a rockshelter in eastern Brazil, has 11 dates ranging from 12,070 to 9350 RCYBP (Prous and Fogaça 1999). Pubenza in the highlands of Colombia has four dates ranging from 17,790 to 13,280 RCYBP. Lt Norte Sul produced lithics in 326 locations along the 516-km transect with 24 dates ranging from 11,300 to 1150 RCYBP. The Lt Brazil–Venezuela transect recorded lithics in 230 locations with 13 dates ranging from 13,660 to 1170 RCYBP (Meggers and Miller 2003). The excavation in the rockshelter at Pedra Pintada on the middle Amazon produced more than 30,000 lithics from levels with 56 dates ranging from 11,110 to 10,000 RCYBP (Roosevelt et al. 1996). In short, there is no doubt that South America was populated by at least 15,000 RCYBP. The reliability of isolated dates several millennia earlier that have been reported from numerous locations, notably Pedra Furada (Guidon et al. 1994), Sand Elina (Vilhena et al. 1993), and Monte Verde (Dillehay 2000b), remains to be established.

The credibility of the early radiocarbon dates is enhanced by

MRS. MARCIA BAKRY, SCIENTIFIC ILLUSTRATOR, DEPT. OF ANTHROPOLOGY, NMNH, SMITHSONIAN INSTITUTION

the presence of mastodon, horse, glyptodon, and ground sloth bones with cut marks in Patagonian and Chilean sites (Borrero 1999; Núñez et al. 1994) and mastodon remains at Taima-Taima and El Jobo in Venezuela (Gruhn, this volume), and at Tibitó and Pubenza in highland Colombia (Ardila et al. 1989; Van der Hammen and Correal 2001), among other locations.

Early dates on human skeletal remains also support the existence of a pre-Clovis occupation of South America (Neves et al. 2004). Several dozen crania dating prior to 8000 RCYBP encountered in rockshelters in Minas Gerais share with Kennewick Man a long narrow skull, low narrow face, subnasal prognathism, and low, wide orbits and noses. Statistical comparison of four female and five male crania from Cerca Grande in the Lagoa Santa region employing principal components analysis and multidimensional scaling verified the strong relationship with Africans, Australians, and Polynesians obtained based on the earliest North American individuals (Neves et al. 2004).

Lithic Complexes

A striking feature of the earliest South American complexes is the predominance of unifacial scrapers, choppers, knives, and expedient tools, cores, and flakes. Shaped artifacts and bifacial projectile points are rare prior to ca. 11,000 RCYBP and become significant only in parts of the Andean region and the southern lowlands during later times (Miotti 2003, Table 1). This situation suggests that part of the problem of identifying pre-Clovis sites in North America is incorrect expectations of the characteristics of early lithic complexes. Rather than tracing early immigrants from north to south, we should be looking in South America for clues to early North American sites.

Projectile Points

Although El Jobo and fishtail projectile points (Figure 13A, B) have become hallmarks of paleo South American complexes, they give a misleading impression of the lithic inventory of the initial immigrants. Fishtail points, described by Junius Bird from Fell's Cave in Patagonia, have attracted attention because the occasional occurrence of fluting suggested they derived from Folsom or Clovis (Bird 1938). Fluted fishtail points were subsequently identified at El Inga in the highlands of Ecuador and now are known to have a distribution from the Southern Cone of South America northward along the Andes into Central America and as far north as the Los Grifos site in Chiapas, Mexico (García-Bárcena 1979; Santamaria 1981). The Los Tapiales site in Guatemala has an accepted date of 10,700 RCYBP (Gruhn and Bryan 1977). This date is compatible with models suggesting a Clovis origin for the South American fishtail fluting technology (see Pearson 2004). However, recent investigations in Patagonia have produced earlier radiocarbon dates of 10,800 RCYBP at Cerro la China (Flegenheimer and Zárate 1997) and 12,390 RCYBP at Cueva del Medio (Borrero et al. 1998; Nami 1996), suggesting that the situation is much more complex and may favor an independent origin of fluting in South America. An alternative hypothesis is that fluting originated in Latin America and diffused from south to north (Borrero and Franco 1997; Martinez 2001, Miotti 2003).

Figure 11 (opposite). Goshen and Plainview site locations.

Lanceolate El Jobo points, identified by José Cruxent (Cruxent and Rouse 1956) at El Jobo, Taima-Taima, Muaco, and other sites on the northwest coast of Venezuela during the 1950s (Ochsenius and Gruhn 1979), have been encountered in a few early sites in the southern Andes, including Monte Verde. They seem to precede the fishtail type, but apparently never became popular. Indeed, a striking feature of South American lithic complexes is the rarity of projectile points, not only in early sites, but throughout the prehistoric record.

Co-traditions

The probability that the initial lithic complexes in South America derive from North America makes their characteristics relevant for identifying pre-Clovis North American sites. Two regional traditions that had emerged by 11,000 RCYBP are potential candidates, the unifacial expedient tool Itaparica tradition of eastern Brazil and the bifacial projectile point Paiján tradition of north coastal Peru.

1. The Itaparica Tradition The Itaparica tradition is a regional expression of the unifacial technology that predominated in South America throughout the prehistoric period. It extended across eastern Brazil from Piauí to Minas Gerais and Goiás, where vertical escarpments with numerous rockshelters, undulating uplands, and river valleys with mixed forest and savanna vegetation (*cerrado*) provided both shelter and a wide variety of terrestrial and aquatic plant and animal resources for human exploitation (Barbosa 2002). Detailed survey of a 600-km² area in the Serranopolis region of southern Goiás identified 40 rockshelters clustered in different watersheds, but no open sites (Schmitz et al. 2004). Stratigraphic excavations produced more than 164,612 lithics consisting mainly of flakes and expedient tools. Rare objects showing modification included choppers, ovoid bifaces, unifacial limaces, and scrapers. Only seven fragments of bifacial projectile points were encountered, representing the upper part of six triangular blades and one stem with a rounded base.

An intensive survey in the Peruçu Valley in northern Minas Gerais recorded at least 60 sites (Fogaça and Lima 1991, Prous and Fogaça 1999). Stratigraphic excavations in several rockshelters identified assemblages of expedient tools and unifacial end- and sidescrapers, but only rare bifacial projectile points. Similar results have been reported at rockshelters and open sites elsewhere in the region. Survey and excavation along the Lt Norte-Sul transect encountered a single broken stemmed projectile point among thousands of cores, flakes, and expedient tools (Meggers and Miller 2003).

Although sites in Amazonia have not been formally assigned to the Itaparica tradition, expedient tools also dominate them. No projectile points were encountered in excavations at the Pedra Pintada rockshelter (Roosevelt et al. 1996; the illustrated examples of points are from the Tapajós River), or during survey in Roraima, Amapá (Meggers and Miller 2003), or French Guiana (Vacher et al. 1998).

MRS. MARCIA BAKRY, SCIENTIFIC ILLUSTRATOR, DEPT. OF ANTHROPOLOGY, NMNH, SMITHSONIAN INSTITUTION

In spite of the early occurrence of El Jobo and Fishtail projectile points in the Andean area and the Argentine pampa, unifacial technology remained more common even there (Miotti 2003).

2. The Paiján Tradition A notable exception to the scant evidence of projectile points in the lithic industry of early South American cultures (Figure 13C) is found in the Paiján tradition that dominated the coastal plains and foothills of the Zaña, Jequetepeque, Moche, Cupisnique, and Chicama valleys of north coastal Peru between ca. 12,000 and 8000 RCYBP. More than 300 sites have been recorded in the Cupisnique and Zaña valleys, mainly small camps and quarries, but also including rare rockshelters (Dillehay et al. 2003). Faunal remains are restricted to modern species, especially lizards, snails, deer, birds, and fish. Milling stones imply processing of wild plants. In addition to unifacial tools, scrapers, cores, and limaces, the lithic inventory consists of a variety of bifacial stemmed projectile points with curved or straight sides sometimes tapering to "needle-nosed" tips, horizontal to downward projecting shoulders, and short narrow stems with straight (sometimes fluted), tapering, or concave sides and horizontal, bifurcated, or rounded bases. The existence of several varieties on the Sabana de Bogotá, the middle Magdalena, and other locations in central and northern Colombia, where they are known as the Restrepo type, extends their distribution to the north (Ardila and Gerardo 1991). The similarity in morphology and chronology between the projectile points of the Paiján tradition and the North American Western Stemmed series seems unlikely to be coincidental.

Although no stone projectile points have been encountered in archaeological sites in Amazonia, complete examples of most of the Paiján types have been found in isolated locations along the Xingu, Tapajós, Negro, Branco, and Essequibo rivers in various parts of the lowlands. They show no evidence of use, suggesting they may have had symbolic significance (Meggers and Miller 2003, Fig. 10.2).

Summary and Issues of Diffusion

Archaeologists have produced detailed studies of local and regional cultural developments throughout the Americas (Bonnichsen and Turnmire 1999; Bonnichsen et al. this volume; Dillehay 2000a; Haynes 2002; Lavallée 2003; Meggers and Miller 2003; Miotti et al. 2003; West 1996). The emerging picture is that by Clovis times, 12,000–10,700 RCYBP, there were several well-developed projectile point and expedient tool complexes in North and South America. The most prominent of these include the Nenana complex in Alaska, the Western Stemmed point complex in the western United States, Goshen/Plainview in the High Plains and Southern Rocky Mountains, and El Jobo, Paiján, Fishtail, and Itaparica in South America. The origins of these patterns and their regional adaptations remain ambiguous.

Figure 12 (opposite). Selected Paleoamerican sites and the extent of exposed continental shelf and glaciers during the Last Glacial Maximum of Latin America.

There are several ways to account for the spread of these complexes, either by direct population expansion or by idea-diffusion, and of course by a combination of these processes. It has been argued that the spread of fluted points by idea-diffusion is improbable because there is no reason for widely separated people to adopt the Clovis complex of traits simultaneously (Haynes 2002). Moreover, if there were no earlier populations, as in the Clovis-first model, there would be no one to adopt the fluting technology. Hence, the rapid spread of Clovis is attributed as a colonizing event and variations in technology represent region adaptations through time. However, the contrary point of view suggests there were pre-Clovis peoples in the Americas and that there may have been practical rational reasons for a widespread shift in adaptive systems by different peoples; and global climate change at the end of the last Ice Age may have been the catalyst. During this period of rapid environmental change and biotic reorganization, humans had to cope with changing resource distributions and respond to new local conditions by adopting technological innovations and creating new social organizational patterns that facilitated communal hunting and gathering. Future research and more reliable radiocarbon dates will no doubt clarify these issues.

Paleobiology

In contrast to archaeology's focus on modeling regional patterns and routes, paleobiologists have used the complex and variable features of the cranium and teeth to distinguish one population from another. These studies are based upon two premises. The first is that skeletal and dental features, while influenced by developmental and environmental conditions, are based on genes. The second premise is that the expressed genetic components of the traits can be used to document the affinity of one population to others.

The use of skeletal evidence to infer ancestral/descendent relationships has long been paleontology's fundamental tool to trace the evolutionary history of all vertebrate animals. It was W. W. Howells's (1973, 1989) monumental study of the size and shape of human crania from populations throughout the world that has served as the foundation for current human craniometric studies. His work clearly demonstrated that human crania from populations in close proximity to one another were more similar morphologically than crania from distant populations, a reflection of their genetic similarity.

Inferences about ancestral relationships based on comparisons of recent populations alone, however, cannot predict the presence of extinct populations, nor can the role that these extinct populations played in shaping intermediate populations be understood. The great research value of analyzing the skeletons, however, is their potential for preservation. Analyzing well-dated ancient cranial remains provides that crucial window for viewing the past.

Using well-dated ancient remains from North America and the eastern Pacific rim, researchers including Brace and colleagues (1989, 1990, 2001, 1992), Jantz and Owsley (1999, 2001), and Steele and Powell (1999, 2002) have consistently documented

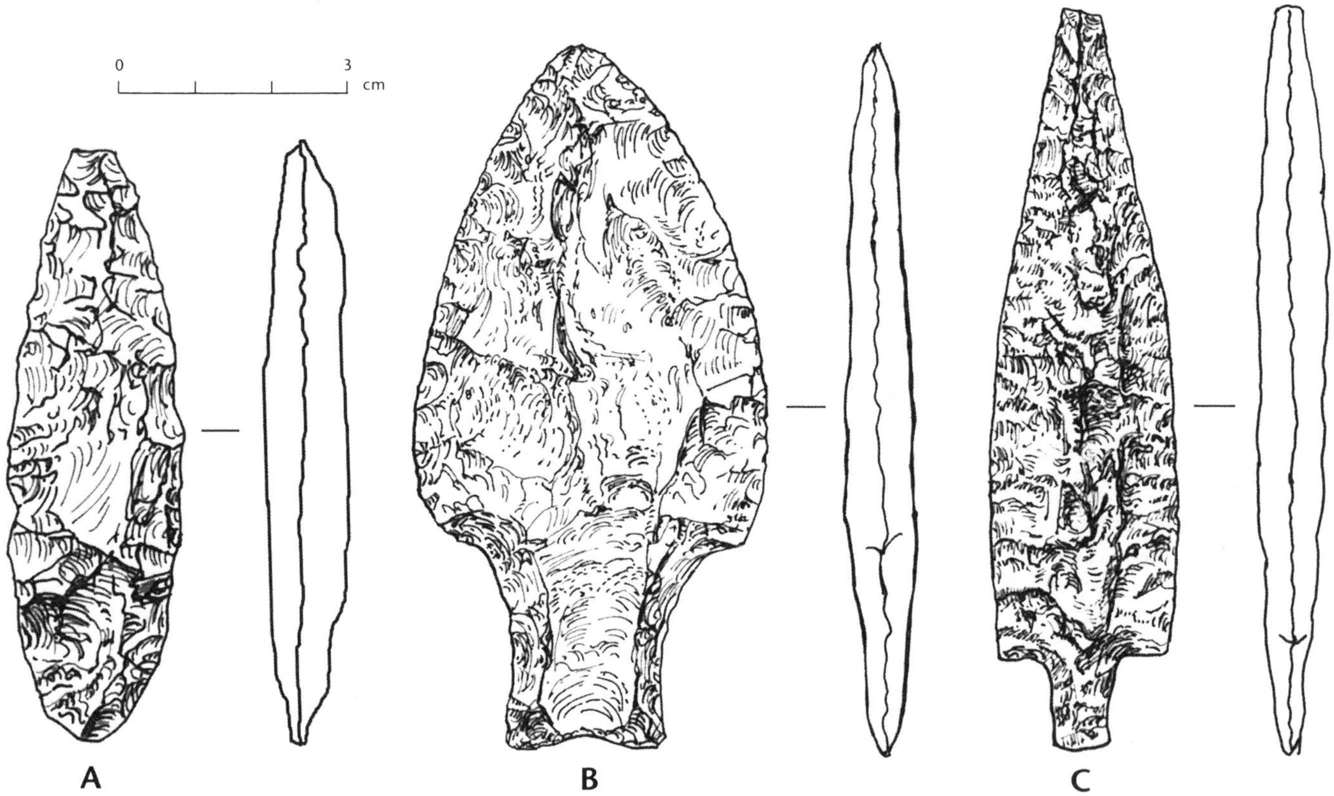

Figure 13. Early American bifacial projectile points. **A,** El Jobo; **B,** fluted fishtail; **C,** Paiján.

that: 1) the earliest North American are more morphologically similar to one another than to extant Native Americans; 2) they more closely resemble southern Asian populations than do Native Americans; and 3) they do not resemble recent northern Asian populations as closely as Native Americans do. Neves and colleagues (1991, 1993, 1996, 1999, 2000, 2004) have reached similar conclusions about the distinctiveness of ancient South American remains. Recent studies of the oldest well-dated human dental remains have documented the morphological distinctiveness of these samples from recent Native Americans and northern Asians as well (Powell 1994, 1995).

While the early North American Skeletal remains appear to differ from more recent Native American, the extent of these differences has been questioned (Powell and Neves 1999) and the cause of these differences remains controversial. One viewpoint is that the differences indicate that the earliest North American remains represent a population primarily descended from a northern Asian population while more recent Native Americans are primarily descended from northern Asian population. An alternative view is that the differences are the result of genetic drift and the fission and fusion of populations after their arrival in the New World, rather than reflecting different founding populations. Although the controversy has yet to be resolved, it is clear that human habitation of the New World was undoubtedly a more complex process than previously envisioned. It is also clear that we have not reached an understanding of who were the first immigrants to the New World.

Beginning in the 1960s the differences among living populations began to be assessed using an expanded number of biochemical features emphasizing blood types and protein variants. This approach soon was followed by comparing peoples using short segments of nuclear DNA and mitochondrial DNA. Schurr (this volume) and Smith et al. (this volume) have reviewed the history, documenting the growing value of this line of research. Both researchers have emphasized that the history of the New World peoples was more complex than previous data suggested and that interpreting the evidence is more difficult and equivocal than anticipated. Both have also confirmed that the preponderance of these studies support the Asian origin for Native Americans and have recognized that two lineages, the Eskimo/Aleut and the Na-Dene, were probably populations founded in Northeast Asia. How many other populations, if any, inhabited the Americas has not been clarified, nor has their homeland if different from Northeast Asia been clearly demonstrated using these lines of evidence. Both Schurr and Smith et al. have provided some evidence for at least one additional founding population for the New World, but the evidence is suggestive only and has not been universally accepted. Schurr (this volume) also proposes a much greater time depth for the presence of humans in the New World.

During the past decade First Americans specialists have focused their efforts in several areas. These include: 1) defining regional adaptive patterns and associated signature artifacts; 2) developing models of general routes that could have been used to get to the Americas; 3) creating statistical models for linking ancient remains with modern and other ancient human populations; 4) identifying and using genetic markers in modern and ancient populations to trace colonization events; and 5) constructing descent models to link ancient peoples with modern populations. Each has enhanced our understanding of the strengths and weaknesses of the available evidence and the possibilities for constructing more adequate models for the peopling of the Americas. None has provided a detailed synthetic model based on multiple lines of evidence to support a colonization event.

Where We Should Go

The interrelated tasks to be addressed in First Americans studies are constructing better descent models, improving our dating techniques, standardizing the reporting of temporal evidence, and improving quality of biological and archaeological evidence.

Constructing Better Models

The central problem confronting scholars is how to construct better descent models that will improve our understanding of Paleoamerican prehistory. With the development of the new archaeology, the focus on historical problems fell out of favor among many American archaeologists. The Kennewick Man lawsuit brought to the forefront many of the historical reconstruction methods that scholars were working on during the 1940s and 1950s. In challenging the government's case, eminent specialists developed a series of seminal affidavits that critically evaluated the rules used to interpret different lines of evidence. These affidavits not only underscored strengths, they also pointed out the weaknesses and exposed flaws in the government's interpretation of the evidentiary record. These documents provide a new standard for evaluating historical evidence.

Before considering rules of evidence governing the interpretation of the evidentiary record, we will consider the variable of time, which provides the framework for correlating multiple lines of evidence.

Time

Modelers interested in linking various lines of evidence must calibrate all time scales to a single standard, using either uncalibrated radiocarbon years or dates calibrated to calendar years before present. If data cannot be placed in the same time framework, they cannot be linked.

Furthermore, in order to correlate paleobiological and archaeological remains one must rely on accurate and reliable radiometric dates and associations between dates and various types of evidence relevant to supporting model propositions. Examples include but are not limited to geological, paleontological, paleoecological, paleobiological, ancient genetic, and archaeological evidence.

Relative time scales are commonly used to order modern-day behaviors that lack empirical antecedents. In formulating relative time scales, analysts survey the range of variability and estimate how much time would have been required to account for the observed variation. Often such calculations are based on untested assumptions about constant rates of change through time. The value of relative time scales is that they organize data into temporal sequences relative to the system of data/evidence under consideration. Unfortunately, potentially important information dated by relative time scales cannot be linked to either calibrated or calendrical time scales.

The success or failure of models seeking to explain the peopling of the Americas must be based on dated empirical evidence. Use of absolute time scales for constructing models is the only reliable approach for linking diverse lines of fossil evidence. By dating ancient cultural and human remains using methods independent of the data themselves, it is possible to assign cultural and paleobiological remains to the time period in which they occurred.

Time is the critical dimension used in constructing models that tie all lines of evidence together in a temporal framework. If evidence cannot be precisely dated or unequivocally associated with other lines of dated evidence, then such evidence will fail the "test of temporal relevance" and must be excluded from the model.

Evidence

In addition to the "test of temporal relevance," there are substantive issues to consider when assessing the interpretive value of various lines of evidence used to reconstruct the natural histories of ancient populations and their biological relationship to living peoples.

1. Biological Evidence Various avenues of biological research have provided significant new information and corroborated evidence gleaned from other research efforts. Furthermore, biological research has even greater potential as new remains are found, new techniques of analysis are developed, and new models are conceived and evaluated. However, as in other lines of inquiry, the strengths of the inferences depend upon the reliability of the evidence, and the evidence must be evaluated before the inference can be accepted and incorporated into higher-order synthetic models.

The greatest strength of the biological research about the New World is that it is scientific inspection, based upon the principles of open inquiry and verification. Without the opportunity to verify or nullify the lines of evidence, the inferences are nothing more than speculation.

Another fundamental strength of some biological data is that they are recoverable from remains that have been preserved under natural environmental conditions. When the research material is accurately dated, it provides the essential opportunity to compare past and present populations, rather than attempting to reconstruct past populations.

While biological inferences have great potential, it is necessary to evaluate the evidence carefully before accepting the inferences. When the research material is accurately dated, it provides the opportunity to compare present with past populations, rather than infer the characteristics of past populations. The first question to address when using biochemical data, genetic data, or morphometric data to reconstruct the descendent relationships of populations is, How accurately do the study samples represent the population being evaluated? There has been a significant collapse and reorganization of many Native American populations during historic times, and with this reorganization a significant amount of gene flow has occurred. Consequently the biological distinctiveness of regional Native American populations has decreased and variation within each of these populations has increased. If the consequences of these historical events are not considered, apparent similarities between populations may give distorted answers to the questions of how long North American regional populations have been separated from one another and when they arrived in the New World. Gene flow during historic times between Native Americans and peoples of European, African, and Asian descent would have a similar effect of increasing the heterogeneity of subpopulations and reducing their distinctiveness from one another. Biochemical and genetic markers traceable to other geographical populations found in Native American samples could also be mistakenly considered evidence that they were part of the ancestral population, or they could create doubt about the authenticity of markers that may truly have been present in the founding populations. Concerns about how accurately fossils represent a population revolve around the scarcity of the fossil record. Are the samples representative of the population they came from?

Other difficulties lie in recovering preserved remains containing soft tissue for analysis. Avoiding contamination of the biochemical and genetic samples during recovery and analysis is also a challenge.

A major issue concerning the interpretation of the skeletal and dental record is the extent that the observable morphology reflects its genetic foundation.

2. Archaeological Evidence Attempts to relate modern tribal groups to the 9200 CALYBP remains of Kennewick Man are an important precedent in exploring the strengths and limits of archaeological methods as a basis for making NAGPRA decisions and for considering cultural affiliations of considerable time depth.

The Ohio Historical Society's (Baker 2003) brief document states that NAGPRA requires appropriate evidence to support claims of cultural affiliation, including the following: 1) existence of an identifiable present-day Indian tribe; 2) existence of an identifiable earlier group, including: a) establishing the biological identity and cultural characteristics of the earlier group; and b) documenting distinct patterns of material culture manufacture and distribution for the earlier group; and 3) existence of shared group identity that may be reasonably traced historically or prehistorically between the present-day Indian Tribe and an earlier identifiable group. The standard for sustaining a claim is a "preponderance of evidence."

The Department of Interior (DOI) identified the "earlier group" to which Kennewick Man belonged as the Windust and early Cascade archaeological phases. In making this claim, the DOI made three assumptions: 1) Windust and/or early Cascade archaeological phase constitutes a valid "group" for the purposes of NAGPRA; 2) Kennewick Man belonged to a group associated with one or the other of these phases; and 3) there is an unbroken chain of evidence that can be used to demonstrate "shared group identity that can be reasonably traced between the Claimant Tribes and Kennewick Man's earlier group."

Based on evidence provided by the DOI's expert consultants, the Windust and Cascade archaeological phases are not valid "Groups" for NAGPRA. Archaeological phases are heuristic devices defined by archaeologists for purposes of classifying, studying, and comparing elements of the archaeological record. Most archaeologists regard a phase as "the largest archaeological unit" for comparing sets of traits limited in time and space (Willey and Phillips 1958). Phases, unless narrowly defined by time, space, and form, cannot be equated with ethnic groups, socio-political entities, or biological populations (Hollinger and Sieg 2002). Phases are not analytical units of comparable scale to modern societies, and chances are poor that they have any social reality. Furthermore, a succession of phases in a regional sequence does not imply either continuity or discontinuity. In sum, the Cascade and Windust phases are not equivalent to modern-day tribes such as "Nez Perce," "Yakama," "Colville," or "Umatilla." It is not scientifically accurate to attempt to link any present-day group with any earlier ancient group using the phase concept.

The DOI assertion that Kennewick Man belonged to a group or community associated with the Windust or Cascade phase is based on the broken projectile point embedded in Kennewick Man's pelvis, but its stylistic and morphological properties are too ambiguous to assign it to either phase. Furthermore, it could have been produced by an adversary group and would thus represent an enemy. K. M. Ames (2000), who prepared the "Kennewick Man Cultural Affiliation Report" for the National Park Service, observed that Kennewick Man might have belonged to any number of groups or societies that lived in, near, or traveled through this region. It is impossible on the basis of available data to affiliate his remains with any particular group.

Ames (2000) also indicates that sometime after 9000 B.C. the central Columbia Plateau was abandoned by the resident populations, a move he attributes to increasingly warm and dry conditions. He identifies a period of significant culture change and reorganization after 4000 B.C. and concludes that the empirical record precludes establishing continuities or discontinuities over increasingly distant time periods. The combined evidence presented by Ames makes it statistically unlikely that an unbroken chain of shared group identity connects Kennewick Man with modern American tribes.

The failed effort to relate Kennewick Man's remains to specific tribal claimants underscores an important issue for all such future efforts. Unlike oral history/folklore, linguistics, and modern DNA studies, archaeological evidence can be dated to specific time periods. However, archaeology has yet to develop a consistent and reliable method for identifying prehistoric social groups. In fact, there may be no sociocultural equivalents to modern tribal groups. Federally recognized tribal groups are unique human organizations created by the federal government for managing Indian peoples. Tribal organization, as it is now known, did not exist during prehistoric times. NAGPRA gives no help in defining what Congress meant by the term group.

In sum, the limits of archaeological evidence prevent doing what Congress mandated archaeologists to do with the passage of NAGPRA. We cannot isolate prehistoric groups in time and space based on the early archaeological record. Efforts to construct a descent model relating Kennewick Man to modern tribal groups and the failure of these efforts demonstrate that much remains to be done in refining our understanding of the evidentiary record and how it can be used for tracing the spread of humans across the face of the earth.

Advancing Paleoamerican Prehistory

During the past decade, First Americans specialists focused their efforts in several areas, including defining regional adaptive patterns and associated signature artifacts, developing general route models that could have been used to get to the Americas, creating statistical models for linking ancient remains with modern and ancient human populations, identifying and using genetic markers among modern and ancient populations to trace colonization events, and attempting to construct descent models to link ancient peoples, such as Kennewick Man's group, with modern populations. Each of these foci has enhanced our understanding of the strengths and weaknesses of available evidence and the possibilities for constructing better models for the peopling of the Americas. Several additional steps that need to be taken to develop a more comprehensive understanding of the existing Paleoamerican evidentiary record include: 1) addressing the lingering doubts surrounding pre-Clovis-age sites; 2) initiating comparative studies of early artifact assemblages; and 3) explaining the origins of regional adaptive patterns.

More than 50 archaeological sites have produced radiocarbon ages greater than 11,500 RCYBP in North and South America (Bonnichsen and Turnmire 1999; Bonnichsen and Schneider 1999; Dillehay 2000b). Even though the Clovis-first model no longer enjoys the widespread support it once did, some specialists still have lingering doubts about claims for pre-Clovis sites.

Paleoamerican sites previously studied need to be reinvestigated, either by new analyses of previously collected samples or by renewed fieldwork using state-of-the-art scientific standards. New methods and approaches have "dated" some aspects of early site reports, especially chronology. One of the most dramatic examples is recent work at the Ushki Lake site in Kamchatka, Siberia (Goebel et al. 2003). Dikov (1968) originally reported a radiocarbon age of 14,000 RCYBP associated with the lowest stratigraphic horizon, Unit VII, whereas new radiocarbon samples from this unit yielded an age of 11,000 RCYBP. The new dates clearly demonstrate that the people who lived at Ushki Lake can no longer be considered possible predecessors of Clovis, as the site falls within the Clovis time period.

In addition to site-level chronological investigations, there is a need to refine the chronology of large-scale regional adaptive patterns such as Nenana, Clovis, Western Stemmed, El Jobo, Fishtail, and others. Developing accurate and reliable beginning and ending dates for regional adaptive patterns will provide a framework for inferring processes responsible for the rise, spread, and termination of cultural and biological developments. The framework will also better define the processes responsible for transitions in specific cultural/biological sequences. At present, many older radiocarbon dates have large plus-and-minus errors or are based on unreliable bone samples.

There have been significant advances in the instruments used for analyzing and reporting artifacts. Historically, specialists produced site reports by describing individual specimens and preparing photographic plates. Specimens can now be measured using electronic calipers, and metric and discrete characteristics can be compared in databases with an open-ended architecture that can accommodate thousands of specimens. Sophisticated instrumentation is available capable of detecting trace elements in parts per billion. Light and SEM microscopes allow investigators to make detailed morphological inspections of specimens at very high magnification. Information produced by a wide variety of methods can now be integrated on personal computers. Color images of important specimens can be posted on the Web, opening a new forum for making results available.

In spite of these technological advances, many of the traditional problems that faced early investigators remain. There are "lingering doubts" about whether specimens recovered from some of the earliest sites, e.g., Blue Fish Caves, Pedra Furada, and Monte Verde I, are artifacts. There continues to be a lack of consensus on how to distinguish stone artifacts from geofacts and bones broken by humans from those altered by natural processes. Verified artifacts have potential for accurately dating the entry of people to the Americas. Innovative experimental work coupled with reinvestigating such localities can enhance our understanding of when the Americas were initially peopled.

First Americans specialists have traditionally worked alone. Although the trend is to rely on interdisciplinary teams to collect and analyze evidence from early sites, comparative artifact studies remain rare. Artifact studies can provide important information about raw material selection, style, technology, and use, as well as issues associated with reconstructing descent patterns. We lack keys or standards for identifying raw material, shape, technology, and use-wear patterns. Furthermore, a generally accepted body of rules is needed that can address middle-range theory questions and thus make it possible to transform process/pattern statements generated in low-range theory into a society-level statement. Finally, we lack a body of rules that will allow us to distinguish consistently between migration, diffusion, and in situ development in early hunter-gatherer tool assemblages.

Although the above statements may sound like a series of condemnations, this is not the case. The field of First Americans studies is beginning to mature. A significant body of archaeological and skeletal evidence has now been recovered from North and South America. Large-scale regional adaptive patterns, including pre-Clovis, Clovis, Nenana, Western Stemmed, Goshen/Plainview, El Jobo, and Fishtail, and at least two skeletal populations have been recognized. The challenge before us is to refine and build on what has learned from previous efforts.

Future research should emphasize comparative studies within and between regions, focusing on the fundamental question of how and why different cultural and biological developments occurred. There is need for comparative studies among American, Asian, and European assemblages to test origin hypotheses. To address cross-regional or intercontinental colonization problems will require cleaning up the evidentiary record collected in the past. Using new technologies, particularly imagery and relational databases, we can compare evidence collected from two or hundreds of sites. Developing imagery databases capable of accommodating significant numbers of specimens will give scholars worldwide access to scattered and otherwise inaccessible collections.

In conclusion, the field of First Americans studies is in transition from domination by the Clovis-first model to awareness of the complexities of the American evidentiary record. New paleobiological and archaeological evidence suggests that the peopling of the Americas was more complicated than previously anticipated and that multiple colonization events were probably involved. Archaeological, craniometric, skeletal, and genetic evidence is illuminating different facets of the process and sending different messages. Paleobiological evidence from genetic and skeletal studies is giving us a glimpse into ties between Paleoamerican and Eurasian populations, and possibly others. The message coming from the archaeological record is that sometime between 12,000 and 11,500 radiocarbon years ago, a series of different regional adaptive patterns emerged using different strategies. Given the resolution of the available chronological record and level of artifact analysis, it is not possible to determine whether these represent independent regional developments or more than one early colonization event. Until there is a prevailing theory or model that integrates all potential lines of archaeological, paleobiological, and genetic evidence into a unified framework, we will not be able to address the general goals of where early populations may have come from, when and where they arrived in the Americas, and how they evolved once they were here.

Acknowledgments

The authors would like to express their appreciation of Marcia Bakry, Senior Illustrator for the Department of Anthropology at the Smithsonian Institution, who produced the excellent line drawings of artifacts and the map of the hypothesized routes for the peopling of the Americas. Marcia also researched and produced the excellent color maps of North America and South America, based on Clapperton (1993), Dyke and Prest (1986) and Manley (2002), showing the extent of glaciers and exposed continental shelves. Site locations for the North American maps and the bibliography were researched by Ashley Smallwood and Robert Lasen in the Department of Anthropology at Texas A&M University. We would also like to thank Jim Chandler for his excellent work in editing the manuscript and Laurie Lind and Mike Waters for their help in coordinating various aspects of the production of this contribution.

References Cited

Ackerman, R. E.
1996 Early Maritime Culture Complexes of the Northern North West Coast. In *Early Human Occupation in British Columbia*, edited by R. L. Carlson and L. D. Bona, pp.123–32. University of British Columbia Press, Vancouver.

Ackerman, R. E., T. D. Hamilton, and R. Stuckenrath
1979 Early Culture Complexes on the Northern Northwest Coast. *Canadian Journal of Archaeology* 3:195–209.

Adovasio, J. M.,
1974 *Prehistoric North American Basketry*. The Nevada State Museum Anthropological and Ethnological Papers No. 16.

Adovasio, J. M., and J. Page
2002 *The First Americans: In Pursuit of Archaeology's Greatest Mystery*. Random House, New York.

Agenbroad, L., D. Morris, and V. L. Roth
1999 Pygmy Mammoth (*M. exilis*) from Santa Rosa Island, Channel Islands National Park, California, USA. In *Mammoths and Mammoth Fauna: Studies of an Extinct Ecosystem: Proceeding of the First International Mammoth Conference at St. Petersburg, Russia*, edited by G. Haynes, J. Klimowicz, and J. W. F. Reumer. Rotterdam: Deinsea. Natural History Museum of Rotterdam.

Akazawa, T.
1999 Pleistocene Peoples of Japan and the Peopling of the Americas. In *Ice Age Peoples of North America; Envoronment, Origins, and Adaptations of the First Americans*, edited by R. Bonnichsen and K. L. Turnmire, pp. 95–104. Oregon State University Press, Center for the Study of the First Americans, Corvallis.

Alexander, H.
1987 *Putu, a Fluted Point Site in Alaska*. Publication 17, Department of Archaeology, Simon Fraser University, Burnaby, B.C.

Ames, K.
2000 Review of Traditional Historical and Ethnographic Information. Chapter 2 in *Cultural Affiliation Report DOI/NPS Kennewick Man Scientific Investigations*. www.cr/nps.gov/aad/Kennewick

Ardila C., and I. Gerardo
1991 The Peopling of Northern South America. In *Clovis: Origins and Adaptations*, edited by R. Bonnichsen, and K. L. Turnmire, pp. 261–82. Oregon State University, Center for the Study of the First Americans, Corvallis.

Ardila C., G. I. Politis, and G. G. Politis
1989 Nuevos Datos para un Viejo Problema: Investigación y Discusiones en Torno del Poblamiento de América del Sur. *Boletín del Museo del Oro* 23:3–45. Bogotá, Banco de la República.

Baker, B.
2003 *Amicus Brief of the Ohio Archaeological Council in Support of Plaintiffs-Appellee for Affirmance of the District Court's Decision*. Case No. 02-35996, District Court No. 96-1481 JE (D.Or.).

Barbosa, A. S.
2002 *Andarilhos daClaridade: Os Primeiros Habitantes do Cerrado*. Goiânia, Universidade Católica de Goiás.

Beaton, J. M.
1991 Paleoindian Occupation Greater than 11,000 yr B.P. at Tule Lake, Northern California. *Current Research in the Pleistocene* 8:5–7.

Beck, C., and G. T. Jones
1997 The Terminal Pleistocene/Early Holocene Archaeology of the Great Basin. *Journal of World Prehistory* 11(2):161–236.

Bedwell, S. F.
1973 *Fort Rock Basin Prehistory and Environment*. University of Oregon Books, Eugene.

Bedwell, S. F., and L. S. Cressman
1971 Fort Rock Report: Prehistory and Environment of the Pluvial Fort Rock Lake Area of South-central Oregon. In *Great Basin*

Anthropological Conference 1970 Selected Papers, edited by C. M. Aikens, pp. 1–25. Anthropological Papers No. 1. University of Oregon, Eugene.

Bird, J.
1938 Antiquity and Migrations of the Early Inhabitants of Patagonia. *Geographic Review* 28:250–75.

Bonnichsen, R., M. Full, and M. Reken
2004 A Preliminary Report of the McMinnville Mammoth, Oregon: A Taphonomic Approach. Paper prepared for inclusion in *Primer Simposio Internacional el Hombre Temprano en Americas y sus Implicaciones en el Problamiento de la Cuncia de Mexico*, edited by J. Conception and S. Gonzales. INHA, Mexico City. In press.

Bonnichsen, R., and A. L. Schneider
1999 Breaking the Impasse on the Peopling of the Americas. In *Ice Age People of North America: Environments, Origins, and Adaptations*, edited by R. Bonnichsen and K. L. Turnmire, pp. 1–26. Oregon State University Press, Corvallis.

Bonnichsen, R., D. Stanford, and J. L. Fasthook
1987 Environmental Change and Developmental History of Human Adaptive Patterns, the Paleoindian Case. In *North America and Adjacent oceans During the last Deglaciation: the Geology of North America, Vol. K-3*, edited by W. F. Ruddiman and H. E. Wright, Jr., pp. 403–24. Geological Society of America, Boulder.

Bonnichsen, R. and K. Turnmire
1999 An Introduction to the Peopling of the Americas. In *Ice Age Peoples of North America: Environments, Origins, and Adaptations*, edited by R. Bonnichsen and K. L. Turnmire, pp. 1–26. Oregon State University Press, Center for the Study of the First Americans, Corvallis.

Bonnichsen, R., and R. W. Will
1999 Radiocarbon Chronology of Northeastern Paleoamerican Sites: Discriminating Natural and Human Burn Features. In *Ice Age Peoples of North America: Environments, Origins, and Adaptations*, edited by R. Bonnichsen and K. L. Turnmire, pp. 395–415. Oregon State University Press, Center for the Study of the First Americans, Corvallis.

Borrero, L. A.
1999 The Prehistoric Exploration and Colonization of Fuego-Patagonia. *Journal of World Prehistory* 13:321–55.

Borrero, L. A., and N. V. Franco
1997 Early Patagonian Hunter-Gatherers: Subsistence and Technology. *Journal of Anthropological Research* 53:219–39.

Borrero, L. A., M. Zárate, L. Miotti, and M. Masone
1998 The Pleistocene-Holocene Transition and Human Occupation in the Southern Cone of South America. *Quaternary International* 49/50:191–99

Brace, C. L., M. L. Brace, and W. R. Leonard
1989 Reflections on the Face of Japan: A Multivariate Craniofacial and Odontometric Perspective. *American Journal of Physical Anthropology* 78:93–113.

Brace, C. L., and K. D. Hunt
1990 A Nonracial Craniofacial Perspective on Human Variation: A(ustralia) to Z(uni). *American Journal of Physical Anthropology* 82:341–60.

Brace, C. L., A. R. Nelson, N. Seguchi, H. Oe, L. Sering, P. Qifeng, L. Yongyi, and D. Tumen
2001 Old World Sources of the First New World Human Inhabitants: A Comparative Craniofacial View. *PNAS* 98:10017–22.

Brace, C. L., and D. P. Tracer
1992 Craniofacial Continuity and Change: A Comparison of Late Pleistocene and Recent Europe and Asia. In *The Evolution and Dispersal of Modern Humans in Asia*, edited by T. Akazawa, K. Aoki, and T. Kimura, pp. 439–79. Hukusen–Sha Publishing Co. , Tokyo.

Bradley, B., and D. Stanford
2004 The North Atlantic Ice-edge Corridor: A Possible Paleolithic Route to the New World. *World Archaeology* 36:459–78.

Breitburg, E., J. B. Broster, A. L. Reesman, and R. G. Strearns
1996 The Coats-Hines Site: Tennessee's First Paleoindian-Mastodon Association. *Current Research in the Pleistocene* 13:6–8.

Broster, J. B., D. P. Johnson, and M. R. Norton
1991 The Johnson Site: A Dated Clovis-Cumberland Occupation in Tennessee. *Current Research in the Pleistocene* 8:8–10.

Broster, J. B., M. R. Norton, D. J. Stanford, C. V. Haynes Jr., and M. A. Jodry
1994 Eastern Clovis Adaptations in the Tennessee River Valley. *Current Research in the Pleistocene* 11:12–14.

Browman, D. L., and D. A. Munsell
1969 Columbia Plateau Prehistory: Cultural Development and Impinging Influences. *American Antiquity* 34:249–64.

Brush, N.
1993 Twelve Thousand Years of Human Occupation at the Eppley Rockshelter. *Current Research in the Pleistocene* 10:5–7.

Bryan, A. L.
1979 Smith Creek Cave. In *The Archaeology of Smith Creek Canyon*, edited by D. R. Touhy and D. L. Rendall, pp. 164–251. Nevada State Museum Anthropological Papers No. 17, Carson City.
1988 The Relationship of the Stemmed Point and Fluted Point Traditions in the Great Basin. In *Early Human Occupation in Far Western North America: The Clovis-Archaic Interface*, edited by J. A. Willig, C. M. Aikens, and J. L. Fagan, pp. 53–74. Nevada State Museum Anthropological Papers 21, Carson City.

Bryan, A. L. (editor)
1986 *New Evidence for the Pleistocene Peopling of the Americas*. Center for the Study of Early Man, University of Maine, Orono.

Bryan, A. L., and R. Gruhn
2003 Some Difficulties in Modeling the Original Peopling of the Americas. *Quaternary International* 109–10:175–79.

Bryan, A. L., and D. R. Touhy
1999 Prehistory of the Great Basin/Snake River Plain to About 8,500 Years Ago. In *Ice Age Peoples of North America: Environments, Origins, and Adaptations*, edited by R. Bonnichsen and K. L. Turnmire, pp. 249–63. Oregon State University Press, Center for the Study of the First Americans, Corvallis.

Campbell, T. N.
1961 A List of Radiocarbon Dates from Archaeological Sites in Texas. *Bulletin of the Texas Archaeological Society* 30:311–20.

Carlson, D. L., and D. G. Steele
1992 Human Mammoth Sites—Problems and Prospects. In *Proboscidian and Paleoindian Interactions*, edited by R. Fox, C. Smith and K. Wilkins, pp.123–48. Baylor University Press, Waco, Texas.

Chauchat, C., and J. Pelegrin
2004 *Projectile Point Technology and Economy: A Case Study from Paiján, North Coastal Peru*. Texas A&M University Press, Center for the Study of the First Americans, College Station.

Cinq-Mars, J.
1979 Bluefish Cave I: A Late Pleistocene Eastern Beringian Cave Deposit in the Northern Yukon. *Canadian Journal of Archaeology* 3:1–32.

1990 La Place des Grottes du Poisson-Bleu dans la Préhistorie Béringienne. *Revista de Arqueologia Americana* 1:9–32.

Cinq-Mars, J., and R. E. Morlan
1999 Bluefish Caves and Old Crow Basin: A New Rapport. In *Ice Age Peoples of North America: Environments, Origins, and Adaptations,* edited by R. Bonnichsen and K. L. Turnmire, pp. 200–12. Oregon State University, Center for the Study of the First Americans, Corvallis.

Clapperton, C.
1993 *Quaternary Geology and Geomorphology of South America.* Elsevier, Amsterdam.

Clark, D. W.
1984 Some Practical Applications of Obsidian Hydration Dating in the Subarctic. *Arctic* 37:91–109.

1991 The Northern (Alaska-Yukon) Fluted Points. In *Clovis: Origins and Adaptations,* edited by R. Bonnichsen and K. L. Turnmire, pp. 35–48. Center for the Study of the First Americans, Oregon State University, Corvallis.

Clark, D. W., and A. M. Clark
1993 *Batza Téna: Trail to Obsidian (Archaeology at an Alaskan Obsidian Source).* Archaeological Survey of Canada, Mercury Series Paper 147, Canadian Museum of Civilization, Ottawa.

Clarke, D. L.
1968 *Analytical Archaeology.* Methuen & Co. Ltd, London.

Collins, M. B.
1999 *Clovis Blade Technology.* University of Texas Press, Austin, Texas.

Collins, M. B., T. R. Hester, D. Olmstead, and P. J. Headrick
1991 Engraved Cobbles from Early Archaeological Contexts in Central Texas. *Current Research in the Pleistocene* 8:13–15.

Cook, J. P.
1996 Healy Lake. In *American Beginnings: Prehistory and Paleoecology,* edited by F. H. West, pp. 363–66. University of Chicago Press, Chicago.

Cowles, J.
1959 *Cougar Mountain Cave in South Central Oregon.* Private Publication. Rainier, Oregon.

Crane, H. R.
1956 University of Michigan Radiocarbon Dates I. *Science* 124:664–72.

Crass, B., and C. Holmes
2004 Swan Point: A Case for Land Bridge Migration in the Peopling of North America. *Abstracts of the 69th Annual Meeting, Society for American Archaeology,* p. 90.

Cruxent, J. M., and I. Rouse
1956 A Lithic Industry of Paleo-Indian Type in Venezuela. *American Antiquity,* 22:172–79.

Curran, M. L.
1984 The Whipple Site Tool Assemblage Variation: A comparison of Intrasite Structuring. *Archaeology of Eastern North America* 12:5-40.

Daugherty, R. D.
1956 The Archaeology of the Lind Coulee Site, Washington. *Proceedings of the American Philosophical Society,* 100:223–78.

Davis, L. G.
2001 *The Coevolution of Early Hunter-Gatherer Culture and Riparian Ecosystems in the Southern Columbia River Plateau.* Cottonwood Field Office, Upper Columbia–Salmon Clearwater District Bureau of Land Management, Idaho.

Deller, D. B., and C. J. Ellis
1988 Early Paleo-Indian Complexes in Southwestern Ontario. In *Late Pleistocene and Early Holocene Paleoecology and Archaeology of the Eastern Great Lakes,* edited by R. S. Laub, N. Miller, and D. Steadman, pp. 251–63. Bulletin of the Buffalo Society of Natural Sciences 33. Buffalo, New York.

1992 *Thedford II: A Paleo-Indian Site in the Ausable River Watershed of Southwestern Ontario.* Memoirs No. 24, Museum of Anthropology, University of Michigan, Ann Arbor.

Dikov, N. N.
1968 The Discovery of the Paleolithic in Kamchatka and the Problem of the Initial Occupation of America. *Arctic Anthropology* 5:191–203.

Dillehay, T. D.
2000a *The Settlement of the Americas: A New Prehistory.* Basic Books, New York.

Dillehay, T. D. (editor)
2000b *Monte Verde: A Late Pleistocene Settlement in Chile.* 2 vols. Smithsonian Institution Press, Washington, D.C.

Dillehay, T. D., J. Rossen, G. Maggard, K. Stackelbeck, and P. Netherly
2003 Localization and Possible Social Aggregation in the Late Pleistocene and Early Holocene on the North Coast of Perú. *Quaternary International* 109-110:3–11.

Dixon, J. E.
1999 *Bones, Boats, and Bison: Archeology and the First Colonization of Western North America.* University of New Mexico Press, Albuquerque.

2001 Human Colonization of the Americas: Timing, Technology and Process. *Quaternary Science Reviews* 20:277–99.

Donohue, J., and F. Sellet
2000 The Chronology of the Goshen Bone Bed at the Jim Pitts Site. *Current Research in the Pleistocene* 18:20–21.

Douglas, C. L., D. l. Jenkins, and C. N. Warren
1988 Spatial and Temporal Variability in Faunal Remains from Four Lake Mojave-Pinto Period Sites in the Mojave Desert. In *Early Human Occupation in Far Western North America; The Clovis-Archaic Interface,* edited by J. A. Willig, C. M. Aikens, and J. L. Fagan, pp. 273–301. Nevada State Museum Anthropological Papers No. 21. Carson City.

Dyke, A. S., and V. K. Prest
1986 *Paleogeography of Northern North America 18,000–5,000 Years Ago.* Geological Survey of Canada, Map 1703A.

Easton, N. A.
2004 Nenana in Canada—2003 Excavation of a Multi–Component Site (KdVo–6) in Mirror Creek Valley, Yukon Territory, Canada. In *Abstracts of the 31st Annual Conference of the Alaska Anthropological Association,* Whitehorse, Yukon, Canada.

Erlandson, J. M.
2002 Anatomically Modern Humans, Maritime Voyaging, and the Pleistocene Colonization of the Americas. In *The First Americans: The Pleistocene Colonization of the New World,* edited by N. G. Jablonski, pp. 59–92. California Academy of Sciences, San Francisco.

Feibleman, J. K.
1954 Theory of Integrative Levels. The British Journal for the Philosophy of Science 5:59–66.

Fiedel, S. J.
2002 Initial Human Colonization of the Americas: An Overview of the Issues and the Evidence. *Radiocarbon* 44(2):407–36.

2004 The Kennewick Follies: "New" Theories About the Peopling of the Americas. *Journal of Anthropological Research* 60(1):75–111.

Figgins, J. D.
1927 The Antiquity of Man in America. *Natural History*, Vol. XXVII (3): 229–39.

Fisher, J. W., Jr.
1992 Observations on the Late Pleistocene Bone Assemblage from the Lamb Spring Site, Colorado. In *Ice Age Hunters of the Rockies*, edited by D. J. Stanford and J. S. Day, pp. 51–81. Denver Museum of Natural History and University Press of Colorado, Denver.

Fladmark, K. R.
1978 The Feasibility of the Northwest Coast as a Migration Route for Early Man. In *Early Man in America from a Circum-Pacific Perspective*, edited by A. L. Bryan, pp. 119–28. Occasional Papers No. 1. Department of Anthropology, University of Alberta, Edmonton.

1979 Routes: Alternate Migration Corridors for Early Man in North America. *American Antiquity* 44(1):55–69.

1994 The Prehistory of Charlie Lake Cave. In *Early Human Occupation in British Columbia*, edited by R. L. Carlson and L. Dalla Bona, pp. 11–20. University of British Columbia Press, Vancouver.

Fladmark, K. R., J. C. Driver, and D. Alexander
1988 The Paleoindian Component at Charlie Lake Cave (HbRf-39), British Columbia. *American Antiquity* 53(2):371–84.

Flegenheimer, N., and M. Zárate
1997 Consideration on Radiocarbon and Calibrated Dates from Cerro la China and Cerro El Sombrero, Argentina. *Current Research in the Pleistocene* 14: 27–28.

Fogaça, E., and M. A. Lima
1991 L'abri du Boquete (Brésil): Les Premières Industries Lithiques de l'Holocène. *Journal de la Société des Américanistes* 77:111–23.

Forester, P., R. Harding, A. Torroni, and H.- J. Bandelt
1996 Origin and Evolution of Native American mtDNA Variation: A Reappraisal. *American Journal of Human Genetics* 59:935–45.

Frison, G. C. (editor)
1996 *The Mill Iron Site*. University of New Mexico Press, Albuquerque.

Frison, G. C., and R. Bonnichsen
1996 The Pleistocene-Holocene Transition on the Plains and Rocky Mountains of North America. In *Humans at the end of the Ice Age: The Archaeology of the Pleistocene-Holocene Transition*, edited by L. G. Straus, B. V. Eriksen, J. M. Erlandson, and D. R. Yesner, pp. 303–318. Plenum Press, New York.

Frison, G. C., and B. Bradley
1999 *The Fenn Cache: Clovis Weapons and Tools*. One Horse Land & Cattle Company, Santa Fe, New Mexico.

Frison, G. C., C. V. Haynes, Jr., and M. L. Larson
1996 Discussion and Conclusions. In *The Mill Iron Site*, edited by G. C. Frison, pp. 205–24. University of New Mexico Press, Albuquerque.

Fryxell, R., T. Bielicki, R. D. Daugherty, C. E. Gustafson, H. T. Irwin, and B. C. Keel
1968 A Human Skeleton from Sediments of Mid-Pinedale Age in Southeastern Washington. *American Antiquity* 33:511–14.

García-Bárcena, J.
1979 *Una Punta Acanalada de la Cueva Los Grifos, Ocozocoautla, Chiapas*. Cuadernos de Trabajo, No. 17, Departamento de Prehstoria, Instituto Nacional de Antropología e Historia, Mexico City.

Gardner, W. M.
1977 The Flint Run Complex and Its Implications for Eastern North American Prehistory. In *Amerinds and Their Early Environments in Northeastern North America*, Proceedings of the New York Academy of Sciences 288:257–63.

Goebel, T.
1999 Pleistocene Human Colonization of Siberia and the Peopling of the Americas: An Ecological Approach. *Evolutionary Anthropology* 18(6):208–27.

2004 The Search for a Clovis Progenitor in Subarctic Siberia. In *Entering America: Northeast Asia and Beringia before the Last Glacial Maximum*, edited by D. B. Madsen, pp. 311–56. The University of Utah Press, Salt Lake City.

Goebel, T., R. W. Powers, N. H. Bigelow, and A. S. Higgs
1996 Walker Road. In *American Beginnings: Prehistory and Paleoecology*, edited by F. H. West, pp. 356–63. University of Chicago Press, Chicago.

Goebel, T., M. R. Waters, and M. Dikova
2003 The Archaeology of Ushki Lake, Kamchatka, and the Pleistocene Peopling of the Americas. *Science* 301:501–05.

Goodyear, A. C.
1999 The Early Holocene Occupation of the Southeastern United States: A Geoarchaeological Summary. In *Ice Age Peoples of North America: Environments, Origins, and Adaptations*, edited by R. Bonnichsen and K. L. Turnmire, pp. 432–81. Oregon State University Press, Center for the Study of the First Americans, Corvallis.

Goodyear, A. C., and K. Steffy
2003 Evidence of a Clovis Occupation at the Topper Site, 38AL23, Allendale County, South Carolina. *Current Research in the Pleistocene* 20:23–25.

Graham, R. W., C. V. Haynes, D. L. Johnson, and M. Kay
1981 Kimmswick: A Clovis-Mastodon Association in Eastern Missouri. *Science* 213:1115–17

Graham, R. W., and E. Lundelius Jr.
1984 Coevolutionary Disequilibrium and Pleistocene Extinctions. In *Quaternary Extinction: A Prehistoric Revolution*, edited by P. S. Martin and R. Klein, pp. 223–49. University of Arizona Press, Tucson.

Gramley, R. M.
1993 *The Richey Clovis Cache*. Persimmon Press, New York.

Grayson, D. K.
1993 *The Desert's Past: A Natural Prehistory of the Great Basin*. Smithsonian Institution Press, Washington, D.C.

Grayson, D. K., and D. J. Meltzer
2002 Clovis Hunting and Large Mammal Extinction: Critical Review of the Evidence. *Journal of World Prehistory* 16:313–59.

Green, T. J., B. Cochran, T. W. Fenton, J. C. Woods, G. L. Titmus, L. Tieszen, M. A. Davis, and S. J. Miller
1998 The Buhl Burial: A Paleoindian Woman from Southern Idaho. *American Antiquity* 63(3):437–456.

Gruhn, R.
1961 *The Archaeology of Wilson Butte Cave, South-Central Idaho*. Occasional Papers of the Idaho State University Museum No. 6, Pocatello.

1994 The Pacific Coast Route of Initial Entry: An Overview. In *Method and Theory for Investigating the Peopling of the Americas*, edited by R. Bonnichsen and D. G. Steele, pp. 249–56.

Oregon State University, Center for the Study of the First Americans, Corvallis.

1995 Results of New Excavations at Wilson Butte Cave, Idaho. *Current Research in the Pleistocene* 12:16–17.

Gruhn, R., and A. L. Bryan
1977 Los Tapiales: A Paleo-Indian Campsite in the Guatemalan Highlands. *Proceedings of the American Philosophical Society* 121:235–73.

Guidon, N., F. Parenti, M. de Fatima da Luz, C. Guerin, and M. Faure
1994 Le Plus Ancien Peuplement de l'Amérique: Le Paleolithique du Nordeste Brésilien. *Bul. Société Préhistorique Française* 91:246–50.

Hamilton, T. D., and T. Goebel
1999 Late Pleistocene Peopling of Alaska. In *Ice Age Peoples of North America: Environments, Origins, and Adapations*, edited by R. Bonnichsen and K. L. Turnmire, pp. 156–99. Oregon State University Press, Center for the Study of the First Americans, Corvallis.

Hanes, R. C.
1988a *Lithic Assemblages of Dirty Shame Rockshelter: Changing Traditions in the Northern Intermontane.* University of Oregon Anthropological Papers No. 40.

1988b Early Cultural Traditions of the Owyhee Uplands as Seen from Dirty Shame Rockshelter. In *Early Human Occupation in Far Western North America: The Clovis-Archaic Interface*, edited by J. A. Willig, C. M. Aikens and J. L. Fagan. Nevada State Museum Anthropological Papers 21, Carson City.

Hannus, L. A.
1989 Flaked Mammoth Bone from the Lange/Ferguson Site, White River Badlands Area, South Dakota. In *Bone Modification*, edited by R. Bonnichsen and M. H. Sorg, pp. 395–412. University of Maine Press, Center for the Study of the First Americans, Orono.

Harrington, M. R.
1948 *An Ancient Site at Borax Lake, California*. Southwest Museum Papers, no. 16.

Haury, E. W.
1953 Artifacts with Mammoth Remains, Naco, Arizona, I: Discovery of the Naco Mammoth and Associated Projectile Points. *American Antiquity* 19:1–14.

1956 The Lehner Mammoth Site, *The Kiva* 21:23–24

1975 *The Stratigraphy and Archaeology of Ventana Cave.* University of Arizona Press, Tucson.

Haynes, C. V., Jr.
1981 Geochronology and Paleoenviroments of the Murray Springs Clovis Site, Arizona. *National Geographic Research Reports* 13:243–51.

Haynes, C. V., Jr., D. J. Donahue, A .J. T. Jull, and T. H. Zabel
1984 Application of Accelerator Dating to Fluted Point Paleoindian Sites. *Archaeology of Eastern North America* 12:184–91.

Haynes, C. V., Jr., D. C. Grey, and A. Long
1971 Arizona Radiocarbon Dates VIII. *Radiocarbon* 13(1):1–18.

Haynes, C. V., Jr., and E. T. Hemmings
1968 Mammoth-Bone Shaft Wrench from Murray Springs, Arizona. *Science* 159:186–87.

Haynes, G.
2002 *The Early Settlement of North America: The Clovis Era.* Cambridge University Press, Cambridge, England.

Heusser, C. J.
1960 *Late Pleistocene Environments of the North Pacific North America.* American Geographical Society Publication 35.

Hester, J. J.
1972 *Blackwater Locality No. 1: A Stratified, Early Man Site in Eastern New Mexico.* Fort Burgwin Research Center Publication No. 8, Ranchos de Taos, New Mexico.

Hollinger, R. E., and L. Sieg
2002 *The Problem of Hopewell Taxonomy: A Review.* Paper presented at the 2002 Midwest Archaeological Conference, Columbus, Ohio.

Hoffecker, J .F.
1996 Moose Creek. In *American Beginnings: Prehistory and Paleoecology*, edited by F. H. West, pp. 363–66. University of Chicago Press, Chicago.

Hoffecker, J. F., W. R. Powers, and N. H. Bigelow
1996 Dry Creek. In *American Beginnings: Prehistory and Paleoecology*, edited by F. H. West, pp. 343–52. University of Chicago Press, Chicago.

Hoffecker, J. F., W. R. Powers, and P. G. Phippen
1996 Owl Ridge. In *American Beginnings: Prehistory and Paleoecology*, edited by F. H. West, pp. 353–56. University of Chicago Press, Chicago.

Holen, S. R.
1996 The Lovewell Mammoth: A Late Wisconsinan Site in North-Central Kansas. *Current Research in the Pleistocene* 13:69–70.

1997 *The Lovewell Mammoth: A Late Wisconsinan Mammoth Processing Site in North-Central Kansas.* Nebraska Archaeological Survey, Technical Report 97-01, University of Nebraska State Museum, Lincoln.

Holen, S. R., and D. W. May
2002 The La Sena and Shaffert Mammoth Sites: History of Investigations, 1987–1998. In *Medicine Creek: Seventy Years of Archaeological Research*, edited by D. Roper, pp. 20–36. The University of Alabama Press, Tuscaloosa.

Holmer, R. N.
1986 Common Projectile Points of the Intermountain West. In *Anthropology of the Desert West: Essays in Honor of Jesse D. Jennings*, edited by C. J. Condie and D. D. Fowler, pp. 89–115. University of Utah Anthropology papers No. 110. Salt Lake City.

Holmes, C. E.
1996 Broken Mammoth. In *American Beginnings: Prehistory and Paleoecology*, edited by F. H. West, pp. 312–18. University of Chicago Press, Chicago.

2001 Tanana River Valley Archaeology Circa 14,000 to 9000 B.P. *Arctic Anthropology* 38(2):154–70.

Holmes, C. E., R. VanderHock, T. E. Dilley
1996 Swan Point. In *American Beginnings: Prehistory and Paleoecology*, edited by F. H. West, pp. 525–36. University of Chicago Press, Chicago.

Howells, W. W.
1967 *Mankind in the Making.* Doubleday, Garden City, New York.

1973 *Cranial Variation in Man: A Study by Multivariate Analysis of Patterns of Differences Among Recent Human Populations.* Papers of the Peabody Museum of Archaeology and Ethnology 67. Harvard University, Cambridge.

1989 *Skull Shapes and the Map: Craniometric Analyses in the Dispersion of Modern Homo.* Papers of the Peabody Museum of Archeology and Ethnology 79. Harvard University, Cambridge.

Hrdlička, A.
1923 The Origin and Antiquity of the American Indian. In *Annual Report of the Board of Regents of the Smithsonian Institution*, pp. 481–93. U.S. Government Printing Office, Washington, D.C.

Irving, W. M., J. T. Mayhall, F. J. Melbye, and B. F. Beebe
1977 A Human Mandible in Probable Association with a Pleistocene

Faunal Assemblage in Eastern Beringia: A Preliminary Report. *Canadian Journal of Archaeology* 1:81–93.

Irwin, H. T.
1967 *The Itama: Late Pleistocene Inhabitants of the Plains of the United States and Canada and the American Southwest.* Unpublished Ph.D. dissertation, Department of Anthropology, Harvard University, Cambridge.

1971 Developments in Early Man Studies in Western North America, 1960–1970. *Arctic Anthropology* 8(2):42–67.

Irwin-Williams, C.
1961 Comments on the Associations of Archaeological Materials and Extinct Fauna in the Valsequillo Region, Puebla, Mexico. *American Antiquity* 34:82–83.

Jackson, L. E., Jr., and A. Duk-Rodkin
1997 Quaternary Geology of the Ice-Free Corridor: Glacial Control on the Peopling of the New World. In *Prehistoric Mongoloid Dispersals*, edited by T. Akazawa and E. J. E. Szathmary, pp. 214–27. Oxford University Press, Oxford.

Jantz, R. L., and D. W. Owsley
1999 Circumpacific Populations and the Peopling of the New World. Paper presented at the *Clovis and Beyond Conference*, Santa Fe, New Mexico, October 30.

2001 Variation among Early North American Crania. *American Journal of Physical Anthropology* 114(2):146–55.

Jennings, J. D.
1957 *Danger Cave.* University of Utah Anthropological Papers No. 27.

Johnson, E., ed.
1987 *Lubbock Lake: Late Quaternary Studies on the Southern High Plains.* Texas A&M Press, College Station.

Johnson, E., L. Litwinionek, and V. T. Holliday
1994 The Sand Creek Mammoth Site, Llano Estacado of Texas. *Current Research in the Pleistocene* 11:70–72.

Jones, G. T., C. Beck and P. D. LeTourneau
1995 A Possible Association between Camelop*s cf. Hesternus* and Lithic Artifacts from the Sunshine Locality in Eastern Nevada. *Current Research in the Pleistocene* 13:27–29.

Jopling, A. V., W. N. Irving, and B. F. Beebe
1981 Stratigraphic, Sedimentological and Faunal Evidence for the Occurrence of Pre-Sangamonian Artifacts in Northern Yukon. *Arctic* 34:3–33.

Keefer, D. K., S. D. de France, M. E. Moseley, J. B. Richardson, D. R. Satterlee, and A. Day-Lewis
1998 Early Maritime Economy and El Niño Events at Quebrada Tacahuay. *Science* 281:1833–35.

Kelly, R. L., and L. C. Todd
1988 Coming into the Country: Early Paleoindian Hunting and Mobility. *American Antiquity* 53(2):231–44.

Kipnis, R.
1998 Early Hunter-Gatherers in the Americas: Perspectives from Central Brazil. *Antiquity* 72:581–92.

Kornfeld, M., and G. C. Frison
2000 Paleoindian Occupation of the High Country: The Case of Middle Park, Colorado. *Plains Anthropologist* 45(172):129–53.

Kornfeld, M., G. C. Frison, M. L. Larson, J. C. Miller, and J. Saysette
1999 Paleoindian Bison Procurement and Paleoenvironments in Middle Park, Colorado. *Geochronology* 14:655–74.

Lahren, L. A., and R. Bonnichsen
1974 Bone Foreshafts from a Clovis Burial in Southwestern Montana. *Science* 186:147–50.

Lanata, J. L., and L. A. Borrero
1999 The Archaeology of Hunter-Gatherers in South America. *Archaeology in Latin America*, edited by G. G. Politis and B. Alberti, pp. 76–89. New York, Routledge.

Laub, R. S., and G. Haynes
1998 Fluted Points, Mastodons, and Evidence of Late-Pleistocene Drought at the Hiscock Site, Western New York State. *Current Research in the Pleistocene* 15:32–34.

Lavallée, D.
2003 The First Peopling of the South American Coast during the Pleistocene/Holocene Transition—A Case Study: The Prehistoric Campsite of Quebrada de los Burros. In *Where the South Winds Blow: Ancient Evidence of Paleo South Americans*, edited by L. Miotti, M. Salemme, and N. Flegenheimer, pp.17–20. Texas A&M University, Center for the Study of the First Americans, College Station.

Lowery, D.
2003 Investigating Clovis on the Delmarva Peninsula. *Mammoth Trumpet* 19(1):10–14.

McAvoy, J. M., and L. D. McAvoy
1997 *Archaeological Investigations of the Site 44SX202, Cactus Hill, Sussex County Virginia.* Research Report Series No. 8, Commonwealth of Virginia, Department of Historic Resources.

MacDonald, G. F.
1985 *Debert: A Paleo-Indian Site in Central Nova Scotia*, 2nd edition. Persimmon Press, Buffalo, New York.

McNett, C. W., Jr.
1985 *Shawnee Minisink: A Stratified Paleoindian-Archaic Site in the Upper Delaware Valley of Pennsylvania.* Academic Press, Inc., Orlando.

Madsen, D. B., and D. Rhode
1990 Early Holocene Pinyon (*Pinus monophylla*) in the Northeastern Great Basin. *Quaternary Research* 33(1):94–101.

Mandryk, C. A. S., and N. Rutter (editors)
1996 The Ice-Free Corridor Revisited. *Quaternary International* 32.

Manley, W. F.
2002 Postglacial Flooding of the Bering Land Bridge: A Geospatial Animation: INSTAAR, University of Colorado, v1, *http://instaar.colorado.edu/QGISL/beringlandbridge*

Martin, P. S.
1973 The Discovery of America. *Science* 117:969–73.

Martínez, G. A.
2001 Fish-Tail Projectile Points and Megamammals: New Evidence from Paso Otero 5 (Argentina). *Antiquity* 75:523–28.

May, D. W., and S. R. Holen
1993 Radiocarbon Ages of Soils and Charcoal in Late Wisconsinan Loess, South-central Nebraska. *Quaternary Research* 39:55–58.

2005 Stratigraphy at the Jensen Mammoth Site in the Central Platte River Valley, Dawson County, Nebraska. *Current Research in the Pleistocene* 22, in press.

Mead, J. I., and D. J. Meltzer (editors)
1985 *Environments and Extinctions: Man in Late Glacial North America.* Center for the Study of Early Man, University of Maine, Orono.

Meggers, B. J., and E. T. Miller
2003 Hunter-Gatherers in Amazonia During the Pleistocene-Ho-

locene Transition. In *Under the Canopy*, edited by J. Mercader, pp. 291–316. New Brunswick, Rutgers University Press.

Meltzer, D. J.
2001 What Do You Do When No One's Been There Before? Thoughts on the Exploration and Colonization of New Lands. In *The First Americans: The Pleistocene Colonization of the New World*, edited by N. G. Jablonski, pp. 9–58. Memoirs of the California Academy of Sciences No. 27. San Francisco.

2004 On Possibilities, Prospecting, and Patterns: Thinking about a Pre-LGM Presence in the Americas. In *Entering America: Northeast Asia and Beringia before the Last Glacial Maximum*, edited by D. B. Madsen, pp. 359–78. University of Utah Press. Salt Lake City.

Merriwether, D. A.
2002 A Mitochondrial Perspective of the Peopling of the New World, In *The First Americans, the Pleistocene Colonization of the New World*, edited by N. G. Jablonski, pp. 295–310. Memoirs of the California Academy of Sciences.

Merriwether, D. A., F. Rothhammer, and R. E. Ferrell
1995 Distribution of the Four Founding Lineage Haplotypes in Native Americans Suggests a Single Wave of Migration for the New World. *American Journal of Physical Anthropology* 98:411–30.

Miller, E. T.
1987 Pesquisas Arqueológicas Paleoindígenas no Brasil Ocidental. In *Investigaciones Paleoindias al Sur de la Línea Ecuatorial*, edited by L. Núñez and B. J. Meggers, pp. 37–61. Estudios Atacameños 8.

Miller, S. J.
1989 Characteristics of Mammoth Bone Reduction at Owl Cave, the Wasden Site, Idaho. In *Bone Modification*, edited by R. Bonnichsen and M. H. Sorg, pp. 381–93. University of Maine, Center for the Study of the First Americans, Orono.

Miller, S. J., and W. Dort, Jr.
1978 Early Man at Owl Cave: Current Investigations at the Wasden Site, Eastern Snake River Plains, Idaho. In *Early Man in America from a Circum-Pacific Perspective*, edited by A. L. Bryan, pp. 129–39. Department of Anthropology, University of Alberta Occasional Papers No. 1, Edmonton.

Miotti, L.
2003 Patagonia: A Paradox for Building Images of the First Americans during the Pleistocene/Holocene Transition. *Quaternary International* 109-110:147–73.

Miotti, L., M. Salemme, and N. Flegenheimer (editors)
2003 *Where the South Winds Blow: Ancient Evidence of Paleo South Americans.* Texas A&M University, Center for the Study of the First Americans, College Station.

Mochanov, Y. A., and S. A. Fedoseeva
1996 Aldansk: Aldan River Valley, Sakha Republic (Chapter 3). In *American Beginnings: The Prehistory and Paleoecology of Beringia*, edited by F. H. West, pp. 157–215. University of Chicago Press, Chicago.

Moeller, R. W.
1980 *6LF21: A Paleo-Indian Site in Western Connecticut.* American Indian Archaeological Institute Occasional Paper 2, Washington, Connecticut.

Morett, A., L. Joaquin Arroyo-Cabrales, and O. Polaco
1998 Tocuila, a Remarkable Mammoth Site in the Basin of Mexico. *Current Research in the Pleistocene* 15:118–20.

Morlan, R. E.
1980 *Taphonomy and Archaeology in the Upper Pleistocene of the Northern Yukon Territory: A Glimpse of the Peopling of the New World.* National Museum of Man, Archaeological Survey of Canada, Mercury Series Paper No. 94, Ottawa.

1986 Pleistocene Archaeology in Old Crow Basin: A Critical Reappraisal. In *New Evidence for the Pleistocene Peopling of the Americas*, edited by A. L. Bryan, pp. 27–48. University of Maine, Center for the Study of Early Man, Orono.

2003 Current Perspectives on the Pleistocene Archaeology of Eastern Beringia. *Quaternary Research* 60:123–32.

Morlan, R.E., and J. Cinq-Mars
1982 Ancient Beringians: Human Occupation in the Late Pleistocene of Alaska and the Yukon Territory. In *Paleoecology of Beringia*, edited by D. M. Hopkins, J. V. Matthews, Jr., C. E. Schweger, and S. B. Young, pp. 353–81. Academic Press, New York.

Morlan, R. E., D. E. Nelson, T. A. Brown, J. S. Vogel, and J. R. Southon
1989 Accelerator Mass Spectrometry Dates on Bones from Old Crow Basin, Northern Yukon Territory. *Canadian Journal of Archaeology* 14:75–92.

Nami, H.
1996 New Assessments of the Early Human Occupations in the Southern Cone. In *Prehistoric Mongoloid Dispersal*, edited by T. Akazawa and E. J. E. Szathmáry, pp. 254–69. Oxford University Press, Oxford.

Neves, W. A.
1996 Cranial Morphological Variation and the Colonization of the New World: Towards a Four Migration Model. *American Journal of Physical Anthropology*, Supplement 22:176.

Neves, W. A., and M. Blum
2000 The Buhl Burial: A Comment on Green et al. *American Antiquity* 65(1):191–93.

Neves, W. A., R. González-José, M, Hubbe, R. Kipnis, A.G.M. Araujo, and O. Blasi
2004 Early Holocene Human Skeletal Remains from Cerca Grande, Lagoa Santa, Central Brazil, and the Origins of the First Americans. *World Archaeology* 36:479–501.

Neves, W. A., J. F. Powell, and E. G. Ozolins
1999 Modern Human Origins as Seen from the Peripheries. *Journal of Human Evolution* 34:96–105.

Neves, W. A., and H. M. Pucciarelli
1991 Morphological Affinities of the First Americans: An Exploratory Analysis Based on Early South American Human Remains. *Journal of Human Evolution* 21:261–73.

Neves, W.A., H. M. Puccarelli, and D. Meyer
1993 The Contribution of the Morphology of the Early South and North American Skeletal Remains to the Understanding of the Peopling of the Americas. *American Journal of Physical Anthropology* 16:150–51.

Núñez, L., J. Varela, R. Casamiquela, and C. Villagrán
1994 Reconstrucción Multidisciplinaria de la Ocupación Prehistórica de Quereo, Centro de Chile. *Latin American Antiquity* 5:99–118.

Ochsenius, C., and R. Gruhn (editors)
1979 *Taima-Taima: A Late Pleistocene Paleo-Indian Kill Site in Northernmost South America.* CIPICS, South American Quaternary Documentation Program.

Orr, P. C.
1956 Pleistocene Man in Fishbone Cave, Pershing County, Nevada. *Nevada State Museum Department of Archaeology Bulletin* 2:1–20. Carson City.

Overstreet, D. F.
2004 Pre-Clovis Occupation in Southeastern Wisconsin. In *New Perspective on the First Americans*, edited by B. T. Lepper and

R. Bonnichsen, pp. 41–49. Texas A&M Press, Center for the Study of the First Americans, College Station.

Overstreet, D. F., and T. W. Stafford, Jr.
1997 Additions to a Revised Chronology for Cultural and Non-Cultural Mammoth and Mastodon Fossils in the Southwestern Lake Michigan Basin. *Current Research in the Pleistocene* 14:70–72.

Peak, A. S., and H. L. Crew
1990 *An Archeological Data Recovery Project at Ca-Cal-S342, Clarks Flat, Calaveras County, California*. Cultural Resource Studies North Fork Stanislaus River Hydroelectric Development Project (volume II, Part I).

Pearson, F. J., Jr., E. M. Davis, M. A. Tamers, and R. W. Johnstone
1965 University of Texas Radiocarbon Dates III. *Radiocarbon* 7:296–314.

Pearson, G. A.
2004 Pan-American Paleoindian Dispersals and the Origins of Fishtail Projectile Points as Seen through the Lithic Raw-Material Reduction Strategies and Tool-Manufacturing Techniques at the Guardiría Site Turrialba Valley, Coasta Rica. In *The Settlement of the American Continents: A Multidisciplinary Approach to Human Biogeography*, edited by C. M. Barton, G. A. Clark, D. R. Yesner, and G. A. Pearson.University, pp. 85–102. Arizona Press, Tucson.

Peck, R. M.
1985 *The Williamson Site, Dinwiddie County, Virginia*. Edited and published by Rodney Peck, Harrisonburg, North Carolina.

Powell, J. F.
1994 Dental Evidence for the Peopling of the New World: Some Methodological Considerations. *Human Biology*. 65:799–815.
1995 Dental Variation and Biological Affinity among Middle Holocene Human Populations in North America. Ph.D. dissertation, Department of Anthropology, Texas A&M University, College Station.

Powell, J. F., and W. A. Neves
1999 Craniofacial Morphology of the First Americans: Pattern and Processes in the Peopling of the New World. *Yearbook of Physical Anthropology* 42:153–98.

Powell, J. F., and J. C. Rose
1999 Report on the Osteological Assessment of the "Kennewick Man" Skeleton (CENWW.97.Kennewick) *http://www.cr.nps.gov/aad/Kennewick.powell_rose,htm*

Powell, J. F., and D. G. Steele
1992 A Multivariate Craniometric Analysis of North American Paleoindian Remains. *Current Research in the Pleistocene* 9:59–62.
1993 Paleobiology of the First American. *Evolutionary Anthropology* 2:138–46.
1994 Paleobiological Evidence of the Peopling of the Americas: A Morphometric View. In *Method and Theory for Investigating the Peopling of the Americas,* edited by R. Bonnichsen and D. G. Steele, pp.141–63. Oregon State University Press, Center for the Study of the First Americans, Corvallis.
1999 Peopling of the Americas: A Historical and Comparative Perspective. In *Who Were the First Americans?* edited by R. Bonnichsen and R. Gruhn, pp. 91–120. Oregon State University Press, Center for the Study of the First Americans, Corvallis.

Powers, W. R., and J. F. Hoffecker
1989 Late Pleistocene Settlement in the Nenana Valley, Central Alaska. *American Antiquity* 54(2):263–87.

Prous, A., and E. Fogaça
1999 Archaeology of the Pleistocene-Holocene Boundary in Brazil. Quaternary *International* 53/54:21–41.

Prufer, O. H., and R. S. Baby
1963 *Palaeo-Indians of Ohio*. The Ohio Historical Society, Columbus.

Reanier, R. E.
1995 The Antiquity of Paleoindian Materials in Northern Alaska. *Arctic Anthropology* 32:31–50.

Rice, D. G.
1972 *The Windust Phase in Lower Snake River Region Prehistory.* Laboratory of Anthropology Reports of Investigations No. 50, Washington State University, Pullman.

Richardson III, J. B.
1978 Early Man on the Peruvian North Coast, Early Maritime Exploitation, and the Pleistocene and Holocene Environment. *Early Man in America from a Circum-Pacific Perspective*, edited by A. L. Bryan, pp. 274–89. University of Alberta, Edmonton.

Rivet, P.
1943 *Les origins de l'homme americain*. L'Arbre, Montreal.

Roosevelt, A. C., M. Lima da Costa, C. Lopes Machado, M. Michab, N. Mercier, H. Valladas, J. Feathers, W. Barnett, M. Imazio da Silveira, A. Henderson, J. Sliva, B. Chernoff, D. S.Reese, J. A.Holman, N.Toth, and K. Schick
1996 Paleoindian Cave Dwellers in the Amazon: The Peopling of the Americas. *Science* 272:373–84.

Sandweiss, D. H., H. McInnis, R. L. Burger, A. Cano, B. Ojeda, R. Paredes, M. del Carmen Sandweiss, and M. D. Glascock
1998 Quebrada Jaguay: Early South American Maritime Adaptations. *Science* 281:1830–32.

Saunders, J. J., G. A. Agogino, A. T. Boldurian, and C. V. Haynes
1991 A Mammoth-Ivory Burnisher-Billet from the Clovis Level, Blackwater Locality No.1, New Mexico. *Plains Anthropologist* 36:359–63.

Saunders, J. J., C. V. Haynes, Jr., D. J. Stanford, and G. A Agogino
1990 A Mammoth-Ivory Semifabricate from Blackwater Locality No. 1, New Mexico. *American Antiquity* 55:112–20.

Scheinsohn, V.
2003 Hunter-Gatherer Archaeology in South America. *Annual Review of Anthropology* 32:339–361.

Schmitz, P. I., A. O. Rosa and A. L. Vietti Bitencourt
2004 *Arqueologia nos Cerrados do Brasil Central: Serranopólis III*. Pesquisas, Antropologia No. 60. São Leopoldo, Instituto Anchietano de Pesquisas.

Schurr, T. G., and D. C. Wallace
1999 mtDNA Variation in Native Americans and Siberians and Its Implications for the Peopling of the New World. In *Who Were the First Americans?* edited by R. Bonnichsen, pp. 41–77. Oregon State University, Center for the Study of the First Americans, Corvallis.

Schweger, C. E.
1969 Comments on "Early Man in America and the Late Pleistocene Chronology of Western Canada and Alaska," edited by A. L. Bryan. *Current Anthropology* 10:359.

Seguchi, N., A. R. Nelson, and C. L. Brace
2003 *New World Paleoindians in Craniometric Perspective: New Look at Old Faces.* Paper presented at 72nd Annual Meeting of the American Association of Physical Anthropologists, April 23–27, Tempe, Arizona.

Sellards, E. H., G. L. Evans, and G. E. Meade
1947 Fossil Bison and Associated Artifacts from Plainview, Texas, with Description of Artifacts by Alex D. Krieger. *Bulletin Geological Society of America*. 58:927–54.

Sellet, F.
2001 A Changing Perspective on Paleoindian Chronology and Typology: A View from the Northwestern Plains. *Arctic Anthropology* 38:48–63

Simons, D. B., M., J. Schott, and H. T. Wright
1984 The Gainey Site: Variability in Great Lakes PaleoIndian Assemblage. *Archaeology of Eastern North America* 12:266–79.

Spiess, A. E., M. L. Curran, and J. R. Grime
1984 Caribou (*Rangifer tarandus L.*) Bones from New England Paleoindian Sites. *North American Archaeologist* 6(2):145–59.

Suárez, R.
2000 Paleoindian Occupations in Uruguay. *Current Research in the Pleistocene* 17: 78–80.

Stafford, T. W., Jr.
1988 Accelerator ^{14}C Dating of Late Pleistocene Megafauna. *Current Research in the Pleistocene* 5:41–43.

Stanford, D.
1978 Some Clovis Points. In *Abstracts of the Fifth Biennial Meeting of the American Quaternary Association*, pp. 140–41. Edmonton.

1979 The Selby and Dutton Sites: Evidence for a Possible Pre-Clovis Occupation of the High Plains. In *Pre-Llano Cultures of the Americas: Paradoxes and Possibilities*, edited by R.L. Humphrey and D. Stanford, pp. 101–23. The Anthropological Society of Washington, Washington, D.C.

1999 Paleoindian Archaeology and Late Pleistocene Environments in the Plains and Southwestern United States. In *Ice Age Peoples of North America: Environments, Origins, and Adaptations*, edited by R. Bonnichsen and K. L. Turnmire, pp. 281–339. Oregon State University Press, Center for the Study of the First Americans, Corvallis.

Stanford, D.J., R. Bonnichsen, and R. E. Morlan
1981 The Ginsberg Experiment: Modern and Prehistoric Evidence of a Bone-Flaking Technology. *Science* 212:438–40.

Stanford, D., and B. Bradley
2002 Ocean Trails and Prairie Paths? Thoughts about Clovis Origins. In *The First Americans: The Pleistocene Colonization of the New World*, edited by N. G. Jablonski, pp. 255–71. California Academy of Sciences, San Francisco.

Stanford, D. J., and M. A. Jodry
1988 The Drake Clovis Cache. *Current Research in the Pleistocene* 5: 21–23.

Stanford, D. J., W. R. Wedel, and G. R. Scott
1984 Archaeological Investigations of the Lamb Spring Site. *Southwestern Lore* 47(2):14–27.

Starikovskaya, Y. B., R. I. Sukernick, T. G. Schurr, A. M. Kogelnik, and D. C. Wallace
1998 mtDNA Diversity in Chukchi and Siberian Eskimos: Implications for the Genetic History of Ancient Beringia and the Peopling of the New World. *American Journal of Human Genetics* 63(5):1473–91.

Steele, D. G., and D. L. Carlson
1990 Excavation and Taphonomy of Mammoth Remains from the Duewall Newberry site, Brazos County, Texas. In *Bone Modification*, edited by R. Bonnichsen, pp. 413–29. Oregon State University, Center for the Study of the First Americans, Corvallis.

Steele, D. G., and J. F. Powell
1999 Peopling of the Americas: A Historical and Comparative Perspective. In *Who Were the First Americans?*, edited by R. Bonnichsen, pp. 97–126. Oregon State University, Center for the Study of the First Americans, Corvallis.

2002 Facing the Past: A View of the North American Human Fossil Record. In *The First Americans: The Pleistocene Colonization of the New World*, edited by N. G. Jablonski, pp. 93–122. Memoirs of the California Academy of Sciences 27, San Francisco.

Stothert, K.
1985 The Pre-Ceramic Las Vegas Culture of Coastal Ecuador. *American Antiquity* 50:613–37.

Swanson, E. H., Jr.
1972 *Birch Creek: Human Ecology in the Cool Desert of the Northern Rocky Mountains 9,000 B.C.–A.D. 1850*. Idaho State University Press, Pocatello.

Tankersley, K. B.
2002 *In Search of Ice Age Americans*. Gibbs Smith, Salt Lake City.

Vacher, S., S. Jérémie and J. Briand (editors)
1998 *Amérindiens du Sinnamary (Guyane): Archéologie en Forêt Équatoriale*. Paris, Editions de la Maison des Sciences de l'Homme.

Van der Hammen, T., and G. Correal Urrego
2001 Mastodontes en un Humedal Pleistocénico en el Valle del Magdalena (Colombia) con Evidencias de la Presencia del Hombre en el Pleniglacial. *Boletín de Arqueología* 16:4–36.

Vilhena V., A. Vialou, and D. Vialou
1993 Les Premiers Peuplements Préhistoriques du Mato Grosso. *Bul. de la Société Préhistorique Française* 91(4–5):257–63.

Wagner, D. P., and J. M. McAvoy
2004 Pedoarchaeology of Cactus Hill, a Sandy Paleoindian Site in Southeastern Virginia, U.S.A. *Geoarchaeology* 19(4):297–322.

Wallace, W. J., and F. A. Riddell
1988 Archaeological Background of Tulare Lake, California. In *Early Human Occupation in Far Western North America: The Clovis-Archaic Interface*, edited by J. A. Willig, C. M. Aikens, and J. L. Fagan, pp. 87–101. Nevada State Museum Anthropological Papers No. 21. Carson City.

Wenner, A. M., J. Cushing, E. Noble and M. Daly
1991 Mammoth Radiocarbon Dates from the Northern Channel Islands, California. *Proceedings of the Society for California Archaeology* 4:1–6.

West, F. H
1981 *The Archaeology of Beringia*. Columbia University Press, New York.

1996 The Archaeological Evidence. In *American Beginnings: The Prehistory and Palaeoecology of Beringia*, edited by F. H. West, pp. 537–59. University of Chicago Press, Chicago.

Willig, J. A
1990 Western Clovis Occupation at the Dietz site, Northern Alkali Lake Basin, Oregon. *Current Research in the Pleistocene* 7:52–56.

1988 Paleo-Archaic Adaptations and Lakeside Settlement Patterns in the Northern Alkali Basin. In *Early Human Occupation in Far Western North America: The Clovis-Archaic Interface*, edited by J. A. Willig, C. M. Aikens, and J. L. Fagan, pp. 417–82. Nevada State Museum Anthropological Papers 21, Carson City.

Willig, J. A., and C. M. Aikens
1988 The Clovis-Archaic Interface in Far Western North America. In *Early Human Occupation in Far Western North America: The Clovis-Archaic Interface*, edited by J. A. Willig, C. M. Aikens, and J. L. Fagan, pp. 1–40. Nevada State Museum Anthropological Papers 21, Carson City.

Wilson, M. C., and J. A. Burns
1999 Searching for the Earliest Canadians: Wide Corridors, Narrow Doorways, Small Windows. In *Ice Age Peoples of North*

America, edited by R.Bonnichsen and K.L. Turmire, pp. 213–48. Oregon State University Press, Center for the Study of the First Americans, Corvallis.

Wisner, G.
1997 Cooper's Ferry Spear Cache: One of NW's Oldest Sites. *Mammoth Trumpet* 13(4):1, 4–6.

Woods, J. C., and G. L. Titmus
1985 A Review of the Simon Clovis Collection. *Idaho Archaeologist* 8:3-8.

Yesner, D. R.
1996 Human Adapation at the Pleistocene Holocene Boundary (circa 13,000 to 8,000 BP) in Eastern Beringia. In *Humans at the End of the Ice Age: The Archaeology of the Pleistocene-Holocene Transition*, edited by L. G. Strauss, B. V. Eriksen, J. M. Erlandson, and D. R. Yesner, pp. 255–72. Plenum Press, New York.

2001 Human Colonization of Eastern Beringia and the Question of Mammoth Hunting. In *Proceedings of the International Conference on Mammoth Site Studies*, edited by D. West, pp. 69–84, University of Kansas Publications in Anthropology 22, Lawrence.

Young, D. E., and R. Bonnichsen
1984 *Understanding Stone Tools: A Cognitive Approach*. University of Maine, Center for the Study of Early Man, Orono.

Zárate, M. A., ed.
1999 The Pleistocene/Holocene Transition and Human Occupation in South America. *Quaternary International 53/54* (whole issue)

Index